NETWORK
PROTOCOLS

McGRAW-HILL SIGNATURE SERIES

Bates, Regis J. "Bud," and Gregory, Donald. *Voice and Data Communications Handbook.* 0-07-006396-6

Black, Uyless. *Frame Relay Networks.* 0-07-006890-9

Feit, Sidnie. *TCP/IP.* 0-07-022069-7

Kessler, Gary C., and Peter V. Southwick. *ISDN: Concepts, Facilities, and Services.* 0-07-034837-X

Kumar, Balaji. *Broadband Communications.* 0-07-038293-X

McDysan, David, and Spohn, Darren. *ATM: Theory and Applications.* 0-07-045346-2

To order or receive additional information on these or any other McGraw-Hill titles, in the United States please call 1-800-722-4726, or visit us at www.computing.mcgraw-hill.com. In other countries, contact your local McGraw-Hill representative.

Network Protocols

Matthew G. Naugle

Signature Edition

McGraw-Hill

New York • San Francisco • Washington, D.C. • Auckland • Bogotá
Caracas • Lisbon • London • Madrid • Mexico City • Milan
Montreal • New Delhi • San Juan • Singapore
Sydney • Tokyo • Toronto

Library of Congress Cataloging-in-Publication Data

Naugle, Matthew G.
    Network protocols / Matthew Naugle. — Signature ed.
       p.   cm.
    Prev. ed. published under title: Network protocol handbook. 1994.
    Includes index.
    ISBN 0-07-046603-3
    1. Computer network protocols. I. Naugle, Matthew G. Network
protocol handbook. II. Title.
TK5105.55.N38   1998
004.6′2—dc21   98-20249
                                            CIP

McGraw-Hill

*A Division of The **McGraw·Hill** Companies*

ISBN 0-07-046603-3

*The sponsoring editor for this book was Simon Yates, the editing supervisor was Ruth
W. Mannino, and the production supervisor was Claire Stanley. It was set in Vendome
ICG by Don Feldman of the McGraw-Hill Professional Book Group composition unit.*

Printed and bound by Quebecor/Fairfield.

McGraw-Hill books are available at special quantity discounts to use as pre-
miums and sales promotions, or for use in corporate training programs. For
more information, please write to the Director of Special Sales, McGraw-Hill,
11 West 19 Street, New York, NY 10011. Or contact your local bookstore.

This book is dedicated to my wife, Regina. Her constant reassurance and, most of all, her unrelenting patience made it possible to write this book. Also many thanks to my three children: Bryan, Courtney, and Lauren, who have proven that life is most enjoyable when looking at it through simplistic, innocent, and trusting eyes.

CONTENTS

Contents

Contents

Contents

Contents

Contents

Contents

NETWORK
PROTOCOLS

The Physical Layer

Introduction

Before the connection of computers to local area networks (LANs), the user would make a copy of the work accomplished on the computer and place it on a diskette. This diskette would then be brought to other users for them to use. For distant sites, the user would mail the diskette. This was affectionately known as "sneakernet."

The LAN was introduced commercially in 1980, enhancing the capabilities of the personal computer (PC). A PC can connect to a LAN to enable the user to communicate with other PC users connected to the same LAN. Originally, a mainframe computer provided the connection to a LAN, and this created two logical connections for the user: mainframe access and the ability to communicate with other PC users directly on the LAN. Figure 1-1 shows personal computers and host computer access on a LAN. Users have the ability to communicate with both personal computers and host computers connected to the LAN.

When LANs were introduced, their connections allowed asynchronous terminal connection to the LAN through the use of a communication server (also known as a terminal server). The reason behind this is that before the LAN, most devices (with the exception of SNA) were connected to their host computer via asynchronous protocols running over RS-232 cables. In order to provide network capabilities like those promoted by SNA, asynchronous terminal servers were invented to allow this connectivity. There was no reason to discard all the asynchronous devices just because the LAN provided a new connectivity.

The asynchronous terminal server allowed multiple asynchronous terminals to connect to it, and the terminal server would connect to the

Figure 1-1
Host to terminal.

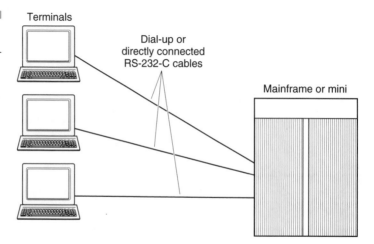

Terminals

Dial-up or
directly connected
RS-232-C cables

Mainframe or mini

LAN. The other end of the connection (before the advent of host LAN boards) was a communication server, which also had a connection to the LAN and multiple asynchronous connections to the host. This allowed a LAN to be installed into a business so that the existing equipment* could be used. This enabled multiple host access and the distribution of printers and modems, while completely eliminating the single-cable-to-host attachment required in centralized computing.

Overview

It is difficult to define exactly what a LAN is. LANs provide many capabilities, but a one-line sentence to sum it up is: *A LAN permits information and peripherals to be shared efficiently and economically*—economically in the sense that LANs can save companies money. This is truly what networks are all about: to allow information to be exchanged in any manner, efficiently and economically.

Given the wide variety of LANs, it is difficult to establish clear advantages and disadvantages of one type of network over another. There are many choices out there, including simple LANs for small environments as well as networks that support very high speeds and large numbers of attachments. The particular choice of a network clearly depends on the requirements that are to be placed on the LAN. Each networking scheme has its own advantages and disadvantages. But by far the most commonly employed LAN type is Ethernet (discussed later).

Networks have been called *distributed* or *decentralized* computing. This is good term for networks, as they allow the processing and application power of the computer to be distributed through the LAN. The computer is no longer the big, complex piece of equipment located somewhere in the building; computers and their attachments are now located throughout a building and sometimes throughout countries. The LAN is the device that allows the connection for distributed computing. Centralized computing is on the decline, but mainframes are far from obsolete. Today more mainframe types than ever are being produced by IBM and other companies, and they are smaller, more efficient, and more powerful than their predecessors. Today the LAN is the "centralized computing" environment.

Mainframes and PCs peacefully coexist on a LAN (see Fig. 1-2). It should be noted that the original methods for accessing the mainframe were not replaced per se; only the physical method of connection was replaced. Previously, each user's terminal was connected to the mainframe with a single cable. Applications were run on the mainframe, and

Figure 1-2
PC LAN with host
connection.

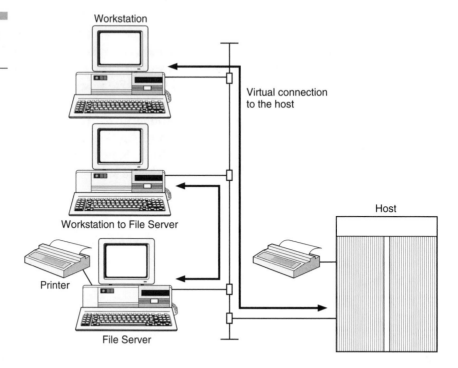

Figure 1-2
PC LAN with host
connection.

the data were output to the terminal. Today, with LANs and PCs, the
mainframe application is still run on the mainframe and the data are
still output to the user's screen. Only the connection to the mainframe
has changed. While applications still run on the mainframe computer,
many of these applications have been ported to run on personal computers. PC users can run their own applications, communicate with others on the LAN, and access the mainframe as if they were still directly
connected to it—all using a standard cabling and protocol scheme for
the attachments to the LAN.

With LANs came many never-before-heard-of technologies, which
will be explained in the following chapters. These technologies were
invented to allow decentralized computing. At first LANs and the software required to operate them seemed extremely foreign to most users,
and the software and hardware are still foreign to most users today. Simply replacing the connection to a mainframe and explaining to users
what happened is not enough. What exactly is a LAN, and how does the
software it uses enable users to communicate over it?

Before a full discussion on LANs is begun, an understanding of the
architecture that all LANs follow is necessary. This architecture is
known as the International Standards Organization's Open Systems

Interconnect model (the OSI model). You will see this model discussed time and time again, so it provides the best introduction to this book.

This book is separated into two parts: access methods and software protocols. Under the discussion of access methods, the workings of Ethernet, Token Ring, and FDDI are fully explained.

Included is a simple introduction to bridging using two methods, transparent and Source Route. The software protocols section explains the protocols that enable computers to communicate with one another. These discussions have been separated because each computer will have both a hardware interface (the network connection) and the software to run the network.

The Open Systems Interconnect Model

The Open Systems Interconnect (OSI) model was developed in 1974 by the International Standards Organization. It divides a LAN into seven processing modules, or layers. Each layer performs specific functions as part of the overall task of allowing application programs on different systems to communicate as if they were operating on the same system.

The OSI model is an architectural model based on modularity. The model is not specific to any software or hardware. OSI defines the functions of each layer but does not provide the software or hardware design to allow compliance to this model. The model's ultimate goal is interoperability between multiple vendors' communication products.

Any communications equipment may be designed after this model. Although the model is mentioned more often in reference to LANs, many data and telephone communications are designed with the OSI model in mind. All the LAN hardware and software explained in this book relate to this model.

This model comprises seven and only seven layers. The modules in sequence of bottom to top are physical, data link, network, transport, session, presentation, and application (see Fig. 1-3 and Table 1-1). Each layer has a specific purpose and functions independently of the other layers. However, each layer is "aware" of its immediate upper and lower layers.

1. *Physical layer.* This layer defines the methods used to transmit and receive data on the network. It consists of the wiring, the devices used to connect a station's network interface controller to the wiring, the signaling involved to transmit/receive data, and the ability to detect signaling errors on the network media (the cable plant).

Figure 1-3
(a) OSI model and
functions. (b) OSI
model filled in.

OSI Layer	Function Provided
Application	Network applications such as file transfer and terminal emulation
Presentation	Data formatting and encryption
Session	Negotiation and establishment of sessions
Transport	Provision for end to reliable delivery
Network	Routing of packets of information across multiple networks
Data link	Transfer of units of information, framing, and error checking
Physical	Transmission of binary data of a medium

(a)

OSI Layer	Protocol
Application	File Server concepts
Presentation	
Session	Courier NetBIOS
Transport	TCP SPP
Network	IP IDP Routers
Data Link	Ethernet, Token Ring bridges
Physical	Wiring systems

(b)

TABLE 1-1

Layers of the OSI
Model

Layer	Function
Application	Specialized network functions such as file transfer, virtual terminal, electronic mail, and file servers
Presentation	Data formatting and character code conversion and data encryption
Session	Negotiation and establishment of a connection with another device (such as a PC)
Transport	Provision for reliable and unreliable end-to-end delivery of data
Network	Routing of packets of information across multiple networks
Data link	Transfer of addressable units of information, frames, and error checking
Physical	Transmission of binary data over a communications medium

2. *Data-link layer.* This layer synchronizes transmission and handles frame-level error control and recovery so that information can be transmitted over the physical layer. The frame formatting and the CRC (cyclic redundancy check, which checks for errors in the whole frame) are accomplished at this layer. This layer performs the access methods known as Ethernet and Token Ring. It also provides the physical layer addressing for transmitted frames.

3. *Network layer.* This layer controls the forwarding of messages between stations. On the basis of certain information, this layer will allow data to flow sequentially between two stations using the most economical path, both logically and physically. This layer allows units of data to be transmitted to other networks through the use of special devices known as *routers.* Routers are defined in this layer.

4. *Transport layer.* This layer provides for end-to-end (origination station to destination station) transmission of data. It allows data to be transferred reliably (i.e., with a guarantee that it will be delivered in the same order that it was sent). It ensures that data are transmitted or received without error, in the correct order (received in the same order as the data were sent), and in a timely manner. Included in this layer is the provision for end-to-end communication, but over an unreliable protocol. This is used for mail programs, network management, host name resolving, and other protocols that do not need the robustness of a reliable transport.

5. *Session layer.* This layer establishes, maintains, and disconnects a communications link between two stations on a network. This layer is also responsible for name-to-station address translation. (This is the same as placing a call to someone on the phone knowing only their name; you must find their phone number if you want to create a connection.)

6. *Presentation layer.* This layer is responsible for data translation (format of the data) and data encryption (scrambling and descrambling the data as they are transmitted and received). It is not always implemented in a network protocol.

7. *Application layer.* This layer is used for those applications that were specifically written to run over the network. Applications such as file transfer, terminal emulation, electronic mail (email), and NetBIOS-based applications are examples.

That is what the OSI model accomplishes. It is not to difficult to understand, and it will become more clear as you read further in this book. Each LAN protocol discussed will be related to the OSI model and will also be identified in terms of the layer in which it resides.

This model will be used throughout the text. Each of the protocols that will be explained has different modules that, when taken as a whole (all seven layers together), constitute a network protocol. Each of these modules fits into a certain layer of the OSI model. Since each layer of the OSI model defines a certain subset of a protocol, we will break down each protocol studied in this book to its layers. In essence, each layer of the model will be shown and explained in each chapter. It is enough now just to introduce the model and to explain each of the layer functions. As you read a chapter on a certain protocol, the sections in that chapter will cover an implemented technology that conforms to the OSI layers.

Topologies

Before moving forward, it is important to understand topologies. Networks are built based on topologies. Topologies are architectural drawings that represent the cable layout and methodologies used for a LAN or wide area network (WAN). Topologies can be hardware dependent; i.e., when a particular LAN is chosen, a specific topology must be followed to implement the LAN. Some LANs have the capability of representing many types of topologies. All of the topologies will become more fully understood as you read through the Ethernet, Token Ring, and FDDI sections of the book.

There are two types of topologies: physical and logical. A physical topology includes the cable plants, for example. A logical topology covers the access method, such as Ethernet, Token Ring, or FDDI. Ethernet has evolved from a bus topology both physically and logically, to a physical star topology and a logical bus topology. This means that for 10BaseT networks, the wiring changed from having all stations connected to a single cable segment to having each station basically having its physical cable connected to a wiring concentrator. Logically, however, all stations still see all packets transmitted on the individual cable plants. This will change if you use a device known a switch. But for now we will start slow and talk about the earlier days of Ethernet.

Star Topology

The *star topology* is probably the oldest topology used for communications. (See Fig. 1-4 where a node is any station that can directly connect

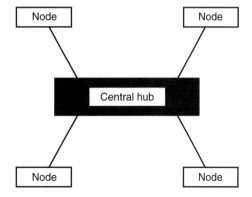

Figure 1-4
Star topology.

to a network.) It was introduced with analog and digital switching devices known as Private Branch Exchanges (PBXs).

In the star topology stations are attached to a common point. As shown in Fig. 1-4, this common point is usually a wiring hub with stations attached via cables that extend from it. Since each of the stations is on a point-to-point link (basically, each one has its own cable plant) with the central wiring hub, the cost and the amount of cable may increase. Considering the type of cable implemented, however, the overall cost is about equal to that of other topologies.

There are two primary advantages to this topology. First, there is no single point of failure in the cabling that would affect the whole network. If one of the cables should develop a problem, the hubs are smart enough to let it affect only the station directly using that cable. The example of the telephone can be used here. If the wire that connects your telephone were broken, only your telephone would be disabled; all other phones would remain operational. Similarly, all other network stations would remain operational. Second, the star topology allows for enhanced network management.

The disadvantage to this topology is the centralized hub. If the hub fails, all connections to it are disabled. However, this is not usually the case because most centralized hubs have multiple passive backplanes, dual power supplies, and other fault-resilient or fault-redundant items to enable high availability and low downtimes.

With the advent of a new wiring system for Ethernet (the access method of Ethernet will be discussed below), known as unshielded twisted pair (UTP), the star topology is now the most common physical topology used for Ethernet networks—replacing the bus topology. The logical topology for Ethernet is still the bus (based on its algorithm, which did not change with any of the changes in wiring methods).

Figure 1-5
Ring topology.

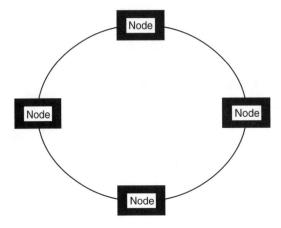

Ring Topology

Figure 1-5 shows the general topology of the ring. In the ring topology all stations attached to the ring are considered repeaters on the LAN that is enclosed in the loop. (A node is any station that can attach directly to the network.) There are no endpoints to this cable topology as there are in the bus topology. The repeater, for our purposes, is the controller board in the station that is attached to the LAN. Each station will receive a transmission on one end of the repeater and will repeat the transmission, bit by bit, with no buffering, on the other end of the repeater. Data are transmitted in only one direction and received by the next repeater in the loop.

The most common cable design for this topology is the star-wired ring. The physical topology in this design is a star while the logical topology is a ring. FDDI is also an example of the ring topology. Physically, FDDI is a point-to-point topology but logically it is a ring topology.

Bus Topology

The *bus topology* is sometimes known as the *linear-bus* topology. It is a simple design, as shown in Fig. 1-6. It uses a single length of cable (also known as the *medium*), with network stations attached to this cable. All stations share this cable, and transmissions from the stations can be received by any station attached to the cable (a *broadcast medium*). There are endpoints to the cable segment commonly known as *terminating points*.

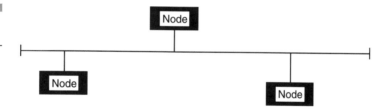

Figure 1-6
Bus topology.

Given the simplicity of this topology, the cost of implementing it is usually low. The management costs, however, are high. No single station on the LAN can be easily administered individually compared with a star topology. There are no management designs in stations that are attached to a single cable. The use of single cable also can lead to a major problem: it contains a single point of failure. If the cable breaks, then no station will have the ability to transmit. The LAN that best represents this topology is the Ethernet access method.

Wiring Systems for Ethernet, Token Ring, and FDDI

This section provides the reader with the cabling systems used for the access protocols of Ethernet, Token Ring, and FDDI. Although the access protocols are discussed later, their terms will be used here. The reader should bear with the terms until the cabling systems are discussed. It is important to understand the cabling system of each protocol before trying to comprehend the protocol. The access methods of Ethernet, Token Ring, and FDDI will be explained in detail in Chap. 2.

The physical layer of the OSI model is the bottom layer and is represented by the wiring systems and associated physical components (connectors) used in LANs. The role of the physical layer is to allow transmission and reception of raw data (a stream of data) over the communications medium (the cable plant). This means that all data being transmitted and received over the network will pass through this layer. For LANs, the layer is implemented fully in hardware.

The first protocol to study is Ethernet. The actual Ethernet protocol will be studied in a moment. First, the wiring systems will be discussed. There are four types of wiring systems that may be used for Ethernet networks:

1. Thick coaxial cable
2. Thin coaxial cable

3. Unshielded twisted pair

4. Fiber

These four types of wiring are also known by many other names. The column headings in Table 1-2 are the most commonly used names. Only the first three cable types will be discussed here. The fiber standard for Ethernet will not be discussed here.

Despite the changes in wiring, the access method defined for Ethernet remains unchanged; that is, the method defined in layer 2 of the OSI model has not been revised. The only thing that changes is the wiring methodology (layer 1 of the OSI model). This is a perfect example of why data communications systems rely on the OSI model for architecture. We can replace the physical layer of the model with something different and the remainder of the layers remain unchanged.

There may be confusion about which type of wiring system is best for Ethernet. A brief summary of the components used in each type of wiring system follows and should be referred to throughout the text. The Ethernet physical components are:

1. Thick coaxial cable—standardized in 1980

   a. Thick coaxial cable

   b. Transceivers

   c. Transceiver cables

   d. 50-ohm terminators

   e. Coring tool

   f. Wiring concentrators (not always required—depends on the size of the network)

TABLE 1-2

Commonly Used Names for Ethernet Wiring

Thick Coaxial Cable	Thin Coaxial Cable	Unshielded Twisted Pair	Fiber
RG-8*	RG-58 A/U or C/U	22–26 AWG† telephone cable	62.5/125 μm
10Base5	10Base2	10BaseT	10BaseF
IEEE 802.3	IEEE 802.3a	IEEE 802.3i	
Thicknet	Cheapernet or Thinnet	UTP	

*Not actually RG-8 cable but characteristic of it.
†American Wire Gauge.

    g. Connectors on cables and network controllers (N-series connectors)

2. Thin coaxial cable—standardized in 1985

    a. RG-58 A/U thin coaxial cable

    b. T connectors

    c. 50-ohm terminators

    d. Wiring concentrators, repeaters (not always required—depends on the size of the network)

    e. Connectors on cables and controllers (BNC connectors)

3. Unshielded twisted pair—standardized in 1990

    a. 22–26 AWG two-pair wire (telephone cable)

    b. Repeater modules and wiring concentrator (always required)

    c. RJ-45 connectors and plugs

Thick Coaxial Cable

This cabling scheme is representative of a bus topology. Thick coaxial cable is characteristic of type RG-8 (RG stands for *radio grade*) cable and was the original cabling scheme used when Ethernet was standardized in 1980. This cable is used on 10Base5 networks. 10Base5 is the standard term applied by the standards body of IEEE (Institute of Electrical and Electronics Engineers, pronounced "I triple E"). It represents the primary characteristics of the cabling scheme, using a kind of shorthand. The 10 represents the transmission speed in megabits per second (Mbps). The middle term represents the transmission signaling method used, and the last number is the longest cable segment used multiplied by 100 m (note that the unit of measurement is meters). Therefore, 10Base5 is a LAN running at 10 Mbps, using baseband (access method) technology (as opposed to broadband), and the longest cable run is 500 m (after 500 m a repeater is needed, which is explained later).

    This type of cable is usually colored yellow, although other colors are available. Every 2.5 m there is a painted black mark to show the placement of a network attachment (a PC, a host computer, a repeater, etc.). Devices may not be placed any closer together than this due to a phenomenon known as *reflection*. Without going into great detail, piercing the cable (placing devices on the cable plant) causes electrical reflections in the cable, which can incorrectly be interpreted by other devices as errors on the network. It is best to follow the markings on the cable.

A network attachment is accomplished through the use of a transceiver cable and a transceiver (see Fig. 1-7). The Ethernet controller (the controller is the integral hardware device that allows a device to attach to and work on the network) cannot attach to the cable plant without the use of these devices.

A *transceiver* (transmitter-receiver) is the intermediate device that transmits and receives the data from the Ethernet controller onto or from the cable plant. The transceiver couples the network device to the coaxial cable and is the most important part of the transmission system. At the time of Ethernet's introduction, around 1980, the components that made up a transceiver were expensive, and the components that allowed for a 10-Mbps transmission were not readily available. But this device provided the ability to transmit a signal up to 500 m from its origin and to detect when two or more network attachments were transmitting at the same time (error detection). This device allowed the transmission speed to be mandated at 10 Mbps, a speed that was unheard of in the data communications industry at the time. The designers of Ethernet could have provided for faster speeds and longer distances, but the cost would have been prohibitive. Initial costs of adding an Ethernet connection to a host computer were estimated to be $1200 to $1500 per connection just for the hardware.

Figure 1-7
(a) Thick coaxial connection.

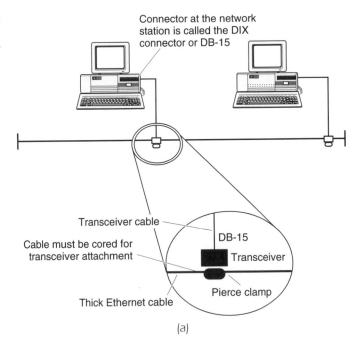

(a)

Figure 1-7
(b)

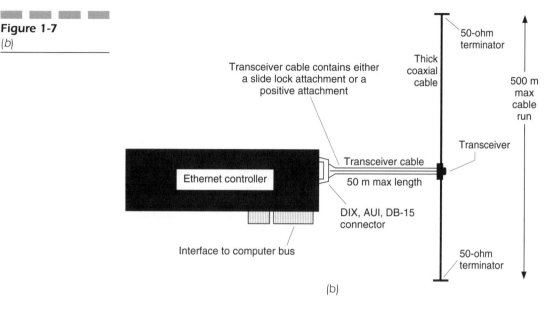

(b)

The transceiver cable connects transceivers to the Ethernet controller itself. It is not a coaxial cable of any type. The cable contains nine individual wires used to transmit and receive data and to relay errors in the transmission system to the controller card. This cable may be up to 50 m long.

Before the transceiver can be physically attached to the cable, the site on the thick coaxial cable must be prepared. A small hole must be cored into the cable to expose the center conductor. This is accomplished with a special piercing (coring) tool, which is a drill bit in a specially designed handle that drills a hole in the cable at a specified depth. The transceiver is then placed on the cable, and the transceiver cable is attached to the transceiver and to the Ethernet controller card of the network attachment.

Each end of a single cable segment is terminated using a device known as a 50-Ω terminator. There are only two terminators per cable segment. The terminators must be used on every cable segment.

Installing a network attachment on thick Ethernet may take five minutes or more to ensure a good connection. This cable type is not commonly used in today's LANs. Its primary use was for factory installations and backbone connections. (Backbone connections will be discussed later.) The advantage to using this type of cable is that it is not as susceptible to noise and can run for longer distances, because of the extended shielding, than the other types of cables used for Ethernet.

Each cable segment of thick Ethernet cable may be up to 500 m long. This length allows 100 network attachments. *A network attachment is any device that may attach to a cable plant.* This includes PCs, host computers, communication servers, repeaters, bridges, and routers. This is important for all cable schemes since networks have limitations on connections. So, on a thick Ethernet cable plant, if there are 100 PCs attached to the cable, you cannot connect another attachment. You must add a repeater to extend the cable plant if more network attachments are needed. If you add a repeater to the cable plant, one attachment (one PC) must be taken off, because the repeater is included in the 100-attachment limitation. There are other ways around this limitation, however. These devices will be fully explained in a later section of this chapter.

For purposes of the present discussion, repeaters are devices that will extend a single cable segment beyond its maximum length. Refer to Table 1-3 for exact specifications on each cable type. For 10BaseX (thick, thin, or UTP) Ethernet, a maximum of four repeaters may be placed between two communicating stations. This does not mean that no more than four repeaters may exist on an Ethernet cable plant. As many repeaters as are needed may exist on any cable plant, but a data path between two communicating network stations may not have more than four repeaters.

No matter what the cable type (thick or thin coaxial, UTP), the maximum number of network stations that may exist on an Ethernet cable plant is 1024. This assumes that devices such as bridges or repeaters are not in use.

Thin Coaxial Cable

The thin coaxial cable scheme also represents a bus topology. Thin coaxial cabling (used on 10Base2 networks) was standardized in 1985. Besides being lower in cost, this cabling scheme offers many other advantages over the thick coaxial cabling scheme. A major advantage is that this cable does not have to be pierced in order to place the transceiver on the cable plant. The external transceiver has been moved to the Ethernet controller itself (see Fig. 1-8). This move represents a cost savings of more than $200 per connection and eliminates the transceiver cable. The cable attaches directly to the back of the Ethernet controller card. RG-58 cable is low-cost cable and has been around for a long time (it is used for other things besides LANs).

TABLE 1-3

Cable Specifications for 10 Mbps Ethernet

Parameter	Value or Specification
Thick coaxial cable (impedance of 50 Ω)	
Signaling techniques	Baseband (Manchester)
Data rate	$\times\ 10^6$ bps
Maximum cable segment length	500 m
Maximum network length (with repeaters)	2500 m
Attachments per segment	100
Attachment spacing	2.5 m, minimum
Connector type	DB-15
Topology	Linear bus
Maximum number of stations per network	1024
Impedance rating of cable	50 Ω (cable ends terminated)
Thin coaxial cable	
Signaling techniques	Baseband (Manchester)
Data rate	$\times\ 10^6$ bps
Maximum cable segment length	185 m
Maximum network length (with repeaters)	1000 m
AttachFgy	Bus
Maximum number of stations per network	1024
Impedance of cable	50 Ω (cable ends terminated)
Unshielded twisted pair	
Signaling techniques	Baseband (Manchester)
Data rate	$\times\ 10^6$ bps
Maximum cable segment length	100 m
Maximum network length (with repeaters)	2500 m using thick coaxial backbone
Attachments per segment	1
Attachment spacing	N/A
Connector type	RJ-45 (8-pin connector) 1, 2, 3, 6 straight through
Topology	Star
Maximum number of stations per network	1024
Impedance of cable	75–150 Ω

Stations are attached to this cable through the use of a BNC connectors and T connectors (see Fig. 1-8*a*). Each cable segment endpoint contains a male BNC connector, and this connector attaches to a T connector. The T connector is placed on the Ethernet card itself. The cable is simply twisted together to form a cable plant.

Like thick coaxial cable, each end of a single cable segment is terminated by a 50-Ω terminator. Usually, the terminator is attached to one end of the T connector on the final attachment at each end of the cable segment.

There are also limitations, not disadvantages, to this cable plant. Compared to the thick Ethernet standard, the maximum cable length is 200 (actually 185) m. Only 30 network attachments are allowed per cable plant. The amount of shielding in this cable has been reduced, although it is more than enough for most applications. Repeaters are used to extend the cable length, thereby allowing more than 30 network attachments to communicate.

Changing the cable plant does not change the repeater rule. No more than four repeaters may separate two communicating devices on an Ethernet network.

Unshielded Twisted Pair

Until the Unshield twisted pair (UTP) cabling scheme was devised, thin coaxial cable was the most commonly used cable type in Ethernet network environments. Unshielded twisted pair (10BaseT) was standardized by the IEEE in October 1990 (see Fig. 1-9). The advantages of this cable scheme are many, and it has become the most popular cabling scheme for Ethernet networks. It is representative of the star topology. The term *twisted* comes from the fact that the wires are twisted together. There must be at least two twists per foot of wire. If the wire were not twisted, the signal strength would be greatly attenuated (greater signal loss).

There are three types of UTP cable used for data as standardized by the EIA (cable standards are not discussed in this book):

Category 3—for use in networks up to 10 Mbps

Category 4—for use in networks up to 20 Mbps

Category 5—for use in networks up to 100 Mbps

Unlike the coaxial cabling schemes mentioned before, UTP is a point-to-point cabling scheme, and its topology represents a star. The cable is

Figure 1-8
Thin coaxial connec-
tions.

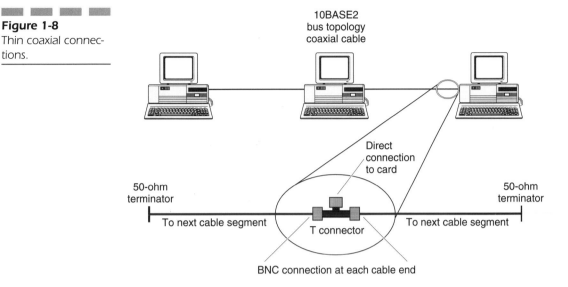

10BASE2
bus topology
coaxial cable

Direct
connection
to card

50-ohm
terminator

50-ohm
terminator

To next cable segment

To next cable segment

T connector

BNC connection at each cable end

(a)

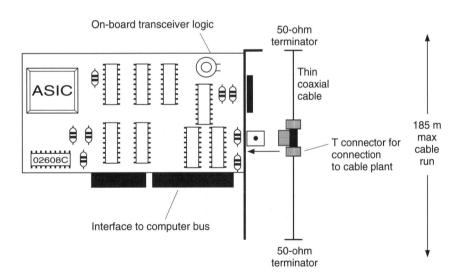

On-board transceiver logic

50-ohm
terminator

ASIC

Thin
coaxial
cable

02608C

185 m
max
cable
run

T connector for
connection
to cable plant

Interface to computer bus

50-ohm
terminator

(b)

Figure 1-9
(a) Unshielded twist-
ed pair. (b) Unshield-
ed twisted pair con-
troller.

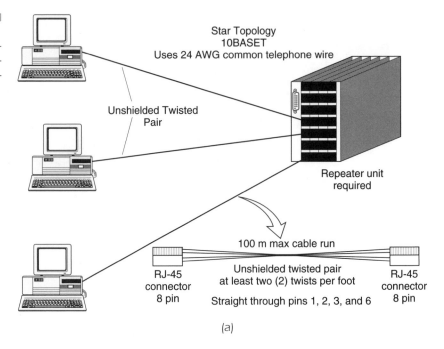

Star Topology
10BASET
Uses 24 AWG common telephone wire

Unshielded Twisted
Pair

Repeater unit
required

100 m max cable run

RJ-45
connector
8 pin

Unshielded twisted pair
at least two (2) twists per foot

Straight through pins 1, 2, 3, and 6

RJ-45
connector
8 pin

(a)

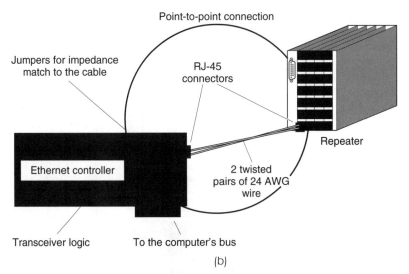

Point-to-point connection

Jumpers for impedance
match to the cable

RJ-45
connectors

Ethernet controller

Repeater

2 twisted
pairs of 24 AWG
wire

Transceiver logic   To the computer's bus

(b)

not a coaxial cable. It consists of four strands of 22–26 AWG (American
Wire Gauge) wire (basically standard telephone wire). Although standard
telephone wire may be used for this cable scheme, there is a UTP cable
standard commonly used for transmitting digital data (LAN data) over
UTP. Consult a cable company for this standard.

UTP may be run for 100 m between the Ethernet controller and the repeater hub. This type of scheme requires the use of a wiring hub. Each network connection is terminated at this wiring hub (which is a repeater; see Fig. 1-10).

With this cabling scheme, there are no external terminators. The cable is terminated on the Ethernet controller and the repeater (at each end of the physical connection). This cable type follows the Ethernet specification, with no more than four repeaters allowed (one 10BaseT concentrator constitutes one repeater, whether or not the two communicating stations are connected to the same repeater) and with a maximum of 1024 stations connected to a physical cable plant (all cable segments combine to form a single cable plant). Although two UTP stations may be attached back to back, each station should run through a repeater to communicate with the other. If two stations are connected back to back, then no other stations can be connected. Two repeaters hubs, though, may be connected through UTP wire. This is not generally recommended, however.

With this cabling scheme, the transceiver is physically located on the Ethernet card, as are thin coaxial controllers. The only thing that is changed is the wiring, the interface, and the connector for the wire. Stations are not concatenated with each other, as is the case with the coaxial cabling schemes.

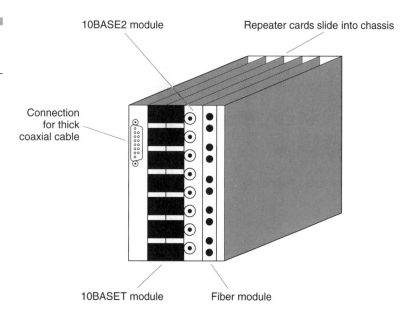

Figure 1-10
Wiring concentrators—hubs.

10BASE2 module

Repeater cards slide into chassis

Connection for thick coaxial cable

10BASET module

Fiber module

A network that utilizes UTP is far more manageable than the aforementioned bus methodologies. For example, if the cable that links a station with its repeater hub is damaged in any way, only that station will
shut down. With the coaxial cabling schemes, if the cable is damaged, all
network stations on that physical cable go down.

The reasons for developing UTP as a cable alternative for Ethernet
were many. This type of cable is very inexpensive. Although some people have indicated that common telephone cable can be used for UTP
installation, it is not recommended, and it is better to order certified
cable. Another reason for using UTP cable is that managing a UTP network is easily accomplished through smart hubs and network management software. UTP topologies also allow for easy installation of or
upgrading to switch networks.

Repeaters

There are many different types of repeaters, but the one most commonly
employed in Ethernet environments is known as a *wiring concentrator,*
which can be integrated into a device known as a hub. Originally, Ethernet repeaters simply attached to thick Ethernet cable segments and
allowed the extension of thick Ethernet to 2500 m. This repeater evolved
into the wiring concentrator, and the wiring concentrator evolved into
the hub. The difference between a concentrator and a hub is that the
hub allows many different types of network devices (routers, bridges, and
repeaters) to be enclosed in one container, and the concentrator simply
allows multiple types of Ethernet or Token Ring repeaters to be
enclosed in one container. Otherwise, the terms can be used inter-
changeably. As shown in Fig. 1-10, wiring concentrators can connect
thick, thin, and UTP networks into one box.

Repeaters allow for cable extensions and thus more stations per network segment. For example, with thin coaxial cable, the maximum number of network attachments to a single cable segment is 30. Repeaters
allow the connection of multiple cable segments, thus enabling more
than 30 stations to attach to a network by repeating the signals over multiple cable plants. With the use of repeaters, the number of stations still
cannot exceed the restriction of 1024 maximum for all cable segments
not separated by a bridge or a router. This is known as a *broadcast
domain,* and repeaters extend the broadcast domain. A broadcast domain
(for our purposes in this chapter) is simply the group of stations that can
"hear" all the information that is transmitted on a cable plant. Repeaters

repeat every signal they see, including error signals. Simple two-port repeaters (connecting two cable segments) have been concentrated into a single device, known as the concentrator. The concentrator houses all repeaters, and all wiring is terminated at this single box. Concentrators have card slots so that the repeaters (in card form) literally slide into the concentrator. A connector in the back of the concentrator connects all the cards. The wiring concentrator is required in a UTP environment. It is this device that terminates every UTP Ethernet connection.

The original specification for repeater units is for the extension of the 10Base5 cabling scheme. Each cable segment is allowed to be 500 m in length, but a single Ethernet cable plant is allowed to stretch to 2500 m in length due to the repeater specification. Each cable segment may attach to a repeater. For 10BaseX networks, there may be no more than four repeaters in a path between two communicating stations in a single broadcast domain (those networks not separated by a bridge, switch, or router). Refer to Fig. 1-11, which shows five Ethernet segments (the cable may be 10Base5 or 10Base2) separated by four repeaters. In this configuration, two of the segments may have only repeater connections; they may not have any other attachments. This is only for extended Ethernet cable segments up to the theoretical maximum limit. Most network installations will not run into this limitation with proper design and the use of wiring hubs.

As Ethernet was upgraded in speed, the repeater limit changed. For 100BaseX (Fast Ethernet) networks, there are two types of repeaters, and

Figure 1-11
Repeater limits.

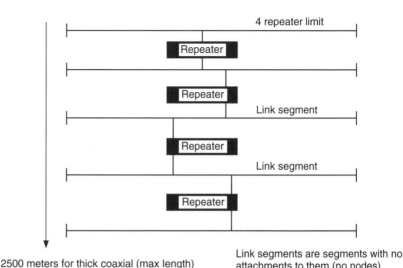

4 repeater limit

Repeater

Repeater

Link segment

Repeater

Link segment

Repeater

2500 meters for thick coaxial (max length)
1000 meters for thin coaxial (max length)

Link segments are segments with no attachments to them (no nodes)

the limit is two between communicating stations. For 1000BaseX (Gigabit Ethernet) networks, the repeater limit is one, and buffered repeaters may be used. These concepts are explained in upcoming sections.

This limitation does not mean that no more than four repeaters may be used on a network. There may be many repeaters on the cable plant—just no more than four repeaters between two communicating stations. One last rule is that if there are five Ethernet cable segments separated by four repeaters, two of the links (usually the ones in the middle) are not allowed to have a network station attachment. Thus the only connections on these two cable segments are the repeaters themselves. What this means is that in any LAN there may be more than four repeaters constituting the network. However, the network must be designed so that station A, when communicating with station B, must not pass through more than four repeaters. This is generally not a problem today with network designs.

Repeaters come in various sizes and shapes from many vendors. The one specification that they must all conform to is the IEEE 802.3c repeater specification. For backward compatibility, information on older-style repeaters as well as the new concentrator hub is included here. There are three types of repeaters:

1. Multiport transceiver units (10Base5 networks)

2. Multiport repeater units (10Base2 networks)

3. Wiring concentrators (10Base5, 10Base2, and 10BaseT networks)

The multiport transceiver unit (MTU) (see Fig. 1-12) is a very old but still implemented style of repeater. It was created to allow easier connection to a thick coaxial Ethernet network. The best-known example is the DEC DELNI (Digital Equipment Local Network Interface). Basically, this is a cable plant in a box complete with transceivers.

The MTU also allowed the connection of Ethernet controllers with or without the use of a cable plant. When Ethernet was introduced, most hosts in a site were minis (VAX computers) and mainframes (SNA hosts, e.g.). Therefore, instead of wiring up the building for Ethernet (thick coaxial cable is hard to install and was once very expensive), a site could install this type of a unit and connect all computers on a floor without having to install a cable plant. One simply used transceiver cables and the repeater. The repeater usually had nine connectors, and this allowed eight connections to it. The ninth connector was used for attachment to an external cable plant. If a cable plant was to be installed, it could be a backbone cable (usually run up the risers of a building and

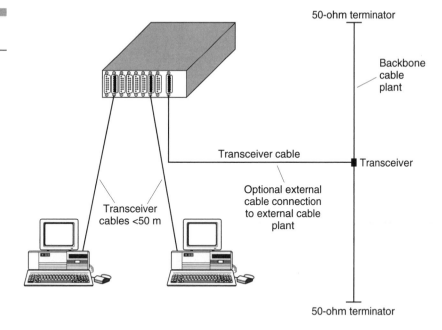

Figure 1-12
MTU connection.

not spread throughout the floor). The MTU provided the connections on the floor, with one port for connection to an external cable plant.

As shown in Fig. 1-13, MTUs can also be cascaded. In this configuration, one MTU acts as the root. All other MTUs have their external cable connection port connected to this MTU. This allows up to 64 stations to be connected without the use of a cable plant.

This repeater is also used when the thick coaxial cable plant is installed throughout a building (not just as a backbone cable). Earlier it was stated that thick Ethernet cable allows for only 100 attachments to a single cable segment, no matter how long that cable is (up to 500 m per cable segment). The MTU overcame this obstacle. Up to eight connections can be made to the MTU, and the MTU is a single attachment to a cable segment. By applying the MTU this way, you can have up to 800 stations on a single cable segment.

As shown in Fig. 1-14, the multiport repeater (MPR) is used with 10Base2 (thin net) cable plants. With the thin coaxial cable came the restriction that only 30 stations may attach to one cable segment, but this did not change the Ethernet maximum of 1024 stations. Therefore, this unit was developed so that up to eight thin cable segments could be attached. Each of the cable segments allowed 29 stations to be attached (the repeater was the thirtieth attachment). There was usually one port

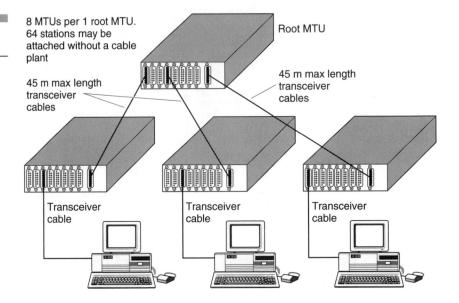

Figure 1-13
Cascaded MTU.

8 MTUs per 1 root MTU. 64 stations may be attached without a cable plant

Root MTU

45 m max length transceiver cables

45 m max length transceiver cables

Transceiver cable

Transceiver cable

Transceiver cable

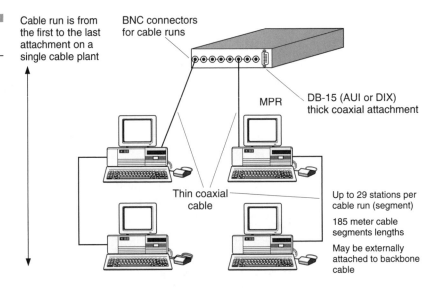

Figure 1-14
10BaseSE2 MPR.

Cable run is from the first to the last attachment on a single cable plant

BNC connectors for cable runs

MPR

DB-15 (AUI or DIX) thick coaxial attachment

Thin coaxial cable

Up to 29 stations per cable run (segment)

185 meter cable segments lengths

May be externally attached to backbone cable

on this unit that provided for connection to an external cable plant. The best-known example of this is the DEC DEMPR (see Fig. 1-15). Over the last few years (especially with the advent of 10BaseT networks), wiring concentrators have become the most popular type of repeater. This is especially true for Token Ring environments. These units have not only 10BaseT connections (a requirement for 10BaseT network stations), but

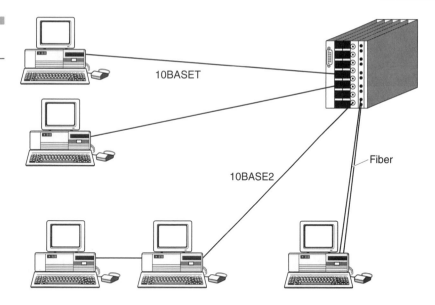

Figure 1-15
Wiring concentrator.

10BASET

10BASE2

Fiber

also 10Base2 and 10Base5 connections, all in one unit known as a con-centrator.

There was no reason why there had to be separate repeater units for each type of cable plant, so all the repeaters types were combined into one unit. This allows the network to be more manageable, for the wiring hubs contain special hardware and software that allow them to be remotely managed over the network (called *in-band management*). This means that a network station located somewhere on the network that contained specialized software could "talk" to the wiring concentrator. In this way, the network administrator could turn on and off ports, get statistics from the unit (collisions, deinsertions, bytes transmitted and received, etc.). This allowed a network management scheme to be brought to Ethernet.

Originally, Ethernet was implemented on a bus topology with all sta-tions attached to it. This made it very hard to manage the individual sta-tions on the network. The Ethernet standard did not mandate any type of management capabilities in the controller card. (In contrast, Token Ring has many management functions built into the IEEE 802.5 specifi-cation.)

The wiring hub made it a little easier to manage the network. It is true that for those stations connected on the 10Base2 ports, the network administrator can get statistics only for the single cable plant or possibly turn off the part attached to the cable plant (not the individual stations on the cable plant). This is extremely useful for those who perform troubleshooting of Ethernet networks. Had it not been for this type of repeater and 10BaseT cabling for Ethernet, Token Ring networks would be the dominant access protocol in use today.

As shown in Fig. 1-16, all three types of repeater units may be inter-mixed on a LAN. Examples of this type of hub are Bay Networks 5000, Cabletron MMAC, and Ungermann-Bass Access One. These hubs are used for much more than simply housing different wiring types for Ethernet. They also house FDDI and Token Ring, and can include abilities to route and switch all stations within the same hub. Each access method has its own backplane, and integral routers and switches contain the capability to communicate between them.

The Physical Layer for Token Ring

The most common type of cable used in Token Ring networks is category 5 (CAT5) UTP. Token Ring started out with a very structured wiring environment. Today it has basically collapsed into a wiring concentrator, CAT5 cable and the DB-9 connector for cable attachment to the NIC card, and the RJ-45 for the end that connects to the wiring concentrator. The most common connector types are the DB-9 on the NIC (network interface controller, the card that is installed on your PC) and the RJ-45

Figure 1-16
Mixing repeater types.

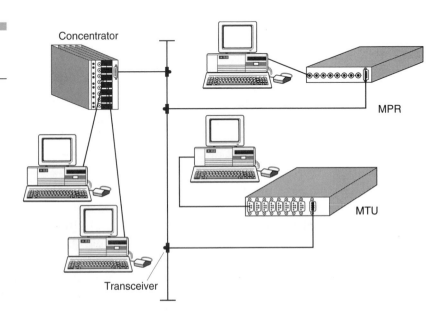

on the concentrator. However, more so than for Ethernet installations, Token Ring networks usually contain a hybrid of concentrator types, wiring schemes, and NICs. Therefore, all types are included in the following discussion.

The cabling scheme use for Token Ring networks is commonly known as the *IBM cabling scheme for Token Ring networks*. It has been designed to provide a structured wiring system that will work with all IBM communicating devices, including the IBM Token Ring.

One chief advantage of the wiring scheme for Token Ring is that it has changed only slightly since its inception. When it was introduced, Ethernet did not have a star topology for wiring. The star wiring for Token Ring was a huge advantage over Ethernet for management purposes. Remember that with Ethernet, a single cable plant is used to connect multiple stations. Any errors that were caused by the physical layer components of the cable affected every station on that cable.

The wiring that was proposed in 1985 is the same type specified today. Token Ring is still a physical star, logical ring topology, the same as it was in 1985. The only new connector type is the RJ-45 for unshielded and shielded twisted pair. This will be fully discussed in the connector section. The types of wire used are described in Table 1-4.

Although there are many different wiring types for Token Ring, types 1, 2, and 3 are the most commonly used. The maximum cable run for type 1 is 300 m for connection between the wiring concentrator and the network attachment. This is for only one wiring concentrator. If there is more than one concentrator on the network, the maximum run of 100 m is recommended. Notice that this is the same maximum cable run recommended for UTP wire for Ethernet.

A survey provided by AT&T shows that most phones are located no more than 100 m from the telephone closet. Usually, where there is a phone, a data jack should be placed there also. Therefore, most cable runs for UTP (Ethernet) and type 3 (Token Ring) extend no more than 100 m from their concentrators. Type 3 wire is the same wire used for UTP in Ethernet.

As shown in Fig. 1-17, network workstations are connected to a concentrator known as a multistation access unit (MAU). This figure shows the original Token Ring concentrator known as the IBM 8228. All Token Ring devices must attach to the MAU. The MAU can use up to eight type 1 wire connections. Today's concentrators are far more advanced than this simple device and contain multiple connections using the RJ-45 connector. The MAU was introduced with IBM's Token Ring. Although IBM still supports the 8228, it has come out with a product

TABLE 1-4

Token Ring Wire
Types

Type	Description
1	An overall, shielded, data-grade cable, with two solid twisted-pair 22 AWG wires. It is available as an indoor version with a braided shield or as an outdoor version with a corrugated metallic shield that is suitable for aerial installation or underground conduit. Type 1 indoor is also available in non-plenum and plenum versions for fire code regulations.
2	A type 1 indoor cable with four solid twisted pairs of telephone-grade (26 AWG) wire added around the outside of the shield. Type 2 is not available in an outdoor version.
3	This is known as the unshielded twisted pair for Token Ring. The original specification stated that this wire can be used where existing or unused phone wire is already in place. This is not recommended. Since this is unshielded cable, a special device known as a *media filter* must be added to the connection on the network attachment to filter out unwanted signals. UTP cable does not have shielding, and the media filter allows only those signals that are data related to pass through to the media card. This media filter is installed on the cable a foot or so ahead of the DB-9 connector.
5	This is a fiber cable used in repeater-to-repeater connections. It uses 100/140-μm cable.
6	This is a data-grade wire of stranded 26 AWG wire used for short runs in patch cables. This cable offers high attenuation and should not be used to connect a network directly to a wiring concentrator over long runs.
8	This is 26 AWG twisted-pair data-grade wire with a plastic ramp covering used in places where the cable must be run over the floor. It is used where cable cannot be run through the walls and floors. Older, historic buildings commonly use this.
9	This is 26 AWG shielded twisted-pair wire in a plenum jacket. It is used in those areas where fire codes restrict the type of wire used in ceilings.

Figure 1-17
A shielded twisted
pair (STP) MAU.

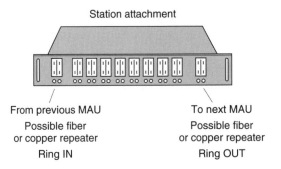

known as the 8230. This concentrator allows up to four MAUs (8228s) to be connected to a fiber or copper repeater unit. This unit is then connected to another 8230 repeater unit that can attach up to four more 8228s. In new installations, in place of these devices concentrators similar to the Ethernet concentrator are used.

The topology represented here is the star-wired ring. As shown in Fig. 1-17, on either side of the MAU are the Ring In and the Ring Out ports. These ports are used to connect to other MAUs (see Fig. 1-18). This is how a single Token Ring network may contain up to 260 stations for type 1 cable (33 eight-port MAUs) and 72 stations for type 3 cable. With the older-style MAUs, the use of more than 72 (9 eight-port MAUs) stations on an UTP cable plant is not allowed because of a phenomenon called *jitter.* This is the ability of an electrical signal to become distorted. UTP allows jitter to become more prevalent compared with type 1 cable. While jitter is still a problem on Token Ring networks, most wiring hubs (MAUs) now use a technology known as *active hub.* This turns each port on a hub into a repeater. With this technology, a UTP network may connect up to 150 stations per UTP ring, although the number varies from hub vendor to hub vendor.

MAUs are connected together through the Ring In and Ring Out ports. One MAU's Ring In port is attached to another MAU's Ring Out port, and that MAU's Ring In port may be connected to the next MAU's Ring Out port. As shown in Fig. 1-18, the last MAU's Ring In port is connected back to the first MAU's Ring Out port. This completes the logical ring. To expand the station limits, a bridge or router is needed; this is explained later.

Figure 1-18
Multiple MAU connection.

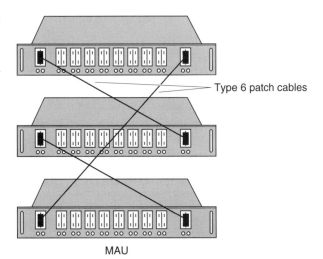

Type 6 patch cables

MAU

The cable that connects a network attachment to the MAU is called the *lobe cable*. When the network attachment wishes to connect to the ring, the attached network station is responsible for providing a phantom voltage (a small, steady voltage applied from the Token Ring controller card to the MAU) that flips a relay in the MAU. This allows the network station to connect to the ring through the MAU. If the network station is powered off, the phantom voltage is gone. The relay will be closed and data will not flow out to the network attachment. This will be covered in more detail later.

Data Connectors

The data connector is the plug that terminates all cable segments in a Token Ring network. There are three types of data connectors used in a Token Ring environment: the hermaphroditic, the DB-9, and the RJ-11 and RJ-45.

Originally, type 1 was the cable type of choice for Token Ring installations. This cable (between the concentrator and the node) consisted of a hermaphroditic connector on one end (connecting to the concentrator) and a DB-9 connection on the end (connecting to the NIC). The hermaphroditic is a large black connector shell specifically designed to be used with type 1 and type 2 wire. It is commonly called the IBM data connector and is used in non–Token Ring SNA environments as well as Token Ring. Normally, hermaphroditic and DB-9 connectors are used only for type 1, type 2 and other types of shielded cable. *Hermaphroditic* means having no gender, and this feature allows two hermaphroditic connectors to connect. This is useful for providing long lengths of cable using multiple shorter cables.

The most common type of cable installed today is CAT5 cable. The pinouts are different for Ethernet and Token Ring, although if you build a cable that uses all eight pins, the same cable can be used for Ethernet and Token Ring. Some installations still use the RJ-11 connector (the same type of connector used in residential phones) on one end of the type 3 cable. This is most commonly found on older NICs to attach to the network station. Remember, with this type of cable, the type 3 media filter has to be used. Type 3 cable is not shielded and allows for unwanted electrical signal (noise, i.e., those signals that do not originate from the ring). The media filter allows only the Token Ring data signal to pass to the controller card. Type 1 and 2 wires have foil shielding surrounding the data cables so that this does not happen. Therefore, media filters are not used with type 1 and 2 cable. Some Token Ring controllers have this filter built in. Check with the Token Ring controller vendor.

The hermaphroditic connector may be used on UTP wire, but it is extremely expensive and there is no need to use it on that cable type. An RJ-45 connector can be used on shielded or unshielded cable wire. Hub vendors use the RJ-45 (eight-pin telephone-type connector) on unshielded or shielded twisted pair wire—but only on the hub. Most Token Ring controller cards still have the RJ-11 or DB-9 connector. The RJ-45 is the same connector recommended for UTP wire in Ethernet installations. The hermaphroditic connector is used on shielded twisted-pair cable to connect to IBM-type MAUs. It is not commonly used on other vendors' hubs.

Multistation Access Unit

Multistation access units (MAUs or MSAUs) are the wiring concentrators for Token Ring. They come in two types: active repeater and nonactive or passive repeater. The active repeater is the type most commonly used in the high-density concentrator hubs. The active repeater is capable of repeating the signal, thereby reducing signal distortion and attenuation. This allows more UTP stations to attach to a single ring. The nonactive or passive repeater does not have this capability and falls under the old rules of network attachment (72 UTP devices per ring).

Physically, there are two types as well: the old hermaphroditic concentrator (the IBM 8228, allowing 8 network attachments, and the IBM 8230, allowing up to 32 network attachments) and the newer high-density concentrator. All in all, these devices are the units that all network attachments connect to. The MAU is the concentrator. The original MAU, the IBM 8228, contained eight hermaphroditic ports. Token Ring concentrators have evolved efficiently to high-density concentrators able to handle over 150 stations per ring. However, the 8228 continues to be used in many installations, so it is explained here.

As Fig. 1-19 shows, the cable that connects the station to the MAU is called a lobe cable and, for most installations, is allowed to be 300 m in total length using shielded twisted-pair (STP) cable. The *Ring In* and *Ring Out* ports are used only for connection to other MAUs. By connecting MAUs together, up to 260 stations (using 33 MAUs) may be combined into one ring (using type 1 cable).

Connecting two MAUs does not create two distinct rings. Only when a bridge or router (explained in the next section) separates the two

Figure 1-19
Controller attach-
ment to a MAU.

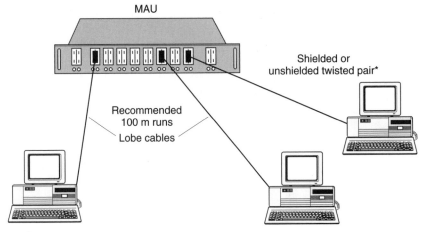

* Different MAU connectors for UTP.

MAUs will there be more than one ring. Although this is not recommended, the maximum number of attachments allowed on one ring
using STP cable is 260. For UTP cable the number is reduced to 72 (using
type 3 cable and a nonactive repeater concentrator).

Each controller on the LAN is responsible for flipping a relay on its
attached port in the MAU, which provides a data path to the ring. The
network attachment applies a phantom voltage to the lobe cable, which
flips the relay to the closed state and allows data to come off the ring
and into the network attachment. It is called a phantom voltage because
it is an electrical signal that uses the same wires as the data signal, but it
will not be mistaken for data. When the Token Ring station is powered
off or brought down for any reason, the phantom voltage is cut off and
the relay is flipped to the open state—in essence taking the network station off the ring.

One important consideration in flipping of the relay is ring length.
Ring length is variable in a Token Ring system. With a network that is
200 nodes large, with all stations powered on, the ring length is the total
length of all cables (including all cables from the users' attachments, the
interconnection of the MAUs, etc.). If half of the stations are powered
off, the ring length is smaller since the stations that are not powered on
do not have their relays activated. Signals passing through the MAU will
bypass the inactive relays and move onto the next activated relay. This
speeds up the network. The lower the number of activated relays, the
shorter the ring length.

Figure 1-20 shows three network stations that are active. Each of the
workstations has an attachment to a MAU. Since each of the workstations

Figure 1-20
MAU operation.

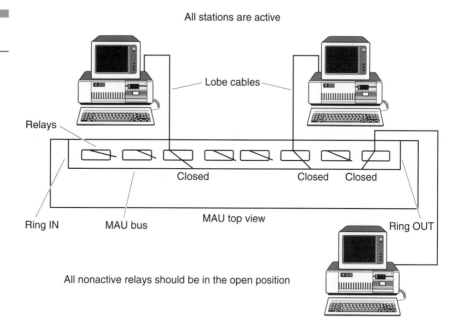

All stations are active

Lobe cables

Relays

Closed   Closed   Closed

Ring IN   MAU bus   MAU top view   Ring OUT

All nonactive relays should be in the open position

is active, the relay is flipped to show attachment to the internal bus of the MAU.

Figure 1-21 shows the same three stations, but one workstation is not active. In this configuration, the workstation that is not active will have its relay in the open state. The data that pass between the other two stations will not be passed to the nonactive workstation.

The new high-density concentrators allow many rings and many stations (usually up to 150) to connect to one concentrator. These concentrators can have multiple backplane rings and cards that have multiple RJ-45 connectors. These new concentrators are usually of the active repeater type and allow for easy network management not only through the star topology, but also through the use of built-in network management capabilities to monitor all data.

FDDI Overview and Physical Layer

Like Ethernet and Token Ring, FDDI (which stands for Fiber Distributed Data Interface) operates at the physical and data-link layers of the OSI model. However, unlike the other two, FDDI's physical layer is split into two sublayers: the physical layer protocol (PHY) and the physical layer medium-dependent protocol (PMD). This is shown in Fig. 1-22.

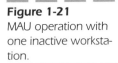

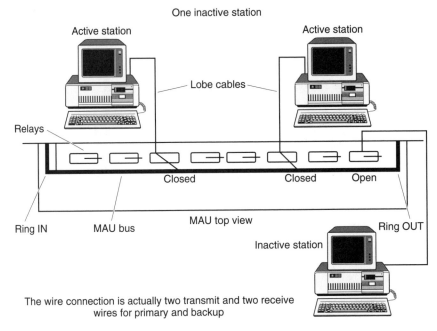

One inactive station

Active station   Active station

Lobe cables

Relays

Closed   Closed   Open

Ring IN   MAU bus   MAU top view   Ring OUT

Inactive station

The wire connection is actually two transmit and two receive
wires for primary and backup

FDDI and the OSI

Similar to other access methods, FDDI provides for LAN access only.
FDDI is defined at the physical layer and at the MAC sublayer of the
data-link layer. At the Logical Link Control (LLC) sublayer of the data-
link layer, FDDI assumes the use of another protocol, called the IEEE
802.2 protocol. The IEEE 802.2 protocol has been adopted to run on all
four LAN standardized protocols. The IEEE 802.1 standard has also been
adopted to run over all four LAN protocols. The most noted of the IEEE
802.1 protocols are the 802.1 standards for bridging.

PHY is specified as the upper sublayer of the physical layer. It is
responsible for symbols (data), line states, encoding/decoding techniques,
clocking requirements, and data-framing requirements.

Figure 1-22
FDDI and the OSI
model.

IEEE 802.1			
IEEE 802.2 Logical Link Control			
MAC	Media Access Control		
PHY	IEEE 802.3 CSMA/CD	IEEE 802.3 Token Bus	IEEE 802.5 Token Ring
PMD			

(SMT shown vertically alongside MAC/PHY/PMD)

PMD is the lower sublayer of the physical layer and is responsible for the transmit/receive power levels, transmitter and receiver interface requirements, error rates, and cable and connector specifications. It is composed of four standards:

1. Physical layer, medium dependent (PMD)
2. Single-mode fiber physical layer, medium dependent (SMF-PMD)
3. Low-cost fiber physical layer, medium dependent (LCF-PMD, under development)
4. Twisted-pair physical layer, medium dependent (TP-PMD)

FDDI is defined at the Media Access Control (MAC) sublayer of the data-link layer. The MAC sublayer is responsible for data-link addressing (the MAC address), media access, error detection, and token handling.

Station Management (SMT) is defined at the physical layer and at the MAC sublayer of the data-link layer. It is responsible for management services, including connection management, node configuration, recovery from error conditions, and the encoding of SMT frames.

The cable most often used with FDDI is fiberoptic cable, though copper cable may also be used. Each ring is allowed to be 100 km in length. In the event of certain physical errors on the ring, wrapping can occur, allowing the total ring length to be a maximum of 200 km.

Untwisted pair cable (UTP) is allowed to run up to 100 m between the network attachment and its concentrator. Only when UTP cable is used for FDDI is a concentrator required. An FDDI ring operates at 100 Mbps.

FDDI stations may be of two types: dual-attachment stations (DAS) and single-attachment stations (SAS). To allow for simpler connection and better fault tolerance of the ring, stations may be attached to a concentrator that is connected to the ring. The use of concentrators is not required but is recommended for most installations. These concentrators are functionally similar to the cable concentrators explained earlier. Stations (both DAS and SAS) may attach to the concentrator, and concentrators offer many efficiencies for the design of an FDDI network.

FDDI Cable

By far the most common cable choice for FDDI is fiberoptic cable. There are two types used: MMF-PMD (multimode) and SMF-PMD (single-

Figure 1-23
FDDI cable makeup.

Cladding

Core

Plastic Outside Covering

mode). As shown in Fig. 1-23, there are three entities that make up fiberoptic cable: the core, the cladding, and the protective coating. The core is a glass cylinder through which the light rays travel. The cladding is a glass tube that surrounds the core; its main purpose is to reflect any stray light rays back into the core. The protective coating is exactly that— a plastic coating that surrounds the core and cladding and protects them. The color of this cable depends on the manufacturer of the cable.

The core and cladding are most commonly referred to by their diameters. The most common are 50/100 μm, 62.5/125 μm, and 100/140 μm. The first number refers to the diameter of the core and the second refers to the diameter of the cladding.

Light is transmitted at one end of the cable and is received at the other end. In essence, the connection of any two attachments is a point-to-point connection. A cable that is capable of handling many different light rays is called *multimode fiber;* a mode is a single light ray. Multimode fiber generally uses inexpensive light-emitting diodes (LEDs) as the light source. There is a high attenuation with multimode fiber. *Attenuation* is the loss of signal strength as the signal travels through the cable. High attenuation means that a signal cannot travel as far down the cable as a signal in a cable that has a low attenuation rate.

Single-mode fiber (SMF) allows only a single light ray to travel through the cable. It uses lasers as its light source and does not have a high attenuation rate. The core of the fiber is usually 8 to 10 μm in diameter, and the cladding diameter is 125 μm. Single-mode fiber comes at a higher cost, partly because of the use of lasers as the light source. Stations may be up to 2 km apart using multimode fiber; with single-mode fiber the separation can be up to 20 km.

Copper cable (UTP) has been approved for FDDI cable. The cable run is allowed to be 100 m between the network attachment and its concentrator, and the network attachment connects to the ring as a SAS. The cable must be CAT5 as rated by the Electronics Industry Association/Telecommunications Industry Association (EIA/TIA). This type of connection requires the use of a wiring concentrator. Copper, however, is generally not used with FDDI. The price of fiber cable has dropped dramatically, although it still does not come close to the price of CAT5 UTP cable. With the advent of 100BaseT (100-Mbps) Ethernet, copper FDDI can be used as an alternative.

Connectors

As shown in Fig. 1-24, two types of connectors are commonly used: the media interface connector (MIC) and the ST connector. The MIC is a flat rectangular connector used to connect multimode fiber to any type of FDDI network attachment. It is constructed with "keys" that are defined by the PMD standard. These are small plastic pieces that are inserted into the top of the connector to ensure that the connector is placed on the correct port type (port types are explained in a moment). Although there are many ways to use these connectors, there are two types of common cable connectors: MIC-to-MIC and MIC-to-ST.

The MIC houses one fiber pair. One fiber pair is the minimum needed to attach to an FDDI network. This is a SAS attachment. Two MICs are used for dual-ring attachment (DAS) and one MIC used for single-ring attachment. Because one fiber pair connects to each MIC, one fiber in this pair is used for transmitting and the other for receiving.

ST connectors are commonly used to connect the fiber to an FDDI patch panel. A patch panel is a passive device that terminates the cable wiring. It allows the manual switching of cables. The panel contains rows of connectors that are terminated for any network attachment. There is another series of rows containing connectors from the concentrator ports. The network attachment cable runs usually terminate at the back of the patch panel. Therefore, there are two series of connectors on the patch panel, one from the network attachments and one from the concentrator ports.

The front of the patch panel allows short cables to connect two network attachment connectors, allowing connectivity between two network attachments. Another way of looking at a patch panel is in relation to the old telephone operator's panel. In front of the operator was a series of plugs and a series of jacks. The operator could use the plugs and

Figure 1-24
FDDI connectors.

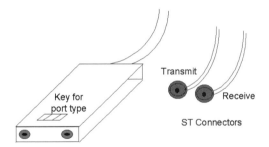

Key for port type

Transmit

Receive

ST Connectors

Media Interface Connector (MIC)

jacks to connect any line with any other line to allow two phones to communicate. A patch panel operates similarly.

ST connectors are not keyed, and care must be taken to ensure the proper connection. One end of this type of cable has the MIC connector, which can be used to plug in to an end station. The other end of the cable has two ST connectors (one for transmitting and one for receiving) and is commonly used for attaching to a patch panel. Other cables in the patch panel will connect to the FDDI ring. ST connectors are less expensive than MIC connectors. The ST connector can be found on single-attachment stations.

Port Types

To guard against illegal physical connections or topologies, the FDDI standard identifies four port types: A, B, M, and S. Ports are the connectors on all FDDI attachments—concentrators, bridges, routers, and network end stations. Any network attachment may use any of the port types. Dual-attachment stations have A and B port types; concentrators have A, B, M, or S port types; and single-attachment stations have only the S port type. The FDDI standard also mandates which port types are allowed to connect to one another. An FDDI network uses a structured cable system; stations are cabled together to form a network by making sure the port types are legal connections. For example, an A port should interface to another attachment's B port. Figure 1-25*a* shows a matrix of port types, legal and illegal. This will be covered in more detail later. Figure 1.25*b* shows some sample port type connections.

Figure 1-25
(*a*) FDDI port matrix.

Port Type

		A	B	M	S
Port Type	A	X	V	VB	X
	B	V	X	VB	X
	M	VB	VB	P	V
	S	X	X	V	V

V - Valid Connection
VB - Valid Connection, PHY B takes precedence
X - Invalid Connection
P - Prohibited Connection

(a)

Figure 1-25
(*b*) Sample port type
configuration.

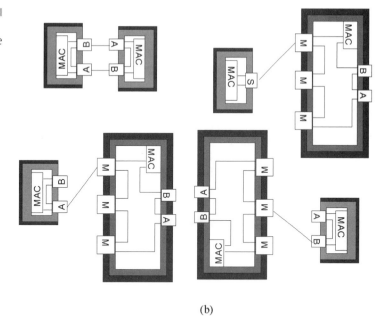

(b)

FDDI Station Classes

A station is any device that may connect to the ring: a concentrator,
bridge, router, or end-user workstation. Even though FDDI is a dual-ring
network, not every station attachment needs to have a connection to both
rings. There are many reasons for this, which will be discussed shortly.

As shown in Fig. 1-26, FDDI has two station classes: the dual-attachment station (DAS) and the single-attachment station (SAS). An SAS provides the least expensive and simplest means of connection to an FDDI
network. It has one S port and connects to the FDDI ring through the
M port of a concentrator. Therefore, SAS connections require the use of
a concentrator. An SAS provides a single attachment to the ring; it does
not have a connection to the dual ring. Therefore, if the SAS connection
is disabled, all ring connectivity is lost for that station. An SAS may be
connected to another SAS, but those stations are the only two that are
allowed on the network. In other words, they form an FDDI ring with
only two stations.

Figure 1-26
Station classes.

SAS connections are used in network stations that may be powered off and on periodically. Because the S port of an SAS is connected to the M port of a concentrator, the M port provides port isolation from the rest of the ring network. This will cause little or no disruption of the main ring, because when the SAS port is inactive, the topology does not change.

The DAS connects to both the primary and secondary rings of an FDDI network. It has two instances of the PHY and PMD, and one or two instances of the MAC. What this means is that the logic required to run the PHY and the PMD sublayers is copied for each port connection, A and B. It takes twice as much logic for a DAS as it does for an SAS. This raises the price.

A DAS may have a connection to an optical bypass relay. An optical bypass relay is an electromechanical device that prevents disruption of the dual ring should the DAS become inactive. It allows the light to reflect to the next station as if the intermediate station were still active. However, the light will not be regenerated as it is reflected to the next station, which can cause attenuation (weakening) of the light signal. As we will see later, disruption of the dual ring can create problems.

The DAS has two ports: A and B. The A port connects to another station's B port, and the B port connects to another station's A port. It is important to note that a DAS does not require a concentrator for attachment to the ring; it is a full-function dual-ring attachment. In case of a failure, the DAS can wrap the ring to isolate the failure.

FDDI networks should not be designed using only DAS connections. This can lead to complications. For instance, if there is a failure, the dual ring will segment itself into separate autonomous rings. The purpose of wrapping is to isolate the failed components and bring the ring back up, even if there are two or more rings when the failed components are isolated. Although DAS is more expensive, it is more resilient than SAS. Figure 1-27 shows the connections of DAS stations. The port types do not depend on the type of attachment. DAS stations can be a router, a PC, a UNIX host, or a concentrator.

FDDI Physical Layer Operation

FDDI stations are connected to a network using a point-to-point, bidirectional physical link (the FDDI cable) between the physical layers of two stations. That is the formal verbiage for the connection. FDDI stations are simply connected directly to each other (in the case of a DAS

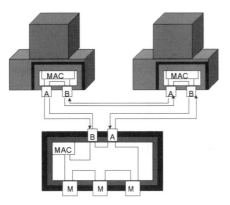

Figure 1-27
FDDI DAS stations.

connection or connections) or may connect to each other through an intermediate device known as a concentrator.

Each attachment to the network has the ability to transmit and receive, but the data move in one direction only: counterclockwise. A single physical connection uses two fibers. A dual attachment uses four fibers, one pair for the connection to each ring. The transmit fiber from one station is connected to the receive fiber of another station, the receive fiber of one station is connected to the transmit fiber of another station, and so on.

In the dual-ring environment, data are allowed to flow in opposite directions on each ring simultaneously. This is not common, for it requires more expensive equipment in the FDDI station attachment. It requires more logic in that two copies of the MAC must be present in the NIC. In single-MAC implementations, data flow only on the primary ring. The secondary ring is idle and is used for backup. Each ring can be up to 100 km long, giving a total length of 200 km. Data on the primary ring flow counterclockwise. Depending on the location of the break in a dual ring, the ring is said to wrap together, meaning that the primary and secondary rings will collapse into one ring (which allows the total length to extend to 200 km).

FDDI Station or Ring Breaks

In normal operation, data will flow only on the primary ring. The secondary ring will remain inactive until there is a failure. In this event, the ring will wrap, and the primary and secondary rings will act as one ring.

Many conditions will cause a ring to wrap. How the ring will wrap is determined by what breaks. An FDDI dual ring will continue to wrap until the failure is isolated. For example, in a network that has five dual-attachment stations, data will normally flow in a counter-rotating fashion (see Fig. 1-28). If there is a break in the fiber between stations B and C, the ring will wrap station B to A and C to D. This will create a single-ring topology and allow the network to continue operating (all stations will continue to communicate, even through the ring has wrapped). Network management will indicate that the ring is wrapped and will identify which stations are in wrap mode. This allows network administrators to find the cable break and fix it. The ring will return to normal operation (two rings) once the FDDI stations have determined that the cable has been restored.

The ring will also wrap if a dual-attachment station becomes disabled. If station B were to become disabled, stations A and C would wrap the ring. Station A would wrap the ring with station E, and station C would wrap the ring with station D. If station B has the optical bypass switch installed, the ring may not wrap; the signal would be reflected to station A just as if station B were still active. If the distance between stations C and A is more than 2 km, the ring may still wrap. This can occur because a multimode fiber run more than 2 km could deplete the strength of the signal; station A would not be able to receive and process it as a valid signal. Station A would lose connectivity with station B, and the ring would wrap.

Figure 1-28
FDDI breaks.

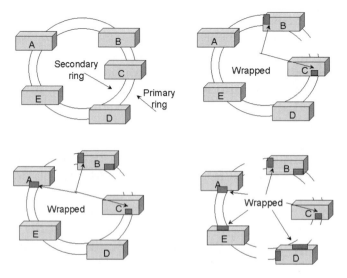

In another scenario, if stations B and D became disabled, station A would wrap with station E, and station C will become isolated. If the fiber broke between stations A and B (not shown in the figure) and the fiber broke between stations E and D, station A would wrap with stations E and B, and C and D would wrap to each other. This will provide two dual-attachment rings. Something as simple as powering off an FDDI attachment will cause this wrapping. This is why stations that attach to the dual ring should be stable devices (devices that are not going up and down frequently). All other devices should attach to the FDDI topology through the use of a concentrator.

FDDI Concentrators

FDDI concentrators provide many benefits. The FDDI concentrator provides functions similar to those of the Token Ring concentrator or the Ethernet concentrator, but it is more technologically advanced. The FDDI concentrator is a device that operates at the physical layer. A concentrator allows an FDDI network to be collapsed into a single node. It is simply a device that provides multiple ports for connection of network stations into the ring. It allows for multiple single-attachment stations, dual-attachment stations, and even other concentrators. Concentrators provide better network management, better topology designs, and greater efficiency. They are not required in certain situations. For example, an FDDI network that contains only DAS attachments does not need a concentrator; DAS attachments can connect to each other to form an FDDI ring. This arrangement is commonly used in FDDI backbone environments using routers. But FDDI concentrators provide a valuable function in reducing the cost of FDDI and increasing manageability.

One of the primary uses of an FDDI concentrator is to provide an FDDI service to those devices that may be powered on and off periodically. This normally would cause a disruption of the ring and may even segment the ring into two or more autonomous rings. The concentrator isolates its network attachments from the dual ring. Concentrators may be cascaded to form one of the most popular FDDI topologies: the dual ring of trees. They provide the root of the tree topology.

A concentrator may use any of the four port types. To provide for this, there are two types of concentrators: dual-attachment concentrators (DACs) and single-attachment concentrators (SACs). These are shown in Fig. 1-29. DACs attach to the ring as a full dual-ring device via its A and

Figure 1-29
FDDI concentrator
topology.

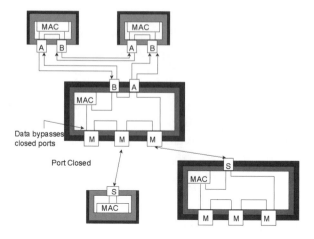

B ports. They provide multiple single-attachment connections to SASs through its M ports. The DAC does not possess any S ports and does not require connection to a dual ring for it to become active. An FDDI network can be as simple as a SAS connection to a standalone DAC.

The SAC attaches to the ring as a single-ring device. It appears as a single-attachment device and cannot connect directly to a dual ring except through a DAC. A SAC will have one S port for connection to another concentrator's M port and provides for multiple SAS connections through its M ports. The SAC does not need any connection on its S port to operate. It can provide FDDI connectivity to SAS stations as a standalone concentrator.

In normal operation, an FDDI network actually consists of two rings. A DAS has two ports: A and B. A DAS that enters the ring uses port A to connect to the incoming primary ring and the outgoing secondary ring. A DAS can be an end station or it can be a DAC.

Port B is the opposite of port A. It connects to the outgoing primary ring and the incoming secondary ring. Port M connects a concentrator port to an SAS port, DAS port, or another concentrator (DAC or SAC) port. This type of port is found only in a DAC or a SAC. Port S connects a SAS or SAC to a concentrator (DAC or SAC). Almost all connections are allowed (whether desired or not), but there is one hard-and-fast rule: No S ports are allowed to connect directly to an A or a B port, and an M port cannot be connected to another M port (there is no use for this anyway).

Concentrators perform two functions: port bypass for inactive stations and port insertion for active stations. Port bypass represents a concentrator port that is closed. This means that the attachment to that port does not have access to the FDDI network. No data from the FDDI network will flow out through this port. The port closing may have been caused by an inactive station (powered off) connected to its port, by network management shutting the port down due to excessive errors, or by another management entity requesting the shutdown of the port. This usually does not affect the normal operation of the ring. Only those ports that have their port shut off will be affected and be denied FDDI services. These services are shown in Fig. 1-30.

Port insertion is the situation in which the network attachment and its concentrator port have a good physical connection and the concentrator port inserts the attachment into the FDDI ring. This provides access to the services of the ring for the network attachment, making the network attachment an active participant in the FDDI ring.

The insertion process causes a momentary disruption of the active ring, but the ring corrects itself quickly and returns to normal operation. This process is accomplished in a very short amount of time.

FDDI Topologies

Like any other LAN, FDDI has topologies that simplify the network and make it more efficient. There are five types of FDDI topologies:

1. Standalone concentrator
2. Dual ring
3. Tree of concentrators
4. Dual ring of trees
5. Dual homing

Figure 1-30
FDDI concentrator functions.

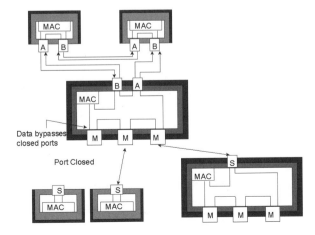

Figure 1-31
Standalone concen-
trator.

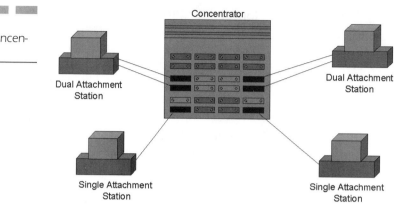

Standalone Concentrator. This topology is the simplest in design. As Fig. 1-31 shows, it consists of a single concentrator with all stations attached to it. The concentrator is at the root of the topology; it acts like the backbone of the FDDI topology. All attachments to this concentrator are SAS. Stations of any type (DAS or SAS) may be connected to the concentrator, but they function as SASs.

This is a good design for small sites that need little network management and a lot of reliability. Any station may be powered off, and the other stations on the ring are basically unaffected. The concentrator provides fault isolation.

Dual Ring. Figure 1-32 shows a dual-ring installation. This type of topology consists of dual-attachment stations attached directly to a dual

Figure 1-32
Dual-ring topology.

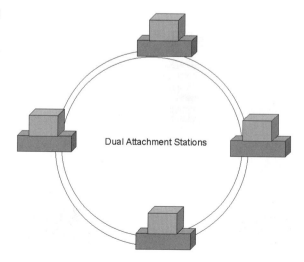

ring that can support users' workstations, bridges, or routers. This is a fault-tolerant topology that is designed primarily for small sites (if the DASs are user workstations). It may act as a backbone for medium to large sites that have a backbone of routers and bridges designed into the network.

This design does have its drawbacks. It does not work well with station attachments that physically move, turn off, or have additions made to them. Each station is attached to the backbone, and it is up to each station to ensure the reliability of the ring. Something as simple as a user powering off his or her workstation will break the ring and cause it to wrap. Worse, multiple failures may cause the ring to wrap multiple times, creating two or more separate FDDI rings. Wrapping is an inherent feature of FDDI, but it can cause connectivity loss between workstations. This type of design should be employed when there is little risk of ring interruption.

Tree of Concentrators. Figure 1-33 shows the tree of concentrators topology. With this design, a concentrator is used as the root of the tree to which network attachments are connected. Attachments may be user workstations (DASs or SASs), other concentrators (DACs or SACs), bridges, and routers. Every attachment will branch from the root.

A better design than the last two mentioned, this topology is used in medium to large networks that tie different groups of network users, and it lends itself well to building these types of environments. With a root concentrator, a second tier of concentrators can be placed for multiple-

Figure 1-33
Tree of concentrators
FDDI topology.

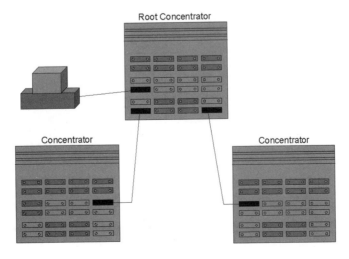

floor connections. This second tier of concentrators would then connect to the root concentrator.

This type of design allows fault isolation to a single station. It also allows greater flexibility in network management, in that individual network attachments or groups of attachments can be singled out for management purposes.

Dual Ring of Trees.   Shown in Fig. 1-34, the dual ring of trees is the most structured and popular type of topology. A dual-ring backbone is the root of the tree. A second tier of concentrators is placed with direct attachments to the root concentrators. All further network attachments—including user workstations, other concentrators, bridges, and routers—are placed on the second tier of concentrators.

The roots are fault-tolerant. Concentrators are simple passive devices that tend to operate for a long time before failing. Usually a cable break will occur before the concentrator fails. This ensures good connectivity no matter how many branches are added to the root. This topology provides fault isolation at any layer of the topology. It also allows easy expansion at any branch of the tree, because network stations should be added to the ends of the branches.

Dual Homing.   Figure 1-35 shows dual homing technology. Although it may not appear so, dual homing is actually an FDDI topology. A topology that contains DACs connected to a dual-ring topology may have an outside DAS with one port connected to one concentrator and

Figure 1-34
Dual ring of trees.

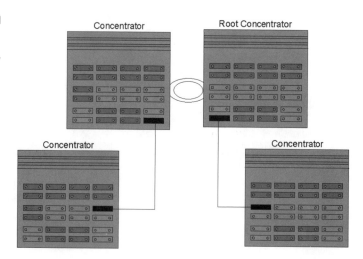

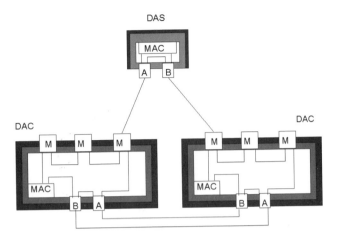

Figure 1-35
Dual homing FDDI
topology.

the other port connected to a different concentrator. This is a dual homing topology.

With this structure a DAS is allowed to connect to two DACs. The A port of the DAS is connected to the first concentrator's M port, and the B port of the DAS is connected to the second concentrator's M port. In this topology, the DAS is not considered part of a dual ring. The only active port on the DAS is the B port (by FDDI rules, the B port has precedence over the A port). If port B fails, the A port is inserted on the other concentrator. This is a dynamic process by which the DAS identifies the B port failure and knows that it can insert its A port. No outside management function is needed.

This is known as a redundant topology, and it is used in environments where uptime is critical in the event of a station or cable failure.

Ethernet, Token Ring, and FDDI

The Data-Link Layer

Ethernet

Since the first edition of this book, Ethernet has undergone one revision (Fast Ethernet) and is currently undergoing another (Gigabit Ethernet). Also, new "wiring center" changes have taken place, including Ethernet switching and full-duplex Ethernet (IEEE 802.3x). This chapter will start out with basic Ethernet (10-Mbps and bus topologies) and proceed to the more advanced aspects of Ethernet.

In a transmitting and receiving communication system, if there is only one cable to use and multiple stations need access to it, there must be a control mechanism in place to allow a fair system for stations to share the cable plant. In an Ethernet LAN system, the control mechanism is an access method known as Carrier Sense Multiple Access with Collision Detection (CSMA/CD). Ethernet applies the functions of the CSMA/CD algorithm. (It is easier to say *Ethernet* than it is to say *CSMA/CD*.) Ethernet basically performs three functions:

1. Transmitting and receiving formatted data or packets

2. Decoding the packets and checking for valid addresses before informing upper-layer software

3. Detecting errors within the data packet (as a whole packet) or on the network

Packets and Frames. Throughout this book, the terms *packets* and *frames* will be used. It is not unusual to say *packet* when you mean *frame*. The difference is about as noticeable as the old problem of bps (bits per second) versus baud. Those who want to argue this point can do so, and the rest of us will continue to learn. For the purists, however, I will explain the difference. Refer to Fig. 2-1. A frame is an encapsulated packet. A packet is an encapsulated datagram (a datagram is fully explained in the respective protocol chapters). For those interested in trivia, the term *datagram* was invented to mimic the term *telegram. Packet* was invented to refer to a small package. Interesting, but not very useful—back to the subject at hand. Framing takes layer 2 information and places layer 1 signaling around it. For example, in Ethernet, framing consists simply of the preamble (and Start of the frame delimiter for IEEE 802.3). For WANs (wide area networks), framing consists of flag bits (usually 7E) at the front of the transmission and flag bits at the end of the transmission.

Figure 2-1
Frames, packets, and datagrams.

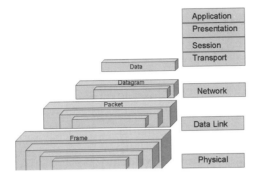

Packets consist of headers around the information that layer 3 has pushed down to the data-link layer. For Ethernet, the packet headers consist of the destination address (DA), the source address (SA), the type field, and the cyclic redundancy check (CRC). This is all fully explained later. Therefore, frames correspond to the physical layer, packets are formed at the data-link layer, and datagrams are formed at the network layer. For this book, the basic unit of information transmitted on any LAN is structured into an envelope called the *packet.*

All data that are transferred (for example, inside of a PC) will be transmitted as units of information on the memory bus between the CPU and memory for processing. When these data have to be transferred between two stations that are separated by a LAN, some special formatting needs to be done to the data. This is the purpose of the packet. This formatting is called *data encapsulation.* Each layer of the OSI model will encapsulate the data received from the next higher level with its own information. There are many encapsulation techniques, and they will be discussed under each protocol. The original data are not changed; information is just appended to the data so the network knows what to do with them.

Figure 2-2 shows the internals of two network stations: a workstation and a file server. In order for the two to transmit and receive data from each other, the data to be transmitted must be formatted for transmission over the network. In this figure the workstation is requesting the file server to provide a directory listing. As the application responds to the request, each layer of the network software places information at the front or end of the data.

For example, the application layer passes information down to the presentation layer. This layer adds presentation layer information to the data and passes it on to the session layer. The session layer adds its information and passes it on to the transport layer. The process is repeated

Figure 2-2

Packet terminating.

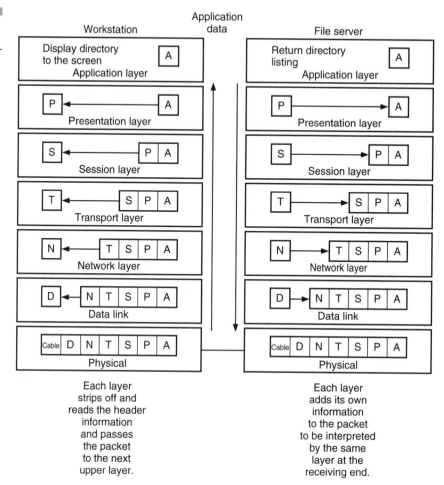

down through the network stack. This is indicated by the letters A, P, S, T, N, and D in the figure, which stand for the application, presentation, session, transport, network, and data-link layers, respectively.

When this transmission is received by the server, each layer of the protocol stack reads the associated information in the received packet. For example, the data-link layer reads the data-link (D) information, strips off the data-link header, and then passes the rest of the packet to the network layer. The network layer reads the network-layer header, strips it off, and passes the rest of the packet to the transport layer. The process is repeated until the rest of the packet reaches the application layer. The application layer then reads the data (in this case, it is a request for read of the hard disk).

When the server responds to this request, it passes the data back down the network stack, and network headers and trailers are added to the applications data. The packet is again formatted as stated previously and sent back to the workstation.

Data that would normally be transferred between the CPU and memory inside of a PC are now encapsulated into a LAN packet for transmission and reception on a LAN. This encapsulation is called a packet and contains information that only the LAN will use to interpret what kind of data it is and where it is to be transferred to. Figure 2-3 shows the packetized information being sent between a workstation and file server in a more simplistic view.

You can think of packets as sentences. When carrying on a conversation with someone, you speak in sentences. These sentences contain a starting point and a stopping point and are directed toward someone who is listening to you. Carried in these sentences may be commands or simply information. Commands could include asking someone to sit down and retrieve something for you. Information could be ideas or data that you have collected and are explaining.

Ethernet is also known as IEEE 802.3. With the exception of the packet format (discussed later under the section on addressing), the algorithm for Ethernet and IEEE 802.3 is the same. Ethernet was invented by Bob Metcalf at Xerox Corporation and was standardized by a consortium

Figure 2-3
Packets.

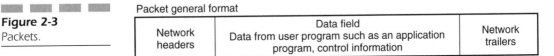

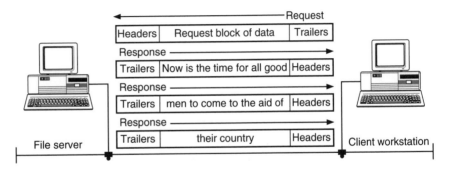

consisting of Digital Equipment Corporation (DEC), Intel, and Xerox (thereby giving the name DIX standard).

The Ethernet algorithm: The following algorithm is for shared cable plants, not switched. The original media are referred to as *shared media*. The current topology for Ethernet is physically a star, logically a bus. The original Ethernet involved a single cable that all stations attached to. We know Ethernet today as a star topology. In a star topology, each host has its own cable that connects to a wiring hub. The connection is known as point to point. This part of the book will explain the Ethernet algorithm in terms of the bus topology.

Refer to Fig. 2-4. A station wishing to transmit is said to contend for sole use of the cable plant. Once the cable is acquired, the station uses it to transmit a packet. All other stations on the cable plant listen for incoming packets or enter defer mode (i.e., a station defers transmission of any information that it wants to send). Only one station may transmit on the cable plant at a time.

To contend for sole use of the cable plant (acquire the cable plant), any station wishing to transmit checks whether the cable plant is busy (i.e., uses carrier sense). It checks to see if any signals are on the cable. If the cable plant has not been busy for a specified amount of time, the station immediately begins to transmit its data. During the transmission, the station listens. This process of listening while transmitting ensures that no other stations transmit data to the cable plant while it is transmitting.

Figure 2-4
Normal Ethernet operation.

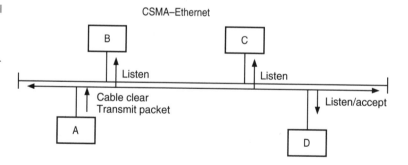

1. Node A receives data to transmit to Node D.

2. Builds a packet.

3. Checks to see if the cable plant is clear (no one else is currently transmitting).

4. Transmits packet while listening to the cable.

5. If there were no collisions, returns to listen mode.

The first 64 bytes of the transmission are crucial. Ethernet based on thick coaxial cable is allowed to extend up to 2500 m (8250 ft, or about 1½ mi) end to end. The rule of 64 bytes (which times 8 bits yields 512 bits, also known as the *slot time*) is important, for it takes 64 bytes transmitted at 10 Mbps to reach all stations possible in a full run of 2500 m.

If no other station transmits during that time and the station transmits all of its data, the station resumes listening to the cable. The transmission is said to have been successful. If any other station transmitted during that time, a *collision* is said to have occurred. Any station transmitting will know a collision has occurred by the structure of the signals on the cable plant. For example, if the strength of the signal doubles, then a station knows that another station has begun transmitting, and the algorithm known as *collision detection* is invoked to enable the network to recover from this error.

Figure 2-5 shows two stations transmitting at the same time, with a collision occurring. When a collision occurs, each of the stations that are transmitting simultaneously (i.e., that are involved in the collision) will continue to transmit for a small length of time (4 to 6 bytes more). This is to ensure that all stations have seen the collision. All stations on the

Figure 2-5
Ethernet with collision.

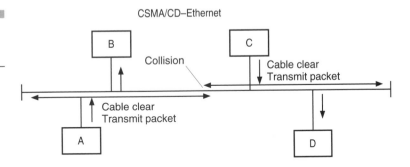

1. Node A receives data to transmit to Node D.

2. Builds a packet.

3. Checks to see if the cable plant is clear (no one else is currently transmitting).

4. Transmits packet while listening to the cable.

5. Node C accomplishes steps 1–4 and starts to transmit.

6. Collision between stations A and C.

7. All stations invoke the backoff algorithm.

8. All stations are free to gain control of the cable plant.

network will then invoke the collision backoff algorithm. The algorithm will generate a random number that will be used to indicate the amount of time to defer any further transmissions. This generated time should be different for each station on the network. Although it is remotely possible, no two stations should generate the same number. To ensure this, the algorithm takes into account the number of previous collisions and other factors. Therefore, all stations should defer transmission for different amounts of time. This prevents two stations from deferring transmission for the same amount of time, thereby reducing the possibility of another collision.

It is important to note that after the collision, all stations will back off. Thereafter, any station may try to gain access to the cable plant. The stations involved in the collision do not have priority. All stations have equal access to the cable plant after a collision.

Ethernet Reception.   Packet reception by Ethernet is just as simple. The receiving station will retrieve all the data being transmitted on the cable plant. Since Ethernet is a broadcast transmission medium, all stations on the cable plant (those not separated by a bridge or a router) will receive all packets transmitted on the cable. This does not mean that all stations will process each packet. The receiver will check to verify a minimum packet size (64 bytes). It will then check to see if the address (addressing is discussed later) from the received packet matches that of its controller board. If the addresses do not match, the packet is discarded and the station will wait for the next packet. The user will not see this process.

If the addresses match, the receiver will check the packet for errors by checking the cyclic redundancy check (CRC) field of the packet. This is a field used to check the validity of the data packet. If there are any errors within the packet, the whole packet will be discarded. If there are no errors, the receiver will check to make sure it is not over the maximum length (1518 bytes, which includes the Ethernet headers and the CRC). If this packet is okay, it is moved off the NIC and into the host computer's memory. The upper-layer software is notified that information for it is waiting to be retrieved. Only the data portion of the packet is handed to the upper-layer protocols (network layer and above). Ethernet will strip off its headers (the addressing) and trailers (the CRC field) and submit the data portion of the packet to the upper-layer software for processing.

If any errors were detected at the data-link layer, the packet is discarded. When a packet is discarded, the receiver will not notify the sender that it has discarded the packet. For that matter, the receiver will not inform the sender that the packet was received in good condition. Once the transmission is completed, Ethernet's job is done, and it does not care what happened to the transmission. Ethernet is known as a *connectionless* protocol. Its main function is to deliver packets to the network and to retrieve packets from the network. It cares only about transmitting and receiving packets. Packets sent and received are not acknowledged at the data-link layer. For this reason, Ethernet is also known as best-effort delivery system or a *probabilistic network.*

The upper-layer software has the responsibility to ensure that the data are received in good condition with no errors and in the same order as sent. These functions will be discussed fully in the following chapters.

Figures 2-6 and 2-7 show the algorithm flow for Ethernet packet transmission and reception. The flow is the same for Ethernet, Fast Ethernet, and Gigabit Ethernet.

This is the simplest way to describe Ethernet. Ethernet has undergone many enhancements over the last 17 years, but the algorithm is still the same. Check the cable plant, transmit if the cable plant is quiet, check for collisions during the first 64 bytes of transmission and for collisions after 64 bytes (known as late collisions), and return to the listening state after the transmission is completed.

All the other stations that were in deferred mode will also be in listen mode. After the cable plant has been quiet for a specified amount of time (9.6 μs), the cable plant is completely free for any station to try to transmit (the algorithm is started over again for all stations).

Fast Ethernet

When Ethernet was invented, the types of devices that were considered for this environment were minicomputers connected together along with terminal servers. The traffic load was very light, because the packets contained primarily one character that was being transferred between the terminal server and the minicomputer.

Today, most user workstations are high-speed personal computers capable of running network-intensive types of programs between multiple devices on the network. The personal computer now uses an internal bus (the pathway used to transfer information between the different components of the PC), which is a very high speed bus. Not only are standard data being transferred, but video and graphical data are being transferred as well. Multimedia applications, high-powered database

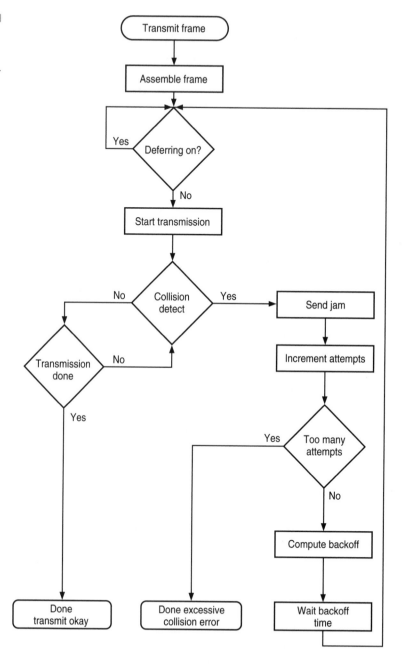

Figure 2-6
Ethernet transmission
flowchart.

applications, and information service connections to such networks as the World Wide Web (WWW) are more commonplace than ever before. Because of this, the Ethernet is easily overwhelmed even with proper network design and segmentation. It was estimated that in 1994 more

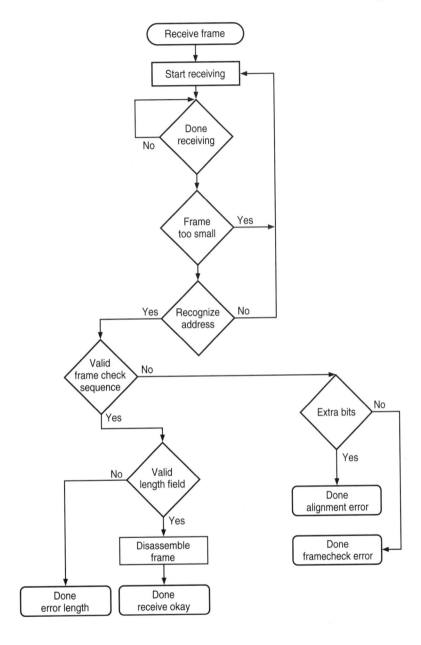

Figure 2-7
Ethernet reception
flowchart.

than 40 percent of all personal computers in the world were connected to a LAN. Furthermore, the average number of users per LAN is estimated to be 21 in 1994 compared to 12 in 1990.[1] Personal computer technology now includes graphical devices such as the Sun SPARC workstation, the PowerPC from Apple Computer, and Pentium-based personal computers. The network technology must be able to keep pace with the types of stations that are being attached. The technology behind Fast Ethernet easily allows for this migration without the high cost and tremendous work involved with network design. Thus, 100BaseT was developed, which provides 10 times the speed of Ethernet at less than twice the cost.

Several other performance upgrade technologies compete for the LAN owner's attention, such as FDDI over UTP, switched Ethernet, and full-duplex and full-duplex switched Ethernet. FDDI and its unshielded twisted-pair version, FDDI over UTP, conform to today's standards. Many FDDI and FDDI-over-UTP products are available. They are suitable for file servers, high-end graphics workstations, and backbone connections, but they provide more capability than most desktops require. Most desktops do not require large, high-powered workstations such as file servers and engineering stations (e.g., Sun SPARC).

The idea that Ethernet could be enhanced started in the early 1990s, with proposals being presented around late 1992. A new group evolved out of the IEEE 802.3 committee, called the IEEE 802.3u working group. The focus of the working group was directed at standards efforts: 100BaseT, which is based on the CSMA/CD algorithm, and 100VG_Any-LAN, which is a demand priority approach. 100VG-AnyLan was moved to the jurisdiction of the IEEE 802.12 subcommittee for further development. 100VG-AnyLAN completely changes the functions of the Ethernet MAC—so much so that the final result is not Ethernet, but a form of "token Ethernet." As of this writing, only one vendor is supporting this standard. Basically, it has been pushed aside in favor of the 100BaseT standard.

100BaseT (also known as Fast Ethernet) is a 100-Mbps version of 10BaseT. It was standardized by the IEEE 802.3 committee in June 1995 and uses the same algorithm as Ethernet (CSMA/CD). It is supported by over 60 Ethernet vendors through a consortium known as the Fast Ethernet Alliance. The packet format, packet length, error control, and management (a collection of functions known as MAC, or Media Access Control) is the same as for 10BaseT. The topology design for 100BaseT is the same as for 10BaseT, the star topology. In fact, it can use all the same wire types as 10BaseT (CAT3, 4, and 5). Furthermore, there is a standard for using 100BaseT over fiber, which is known as 100BaseFX. There are some differences between the two, which are described later.

International Data Corporation.

The controller cards that are currently being developed are dual-speed Ethernet controllers. These controllers can operate at either 10 or 100 Mbps. Some of the controllers are self-sensing, meaning that they can detect whether they are connected at 100 or 10 Mbps. This is known as autonegotiation. This occurs during powering up of the card and is completely transparent to the 10-Mbps system.

In 10BaseT, there is a function known as pulse signaling. Link pulses between the Ethernet controller and the hub port are transmitted every 16 ms when there is no transmission. They are used to signal link integrity. 100BaseT systems can sense this signal and convert operation to 10BaseT mode.

100BaseT builds on this concept by providing fast link pulse (FLP) bursts. A device that supports this feature bursts fast link pulses during startup, device reset, or through a network management command. Each side of the connection must support FLP in order for it to work. Encoded in the FLP is a word that describes to the other end the highest mode of operation that the link partner can operate in. The two sides of the connection must match on the encoded word before the link comes up. The modes of operation are as follows:

- 10BaseT two-pair category 3
- 10BaseT full-duplex two-pair category 3 (full duplex follows the IEEE 802.3x recommendation)
- 100BaseTX two-pair category 5
- 100BaseTX full-duplex two-pair category 5
- 100BaseT4 four-pair category 5

To summarize, 100BaseT systems provide transparent interoperability with 10BaseT systems through the process of autonegotiation. A 100BaseT system starts with the FLP to negotiate with its link partner a common mode of operation. If the remote end does not respond to the FLP or responds with the 10BaseT link pulse test, the partner is assumed to be a 10-Mbps interface and the 100BaseT controller will operate as a 10BaseT interface as well. Self-sensing is a very important feature, especially when moves, adds, and changes must be made to a network. With self-sensing you can simply put the 10/100 hub or switch in the closet and put 10, 100, or 10/100 cards anywhere on the network; the devices in the wiring closet will figure out what is being connected to them and select the appropriate speed. Without this capability, manual intervention is required for moves, adds, and changes.

The cards that do not support self-sensing offer a jumper or software setting. Implementing the 10BaseT standard on the same card that supports 100BaseT allows the controllers to be connected to existing 10BaseT hubs. This provides a migratory path to Fast Ethernet if the network currently uses 10BaseT but in the future will use Fast Ethernet.

The differences between Fast Ethernet and Ethernet exist primarily in the physical layer technology (including repeater hubs) and are noted as follows:

- 100BaseTX—specified for use with CAT5 UTP cable and type 1 STP cable

- 100BaseT4—specified for use with four-pair CAT3, 4, or 5

- 100BaseFX—for use with fiberoptic cable and CAT5 UTP

100BaseTX uses the same signal encoding scheme used by FDDI PMD, standardized by the American National Standards Institute (ANSI X3T9.5). It is based on the ANSI FDDI physical media—dependent (PMD) sublayer, and it provides support for the twisted-pair PMD (TP_PMD) specification (which uses two-pair CAT5 wire) and the fiber specification as well. The encoding scheme is the same as that of FDDI, which is the 4B/5B scheme. The coding scheme runs at 125 Mbps (which yields 100 Mbps). This is in contrast to 10BaseT encoding, which is 20 MHz Manchester and allows for 10 Mbps.

To address the needs of the large installed base of CAT3 (voice grade) wire, the 100BaseT4 standard was created. It uses four pairs of wire in the cable and provides those networks that have previously installed CAT3 wire with the capability of upgrading to Fast Ethernet. This specification also changed the signal coding scheme from Ethernet's Manchester encoding to 8B/6T. Three of the wire pairs are dedicated to transmission/reception, and the fourth wire pair is used for collision detection. 100BaseT4 does not support full duplex, but it does support autonegotiation.

The schematic for 100BaseT4 (see Fig. 2-8) should look familiar. It still uses pins 1, 2, 3, and 6, with pins 1 and 2 used for transmission and pins 3 and 6 used for reception. However, pins 4/5 and 7/8 have been added. These are bidirectional pins, and they can operate in either receive or transmit mode but not both at the same time. You should now see why

Figure 2-8
100BaseT4 pinout.

100BaseT4 cannot operate in full duplex. Either it is using three pairs to transmit or three pairs to receive. 100BaseT4 has added more wire pairs and a new encoding scheme, and it has increased the clock speed from Ethernet's 20 MHz to 25 MHz, all of which now allows for 100-Mbps transmission and reception.

100BaseFX is used primarily for fiber but it is specified for CAT5 cable as well. It is very similar to 100BaseTX. It supports half- and full-duplex modes of operation. It does not support autonegotiation. The difference is that 100BaseFX continually sends a stream of idle characters when there is not data to be sent. When data are to be sent, the idle characters are replaced with a start-of-frame delimiter (SFD). When the transmission is completed, idle characters are again transmitted. This allows of for continuous monitoring of the link, but it disables the autonegotiation. 100BaseFX has the added advantage of allowing 2 km between DTE stations (between two 100BaseT switches, for example) when the mode is full duplex.

The characteristics of the three standards are summarized in Table 2-1.

Hub Support. Like 10BaseT, 100BaseT requires the support of a wiring concentrator (a back-to-back connection between two devices works for two stations only). The hub can accommodate 10BaseT interfaces but not directly. Most hub vendors supply a separate backplane for 100BaseT interfaces. There are two types of repeater interfaces:

- Class I—translational repeaters. Class 1 repeaters support all media types for 100BaseT. This type of repeater produces an inherent delay to support the possible translation between media types. This delay restricts some of the topology diameters (shown in Table 2-2). It also restricts the number of repeaters allowed. Only one class 1 repeater is allowed in a single collision

TABLE 2-1 Characteristics of Fast Ethernet Standards

Type	Medium	Autonegotiation Support?	Continuous Signaling	Coding Scheme	Full Duplex Support?
100BaseFX	Fiber	No	Yes	4B/5B	Yes
100BaseTX	2-pair copper	Yes	No	4B/5B	Yes
100BaseT4	4-pair copper	Yes	No	8B/6T	No

TABLE 2-2

Network Length
Limits for 100
BaseT

100BaseT Medium Type	Maximum Network Diameter, m	Maximum Segment Length, m
Copper only (TX)	200	100
Fiber only (FX)	272	136
Multiple copper (TX) and one fiber FX	260	100 (TX)
		160 (FX)
Multiple copper (TX) and multiple fiber (FX)	272	100 (TX)
		136 (FX)

domain. However, most hub manufacturers allow their hubs to be stacked or multiple cards to be inserted into one hub, which still represents a single class 1 repeater.

■ Class 2—transparent repeaters. Class 2 repeaters support only one media type. It can be any of the 100BaseT types, but only one is allowed per repeater. Class 2 repeaters have low delays because they do not have to translate between media types. Therefore two repeaters are allowed in a single 100BaseT collision domain.

Table 2-3 provides a comparison between Ethernet and Fast Ethernet. Figure 2-9 summarizes the 100BaseT topology.

TABLE 2-3

Comparison of
Ethernet and Fast
Ethernet

Characteristic	Ethernet	Fast Ethernet
Speed	10 Mbps	100 Mbps
IEEE standard	IEEE 802.3	IEEE 802.3u
Algorithm	CSMA/CD	CSMA/CD
Physical topology	Bus or star	Star
Cable type	Coaxial, UTP, or fiber	UTP, STP, or fiber
UTP cable type	CAT3, 4, or 5	CAT3, 4, or 5
Interface type	AUI	MII
Full-duplex support	IEEE 802.3x	On all but 100BaseT4

Figure 2-9
100BaseT topology.

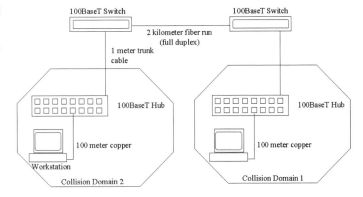

Gigabit Ethernet Physical Layer

Copper is not standardized for Gigabit Ethernet. In the standard finalized in the summer of 1998, fiber is the only cable recommended. (See Table 2-4.)

Token Ring

Token Ring is probably the oldest ring access technique. Supposedly it was originally proposed by Olaf Soderblum in 1969. The IEEE version of the Token Ring access method has become the most popular ring access technique. There are proprietary ring access techniques used by different manufacturers. None of these have been adopted by the IEEE.

TABLE 2-4 Topolgy comparison for Ethernet. *(Courtesy of Gigabit Ethernet Alliance)*

	Ethernet 10BaseT	Fast Ethernet 100BaseT	Gigabit Ethernet Goals 1000BaseX
Data Rate	10 Mbps	100 Mbps	1000 Mbps
Cat 5 UTP	100 m	100 m	100 m
STP/coax	500 m	100 m	25 m
Mulitmode fiber	2 km	412 m (half-duplex) 2 km (full-duplex)	500 m
Single-mode fiber	25 km	20 km	3 km

Although there is still one cable plant and multiple stations needing access to this cable plant, operation of the Token Ring access method is completely different from that of the CSMA/CD algorithm used for Ethernet (IEEE 802.3). On a Token Ring network, a formatted 24-bit (3-byte) packet (shown in Fig. 2-10) is continuously transmitted on the ring. This packet is known as the *token*. The packet contains three 8-bit fields: the starting delimiter (SD), access control (AC), and the ending delimiter (ED). These fields will be explained shortly. With a few exceptions, any station that has received this token and has data to transmit may transmit onto the ring (the cable plant). That station first captures the token and then transmits data to the cable plant. All other stations must wait for the token. When the station has received its original transmission back, it builds a new token and releases it to the network. No station on the ring may transmit unless it has captured the token.

Token Ring provides a tremendous array of built-in management techniques that constantly monitor the controller and the ring. When power is applied to a Token Ring controller, it begins a five-phase initialization routine. Any error in this process will disable the controller from entering the ring.

PHASE 0. A lobe test is performed in which the controller board submits packets to the cable (known as the lobe cable) attached to it to see if it receives frames back. The controller does not insert itself onto the

Figure 2-10

Token Ring frames.

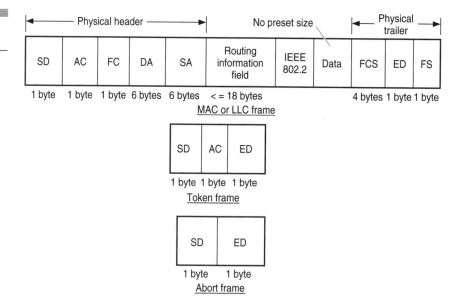

ring (flip the relay in the MAU). The cable is looped back at the MAU, and any packet transmitted should be immediately returned in the same format as the transmission. If this test succeeds, the controller enters phase 1.

PHASE 1. The controller produces the necessary signal (the phantom voltage) to flip the relay in the MAU to insert itself onto the ring. Flipping the relay causes an interruption and produces electrical noise on the ring. This causes an error on the ring that causes the token or any transmitting station's data to be lost. In other words, any station that inserts itself into the ring will cause an interruption on the ring.

A special station on the ring known as the active monitor enables the ring to recover from this error and puts a new token on the ring. Once this is accomplished, the controller waits for a special frame that it knows must be present on the ring. Once the frame has been found, it knows that the ring is active. In the case that no such frame is found on the ring, the controller will assume that it is the first station on the ring and will insert the frame itself and wait for it to return. If this test succeeds, the controller enters phase 2.

PHASE 2. The controller board will transmit one or two frames with the source and destination addresses set to its address. This is called the *duplicate address test* and it is used to check if any other controller has its address. If the packet returns with the address-recognized bit set, the controller will remove itself from the ring. If the frame returns with the address-recognized bit not set, the controller enters phase 3.

PHASE 3. The controller tries to find its neighbor by waiting for certain control frames to pass by. It will also identify itself to its downstream neighbor. In a ring environment, each active station repeats signals it received to the next controller on the ring. In this way a network station can identify who is downstream of it. Likewise, the new station will be identified to its upstream neighbor, for it will repeat data to it. Keeping track of the downstream neighbor is an important network management facility for Token Ring. When a station is added to or deleted from the ring, any station can report this occurrence to the network management on the ring. If this test succeeds, the controller enters phase 4.

PHASE 4. The controller requests its initialization parameters from a station on the network known as the ring parameter server (RPS). The

RPS resides on each ring. It sends initialization information to new stations attaching to the ring and ensures that stations on the ring have consistent values for operational parameters. In this request packet to the RPS will be registration information from the newly attached station. This information is the individual address of the ring station's next available upstream neighbor (NAUN), the product instance ID of the attached product, and the ring station's microcode level. The RPS parameters include the ring number of the attached ring. If there is not a server present, the initializing controller will use its default parameters. The servers on Token Ring are discussed at the end of the next chapter.

Provided there were no errors during this initialization process, the network station is now active on the ring and may transmit and receive data as discussed in the following paragraphs. Even after initialization, the controller has the capability to take itself off of the ring if there were too many errors.

Operation.   Basically, a Token Ring controller may be in one of three states: repeat mode, transmit mode, and copy mode. When a Token Ring controller does not contain any data that need to be transmitted on the network, it stays in a mode known as *normal repeat mode*. This allows the controller to repeat any signals to the next active station on the network.

When the controller has data that need to be transmitted on the network, it must wait until the token frame comes around to it. Once the token is presented, the controller makes sure that the token has not been reserved by another station. If it has not been reserved, the controller captures the token (takes it off the ring) and transmits its data to the ring. With no token on the ring, no other station can transmit.

Each station on the ring receives this newly transmitted information. Upon receipt of the SD field, the network station looks further into the packet to find an indicator that the receiving station either sent the packet or should receive the packet. If the network station did not originally send the packet but the packet is destined for it (i.e., the destination address in the packet indicates that the network station is the intended recipient), the network station continues to copy the frame. If the frame is not destined for it, it simply repeats the frame back onto the ring, unaltered, without continuing to copy.

The station that originally submitted the packet is the *only* station that may take the packet off the ring. This is called *stripping*. The destination station merely copies the frame as it repeats it back onto the ring. In case of an error that prevents the originating station from taking the packet off the ring, a special monitor on the ring known as the active monitor notices that this packet has traversed the ring more than once and takes the packet off the ring.

After the originating station takes the packet off the ring, it submits the token back on the ring. The token frame then circulates around the ring for the next station that has data to transmit.

Whereas the Ethernet specification states that the network will operate at 10 Mbps, there is no specification for clocking on Token Ring. Currently, IBM is supporting two speeds for its Token Ring network. These are 4 Mbps and 16 Mbps, commonly known as the 16/4 standard. The original Token Ring network ran at 4 Mbps. In 1989 IBM began to ship the 16-Mbps controllers. Some words of caution are necessary here. A 16-Mbps station will not operate on a ring where the stations are running at 4 Mbps (called the ring speed). The 16/4 controller can be switched and will operate at 4 Mbps. The switch from 16 or 4 Mbps is usually software selectable with a utility program shipped with any Token Ring controller. The 4-Mbps controllers will not operate on 16-Mbps rings. The 16/4 controller is the only controller that may operate at either 16 or 4 Mbps. The 16- and 4-Mbps rings may be intermixed (separated by individual rings running at the same speeds) through a special device known as a *bridge* or a *router,* which will be covered later. Token Ring frames are shown in Figs. 2-10 and 2-11.

Figure 2-11
Token Ring frame fields.

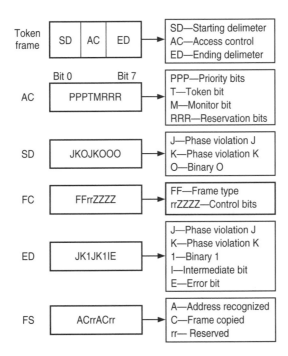

Essentially, there are four types of frames (IBM uses the term *frame* instead of *packet*). The four types are LLC (logical link control, data), MAC, token, and abort frames. A MAC frame is used to maintain the ring. The MAC frame performs housekeeping chores on the ring such as purging the ring, sending information to one of the monitors, detecting errors, and finding a neighbor. MAC frames are generated by the MAC on the Token Ring NIC. The abort frame is used by a station to indicate to all other stations that its last transmission should be aborted. The data frame is used to transmit and receive application data on the ring.

When a Token Ring network station receives the SD field of an incoming frame, it knows that a frame is approaching, but it has no idea what type of frame. The receiving station continues to read the frame. The next field is the AC field, shown in Fig. 2-11, which contains some bits that provide more information to the controller. If the T bit is set, this indicates to the network station that the frame is a token frame (no data is associated with a token frame). Notice also in this figure the priority (P) and reservation (R) bits. If the network station wants to transmit a packet, it reads the P bits, which indicate the current priority of a token frame. If the priority bits are equal to or less than the network station's priority bits, a frame may be transmitted. If the P bits are higher, the network station may "reserve" the token by setting the R bits, indicating that it has the next shot at it. The M bit is set by the active monitor to indicate that this frame has passed by an active monitor.

If the T bit is not set (indicating that the frame is a LLC/data frame), the next field is the frame control (FC) field. If the T bit is set (indicating that the approaching frame was a token frame), the next field is the ED field. The FC field indicates whether the frame is a data frame (LLC frame) or a MAC frame. A MAC frame is used for network management purposes such as reporting errors and duplicate address testing. The MAC frame types are represented by the Z bits. The r bits are reserved.

The next fields are the destination and source address fields, which indicate the 16- or 48-bit address of the sender and the 16- or 48-bit address of the intended recipient of the frame. The receiving controller reads the destination field to determine if the packet is intended for it. If the destination address is not that of the receiving station, it simply repeats the rest of the packet back to the ring. If the destination address is its address, it copies the rest of the frame. The format of the source and destination address fields is discussed later.

The next field is the routing information field (RIF). This is used for source routing information and is explained in detail later. The

DSAP, SSAP, and control fields are used by the data link for controlling the frames on the LAN. This is fully discussed in Chap. 3 of this book.

The data field contains user data for a LLC frame and contains management information if it is a MAC frame. To distinguish between the two types of frames, the controller reads the FC field. If the FC field indicates a MAC frame, it will read the Z bits; otherwise, those bits are ignored.

The FCS is a 32-bit cyclic redundancy check and is used to maintain accuracy for the whole frame (minus the frame status field).

The ending delimiter (ED) field has special symbols, J and K, to indicate to the controller that the end of the frame is arriving. It has two other important bits: the I and the E bits. The I bit indicates to the recipient of this frame that it is an intermediate frame, and more data will immediately follow in another packet. In other words, the transmitting station is transmitting multiple frames while holding the token. I have yet to see this bit used. The E bit indicates that another station has found an error in the frame and that it should not be copied. This type of error involves an electrical signal. This frame should return to the sender to be "stripped." Stripping means that the station takes the packet off the ring and puts a token back on the ring.

The frame status (FS) field has two bits of importance. These are the address-recognized (A) and the frame-copied (C) bits. The A bit is used to indicate that a destination station recognized its address. The destination will set this bit to indicate to the sender that the address was recognized. This bit is also used during Token Ring controller initialization. When an initializing controller starts, it sends out a duplicate address frame. With this frame, the controller sets both the destination and source address to its physical address. If the frame returns with the A bit set, the initializing controller knows that another station on the ring has the same address. The initializing controller removes itself from the ring after notifying the ring error monitor.

If the C bit is set, a destination station has copied the frame. When the frame returns to the originator, it will know that the destination has copied the frame and, therefore, that it does not have to resend the frame. If the E bit is set, the A and C bits should not be set. The A bit can be set without the C bit, indicating that the destination station recognizes its address but could not copy the frame for some reason. The frame may also be returned with both the A and C bits set to a 0, indicating that no station is out there with that address, and therefore the frame was not copied.

Notice that this field has two A bits and two C bits. The reason behind this is that the FCS does not cover the frame status field. The FCS is set by the originator of a packet to be used by the destination station to ensure the integrity of the packet. Since the A and C bits are set by the destination station, the A and C bits are not covered (with a few exceptions, those bits should always be set to zero by the sending station). If the C bit is set and the A bit is not set, the frame is in error. The receiving station must recognize its address before copying the frame.

A ring topology operates as a closed loop. Therefore, maintenance of the ring is different from that in a bus topology such as Ethernet. A few errors can occur that require special attention. Maintenance to the ring is accomplished through the use of monitors. There are many monitors that run on a Token Ring network. All are used for configuration and status report information. The monitors may be separate devices on the network, but most work as an application of the NIC. The end user will not even know that the monitor is operating on the workstation's NIC. The following is a short listing and definition of the monitors.

ACTIVE MONITOR.   The active monitor resolves the following conditions:

Lost tokens

Frames and priority tokens the circle the ring more than once

Other active monitors on the ring (only one may be active at a time)

Short ring (a ring with such a low bit delay that it cannot hold a token)

Master clock control

RING ERROR MONITOR.   The ring error monitor observes, collects, and analyzes hard-error and soft-error reports sent by ring stations and assists in fault isolation and correction.

CONFIGURATION REPORT SERVER.   The configuration report server accepts commands from the LAN Network Manager to get station information, set station parameters, and remove stations on its ring.

RING PARAMETER SERVER.   The ring parameter server resides on each ring in a multiple-ring environment. It sends initialization information to new stations that are attaching to the ring, ensures that stations on the ring have consistent values for operational parameters, and forwards registration information to LAN Network Manager from stations attaching to the ring.

LAN BRIDGE SERVER. The LAN bridge server keeps statistical information about frames forwarded through a bridge and provides bridge reconfiguration capabilities.

Some of the servers operate on the Token Ring NIC, and some operate on devices such as a Token Ring bridge or router. All of the devices collect and store information. What retrieves this information is the IBM LAN Network Manager. This is a PC application that communicates optionally with NetView, IBM's host-based network management product. It can collect information from the NIC cards and it can communicate with the NIC cards as well. It may want to communicate with a NIC card to tell it to remove itself from the ring. It allows a network administrator to manage multisegment IBM Token Ring networks. It also provides facilities for managing the LAN media and LAN adapters in the network and for managing the bridges that interconnect the networks. Figure 2-12 shows the relationship between the servers, the ring stations, LAN Network Manager, and the SNA control point (NetView). You can think of this application as the server portion of the SNMP management system.

Figure 2-12
LAN reporting mech-
anism for Token Ring.

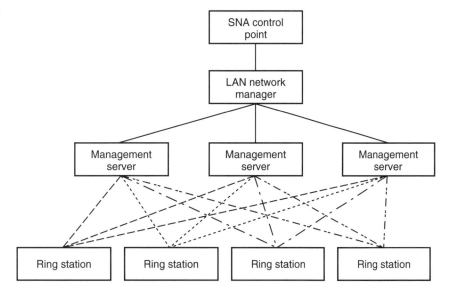

FDDI

The FDDI standard is actually of a set of standards as established by the ANSI FDDI is a token-based ring access method that allows stations to access a cable plant operating at 100 Mbps. It can connect up to 500 dual-attachment stations in a 100-km network. It differs from other ring access methods in that it is a timed-token protocol; that is, each station is guaranteed network access for a certain time period that is negotiated between all active stations upon startup and when a new station joins the ring.

The topology for FDDI is mixed. Logically, FDDI is a ring, but physically it is a point-to-point or a star connection. Without the use of a wiring concentrator, the physical connection is point-to-point. Each network attachment has a physical connection with another attachment. With the use of a concentrator, the physical topology turns into a star configuration, much like Token Ring. The main difference is that Token Ring requires the wiring concentrator whereas FDDI does not. Network stations in FDDI can be connected directly to each other. FDDI operates at 100 Mbps. Token Ring runs at 4 and 16 Mbps.

As shown in Fig. 2-13, an FDDI dual ring is composed of two rings (known as the primary and the secondary ring) that normally operate independently. Data can travel on both rings in opposite directions. Although data are allowed to travel on both rings, commonly data are transmitted only on the primary ring until a certain fault occurs, in

Figure 2-13
FDDI dual ring.

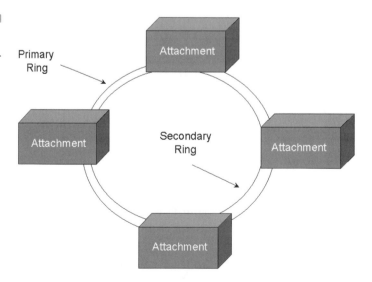

which the two rings may become one. When a fault does occur, such as a DAS station powering down or a cable break, and the two rings combine, this is known as wrapping.

FDDI allows stations to communicate over a dual-ring topology by guaranteeing access to the cable plant at timed intervals using a token. This may seem similar to the Token Ring architecture, but the operation of FDDI is different. Both are ring architectures, but the algorithms for the access methods are different.

A network station must wait for the arrival of the token before transmission may begin. Upon arrival of the token, the network station captures the token, and this stops the token repeat process (no token will now be on the ring, thus guaranteeing that no other station has access to the cable plant). A station transmits a series of frames to the next active station on the network (data flow in one direction only). Frame transmission continues (meaning that frames may be transmitted) until the token-holding timer expires or the station has no more frames to transmit. The station then releases the token to the ring.

The downstream neighbor receives these symbols, regenerates them, and repeats them to its downstream neighbor. When the frame returns to the originator, it is "stripped" from the ring. The destination station does not strip the frame. Only the originator of a frame may take it back off the ring.

FDDI is often compared to Token Ring, and sometimes it is called a faster Token Ring. This is not true at all, for FDDI is a timed-token protocol. The following discussion on timers will show the difference.

FDDI Ring Timers.   Proper ring operation requires

1. Physical connection establishment between two network stations
2. Ring initialization (the ability of the ring to continuously circulate a token without error)
3. Steady-state operation
4. Ring maintenance (dynamic ring management capability on the part of the stations on the ring)

In order to implement the timed-token protocol, a station must use certain timers. Each of the timers is used to control the scheduling process of transmitting frames. For the most part, they simply control the length of time during which a station may transmit before the token is released. They are also used to register the total time it takes for the token to travel around the network. Stations are locally administered; that is, each station has built-in management capabilities, and these enable the network station to dynamically manage the ring. Various timers in an FDDI network are used to regulate the operation of the ring.

TARGET TOKEN ROTATIONAL TIMER (TTRT). This is a ring latency parameter that sets the delay for the ring. This timer is set only during ring initialization. Special MAC frames are passed between the network stations called *T_Req* frames; these are requests for a token rotation time value. Each station submits a request for how fast it wants the token to rotate. That is, if a station wants to see the token once in X milliseconds, it will put a request in for T_Req = X ms. The station requesting the fastest T-Req frames will win, as it satisfies the requests of all other stations. The station requesting the lowest time wins because it is assumed that the application running on that station requires the token to circulate at the specified rate. Once this process is accomplished, all stations on the ring set the value to the lowest bid.

TOKEN ROTATIONAL TIMER (TRT). This timer measures the time between successive arrivals of the token—that is, how long it takes for the token to return once it has come through. It is initialized to the TTRT time. Every station has this timer, and it is locally maintained. If the token arrives within the TRT time, this timer is reset back to the TRT. If the token does not arrive within the TRT time, the network station increments a late counter and reloads TRT with the value of the TTRT. If this timer expires twice, the network station tries to reinitialize the ring.

TOKEN HOLDING TIMER (THT). This timer determines the amount of time that a station may hold the token after capturing it. This timer is loaded with the TRT time when a station captures the token. The TRT timer is reloaded with the TTRT to start timing the next token rotation even though this station has the token. When a station is holding the token, the network station may transmit as many frames as it can without letting the THT expire. The station must release the token before the expiration of this timer. The THT is loaded to the time remaining in the TRT every time the token arrives. Therefore, if a station receives a token with 10 ms left in the THT, it may transmit data for 10 ms.

VALID TRANSMISSION TIMER (TVX). This timer allows an FDDI ring to recover more quickly from the loss of a token. Normally, when twice the TTRT time elapses, the ring should reinitialize itself. Because the TTRT maximum value is 165 ms, in the worst case this could take up to 330 ms.

This timer allows the ring to initialize much more quickly, normally within 2.4 ms. Each station has this timer. It times the duration between valid frames and is reset upon receipt of a valid frame. This time is calculated from the time it would take to transmit a 4500-byte frame around a maximum-size ring of 200 km. This equals 2.5 ms.

The net effect of all the timers is that the token should never take more than twice the TTRT time to completely circulate the ring. If the TRT timer should exceed this, the ring will reinitialize because it will be assumed that the ring is no longer operational.

FDDI Frames. As indicated in Fig. 2-14, data to be transmitted on the FDDI ring must be encapsulated in frames, which contain very distinct fields. Unlike the case for Token Ring, there is not a master clock on the FDDI ring. Each station manages its own clock when transmitting frames. Each station's clock is independently generated by its hardware and has the possibility of being slightly different from the others on the ring. Therefore, for each station to adjust to the transmitting station's clock, FDDI uses a preamble field at the beginning of each transmitted frame.

The starting delimiter (SD) is a specially formatted field that contains indications of signal phase violations. These are mistimed signals that would normally cause an error. But the FDDI chipset has been

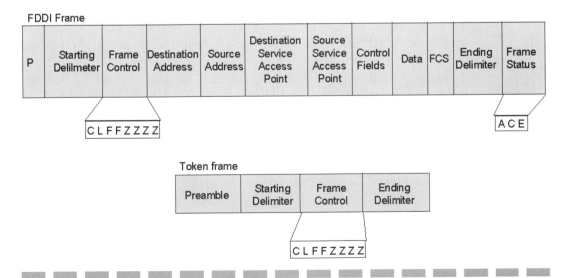

Figure 2-14 Transmitting data on the FDDI ring.

programmed to recognize two special violation flag patterns. The format of the SD field is static, and it contains the violation flags. It is set this way so that the receipt of the SD indicates the presence of a frame approaching.

The frame control (FC) file indicates what kind of frame is approaching. For instance, it could be a token frame, SMT frame, beacon, claim-token, or LLC (data) frame. This field's function is similar to that of the Token Ring FC field.

The destination address (DA) and source address (SA) fields can be set to 16 or 48 bits. This is indicated by the L (length) bits in the FC field. All stations on the ring must be set to 16 or 48. All FDDI implementations use 48-bit addresses. The 16-bit address is not used.

The ending delimiter (ED) indicates the end of a frame. Like the starting delimiter, it has signal phase violation indicators placed at certain locations so that the controller will recognize this field as the end of the frame.

The frame status (FS) field contains three important bits that are set by other stations on the ring. They are the address-recognized (A) bit, the frame-copied (C) bit, and the error (E) bit. The address-recognized bit is set by a station for one of two reasons: the station that received the frame recognized its address in a duplicated-address SMT (station management) frame check, or it recognized its address in a data LLC frame. But just because a station recognized its address does not mean that it copied the frame; it could have been too busy to do so. The C bit indicates that a destination station was able to copy the frame. This bit will not be set unless the A bit is set. The E bit indicates that some station (not necessarily the destination station) found a signaling error in the frame. If this bit is set, the C bit can never be set, because any frame found to be in error will not be copied. The A bit can be set without the C bit being set.

Connection Establishment. Upon powering up an FDDI controller, certain checks are conducted at the physical layer to ensure proper connection. The connection management (CMT) portion of station management (SMT) controls the physical connection process by which stations find out about their neighbors. Even though FDDI is a ring topology, the connections are really point-to-point. FDDI ports are physically connected in a one-on-one relationship with other attachments.

Confirmation of a good connection is accomplished by stations transmitting and acknowledging defined line state sequences. During this process, network stations exchange information about port types, connection rules, and the quality of the links between them. If the connection type is accepted by each end of the link and the quality link test passes, the physical connection is considered established. This connection test is accomplished by each point-to-point connection on the FDDI ring. Each station on the ring is considered to have a point-to-point connection with its directly connected neighbor. Each connection between two points accomplishes these checks. Once these checks are completed, the ring will initialize.

Ring Initialization. Figure 2-15 illustrates that after the station physical attachments have achieved a good physical connection, the ring must be initialized. Since FDDI uses a token to grant a station access to the ring, one of the stations must place the token onto the ring. This process is known as the claim process and is accomplished when

- A new station is attached to the ring.
- There are no other active stations on the ring.
- The token gets lost.

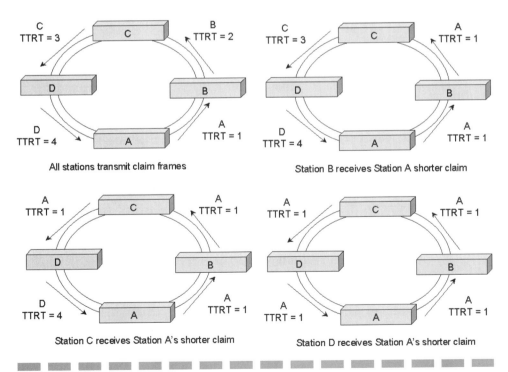

All stations transmit claim frames

Station B receives Station A shorter claim

Station C receives Station A's shorter claim

Station D receives Station A's shorter claim

Figure 2-15 FDDI token claim process.

Any station on the ring has the capability to initialize the ring, but a bidding process ensures that only one station initializes the TTRT timers. The station with the lowest TTRT wins the right to initialize the ring.

This process starts when the MAC entity in one or more stations enters into the claim state, in which it continually transmits MAC claim frames containing the station's MAC address and its TTRT bid. When another station on the ring compares a received claim frame with its own bid for TTRT, it may take one of two actions. If the time bid for TTRT in the received frame is shorter, the station quits transmitting its own claim frames and starts repeating the received claim frame. If the time bid for TTRT in the received frame is longer, the station continues transmitting its own claim rather than repeating the received claim frame.

The token claim process stops when a station receives its own claim frame back. This means that all other stations on the ring have conceded their right to initialize the ring to that station. The winning station issues a new token to the ring. If two stations bid the same TTRT, the station with the highest address wins.

Neighbor Notification and Duplicate Address Check. When an FDDI controller is initialized or the port on the concentrator is started or restarted, each network station must find out about its next available upstream neighbor (NAUN). This process, called the neighbor notification protocol (NNP), is similar to the active monitor present and standby monitor present algorithm used in Token Ring.

There are two times when this algorithm is invoked: when a station first becomes operational on the ring and every 30 seconds thereafter. The algorithm is as follows. A station transmits a management frame, called a neighborhood information frame (NIF). Inside the NIF are two fields called the upstream neighbor address (UNA) and the downstream neighbor address (DNA). Using these frames, a network station not only can find the addresses of the stations on its ring, but can check for a duplicate of its own address. The other purpose of the protocol is to verify that the transmit and receive paths are functional.

Say, for example, that there are three stations on the ring: X, Y, and Z. Station X transmits the NIF NSA (next station addressing) request with the UNA and DNA fields set to unknown. Station X's downstream neighbor, station Y, receives this NIF request, sets the UNA field to X, and transmits a NIF NSA response. Station Y also transmits a NIF NSA request, with UNA set to A and DNA set to unknown. Station Y now knows that its upstream neighbor is X.

Station Z receives station Y's request, sets the UNA to Y, sets the DNA to unknown, and transmits this as a response NIF NSA. Station Z now knows that its UNA is Y. Immediately station Z transmits a NIF request with UNA set to Y and DNA set to unknown.

Station X receives station Y's response and now knows that its downstream neighbor is Y. Station X also receives station Z's NIF request and now knows that its UNA is Z. Station X transmits a NIF response with the DNA field set to Y and the UNA field set to Z. Station Y receives station Z's NIF response and now knows that its DNA is Z. Station Z receives station X's NIF response and now knows that its DNA is station X. At this point, each station knows its UNA and DNA.

During this operation, all stations check for the A bit setting. This indicates whether a station recognized its address. Because these types of frames have the DA and SA set to the originator, the A bit should never get set. If it does, there is another station out there with the same address. If a duplicate address is detected, the LLC services in the station that recognized the duplicate are disabled. The SMT services are still active. The LLC services will be reenabled once the duplicate address problem has been resolved by manual intervention.

Now that the ring has been initialized, the first time a new token circulates the ring, it cannot be captured by any station. Instead, each station that receives the token will set its own TTRT to match the TTRT of the winning station. On the second pass of the token, network stations may transmit synchronous traffic. On the third pass of the token, station may pass asynchronous traffic.

Synchronous and Asynchronous Data Transmission. There are two types of transmission on FDDI: synchronous and asynchronous. Synchronous traffic is reserved bandwidth that is guaranteed to a network station holding the token. This service is used primarily for voice and video applications. These types of frames are always transmitted before any asynchronous frames. Support for synchronous traffic is optional, and it is generally not used.

Asynchronous traffic is the most common mode of operation available to network stations. It is a service class in which unreserved bandwidth is available to the station that has captured the token. Asynchronous traffic can be further classified as restricted and nonrestricted. Nonrestricted asynchronous allocation supports eight priority levels. Asynchronous transmission allows a ring station to use a timed-token protocol. This allows for dynamic bandwidth allocation. Restricted asynchronous transmission allows extended transmissions to exist between two or more stations on the FDDI ring. It allows ring stations to "hog" the cable for their transmissions. This requires a station to see the token to indicate that the ring is restricted for its use. Restricted asynchronous mode is optionally supported on FDDI.

These two service classes, synchronous and asynchronous, should not be confused with the asynchronous and synchronous services that mainframe or minicomputers and their terminals use to communicate.

Normal Operation. Once all the verifications and tests have been successfully completed, an FDDI ring may start normal data transmission, or steady-state operation. Stations may transmit frames to each other using the rules of the FDDI protocol. The ring will remain in steady state until a new claim process is initiated.

Frames are transmitted on a ring using the timed-token protocol. This protocol is a series of steps that the network station must perform before data transmission is allowed on the ring. A network station wishing to transmit must wait for the arrival of the token. Upon the token's arrival, the network station captures the token, stopping the token repeat process. The TRT timer is loaded into the THT timer, and the TRT is reset to the value contained in TTRT.

A station transmits a series of symbols (data bytes), which when combined form frames, to the next active station on the network. Frame transmission continues until the THT expires or the station has no more frames to transmit. For example, if a station receives the token and its THT has 10 ms left, the station will transmit asynchronous frames, in order of priority, until the THT expires. The station then releases the token to the ring and resets the THT.

Every other active station on the ring receives the transmitted frame and checks it for the destination address. If the destination address of the received frame is not its own, it will check for signal or FCS errors and repeat the frame to its downstream neighbor. If the destination address is its own, it copies the frame into its receive buffer, sets certain control bits in the frame to indicate to the sender that it has copied the frame and recognized its address (the A and C bits in the FS field), and repeats the frame to its downstream neighbor.

Any other downstream station that receives this frame checks it for errors and repeats the frame. If any station detects an error (usually a signal or framing error), it can set the error bit (the E bit in the FS field) in the frame. If the destination station has not copied the frame and detects the error bit set, it may set the address-recognized bit and the E bit and repeat the frame.

This process continues until the frame circles the ring and the station that originally transmitted the frame removes it from the ring, a process known as stripping.

Stripping and Scrubbing FDDI. To minimize delays for data to pass through the controller, each network station reads and repeats each frame, bit by bit, as it receives the frame. As shown before, the frame contains destination and source MAC addresses to identify what station set the frame and what station the frame is intended for. Under normal circumstances, the station that transmits a frame is the only one that is allowed to remove it from the ring. When a station recognizes its own address in the source address of the frame, it knows that it must strip the frame from the ring. But by the time it recognizes the frame as its own, the frame header and destination address have already been repeated to its downstream neighbor, leaving a fragment of the original frame on the ring.

To ensure that the ring does not deteriorate because of this frame fragmentation, each FDDI station contains the ability to remove these fragments from the ring through SMT. They are either removed by the downstream neighbor or by the operation of the repeat filter in each active station's PHY. This process is called *scrubbing*.

There is no standard procedure for scrubbing the ring of unwanted or orphaned frames. The process of scrubbing is enacted to keep the ring from deteriorating due to stray frames. This can happen when a new address joins the ring or leaves the ring. If a station has transmitted a frame and leaves the ring before it can strip it, it is the responsibility of another station to remove this stray frame. To do this, a station will send a series of idle frames to the ring. At the same time, the MAC entity of the FDDI controller card will strip the ring of frames and token. This will force other stations to enter into the claim process. The scrubbing process ensures that any frames on the ring originated after the ring was scrubbed. It eliminates the possibility of stray frames continually circulating the ring.

FDDI Station Management (SMT). This section introduces some of the management techniques imposed by the FDDI specification. It is not intended to provide a complete understanding of all the management functions; its purpose here is only to expose you to FDDI management.

Management within FDDI is called station management (SMT). FDDI provides standardized management functions that are present in every network attachment on a ring. These cover a broad range of items, from signaling errors to lost tokens. The paramount idea behind the management functions is to enforce a standard FDDI management platform. All FDDI controllers contain the same management entities and all controllers function the same (with respect to management). SMT is used only on a ring-by-ring basis. That is, the management functions are performed locally. The management frames do not cross bridges or routers. SMT provides for the integrity of the ring as well as access to the network management services.

SMT not only manages a network attachment internally, but produces frames on the network as necessary. Neighbor notification, duplicate address flags, and invalid port type connection notifications are examples of the SMT frames that may be generated on an FDDI network.

1. Connection management (CMT) operates at the link level (the physical link between two stations) and consists of three components:

    Physical connection management (PCM)

    Configuration management (CFM)

    Link error monitoring (LEM)

2. The network attachment level consists of two components:

    Configuration management (CFM)

    Entity coordination management (ECM)

3. Ring-level management consists only of ring management (RMT).

4. Frame-based management consists of a series of frames that allow remote management of the ring stations over the network.

PHYSICAL CONNECTION MANAGEMENT (PCM). PCM initializes the duplex connection between two stations. Each side negotiates with the other until both ports agree that a connection will be allowed. PCM has four functions:

1. It performs synchronization of the link between two stations. Certain signals are passed between the ends of the connection and are used to ensure that each end is willing to have a connection established based on a negotiated sequence of requests.

2. One of these requests is to identify the port types on each end of the connection. This is useful for removing undesirable links. For example, the connection from port A to port A is allowed, but it can produce unknown results in the ring.

3. Another request is to perform a link confidence test (LCT) and ask for the time period of this test. The two parties in the connection pass a series of frames between each other and monitor the results. This test is performed prior to connection establishment to isolate faulty physical links. This test is also accomplished after the connection has been established as a postconnection test, to ensure that the connection does not degrade over time. A faulty link could be a link that, when tested, produces too many errors.

4. PCM finds the identity of the downstream and upstream neighbors (it identifies the MAC address of each end of the link). The two sides of the connection will pass tokens and neighborhood information frames (NIFs) before the port is placed onto the operational ring. By finding the MAC addresses of the UNA and DNA, this test can be used for security reasons to enable only certain stations to enter the ring.

CONFIGURATION MANAGEMENT (CFM). CFM is used for monitoring the port or the MAC. Each port contains a configuration control element that controls its state. It controls the insertion or removal of a port or MAC on the active ring. It not only monitors the link at initialization, but provides continuous monitoring during port operation. It performs the interconnection of the PHYs and the MACs within a node. Remember that FDDI operates on a primary and secondary ring, and it also operates in a mode known as *local* (basically, a port in a loopback that is issued for diagnostics and monitoring of the port before it is placed on the active ring). CFM is the entity that places a station on a requested path. CFM also configures the node in the event of a fault and can place a port in the wrap state.

LINK ERROR MONITORING (LEM). Link error monitoring is part of PCM and its functions are spread over PMD, PHY, MAC, and SMT in an FDDI network attachment. It is provided by the FDDI hardware. For example, PMD detects a loss of electrical signal, PHY detects and accumulates line state violations and clock rate problems, and MAC detects FCS errors. Each entity of the FDDI station has the responsibility to monitor and detect errors on its active links. Information gathered by these monitors is compared to set thresholds.

Depending on the accumulated information and when the errors occur (for example, during initialization or when the ring is considered up and running), the link may be determined to be faulty and could be disabled, leading to a wrapped condition. The accumulated information might not take a link down but may consider it marginal and notify the ring of this station's condition through the use of SMT MAC frames.

ENTITY COORDINATION MANAGEMENT (ECM).   The entity coordination management function can start and stop any or all PCM functions. For example, it is this entity that will allow a PCM to initialize a connection setup between two stations. Its purpose is to act as an interface to SMTs requests of the individual PCMs.

In the event of a beacon condition, ECM on a network station stops all activity to the PCMs and activates the trace signal to detect errors. An internal test occurs on this network station. If the station passes its test, then after the trace signal is sent ECM reinitializes all other SMT components in the network station.

ECM controls the optical bypass switch, which is used when the network station's ports are shut down for any reason. Because the node is not inserted as an active repeater in the ring, the optical bypass is functionally a mirror that reflects an incoming signal to the disabled network station's downstream neighbor without disrupting the dual ring.

RING MANAGEMENT (RMT).   The purpose of ring management is to control the MAC logic. The MAC standard specifies how the various frame types are built, including normal data frames, the token frame, management-specific frames, and so on. Maximum and minimum lengths for the frames are defined, along with methods of error checking, the timed token access protocol, destination/source addressing, and the various operational modes and conditions. Initialization of the operating parameters using the claim and beaconing mechanisms is also handled by the MAC.

The MAC is designed to control initialization, error detection and recovery, and duplicated address detection. RMT receives signals for the MAC chip indicating a condition change. RMT has four functions:

1. Upon receiving a signal from the CAM (which controls the state of the port) that a port has inserted itself into the ring, RMT causes the MAC chip to start the claim service or, if inserting caused the ring break, to beacon the ring.

2. After this, RMT checks for a duplicate address on the ring. MAC addressing on FDDI allows the MAC address to be supplied by a management function of the network station. Normally, the

FDDI controller uses the PROM address. To detect any other station that may be using the same address, this function is performed using the SMT NIF (neighborhood information frame).

3. During a hard error on the ring (an error that causes the ring to become disabled), stations send out beacon frames. This helps stations determine where the fault is (the MAC address of the station that is transmitting the beacon and its UNA are embedded into the beacon frame). If a station becomes stuck in this condition, it is RMT that tries to recover the network from this condition. RMT notifies the ring network manager (using a special multicast address) of the condition, using the trace feature.

4. The MAC monitors restricted tokens. A token may be restricted so that only certain stations may use it. This effectively stops all normal transmission of data. This option is rarely used, but the possibility exists. When the MAC on a network attachment receives a restricted token, it notifies RMT of this. RMT knows that a token can be restricted only for a certain amount of time. If this time limit is exceeded, RMT transmits claim-token beacon SMT frames that interrupt the restricted token and force the ring back to normal service. The default for the restricted token timer is 0. This means that no restricted token is allowed on the ring.

Frame-Based Management.   The aforementioned management entities are primarily for point-to-point management of the network attachments. Some portions of the FDDI management scheme can be performed either locally on the network attachment or remotely through the use of MAC frames to control the management entities. Not only do these SMT entities control the FDDI network, but a network administrator can also control an FDDI network. For example, a network administrator can force a port to insert itself into the ring or remove itself from the ring. A network administrator can query for statistical information from the network attachment. This type of remote management is accomplished by passing SMT frames across the ring from the network administrator's network station.

Frame-based management (local or remote) requires the use of the FDDI network and special MAC frames to perform the management. There are many network management types available. In order to prevent any one nonstandard network management protocol from taking over the management procedures of FDDI, a standardized remote management scheme was developed. This can be used in conjunction with a popular standard known as Simple Network Management Protocol (SNMP).

There is a process on every FDDI network attachment called the SMT agent. As the name states, it acts as an agent for SMT and provides for SMT frame services. When a SMT MAC frame is received and the FC field is set to 41 (hex) or 4f (hex), the frame is handed to the SMT agent. The SMT agent reads the frame and generates a response if needed. Eight frame types are supported:

1. Neighborhood information frames (NIF)

2. Status information frames (SIF)

3. Request denied frames (RDF)

4. Echo frames

5. Status report frames (SRF)

6. Parameter management frames (PMF)

7. Extended service frames (ESF)

8. Resource allocation frames (RAF)

NEIGHBORHOOD INFORMATION FRAMES (NIF). These frames perform a "keep-alive" function. Inside this frame is a description of the station, the state of the station, and the capabilities of the MAC. The NIF allows a station to send a frame out to notify its downstream neighbor of its existence. The downstream neighbor's response is addressed directly to its upstream neighbor. All stations attached to the ring perform this operation every 30 seconds. Through this process, all stations know who their upstream neighbors and downstream neighbors are.

STATUS INFORMATION FRAMES (SIF). These frames are used to request more information about a station, such as the number of ports, the number of MACs, the neighbors of each MAC, and how the station is connected to the ring (DAS, SAS, etc.). Also included in these frames could be the LEM or any of the frame counters in the MAC.

REQUEST DENIED FRAMES (RDF). These frames are used to deny requests received from the ring. An example is an SMT request from an unsupported SMT version.

ECHO FRAMES. These frames are used to test a path to a ring station using a simple echo request/response protocol. A ring station sends a request to another ring station, which should respond with an echo response frame. The information in the echo frame's information field is copied by the recipient and echoed back to the requester.

STATUS REPORT FRAMES (SRF). A management station must have a reliable image of the network so that it may properly perform management functions. This frame is sent with a special (reserved) multicast address that allows stations to report their status to the management station. Status report frames are used to indicate that certain events or conditions have taken place—for example, the ring has wrapped, a duplicate address was detected, the LEM thresholds have been exceeded (because of excessive line errors, for example), or an illegal configuration (port type mismatch) has been found.

PARAMETER MANAGEMENT FRAMES (PMF). The PMF protocol allows a network manager to query information frames of SMT on a ring station or make changes to any of the entities in SMT. There are five functions of PMF frames: to get information from a remote ring station, change an attribute value, delete an attribute value, add an attribute value, and retrieve information.

EXTENDED SERVICE FRAMES (ESF). These frames are used to allow FDDI vendors to define their own SMT frames and services as addenda to the established ones. They can be used by an FDDI vendor to place in the SMT vendor-specific information that is not supplied by the standard SMT.

RESOURCE ALLOCATION FRAMES (RAF). These frames are used to request FDDI allocations from a network resource. They are not implemented in version 6.2 of SMT.

Addressing the Packet

Throughout the previous section MAC or physical-layer addressing was discussed. These addresses are very important in a LAN environment. But what exactly are these addresses? We must be careful to differentiate these addresses from other types of addresses—IP addressing, for example. Addressing is strictly in the data-link layer and only identifies a station on the Ethernet from a NIC point of view. An IP address is a software address that identifies a particular station running the IP protocol. This may sound confusing, but simply remember that the MAC address is like the number on your house. It does not identify the contents of your house or what language the people in the house speak, only the physical location of the house on your street.

Local area networks started out in a single-cable, multiple-attachment environment. With this setup all stations can see and process any of the packets sent along the single cable. Instead of each attachment receiving all packets and letting its upper-layer software determine if the packet is meant for the attachment, each station is assigned a unique address on the cable segment. In this way all stations can receive any packet, but each station compares its address to the destination address in the received packet. If there is not a match, the attachment will discard the packet without disturbing its upper-layer software. Each station has a unique address, and this is a universally unique address (Token Ring allows us to get around this through the use of locally administered addresses). Blocks of addresses are handed out to NIC card manufacturers.

Each station on a LAN is identified by a special address. Known as the MAC-layer address or the physical-layer address, this address is used to "physically" identify a station on a LAN. It is different from the address used to identify the station via the network software that it is running. (Software or protocol addresses are discussed in the following chapters.) It is called a MAC address, for it is at this sublayer (the media access control, or MAC) of the data-link layer that addressing is defined.

Identifying a host by its physical address is done by placing its location on the network. Like using the phone number in the telephone system, addressing in either Token Ring, Ethernet, or FDDI is the most basic way two stations communicate with each other over a LAN. Once the connection is established, all data will be transferred between the source and destination addresses of the two stations.

Ethernet V2.0, IEEE 802.3, and Token Ring addressing will be discussed in this section. All address numbers will be specified in hexadecimal unless otherwise noted.

Frame Identification (Ethernet)

Refer to Fig. 2-16. The Ethernet packet, reading from left to right (byte 0 to byte 5), consists of the destination address, the source address, the type field, the data field, and the cyclic redundancy check (CRC).

1. The *destination address* is the address of the immediate recipient of the packet (it does not have to be the final destination). Included

Figure 2-16

M—multicast bit.

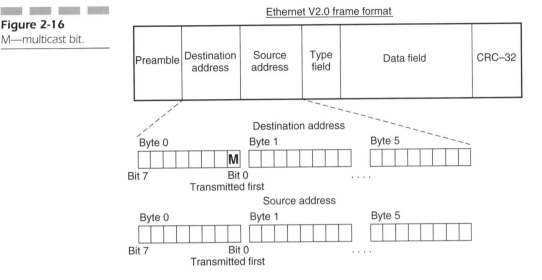

in this field is the M bit, which indicates if the packet is a multi-cast or broadcast address (explained shortly).

2. The *source address* is the address of the sender of the packet.

3. The *type field* indicates what type of data is in the packet: TCP, XNS, AppleTalk, etc.

4. The *data field* contains upper-layer software headers and user data.

5. *CRC-32* is an error-checking algorithm to ensure the integrity of the packet.

Although the data encapsulation techniques (Ethernet, IEEE 802.3, and Token Ring) are all different, they use something similar. This is the addressing portion of the packet. Each packet that is transmitted onto a network must have a physical address associated with it. All physical addresses for Ethernet, IEEE 802.3, and Token Ring are 48 bits long. There used to be an allotment for 16-bit addresses, but this addressing length is no longer used. There are two physical addresses in each packet: the source and the destination address. Each is 48 bits long and is expressed as 6 bytes. There is a 6-byte source address and a 6-byte destination address.

For LAN interfaces, there are three types of physical addresses:

1. *Unique station address.* Much like a telephone number, there is one number assigned per machine. When a packet is transmitted onto the network, each station will receive the packet and look at the destination address field. The LAN interface card will make a comparison with the

address that was loaded into its memory at startup time. If there is a match, the adapter passes the packet on to the upper-layer software (TCP/IP, XNS, IPX, etc.) to determine the outcome. If there is not a match on the address, the LAN interface card simply discards the packet. These are known as *unicast packets*.

2. *Multicast address.*   A special type of address is used to identify a group of network stations. Each network station has a multicast address assigned to it. In this way, a single station may transmit a packet with the destination address set to multicast and it can be received by more than one network station on the network. The easiest way to identify an Ethernet or IEEE 802.3 multicast address is by the first byte. If the first byte is an odd number, the address is a multicast address. A multicast address example is the spanning tree algorithm used in bridges. With this address, all bridges may receive information by one station transmitting it in multicast mode. How a station indicates a multicast destination address will be shown later. Source addresses are never multicast. Multicast addresses are also used with the RIP version 2 protocol and with the Open Shortest Path First protocol (a routing protocol for IP).

3. *Broadcast address.*   This is a special form of multicast address. The term *broadcast* indicates that the packet is destined for all stations on the network. Each station on the network will pick up this packet and will automatically send it to its upper-layer software. The physical address for broadcast is FF-FF-FF-FF-FF-FF. IBM Token Ring also includes C0-00-FF-FF-FF-FF as a broadcast address. This type of broadcast is usually used on the local ring.

Since Token Ring has no multicast system, there is one more type of address used in Token Ring networks. It is called the *functional address*. If the first two bits are set to 11 (hex C) and bit 0 of byte 2 is set to 0, then the address is a functional address. A packet is either addressed with a unique address or a broadcast address. The functional addresses in Table 2-5 are reserved, except for the user-defined range. Functional addresses have a special meaning in Token Ring.

Physical (MAC) Addressing

The physical address is placed by the controller manufacturer into a PROM (programmable read-only memory), a chip on the card that can be programmed with information and hold this information even during a power cycle (on/off) on the LAN card. Upon startup of the net-

TABLE 2-5

Functional
Addresses

Function Name	Functional Address (Hex)
Active monitor	C00000000001
Ring parameter server	C00000000002
Ring error monitor	C00000000008
Configuration report server	C00000000010
NetBIOS	C00000000080
Bridge	C00000000100
LAN network manager	C00000002000
User-defined	C0000008000 through C00040000000

work station software, the network controller software will read its physical address from the PROM and assign this number to its software. For example, when the Ethernet drivers (the software that is loaded during your workstation power-up to run the NIC) are loaded into a network station, the drivers read the address from the PROM and use the address in its software loading. It will use this number as the physical address for all packets that are transmitted or received by this network station. Token Ring controllers have the capability of allowing the network administrator to overwrite this PROM address with a private address known as the locally administered address or LAA. To indicate this type of address, bit 1 of byte 0 of a Token Ring frame must be set to 1. This will be more apparent when we discuss the addressing of the Token Ring frame. But for now, any Token Ring address that starts out with a 4 is a locally administered unique address.

As shown in Figs. 2-17 and 2-18, there are distinct entities in the addressing portion of a frame. Figure 2-17 shows the Ethernet frame, and Fig. 2-18 shows an IEEE 802.3 frame with IEEE 802.2 headers. For now, pay attention to the address headers of each of the frames. The destination address is 48 bits long and is broken down into six 8-bit fields. The first byte is on the leftmost side of the packet and is labeled byte 0. The last byte is on the rightmost side and is labeled byte 5.

The first three bytes of the physical address, either the source or the destination address, indicates the vendor. It was not intended to indicate the vendor; it just ended up this way. Originally, Xerox handed out these addresses. It is now the responsibility of the IEEE. The purpose of having the addresses assigned by a central authority is simple. It ensures that there will not be a duplicated address in any network interface card.

Figure 2-17
Generally defining
the address fields.

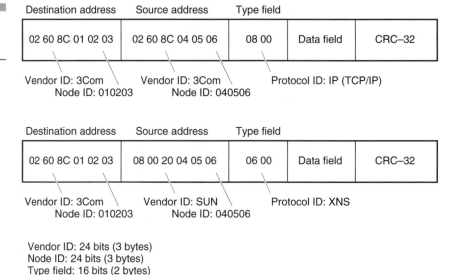

Figure 2-18
IEEE fields.

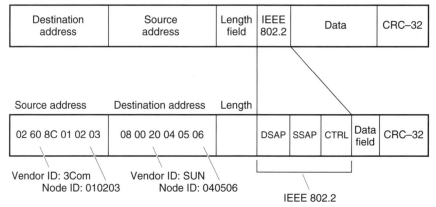

Each vendor of a LAN interface card must register its use with the IEEE. IEEE assigns the vendor the first three bytes of the six-byte total address. For example, 00-00-A2-00-00-00 is assigned to Bay Networks, 02-60-8C-00-00-00 is assigned to 3Com Corporation, and 08-00-20-00-00-00 is assigned to Sun Microsystems.

The last three bytes of the address can be assigned by the manufacturer. For example, Bay Networks (00-00-A2) is allowed to assign the last three bytes to the LAN cards for their routers. This means they are allowed to assign up to 2^{24} (16,777,215) individual addresses to their LAN cards, one per card. Vendors are allowed to assign the last three bytes using whatever numbers they like. Normally, these LAN cards are assembled in assembly-line fashion, and each card produced is incremented by one in the last three bytes.

For example, in Fig. 2-17, the destination address on both of the frames is 02-60-8C. This indicates that the LAN interface card was manufactured by 3Com Corporation. These three bytes were assigned to 3Com by the IEEE. The following three bytes are assigned by 3Com. The normal way to do this is to increment the node ID portion by one for each card manufactured.

Notice in Fig. 2-17 that the source address is different. The address is 08-00-20, which is the address assigned to Sun Microsystems. LAN controllers do not differentiate between vendors. The controllers read the whole 6-byte address as an address. The controllers do not care who manufactured the card.

The differences between an Ethernet packet and an IEEE 802.3 packet should also be noted here. Ethernet packets use a type field, and IEEE 802.3 packets use a length field. How do these two fields correlate? They don't, except in their position in the packet. The type field is used to indicate the network protocol process that sent the packet. For example, in Fig. 2-17, 0800 indicates that the packet originated from a TCP/IP process. The setting 0600 indicates that the XNS process was used. If the field has 8137, H indicates that the packet was sent from a Novell NetWare station.

Now refer to Fig. 2-18. IEEE 802.3 frames have the length field in place of the type field. This indicates the length of the data field, not the total length of the packet... How does the packet indicate what process was used to send it? IEEE 802.3 frames (with the exception of proprietary NetWare frames) should use IEEE 802.2 headers. This includes the destination service access point (DSAP) field, the source service access point (SSAP) field, and the control field. The DSAP field indicates which destination process the packet is intended for. The SSAP field indicates the source process that sent the packet. The control field is used for IEEE 802.2 link control. For example, if the DSAP and SSAP fields are both filled with an E0 and the control field is an 03, the packet is a nonproprietary Novell NetWare packet. These fields are roughly analogous to the type field in an Ethernet packet.

An IEEE 802.3 Ethernet controller can differentiate between the two types of packets by reading the twelfth byte of a packet. This byte will be either a length field or a type field. Ethernet type fields begin at 0600 (hex). The largest packet size of Ethernet is 1518 bytes, which is 05FE (hex). Therefore, if the number at byte 12 is equal to or greater than 0600 (hex), it is an Ethernet framed packet.

Bit-Order Transmission

The bit order for addressing the packet for Token Ring and FDDI is the same. Therefore, Token Ring only will be shown. First refer to Fig. 2-19. This is an IEEE 802.3 frame. Most of the fields have been explained in the previous text. The purpose of this figure is to explain the way the bits in the bytes are transmitted. This is very important, for IEEE 802.3/Ethernet and Token Ring do it differently. This leads to many incompatibilities between the two architectures.

First, the IEEE 802.3 address fields are divided into six bytes. In this figure, the whole frame is transmitted left to right, but the bits in each of the fields are transmitted right to left. Refer to the bottom part of Fig. 2-19, which shows an exploded view of the address fields. Notice that bit 0 is the rightmost bit. This is called the least significant bit of the byte and is transmitted first. The next bit transmitted is the bit to the left of bit 0. Once all the bits of the byte are transmitted, the next byte (byte 1) is transmitted. Once again, bit 0 is transmitted first.

For a better view, refer to Fig. 2-20. This figure shows a destination address of 02608C as the first three bytes. For convenience, it is converted

Figure 2-19
IEEE 802.3 CSMA/CD frame.

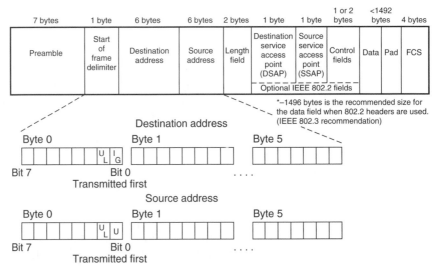

Figure 2-20
Ethernet/IEEE 802.3
bit order and trans-
mission.

Destination address	Source address
02608C010203	02608C040506

The first three bytes in binary
00000010   01100000   10001100. . .

The first three bytes as transmitted on the cable plant
01000000   00000110   00110001. . .

to binary. Notice, though, that when the bits are transmitted, each byte appears to be reversed. This is the way Ethernet/IEEE 802.3 frames are transmitted on a cable plant.

For IEEE 802.3 addressing, two other bits in the address are important. Refer again to Fig. 2-19. Bit 0 of byte 0 is on the right side of the first byte of the packet. If this bit is set, the packet is a multicast packet. IEEE 802.3 frames call this the individual/group address bit. It is the same as the multicast bit in the Ethernet V2.0 frame. This is the first bit trans-mitted on the LAN. Therefore, if the first bit transmitted on an Ether-net or IEEE 802.3 LAN is a 0, it is an individual address. If the first bit is a 1, it is a multicast address. For example, for the address 02-60-8C-00-01-02, the first bit transmitted would be a 0. But for 03-60-8C-01-02-04, the first bit transmitted would be a 1. Therefore, the latter address is a multicast address. The second bit indicates whether it is a universally assigned (IEEE) address or a locally assigned address (the local site). This bit is rarely used in Ethernet networks.

Token Ring is completely different. Like Ethernet/IEEE 802.3, bit 0 is the first bit transmitted on a Token Ring LAN, but for Token Ring *bit 0 is the leftmost bit of the byte*. Refer to Fig. 2-21. As with IEEE 802.3, this is the individual/group address bit. It is the same as the multicast bit in the Ethernet frame and is the first bit transmitted on the LAN. There-fore, if the leftmost bit (bit 0) of byte 0 is set to a 1, the packet is a multi-cast packet. The next bit (bit 1) of byte 0 is the universal/local adminis-tration bit. If this bit is set, then the address has been locally assigned. This means that the address PROM is not used to determine the MAC address of the card; it has been assigned by a network administrator of the LAN. This can have many advantages. Some LAAs have the phone number of the person that is using the address. Some LAAs are assigned by building, floor, and cubicle number. There are many way to assign these addresses, but bit 1 will always be set. Therefore, LAAs always start with 4 as the first digit (40 as the first byte in hex).

Figure 2-21
IEEE 802.5 frame
format.

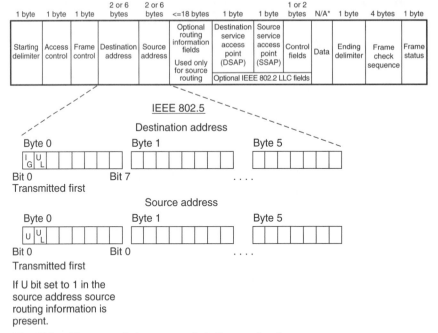

The disadvantage of these addresses is that every address on the LAN must be unique. No two MAC addresses should be the same; therefore, the addresses are usually assigned by a central authority, usually the LAN administrator's office. In other words, each company that uses LAAs is free to assign their own addresses, but the assignment should be managed by one group of people. This can become an administrative nightmare. Users may also be allowed to assign their own addresses, which can lead to a security breach. A lot of companies are applying filtering capabilities to their forwarding devices, known as bridges and routers, so that certain addresses are allowed on certain LANs and other addresses are "filtered" from entering some LANs. Token Ring controllers check for duplicate addresses upon startup.

In summary, Token Ring and Ethernet/IEEE 802.3 both signify the left-hand byte as byte 0. They both state that bit 0 of byte 0 should be the first bit transmitted on the LAN. The problem is that Ethernet/IEEE 802.3 has the rightmost bit as bit 0 and Token Ring has the leftmost bit as bit 0. This leads to a complication. If Ethernet/IEEE 802.3 transmits and receives MAC addresses one way and Token Ring transmits and receives them in another way, how can network stations located on different LANs (one on Ethernet and one on Token Ring) communicate with each other? Obviously, they use completely different access methods, and an Ethernet controller cannot simply be placed on a Token Ring LAN. But even if they could, how would you make up for the address format differences? How would a Token Ring controller know which way to read an Ethernet formatted packet (at the MAC level of headers anyway)? The answer is that they cannot without the use of a device known as a router or a bridge. These devices are explained in the next chapter. Routers fully understand their specific protocol and what in their packet needs to be translated. Therefore, all routers and some vendors' bridges understand how to translate between the two LANs.

CHAPTER 3

Bridges, Routers, and Basic Network Information

Introduction

This chapter provides a brief introduction to transparent and source route bridges, multiprotocol routers, and network protocols. In environments with a variety of computer resources (PCs, graphics workstations, mini, and mainframes), there are varying requirements for network bandwidth. For instance, clustered minicomputers, diskless workstations, or PCs sharing a file server place a great burden on a network because of numerous data transfers. When such traffic begins to seriously affect performance, the most efficient way to deal with the problem is to divide the network into separate networks. These separate networks are then brought together to form an internet by means of bridges or routers.

Transparent Bridges

There are three types of bridges:

- *Transparent* bridges operate transparently to any attachment or protocol running on the LAN. This bridge type can be used in Ethernet, Token Ring, or FDDI environments.

- *Translation* bridges can convert between two LAN types (Ethernet to Token Ring, etc.).

- *Source route* is used in Token Ring networks (available for FDDI as well but rarely used). A source station sends discovery packets to the network to find the destination. The source station picks the path to a destination by receiving and reading discovery packets that are returned to it.

All three will be explained in the following paragraphs. However, a point to make here is not to confuse the protocol known as the *spanning tree protocol* with any of the above bridge types. STA operates independently of the three, and in fact STA works on all three bridge types. STA is the protocol that disables loops in a bridged network to provide a single path between any two communicating stations. STA is fully covered later.

Introduction

Shared environments led to the haphazard placement of network stations throughout the network. Before the advent of bridges, everything that could be placed on one cable plant was placed there, regardless of the type of application. Refer to Fig. 3-1. Graphics workstations were on the same cable plants with equipment that used very little cable bandwidth (i.e., terminal servers). The stations that did not generate much network traffic had to contend for use of the cable plant equally with the equipment that generated a lot of network traffic. Soon, people found that the network response was extremely slow. Some even went so far as to say that Ethernet had found its true limitations. The year was 1985.

Before an uproar was started, a new networking concept arrived. A device was needed that would allow us to segment the cable plant based on network utilization, yet give us complete transparency. Refer to Fig. 3-2. The answer came in the form of a device called a *bridge*. The true name for the algorithm is *transparent bridging*. It was first introduced for Ethernet networks, and at that time it was commonly referred to as an Ethernet bridge. However, this was a misnomer, for the bridge could operate on any LAN type; it would be a few years before Token Ring supported transparent bridging instead of source route. In order for transparent bridging to work properly, the interface on the bridge must support a special receive option known as *promiscuous mode*. This enables the receiver to receive all packets, regardless of address, and process them (hand them to the upper layer software).

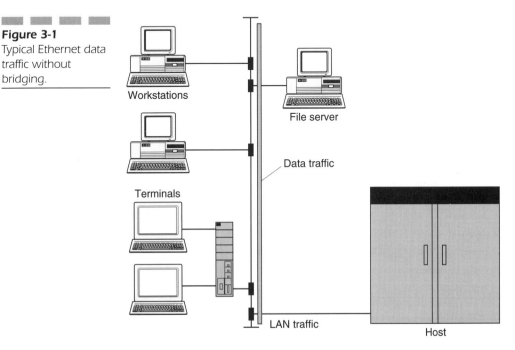

Figure 3-1
Typical Ethernet data traffic without bridging.

Workstations

File server

Data traffic

Terminals

LAN traffic

Host

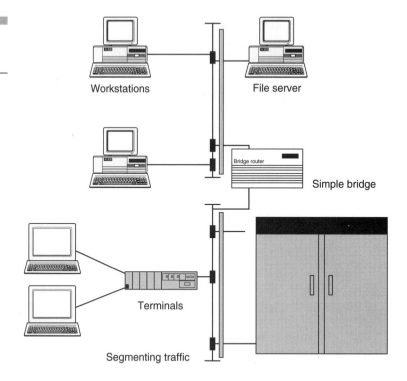

Figure 3-2
Bridged Ethernet
environment.

Workstations

File server

Bridge router

Simple bridge

Terminals

Segmenting traffic

Refer to Fig. 3-3, in which a bridge interconnects two Ethernet cable segments and watches all packets that traverse the cable. The bridge operates by inspecting all packets on the network for their addresses. An Ethernet bridge operates in a promiscuous mode, receiving all packets, not just the packets with the destination address assigned to them. Basically, a bridge performs three functions: *learning, filtering,* and *forwarding.*

As a bridge receives each packet, it notes the source address of the packet and the port number that it received the packet on. The source address is extracted and matched to what is known as a forwarding table. If the address is not in the table, it is added to the table; the port number that the bridge received the packet on is also added. If the source address is in the table, the packet is simply discarded. This is known as *learning.* The bridge keeps a table of source addresses for each cable segment that it is attached to. It is a single table that consists of MAC addresses and the physical bridge port that it was learned from. Any transmitted packet on the cable segment will suffice for the learning process. No special packets are transmitted on the cable plant for the bridge to learn. Remember, this is transparent bridging. An end station addresses the packet to the destination station with which it wishes to communicate.

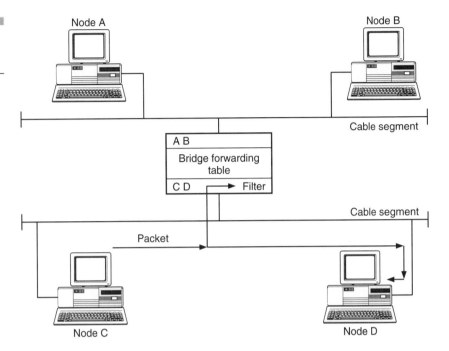

Figure 3-3
Bridge routing
table—filtering.

The end station does not know the bridge exists. In contrast to routing technology, end-station transmitted packets are not directed toward the bridge. As far as the transmitting source is concerned, the destination is on the same cable segment as it is.

The destination address of a received packet is also extracted and compared to the table. A couple of actions can result from this comparison:

1. *Filtering.*   After receiving a packet and processing the source address, the bridge will look at the destination address of the packet. It will perform a lookup in its address table to find a match for that address. If a match is found, the bridge will check to see which port is associated with the match. If the port in the match is the same port that the packet was received on, the bridge will discard the packet. The packet will not be forwarded to any cable plant attached to this bridge. A port number match indicates that the destination is on the same cable plant as the source of the packet.

2. *Forwarding.*   Refer to Fig. 3-3. If network station C transmits a packet to network station D, the bridge will receive the packet (not because it was specifically transmitted to it, but because it listens and receives all packets that are on a cable segment). It will perform a filter operation for the destination D is on the same cable segment as source C. Therefore, the bridge will not forward the packet to cable segment 2. It will discard that packet. Again, since Ethernet is a broadcast network, station D will receive that packet as if the bridge had not even been there, but this reduces the amount of traffic on cable segment 2, allowing for better utilization of that network.

If the port number in the table did not match the port number that the packet was received on, the bridge finds the port number associated with the destination address and forwards the packet out that port only. Refer to Fig. 3-4. Node C transmits a packet to node A. If a match is found in one of those tables, the bridge will forward the packet onto that cable segment and only that cable segment. In this case, the packet was forwarded to the LAN that has node A.

Under certain circumstances, the destination address will not be found in the bridge's table. If the bridge cannot find a match in any of the tables, it will forward the packet to all cable plants attached to it except for the one that it received the packet on. When would this happen? There may be times when a bridge has not yet learned the address and made an entry into one of its tables. When the bridge cannot find the address indicated by the destination address field, it must forward the packet to all active ports that the bridge has (again, except the received port). This is known as *flooding*. Flooding the packet to all ports will not cause any errors on the network, and the table will be quickly updated, because when the destination station transmits a response to a packet, the bridge will then learn the address of that packet and place the address in the correct table, and it will not have to forward the packet to all cable plants after that. This is known as forwarding an unknown packet. It should not happen frequently. In most cases, unknown addresses are usually learned within one minute of bridge initialization. A bridge must forward to all active ports every packet for which it does not know the destination address.

Some protocols must use multicast and broadcast addresses in order to properly operate with other stations on the network. With addresses such broadcast and multicast, the bridge will always forward these packets to every active cable segment that is attached to the bridge (flood the packet to all attached cable segments). The bridge can be configured not to forward these types of packets, but this is not usually the case. Bridges will flood multicast and broadcast packets. But this creates a problem. What if redundant paths are built between source and destination sta-

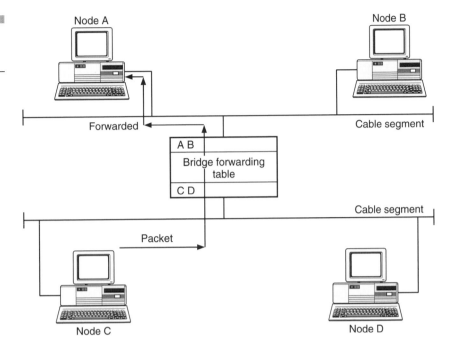

Figure 3-4
Bridge routing
table—forwarding.

tions? Wouldn't this create a loop situation for broadcast and multicast packets? That is why the spanning tree algorithm was invented.

Redundancy and the Spanning Tree Algorithm

It is possible to place two bridges in parallel—that is, two bridges can interconnect the same two or more cable segments. In Fig. 3-5a, two bridges interconnect the same two Ethernet LANs. An active loop configuration is not allowed in Ethernet bridging. There is a protocol that allows the physical loop to exist, but not a logical loop. This protocol is the spanning tree algorithm, or STA. However, redundancy is a major consideration in network design.

You might think that since the bridges interconnect the same cable plants, each bridge will forward a copy of a packet destined for a remote cable plant. That is, if one station transmits a packet and neither bridge knew about the other, both bridges would forward the same packet to the other cable segment. In such a case one packet would be transmitted and both bridges would forward the packet, resulting in two or more

Figure 3-5
Redundancy: (a)
Loops. (b) Blocked
loops.

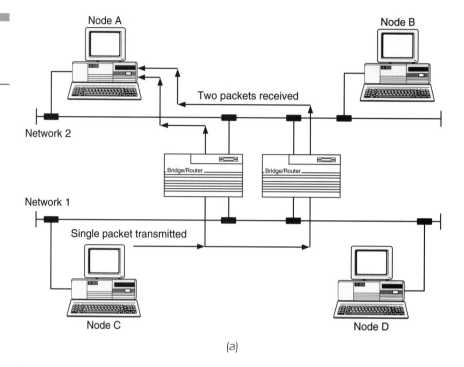

(a)

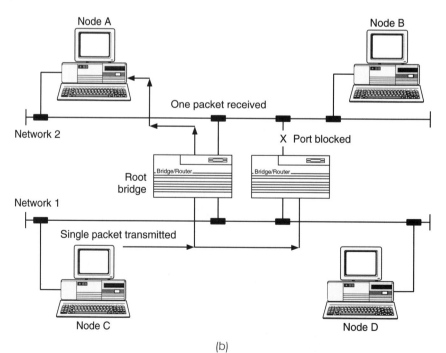

(b)

packets on the forwarded LAN. This can produce loops and fatal errors on the network. In fact, these environments will cause tremendous problems in bridged networks. For example, packets that are sent out in broadcast mode will be picked up by both bridges. Each bridge will forward a copy of the packet to the next LAN. Each bridge will receive the other's broadcast on the forwarded LAN. Since the destination address is broadcast, each bridge will forward the packet back to the originating LAN. This loop will continue until one of the bridges stops forwarding.

Redundancy has to be allowed. Redundancy is the biggest use for parallel or looped bridges. For Ethernet, the spanning tree protocol enables this redundancy without the duplication of packets.

Spanning tree is the protocol, approved by the IEEE 802.1d committee, that enables bridges to interconnect two or more of the same cable plants or to allow for a redundancy in the network. The protocol enables bridges to talk to one another using reserved multicast packets known as bridge protocol data units (BPDUs). These packets allow bridges to learn about one another. If there is a loop (indicating parallel bridges), the best path will be chosen and the other one placed in a logical blocking state (unable to forward packets). One bridge will become the ruler of the spanning tree LAN (known as the root bridge), and it will place all of its ports in the forwarding state. Other bridges, for looped paths, will have some ports placed in a blocking state.

In the first stage in creating a loop-redundancy network, all the network bridges select a root bridge. The bridge that wins the right to become the root bridge will have the highest priority set (a priority number is given to the bridge when it is installed) in its transmitted BPDU. Upon startup each bridge thinks that it is the root bridge and immediately starts to transmit root BPDUs. Upon receiving a root BPDU and noticing that its address is different (indicating another bridge on the network), the bridge compares priority numbers. A high priority is a low numerical value, and a low priority is a high numerical value. If there is a conflict between two or more bridges with the same priority, the bridge with MAC address having the lowest numerical value will win. Other bridges quit transmitting root BPDUs when they notice another's higher priority. Since each MAC address on an Ethernet LAN is unique, there will not be a tie again. Each bridge that receives a root BPDU compares it to its own root BPDU that it is transmitting. If it loses via the priority, it quits transmitting its own root BPDU and forwards the root BPDU that won. However, it will add the cost of the port that it received the root BPDU on. In this way, another bridge can determine if it has a better (lower cost) back to the root, explained next.

Once a root bridge is selected, that root bridge then propagates the root BPDUs to find loops, and certain bridges will dynamically shut down one or more of their ports (known as a *port in the blocking state*) to disable any loops in the bridged network.

Bridges running STA determine how to shut down their ports that have a looped configuration, for each of their ports will have a number assigned to it known as a *cost*. This can be assigned as a default number or assigned by the network administrator at installation time. These cost numbers are propagated throughout the network through the BPDUs. The goal is for a port to have the lowest cost to the root bridge (in the direction of the root bridge). The cost of the received port is added to the received root BPDU, and the bridge forwards this information out its ports. In this way, the cost field of the root BPDU is cumulative; as the root BPDU is propagated (forwarded), the cost is more expensive to the root, because the packet is continually farther and farther away for the root bridge.

Therefore, the bridges that put their ports in the blocking state are the bridge ports that have the highest cost associated to the path to the root bridge. If there is a loop between two bridges, each bridge will determine which has the highest cost to the root bridge. The bridge that has the highest-cost port to the root bridge will put its port into the blocking state. The other bridge will keep its port in the forwarding state, allowing the forwarding of data frames.

The cost is set during the configuration of the bridge. It should be set so that the highest cost is inversely proportional to the line speed. For example, a T1 line that runs at 1.544 Mbps will have a higher cost than an Ethernet segment that runs at 10 Mbps. In other words, once the root bridge is selected, bridges that have the highest cost to get to the root bridge will shut down their ports. If two bridges have the same cost to the root bridge, a series of tie breakers become involved and eventually one port will enter into the blocking state.

This process of providing a loop-free topology usually takes about 30 to 35 seconds, depending on how the network administrator has set up on the bridge. When the bridges have invoked the STA, all forwarding of data will be stopped and bridge tables will be flushed (erased). Once a loop-free topology is established, the root bridge will transmit root hello BPDUs to the network (default every two seconds). Every bridge on the network waits for these hello BPDUs, whether or not they are in blocking or forwarding mode. If a bridge receives them, it assumes the root bridge is alive and the path to get there is good. Each packet is examined as well to ensure that the current state is the most correct one.

If a bridge receives a root BPDU and believes it has a lower cost to the root, it will place its port in the forwarding state. In the event that a bridge does not receive the hello BPDU (the default maximum wait timeout is 20 s), the bridge that discovered this will issue a topology change configuration BPDU, and the bridged network will reconfigure to establish another loop-free topology.

Refer to Fig. 3-5b. In this very simple configuration, the bridge on the left was selected as the root bridge because of its priority. Therefore, the bridge on the right will put one of its ports into the blocking mode. In this picture, both ports have equal access to the remote bridge. But this one is a special case. Since both sides of the root bridge must be kept in the forward state, the bridge on the right must determine internally which port to keep active. This becomes an internal decision on the bridge. This can be accomplished via a port ID or the MAC address of the port. The lower the MAC address, the better the chance you have of keeping the port in the forward state. Therefore, this tie breaker assumes that on the bridge on the right, the upper port had a higher MAC address than the port on the lower side of the bridge.

Being in the blocking state does not mean the bridge is disabled. Although the bridge will not forward any data packets or learn any addresses while in the blocking state, it will continue to listen for hello packets submitted by the root bridge. These hello packets are special packets submitted by the bridge to let all other bridges (active or blocking state) know that the root bridge is still alive. If a bridge with a port in the blocking state does not receive these hello packets for a certain amount of time, the LAN bridges will reconfigure and create a new tree topology.

NOTE: *The spanning tree algorithm allows only one path, loop free, to any destination in a bridged environment.*

One last comment on Ethernet (transparent) bridges. With all costs equal, the total amount of bridges in a linear path (in a row) between two stations is limited to seven. This is not a strict standard (i.e., defined in the STA), but one that is generally adhered to by most network designers. This is due to latencies that may start to occur after a packet traverses eight bridges. This is for the transparent bridges only. With unequal costs (different transmission media, 56k serial lines FDDI, etc.) the number becomes a variable depending on the protocol that is being bridged.

Source routing bridges (discussed next) allow only seven linear bridges or hops to a destination. IBM has recently revised this standard to allow for 13 hops. There are very few implementations of this. The foregoing explanation always mentioned Ethernet as the cable access method for these bridges. Since transparent bridges make forwarding decisions based on the 48-bit MAC address, transparent bridging could work on Token Ring networks. In fact, Token Ring switches most often operate in transparent mode.

What do switches (L2, or Layer 2, switches) do? Switches are really bridges with lots of ports. The cost of the hardware to produce a bridge has dropped dramatically, and we can now put multiple ports in a single box, whereas before with bridges we usually put two or four ports per box and attached whole cable segments (with multiple stations attached) to the port. The price of the port (5–10 years ago) was about $5000 per port. The switch operates today exactly like the bridge, but each port can have its own network device attached to it.

A switch can still have whole cable segments attached to it; this is known as segment switching. Remember, a switch is a bridge. The difference is that each port can have its own attachment to a single computer whereas the bridge connects cable segments.

Token Ring Bridges

Originally, Token Ring bridging was based on an algorithm known as source routing. For a few years we had two standards for bridging, transparent (the algorithm used in Ethernet) and source routing. Making the two talk to each other required the use of another type of bridge, known as a translational bridge. Tough, you bet! Today, Token Ring bridging incorporates both transparent and source route standards. IBM no longer holds the iron fist for Token Ring environments, and many vendors today support transparent bridging for Token Ring. In fact there is an IEEE 802.5m standard that allows for operation of both based on one bit in the source address. This bit is known as the Routing Information Indicator and tells the receiver that source route information is contained in the packet. If this bit is not set, then source route information is not contained in the packet and the packet can be transparently bridged. Although source route networks can be switched, the ability to transparently bridge a Token Ring network allows for a true Token Ring switched environment to exist, one token ring station per switched port.

The following contains information on the source route standard. This method of bridging Token Ring LANs can exist only on Token Ring LANs. It will not work for Ethernet LANs. The algorithm for source routing is much more detailed than what will be described here. It will be only briefly described in the following paragraphs.

First we will show an example, and then we will explain the entries in the Token Ring packet. Refer to Fig. 3-6a. Token Ring environments have a number assigned to each ring. This number is assigned when configuring the bridge ports. The ring on the left side of the picture has been assigned ring number 4. The ring on the right side has been assigned ring number 3. These numbers are not assigned in the network station. They are assigned when configuring the bridge for operation on the ring. The bridge on top is assigned number 5, and the bridge on the bottom is assigned number 6. These are bridge numbers not ring numbers, so in order to operate in a source route environment you must assign ring numbers to the ports of a bridge and a bridge number that is configured as a global (on that bridge) parameter.

Remember from Chap. 2 that the Token Ring frame has a field known as the routing information field, or RIF. Besides the routing control field, there is a variable number of route descriptors (RDs). The combination of the ring and bridge number creates a two-byte field (used in the route descriptor field of the RIF field) with the ring number being 12 bits in length (leftmost bits) and the bridge number being 4 (the rightmost bits). This allows for 4095 rings and virtually an unlimited

Figure 3-6*a*
Source routing.

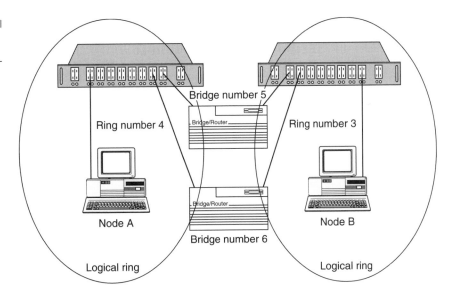

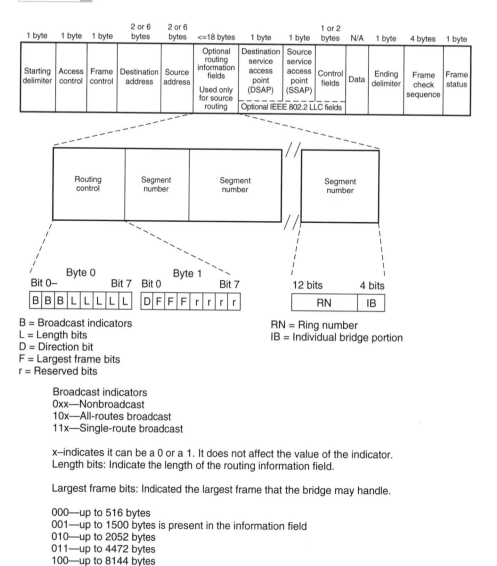

B = Broadcast indicators
L = Length bits
D = Direction bit
F = Largest frame bits
r = Reserved bits

RN = Ring number
IB = Individual bridge portion

Broadcast indicators
0xx—Nonbroadcast
10x—All-routes broadcast
11x—Single-route broadcast

x–indicates it can be a 0 or a 1. It does not affect the value of the indicator.
Length bits: Indicate the length of the routing information field.

Largest frame bits: Indicated the largest frame that the bridge may handle.

000—up to 516 bytes
001—up to 1500 bytes is present in the information field
010—up to 2052 bytes
011—up to 4472 bytes
100—up to 8144 bytes
101—up to 11407 bytes
110—up to 17800 bytes
111—Used in an all-routes broadcast

Direction bit: Indicates to the bridge the order in which to read the information in the routing information fields.
Ring number: Indicates the number for a particular ring.
IB: Indicates the number for the bridge.

Figure 3-6b IEEE 802.5 frame format with defined RIF field.

number of bridges, as explained next. The ring numbers must be unique for each ring on the network. The bridge number does not have to be a unique number. If there are no parallel bridges (those bridges interconnecting the same two rings), all bridge numbers may be the same, allowing virtually a limitless number of bridges on a ring. The following text will show how a packet is forwarded through a Token Ring bridge.

If node A needs to communicate with another device, node B, it transmits certain packets on the ring in order to find the other station. It first transmits a packet called the *all stations broadcast packet* in an attempt to find the destination. If the destination station is on the same ring, it should respond and the originator and the destination can set up a local session.

If there is no response on the local ring, the originating station assumes that it may be on a remote ring. The originating station will then broadcast a *dynamic route discovery packet* in an attempt to find the destination on a remote ring. This packet will have routing indicators— i.e., the optional routing field—built into it that any bridge attached to the ring will know what to do with. The originating station can submit an all-routes discovery packet (the actual name for this is the all-routes explorer, ARE, frame) or a single-route discovery packet. Refer to Fig. 3-6*b*, which shows the complete Token Ring packet format with the full RIF field, which will be explained in the following text.

Depending on the programmer of an application, dynamic route discovery packets can be sent by the end station in one of two modes:

1. *Single-route explorer with an all-routes return:* Send out an explorer packet so that only one copy of the packet is forwarded to any ring. When the destination station returns the packet, all bridges forward the packet.

2. *All-routes explorer with a single-route return:* Every bridge forwards this packet to the destination, and the destination station flips a direction bit in the routing control field of the packet so that each packet sent back by the destination station will follow the exact path that it took to get to the destination.

Note the broadcast indicators of the routing control field in Fig. 3-6*b*.

Again, the specific method is up to the application developer. For example, Novell NetWare uses a single-route explorer with an all-routes return. Most TCP/IP implementations use an all-routes explorer with a single-route return. It is all up to the developer.

For example, in Fig. 3-6a, if node A transmits an all-routes discovery packet to find node B, each bridge attached to the ring (both bridges 5 and 6) will forward the packet to all rings connected to it. In the process of forwarding the packet, the bridge will stuff into the RIF segment number field three things:

1. The ring number of the ring from which the packet came (ring 4)

2. The bridge number assigned to that bridge (bridge 5 or bridge 6)

3. The ring number the packet is being forwarded to (ring 3)

How is all this accomplished via the Token Ring packet? Again, refer to Fig. 3-6b. These three things are put into a special field in the Token Ring packet called the routing information field (RIF) segment number. This field is present only when the packet must traverse a bridge. If the packet does not have to traverse a bridge to get to its final destination, this field will not be present in the packet. After this, the bridge will forward the packet onto the next ring, ring 3. Therefore, when the packet reaches the destination, the packet from bridge 5 would look like Fig. 3-6c. Since bridge 6 would also forward the packet, it would be similar, except there would be a 6 in place of the 5 for the bridge number in the RIF field.

For simplicity, I have not set any bits in the routing control (RC) field in Fig. 3-6c. The RC field contains:

- Broadcast indicators, which specify which of three types of source route packets this one is

- Length bits, which indicate the length of the RIF including the RC field

- Direction bit, which tells the bridge which way to read the RIF. A 1 means read the RIF from right to left, and a 0 means read the RIF from left to right.

- The largest frame bits, which tell other bridges how large a frame a bridge can forward

- r bits, which are reserved. IBM has modified this field to allow for more granularity in the largest frame indicator.

Figure 3-6c
IEEE 802.5 packet
with filled-in RIF field.

DA	SA	Data-link header	RC	0045	0030	Data	Packet trailers

RIF field

The broadcast type bits are set by the transmitter (not necessarily the source) of the packet. When a node wishes to find another node, it must first transmit an *dynamic discovery* packet. There are two types of discovery.

■ All-routes explorer
■ Single-route explorer

The first is by means of an all-routes explorer with a single-route return. This discovery packet is transmitted out to find a destination station no matter what the topology. Every bridge forwards this packet to all of its active ports. Upon receipt the destination will set the direction bit to 1 and echo the packet (setting the source and destination addresses appropriately) out as a specific route return (setting the broadcast type to a specific route). Upon receiving the packet, the bridge will read the prebuilt RIF and forward the packet out the port indicated in the RIF. A bridge does not add any information to the RIF with this bit set, and the broadcast indicators show that it is a specifically routed frame (single-route return). If there are multiple paths to the destination, the originator of the packet receives multiple return packets. The source chooses which one it wants to use, and from that point on the source and destination stations communicate using the prebuilt RIF from one of the returned dynamic discovery packets. Usually, the originator chooses the first packet that returns, even if the route is longer, because it assumes that that path is less congested.

A second type of discovery is by means of a single-route explorer with an all-routes return. If the originator of the dynamic discovery packet sends this frame out, it will be forwarded only by those bridges whose ports are not in blocking mode. In other words, this type of frame assumes that the spanning tree algorithm is running on the bridge topology and that loops in the network have been taken out. There is only one path to the destination. The destination station should receive only one discovery packet. The destination station sends the packet back to the originator, but it sets the broadcast type to all-routes explorer. It basically builds a new packet to send back to the originator. All paths to the originator are taken, and the RIF field gets filled in along the way. The originator usually chooses the first response that it gets, and from that point on the source and destination communicate using a *specifically routed frame.*

The specifically routed frame is a type of source route frame that does not have anything to do with a discovery frame. The specifically routed frame type indicates that a source and a destination are communicating via a path described in the RIF field. This type is used after the source has discovered and chosen a path to the destination station.

As the packet traverses each bridge, the bridge changes the RIF length field as it adds an RD. That bridge also sets bits as to the largest frame that it can transmit. The bits are prebuilt as to frame size. The possible entries are given in Fig. 3-6*b*. If a bridge receives a packet having the field set to 2052 and can transmit only 1500 bytes, that bridge will reset the field to 1500. If the bridge can transmit larger packets, it will not enter anything into the field.

The direction bit indicates to a bridge which way to read the RIF field: right to left for a packet being transmitted from the destination to the source station, or left to right for one being transmitted from the source station to the destination station. It is used on specifically routed frames and responses to all-routes explorer frames. Instead of having a destination station rebuild the RIF field, it can set the direction bit and send the packet back. In this way, bridges will read the information from last to first.

Single-route explorer types are responded to by the destination station as all-routes explorers. If the destination station receives a frame that was sent to it as a single-route explorer, it does not set the direction bit and send the packet back to the originator. It builds a new frame and sends it as an all-routes response. See the preceding rules for what type of response packet is generated for single-route and all-routes explorer packets.

The rrrr bits are reserved, but IBM is now allowing their use to indicate more largest-frame sizes.

In Fig. 3-6*c* the 0045 indicates that the packet traversed ring 4 through bridge 5 and then onto ring 3, which is where the destination station resides. The 0 following the ring number 3 was put in by the bridge to fill out the field to a 16-bit boundary. If there were another bridge in the path to the destination, the next bridge would strip off that 0, put in its bridge number, and then add the next ring number followed by a 0 into the RIF field. Figure 3-6*c* shows the RIF field as if the packet traversed ring 4, bridge 5, ring 3.

Each bridge the packet traverses inserts that type of information into the packet. As the packet traverses the rings and bridges in the network, the path (ring received from, bridge number traversed, ring forwarded to) that it took is inserted into the packet. Each bridge inserts its own information into the packet.

There is one exception to the RD stuffing that a bridge has to do. The first bridge to receive an explorer (indicated by the RIF field set to 2 bytes), must add two RDs. The first RD indicates the ring number that the bridge received the packet on. It also adds its bridge number to the first RD. The second RD only has the ring number that the bridge forwarded the packet to. The bridge number of the second RD is set to 0, simply to fill out the field on a 16-bit boundary. Each subsequent bridge only manipulates the last RD in the RIF, placing its bridge number and one new RD in the field. In this new RD, the bridge places the ring number that the packet was forwarded to. It, too, will use 0s to align the packet on a 16-bit boundary. Therefore, as you have probably figured out, the last RD will always have a 0 in the bridge field.

When the packet reaches the destination, the destination responds with a response packet. Depending on whether the packet was originally sent as an all-routes explorer or a single-route explorer, the destination responds with the opposite of what it received. See the preceding rules on route explorers.

As shown in Fig. 3-6a, there is more than one route to the destination. Therefore, more than one all-routes explorer packet would make it to the destination. The destination will respond to each of the packets. Each packet will take the reverse path back to the originator.

Multiple packets will therefore arrive back to the originator of the packet. Which path does the originator decide to take? The most followed rule is that the first response packet that makes it back to the originator wins. The station uses the route of the first response packet when the other packets arrive; the originating station may discard the follow-on packets.

The foregoing text mentioned that a multiple-response packet may arrive back to the originator of a route discovery packet. This may happen as a result of two of more bridges interconnecting two or more of the same LANs. In transparent bridging for Ethernet, one or more of the bridge ports would be set to blocking, allowing only one path to be active at a time. In source routing for Token Ring, all bridge ports can be and usually are active. This allows for forwarding of all-routes explorers, so multiple packets can arrive at the destination. This is okay and is expected in Token Ring. Every path to a destination should be found, and may the fastest response win. Each packet has different routing entries, for each ring number in a Token Ring LAN is always unique. Only bridge numbers may be reused, because the RIF field is always read as the ring number in, the bridge number, and the ring number out. However, to allow for efficiencies, single-route explorers are also used. They operate with the spanning tree protocol for Token Ring, which shuts down (blocks) ports, enabling only one path for a discovery packet to the destination.

One final note on source route bridges. The spanning tree protocol has been adapted to run with source route bridges. Therefore, STA will shut down redundant loops only for SRE frames. This is only useful when the end station uses the single-route explorer for route discovery. Bridges that have their ports enabled will forward this SRE packet. Bridges with the ports in the blocking mode will not forward the frame.

An end station that submits an all-routes explorer frame for route discovery will still have the frame forwarded by all ports on a bridge, whether it is in blocking mode or not. In this way, source route bridges are using the STA to use the cable plant more efficiently.

Differences

The differences between the two types of bridges should be apparent. Transparent bridges operate completely transparently to any of the stations on the network. Transparent bridges communicate only with other transparent bridges. Using transparent bridges, end stations are not aware of any bridges on the network. Ethernet network stations operate on the network as if the bridges were not there. Therefore, no modifications have to be made to the Ethernet algorithm or the packet format. With a few exceptions, you simply place the bridge on the network and let it start to work.

Just the opposite, Token Ring bridges do not contain any intelligence, except to stuff the packet with ring and bridge numbers. Each one of the bridges must be assigned a ring number for each of its ports, and internally it must be assigned a bridge number. The end station on a Token Ring network has no idea what its ring number is. The bridge handles this for it.

The Source Route algorithm forces the end stations to find each other, producing a lot of overhead on the rings. (*Overhead* is defined as necessary protocol packets that do not contain any user data.) While these packets are being transmitted, no other station may transmit. The packet format for Token Ring has to be modified to allow for the route information. End stations on a Token Ring network do know about the bridges on their network. In order for source route to work, the Token Ring frame has to be modified to allow for the RIF.

One clear advantage for transparent bridges is that they may operate in either Token Ring or Ethernet environments, although they are rarely found on Token Ring LANs. IBM supports Token Ring and source routing for its implementation, and most vendors have followed. Source routing bridges operate only in Token Ring environments.

The IEEE committee that sets standards for LANs has adopted the transparent bridge algorithm for all its LANs (802.3, 802.4, and 802.5). Source routing has been adopted only for IEEE 802.5 and FDDI (but not often implemented). It has not been adopted for any of the other LAN standards.

One big advantage of bridges (no matter what kind) is that they are protocol transparent. This means that any protocol may run on top of them: XNS, TCP/IP, DECnet, AppleTalk, etc., may all run through a bridge, whether it is transparent or source route. This represents a big advantage for bridges. At the time of their introduction, there was one router for every protocol on the network. Multiprotocol routers were in their infancy, and it would be years before they were easily available and reliable. Bridges offered this first off in 1985.

One last thing that may be noted here. What if you want to transfer data between Token and Ethernet networks? The best performance and smallest problems are obtained by routing data between the two types of networks.

Address Conversion

Some protocols are not routable protocols. When a protocol cannot be routed (NetBIOS or LAT protocols, for example), there is a standard adopted by the IEEE committee that will allow for seamless transmission between Ethernet and Token Ring. It is modeled after the IBM 8209 bridge, and it is called a translation bridge because it translates between Ethernet and Token Ring MAC address format. Recall from the discussion of address formats the way Token Ring and Ethernet addresses are formatted and transmitted onto the cable. Ethernet and Token Ring are at opposite starting points for transmissions of bit 0. FDDI follows Token Ring address formats. Therefore we must be able to bridge Ethernet and Token Ring and translate the addresses between the two. The translation bridge does this. For a received Ethernet or Token Ring packet, the bridge reverses the bits in each byte and rebuilds the MAC header. For example, an address of 02-60-80-01-02-03 on Ethernet will translate to 40-06-31-80-40-C0 on Token Ring. This is the first process: MAC address conversion (see the following example).

The second action is that the bridge must be informed of any protocol that may use the MAC address inside the packet somewhere. The most common protocol is IP-ARP. The ARP packet finds MAC addresses that are assigned to an IP address. Note that the bridge must be made aware that it has received an IP-ARP packet and that it must convert the address in the MAC header as well as inside the IP-ARP packet.

Third, the bridge must be told that if a token packet needs to be translated to Ethernet, it must use either the Ethernet or the IEEE 802.3 frame format.

Table 3-1 allows for easy translation of hex strings and should be used to assist you in bit reversing.

EXAMPLE   The following MAC address needs to be converted: 02-C4-20-01-02-03. Let us try the first byte, 02. The left nibble (left-four bits, or 0) are applied to the left-hand column. The right-four bits (the 2) are applied to the top row. Follow the two numbers across and down until they meet in the table. That number, 40, is the bit reversal of the column/row combination. After applying the whole address to the conversion table, the converted adress is 40-23-04-80-40-C0. From Table 3-1, it does not matter whether the address is Token Ring or Ethernet. Each type's MAC address may be applied to the table in the same manner specified to see what it would translate to on the other type of LAN. The disadvantage of this type of bridging is that addresses may be duplicated after the bit reversal process, when forwarded to the next LAN.

Many different frame conversions can be accomplished, depending on the type of received packet. This is beyond the scope of this book. It is fully detailed in my book *Local Area Networking*, published by McGraw-Hill (ISBN 0-07-912256-6).

Routers

Theory

General router information will be covered here. As protocols are discussed in subsequent chapters, routers will be discussed in detail for each specific protocol.

Networks running the appropriate software are usually divided logically by a device known as a *router.* At one time (late 1984 to early 1985) routers were expensive and were very slow for packet forwarding (300 to

TABLE 3-1 Bit Reversal Conversion Chart

First Four Bits (Below)	Second Four Bits (Across)															
	00	1	2	3	4	5	6	7	8	9	A	B	C	D	E	F
00	00	80	40	C0	20	A0	60	E0	10	90	50	D0	30	B0	70	F0
1	08	88	48	C8	28	A8	68	E8	18	98	58	D8	38	B8	78	F8
2	04	84	44	C4	24	A4	64	E4	14	94	54	D4	34	B4	74	F4
3	0C	8C	4C	CC	2C	AC	6C	EC	1C	9C	5C	DC	3C	BC	7C	FC
4	02	82	42	C2	22	A2	62	E2	12	92	52	D2	32	B2	72	F2
5	0A	8A	4A	CA	2A	AA	6A	EA	1A	9A	5A	DA	3A	BA	7A	FA
6	06	86	46	C6	26	A6	66	E6	16	96	56	D6	36	B6	76	F6
7	0E	8E	4E	CE	2E	AE	6E	EE	1E	9E	5E	DE	3E	BE	7E	FE
8	01	81	41	C1	21	A1	61	E1	11	91	51	D1	31	B1	71	F1
9	09	89	49	C9	29	A9	69	E9	19	99	59	D9	39	B9	79	F9
A	05	85	45	C5	25	A5	65	E5	15	95	55	D5	35	B5	75	F5
B	0D	8D	4D	CD	2D	AD	6D	ED	1D	9D	5D	DD	3D	BD	7D	FD
C	03	83	43	C3	23	A3	63	E3	13	93	53	D3	33	B3	73	F3
D	0B	8B	4B	CB	2B	AB	6B	EB	1B	9B	5B	DB	3B	BB	7B	FB
E	07	87	47	C7	27	A7	67	E7	17	97	57	D7	37	B7	77	F7
F	0F	8F	4F	CF	2F	AF	6F	EF	1F	9F	5F	DF	3F	BF	7F	FF

1000 packets per second, or pps). Not only that, some commercial router configuration tables were manually configured, and this was a network management nightmare for medium and large networks. Bridges not only provided self-learning capabilities but could also make forwarding decisions within the cable speed of Ethernet (actually 10,000 pps in 1986 and 14,800 pps today.) But bridges have the disadvantage of building flat networks, which are networks that have no hierarchy. Every node is on an even level with every other node. This means that all nodes have the capability of seeing traffic they do not need to see, namely, broadcast traffic. The protocols that run networks are broadcast oriented. A bridge must forward all broadcast packets to every port but the one that it received it on. Therefore, all broadcasts are forwarded. Simply stated, bridges do not allow for scaling.

But routers do! Routers build hierarchical networks. Routers are devices that interconnect multiple networks, primarily running at the same high-level protocols (TCP/IP, XNS, IPX, etc.). They make forwarding decisions at the network layer of the OSI model. With more software intelligence than simple bridges, routers are well suited for complex environments or large internetworks. In particular, they support active, redundant paths (loops) and allow logical separation of network segments. (Each separate network in a routed environment is given a network number. This is analogous to the area code numbering system for the telephone system.)

Refer to Fig. 3-7. Unlike transparent bridges, which operate at the data-link layer and operate only on the MAC addresses in the packets, routers work with network numbers that are embedded into the data portion of the packet and determine the forwarding of each packet based on this network number. These network numbers are similar to geographic area code numbers. For local calls, you simply dial the seven-digit number. The switching offices of the telephone system will automatically route the call locally to its destination. For a long-distance call (i.e., a call to a number in another area code, say) you must enter the area code before the number in order to complete the call. If you call from New York to Virginia, for example, the telephone switching office in New York will look at the area code and notice that it is assigned to Virginia. It will route the call to Virginia, where the local switching offices there will route the call to its final destination. This is similar to the routing functions in a WAN-LAN connection.

Packets that are bound for nonlocal networks are sent specifically addressed to a router by the software that resides in the network layer of the OSI/ISO model. Depending on the type of network operating soft-

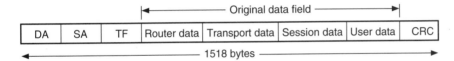

The Ethernet packet is still possible in 1518-byte length. Routing information is entered into part of the data field. Transport-layer software (if there is any) is also placed in the data field.

The amount of space available for data is reduced, but with 1500 bytes there is still plenty of room for user data. Router information length is from 20 to 60 bytes.

The receiving station will know where the router information starts and where it ends. This information will not be mistaken for data.

Everything, whether it is user data or control information for the network stations, is stuffed into the packet and transmitted to the network.

Figure 3-7 Router information in packet.

ware (TCP/IP, XNS, IPX, etc.), there are many different ways of traversing the routers. The following example closely resembles the Internet Datagram Protocol (IDP) of the XNS (Xerox Network System) environment. Other routing protocols are similar and will be explained fully in the succeeding chapters on protocols. XNS was specifically designed to run on a network, and it is the easiest to comprehend. It will be briefly explained here, and will be further explained in the routing sections of all subsequent chapters. Each protocol implements routing a little differently.

Operation

Networks that allow the use of routers must employ what is known as a *network number*. This network number is not the same as a physical address that identifies a station on the network. The network number is the number assigned to identify a particular logical network. It is usually a single entity grouping multiple network workstations. Network stations are grouped together by a single network number. This constitutes a single LAN—a zip code, if you will. The combination of the network number and the physical address of the station on the network uniquely identifies any station on the internet. The term *internet* is used when multiple single networks are grouped together to form a larger network though the use of routers.

If a network station has the network-layer software present, thereby making it fully aware of a internetworked environment, it is commonly used in the following manner. The subsequent text is a general overview for the succeeding chapters, which will explain in detail how each protocol's network layer operates. Most network layers function in a similar manner.

Refer to Fig. 3-8. When the network-layer software of the network operating system receives data from its upper-layer software that are bound for a station that is across a router, the normal process is to look at the network address (not the physical address) of the destination packet and compare it to its local network address. If the network number is different from its own, the network layer software knows that the packet

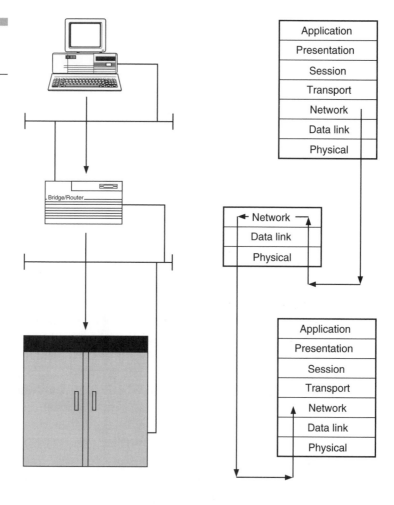

Figure 3-8
A routed environment.

is bound for a remote network. It then attempts to find a router that can process this packet via the shortest route. Routers are the only device that may forward a packet to a remote device on an internet. Network numbers are assigned to a workstation in various ways (protocol dependent).

The network-layer software that resides in an end station may hold a table of the network numbers and associated routers that it knows about. These entries contain the router's physical address, whose network numbers the router is associated with, and the distance to the final destination. If the network number is not in the table, or the end station does not support holding a table, it will request the routers on the network to send it information about a particular destination. In addition, an end station may be configured with a default router. This is an address of a router that the station will send all remote packets to.

If the router returns information about the remote network, the station uses this returned information to address its packet to be handled by a router or a series of routers to enable the packet to reach its final destination. Once formatted, a packet is submitted to the network with the destination MAC address set, not to its final destination, but to a particular router. The final destination network number and host number are embedded somewhere in the network-layer header of the packet. Each router in a path to the destination determines where to forward the packet by reading the network header information in the packet. *This is the basis of packet switching.*

When a router receives a packet, it reads the fields of the packet's network-layer header that contain the routing information, particularly the destination network number. If the destination network number is directly attached to that router, it simply forwards the packet, physically addressed to the final destination. If the destination network is not directly attached to that router, the router will have a list of other routers that it knows about. This listing is known as a *routing table.* The table should have the address of the router next in line to the final destination network. The router then physically addresses the packet directly to that router. When the next router receives that packet, it invokes the same algorithm as the previous router to get the packet to its final destination.

When the packet finally reaches the router with the destination network attached, that router physically addresses the packet to the final destination. The physical source address of the packet will be that of the router, so that the destination will know how to address any response packets. Any return packets by the destination network station are addressed directly to that router and are forwarded in the same manner to the originator of the packet.

It is important to remember that the originating station, knowing the packet is destined for another network, will physically address the packet to the router and not to the final destination. This enables the router to receive and process the packet. Remember that routers only process packets that are physically addressed to them. The final destination network and destination network attachment are embedded in the network-layer header of a packet (refer to Fig. 3-7). The router uses this network-layer information to determine where to route the packet.

With bridging, an end station physically addresses the MAC destination address to the final destination. In a bridged environment, all stations are assigned the same network number.

Another bridge is known as a source route transparent (SRT) bridge. This function is primarily found in Token Ring switches. It follows the IEEE 802.5m standard. The bridge reads the first bit of the source address field of the received packet (this is known as the routing information indicator, or RII, bit). If it is set to a binary 1, then there is source routing information in the packet. If the bit is set to a 0, there is no source routing information in the packet and the packet should be transparently bridged. This is an SRT bridge capability, not a translating capability.

Conclusion

Routing is covered in much greater detail in each of the following chapters. The point of the previous text was to illustrate routing conventions. The next chapter will show the details of each protocol.

The following is a comparison between a router and a bridge.

Routers

1. Routers build routing tables from dynamically discovered network numbers in their domain.

2. Routers automatically allow for redundant paths. Some routing protocols offer the routing of data to a destination over multiple paths (OSPF = equal cost multipath).

3. Routers offer some intelligence in that they will respond to any originating station when the destination is unreachable or if there is a better route to take.

4. Some routers allow for different-size packets on the network. If one segment of the network allows for only 1518 bytes (the maximum Ethernet packet size) and the other segment allows for 4472

(Token Ring maximum per 4 Mbps), most 16-Mbps networks will use the 4-Mbps packet size of 4472. Routers will fragment the original packet and then reassemble it at the destination network.

5. Routers allow for multiple paths to the same destination station. One network station could use one path to reach a destination, and another network station might use an entirely different path but still reach the same destination.

6. Routers segment the network into logical subnets by assigning an ID number to an individual network. This allows for better network management, among other things.

7. Routers do not forward local broadcast packets. This can eliminate what are known as *broadcast storms*.

8. Routers operate at the OSI network layer (layer 3).

9. Routers offer efficiency through the building of hierarchical networks.

10. Routers offer better Ethernet to Token Ring or FDDI connectivity. A nightmare can occur when trying to connect FDDI, Token Ring, and Ethernet via bridging or switching. It works, but the problems that can occur are not worth the risk. Route when traversing multiple access methods.

The reason routers easily provide the capability in item 10 is that routers fully understand their protocols and the LANs (Ethernet, Token Ring, etc.) they are connected to. End stations send their packets directly to the router. Upon receipt of a packet, the router strips off the MAC header of the received packet. It does a routing table lookup, and based on the information found in the table, it builds a new MAC header appropriate for the port type that it will forward the packet out to. This allows a router to build a new packet header when the packet is forwarded. Routers understand which part of their packet needs translation. Therefore, routers easily provide this address translation for packets that must be transmitted between workstations on different types of LANs.

Bridges

1. Bridges (and data-link layer switches) are less expensive than traditional routers. However, an advancement known as layer 3 switching is driving the cost of routers, on a per port basis, down to the level of layer 2 switches and bridges.

2. Transparent bridges and switches allow for loops, but only one path to a destination network can be active at any one time.

3. Bridges and switches are less complicated devices.

4. Transparent bridges are invisible, or transparent, to any other device on the network (except other bridges). Network stations do not need specialized software to operate with a bridge. This can reduce the network software executable size in a network workstation.

5. Bridges increase the available bandwidth of a LAN by physically segmenting network stations to their respective LANs.

6. Bridges have slowly but surely turned themselves into devices known as switches. Basically, a switch is a bridge with lots of ports. A switch is the combination of a wiring concentrator and a bridge. The technology to build a bridge using application-specific integrated circuits (ASICs) has greatly reduced the cost per port of a bridge. Therefore, instead of sharing a whole Ethernet segment to a single port on a bridge, the switch port connects directly to the network station. A switch can connect to shared networks, or it can connect directly to a network attachment. It can forward at wire speeds. Switches can commonly be found for $100 per port (whereas in 1985 a two-port bridge sold for $5000 per port).

Multiprotocol Routers

Bridges are rarely sold standalone any more; their functions have been incorporated into routers. When a router can support more than one protocol (either multiple network layers or an integrated bridge function), it is called a *multiprotocol router*. Multiprotocol routers not only bridge but also route many different types of network software—all out of the same box.

The operation of the multiprotocol router is straightforward, as shown in Fig. 3-9 and flowcharted in Fig. 3-10. With few exceptions, when the packet arrives at the router, the physical source address of the packet is checked and, if needed, added to the bridge table. Next, the router checks the destination address of the packet. If the destination address is that of the multiprotocol router, the multiprotocol router attempts to route the packet (remember, in a routing environment all packets to be routed are addressed to a router and not to the final desti-

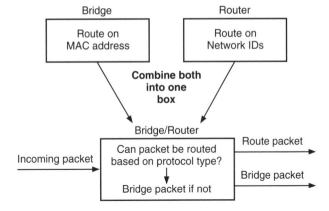

Figure 3-9
Multiprotocol
bridge/routers.

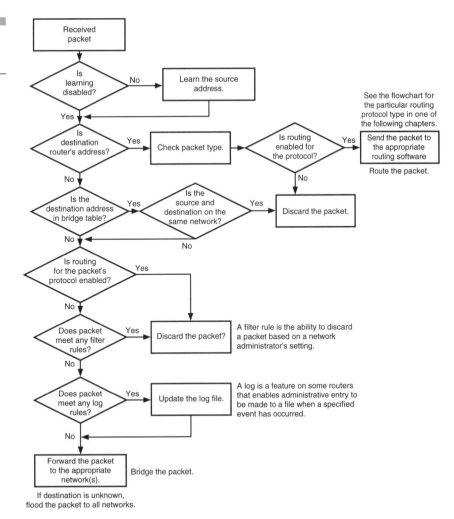

Figure 3-10
Multiprotocol router
flowchart.

nation). If the address is not that of the multiprotocol router, the router bridges the packet. Broadcast packets are checked also, and forwarded if necessary.

Again, it is best if the bridging and routing functions are allowed to reside in the same unit. Such a unit is called a *multiprotocol bridge/router.* This type of router knows when to bridge the packet and when to route it. The two flowcharts in Figs. 3-9 and 3-10 show data flow in a multiprotocol environment and in a pure routing environment.

Translating Bridges

Now that bridges and routers have been explained, we come back to the question of how Ethernet/IEEE 802.3 networks converse with Token Ring networks. The answer to this is translation bridging, and the most notable device that accomplishes it is the IBM 8209 Ethernet to Token Ring bridge. A question arises here: Why should we bridge the two, and not simply route wherever we can? The answer is that not all protocols are routable. NetBIOS, LAT (a DEC terminal and printer protocol), and SNA cannot be routed and must therefore be bridged. If there are devices that reside on both Ethernet and Token Ring, we must provide a translation bridge to facilitate communication.

The 8209 builds two separate database tables. On the Ethernet port, it operates in promiscuous mode, watching all data traffic on its Ethernet. It builds a forwarding table based on all source addresses it has seen. Along with this, it creates an entry for the address to indicate whether it is an Ethernet or IEEE 802.3 formatted frame.

The 8209 also builds a forwarding table for the Token Ring port. The 8209 caches the source address of a received source route discovery frame into that table. Mated with this source address is the received RIF associated with the source MAC address. Therefore, when the 8209 places the source address and its RIF into the table, it translates the frame and forwards the frame to the Ethernet port. One of the key functions of any translation bridge is the bit reversal process that must take place when the MAC address of the received packet is converted between Ethernet and Token ring frames. Ethernet uses the canonical format, and Token ring and FDDI use the noncanonical format; in other words, Ethernet holds the first bit transmitted as the rightmost bit of the byte, and Token ring holds it as the leftmost bit of the byte, as explained before in this chapter.

The interesting thing about the 8209 is that it sees the Ethernet as a Token Ring. Therefore, when translating from Ethernet to Token Ring, it places a bridge and ring number of the Ethernet segment in the RIF. This is never seen on the Ethernet cable, but it is on the Token Ring cable. In this way, the source route of a specifically routed frame can be forwarded to the Ethernet cable.

What happens when a source route frame is received and it needs to be forwarded to the Ethernet? How does the bridge know which frame type to use? The 8209 is not perfect, and there is a default configuration setting to be used for unknown frame types. This means that you have to know a little bit about your network before setting up one of these devices.

Gigabit Ethernet—
The Need for Speed

Many types of networks have entered our communications world. Token Ring claims to have its roots tied to the 1960s (which would explain a lot about Token Ring). DECnet (over serial lines) came to us in the late 1970s. The experimental Ethernet that ran at 2.94 Mbps started around 1974. In 1980, the adoption of version 1.0 and version 2.0 (1982) of Ethernet by Digital, Intel, and Xerox (the DIX standard) pushed up the speed and changed a few other things in the MAC header and allowed for 10 Mbps transmission. This speed seemed to be more than enough for years to come, but this was 1980, and the personal computer as a fully productive business tool was only a gleam in an eye. As advances in microprocessor technology pushed the speeds of personal computers ever higher, prices of personal computers decreased and the desire to connect every computer together (whether it is a mainframe, mini, or personal computer) created congestion and has increased the need for bandwidth.

Topology designs require that the backbone of a network be 10 times faster than the network connecting to it. FDDI satisfied this requirement in the early 1990s as a 100-Mbps system. FDDI never emerged as a desktop solution and has remained primarily as a backbone solution. But this creates a real problem with major differences between Ethernet and FDDI. An FDDI station cannot simply attach to an Ethernet network, and vice versa. Bridges and routers accomplished this.

Advancements in chip technology (application-specific integrated circuits) increased the speed of Ethernet without changing the basic operations (and frame format). Fast Ethernet, at 100 Mbps, soon became common (not yet pushing out FDDI, but competing with it). Fast Ethernet has some limitations, such as the number of repeaters possible, but the basic MAC frame format and topology do not. Fast Ethernet can run full duplex with ease and has many other network management advances as well. But the key function of Fast Ethernet was that there were no changes to the frame format. Therefore, no translations had to take place, and this allowed for an advancement in frame-to-frame switching. With some simple buffering techniques, the forwarding of data between Fast Ethernet and Ethernet is accomplished transparently.

Now comes one more advancement: Gigabit Ethernet. It is coming out of the same IEEE 802 working group as did all previous generations of Ethernet. The Gigabit High-Speed Study Group was originally formed in November 1995. The group's mission was to decide if Gigabit Ethernet warranted development of a standard, and it resulted in the writing of a project authorization request (PAR). The PAR was accepted at the July 1996 plenary meeting, at which time the Gigabit Ethernet task force was formed as the IEEE 802.3z working group.

Technical submissions were received until the November 1996 meeting, at which time the committee established final objectives and work items that would encompass the Gigabit Ethernet standard. The objectives are:

- To develop a 1000-Mbps Ethernet that allows for half- and full-duplex operation
- To continue to use the IEEE 802.3 Ethernet frame format
- To continue to use the CSMA/CD access method and allow up to one repeater per collision domain
- To develop backward compatibility to 100BaseT and 10BaseT 802.3 Ethernet standards

The future plan of the task force is to prepare a draft standard by the July 1997 meeting, at which time the draft standard will go to letter ballot. Following that it will go through the standard comment and editorial review process, which can take another eight months before it becomes an official IEEE 802 standard.

Another group dedicated to Gigabit Ethernet is the Gigabit Ethernet Alliance. The GEA is an open industry forum (made up of companies such as Bay Networks, Cisco, Madge, etc.) whose mission is to accelerate the development of the IEEE standard for Gigabit Ethernet, and to promote this new technology in the marketplace through participation in trade shows, technology seminars, and interoperability testing. The GEA has a Web site at *www.gigabit-ethernet.org*. GEA member companies are grouped into steering members and participating members. This organization works in conjunction with the IEEE process by providing a more frequent technical forum for multivendor development and testing of the Gigabit Ethernet technology in development. Ultimately the work of the alliance is fed into the IEEE as submissions, and there it undergoes further scrutiny.

Gigabit Ethernet should be ratified by June 1998.

Ethernet Technical Review

The topology of a shared 10-Mbps Ethernet was dictated by the CSMA/CD *collision domain*. A collision domain is the visibility distance between stations on a network—that is, the round-trip travel time between two end stations on the network. This is necessary because two stations can begin transmission at the same instant on a quiet network, and they need to be able to detect this fact within the period of time known as the *collision window* in order for the network to operate correctly. In other words, the transmission must begin, propagate to the farthest reaches of the network, and return within a defined period of time.

Refer to Fig. 3-11. In this case the collision domain is a network collision domain, or a diameter of roughly 3 km. This means that a minimum packet size is 64 bytes or 512 bits (called a slot time). This is the maximum amount of time (in bits) to travel this distance. At a 10-Mbps transmission rate each bit requires 100 ns (nanoseconds) of time to transmit. Ethernet's origins are based on a bus topology, which is a shared medium that allows for up to 1024 stations to attach to single collision domain.

This network topology can be extended by using bridges (commonly known as switches today), or routers. Both of these devices terminate

Figure 3-11
10BaseT configuration.

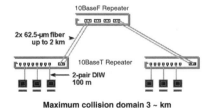

collision domains. For example, if you have two network segments separated by a single bridge you now have two collision domains.

Refer to Fig. 3-12. When Ethernet is scaled to 100 MB/s the collision domain is still 512 bit times. However the length of each bit is only 10 ns, so the total distance is reduced by a factor of 10. This worked for Fast Ethernet because 100 m is the maximum distance of a UTP horizontal link installed in conformance with ISO 11801 (EIA/TIA 568), and it even allows for a two-repeater configuration.

To run distances longer than 100 m, point-to-point full duplex links are allowed. Full duplex effectively turns off the CD portion of the CSMA/CD protocol. When you think about it, if there are only two nodes on the cable plant, why do you need collision detection?

Fast Ethernet topologies extend to more than 10 km using Fast Ethernet switches that run full duplex and single-mode fiber. This backbone type of network can then be connected to other types of shared and shared/switched Ethernet networks to complete the topology.

Fast Ethernet as a trunk or backbone protocol was short lived. New NICs known as 10/100 cards were installed, because the price difference between the 10 and 10/100 was minimal. But if the edge devices are running at the same speed as the backbone or trunks, what advantages did Fast Ethernet give us on the backbone? Even faster backbones needed to be created. Asynchronous Transfer Mode became the buzzword of choice in 1994 because its scalable properties were outstanding. However, ATM is beyond the scope of this book.

Proponents of Ethernet were adamant that the speed of Ethernet could be increased again by a factor of 10, thus, producing 1000-Mbps Ethernet. But remember, speed is measured in bit times (now it is 1 ns in length), and if we increase the speed, we reduce the collision domain size. Therefore, without modifying the collision domain, the maximum length is about 20 m. ATM proponents thought this absurd, and many stated that Gigabit Ethernet was doomed. But to show you how strong

Figure 3-12
100BaseTX-FX topology.

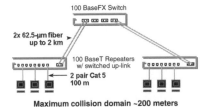

the Ethernet name was, Gigabit Ethernet is alive and well and products for it are already available.

Figure 3-13 shows a comparison of the different Ethernet versions. The original Ethernet was based on coaxial cable (thick and thin versions); later, development of additional physical media options, including fiber optics and UTP, soon came about as the advantages of the star topology soon became apparent (Token Ring's physical topology is based on the star topology). Ethernet advocates were determined to not allow Token Ring to undermine the advantages of Ethernet, and hence Ethernet based on UTP cabling was born.

Starting Up Gigabit Ethernet

One of the most difficult tasks in building a new higher-speed communications technology is development of the new physical layer. Fast Ethernet began by taking two existing technologies and creating a new standard in a very short period of time. It took the Ethernet MAC standard (frame definition and CSMA/CD protocol), and it defined how to marry that with the FDDI physical layer from TP-PMD running at 100 Mbps. By using the FDDI physical layer, Fast Ethernet avoided this time-consuming part of the process.

Later physical layers for Fast Ethernet have in fact taken longer to develop and deploy (100BaseT4 and 100BaseT2). While they enrich the overall capabilities of Fast Ethernet, it is good that they did not impact the initial deployment of Fast Ethernet products.

Similarly, Gigabit Ethernet is taking existing technology and creating a new standard. It is taking the Ethernet MAC and modifying it as nec-

Figure 3-13
Different Ethernet technologies.

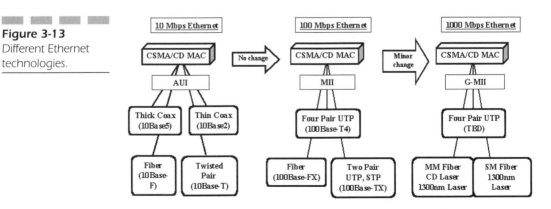

essary to operate in a gigabit LAN. In addition, it is taking an existing physical layer, this time from Fiber Channel, and modifying it to support Gigabit Ethernet. In this case the modification is necessary to change the operating data rate of the Fiber Channel components from 800 Mbps to 1000 Mbps. While this involves a significant amount of effort, it should not require the amount of work an entirely new physical layer would require.

It is a goal of Gigabit Ethernet to support a 200-m topology single-repeater network. This allows for 100-meter links between end stations and servers operating over either copper or fiberoptic networks. Fiberoptic networks will come first, and Gigabit Ethernet over copper cabling may come next.

The objective of the Gigabit Ethernet standards activity is to standardize a technology capable of supporting 100-m links with CSMA/CD. Basically it is an extension to the existing 10- and 100-Mbps versions of Ethernet. It is an IEEE standard (draft standard as of this writing). It will support both CSMA/CD and full-duplex (IEEE 802.3x) modes of operation. Additionally, full duplex includes an optional standard technology for flow control, so future switching products based on Fast Ethernet and Gigabit Ethernet will include this enhancement. Full-duplex technology has the additional benefit of eliminating Ethernet collisions, and it can provide a better option for high-performance devices such as servers.

The essential element is the frame format. This allows Ethernet, Fast Ethernet, and Gigabit Ethernet switches to be simple and low-cost. Except for design considerations, it also allows companies to fully support the standard from day one. A company will have to upgrade its protocol analyzer, but the operation of Ethernet is well known. Furthermore, changes in protocol software are minimal. A new driver to support the gigabit controller card is all that is needed.

Gigabit Ethernet is still a star-wired standard, and it is targeting copper (probably won't happen) and fiberoptic transceivers for its media. The topology rules (CSMA/CD mode) are essentially the same as for Fast Ethernet, except that it supports only a single repeater, not a two-repeater configuration. Ethernet allows for four repeaters. There is in Gigabit Ethernet a new concept, that of a full-duplex buffered repeater. This is still being defined; it will possibly provide for full-duplex operation over a repeater-based network, with a better cost/performance ratio than existing repeaters.

In a similar way Fast Ethernet (100BaseT) used the existing media access control layer, but deployed a new interface specification (media-independent interface, MII) to allow for physical layer flexibility. Over the last three years Fast Ethernet has seen the development of four different physical layers. This was explained in the previous section on Fast Ethernet.

Gigabit Ethernet will support the same strategy. Through this standard interface (something similar to the MII of Fast Ethernet) Gigabit Ethernet intends to support a wide variety of media options, such as CD laser (short distance) and "long wavelength" laser (long-distance multimode and single-mode fiberoptic) transmission. It does not appear that Gigabit Ethernet will be running over copper for its first version. Later versions may support this.

Refer to Fig. 3-13. Gigabit Ethernet networks that operate on shared environments have a big problem: the collision window. Using current methods of CSMA/CD, the collision domain can extend 20 m. This is unacceptable. Again, full-duplex switched Gigabit Ethernet environments do not have this problem, for the CSMA/CD algorithm is turned off (there are no other devices on their cables to contend with).

Note that even though we are running at 1000 Mbps, the Gigabit Ethernet MAC is essentially the same, although it is being modified slightly to allow for 100-m repeater links. The rules that have changed are on the physical layer. This provides a great advantage in that deploying a Gigabit Ethernet backbone network does not require a great deal of new training, replacement of existing equipment, new test tools, or software revisions for the switches or routers that you may have deployed.

There are a couple of approaches to this problem. The current approach the IEEE committee is taking is to extend the collision window from the current 512 bits to 512 bytes. This will require padding each short frame in a CSMA/CD network to a minimum length of 512 bytes. This has an impact on network efficiency (discussed later).

The problem with this is that the existing Ethernet/Fast Ethernet MAC requires that the distance of the network must be no larger than the amount of time it take 512 bits of data to propagate across the network. In the case of Fast Ethernet this time is $512 \cdot 10$ ns $= 5.12$ μs. In the case of Gigabit Ethernet this time is only 512 ns, which would translate into a network diameter of roughly 20 m.

Refer to Fig. 3-13. In order to meet the goal of 200 m, the IEEE 802.3z committee is considering extending the collision window from the current 512 bits to 512 bytes. This means that the minimum Ethernet frame will be extended from the current 64 bytes to 512 bytes as well. The committee contemplates accomplishing this by padding existing Ethernet frames with special codes to indicate an extended carrier.

Figure 3-13 shows how carrier extension works. The minimum Ethernet frame is 64 bytes. Carrier extension is a special code that is added to the frame after the FCS, which provides the time necessary for the CSMA/CD protocol to detect a collision event on a gigabit network. This results in the loss of efficiency for shared networks, as shown in the following.

Since data are transmitted during the extended carrier, performance is limited to that of a 100-Mbps network. The advantages that occur when using this method are noticed when the network is carrying large frames.

Full duplex does not have this problem, for the CSMA/CD protocol is disabled and carrier extension is turned off. This means that full-duplex Gigabit Ethernet links actually have 10 times the throughput of Fast Ethernet links.

Figure 3-14a illustrates the relative throughput (in frames per second) for Gigabit Ethernet (both GE with and without carrier extension) and for Fast Ethernet (FE). It can be noted that at small frame sizes Gigabit Ethernet with carrier extension has marginally faster throughput than Fast Ethernet. However, at larger frame sizes there is no carrier extension, and Gigabit Ethernet achieves 10 times the throughput of Fast Ethernet.

Figure 3-14b depicts the actual throughput of information on Gigabit Ethernet V's Fast Ethernet in Mbps. As this chart demonstrates, Fast Ethernet follows the classic curve dependent on frame size. (This does not factor in the effects of congestion on efficiency). Gigabit Ethernet without carrier extension follows a similar curve, but with 10 times the speed. However, Gigabit Ethernet with carrier extension has lower performance until the frame size gets to the minimum carrier length of 512 bytes. These theoretical transmission rates can actually be achieved in a collision-free environment using full-duplex mode.

Another way to look at it is shown in Fig. 3-14c. It shows that Gigabit Ethernet with carrier extension has reduced efficiency at small frame sizes and then achieves the characteristic Ethernet curve when it crosses the 512-byte threshold. This may have an impact on performance for smaller Ethernet packets. To compensate, a new feature called *packet bursting* has been incorporated. This is the buffering of multiple packets and their transmission as one frame of 512 or more bytes. Another method that is being investigated is the use of buffered repeaters, in which packets are buffered at the repeater and then repeated as a single transmission.

Figure 3-14
Throughput in frames
per second of Gigabit
Ethernet (GE) with
and without carrier
extension and of Fast
Ethernet (FE).

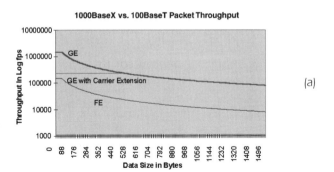

(a)

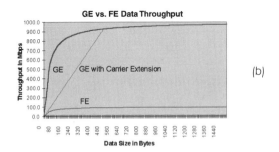

(b)

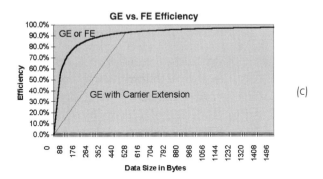

(c)

Gigabit Ethernet Transmission Methods

Another change in Gigabit Ethernet is standardizing on two different
fiberoptic systems. One is designed for short-haul applications, and the
other is designed for longer-distance applications. Refer to Fig. 3-15. Gigabit Ethernet supports two different physical encoding technologies and
four different media transmission options. Most of Gigabit Ethernet's

Figure 3-15
1000BaseX topology.

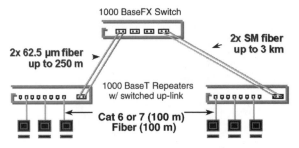

applications will use the standard Fiber Channel encoding, which is an 8B/10B code (8 bits of data are converted into 10 baud along with clock on the cable).

The four transmission options are:

1. CD lasers (using the 850-nm laser wavelength) for low-cost short-distance fiberoptic transmission (550 m)
2. Long-wavelength (1300-nm) lasers for long-distance fiberoptic transmission (3 km)
3. Copper transceivers for simple 25-m coax links (to be determined)
4. Copper transceivers for 100-m CAT5 UTP links (to be determined)

The task force is also including a specification for a media-independent interface just like the AUI connector on 10BaseT and the MII on 100BaseT. This one has the name of *gigabit media-independent interface* (GMII).

Figure 3-16 shows the relationships between the different components of the Gigabit Ethernet standard. At the top of the figure is the Gigabit MAC, which implements the carrier extension necessary for CSMA/CD operation. The Gigabit Ethernet MAC is designed to connect to the physical encoder/decoder through the gigabit media-independent inter-

Figure 3-16
Gigabit Ethernet physical layer.

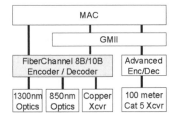

face (GMII). Two different encoder/decoder options are shown, one using the standard Fiber Channel 8B/10B encoding scheme, and a more advanced one for category 5 UTP applications for 100-m links.

This encoder is designed to support 1300-nm optics for long-distance operation over 10-μm single-mode and 62.5-μm multimode fiberoptic cable. It also supports 850-nm optics (based on low-cost CD lasers) for short-distance operation over 50-μm and 62.5-μm fiberoptic cable. Finally, this encoder/decoder can drive coax-based copper cables for short-distance interconnect applications.

A new advanced encoder/decoder is being defined to support transmission over category 5 UTP cable plants at a distance of up to 100 m. The speed of the 8B/10B Fiber Channel encoder/decoder is increased to 125 MHz to support 100-Mbps Gigabit Ethernet.

The current Fiber Channel Interface operates at 1.063 gigabits per second (Gbps), and it will be enhanced to 1.250 Gbps as soon as technology allows (application-specific integrated circuit development). Using 8B/10B and 1.250 Gbps speeds will allow Gigabit Ethernet to run at the proposed 1000 Mbps.

Two different kinds of fiberoptic transceivers are defined for use in Gigabit Ethernet applications: 850-nm CD laser transceivers (known as short-wavelength transceivers) are defined for short-distance transmission over 50-μm and 62.5-μm multimode fiberoptic cable, and 1300-nm laser transceivers (known as long-wavelength transceivers) are defined for use with 62.5-μm multimode fiberoptic cable and 10-μm single-mode fiberoptic cable. Both of these kinds of laser optics uses the Fiber Channel 8B/10B encoder.

Table 3-2 summarizes the various encoding, transceiver, media, and distance options defined for Gigabit Ethernet. Interesting to note is that 50-μm multimode fiber provides for 550-m links with Gigabit Ethernet, while 62.5-μm multimode fiber only supports 260-m links. Laser optics at 1300 nm are really 10-μm optical transmitters that can drive either 10-μm single-mode fiber or 62.5-μm multimode fiber. Single-mode fiber distances of up to 3 km are supported, making this a viable technology for small campus networks (traditional 10-μm distances are 10 km or more for other communications standards). Campus multimode 62.5-μm fiber has more significant limitations making it usable up to only 800 m.

Figure 3-17 illustrates the standard horizontal cable plant that is used in most structured wiring installations today. The total distance from the desktop divide to the wiring closet hub is 100 m. The international standard for this architecture is ISO 11801, which was previously commonly known in the United States as EIA/TIA 568. Gigabit Ethernet sup-

TABLE 3-2 Gigabit Ethernet Physical Distances (*Courtesy of the Gigabit-Ethernet Alliance*)

Standard	Fiber Type	Diameter, μm	Modal Bandwidth, MHz · km	Minimum Range, m
1000BaseSX	MM	62.5	160	2—220*
1000BaseSX	MM	62.5	200	2—275†
1000BaseSX	MM	50	400	2—500
1000BaseSX	MM	50	500	2—550‡
1000BaseLX	MM	62.5	500	2—550
1000BaseLX	MM	50	400	2—550
1000BaseLX	MM	50	500	2—550
1000BaseLX	SM	9	NA	2—5000

*The TIA 568 building wiring standard specifies 160/500 MHz · km multimode fiber.

†The international ISO/IEC 11801 building wiring standard specifies 200/500 MHz · km multimode fiber

‡The ANSI Fiber Channel specification specifies 500/500 MHz · km 50-μm multimode fiber and 500/500 MHz · km fiber has been proposed for addition to ISO/IEC 11801.

Figure 3-17
ISO 11801 horizontal cable plant.

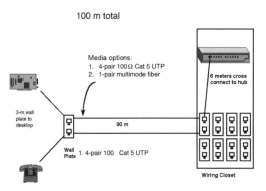

ports two of the many cable types defined in ISO 11801, namely one-pair multimode fiberoptic (50-μm or 62.5-μm), or four-pair category 5 UTP (UTP requires development of a new physical layer encoding and transmission technology).

Tables 3-3 and 2-4 show comparisons between Ethernet, Fast Ethernet, and Gigabit Ethernet.

TABLE 3-3

Gigabit Ethernet
Summary

	Ethernet	Fast Ethernet	Gigabit Ethernet
Speed	10 Mbps	100 Mbps	1000 Mbps
Cost	X	2X 10 BT	2–4X 100 BT
IEEE standard	802.3	802.3u	802.3z
Media access protocol	CSMA/CD	CSMA/CD	CSMA/CD
Frame format	IEEE 802.3	IEEE 802.3	IEEE 802.3
Topology	Bus or star	Star	Star
Cable support	Coax, UTP, fiber	UTP, fiber	UTP*, fiber
Network diameter (max.)	2500 m	210 m	3 km
UTP link distance (max.)	100 m	100 m	100 m†
Media-independent interface	Yes (AUI)	Yes (MII)	Yes (GMII)
Full duplex capability	Yes	Yes	Yes
Broad multivendor support	Yes	Yes	Yes
Multivendor availability	Now	Now	1998

*Standard not available
†Not adopted.

Products

What kind of products can you expect from this technology? Basically, whatever the market demands and will pay for. This includes buffered repeaters, switches, and router interfaces. Combinations of these products will also appear as the topology demands. Products such as combined hub repeaters are expected.

Gigabit Ethernet is also expected to run with all available protocols as well, including Novell NetWare (IPX/SPX), TCP/IP, and many others. Again, this is a change at the physical layers. Upgraded drivers will be required to support this in desktop systems, but this is a very minor change. The real advantage, besides speed, is that CSMA/CD is a known and trusted technology. Major upgrades to the customer's networks are not necessary, and support staffs do not have to be retrained. New protocol decoders will prevail, but they will be very cheap as well. Even books on Gigabit Ethernet are expected out, although they will be very small. Most will provide topology design examples and troubleshooting techniques.

Quality of Service

This probably causes a bigger debate than carrier extension. Gigabit Ethernet provides a large pipe to send information through, but simply providing a larger pipe does not always alleviate bandwidth problems and other considerations such as video and voice over a network.

Gigabit Ethernet in and of itself does not provide any quality-of-service guarantees. Remember, this is an extension of Ethernet and the CSMA/CD algorithm, which has never provided quality-of-service guarantees. Therefore, the question remains, how do we get quality-of-service guarantees on Ethernet? The answer will be found in external protocols that are currently being developed, such as Resource Reservation Protocol (RSVP) for the TCP/IP protocol. The problem with these add-ons is that they are protocol specific. RSVP provides something for TCP/IP, but what about Novell NetWare? On the IEEE side, there are IEEE 802.1p and IEEE 802.1Q They provide a method of tagging packets and indicating priority in them. But some of these, like IEEE 802.1Q, may require changes to the chip set in many switches and NIC cards, because changes have been made to the MAC format. IEEE 802.1Q provides for a four-byte entry of type VLAN and VLAN ID right after the source address of an Ethernet packet, pushing the maximum frame length to 1522 bytes. But priority protocols (i.e., RSVP) to IEEE 802.1x priorities have not been clearly defined. Therefore, this subject is still under many considerations as of this writing. Today, ATM still has the best quality-of-service guarantees for any protocol that is run over it.

Finally, Table 3-2 summarizes Gigabit Ethernet.

IEEE 802.2

Introduction

IEEE 802.3 frames generally support the use of IEEE 802.2 data-link headers. The IEEE 802.x specification expects the use of IEEE 802.2, in either mode or though a protocol known as the SubNetwork Access Protocol (SNAP). Both of these frame types are discussed in this chapter.

First, let's dive into the functions of the IEEE 802.2 protocol known as logical link control (LLC). LLC is the standard published by the IEEE 802.2 standards body. The protocol of IEEE 802.2 LLC is actually a subset of another protocol, High-level Data Link Control (HDLC), a specification presented by the International Standards Organization (ISO). Specifically, LLC uses a subclass of the HDLC specification. It is formally classified as BA-2,4. It uses the asynchronous balanced mode (ABM) and the functional extensions, options 2 and 4. Well, enough of the techno-history. Suffice it to say that the protocol of LLC allows network stations to operate as peers in that all stations have equal status on the LAN.

It is important to understand that IEEE 802 is a data-link protocol. It controls the link between two stations. It has nothing to do with the upper-layer protocols that may run on top of it. As shown in Fig. 4-1, the IEEE 802 committee divided the data-link layer into two entities: the MAC layer and the LLC layer. Splitting the data-link layer provides some

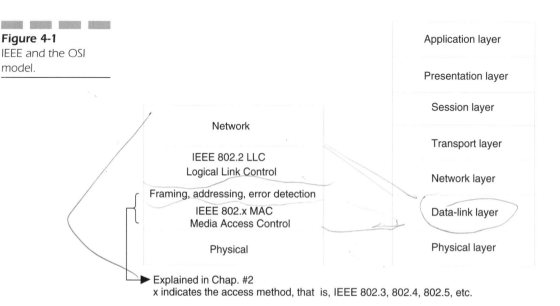

Figure 4-1
IEEE and the OSI model.

Application layer

Presentation layer

Session layer

Network

Transport layer

IEEE 802.2 LLC
Logical Link Control

Network layer

Framing, addressing, error detection

IEEE 802.x MAC
Media Access Control

Data-link layer

Physical

Physical layer

Explained in Chap. #2
x indicates the access method, that is, IEEE 802.3, 802.4, 802.5, etc.

benefits. First, it provides for peer-to-peer connectivity between two network stations, which reduces the LAN's susceptibility to errors. Since it is a subset of the HDLC architecture, it provides a more compatible interface for wide area networks. It is independent of the access method (Ethernet, Token Ring, etc.). The MAC portion of the data-link layer is protocol-specific. This allows a IEEE 802 network more flexibility. LLC is placed between the MAC layer specification of the data-link layer and a network-layer implementation. LLC2 (connection-oriented) is used to link a local area network to a wide area network and to link network stations and SNA hosts. An example is the IBM LAN Server program. This network workgroup operating system is based on a session-layer protocol, NetBIOS, that operates functionally similar to Novell NetWare. IEEE 802.2 provides data-link services for the upper-layer protocols. These include data transfer and link establishment, control, and disconnection between two stations.

When the IEEE started work on the LLC subset, they knew that providing a connection-oriented service only (type 2, or LLC2) would limit the capability of this protocol in the LAN arena. Most applications currently operating on a LAN do not need the data integrity functions provided with the LLC2 protocol. Furthermore, time-sensitive applications could not tolerate the tremendous overhead involved with establishing and maintaining a connection-oriented session. Also, most LAN protocols already provided for this type of functionality in their software.

With this in mind, the IEEE 802 committee provided a connectionless mode of the LLC protocol as well as a connection-oriented specification. Thus, the IEEE 802.2 committee allowed for three types of implementation for LLC:

Type 1. known as LLC1, uses the UI (unsequenced information) frame and sets up communication between two network stations as an unacknowledged connectionless service.

Type 2. commonly known as LLC2, uses the conventional I (information) frame and sets up an acknowledged connection-oriented service between two network stations.

Type 3. using something called AC frames, sets up an acknowledged connectionless service between two network stations.

Types 1 and 2 will be discussed in this chapter.

First, we will present an overview of connection-oriented service versus connectionless service. Connection-oriented service means that two stations that wish to communicate with one another must establish a connection at the data-link layer. When station A wants a connection to station B, it sends control frames to station B to indicate that a connection is desired. Station B may respond with a control frame indicating that a session can be established, or it may respond to Station A that it does not have a connection. If a connection is allowed to be established, station A and station B will exchange a few more control packets to set up sequence numbers and other control parameters. After this is accomplished, data may flow over the connection. The connection is strictly maintained with sequence numbers, acknowledgments, retries, etc.

Connectionless service is the opposite of connection-oriented service. A connection is not established at the data-link level before data are passed. A connection will still be established before data are transmitted, but it will be the responsibility of a particular network protocol (TCP/IP, NetWare, etc.). Connectionless service means that a frame is transmitted on the network without regard to a connection at the data-link layer. It is the responsibility of the upper-layer software of a network operating system.

This method of data transfer does not provide for error recovery, flow, or congestion control. There is no connection established before the transmission of data, and there is no acknowledgment upon receipt of data. This type of functionality requires less overhead.

This protocol was allowed so that existing protocols, such as TCP/IP, XNS, and NetWare, could migrate to the IEEE 802.2/802.3 protocol. The connectionless protocol of LLC specifies a static-frame format and allows network protocols to run on it. Network protocols that fully implement a transport layer generally use type 1 service.

Table 4-1 shows the frame formats, commands, and responses used by the three types of LLC. This table should be referred to throughout the chapter. Entries in this table will be explained in the next few pages.

LLC Type 2 Operation

Connection-Oriented LLC2—Asynchronous Balance Mode

This is the most complicated of the services, so it will be explained first. Connection-oriented services provide the functions necessary for reliable data transfer (similar to a transport-layer function). Connection-oriented services allow for error recovery, flow, and congestion control.

TABLE 4-1

LLC Commands
and Responses

Type of Frame	Format	Command	Response
1		UI	
		XID	XID
		TEST	TEST
2	I	I	I
	S	RR	RR
		RNR	RNR
		REJ	REJ
	U	SABME	UA
		DISC	UA
			DM
			FRMR
3		AC	AC

They involve the use of specific acknowledgment that the connection is established. They provide for flow control, ensuring that the data arrive in order as they were sent and that the connection is not overloaded. This type of circuit has a tremendous amount of overhead compared to LLC1.

Connection-oriented link methods were originally used for data transfer through serial lines. Serial lines are used to connect networks through the phone system. Just a few years ago, serial lines tended to carry a lot of noise (unwanted electrical signals permeating the copper wire). Part of the HDLC protocol addresses the reliability of data. Although modern data lines from the phone company are conditioned to handle data (mainly through upgrades in the cable plant involving the use of fiber cable), there still are some lines that remain noisy, and the connection-oriented services are used to handle this type of link.

Connection-oriented services are used with LANs today, most commonly for protocols that do not invoke a transport or network layer. A good example of this type of protocol is NetBIOS. The IBM LAN Server program uses this type of connection. However, most LAN protocols have network, transport, and session layers built in and therefore use the connectionless mode of LLC. This type of data-link operation is used on a Token Ring network. Also, most Ethernet operations do not use connection-oriented services at the data-link layer.

Figure 4-2
Class versus opera-
tion matrx.

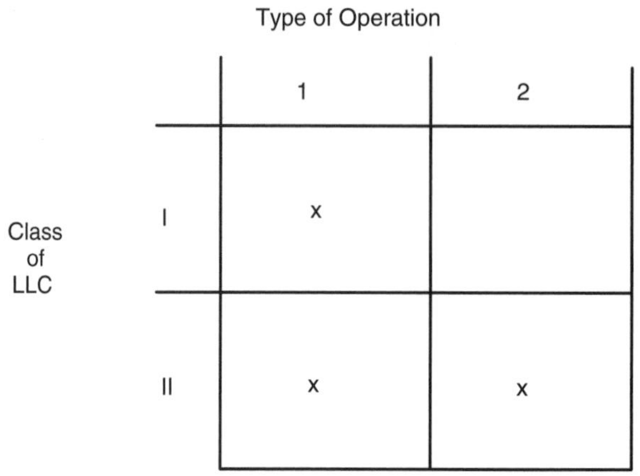

Type of Operation

Figure 4-2 shows the class of service in relation to the LLC type. This figure shows that network stations supporting type 2 can provide both type 1 (connectionless) and type 2 (connection-oriented) services. Type 1 can provide only type 1 service.

Frame Formats

An LLC frame is shown in Fig. 4-3. The following discussion contains information on the frame types for IEEE 802.2. It may be confusing to read, but the end of the chapter gives examples of how they are used.

The destination service access point (DSAP) address identifies one or more service access points to which the LLC information field should be delivered.

The source service access point (SSAP) address identifies the service access point that originated the message.

For LANs implementing LLC, a service access point (SAP) identifies a particular service that resides on a network station. It is used to send or receive a message and is analogous to the Type field in the Ethernet frame. A SAP of FE indicates that the OSI protocol owns the packet. A SAP entry of F0 indicates NetBIOS. For a full listing of SAPs, see the end of this chapter. SAPs are used to tell the network layer which network process is to accept the packet and which network process submitted the packet. Companies must register their protocols with the IEEE to receive a SAP.

SAP Addressing.   There are four types of addresses:

1. Individual address, used by DSAP and SSAP
2. Null address, indicated by all zeros in the DSAP and/or SSAP
3. Group address, which is used only by the DSAP
4. Global DSAP, indicated by all 1s in the DSAP field, used to designate the group of all active DSAPs on the network.

Each SAP address consumes exactly one octet. The DSAP address contains seven bits of address (indicated by the D bits in Fig. 4-3) and one bit (the I/G bit) that identifies the address as an individual SAP or a multicast SAP. This is the leftmost bit of the DSAP field as illustrated in Fig. 4-3. This type of address shows that the packet is intended for a group of SAPs at the destination end of the link. If this bit is set to a 0, it is an individual address; if it is set to a 1, it is a group address.

The SSAP contains seven bits of address (indicated by the S bits in Fig. 4-3) and one bit (C/R) to indicate whether the packet is a command or

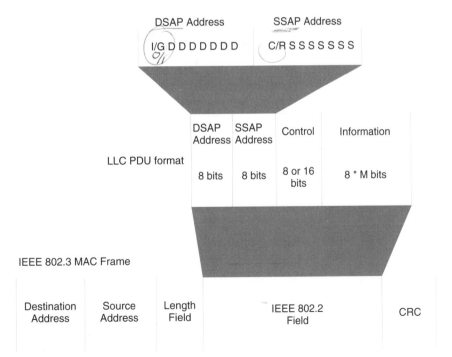

Figure 4-3
IEEE 802.2 fields.
(*Courtesy IEEE*)

Frame shown is IEEE 802.3. The IEEE 802.5 frame also uses the IEEE 802.2 fields.

response type of packet. If the C/R bit is a 0, it is a command packet; if it is set to a 1, it is a response packet. This will be explained further.

The control field is expanded as shown in Fig. 4-3. For supervisory and control frames this field may contain one or two octets that represent functions of the frame. For data frames, this field will contain sequence numbers when needed.

Poll and Final (P/F) Bits.   P/F bits are used between two communicating stations (termed a *primary* and a *secondary* station) to solicit a status response or to indicate a response to that request. It is a P bit when used by the primary station (the requester) and an F bit when used by the secondary station (the responder). In LLC any station can transmit a frame with the P bit or the F bit set. There is no master-slave relationship associated with it.

A frame with the F bit set does not indicate the end of a transmission in asynchronous balance mode (ABM), which is the mode used in LLC2. This bit is used as a housecleaning method between two stations to clear up any ambiguity. It is used to indicate that the frame is an immediate response to a frame that had the P bit set.

For example, when a station wants to set up a connection with another station, it submits a frame known as Set Asynchronous Balance Mode Extended (SABME[1]). In this frame, the P bit is set to 1. The destination station, upon accepting a connection request, will respond with an Unnumbered Acknowledgment (UA) frame, and the F bit will be set. A P bit frame is acknowledged immediately.

In order to effect the preceding functions, three types of frames are transmitted or received in a connection-oriented network:

1. *Information frame.*   This frame is used to transfer user data between the two communicating network stations. Included in the information frame may also be an acknowledged receipt of previous data from the originating station.

2. *Supervisory frame.*   This frame performs control functions between the two communicating stations. These include acknowledgment of frame, request for the retransmission of a frame, and request for control of the flow of frames (rejecting any new data).

3. *Unnumbered frame.*   This frame is used for control functions also. In LLC1 it is used to transfer user data. In LLC2 the frame is

---

[1]SABM and SABME are the same frame. The E asks the destination to set up for extended sequencing, known as modulo 128.

used for session initialization or session disconnect. It can also be used for other control functions.

All three types of frames are shown in Fig. 4-4. The supervisory frame provides for four commands or responses. These commands are receive ready (RR), reject (REJ), receive not ready (RNR), and selective reject (SREJ). Supervisory frames do not contain any user data and are used to perform numbered supervisory functions. Refer to Table 4-1 as you read the following text.

Receive ready (RR) is used by the source or destination station to indicate that it is ready to receive data. It is also used to acknowledge any previously received frames. For example, if the station previously indicated that it could not receive any more data by using receive not ready (RNR), it will send the RR frame to indicate it can again accept new data. When running a protocol over IEEE 802.2 (SNA over Token Ring, for example), these packets will traverse the ring even when there is no

Figure 4-4
Control fields. (Courtesy IEEE)

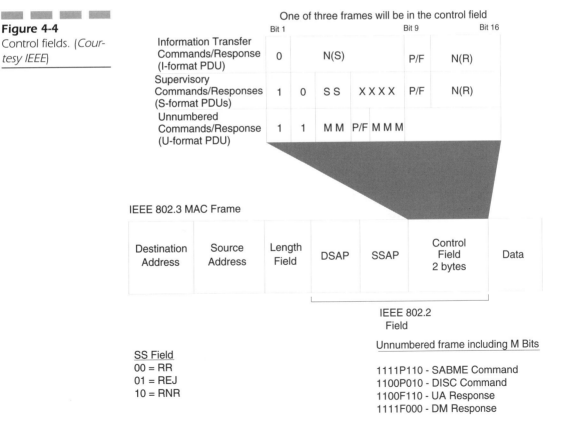

data to send. It is a polling frame to ensure that the link is still good. It is usually sent (as a poll) every few seconds. This will consume bandwidth with nondata frames.

Receive not ready (RNR) is used by a receiving network station to indicate that it saw the data packet but was too busy to accept it. Therefore, a network station will send this frame to indicate to the source not to send any more information until further notice. The receiving network station will indicate that it is again ready to receive data by using the RR frame.

The reject (REJ) frame is used to indicate a request for a retransmission of frames (more than one), starting with the frame indicated by a sequence number in the field known as N(R). Any frames of N(R)-1 were accepted and are acknowledged. Sequencing is explained shortly.

Finally, the unnumbered frame has commands and responses that are used to extend the number of the data-link control functions. These commands are

Set Asynchronous Balance Mode Extended (SABME). This frame is used to establish a data-link connection to a destination network station. It will establish the link in asynchronous balanced mode, which will be explained in a moment. No user data are transferred with this frame. The destination, in response to this packet, will send an Unnumbered Acknowledge (UA) frame back to the originator. All sequence counters are set to 0 upon receipt of a SABME and receipt of the UA.

Disconnect (DISC). This frame is used to disconnect a session between two communicating stations. No user data are sent with this frame. Upon receipt of the frame, the destination station should acknowledge it with a UA frame.

Unnumbered Acknowledge (UA). This frame is sent as an acknowledgment to the SABME and DISC commands.

Frame Reject Response (FRMR). This frame differs from the simple reject frame in that the sender of it is rejected. It is a noncorrectable frame.

Sequencing of Data

Most LAN systems offer some type of sequencing for data delivery. Whether it is simplex, as in Novell's stop-and-wait method, or complex, as used with LLC2 and TCP/IP, some type of sequencing is always used to ensure proper data delivery.

The following text describes a generic method of sequencing—one is that is definitely used in LLC2 operation but that has been around for many decades and whose modes and methods have been copied by many different LAN architectures. Most sequencing follows a method known as automatic return request (ARQ). Examples of the method follow.

Sequencing of transmitted data ensures that the data will be presented to the receiver in good condition and in the same order as sent. Imagine sending data (a lawyer's legal text of a court trial, for example) between a PC and a host, and during the transfer, the data get mixed up and received in the wrong order. Without sequencing of data, the host's application would receive the data as presented to it by the LAN software and would then process the data (save to a file, input into a database, etc.). If the file is saved, it will not be saved in the way it was sent. Needless to say, in any application, misordering LAN data can have catastrophic effects.

During the initiation of a connection, part of the handshaking that goes on is the establishment of a data window. For two communicating network stations A and B, B will establish a window for A and A will establish a window for B. These windows are maintained by state variables. Another name for a state variable is a *counter.* The transmitting station will maintain a send-state variable, V(S), containing the sequence number of the next frame to be transmitted. The receiving station will maintain a receive-state variable, V(R), containing the sequence number that indicates the next frame expected to be received. V(S) is incremented with each frame that is transmitted by a network station. This counter is also placed in a sequence field in a transmitted frame.

When a network station receives a frame, it checks the sequence number, N(S), in the frame received. If this field matches its V(R), it increments its V(R) by one and places this number in frame N(R) of some type of acknowledgment packet that is transmitted back to the originator of the frame.

If V(R) does not match the received sequence number, usually after the expiration of a wait timer the station sends a negative acknowledgment packet to the originator. In this packet is the sequence number of its value in V(R). In LLC2, this type of frame is a REJ or SREJ frame. In other words, the station expected one sequence number but received another. The number it expected will be transmitted back to the sender in hopes of receiving the correct packet and sequence number. Refer to Fig. 4-4 to see N(S) and N(R) in a frame. If extended (7-bit) sequencing is used, the control field is extended to 16 bits. However, if 3-bit addressing (modulo 8) is used, the control field is extended by only one byte.

When this packet is received by the originating station, the station looks at the received sequence number N(R). It knows that it has already sent this frame, but something went wrong in the process. If possible, the station will retransmit the old frame.

What is the order for sequence numbering? Not all protocols operate in the same way. For LLC2, the sequence numbering starts at 0 and ends at 127 (known as modulo 128). This means that sequence numbers may go as high as 127, but then they must return (wrap around) to 0. This permits 127 frames to be outstanding (not acknowledged). It does not permit 128 frames to be outstanding, for the value of V(R) is the next expected sequence number. This scheme also guards against wrapping. If a station has 127 outstanding frames (0–126), a sending station may not use 0 again until it has been acknowledged. Most frames, however, are acknowledged within a few sequence numbers.

One important consideration is the actual size of the window. While modulo 128 is nice in that 127 frames may be transmitted with one acknowledgment, this constitutes a resource constraint. No transmitted frame may be erased in the sending network station's memory until it has been acknowledged. This means that a network station should have enough memory to store that amount of data until it is acknowledged.

The window can be shut down at any time to prevent any more frames from certain stations from being received. This allows for efficient use of resources and also allows a network station to tend to other stations on the network. In other words, it eliminates the possibility of one station hogging another station's time.

If a window of size 7 is opened to another station and six frames are outstanding, the window will be closed. If an acknowledgment is received for six frames, the window will be opened again for six frames. This is called the *sliding window.*

An advantage of a windowing system is *inclusive acknowledgments.* If a network station has five outstanding (unacknowledged) frames (0–4) and it receives an acknowledgment frame of 5, this means that the destination station has received frames 0–4, and the next frame the receiving station sees should have a sequence number of 5. This keeps the receiving station from having to transmit five acknowledgment frames. There will be less overhead on the network and the network stations (only one acknowledgment is transmitted instead of five).

The ability to detect and correct sequence errors is characterized by three types of retransmissions: Go Back to N, Selective Repeat, and Stop and Wait. With Selective Repeat, only the indicated frame needs to be retransmitted. Go Back to N specifies that not only a particular sequence number is to be retransmitted, but also any frames before that and up to the last acknowledged sequence number. Stop and Wait means to send a packet and do not transmit another until that packet has been acknowledged. All three types have their merits.

Selective Repeat offers better bandwidth utilization in that only the out-of-sequence frame needs to be retransmitted. But the receiving station must wait for that frame and, when it does arrive, it must reorder the data in the correct sequence before presenting it to the next layer. This consumes memory and CPU utilization. Go Back to N is a simpler method; however, it uses more network bandwidth and is generally slower than the Selective Repeat. LLC2 uses the Selective Repeat method. Other network protocols use a variation of the two.

The Stop-and-Wait method is said to have a window size of 1, for only one frame may be outstanding at a time. With this method, a transmitting station will transmit a frame and wait for an acknowledgment. It cannot transmit any more frames until it has received an acknowledgment for the previous frame. The sequence numbers used can be of two types: a modulo number and a 0–1 exchange. With a 0–1 exchange, a starting number is established between the two stations, say a 0. When the transmitting station transmits a packet, it sets the sequence number to 0. When the receiver gets the packet, it sets its received sequence number to 0 and then acknowledges the packet. When the response packet is received by the original station, it notes the 0 sequence number and then sets the next transmit sequence number to 1. The sequence number will flip-flop in this way throughout the length of the data transfers; the only two sequence numbers used are 0 and 1. Novell implements a variation of this protocol in NetWare 3.11 and previous releases.

An example of LLC2 sequencing is included at the end of the chapter.

Timer Functions

Timers are used throughout the LLC2 mode of operation. These timers include

1. *Acknowledgment timer.* A data-link connection timer that is used to define the time interval in which a network station expects to see a response to one or more information frames or a response to one unnumbered frame.

2. *P-bit timer.* Defines the amount of time that a network station will wait for a response frame regarding a frame that was sent with the P-bit set.

3. *Reject timer.* Defines the amount of time that a network station will wait for a response to a REJ frame sent.

4. *Busy-state timer.* Defines the amount of time that a network station will wait for a remote network station to exit the busy state.

Three parameters that are used are

N2. The maximum number of times that a PDU (Protocol Data Unit, i.e., pocket) is sent following the expiration of the P-bit timer, the acknowledgment timer, or the reject timer

N1. The maximum number of octets allowed in an I PDU

k. The maximum number of outstanding I PDUs (those that have not been acknowledged)

Connection-Oriented Services of the IEEE 802.2 Protocol

There are two modes of operation for LLC2: operational mode and nonoperational mode.

Nonoperational mode is not completely discussed in this text. It is used to indicate that a network station is logically (not physically) disconnected from the physical cable plant. No information is accepted when a network station has entered this stage. Examples of possible causes for a network station to enter this mode are

1. The power is turned on but the receiver is not active.

2. The data link has been reset.

3. The data-link connection is switched from a local condition to a connected-on-the-data-link (online) condition.

In operational mode, three primary functions are provided:

1. *Link establishment.* A source station will send a special frame to a destination indicating that a connection is wanted. This frame will be responded to with an acceptance or rejection. Once a connection is established, the two network stations provide each other with a series of handshaking protocols to ensure that the other is ready to receive information.

2. *Information transfer.*   Data from the user's applications are transferred from the originating station to the remote station, and the data are checked for possible transmission errors. The remote station will send acknowledgments to the transmitting station. During this phase the two network stations may send control information to each other indicating flow control, missing packets, etc.

3. *Link termination.*   The connection between the two stations is disconnected and no more data are allowed to be transferred between the two stations until the session is reestablished. Usually, the link will remain intact as long as there are data to send between the two stations.

Details of LLC Type 2 Operation

In operational mode, the data link enters into a mode known as asynchronous balance mode (ABM). This type of operation has nothing to do with the asynchronous data transfer commonly found in terminal-to-host connectivity and through the modems on your computer. *Asynchronous* in this case means that a connection may be initialized by any station on the network. In this type of operation a connection at the data-link layer is established between two SAPs. Each end of the connection is able to send commands and responses at any time without receiving any type of permission from the other station. There is no master-slave relationship in this mode. The information exchanged between the two is command information (indicating sequence numbers or that a station is busy and cannot accept any more data). This mode is also used for user data transfer.

ABM in LLC2 has four phases of operation:

1. Data-link connection phase
2. Information transfer phase
3. Data-link resetting phase
4. Data-link disconnection phase

Data-Link Connection Phase.   Refer to Fig. 4-4. Any network station may enter into this state with the intention of establishing a session with another network station. It will use the unnumbered frame SABME. When the SABME frame is sent, an acknowledgment timer is set. If the frame is not responded to, it will be resent up to N2 times. If the frame is not responded to in that amount of time, the connection attempt is aborted. This type of frame has the P bit set.

The two responses that may be received back are the UA and the DM. DM stands for disconnect mode. This enables a network station to indicate that a connection is not allowed. If a connection were already set up and a network station received this response from the other end of the connection, the connection would be broken. The F bit will be set in the response packet. With receipt of the DM frame, the acknowledgment timer is stopped and that network station will not enter into the information transfer stage. The connection attempt is aborted. The station should report this condition to its upper-layer protocols.

Upon receiving a UA response frame back, the connection is established and the station enters into the information transfer phase.

Information Transfer Phase   When a network station has sent a UA response packet or has received a UA packet, that network station immediately enters the information transfer phase. The connection is established and the exchange consists of sending and receiving I (information) and S (supervisory) frames. If a network station receives a SABME frame while in this phase, it will reset (not disconnect) the connection. This frame is used to return the connection to a known state. All outstanding frames are lost and sequence numbers are reset to 0. Retransmission of any outstanding frames will occur between the two network stations, and they will be acknowledged at this time.

When a network station has user information that needs to be sent (or resent as the case may be), it will do so using the I frame. In this sequence, numbers need to be put into the packet. See Fig. 4-4, which shows the I frame format.

Sequencing within LLC2 is accomplished with modulo 128. This means that the sequencing starts at 0 and reaches its upper limit at 127. From there, it returns to 0. (Zero is a higher number than 127.)

Refer to Table 4-2 and Fig. 4-4. When a source sends an I frame to the destination, the N(S) field will be set (in the frame) to its current variable V(S), and the N(R) will be set to its V(R) for that particular connection. At the end of this frame, transmission will be incremented by one. When sending frames, a station may transmit more than one frame without receiving an acknowledgment for any of the frames. For each frame sent it increments V(S) by one. This is an upper limit to the number of outstanding packets that may be unacknowledged. When this limit has been reached, the sending station will not send any more frames, but it can resend an old frame.

TABLE 4-2

LLC2 Sequence
Counter Definitions

V(R)	Receive-state variable. A counter maintained by a network station. This counter indicates the sequence number of the next-in-sequence I PDU to be received on a connection. It is maintained in the network station and not the frame.
N(S)	Sequence number of the frame (called the send sequence number). Located in the transmitted frame, this field will only be set in information (I) packets. Prior to sending an I frame, the value of N(S) is set to the value of V(S), the send-state variable. This is located in the frame and not in the network station.
V(S)	Send-state variable. This number indicates the next sequence number expected on the connection. It is incremented by one for each successive I-frame transmission. It is maintained in the network station and not the frame.
N(R)	Receive sequence number. This is an acknowledgment of a previous frame. It is located in the transmitted frame. All information and supervisory frames will contain this. Prior to sending that type of frame, it is set equal to the value of the receive-state variable V(R) for that connection. N(R) indicates to the receiving station that the station that originated this frame accepts all frames up to the N(R) minus 1.

Also, if the transmitting station enters into the busy state, it can still transmit I frames; it just cannot receive any. If the source receives an indication that the destination is busy, it will not send any more I frames until it receives a receive ready (RR) frame from the remote station.

When a transmitting station enters into the FRMR (frame reject) state, it will not transmit any more information for that particular link. When the sending station transmits this frame, it starts an acknowledgment timer in anticipation of receiving a response packet for its transmission.

Data-Link Disconnection Phase. In the information transfer phase, either network station may disconnect the session. This is accomplished by transmitting a DISC frame to the other network station. When it sends this packet, the station will start an acknowledgment timer and wait for a response. When it receives a UA or DM response packet, the timer is stopped and the station enters into the disconnect mode.

If this timer expires before receipt of a response packet, the originator again sends the DISC packet and restarts the timer. When an upper limit of resends is reached, that station enters the disconnect phase and informs the upper-layer protocols that an error has occurred in attempting to disconnect from the remote station.

When a network station receives a DISC packet, it responds to it with a UA packet and enters into the data-link disconnection phase. While in this phase, a network station is able to initiate sessions with other network stations and can respond to session requests. If it receives a DISC packet, it will respond with a DM packet. If it receives any other type 2 command with the P bit set to 1, it will respond with the F bit set to 1.

LLC2 Frame Reception

Frame reception for LLC2 involves many more functions than the transmitting of a frame. When a network station is not in the busy condition and receives an I frame containing the sequence number that matches its V(R) variable, it accepts the frame, and increments the V(R) by one. Then if the receiving station has another I frame to transmit, it will transmit the frame as indicated previously but will set the N(R) variable in the transmitted packet to its V(R) and transmit the packet. LLC2 does not use separate packets to indicate an acknowledgment unless there are no more I frames to be sent. In this case, it will send a receive ready (RR) frame with N(R) set to its V(R).

It may also send a receive not ready (RNR) frame back, indicating that it is now busy and no more data should be sent—but it acknowledges your packet, as indicated by N(R).

If any frame is received as an invalid frame (e.g., wrong SAPs), the frame is completely discarded. Figure 4-5 shows the variable states of N(R), N(S), V(R), and V(S) for a simple transmission between two network stations.

A Live Connection

Refer to Fig. 4-5. To establish a connection, station A sends a UI (unnumbered information, i.e., nonsequenced) frame (a Set Asynchronous Balance Mode frame) with the P bit set. Station B allows the connection to send back a UA frame with the F bit set. The connection is now established. Table 4-3 describes the session in detail.

Figure 4-6 shows how LLC2 uses timers. It also depicts how the P/F bit can be utilized to manage the flow of traffic between two network stations. The exchange is described in Table 4-4.

Figure 4-7 depicts the use of the reject command. The exchange between stations is detailed in Table 4-5.

Figure 4-5
A sample user session.

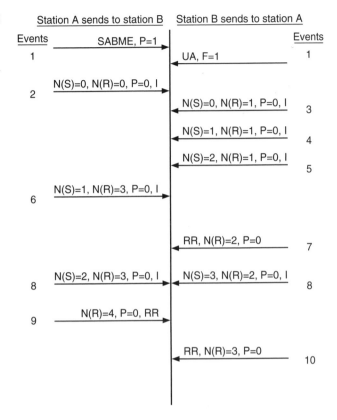

| Station A sends to station B | Station B sends to station A |

If there is no data to send, both sides will continue to send
RR frames to each other (about every 3 seconds)
to make sure each side of the link is active.

The preceding examples were taken from another HDLC exchange. LLC2 follows this flow very closely. The examples were presented mainly to show how a connection is set up and to illustrate the exchange of sequence numbers. It also shows what happens during a frame loss.

The preceding text basically explains how most transport-level protocols work. They do not all follow this exact interpretation of LLC2, but the functions are basically the same. LLC2 is primarily used in transporting SNA traffic across Token Ring, when IBM LAN Server (IBM's alternative to Novell's NetWare) is being used, and with LANs that are implementing the use of serial lines with their network (bridges or routers that are using a telephone line to connect to a remote network). Usually, serial lines are conditioned to handle the digital traffic, but

TABLE 4-3

Description of
Sample Session

Event	Operation
1	Station A requests a link to be established to station B by sending station B a SABME. Station B responds with a UA. Both the P and F bits are set to 1.
2	Station A sends an information frame and sequences the frame with N(R) = 0. N(R) = 0 means that station A is expecting to receive a frame with its field of N(S) = 0. The P bit is set to 0.
3–5	Station B sends information frames (I), number N(S) = 0 through N(S) = 2. Its N(R) field is set to 1, which acknowledges station A's frame sent in event 1 (it had an N(S) value of 0). Remember, the N(R) value states that the station acknowledges all previously transmitted frames. The N(R) value also identifies the next N(S) value that is expected from station A.
6	Station A sends an I frame sequenced with N(S) = 1, which is the value station B expects next. Station A also sets the N(R) field to the value of 3, which inclusively acknowledges station B's previously transmitted frames numbered N(S) = 0, 1, 2.
7	Station B has no data to transmit. However, to prevent station A from "timing out" and resending data, station B sends a receiver ready (RR) frame with N(R) = 2 to acknowledge station A's frame with N(S) = 1 (sent in event 6).
8	The arrows depicting the frame flow from the stations are aligned vertically with each other. This means that the two frames are transmitted from stations A and B at about the same time and are exchanged almost simultaneously across the full-duplex link. The values of the N(R) and N(S) fields reflect the sequencing and acknowledgment frames of the previous events.
9–10	Stations A and B send RR frames to acknowledge the frames transmitted in event 8. If neither side has data to send, each side will continue to send these frames to ensure that the other side is active. The sequencing will remain the same throughout the RRs transmitted.

there may be instances where the serial line is not conditioned; the LLC protocol ensures that data are reliably transferred across these lines.

LLC Type 1 Operation

With the exception of SNA and the IBM LAN Server program, type 1 is the most commonly used class of LLC. It is used by Novell, TCP/IP, OSI, and most other network protocols. There are no specific subsets for the operation of LLC type 1. Type 1 operation consists of only one mode—information transfer.

Figure 4-6

T1 timer. (*Courtesy of Uyless Black*)

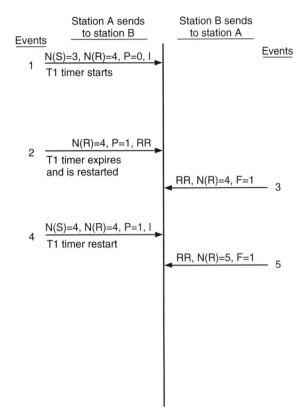

TABLE 4-4

Use of Timers

Event	Operation
1	Station A sends an I frame and sequences it with N(S) = 3.
2	Station B does not respond within the bound of the timer. Station A times out and sends a receive ready (RR) command frame, indicated by the P bit set to 1.
3	Station B responds with F = 1 and acknowledges stations A's frame by setting N(R) = 4.
4	Station A resets its timer and sends another I frame. It will keep the P bit set to 1 to force station B to immediately respond to this frame.
5	Station B responds with an RR frame with N(R) = 5 and the F bit set to 1.

Figure 4-7
Reject. (*Courtesy of Uyless Black*)

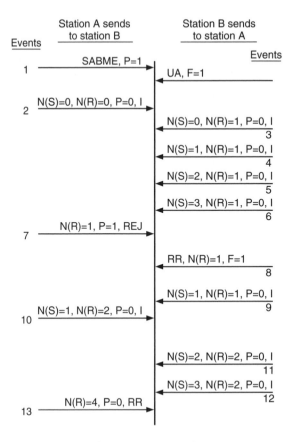

	Station A sends to station B		Station B sends to station A	
Events				Events
1	SABME, P=1		UA, F=1	
2	N(S)=0, N(R)=0, P=0, I			
			N(S)=0, N(R)=1, P=0, I	3
			N(S)=1, N(R)=1, P=0, I	4
			N(S)=2, N(R)=1, P=0, I	5
			N(S)=3, N(R)=1, P=0, I	6
7	N(R)=1, P=1, REJ			
			RR, N(R)=1, F=1	8
			N(S)=1, N(R)=1, P=0, I	9
10	N(S)=1, N(R)=2, P=0, I			
			N(S)=2, N(R)=2, P=0, I	11
			N(S)=3, N(R)=2, P=0, I	12
13	N(R)=4, P=0, RR			

Information Transfer

Type 1 operation does not require any prior connection establishment between the source and destination network stations before data are transferred between them. Once the SAP information field has been sent, information may be transferred.

Two other types of frames, besides information frames, may be transmitted using LLC1 operation: the exchange identification (XID) and the TEST frames.

XID

The XID frame is used for the following functions:

1. An XID frame with a null SAP can be used to retrieve information from a remote network station. It is a form of the keep-alive packet that is used to check the presence of the remote station.

TABLE 4-5

Use of the Reject
Command

Event	Operation
1	Station A sets up a connection with station B by sending a SABME, and station B accepts this request by sending back a UA.
2	Station A sends an I frame and sequences it with N(S) = 0. The N(R) = 0 means that it expects station B to send an I frame with a send sequence number of 0.
3–6	Station B sends four frames numbered N(S) = 0, 1, 2, 3. The N(R) value is set to 1 to acknowledge station A's previous frame. Notice that the N(R) value does not change in any of these frames because station B is indicating that it still expects a frame from station A with a send sequence number of 1. During these transmissions, we assume that the frame with N(S) = 1 is distorted.
7	Station A issues a REJ frame with N(R) = 1 and P = 1. This means that it is rejecting station B's frame that was sequenced with the N(S) = 1, as well as all succeeding frames.
8	Station B must clear the P bit condition by sending a non-I frame with the F bit set to 1.
9–12	Station B retransmits frames 1–3. During this time (in event 9), station A sends an I frame with N(S) = 1. This frame has N(R) = 2 to acknowledge the frame transmitted by station B in event 8.
13	Station A completes the recovery operations by acknowledging the remainder of station B's transmissions.

2. With the LSAP address set to a group SAP, the XID frame can be used to determine group membership.

3. It can be used to test for duplicate addresses (see the Chap. 2 section on Token Ring).

4. If the link between the two stations is operating in LLC2 mode, this frame can be used to request receive-window sizes.

5. It can be used to request or identify a service class from a remote station (type 1 or type 2 LLC operation).

6. It can be used with IBM Token Ring source routing to find a remote station through source route bridges. This frame is the dynamic explorer frame used by source route bridges.

Test

The primary use of the TEST frame is for loop-back functions. A network station will transmit a TEST frame to test a path. It will be transmitted on a certain data path with the information field set to a specific entry. The network station will then wait for the frame to be responded to, and the information field in the frame will be checked for errors.

Although LLC1 operation does not require a response frame to any LLC1 command frame, these two frames do require a response, but responses are made in unnumbered format (no use of the P or F bit).

SNAP

The most common implementation of LLC1 operation is through a special subsection of the IEEE 802.2 specification known as SubNetwork Access Protocol (SNAP). Most LAN vendors implement this protocol, although the switch to true LLC1 operation is becoming more popular. SNAP was introduced with LLC to give network protocols an easy transition to the new frame formats created by the IEEE 802.2 committee. SNAP can be used with Novell, TCP/IP, OSI, and many other full OSI stack protocols. Any protocol may use SNAP and become a pseudo–IEEE-compliant protocol.

As stated previously, IEEE 802.2 defined two fields known as the DSAP and SSAP. For the most part, the SAP fields are reserved for those protocols that implement IEEE 802.x. One SAP has been reserved for all non-IEEE standard protocols. To enable SNAP, the SAP value in both the DSAP and SSAP fields are set to AA (hex). The control field is set to 03 (hex). Many different protocols may be run using this one SAP. Therefore, to differentiate between the various protocols, any packet with the AA SAP address also has a five-byte protocol discriminator following the control field. Refer to Fig. 4-8.

Since this is type 1 operation, DSAP and SSAP will be set to AA, and the control field (only one byte for type 1 operation) will be set to 03 to indicate unnumbered information packets. Following this control field are five bytes called the *protocol discriminator.* This identifies the protocol family to which the packet belongs.

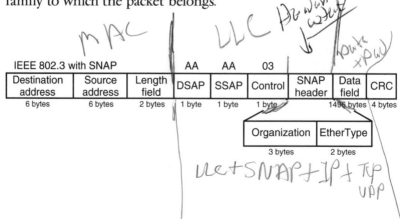

Figure 4-8
SNAP packet.

Figure 4-8 shows an IEEE 802.3 Ethernet packet with a SNAP header. This packet contains four distinct parts:

[handwritten: MAC]

1. The data-link encapsulation headers (destination and source address and the CRC trailer)
2. The three-byte 802.2 headers (set to AA, AA, and 03) *[handwritten: DLC]*
3. The five-byte protocol discriminator immediately following the 802.2 header *[handwritten: SNAP]*
4. The data field of the packet

The important field to notice is the protocol discriminator. The first three bytes of this field indicate the vendor (080007 indicates Apple Computer, for example). The next two bytes identify the type of packet (the EtherType field).

If the first three bytes of the protocol discriminator are set to 00-00-00, this indicates a generic Ethernet packet (not assigned to any particular vendor), and the next two bytes will be the Type field of the Ethernet packet. The use of three bytes of 0s indicates the use of an Ethernet frame. This is useful when the frame traverses different media types. If the frame is transposed to allow passage on another media type, the field of 0s indicates that any time the frame is forwarded to the Ethernet medium, it will be built as an Ethernet frame and not as an IEEE formatted 802.3 frame. If this field contains an entry other than 0s, it indicates that the IEEE 802.3 frame (and not the Ethernet frame) should be used when forwarding to an Ethernet LAN. The use of all 0s simply indicates an encapsulated Ethernet packet.

For example, a TCP/IP packet could have this field set to five bytes of 00-00-00-08-00. This signifies an organization ID (protocol discriminator) of 00-00-00 (which states that it is an Ethernet frame), and the Ethertype is a 08-00, which is the Ethertype for IP messages. Following this is the Ethernet data frame.

SNAP allowed Ethernet vendors to quickly switch their drivers and network protocols over to the IEEE 802.x packet format without rewriting a majority of the code. This format allowed vendors that had drivers written for the Ethernet system to port the network operating code quickly over to the Token Ring data-link frame format. A lot of vendors, in support of Token Ring, use the SNAP method to implement their existing code to run on top of Token Ring. It is quick, simple, and easy, and it enables operation between different vendors.

Today, vendors are switching their code over to the LLC frame format. For example, Novell NetWare used to use SNAP. The format was AA-AA-

F0 OSI	42 Spanning Tree BPDU	F8, FC Remote Program Load
E0 Novell	FF Global LSAP	04, 05, 08, 0C SNA
F0 NetBIOS	F4 IBM Net Mgmt.	AA SNAP
06 TCP.IP	7F ISO 802.2	80 XNS
	00 NULL LSAP	

03-00-00-00-81-37—AA for the DSAP/SSAP/Control fields, three bytes of 0s to indicate encapsulated Ethernet, and then the Ethernet Type field assigned to Novell: 8137. Novell has registered their NetWare operating system with the IEEE, and can now use the SAP address of E0 in their LLC frames (see Table 4-6). They use LLC1 for their transmission. This means that the DSAP and SSAP are set to E0 and the control field is set to 03 to indicate connectionless or LLC1 communication. Following these fields would be the IPX header starting with the checksum of FFFF.

The choice between connection-oriented networks and connection-less networks centers around the functionality that is needed. The connection-oriented system provides for the integrity of data, but with it comes the extreme burden of overhead. Connectionless systems consume less overhead but are prone to error.

Error rates on serial lines tend to be in the 1 in 10^3 range, while errors on a LAN tend to be in the 1 in 10^8 range. Therefore, it does not make a lot of sense to use connection-oriented services for a LAN. Error control is usually provided for by the application as well as the upper-layer software of a LAN (the transport layer). Today, connectionless methods are the more common.

Figure 4-9 summarizes all Ethernet/IEEE 802.3 and IEEE 802.5 with IEEE 802.2 data-link encapsulation types. The figure shows the IEEE 802.5 packet with SNAP. This packet can be used with or without SNAP. SNAP is shown with the IEEE 802.5 packet (and not as a separate packet) for space considerations only.

Ethernet Encapsulation Methods (Data-Link frame encapsulation)

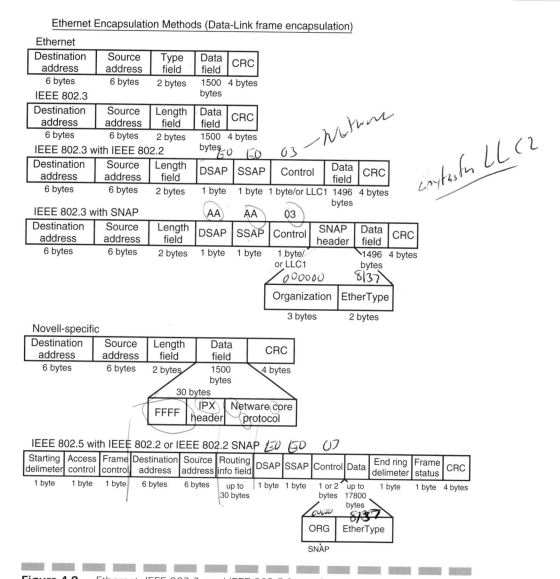

Figure 4-9   Ethernet, IEEE 802.3, and IEEE 802.5 frame formats.

CHAPTER 5

Xerox Network
System

Introduction

A network consists of many parts. Groups of networks are usually combined to form internetworks. There are hardware components, software components, and countless applications that run on a network. The previous chapters were concerned with the hardware components of a network, the NIC card. Those chapters explained the physical and data-link layers of a network. The network through application layers are primarily written in software, and these layers are usually grouped into a single protocol. The current protocol suite of choice (and it probably will be for some time) is the Transmission Control Protocol/Internet Protocol (TCP/IP). TCP/IP can be extremely complicated to understand, so a chapter on the Xerox Network System (XNS) protocol is included first. This protocol is fairly simple to understand and is the basis of the NetWare protocol. Reading this chapter first will provide you with a solid understanding of the network, transport, session, and application layers.

Is XNS actually a protocol? It depends on whom you ask. Yes, it is a protocol, but more so it is a protocol implementation. There are many variations of XNS (Ungermann-Bass, Intergraph, and Novell NetWare have implemented their own proprietary versions, and 3Com had their own network software based on it called Xfile, Xprint, and Xmail; XNS is more of an architecture rather than a strictly adhered-to standard. The protocol itself is about 20 years old. It does provide great insight into how a network protocol should operate. For this reason it is used as an introduction to learning the network protocols. It provides the architecture for the upper-layer protocols (OSI layers 3 through 5).

The protocols behind XNS can give implementers a clear understanding of how to build a network protocol. The unfortunate thing about XNS is that the specification is "loosely" written. It is a protocol that allows for a tremendous variety of vendor-specific implementations. It was implemented by most vendors as an architecture to follow. XNS was derived to operate over a LAN—specifically, Ethernet.

XNS was the first network protocol commercially implemented over Ethernet. It was implemented as an open document to show how a software protocol should run over a LAN. The protocol was implemented by multiple vendors at the start of the network revolution in the early 1980s. All vendors implemented the protocol suite up to layer 4 (the transport layer), the sequence packet protocol, or SPP. The session-layer protocol known as courier, explained later in this chapter, was not used by most LAN vendors. All vendors implemented their own session-layer schemes. Therefore, in the early 1980s, most network architectures were based on XNS, but all were developed as proprietary network systems. This meant that a protocol based on XNS from one LAN vendor would not work with another LAN vendor's XNS equipment.

This created a multitude of problems. While the XNS protocol was a good protocol, it locked a company into one network vendor. For example, if you ordered network equipment from vendor A and decided later to implement vendor B's equipment, the two vendors' equipment would not interoperate if each was using its own implementation of the XNS protocol. This could represent a large investment in one vendor. If a company needed a certain type of networking equipment, and vendor A did not supply it but vendor B did, the company was faced with some very hard choices.

XNS is a good LAN protocol implementation. Unfortunately, no vendor implemented the exact same version. I have chosen XNS as the first protocol to study, for XNS does provide a good implementation from which to learn LAN protocols. It has all the pieces needed to construct a protocol suite for networking.

Other protocols such as TCP/IP proliferated as standard types for all protocols to run, for they allowed multiple LAN and computer vendors to interoperate. Vendor A's equipment would interoperate with vendor B's, for both implemented the TCP/IP protocol. The TCP/IP protocol was built as an open standard, and all vendors that implemented it had to follow one and only one standard. TCP/IP is covered in a later chapter.

Network Definition

Figure 5-1 shows a network (the solid-line box) and an internetwork (the dotted-line box). A network can be defined as a transmission medium that is used to carry internet packets through the network. A network is a collection of hardware and software that, taken together, provide a means to share data. An internet is usually a composition of networks tied together by devices known as *routers*.

There are many entities that constitute a network. For this introduction, a few important entities will be described.

Hosts are classified into two groups: internetwork routers and network stations. Stations are classified as workstations or servers.

A network is defined as a communication system used to carry internet packets that connect stations. The network, in turn, is connected to a

Figure 5-1
XNS network
internet.

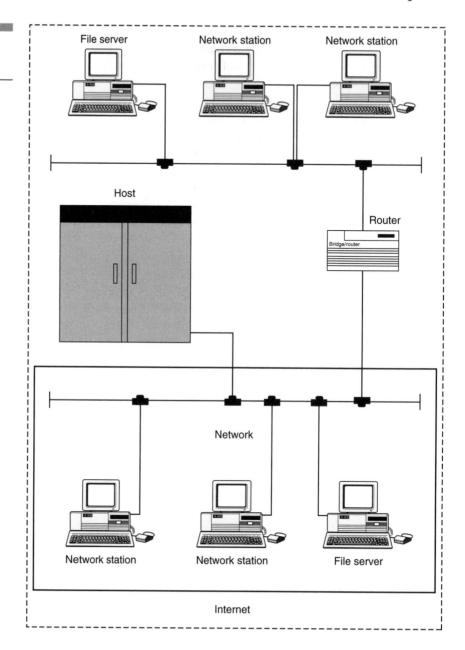

router that connects the internetwork (see Fig. 5-1). With a few exceptions, stations on the same network can communicate among themselves without the use of a router. Every network is assigned a unique number. This, in a crude way, is like the area code of the telephone system. Network numbers are also assigned to routers that are connected together to form a circuit using leased lines (serial lines specially conditioned to handle digital data traffic or used between directly connected networks). Data serial lines are like a static switched circuit and are used only for data.

Individual stations on a network are not assigned individual network numbers (TCP/IP, however, does assign network numbers to individual stations). In XNS, workstations take on the network number assigned to their network. This means that every station on an individual network is assigned the same network number. In this way, a group of stations can be identified by the network number.

During the beginning of the commercial network marketplace, Ethernet was found in most installations. Therefore, XNS became the most popular network protocol scheme implemented. XNS was specifically written as an architecture to use Ethernet as its data-link and physical layers. Interoperability was considered, but companies communicating with other companies through a global Internet was not. In the early 1980s a customer generally bought mainframe or minicomputer systems from one company. All parts associated with that computer system, as well as maintenance, were also purchased from that company. The network system was no different. Buying software and hardware from one company was the norm.

As shown in Fig. 5-2, certain individual protocols make up the architecture of XNS. This figure shows how the protocols of XNS were assigned to levels.

Internet datagram protocol (IDP). This network-level protocol provides the data delivery functions needed to route (local or remote) data in an internet. This protocol provides network-level addressing, routing, datagram packet formatting, etc.

Routing information protocol (RIP). Believe it or not, RIP was invented for XNS and it then migrated to TCP/IP. Similar to other RIP implementations, it provides dynamic updating of routing tables based on a distance vector algorithm. It should be noted that the specification does not include special routing algorithms such as split horizon or poisoned reverse. Most vendors are implementing them, though. Its update interval is 60 s.

Figure 5-2
XNS model. (*Courtesy Xerox Corp.*)

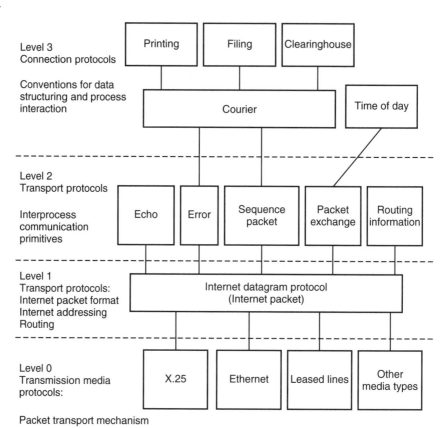

Level 3
Connection protocols

Conventions for data
structuring and process
interaction

Level 2
Transport protocols

Interprocess
communication
primitives

Level 1
Transport protocols:
Internet packet format
Internet addressing
Routing

Level 0
Transmission media
protocols:

Packet transport mechanism

Error Protocol. This protocol allows one network station to notify another network station that an error has occurred in a received packet.

Echo protocol. This protocol allows for testing of a network path or for recording round-trip delay times.

Sequence packet protocol (SPP). A transport-level protocol, this protocol allows for reliable data exchange between two stations. Originator and receiver network stations synchronize and establish a connection between themselves. During this process, a sequence number is established to mark the packets exchanged between the two. This sequencing is accomplished on a packet-level basis.

Packet exchange protocol (PEP). A request and response protocol used in transaction-oriented applications, this is similar to the unreliable protocol of UDP in the TCP/IP arena.

Courier. This is a session-level protocol that provides remote procedural call services.

Clearinghouse. This enables the naming services to exist on XNS.

File service. This is a file transfer service.

Printing service. This is a remote printing service.

Time of day. This provides date and time service.

With the exception of file service, printing service, and time of day, each of the aforementioned protocols will be discussed. XNS provides an excellent example of how a software LAN protocol suite is built. Each of the services provided shows the functions needed to provide for data exchange between two communicating network stations on a LAN. Generally, there is one protocol for every service needed on a LAN. Studying each protocol will lead to a better understanding of any other upper-layer protocol.

The preceding protocols are the suite that make up the XNS protocol stack. The most popular implementation of the XNS protocol is Novell NetWare's IPX. IPX emulates this protocol in the internet and transport layers—IDP and SPP. SPP is used primarily for peer-to-peer services such as remote console (RCONSOLE), SNA gateways, etc. After the transport layer, Novell implements a proprietary protocol stack known as NCP or the NetWare Core Protocol.

XNS segments protocols into levels. This can be confusing, but for completeness, I have included this concept. These levels can be roughly translated to OSI layers as shown in Fig. 5-2.

Level 4 (OSI layer 7)	Application protocols
Level 3 (OSI layers 5–7)	Control protocols: conventions for data structuring and process interaction
Level 2 (OSI layer 4)	Interprocess communications primitives
Level 1 (OSI layer 3)	Transport protocols: internet packet format and internet addressing and routing
Level 0 (OSI layers 1 and 2)	Transmission media protocols: packet transport mechanism

Level 0 protocols were given in the beginning of this book. This discussion on XNS will start with Level 1, or the network-level protocol of the internet datagram protocol (IDP).

The network-level protocol operates much like delivering a letter. If you wish to send a letter to someone, you write down the data in a letter. You then insert this letter into an envelope and put the destination address and return address on the envelope.

From here, you put the envelope in the box, where the mail service will pick it up. The mail service will note the destination address of the letter and route the letter to its final destination. If the mail service cannot find the recipient, it will inform the sender of this problem. The sender may then receive the letter back.

If the letter is deliverable, the recipient will read the letter and, if necessary, send a response. The recipient knows where to send the response by the return address on the letter. This letter is then routed back by the mail service to your address, where you receive it.

This may sound like a connection-oriented protocol. It is not. Most LAN network layer protocols are connectionless or datagram oriented. But most offer checkpoint error detection for host or network address problems in the attempt to deliver a message.

IDP provides, like other network protocols, the network addressing, routing, and datagram packet formatting of data. This layer provides for not only the routing of datagrams in the XNS Internet, but also the network-level addressing.

XNS can run over any data-link protocol, including Token Ring, Ethernet, FDDI, and any WAN protocol. We will start our discussion of the XNS system with the internet transport protocols. Since XNS was built to run specifically on top of Ethernet, this will be the medium of choice for the continuing discussion. XNS's general functions are shown in Fig. 5-3.

Terminology and Definitions

Before the XNS levels are detailed, some terms need to be defined. Every packet transmitted will have a source and a destination address. This may be the MAC-layer address or it may be a protocol address at the network layer. The MAC-layer address is directed at the physical interface to the network. The Ethernet card in a PC that connects to an Ethernet cabling system is identified by this address, for example.

The data flow between the source and destination will always have the sender of the data and the recipient of the data clearly identified in

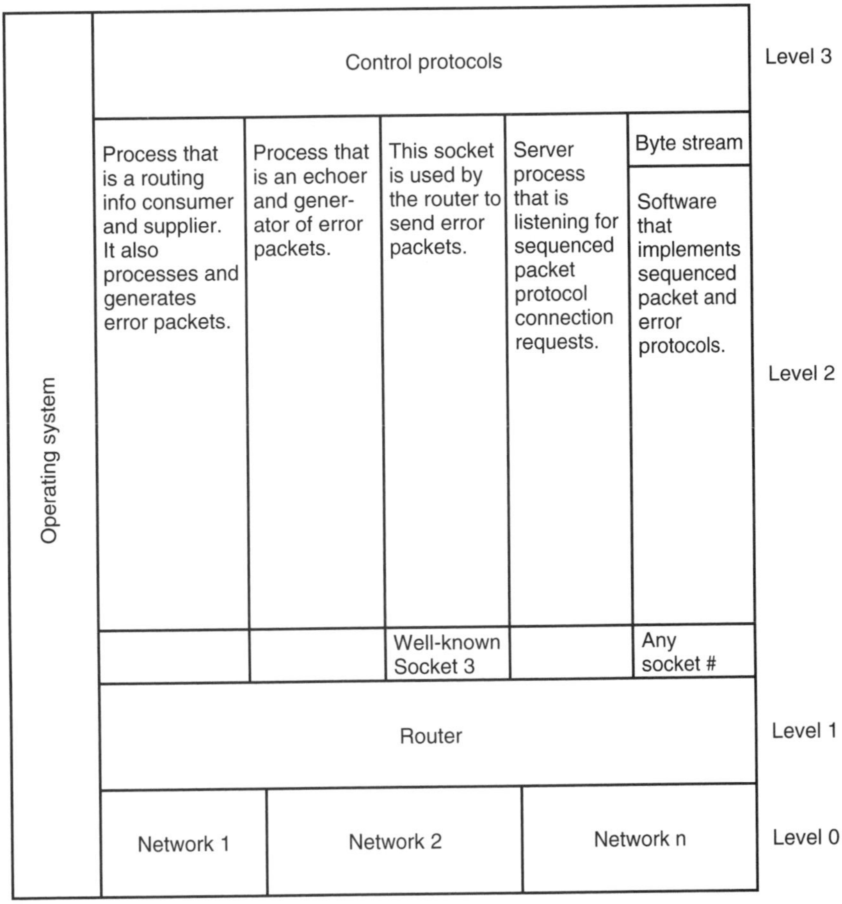

Figure 5-3
Overall XNS model.
(*Courtesy Xerox Corp.*)

						Level
	Control protocols					Level 3
Operating system	Process that is a routing info consumer and supplier. It also processes and generates error packets.	Process that is an echoer and gener-ator of error packets.	This socket is used by the router to send error packets.	Server process that is listening for sequenced packet protocol connection requests.	Byte stream Software that implements sequenced packet and error protocols.	Level 2
			Well-known Socket 3		Any socket #	
	Router					Level 1
	Network 1	Network 2		Network n		Level 0

the packet. They will be identified by their protocol addresses in the IDP header.

A *connection* is an association between two communicating stations (usually the sender and the receiver). After a connection is established, data should flow reliably between the two stations. A connection will usually have a *listener* and an *initiator*. The listener will wait for connection from remote sources, and the initiator will be the station that attempts to connect to other stations on the network.

A system that supplies services to others is called a *server*. Each service (file, print, terminal) in the server has a *supplier* and a *consumer*. Like the connection, the supplier "supplies" information upon request from the consumer. An example of this is a user's workstation that has a connection to a file server.

A *network* is a transmission medium that is configured to carry internet packets. A *transmission medium* is any set of communications equipment that is configured to carry data. This includes the cable, the network interface card, and the software to send and receive packets (data).

A *host* can be a network station as long as it is supplied with a 48-bit address to identify it on the network. The host must supply communication protocols to enable other stations (hosts) on the network to communicate with it.

A *port* is an identified service in a host and may be the source and/or destination of packets. It is an integer number that represents an application on a server on the internet. The service may be a file server application, a terminal service, or some type of communication service. Data originate in and are sent via ports. *Port numbers* are numbers assigned to services in a network station. These services are identified with human-readable names. Network software uses port numbers. They identify which service is being requested and with which service a network station should respond to packets.

Socket numbers are the combination of host address, network address, and port number. The XNS documentation calls port numbers socket numbers. To be consistent with the documentation, this text will use the term *sockets* to mean ports. However, port numbers are used to define a particular service running on a network station. A socket number is used to uniquely define a service by including not only the port number but also the network and host numbers. Therefore a socket number is the combination of the port, host, and network numbers.

A network may be one of three types: a broadcast network, a multicast network, or a point-to-point (nonbroadcast) network. A *broadcast network* is one in which all hosts on the network can receive data. (It is possible to transmit to all hosts on the network with one packet.) Ethernet and Token Ring are examples.

A *multicast network* enables the communications software to transmit data to a subset of hosts on the network. Ethernet, Token Ring, and FDDI support the transmission and reception of multicast packets. An example is the spanning tree algorithm packets for Ethernet bridges or router update messages between routers (OSPF or RIPv2).

A *point-to-point network* is a communication medium between two hosts only (an example is two networks geographically separated but connected to form an internet by telephone serial line communications).

Finally an *internetwork* or *internet* is the interconnection of networks that carry internet packets, as illustrated in Fig. 5-1.

Sockets and Routing

The internet transport protocols support data communication between *sockets* on a network. These sockets may be on the same host, on different hosts on the same network, or on different networks (the Internet). Sockets will be explained in detail later. For now, a socket is an endpoint for communication in a network. It allows the network protocol in a workstation to understand the data's destination.

Switching internet packets between sockets that reside on the same host, between sockets and networks (the same host is the source or the destination of the packet), and between networks is the function of *routing*. The routing function is found on every station on a network or internet. Just because data are transferred between two sockets on the same host, or between two hosts on the same network, does not mean that the routing function is not enabled. The main purpose of the routing function is to deliver data between sockets no matter where the sockets reside.

Routing data between stations on the same LAN is called *direct routing*. Sending data between stations on different networks is called *indirect routing*. In direct routing, the transmitting station knows the network is local, because the destination network is the same as its network number. Therefore, routers are not invoked, and the packet is simply transmitted to the network. For indirect routing, the router function maintains a table, known as a *forwarding table*, that contains two important entries: a network number and a cost to that network from the router. Each router will know every network number on the XNS Internet. The routers will assign a cost, more commonly called a *hop*, to each network number. The *hop count* is the number of routers between a router and the associated network number on the internet. Routers update their tables dynamically using the routing information protocol. All of this is explained in more detail shortly.

XNS at Level 0

The functions provided at level 0 are simply those needed to transport data across the transmission medium (the broadcast, multicast, or point-to-point network). At this level, the functions do not care what the data represent; the data are provided using level 1 protocols for the mere action of being transported over the medium. Examples of broadcast transmission mediums are Ethernet, Token Ring, FDDI, and the WAN protocol (frame relay, leased line using PPP, SMDS, ISDN).

Nonbroadcast or Point-to-Point

Serial lines use the CCITT standard High-Level Data Link Control (HDLC) protocol for data encapsulation. Data to be sent across the serial line are wrapped in an HDLC frame so that the device at the other end of the serial line will understand what to do with the packet. HDLC framing is simply the signal that encapsulates your data for transmission over the serial line. This framing is removed at the remote end of the serial line.

At this level, the packet is treated simply as data. Level 0 packets are interpreted only by level 0 protocols. Refer to Fig. 5-4. Data handed to level 0 by level 1 is encapsulated (enveloped) by level 0 headers and trailers to be transmitted over the medium. The exact encapsulation technique depends on the medium. Ethernet uses one type of encapsulation, Token Ring, and IEEE 802.3 another, etc. The host that receives the data must use the same type of transmission medium for it to properly deencapsulate the packet. A system that is not running the same level 0

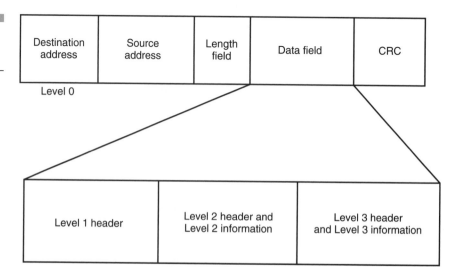

Figure 5-4
IEEE 802.3 frame with IDP headers.

protocols as the sender should not appear on the same transmission medium.

Packet encapsulation is the function of transforming level 1 data (an internet packet) into whatever packet form is supported by the transmission medium, so that it simply appears as data. Generally, this involves the adding of headers (information at the beginning of the packet), trailers (information at the end of the packet), and error correction information, called cyclic redundancy check (CRC). The deencapsulation process reverses this process at the destination host of the packet.

In order for the packet to be transmitted, it must be properly addressed according to the conventions of the transmission medium (for Ethernet and Token Ring, all level 0 addresses are 48 bits in length).

The packet encapsulation for IEEE 802.3 transmission is shown in Fig. 5-4. All upper-layer information is placed in the level 0 packet. All padding (to fill up the packet to the minimum number of bytes for transmission) is done at the encapsulation stage. This extra padding will be removed by the destination station. Packet encapsulation for each layer is shown in Fig. 5-5, which shows each layer of the OSI model and the encapsulation that is used. Each layer will add its own information to the beginning of the encapsulated data. For example, as data are handed to it from the application layer, the presentation layer adds its own special information to it. This special information is known as *header information*. When the packet is received, the network software at each OSI layer reads the header information and then processes the packet accordingly.

Packet Reception at Level 0.   When the internet packet is received, it must be deencapsulated by the level 0 protocols (address and CRC check), and then the rest of the packet is transferred to the level 1 protocols. The process of deencapsulation is simply the reverse of the encapsulation methods. Error checking is usually first done, and if the packet is received in good condition (no errors), it is deencapsulated by level 0 and transferred to the level 1 protocols for processing. The level 1 protocols may give it to a socket or it may reencapsulate it and transmit it to another network (routing). If it is reencapsulated, it must conform to the transmission-medium encapsulation methods of the new network. This is one reason why protocols that incorporate a routing layer can be delivered to different transmission mediums (received as an Ethernet packet and translated and then transmitted to a Token Ring network); bridging a packet cannot do this. Bridging essentially runs at level 0. XNS level 1 encapsulation is shown in Fig. 5-6.

Figure 5-5

Layered encapsulation.

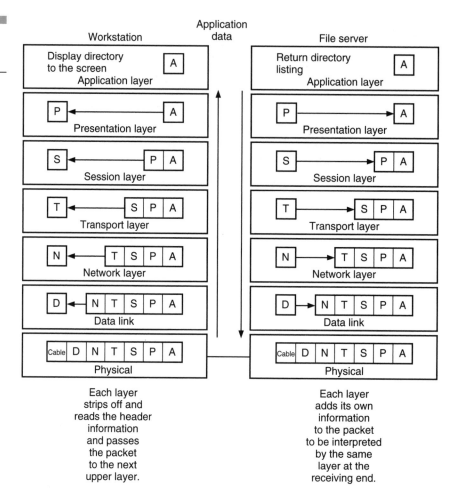

The design of the internet transport protocols is to treat all packets independently. Each packet is a datagram and is transmitted in what is known as connectionless mode. It is transmitted over the transmission medium; an acknowledgment is not expected from the receiver of the packet at this level.

XNS defines only one protocol at level 1. It is called the internet datagram protocol (IDP). The purpose of this protocol is to address, route, and deliver standard internet packets. It provides the delivery system for data transmitted and received on the internet. This process delivers a

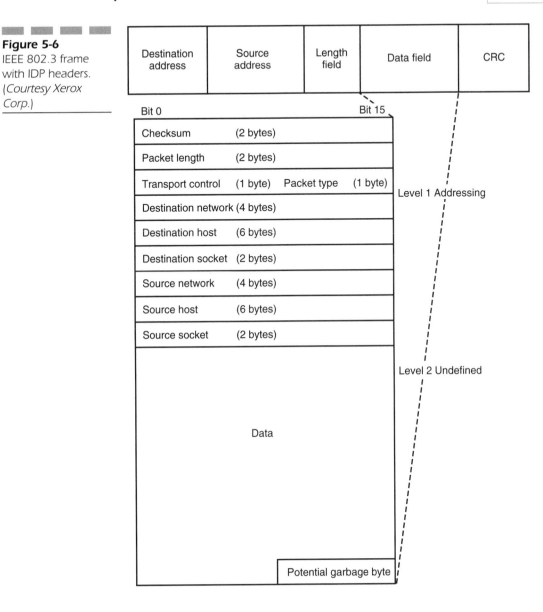

Figure 5-6
IEEE 802.3 frame
with IDP headers.
(*Courtesy Xerox
Corp.*)

connectionless, best-effort delivery service. There is no connection setup for the delivery of packets. The internet packet is transmitted, and the protocol continues with the next packet in the queue. If, for some reason, the packet never makes it to the final destination, another, —higher-layer protocol (usually a transport-layer protocol) will time out and the packet will be retransmitted. This is the function of the sequence packet protocol (SPP), which will be discussed in a later section.

The format of the IDP header and its encapsulation in an IEEE 802.3 packet is shown in Fig. 5-6. It has a source network address and a destination network address. Its primary concern is dealing with the delivery of a datagram (formatted data) onto the network.

The network address totals three distinct fields. The 32-bit network number, which allows the use of 4,294,967,296 network numbers, uniquely identifies a single network on the internet. The 48-bit host number, which allows for 2.814749767107E + 14 hosts, is unique for every network attachment on the internet. Whereas network numbers group network stations into a single entity, host numbers define a single host. The 16-bit port number, which allows for 65,536 services, uniquely identifies a socket within the operating system on the host.

Host Numbers

The host field is identical to the 48-bit physical MAC address of the network station. In the internet header, it identifies the originator of the packet, the source host number, and the final address (the destination host number) to which the packet is destined. When working with an internet, even though the address mimics a MAC address in format, it may or may not be the MAC address of the originator or an intermediate recipient. In the IDP header, this field contains the 48-bit MAC address of the original sender of the datagram and the 48-bit MAC address of the final recipient of the datagram.

This has many advantages compared to other network protocols (protocol address, TCP/IP, DECnet, AppleTalk, etc.). For example, when a host is moved from one network to another, it will not have to change its host address to the new network host address scheme (the network number may change, though). When the internet packet is encapsulated for transmission over Ethernet, there is not a separate process to provide translation between the protocol address and the physical-layer address, as in TCP/IP or AppleTalk, which reduces the overhead needed for transmission. This also allows the elimination of tables in the network station's conserving memory (at the peak of XNS's popularity, memory was very expensive).

Network Numbers

The host number identifies a particular host on the internet, and the network number identifies a group of hosts on the internet—specifically, on one network. The network number is like the area code in the phone system. It serves an important role on the internet when hosts are separated by routers. With network numbers, routers need to know only a path to each network on the internet (a table of network numbers and the path to get there). This reduces the amount of information that a router must contain to perform its functions. Providing for network numbers allows hierarchy to be implemented in a network. Building a network hierarchically allows for more efficient use of resources through better network management, reduced congestion, maintainable broadcast domains, etc. Imagine if the whole United States used one area code. It would not work.

The routing function (direct and indirect) uses only the network number to route packets. It does not care about the host numbers until the packet reaches the final network. If the source and destination are on the same network (same network number), the packet is simply routed locally. If the destination is on a remote network, workstations invoke the use of routers for datagram routing. It is up the final router to determine how to address and deliver the packet to its final destination. Keeping track of network numbers instead of host numbers provides for more efficient routing. Network numbers are the only entries in the routing table (used to determine a forwarding path). DECnet and OSI are the only protocols that are exceptions to this.

An internet packet addressed to a host will contain the network number on which the host resides. Routers will attempt to deliver the datagram to the host based on this network number. All network numbers must be unique on the internet in order for routing to function properly. If two autonomous networks are merged, the network numbers must remain unique throughout the merged network.

Socket Numbers

A socket is an integer number assigned to a service on a host (destination or source). Again, the XNS specification likes to vary the use of this term. This will identify the process to which the originator or packet wishes to communicate (destination socket), and it will identify the process to which the recipient can return a response (source socket). Certain socket numbers are considered "well known." This means they are reserved for a specific use. They are also called *static sockets*. Table 5-1 shows the well-known sockets and their descriptions for XNS.

TABLE 5-1

XNS Socket Types

Function	Well-Known Socket Number, Octal
Routing information	1
Echo	2
Router error	3
Experimental	40–77
Dynamically assigned	4000–6000 (decimal)

Other socket numbers are known as *ephemeral.* This means that they are dynamic and can be reused. These socket numbers do not need to be unique throughout the internet, because a socket number is combined with the host and network numbers to uniquely identify a network, host, and service needed on that host. In other words, the network address, host address, and socket address taken as one number will identify a service on the internet.

For a socket number assigned to a file service (for example, 0451), there may be multiple requests for connections to the server containing that socket number. Connection identifiers are used so that the host providing that file service will be able to differentiate between multiple incoming connections wanting access to the same destination socket number for the file service.

Simply stated, socket numbers identify a service running on a host (such as a file service, print service, or database service). The number enables a workstation to communicate with the service.

Identifying the IDP Fields

Checksum. This is an error-identifying number. It is a software checksum and is used in addition to the transmission media's checksum. This field is optional. A value of FFFF (hex) in this field indicates that checksumming is turned off at this layer. Novell's NetWare usually does not use the checksum field in its proprietary IPX implementation. This works out because the SAP of FF is a global SAP and does not hurt anything..

Length. This field indicates the length of the internet packet measured in bytes. This includes the checksum field. It does not include the transmission-medium encapsulation headers and trailers (level 0). It is equal to the length of the data field plus 30 bytes. The 30 bytes is the length of the network-layer header. It will always be this length. The maximum length of the internet packet is 576 bytes. TCP also states that systems in routed networks should adhere to the maximum datagram size limit of 576 bytes (total length). All systems should be able to handle this maximum datagram size. XNS expects level 0 protocols to provide fragmentation and reassembly (slicing up a packet to fit the transmission size of the transmission medium). Level 0 protocols have never implemented this. The 576-byte packet includes the following:

- 30 bytes for network header information
- 12 bytes for transport-layer information
- 22 bytes for session-layer (level 3) information
- 512 bytes for a typical disk page (or a block of memory) for XNS systems

This does not mean that all XNS packets must be of this size. Two communicating stations may negotiate a different packet size, either larger or smaller. Although the XNS recommendation is 576 bytes, most XNS implementations (Novell NetWare) do not follow this recommendation, even when using routers.

Transport Control. This is used only by routers, to manage the transport of internet packets on the internet. It is initialized to 0 by the creator of a packet. As the routers change this field, the checksum (if used) is recomputed. Each time a router forwards the packet to another network, it adds 1 to this field. This indicates how many routers the packet has traversed. When this field reaches 16, the packet is discarded by the last router. This ensures that packets do not endlessly loop in an internet. Some implementations will send an error packet back to the originator of the packet by using the error protocol (explained later). With a maximum of 16 allowed in this field, only bits 4–7 are used. Bits 0–3 are not used and should be set to 0.

Maximum Packet Lifetime (MPL). This is used to estimate the maximum time an internet packet will remain in the internet as governed by the internet transport protocols. The recommended standard for this entry is 60 s. This was computed by estimating that the most time a router will delay forwarding a packet is 1 s. A packet is allowed to traverse 15 routers before being discarded. This would allow the MPL to reach 15. However, the number is multiplied by 4 for rare cases of extreme delay. There could be low-speed lines in a very large network.

Figure 5-7
IDP demux efforts
based on assigned
packet type. (*Cour-
tesy Xerox Corp.*)

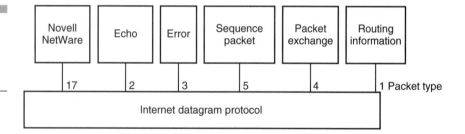

It is recommended that network drivers that cannot transmit a packet within one-tenth of the MPL time discard the packet.

Packet Type.   Refer to Fig. 5-7 and Table 5-2. This field is regulated by Xerox. Anyone following this protocol should register their protocol with Xerox so that Xerox can assign a packet type. This field indicates the format of the data field of the internet packet and is used by Xerox to identify the user of the packet. The types recommended are shown in Table 5-2. Experimental types are used when a vendor is developing a new protocol.

It primarily identifies the level 2 function of the embedded data in the data field of the packet. Novell has a registry number of 17 (decimal) for their XNS implementation called Internet Packet Exchange (IPX).

The level 1 protocol does not interpret this field; it merely reads it and then passes the packet information up to the appropriate socket identi-fied by the packet type. It provides information on which "well-known process" the IDP should hand the packet to.

Source and Destination Network Addresses.   The source and desti-nation network numbers identify two things: the originator of the packet (network and host addresses) and the intended recipient of the packet (net-

TABLE 5-2

Packet Types

Protocol	Packet Type, Octal
Routing information	1
Echo	2
Error	3
Packet exchange	4
Sequence packet	5
Experimental	20–37

work and host addresses). Providing the address, network, and host numbers is enough to identify any network address on the entire network.

Data. The data portion of the packet is nothing more than a sequence of bytes that is completely transparent to the level 1 protocol. The length of the data is derived by subtracting 30 from the length field, and may range from 0 to 546 bytes. It could contain transport, session, and application data, or it may contain commands. Level 1 does not care about this field.

Garbage Byte. At the end of the data field, there may be an odd number of bytes in the packet. This type of packet cannot be transmitted. To produce an even number of bytes, the level 1 protocol may add a garbage byte, which will be included in the checksum but not in the length field.

Any client may interface to the internet datagram protocol by simply acquiring a socket from the network operating system and then sending and receiving data using that socket number. It should be noted that packets are sent using a best-effort delivery service. Packets may be lost even in the highest integrity type of network.

If the checksum is found to be in error, the packet may be thrown away and no error packet sent to the originator to indicate there was a problem with the received packet. Packets may also be discarded due to buffer (RAM) constraints.

Included in the function of the internet datagram protocol is the capability of a routing function. A router is a store-and-forward device that routes packets to different networks based on a network number and a routing table. The routing table must be updated dynamically (must learn of other networks without user intervention). XNS accomplishes this using the routing information protocol (discussed shortly).

Packets may be sent directly to the destination (unique destination address) or they may be sent in multicast or broadcast mode (addressed so that many stations can receive the one packet). IDP packet flow will be shown later.

Level 2

The level 2 protocols consist of echo, error, sequence packet, packet exchange, and the routing information protocol. Each operates independently.

Routing Information Protocol

For a hierarchical network based on network numbers, there must be a method to distribute those network numbers to facilitate routing. Any station must be able to transmit to and receive from any other station on the internet. Therefore, all stations must know all the network numbers on the internet or know who to request information from about a network number.

To enable network numbers and their associated paths to become well known on an internet, a dynamic routing update protocol was invented known as the routing information protocol or RIP. RIP is the most widely implemented routing update protocol today in heterogeneous networks. Although it does have its drawbacks, it provided a good beginning to widespread use of dynamic routing protocols. Variations of this protocol are implemented on AppleTalk, TCP/IP, IPX, and XNS. A high-level overview of how distance vector protocols operate follows. Refer to the specified protocols for more information on how they run the RIP protocol.

The indirect (nonlocal) routing of datagrams is accomplished using a special store-and-forward device known as a router. The router knows of other networks by means of an internal database table that lists the network numbers and the path associated with the network number. The routing information protocol is the means by which the database is dynamically maintained. It is one of the protocols located at level 2.

All XNS routers perform three tasks: supply routing information to any station that requests information, update other routers about the networks that it knows about, and provide a forwarding path for datagrams.

A network workstation (nonrouter) has various methods of obtaining its network number. Some XNS implementations allow an initializing workstation to read information from the cable plant and obtain its network number in this way. Others get their network numbers when they are initialized (preconfigured), while others retrieve this information from a server.

If the workstation does maintain a table of network numbers, this table may be simple (with only one entry) or it may be complex in that it maintains a routing table, just like a router, but does not update the numbers of network stations on the network (called passive RIP implementation). The passive implementation means that the device maintains information about the network (maintains a routing table), but it does not exchange this information with other devices, nor does it respond to requests for routing information. The active implementation means that the protocol maintains a routing table and provides routing updates (exchanges its routing table with other routers.) This implementation can also respond to stations requesting information about a network.

For networks implementing RIP, an assumption is made that the internetwork is no larger than a few hundred networks. Based on this assumption, the network number space is called "flat." With RIP, there are no assigned area numbers separating the networks. It is a flat address space, and all routers are considered equal with each routing table considered trusted (in the sense that all routing information dispersed by any router is considered reliable).

In TCP/IP protocol, a concept known as the default route is used in many PCs. This is the router that the PC is instructed to use when the data it has to transmit are for a destination station not on its local network. All XNS stations implement the passive version of RIP; they do not employ the concept of a default route. Novell NetWare, which is the most popular protocol that uses some of the XNS protocol, allows workstations to find network numbers through another protocol that converses with a router (or router function in a server). In other words, NetWare workstations snoop their network ID from information found on the network. XNS stations do not do this. However, since XNS is more an architecture than a protocol, some vendors may implement different methods of determining the network number.

Implementing RIP. RIP is based on a concept known as *distance vector.* The vector is the network number, and the distance is the cost associated with that network number. The routers exchange network reachability (vector, distance) information through the broadcasting of their routing tables, which contain a listing of distance-vector entries. Each entry in the table consists of a network number (the vector) and the amount of routers (distance) between it and the final network. There are actually more entries in the table associated with a network number, but those entries are held in the router and are not broadcast to other routers or workstations implementing active RIP. These entries include the port number associated with a network number and an age entry indicating how long it has been since the router has seen an update on a network number.

RIP measures distance networks through a metric known as a *hop.* Any time a datagram must traverse a router, it is considered one hop. A maximum diameter (distance) of 15 hops is allowed for any packet to reach the destination network. If a data packet has traversed 15 routers and the sixteenth router receives that packet and knows (a field in every XNS packet [transport control] indicates how many routers the packet has traversed) it must traverse one or more routers beyond it to get to the destination network, the packet will be discarded at the sixteenth router. An error message may be sent to the originator of the packet indicating that the packet was discarded. Again, XNS does not specify it, but some implementations of XNS allow it. IP does not have this function in which a router checks the packet for how many routers it has traversed. It does have another field, called the time to live (TTL) field, that is used to eliminate packets that have been hopping around the network (usually in a loop)

When a new host is initialized and possesses only the requester (passive) RIP process, it will broadcast a request packet on its directly connected network and will build its table from all response packets received. The requester will be able to identify the directly connected network numbers by the RIP response packets that have the internetwork delay (the hop count is called *internetwork delay* in the XNS specification) set to 0. Any nonzero numbers in this field indicate a nonlocal network. Although the XNS specification states this, in reality this procedure is implementation dependent. Novell NetWare does not implement it this way, but AppleTalk does. Again, XNS is an example network architecture.

This is how a workstation finds out its network number without having it assigned by the network administrator. The network administrator only has to assign a network number in the routers, and the workstation will find out about it though the router RIP responses (or other methods in other XNS implementations). For XNS, no more than one network number should be assigned to one network (broadcast domain). Again, depending on how XNS is implemented, different network vendors may or may not employ this.

Using distance-vector algorithms, each router initializes its table starting with entries of those networks that are directly attached to it. The usual cost associated with this is 1 (for one hop). For example, if a router is configured with four interfaces of network numbers 1, 2, 3, and 4, then the first entries in the routing table will be networks 1, 2, 3, and 4 with a cost of 1 (assuming the XNS implementation uses 1 as the starting cost). Their hop counts will be set to 1, and this table will be broadcast immediately out to their directly connected networks. This tells other routers (or stations that are implementing RIP) on the network the network numbers that they have configured as their directly connected networks and to get to those networks, the cost is 1.

All routers on an internet will do this, and this information will be propagated throughout the entire internet. Routing table entries can be deleted only by a time-out process expiring in three minutes. Any router that is gracefully taken down should report this by transmitting a RIP response packet to the network with all internetwork delay (hop count) entries set to 16, which means the network is unreachable. This indicates to other routers that that particular router is no longer able to forward packets to the network.

The protocol of RIP was built on the following assumptions. The network may contain Ethernet (broadcast networks), leased lines, and/or public or private networks (nonbroadcast networks) that use packet or circuit switching. Each of these types of network has different bandwidth characteristics, but the metric (the length) for each delay is generally defined in terms of hops or internetwork router hops. In other words, RIP pays no attention to the medium or its speed. A 56-kbps circuit is considered a single hop as is a 45-Mbps serial line. This is one major drawback of RIP.

Internetwork routers use RIP to inform each other of the topology of the internet. With the predominant network being Ethernet, RIP utilizes the broadcast capabilities of the internet datagram protocol and Ethernet to distribute this routing information. Public data networks may support broadcast addressing, or it may be too expensive, in which case these types of networks must know the identities of the routers with which they will be exchanging information (in other words, they must be statically configured with the other routers' addresses in order to exchange information with them). There are limited dynamic exchanges between routers on public switch networks.

For RIP to work efficiently, it should not be implemented in large, complex internetworks (although it is being used in some). There are many disadvantages in having all internetwork routers maintain complete database tables of all other networks in the internet and then broadcast this information to all other routers on the internet. There are also different bandwidth implementations (different speeds for leased serial lines and fast, reliable networks like Ethernet). In addition, queuing delays (router delays in forwarding the packet) may cause extreme congestion on the internet, and routing updates may be delayed to the farthest routers. Refer to Fig. 5-8.

The fields of the RIP packet are defined as follows.

OPERATION. This field indicates the type of RIP packet. It can be one of two types:

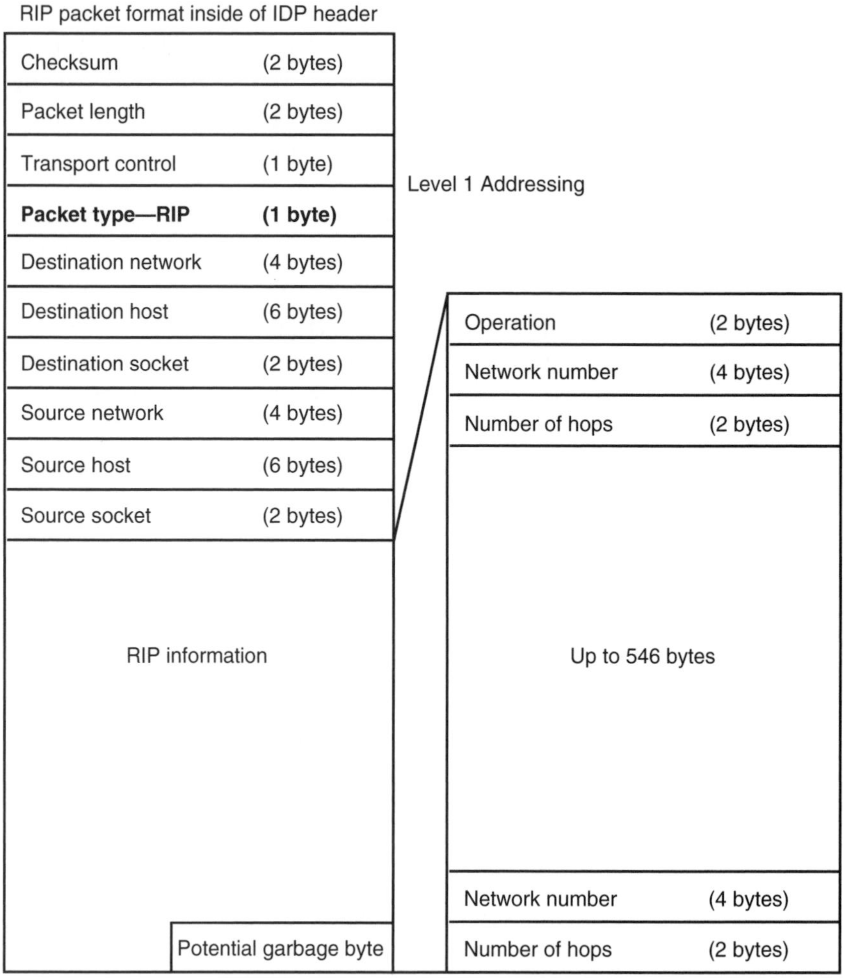

Figure 5-8
RIP packet. (*Courtesy Xerox Corp.*)

RIP packet format inside of IDP header

Checksum	(2 bytes)	
Packet length	(2 bytes)	
Transport control	(1 byte)	Level 1 Addressing
Packet type—RIP	**(1 byte)**	
Destination network	(4 bytes)	
Destination host	(6 bytes)	
Destination socket	(2 bytes)	
Source network	(4 bytes)	
Source host	(6 bytes)	
Source socket	(2 bytes)	

Operation	(2 bytes)
Network number	(4 bytes)
Number of hops	(2 bytes)

RIP information

Up to 546 bytes

Network number	(4 bytes)
Number of hops	(2 bytes)

Potential garbage byte

1. A 1 in this field indicates a request for routing information.

2. A 2 in this field indicates a response to a request for information.

CONTENTS. Each entry (there may be more than one) in the contents field contains one or more tuples that consist of a 32-bit object network (a network number) and a 16-bit internet delay, in hops (how many routers away from the originator of a packet the destination network is).

Depending on whether the packet is a request or response packet, the contents field can be filled in the following ways. If the RIP packet's operation field indicates a request, then specified in each tuple of the RIP packet is the object network number of the network that the station submitting the RIP request seeks information about. The internetwork delay will be set to infinity (in RIP this means 16). If the requester wants information about all the networks that the router knows about, the contents field is set to "all" (all FFs) with internetwork delay set to infinity.

If the operation is a response, this means that the router is responding to a request for information or is providing a periodic update for other routers. The contents field will contain one or more tuples indicating the appropriate network number and the internetwork delay (the cost or hop count) that the router knows about. For example, if a request was received by the router for information on all networks that it knows about, the router would build a response packet in the format of Fig. 5-8 that will contain one or more tuples consisting of the network number and cost for each network number the router knows about.

When a routing information supplier transmits or broadcasts a packet that is a response, the internetwork delay (the cost, measured in hops) is the delay that the packet will experience in being forwarded to the final network number (as indicated by the object number) if the packet is forwarded by the router that generated the response. If the router forwards a packet to a directly attached network (the cable segment on the other side of the router), the internetwork delay is 1. For each router the packet must traverse to reach the final destination, this number is incremented by 1.

A delay of infinity (16) means the network cannot be reached or that the router does not know of the object network. For example, if the RIP request was sent out by a workstation for network X and a router responds to that request with an internetwork delay of 16, this indicates that the router does not have that network number in its database table or that the network number for some reason has been taken off the internetwork (the router that attached that network was taken down, for example). Finally, the delay to reach a router on a directly connected network will always be 0. (Note that other implementations use 1 for directly connected networks.)

Routing Table Maintenance.   RIP is used in every router and network station on the network to maintain a cache (a portion of RAM memory) of network entries called the routing table. RIP is the algorithm that is used to maintain this routing table and serves several purposes. First it permits the quick initialization of a routing table when a network station starts. Second, it ensures that a routing table is up to date with any changes that occur on the network. These changes could involve new networks being added or a router becoming temporarily disabled.

As shown in Table 5-3, each entry in the table contains

1. A network number (known as the object network)

2. A delay (measured in hops, the number of routers between it and the destination network)

3. **a.** The directly connected network (there will be one entry for each attachment the device has; for example, if the device is attached to three Ethernet segments, there will be one entry in the table for each attachment) or

   b. The host number of an internetwork router (only if the delay is not 0; in other words, it is not directly attached), which is used to forward a packet to another network by using another router

4. A timer used to age out old entries—networks that have not been heard from for a specified amount of time

During initialization, a network device will accomplish the following:

1. If the device is a requester only, called the RIP passive service (a user's workstation, for example), its table may contain an entry only for the network it is directly attached to. This will allow it to make a comparison between its network number and a destination station network number. If the two are different, the user's workstation knows that a router is needed to deliver the packet.

Routing tables can consume large amounts of memory. Therefore, in a nonrouting network station (a workstation) only one network number is held in the table. This is the number of the network it is directly

TABLE 5-3

An XNS Routing Table

Network Number	Delay, Hops	Host Number	Timer
1	1	02608c1234565	50
20	2	02608c123456	100
50	0	Local	34
20	0	Local	45
23	0	Local	40

attached to. If it needs to communicate with a remote network station on the internet, it will ask a local router for the information. This saves RAM memory. Some network protocols do not allow a workstation to possess the RIP process. They give the network workstation the address of a local router and make it send all routable data packets to it. The router will figure out the shortest path and, if there is another router that can handle the request with a shorter path, it informs the workstation and the workstation will then use the other router.

2. If the device is capable of being a requester and responder (a router, for example), the table will contain entries for both the directly connected networks and all other networks on the internetwork. Most routers, upon initialization, enter the network numbers of their directly connected networks as the first entries in the table, and the cost is set to 0, indicating direct connection.

Active routers perform two operations: responding to requests for routing information (event driven) and periodically transmitting their routing tables to their directly connected networks. This means that when a router transmits its table to the network, it keeps other routers informed on the state of the internetwork.

Figure 5-9 shows the propagation of routing table information. This figure shows updating without the use of an algorithm known as split horizon. This will be explained later. Also, this figure shows one direction of the update process. In actuality, there is no synchronization between routers in updating other routers with their tables. Each router updates when its time expires.

After initializing, router A adds its directly attached network numbers to its table and marks these entries as local (directly attached). It then transmits this table to network 2 as a RIP response. Each of the hop count entries will be 1 before transmission. They were initialized to a 0, but when the table is sent out, they transmit it as 1, because they are directly attached networks on router A.

Router B receives this packet (it will be addressed to the RIP socket number and physically addressed to broadcast). The RIP cost assigned to the receive port of the router is 1. Router B therefore adds 1 to each entry of the received table and then compares the entries in the received table to its routing table. If there is not an entry for a particular network number, router B will add the network number. If the network number is already in the table, router B compares the cost (hop count) of the received routing table with its own. If the hop count is lower, router B changes that entry in the table. If the received hop count is greater than

Figure 5-9
A routing table
update not imple-
menting split
horizon.

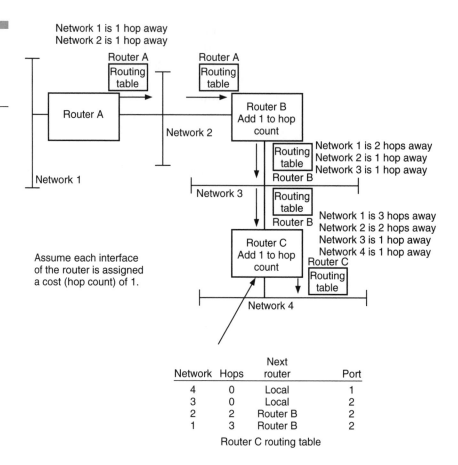

Figure 5-9
A routing table update not implementing split horizon.

Network 1 is 1 hop away
Network 2 is 1 hop away

Router A
Routing table

Router A
Routing table

Router A

Network 2

Router B
Add 1 to hop count

Routing table
Router B

Network 1 is 2 hops away
Network 2 is 1 hop away
Network 3 is 1 hop away

Network 3

Routing table
Router B

Network 1 is 3 hops away
Network 2 is 2 hops away
Network 3 is 1 hop away
Network 4 is 1 hop away

Router C
Add 1 to hop count

Routing table
Router C

Network 4

Assume each interface
of the router is assigned
a cost (hop count) of 1.

Network	Hops	Next router	Port
4	0	Local	1
3	0	Local	2
2	2	Router B	2
1	3	Router B	2

Router C routing table

its internal table, it discards that entry of the received routing table. Once router B has updated its table, it may transmit it to network 3. Router C receives this table and repeats the preceding process.

Reception of a RIP Response.   This type of packet is received in two ways. First, it may be received in return for a previous request that was transmitted requesting information about a network or a set of networks. Second, it could be received by "gleaning." This is the process in which a local router sends out its routing table periodically, and since this response packet is generated in broadcast mode, all stations will receive it. The needed information is gleaned from the packet. The packet was requested by another station, but since it is sent in broadcast mode any network station will receive it and build a table based on the information found in the packet.

Updating the Table. This was explained previously but is covered here in greater detail. All routers perform this routine.

This is the real purpose of the RIP protocol—the ability to dynamically update a router's table periodically or when an event has occurred that forced a RIP response packet to be generated. When a network device receives information about other networks (a RIP response), it must decide whether it should update its table (whether it is a requester only or both a requester and responder, i.e., a router). Some decisions are simple. For example, if the information in the RIP response packet contains a network number of a directly connected network, then no update is necessary. That is, the RIP response packet contains information about a network that is already directly connected to that device. The algorithm will disregard that information and its table will not be updated.

However, table updates are required for the following situations:

1. There is not an existing entry in the table for a network number.

2. The existing entry in the table has not been updated for 90 s, which suggests the entry may no longer be valid.

3. The delay in the received RIP packet for a network number is less than the delay already entered in the table. In other words, a new route was found that offered less delay to the target network.

The RIP process adds new networks, changes an existing entry, or deletes an entry from the table.

When any modifications to the table are done, the host address in the table is set to the source host entry of the packet (see Fig. 5-8). When a packet is received, a 1 is added to each cost entry in the received routing table before any comparisons are made to the table. Located in a RIP response packet are network numbers from a router's table. These indicate networks to which the router can forward packets.

In other words, if they are from another router, it is assumed that to reach those indicated networks, the packet will have to travel across that router. Anytime you traverse a router, the internetwork delay increments by 1. However, some implementations allow you to artificially configure this number, in order to force a particular route to be taken over another. The route with the lowest hop count is taken. This is the reason the router automatically increments each entry by 1 before accomplishing any comparisons. However, not all protocols follow this. Some implementations add 1 to each entry before transmitting a RIP response packet, so that when the RIP response packet is received, a 1 is not added to each entry before the comparison. Either way works. The aforementioned procedure is the method the XNS standard recommends.

If the entry is not new but indicates a better route to an existing network, the old entry is completely replaced with the new entry. The information contained in the RIP packet is placed in the table, and the delay in the table is set to the delay indicated in the RIP packet. If this delay exceeds 15, the object network is said to have become inaccessible. The internet delay is set to the value indicated in the RIP packet plus the internet delay of the router. Usually this means it is set to the internet delay of the RIP packet plus 1.

For a change to be made, a 1 is added to the delay indicated in the RIP response packet and then compared to the delay in the table. If the delay in the RIP response packet is lower, the entry is changed. If it is the same or higher, the entry in the RIP response packet is ignored.

Upon an update, the time-out entry for each update entry is set back to 90. Each entry in the table has its own expiration timer, initialized to 90 s.

Handling Time-Outs.   The XNS standard states that for the network device that possesses only the RIP requester, timing out an entry is done at twice the normal 90 s. The table will be set with an internetwork delay of 16 (infinity) and the entry may be discarded.

Routers may keep this infinity entry longer so that with each routing update, all routers on the internet will know that this network path is no longer valid. During this time-out period, if the requester or a router receive new information about the object network, the table entry may be reinitialized and the network entry will become valid once again.

Timing for Updating the Network.   Routers transmit their complete routing tables once every 30 s to their directly connected networks. The packet is addressed to a MAC broadcast destination address. If the network is point to point (serial connections between routers), the packet will be addressed specifically to those routers. If a router updates a routing table entry, it should immediately notify all other routers with a response packet. This is known as *event-driven updating.* In other words, if a router changes an entry in its table for internetwork delay, it will not wait for the 30-s timer to expire before transmitting a RIP response packet. It will transmit this packet immediately, usually with only that entry in the packet.

Under XNS, a router should gracefully shut itself down by transmitting a RIP response packet with the internetwork delay set to infinity (16) for all network entries in its table. This allows the router to tell other routers that it will no longer be able to forward packets to those networks, even before it cannot do so.

Finally, when a responder replies to a requester and it does not know about the network being requested, it responds with the network entry in a response packet, but with the internetwork delay set to infinity (16).

Complications of the RIP Implementation. RIP exists primarily for those networks that are not large and complex and that experience relatively few changes in topology (for instances of networks going up or down for whatever reason). Since most networks implementing XNS are Ethernet, it could be said that the transmission speed is high and deemed reliable and that the speed is constant. However, negative circumstances can occur using this protocol.

When an internetwork router initializes and brings a shorter route (or the only route) to a network than any other router knows about, all routing tables in all routers and nonrouters will have to be updated. This should occur within $30n$ s, where n is the number of hops (internetwork delay) from the newly initialized router to the router most distant from that router, and 30 represents the normal broadcast interval of RIP routing updates (routers broadcasting their routing tables to their directly attached networks).

When an internetwork router is taken out of operation, and if another route exists that the router used to forward packets, this new router for the replaced network number will be discovered within 90 s. This is the time after which a routing table entry in any router becomes suspect as unreachable, and this entry becomes eligible for replacement with another entry.

If a network goes away and there is not an alternative route to it, the entry in the routing tables in all routers for this network will be purged within $90 + 30 \cdot \max(n, 15)$ s (although it should occur much faster than this). RIP router updates occur every 30 s.

In a single router, an entry that has not been updated for 180 s is considered suspect and will be removed from the table if an update is not received in another 60 s. Therefore, an entry in a table will be deleted in 240 s (if the router does not hear from it).

It is possible during this time of network inaccessibility and the purging of the routing table entries that a packet may get caught in a routing loop (a packet will endlessly be forwarded by routers that are misinformed about a path to a network destination. XNS alleviates this problem, for any packet on the internetwork has a lifetime. Once the packet's lifetime is exceeded, the packet will be discarded.

Every implementation of RIP (TCP/IP, AppleTalk RMTP, Novell NetWare IPX) is affected by the drawbacks of RIP. In order to make RIP operate correctly, the following protocols were implemented:

1. Split horizon
2. Poisoned reverse
3. Hold-down timers
4. Triggered updates

Routers will not broadcast their entire routing tables out of all their active ports. Certain entries in the table will be omitted before the table is sent. This is the purpose of split horizon.

Referring to Fig. 5-10a, with router A directly attached to network 1, it will advertise that route through all its ports as a distance of 1. Router B receives this and updates its table as network 1 with a distance of 2. Router C receives this and updates its table as network 1 with a distance of 3. Notice that all routers broadcast all the information in their tables through all ports (even the ports from which they received the update).

Why does router B broadcast a RIP update of network 1 to router A, when router A already has a direct attachment to it? Wouldn't this confuse router A into thinking that another route existed for network 1? Normally it would, but remember that the only changes that RIP makes to its tables is when then the distance is lower when there is a new entry, or when the next router path taken to a network changes its hop count. Since the received hop count is higher, router A simply ignores that particular entry in the update table.

Using the original algorithm, a serious problem occurs when router A loses it reachability to network 1. It will update its table entry for that network with a distance of 16 (not reachable) but will wait to broadcast this information until the next scheduled RIP update. So far so good, but if router B broadcasts its routing table before router A (not all routers broadcast their tables at the same time), router A will see that router B has a shorter path to network 1 than it does (a distance of 2 for router B versus a distance of 16 for router A), and the new entry will be made. Now router A, on its next RIP update broadcast, will announce that it has a path to network 1 with a distance of 3 (2 from the table entry received from router B and 1 more to reach router B). There is now a loop between routers A and B. A data packet destined for network 1 will be passed between routers A and B until the transport field is 16. This is known as *looping,* and its cause is called *slow convergence.* The RIP protocol works extremely well in a stable environment (an environment where no routers or their networks ever change)—a *stable convergence.* Convergence is the ability of the network to learn about bad destinations and to correctly mark them as unreachable in a timely manner.

Figure 5-10
Routing table
updates (RIP). (a) Not
implementing split
horizon. (b) Imple-
menting split
horizon.

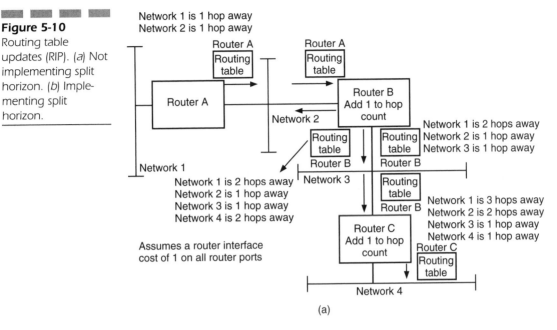

Network 1 is 1 hop away
Network 2 is 1 hop away

Router A
Routing table

Router A
Routing table

Router B
Add 1 to hop count

Network 2

Routing table
Router B

Routing table
Router B

Network 1 is 2 hops away
Network 2 is 1 hop away
Network 3 is 1 hop away

Routing table
Router B

Network 3

Router A

Network 1

Network 1 is 2 hops away
Network 2 is 1 hop away
Network 3 is 1 hop away
Network 4 is 2 hops away

Assumes a router interface
cost of 1 on all router ports

Router C
Add 1 to hop count

Network 1 is 3 hops away
Network 2 is 2 hops away
Network 3 is 1 hop away
Network 4 is 1 hop away
Router C

Routing table

Network 4

(a)

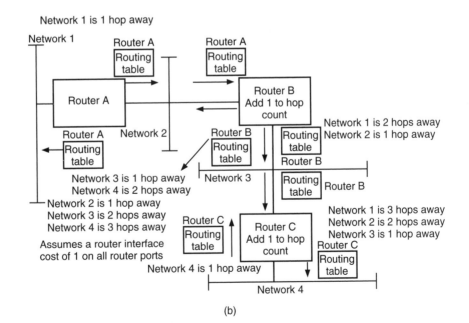

Network 1 is 1 hop away

Network 1

Router A
Routing table

Router A
Routing table

Router B
Add 1 to hop count

Router A

Network 2

Router B
Routing table

Routing table
Router B

Network 1 is 2 hops away
Network 2 is 1 hop away

Router A
Routing table

Network 3 is 1 hop away
Network 4 is 2 hops away

Network 3

Routing table
Router B

Network 2 is 1 hop away
Network 3 is 2 hops away
Network 4 is 3 hops away

Assumes a router interface
cost of 1 on all router ports

Router C
Routing table

Router C
Add 1 to hop count

Network 1 is 3 hops away
Network 2 is 2 hops away
Network 3 is 1 hop away
Router C

Routing table

Network 4 is 1 hop away

Network 4

(b)

Even future RIP updates will not quickly fix the convergence in this case. Each update (every 30 s by default) will add 1 to the table entry, and it will take a few updates to outdate the entry. This is slow convergence and it causes errors in routing tables and leads to routing loops.

To overcome this and other problems, a few rules were added to the RIP algorithm. *Split horizon* states that any router will not broadcast a learned route through a port from which it was received. Thus, router B will not broadcast the entry of network 1 back to router A, since it learned of this route through router A. This keeps router B from broadcasting back to router A the reachability of network 1, thereby eliminating the possibility of a lower hop count being introduced when network 1 went away.

Figure 5-10*b* shows how routers that implement split horizon broadcast their tables. Notice they do not broadcast information about their directly connected networks from the port attached to the directly connected network. In other words, router A does not include information about network 2 when broadcasting a routing table update from the port that is attached to network 2. Router C, in broadcasting a routing update to network 3, will not include information it learned through that port. In other words, information about networks 1, 2, and 3 are not included with the update.

The second protocol is called a *hold-down timer.* This rule states that once a router receives information claiming that a known network is not reachable, it must ignore all future updates that include an entry (a path) to that network, typically for 60 s. Not all vendors support this in their routers. If one vendor does support it and another does not, routing loops may occur.

Other rules that help eliminate the slow convergence problem are *poison reverse* and *triggered updates.* The poison reverse rule states that once the router detects that a network connection is disabled, the router should keep the present entry in its routing table and then broadcast "network unreachable" (metric of 16) in its updates. This rule becomes efficient when all routers in the internet participate using triggered updates. Triggered updates allow a router to broadcast its routing table immediately following receipt of this "network down" information. The XNS specification does call for this to be implemented.

The main advantages of the RIP protocol are that it is a good protocol for networks that are stable, it provides adequate responsiveness, and it is not complicated to implement. With most routing implementations, the user simply sets RIP to on, and RIP will take care of the rest of the work. All updating will occur without user intervention. Remember, though, this is used with a flat network address space, and each router is a trusted router. Each router transmits its entire routing table, even when no change has occurred. Each router takes for granted that any router is reliable and that the routing tables are without error. Propagation of the routing information is based on the previous router's table. If errors occur on any router, then they will be propagated throughout the network. RIP is a good router table update implementation for small networks with equal transmission-medium speeds (all Ethernet or all Token Ring) and where the internetwork is stable (networks do not go up and down). Otherwise, hierarchical routing schemes prove to have many advantages over flat network address routing schemes.

With the hierarchical internet, the internet is divided into areas, all of which are connected to a single "backbone" area. Dividing an internet into areas avoids loops while still allowing for multiple paths and therefore redundancy. It also assists in routing table updates when a change occurs. With a hierarchical internet, a change will affect only the particular nodes in that area. The updates are not passed to those nodes that are not considered "trusted." That is, those routers that do not need to know about a change will not be updated, and the network will still operate efficiently. With this type of network, overhead is drastically reduced.

To operate hierarchical networks, there are level 1 and level 2 routers. Level 1 routers update other level 1 routers within their own area. Level 2 routers update other level 2 routers for information between areas. There must be at least one level 2 router per area.

Refer to Fig. 5-11, which shows three areas. In area 1 there are two level 1 routers that have connections to the Ethernet segments (EN). There is one level 2 in area 1 router that has connections to other level 2 routers in areas 2 and 3. If the packet to be routed is in the same area, the level 1 routers will route the packet locally (it will not leave area 1). If a packet on area 1 is destined for area 2, it must be transmitted to the level 2 router for it to be delivered to the area 2 router. Once the packet has been delivered to area 2, it will be locally routed until it reaches its destination.

Routing across a Serial Link.   Refer to Fig. 5-12. For routing across a serial link (leased telephone line), the data-link and physical layers change, as does the format of a packet. When the network medium changes from Ethernet to Token Ring, the packet headers change for the data-link portion. This holds true for transmitting data across a serial link.

Figure 5-11
Level 1 and level 2
routers. (*Courtesy
Uyless Black*)

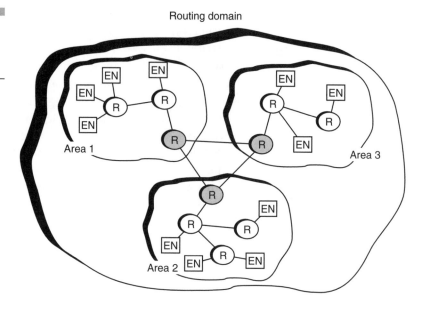

Routing domain

Area 1

Area 2

Area 3

R = Router

EN = End node (End system [ES])

(R) = Level 1 router

(R) = Level 2 router

Figure 5-12
Wide area network-
ing with serial lines.

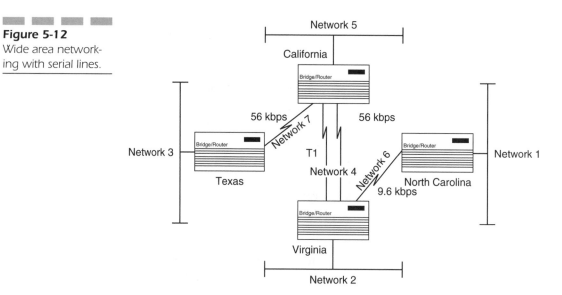

Network 5

California

Network 3

56 kbps

Network 7

56 kbps

Texas

T1

Network 4

Network 6

North Carolina

9.6 kbps

Network 1

Virginia

Network 2

When a packet is routed over a point-to-point link (two routers connected by a serial interface), the MAC headers of the packet received by the router are not stripped off, but serial data-link headers are put on. In other words, if a packet is received on the Ethernet and has to be routed to another router through a serial interface, the Ethernet headers (the address, the Ethertype field, and CRC) are not stripped off, but a serial line header is put on, and the packet is transmitted over the serial interface. When the packet is received at the remote end of the serial link, the Ethernet headers pertaining to the router that received the packet over the serial line are placed in the packet, and the packet is then routed to its destination as if it were received by a LAN interface.

Since the serial link is a point-to-point connection (no intermediate stops along the way), this is the most efficient method. The packet will not be received by anything but the remote end of the link, and the extra header data that need to be transmitted on the link will consume little bandwidth.

This is allowed because the MAC address header keeps getting changed by every router along the way en route to the packet's final destination. So why not strip off the MAC headers before it is transmitted on the serial link? This really does allow for better utilization of the serial link. It will need new MAC addresses when it reaches the remote router, so it is left up to that router to do it.

This includes the case where you are bridging a packet across a remote link. Since the bridging algorithm works with the MAC address, it must be left intact before, during, and after crossing the link. This type of packet is encapsulated in a serial header and then transmitted on the serial link. The remote link will strip off the serial headers and bridge the packet just as if it were received on an Ethernet cable segment.

Notice also in Fig. 5-12 that serial lines are assigned network numbers. You cannot assign one network number for all the serial lines. The one exception to this is shown for the routers between California and Virginia. These two lines are grouped to form one logical connection between the two routers. This is a very common application for routers, and almost all router vendors support this. Data sent down these lines are usually load balanced, so that only one packet is transmitted down one of the links. (A single packet will not be transmitted down both links. This would cause duplication at the other end.) Only in this case of grouping can two or more serial lines be assigned a single network number.

Packet Flow.   Figure 5-13 shows how a packet is formatted and sent to the router. The importance of this figure is to show how the MAC

Figure 5-13
Abbreviated version
of data transfer using
IDP headers.

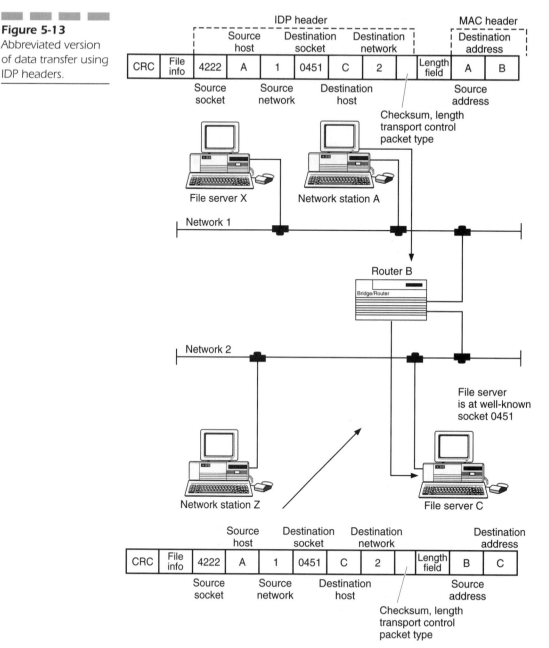

address changes and to show that the internetwork packets are addressed directly to the router. The final destination host and network address are embedded into the network header of the packet.

In this figure network station A wants to transmit data to another station, server C. It is assumed here that the destination network station network and host address were found through previous action on the workstation. This will be explained further in the section on the Clearinghouse Name Service at the end of this chapter. For this example, though, the destination station address is already known.

Next, the workstation will compare the network address of server C with its own network number. If it is the same network number, the network station will send the packet directly on the local LAN. If the network numbers are different, it must find a router to forward the packet. In this case, the network numbers are different, and therefore, workstation A must find a router to forward the packet to the destination.

Workstation A will send a RIP request packet out to network 1 and wait for a response. Inside the RIP request packet is the network number needed. In this case, it is network 2. Router B will receive this packet and perform a routing table lookup. Router B will send a RIP response packet back to workstation A.

Workstation A will receive this router RIP response packet. Workstation A then needs to format a packet to send to router B. It will fill out the network header with the source network number and host number set to its own. It will then ask the IDP layer for a dynamic socket (in the range of 4000 to 6000, decimal). The network station will then set the destination network and host with the address of the final destination. The socket number in this case is set to 0451. This is a static socket (using this protocol) for a file server. Finally, the network station will assign the physical (MAC) address of the packet. The source address will be its own, and the destination will be set to the physical address of the router. It will then transmit the packet to the network.

At the top of Figure 5-13 is the packet format built by workstation A. The MAC header contains two entries: the physical destination address of the router (B) and the source address of the workstation (A).

The IDP header contains the following entries:

Destination network	2
Destination host	C
Destination socket	0451
Source network	1

| Source host | A |
| Source socket | 4222 |

The destination network and host addresses in the IDP header are those of the final destination, not the router. The router uses this information to determine how to forward the packet. The source network is used by the destination station (the file server). When the file server transmits a response to this packet, it will know the network to send a response to.

The router will receive this packet from workstation A and strip the MAC headers, the length field (the physical addresses), and the data-link trailers (the CRC). What remains is the IDP header. It will examine the destination network address and perform a routing table lookup.

From the routing table, the router will know that the destination network number is directly attached on the other side of the router. Thus, the packet will be directly delivered to the network and does not have to be forwarded to another router. Router B will now format the packet and transmit through the physical router port connected to network 2.

Router B will extract the destination host number from the received IDP header and place this in the destination physical address (the MAC address) field. It will place its own physical address in the data-link source address field. With a few exceptions, the rest of the packet will be left alone. It will be appended to the new data header the router has just built. The router will then transmit the packet to the network. This is shown at the bottom of Fig. 5-13.

Notice that the MAC header contains file server C's physical address as the destination address and router B's physical address as the source. The IDP header remains the same, except for the CRC (checksum) and the hop-count fields. (These fields are not shown in this figure, but when the packet traverses the router, the transport control is incremented by 1 and the CRC is recalculated. The packet is then transmitted to network 2..)

File server C will receive the packet, strip off the data-link headers, and then recognize this as a file server call of some type (the file info portion of the packet). It will process the packet according the call information and, if necessary, respond to the source.

The importance of the preceding paragraphs is to illustrate that the destination MAC address of the packet will constantly change. Except for checksum and transport control, the IDP header will not change. The router that receives the packet will determine the best route to the destination network and will address the packet directly to the destination station (if the network is directly attached to the router) or it will physically address the packet to the next router in line to the destination station.

Figure 5-14 shows the preceding sequence as a flowchart.

Level 2 Error Protocol

The error protocol is used by any service on the network (primarily in a network station) that notices an error. If the error packet is generated by a well-known socket number, the source socket in the error packet will be set to that number so the recipient of the packet will know what well-known service noticed the error. If the error was noticed by the IDP layer, it will be sent by the well-known router error socket. The packet is always sent to the source socket of the service that created the error. There is no acknowledgment of this type of packet, nor is one generated in response to a multicast or broadcast packet. The format of the error packet is shown in Fig. 5-15. The packet type is the error protocol specified in Table 5-2.

The format of the packet is quite simple and contains two 16-bit words that indicate the error number and the error parameter. The error number shows the kind of error, and the error parameter is a parameter for certain types of errors. These are listed in Table 5-4. These words are followed by the first portion of the offending packet. This includes all of the IDP header and as much of level 2 and higher levels as the implementor of the XNS protocol desires. Xerox recommends that at least 42 bytes be copied, since that will include the SPP header.

When the receiving process gets the error protocol packet, it will try to determine the cause of the error and report this to the end user. For example, if the error packet received was for a SPP error, indicating that the socket does not exist at the destination host, the process can assume that the socket was deleted at the destination host and the connection was disabled. This would be turned into a descriptive response by the programmer so that the end user may understand what happened to the connection. It is up to the software programmer of the protocol to create a meaningful message. It may not be included at all.

The error protocol may be used by any of the level 2 services.

Level 2 Echo Protocol

The echo protocol transmits an echo packet to a destination station to determine if a remote host is alive and operating, and to determine the

Figure 5-14
XNS routing flow-
chart.

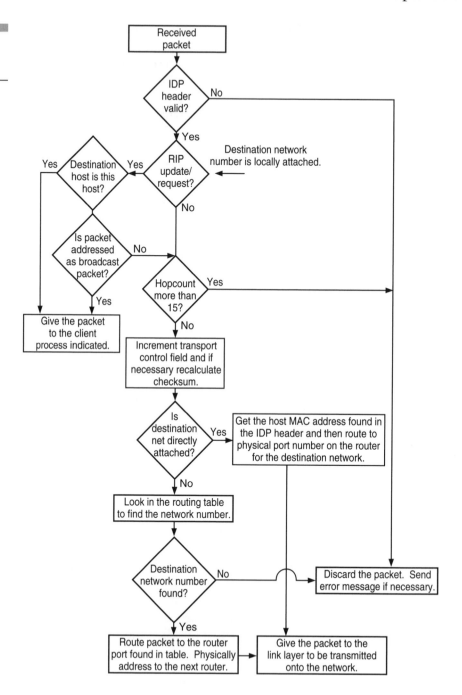

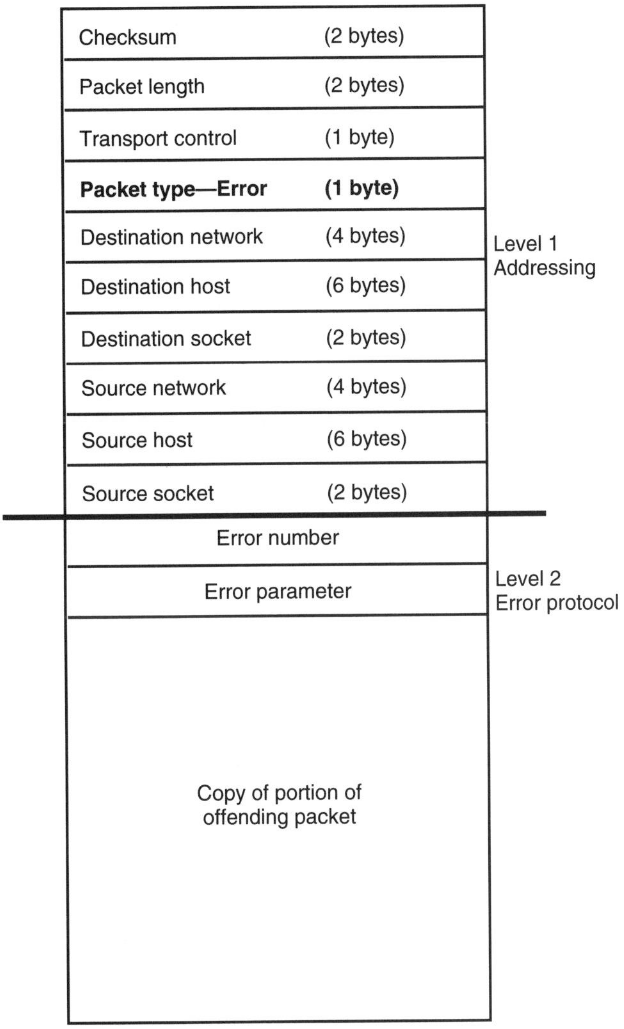

Figure 5-15
IDP error packet.
*(Courtesy Xerox
Corp.)*

Checksum	(2 bytes)
Packet length	(2 bytes)
Transport control	(1 byte)
Packet type—Error	**(1 byte)**
Destination network	(4 bytes)
Destination host	(6 bytes)
Destination socket	(2 bytes)
Source network	(4 bytes)
Source host	(6 bytes)
Source socket	(2 bytes)

Level 1
Addressing

Error number
Error parameter
Copy of portion of offending packet

Level 2
Error protocol

path to that destination. This is a simple protocol in that an echo packet is sent out and the destination should send the packet back to the originator. The packet type is set to the echo packet type, as shown in Fig. 5-16.

The first word of the packet indicates the operation. It is set to 1 for echo request and 2 for echo reply. If checksumming was invoked on the request, it will be checked and recomputed for the reply. If the packet received was an error, the error protocol will be invoked and the error protocol packet will be sent. The echoed data will be the same as the data that were in the echo request packet.

TABLE 5-4

The Error Protocol
Assignments

Error Number, Octal	Description
0	An unspecified error is detected at destination.
1	The checksum is incorrect, or the packet has some other serious inconsistency detected at the destination.
2	The specified socket does not exist at the specified destination host.
3	The destination cannot accept the packet due to resource limitation.
1000	An unspecified error occurred before reaching the destination.
1001	The checksum is incorrect, or the packet has some other serious inconsistency before reaching destination.
1002	The destination host cannot be reached from here.
1003	The packet has passed through 15 internet routers without reaching its destination.
1004	The packet is too large to be forwarded through some intermediate network. The error parameter field contains the length of the largest packet that can be accommodated.

The echo protocol is often used to ensure the proper operation of the internet datagram protocol since it uses this protocol to send the packet. The echo protocol packet can traverse routers.

Level 2 Sequence Packet Protocol

A Transport Layer Implementation.   When working with networks, you will often hear that the upper-layer protocols (the transport layer) provide error correction for packets transmitted and also guarantee packet transmission reliability. While the network layer delivers data with a best-effort service, if anything happens during this process, the transport layer will guarantee retransmission for the reliable delivery of data. All packets sent over the network layer are guarded by the transport layer to ensure proper delivery.

This layer establishes connections for the session layer. The Sequence Packet Protocol (SPP) will establish a unique connection identifier for each connection the network station has. With each connection, the sequence packet protocol provides for reliable transmission of successive

Figure 5-16
IDP echo packet.
(*Courtesy Xerox
Corp.*)

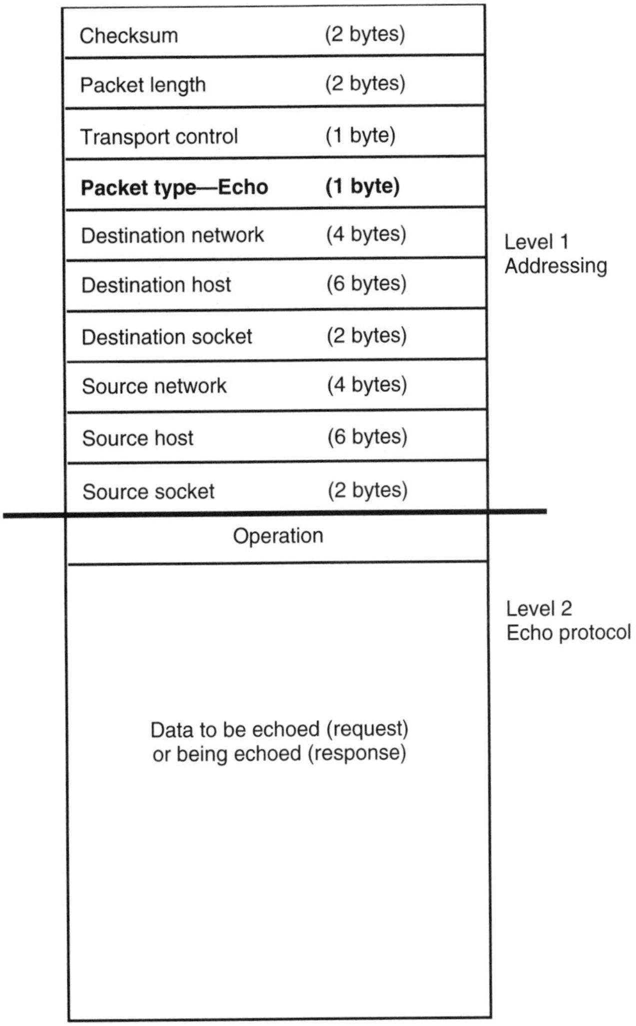

Checksum	(2 bytes)	
Packet length	(2 bytes)	
Transport control	(1 byte)	
Packet type—Echo	**(1 byte)**	
Destination network	(4 bytes)	Level 1 Addressing
Destination host	(6 bytes)	
Destination socket	(2 bytes)	
Source network	(4 bytes)	
Source host	(6 bytes)	
Source socket	(2 bytes)	
Operation		Level 2 Echo protocol
Data to be echoed (request) or being echoed (response)		

internet packets for the client process that requested it. For example, when a client process such as a file transfer program needs to send data to a destination, the SPP will provide the reliability for this transmission.

On a given connection, each packet will be given a sequence number (as part of SPP's header) and it will then given to the IDP layer for transmission onto the network. These sequence numbers are used for the following purposes:

1. *Ordering of packets.* When packets are transmitted, all packets may not be received in the same order as they were sent. If the packet traverses routers, any packet may be dropped. It could have been

lost while on the transmission medium (for example, Ethernet). In order that the packets arrive at the receiving end in the same order in which they were sent, the packets must arrive in ascending sequence number order. Any packet received out of order will be requested again by the destination.

2. *Detect and discard duplicate packets.*   If a packet was transmitted and there was a delay in getting it to the destination, the originator may time out waiting for an acknowledgment and resend the packet. In the meantime, the server may acknowledge the original packet. The duplicate packet is then received. The transport layer software (SPP) will notice that this sequence number has already been acknowledged and will discard the duplicate packet.

3. *The ability to acknowledge receipt of a packet.*   Any sequenced packet that arrives in good condition at the destination station will be acknowledged.

XNS sequencing occurs on a packet basis. Every packet transmitted will have at most one sequence number. This is in contrast to protocols such as TCP/IP, where every byte of the data is assigned a sequence number. This is known as a *byte-oriented protocol.* SPP is a *packet-oriented* protocol.

A source network station may use sequence numbers up to and including the number specified by the recipient of the packet (the destination). There is a preestablished starting sequence number that will be known on both ends of the connection. The initiation of the connection will also initialize the sequence number to start with. The starting sequence number for XNS SPP protocols is 0.

A connection also establishes connection identifiers. These are 16-bit numbers, one specified by each end of the connection. An implementation of this is to read the clock register of the processor and assign this as the connection identifier. This establishes a unique connection identifier at any connection attempt. Therefore, there is no need to maintain a cache table for connections already taken.

The maximum packet size for XNS packets, including SPP headers, is still maintained at 576 bytes, although many XNS implementations do not follow this. For example, Novell's implementation called Internet Packet Exchange (IPX) allows for packet sizes up to the maximum size of the transmission medium. This can be 1500 bytes for Ethernet and 4472 bytes for Token Ring. XNS still recommends 576 bytes, for reasons explained earlier. Upon connection time, two communicating network stations will negotiate for the maximum packet size.

SPP allows for different types of acknowledgments. It can acknowledge a packet when it is received (some versions of Novell implement this). Other implementations allow several packets to be received before an acknowledgment is sent. At connection time, the source station will tell the destination what type of acknowledgment it would like.

The packet format for SPP is shown in Fig. 5-17 and is explained as follows.

Connection Control. The connection control field contains eight bits, of which only four are used (0–3). It is used to identify the control actions of SPP.

System Packet Bit. The system packet bit is used to send probes (to ensure that a connection is still alive, even when there are no data to send), or it may be used to return acknowledgments. The protocol of SPP requires data to be sequenced contiguously. System packets contain no data and are primarily used to maintain the connection. When there are no data to send to the destination, SPP will still ensure that the other side of the connection is active and can receive packets. To do this, it must send a system packet with the next unused sequence number; otherwise, the receiving end will send an error packet with a bad sequence number as the source of the problem. System packets are used to maintain the connection. The XNS standard does not specify how often these are to be sent. This is left to the implementor.

Send-Acknowledgment Bit. This bit is set to ensure that the receiving end immediately sends an acknowledgment upon accepting the packet. If the receiving end has data to send back, the acknowledgment will be in that packet. Otherwise, it will send a system packet back as an acknowledgment. Acknowledgments may be sent at any time, even without this bit being set.

Attention Bit. With this bit set, the source socket is telling the destination socket that this packet needs to be given immediate attention. The attention bit and the system packet bit cannot both be set at the same time. It is a way of immediately getting the destination's attention, even when the destination is extremely busy. These types of packets are sent infrequently and usually under emergency conditions.

End-of-Message Bit. If this bit is set, a single message has been sent and a new message will begin with subsequent packets. To make SPP

Figure 5-17
IDP/SPP packet.
(*Courtesy Xerox
Corp.*)

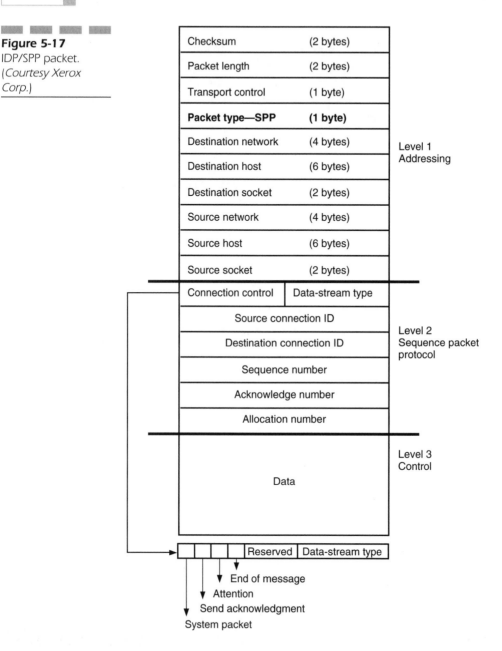

Checksum	(2 bytes)
Packet length	(2 bytes)
Transport control	(1 byte)
Packet type—SPP	**(1 byte)**
Destination network	(4 bytes)
Destination host	(6 bytes)
Destination socket	(2 bytes)
Source network	(4 bytes)
Source host	(6 bytes)
Source socket	(2 bytes)

Level 1
Addressing

Connection control	Data-stream type
Source connection ID	
Destination connection ID	
Sequence number	
Acknowledge number	
Allocation number	

Level 2
Sequence packet
protocol

Data

Level 3
Control

			Reserved	Data-stream type

End of message
Attention
Send acknowledgment
System packet

efficient, incoming packets are viewed as a stream of bytes. There are no identifiers in the packet to indicate the type of data received. The end-of-message bit indicates only that the packet contains the end of that particular message. For example, a service (client process) may request that a block of bytes be read from a file located on a remote file server (somewhere on the network). This bit could be set to indicate that the full block was read and there are no more data to follow. Also, a request may be sent for data information in which the response requires multiple packets to be sent. The final packet in the response to this request will have this bit set. The system packet bit and the end-of-message bit cannot both be set at the same time.

Datastream Type.   This field is actually ignored by SPP. This message is used exclusively for higher-level protocols (client processes) so that they may provide escape sequences to the destination without having to put this in the data portion of the packet. This field could indicate the means to abort the sending of a file. It can also be used to indicate that data are in the data field, to indicate that the end of the data has been reached, or to reply to the originator that the data have been received and processed correctly. It is vendor-specific.

Source and Destination Connection Identifier.   These fields identify the source and destination connection numbers for two communicating stations. Since any station may have more than one connection to a network device, this serves as a way to identify which connection the packet is intended for. This will be discussed in detail later.

Sequence Number.   The basic function of the sequence number is to count the packets sent or received on a connection. Data flow in each direction on a connection is sequenced. Each network station on a connection maintains its own sequencing. Upon connection, the sequence number is set to 0 for all new connections. The count will ascend by 1 for each packet sent. If the count exceeds the number for a 16-bit field (65,535 decimal), the sequence will start again from 0. In this way the destination knows how to order incoming packets (for a connection ID), to discard a duplicate packet, to acknowledge a packet, and to send control (connection maintenance) information. If packets 4, 5, and 7 arrive, for example, the destination knows that packet 6 is missing and will inform the source of this error. The source may then send packet 6 or may send a stream of packets (4, 5, 6, and 7), depending on the control algorithm used for sequencing.

The three types of sequencing most commonly used are:

1. Go back to N

2. Selective

3. Stop and wait

Sequencing enables workstations to synchronize the transmission of packets. For example, as shown in Fig. 5-18a, two stations have established a connection. Workstation A and file server B have also established that sequence numbers will begin with 0. Next, workstation A sends five packets to file server B. File server B responds with an acknowledgment packet. The acknowledgment number will be 5. After this, workstation A sends five more packets to file server B and waits for an acknowledgment response.

If the file server, as shown in Fig. 5-18b, sent back an acknowledgment packet containing a 2, this means that certain packets were not received. The file server is telling the workstation that packets 0 and 1 were received in good shape, but packets 2, 3, and 4 were not. This indicates to the workstation to go back to the third packet and resend packets 2, 3, and 4 (the third, fourth, and fifth packets). After this, the file server sends the acknowledgment back to the workstation indicating that all packets were received in good condition.

Selective reject is a way to indicate that one packet was not received. In this case, the file server could have sent the packet a selective reject of 3, indicating to the workstation that packet 2 was not received.

Stop and wait is also called the *ping-pong sequence method.* When workstation A sends a packet to file server B, it stops and waits for file server B to send an acknowledgment for that one packet. Upon receipt of an acknowledgment from the server, the workstation may then transmit one more packet to the file server. As of this writing, this is the method used by Novell NetWare. XNS does not specify any particular method for sequencing of data.

The originator of a connection expects to receive an allocation number indicating that it may use sequence numbers up to and including a particular value.

Acknowledge Number.   This field indicates the sequence number of the next expected packet. Acknowledgments are cumulative; by specifying the next sequence number expected, all previous packets are said to be acknowledged. This means that with an acknowledge number of 7, all previous packets before 7 have been acknowledged.

Xerox Network System

Figure 5-18
(a) Sequencing.

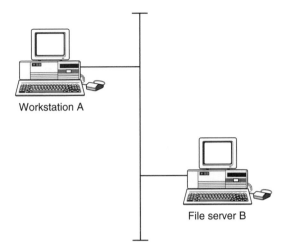

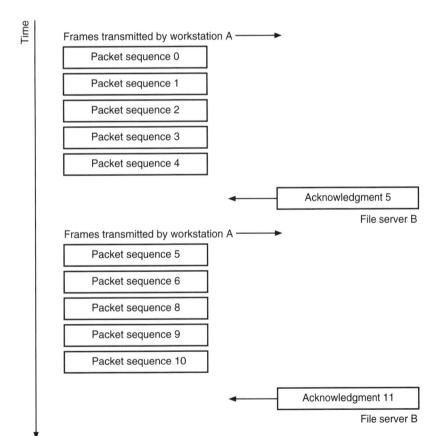

Workstation A

File server B

Time

Frames transmitted by workstation A ⟶

| Packet sequence 0 |
| Packet sequence 1 |
| Packet sequence 2 |
| Packet sequence 3 |
| Packet sequence 4 |

Acknowledgment 5

File server B

Frames transmitted by workstation A ⟶

| Packet sequence 5 |
| Packet sequence 6 |
| Packet sequence 8 |
| Packet sequence 9 |
| Packet sequence 10 |

Acknowledgment 11

File server B

(a)

Figure 5-18
(*b*) Sequencing with
error.

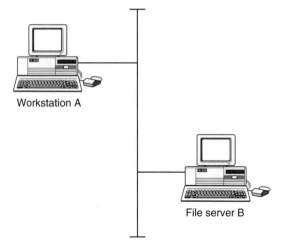

Workstation A

File server B

Time

Frames transmitted by workstation A ⟶

| Packet sequence 0 |
| Packet sequence 1 |
| Packet sequence 2 |
| Packet sequence 3 |
| Packet sequence 4 |

Acknowledgment 2

Frames transmitted by workstation A ⟶

| Packet sequence 2 |
| Packet sequence 3 |
| Packet sequence 4 |

Acknowledgment 5

(b)

Allocation Number. This field specifies the sequence number up to and including which packets will be accepted from the remote end. Finding the difference between the allocation number and the acknowledge number, and adding 1 to this, will indicate the number of outstanding packets in the reverse direction. This is a good congestion control mechanism. It indicates how much data may be sent at one time. In other words, the destination may say, "received packets up to X; have enough space to handle N more packets." This is known as providing a window on the connection for sending information.

Data. This field will be filled with data only when the packet is not a system packet. This field is usually destined for the client process (a file server, for example), and the level 2 protocol will pass it to the client process without looking at it. The length of the data field is calculated by subtracting 42 from the internet packet length (IDP and SPP headers of 30 and 12 bytes, respectively). This field is recommended to have 534 bytes maximum.

Establishing and Opening Connections. SPP is the protocol that establishes, maintains, and disconnects sessions between sockets (client processes) on two communicating stations. SPP does not do this by itself. It provides this as a service to the session layer. The client process (the application or the session layer) must request that these sockets be opened and that connections be established. SPP listens only to the client process and does this according to the commands that are set by it. A connection between two network stations will be established only once the destination network and host numbers, the connection IDs, and the socket numbers are known. Without this information, a connection cannot be established. Figure 5-19 shows the relationship of socket numbers.

A connection is established as follows. There are two ends to a connection: the server and the consumer (the client). Server applications processes can be processing client requests such as file transfer, mail, name service, and terminal emulation. All these client processes must possess the well-known socket number. It will advertise this socket number on the network (or internetwork) so that other network stations will know what socket to connect to. Otherwise, the network station must have this well-known socket number hard-coded into the software.

A server process will open up by establishing a relationship with the SPP protocols. The server process will tell SPP to open up a socket (specified by the client process) and listen to the network for connection attempts to that socket. This is known as the *service listener process.*

Figure 5-19
Socket addressing.

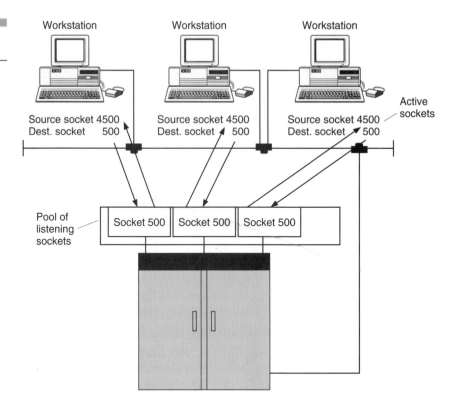

At the other end of the spectrum is the consumer, or client, process. This process (the client end of a file transfer program, for example) asks SPP to open a connection to a specified service listener process. SPP in the workstation will build a packet using a dynamic, unassigned socket (not a static or well-known socket), will assign a connection ID to this connection attempt, and will build a connection type of packet and send it to the destination station. SPP has to create a source socket to use so that the service listener station (the server) will know what entity in the requesting station to respond to when sending back a packet.

Upon receipt of this packet, the destination's service listener will allow the creation of the connection, note the source socket and source connection IDs, and assign a connection ID of its own (the destination ID); the connection is then said to be established at the server end. The destination (the server process) will send a system packet to the originator of the connection (including its socket and connection IDs), and the consumer end now has an established connection that it is awaiting the next packet.

At this time, data may flow between the source and destination stations using the socket numbers and the connection IDs to identify it. Sequence numbers (initiated at 0) will also be incremented.

Terminating Sessions. In order to terminate a session gracefully, three messages need to be sent between the source and destination station. First, a packet must be sent indicating that all data have been sent. Second, a packet must be generated that indicates all data have been received and processed. Third, a message must be sent indicating that the sender comprehends this action. All this is accomplished using the client-layer protocol. It is not up to SPP to terminate a session. A vendor-specific application will terminate a session. The client process terminates a session for XNS protocols. Most other protocols allow the transport layer to create, maintain, and terminate a session using a predescribed set of sequencing that is standardized and not vendor-specific. This action (terminating a session) is part of the reason any vendor that implements XNS makes it proprietary for a lot of features. The way XNS operates is left up to the vendor and not set in the standard.

Some implementations of XNS use the following algorithm to end a session. The datastream type will be set to "end," and the network station that wishes to end this connection will send this packet. It should be replied to by the remote end of the connection with a datastream-type "end reply." With this, the network station that sent the original "end" request will also reply with "end reply." This will terminate the connection. During this exchange, sequence numbers are still used; otherwise, the packet would be in error.

SPP Implementations. The receiving end of a connection controls the flow of data to it. The originator of packets controls the flow of acknowledgments. In other words, the source station controls the acknowledgments and the destination station controls the data flow. With XNS SPP, Xerox gives guidelines only for controlling the request for acknowledgments, the generation of acknowledgments, and the retransmission of packets. There is no standard for it. Different sequence and acknowledgment protocols exist with different implementations of XNS.

Once a connection is established, acknowledgments may be sent at any time (they do not have to be requested). Typically, the sender sets the send-acknowledgment bit in a packet with the sequence number corresponding to that permitted by the receiving end's allocation number (how many packets the receiving end can accept). The originator of a packet may send a probe packet, which is nothing more than a system packet with the send-acknowledgment bit set. This packet asks for an allocation number and/or an acknowledgment.

Acknowledgments and allocation numbers can be "piggybacked" to data packets. For example, if a packet is sent to a destination, and the destination has data to send back to this originator, it will also include in the packet and acknowledgment number and an allocation number. This conserves not only processing cycles on the station but also conserves bandwidth on the network. If there are no data with which to respond, these fields cannot be set, and the receiving end should respond with a system packet whenever one is requested.

The originating end of a connection will time the period since a packet was sent. Upon expiration of this time, if it has not received an acknowledgment, it will assume that the packet was lost, and it has the capability to retransmit the packet. This time is called the *round-trip delay* and is calculated by sending a send-acknowledgment packet and waiting for the response. This time is kept as the retransmit time. It is only a suggestion by Xerox.

SPP does use the error protocol when an error has occurred using SPP.

Level 2 Packet Exchange Protocol

The packet exchange protocol (PEP) (not to be confused with PUP*f*a no-longer-used XNS routing protocol) is used to transmit data between source and destination stations on a connectionless transport service. Unlike SPP, this transport-layer protocol does not establish a reliable session with the destination before data are sent over the link. The protocol is simple in architecture and is "single-packet-oriented." The protocol will submit a retransmission of a packet, but it does not detect duplicate packets at the receiving end. It is primarily used for simplex types of transmissions.

This protocol can be used for network management, name service, time service, or any other service that can deal with the unreliable nature of the protocol. It is basically a transport mechanism for applications (services) that do not require the reliability of SPP. It is also used with protocols that have built into their application the functions provided for by SPP. There are many independent programs out there that are not public user oriented. This provides a transport service for specialized programs as mentioned previously.

The structure of the PEP packet is shown in Fig. 5-20.

Figure 5-20
Packet exchange pro-
tocol packet. (Cour-
tesy Xerox Corp.)

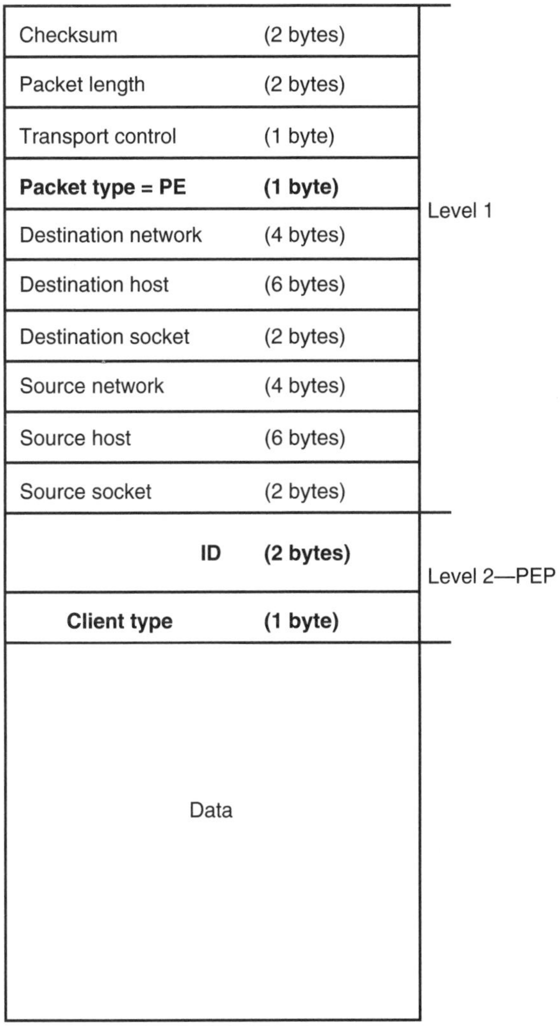

Checksum	(2 bytes)	
Packet length	(2 bytes)	
Transport control	(1 byte)	
Packet type = PE	**(1 byte)**	Level 1
Destination network	(4 bytes)	
Destination host	(6 bytes)	
Destination socket	(2 bytes)	
Source network	(4 bytes)	
Source host	(6 bytes)	
Source socket	(2 bytes)	
ID	**(2 bytes)**	Level 2—PEP
Client type	**(1 byte)**	
Data		

ID. This is a 32-bit field, and the source will set it to a specific number. Upon receiving a reply, the source networks station will look at this field and, if it is the same as in the packet previously sent, the packet exchange is said to be complete. That is, the source sends the destination a packet and places a number in this field and then waits for a reply before any more information is sent. Upon receipt of this packet, the destination station places the same ID back into this field and replies to the source network station. Of course, the destination will send a reply only if the source packet was received in good condition. Otherwise, it will discard the packet.

Client Type. This field contains a field registered with Xerox that will identify the source and destination client (the service that submitted the request and the service intended to receive the request). There will be a unique client type for every service offered under this protocol. Therefore, there will be different ones for time of day, name service, and a network management routine.

As shown in Fig. 5-20, the typical IDP packet header will be formed. This includes the source and destination network/host addresses and the socket address. This time, the socket number will not identify a particular service being requested. The client type will specify this.

This protocol can be faster to use with services that do not require a connection-oriented routine to exchange data. Many application programs have been developed that take advantage of this protocol. Remember, though, that XNS implementations are usually specific to the vendor. One vendor's protocol will usually not work with implementations of another vendor's XNS protocol. Since XNS is a network architecture, it must include a connectionless transport protocol in order to be thorough in the transport-layer offerings.

Courier—The Session-Level Interface. Courier is known as a remote procedural interface that not only establishes, maintains, and disconnects sessions between two communicating sessions, but also allows (as the name implies) services to be invoked remotely. These services could be opening a file, reading a section of the file, closing the file, etc. The following text is provided so that the reader may understand the beginnings of a remote procedural interface. It is not intended as a strict learning tool or to be used to write code. Courier is a different way of providing services on a network.

Other protocols provide network services (file, print, and terminal) that use the session layer to establish and maintain a link with a remote network station, but use socket numbers to identify and invoke the remote service. This type (socket calls) has a client (the requester of a service) and server (the provider of a service) interface that allows the service to use the session layer to maintain the connection while the application layer provides the service.

There are three sublayers to this session-layer protocol:

1. Sublayer 1—the transport
2. Sublayer 2—the data types
3. Sublayer 3—the message types

With other protocols, once the physical through the transport layers are defined, remote applications can be accessed by building a session to the remote network station and giving the remote station a port number with which you wish to communicate. For example, with TCP/IP, if you wish to communicate with the remote terminal application, you build a packet (with the transport layer first establishing a connection to the remote network station) with the destination assigned to port 23. The remote station will spawn a process with your network, host number, and source socket number to the TELNET application. It would then go back, waiting for more connection attempts at that socket number. The file transfer protocol is assigned a different port number, and that application awaits connections only at that port number. Courier works similar to this but is different in many ways.

All calls to courier are made to socket 5. If you wish to communicate with a remote station and wish to access a remote application, all requests go to port number 5. (Again, remember that Xerox calls their ports *sockets*, when in reality a socket is the combination of the network number, host number, and port number.) In your packet will be an application number. This number represents the application that you wish to communicate with.

These program numbers are all unique. No two program numbers may be assigned the same number. Those vendors who wish to use courier at the session layer must register their applications with Xerox. Xerox will assign the company a block of application numbers to use with their applications. This is similar to the process that Xerox used to do with Ethernet addresses and Ethernet type-field assignments.

Each program is also assigned a program name, but this has no significance to the courier protocol. The application name will never be transferred in any packet exchanges and is said to have local significance only. Program names do not have to be unique. The program numbers distinguish the applications.

Furthermore, every application must also contain a version number, which is used to determine the version of the application being used. This is to ensure that both the source and destination stations will communicate using the same version of the application program.

A remote program will have one or more procedures with a full complement of error recovery features and statements that should be used. When a program is used with the courier protocol, it must declare itself with a program name (the identifier) and its version number. All procedures inside the program must be declared with a number assigned to each procedure. No two of these numbers can be used again in the program.

There are four messages defined at the message layer:

1. *Call.* For written software programs, the program name is the name of the executable, and the procedures are nothing more than the functions that the program can call and execute. The following is an example that simply opens a file, reads a block of data, writes it to the screen, and then closes the file.

```
Program name readwrite 15 version 1
begin;
Credentials: Type Record [user, password; string]
Mode: Type {readpage(0), writepage(1), ReadAndOrWritePage(2)};
OpenFile: Procedure [credentials: Credentials, filename: string,
    mode:Mode]
returns [handle:unspecified, pagecount:cardinal]
reports [...NoSuchFile,...] 0;
NoSuchFile:Error 2;
...
End
```

The numbers in the function names are the procedure numbers that courier requires a program to declare.

With that declaration, the user's program builds a call message type to the procedure Openfile and uses the user name Matt, the password Naugle, and the file name book.doc with the mode as readpage. This would be coded into a packet as follows:

```
call[
    transactionID:0
    ProgramNumber : 13, versionNumber : 1, procedureValue:0,
    procedure Arguments:[
        credential: [user, Matt", password: "Naugle"],
        filename : "book.doc",
        mode:readPage]];
```

At the user's workstation this will place a courier call to the remote network station stating that it would like to invoke program number 13, procedure 0 (read), and that the credential to get in is the user name and password. The file name would be transmitted as the string book.doc. This would be formulated into a packet and sent to the remote network station, which would process this and return the information to the requester.

2. *Reject.* This forms a reject message to the originator explaining why the request was rejected. It could be a wrong version number, the program does not exist, the procedure within the program

does not exist, or there was an invalid argument in the request. Otherwise, it will not indicate what the error was.

3. *Return.* The return message is a reply to a previous call. This usually contains some type of user data that the request indicated.

4. *Abort.* This message indicates that an error has occurred that is unspecified. This message will return the original arguments used in the request. These can be vendor-specific error codes.

With this information in hand the following is a typical transaction using the courier protocol:

```
FileAccess: program 13 version 1
begin
—(types and constants written here)
Credential: Type Record[user, password: string]
Mode: Type {readPage(0), writePage(1), readAndOr WritePage(2)};
PageContents: Type Array of 256 of unspecified;
—procedures
OpenFile:Procedure [credentials: Credential, filename: string,
    mode: Mode]
returns [handle: unspecified, pageCount: cardinal] reports
    [NoSuchUser, IncorrectPassword, NoSuchFile,
AccessDenied, FileInUse, InvalidMode] 0;
ReadPage: Procedure [handle: unspecified, pageNumber: cardinal]
    reports [NoSuchUser,
IncorrectPassword, NoSuchFile, AccessDenied, FileInUse, Invalid-
    Mode] 1;
WritePage: Procedure [handle: unspecified, pageCount: cardinal,
pageContents: PageContent] reports (InvalidHandle, IncorrectMode,
    FileTooLarge] 2;
CloseFile:procedure [handle: unspecified] reports [InvalidHan-
    dle] 3;
—errors
NoSuchUser; error 0; —user unrecognized by server; user
    unregistered
IncorrectPassword; error 1; —password specified not that
    of specified user
NoSuchFile; error 2 —filename unrecognized by server, file
    does not exist
AccessDenied; error 3; —user entitled to access file in
    specified mode
FileInUse error [user:string] 4 —file already open for
    specified user
InvalidMode error 5 —invalid mode, not read..., write...,
    or readAndOrWritePage
InvalidHandle error 6 —invalid handle, perhaps obsolete by
    Close File
IncorrectMode error 7 —requested operation inconsistent
    with open mode
NoSuchPageNumber error 8 —requested page unreadable, not
    present in file
FileTooLarge error 9; —requested page unwritable, file
    would be too large
end.
```

Clearinghouse Name Service—An Example Name Server. Any user may use the internet address to connect to any service on the network. But for a user to remember the addresses of all the network servers on the network is an impossible task. Users are more likely to remember names before they remember numbers. This is the purpose of the Clearinghouse Name Service. Without a name service, a user would have to remember the 6-byte MAC address, the 4-byte network number, and the 2-byte port (socket) number in order to get a connection to the remote station.

To easily understand how a network name service operates, think of the "white pages," "yellow pages," and the operator in the telephone service. A network name service operates very similar to this.

When we want to call someone, we do not simply enter the name of the person into the telephone. We enter the number of the person we are trying to reach. If we do not know the number of the person, we look in the telephone book to find the name of the person, and immediately following this is the telephone number. This is a simple example. What if we are trying to reach someone who is outside the area in which we are dialing? We must invoke the services of the telephone operator.

In this case, we ask the operator for the number of a person. This is accomplished by giving the operator the last name of the person. If the last name is Naugle, there is a good chance that the operator will be able to give you the number right away. If the name is Jones or Smith, the operator will usually ask for more information, such as a first name or maybe a street address.

The other type of name lookup is the yellow pages. The yellow pages of the telephone system provide a more generic name-to-number lookup. Using the yellow pages, we look up a number by category rather than by name. For example, if we need some electrical work done on the house, we can look up "electrical" in the yellow pages to get close to the number we are looking for. Once we get to that section of the yellow pages, we should be able to get more specific information. We will be able to flip through a few pages to find the name of the specific electrical company that we want.

Therefore, there are three ways to find a number in the telephone system: by name, by number, or by subject. The telephone system has to have a number before it can make a connection between two users of the telephone. The users do not like numbers and prefer using names. The problems that exist in the telephone system also exist in the network. Therefore, Xerox created the name system for XNS and called it the Clearinghouse Name Service (CNS).

CNS names are divided according to the scheme name@domain@organization. This is a naming service based on organizations; within them are domains, and within the domains are the names. These three parts form a hierarchical naming structure so that two different names with the same value for the organization and domain parts are said to be in the same domain. As long as one part of the three is unique, the whole name is said to be unique. With this system, there may be many domains within one organization, and there may be more than one organization within the name structure.

There are two parts to this service: the client end (the part in the user's workstation) and the server end (a centralized server that contains a database of all the XNS Internet address-to-name mappings).

Name Server Functions.   As shown in Fig. 5-21, there are two entities to the name service. The client is a simple process that requests information from the name server. This request could be a connection attempt to a server. This connection request is accomplished by the user typing in the connection command and the name of the server to which the user wishes to connect. Another example is a mail program with which the user wants to see all the users in a certain domain (to address the mail message). These requests are made at the client side of the name service, and the request is formed into a packet and sent out to the network. It is the responsibility of the server to receive this packet and try to find the internet address of the server or to respond with all the names in the domain requested. The client will first submit a packet to

Figure 5-21
Example name service.

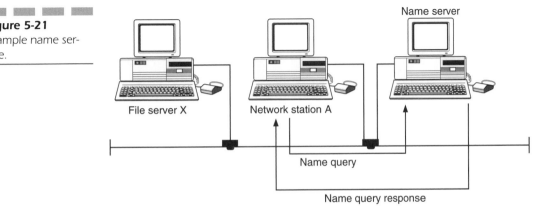

Name server

File server X   Network station A

Name query

Name query response

the name server to find the address of the file server. The name server should respond to this request, and the client will be able to extract the internet address of the file server from this response. A name service may span routers.

Not all protocols invoke this type of name service. Some name services are distributed. This means that the name service runs on multiple network stations and is synchronized. There is another type of name service in which each of the network stations maintain a names directory for the name-to-internet address mappings. AppleTalk and NetBIOS are examples of this type of name service. The response that a name server provides contains the full internet address (network number, host number, socket number) of the requested named service. For XNS, this is how workstations find network servers on the network. A user may request to be connected to Server1@Bldg5@Warehouse. This request would go out to the name server, and the name server should reply with the full internet address (network number, host number, and port address) to the user's workstation. The user's network protocol would then be able to find the requested server by an address and not a name. The name is used only for human intervention. It is easier to remember a name than it is to remember a list of numbers.

The XNS name server is a centralized server that resides on the network somewhere. Although the XNS specification does not call for it, there can be more than one name server on a network. (This is known as a *distributed name service* and requires the name servers to update each other, which is called *synchronization.*)

Upon receipt of a request, the name server reads the type of request. From this the file server responds to the user's request. Once the user has received its response, it may then proceed with whatever actions are necessary to complete its task. The name service is not invoked again until the next request.

Novell NetWare uses a name service, but it is a variation of the distributed name service. NetWare service names are distributed throughout the network with the help of the routers. Routers maintain service advertisement protocol (SAP) tables, which contain the name of the server, the internet address, and the service provided, along with a few other items.

Novell NetWare

Introduction

TCP/IP seems to be the emerging protocol of choice, but, by far, the most frequently installed of the workgroup client-server class of networks is Novell NetWare. As of this writing, even Microsoft is trying to invade the Novell turf with Windows NT. While NT has a lot of advantages, NetWare continues to thrive simply based on installed base and a new suite of products for intranets. This is known as Intranetware. TCP/IP and NT zealots continue to predict the demise of NetWare, but they are off track. Why are there so many requests in the employment section for Novell-certified engineers? Simply stated, it is human nature. Sixty million users currently depend on NetWare. Thousands of home-grown applications are continuing operate using Novell. For the future, Novell still controls 60 percent of the market, and this will help it win. NT does have the advantages of working with multiple protocols simultaneously and includes the ability to allow for an Internet server to be used on the server. But NT is plagued with disadvantages as well. Scaling, a name structure based on domains, etc., will continue to hamper NT installations. But, as we learned from ATM, companies will not replace a network infrastructure simply on the basis of a promising new method or technology. If something works, stick with it. TCP/IP will be the dominant protocol within a couple of years, and corporations know the advantage of maintenance of a single protocol.

Novell has rebounded with a suite of products, most notably IntraNetWare. IntraNetWare is a suite of functions for building a intranet based on Novell. The functions and protocols are TCP/IP based and include directory services, a Web server, IP/IPX gateway, support for native IP, replication services, and a suite of other functions that will enable you to build an intranet without having to pull out your current IPX infrastructure. In fact, IntraNetWare allows for multiple protocols to be run on the intranet for compatibility or conversion. However, as of this writing, release 5 of NetWare is going to quit using the IntraNetWare name and revert back to NetWare. The future remains to be seen. NT is making great inroads, but NetWare is still entrenched.

Novell NetWare's core protocol and operating system are based on a proprietary scheme known as Internet Packet Exchange (IPX) and NetWare Core Protocol (NCP). This is the topic of this chapter. The network operating system allows workstations and their associated file servers to exchange files, send and retrieve mail, and provide an interface to SNA terminal emulation and database programs, among a host of other applications.

Novell NetWare's popularity grew very strong not only because of its ability to provide file and print services but also because of its ability to support multiple manufacturers' network interface cards and many different types of access methods (Ethernet, Token Ring, ARCnet, Proteon Pronet-10, FDDI). Novell's install base primarily includes DOS-based personal computers but also offers connection services to Apple, UNIX, IBM SNA, and OS/2 environments. Because of its low cost during the ramp-up of the LAN environment (early 1980s), the access method of ARCnet was very popular with NetWare environments. NetWare supported the ability to route packets to and from ARCnet, Ethernet, and Token Ring networks directly on the file server and therefore gave users the ability to communicate to file servers and other users no matter what access method they were working with. That is, since NetWare supported a native IPX router, it supported the translation between a variety of access methods.

As Ethernet and Token Ring became the access methods of choice, NetWare aided the migration to these networks from ARCnet. Furthermore, Novell supported almost any manufacturer of network interface cards (NICs). Novell grew a strong operating system from this environment and today is the number one manufacturer of workgroup computing operating systems (claiming 60 to 70 percent of the market).

Finally, Novell's design goal was to be the most competitive as well as the highest-performance LAN operating system.

This chapter will give you an understanding of how Novell NetWare operates on a LAN and WAN. The key topics for discussion are: (1) the user interface (workstations and servers), including the workstation shell, the NetWare File Service Core Program, and the NetWare Core Protocol, and (2) the LAN operating system of Internet Packet Exchange (IPX). This chapter is not designed to teach you how to operate NetWare or how to apply the services provided with NetWare (although some of the services are used as examples). The objective of this chapter is to provide you an insight on what goes on behind the scenes of this network operating system. When you start and log in to the NetWare network, exactly what happens, and how do the file server and workstation communicate over the LAN or an internetwork?

For users in the Novell environment there are two primary or user-identifiable physical entities: a workstation and a server interconnected through a LAN. Refer to Fig. 6-1. The workstation is usually a personal computer (DOS, OS/2, UNIX, or Apple operating system) that makes requests for file and print services from an entity known as the *file server*. The file server runs a proprietary operating system known as Net-

Figure 6-1
Basic Novell internet.

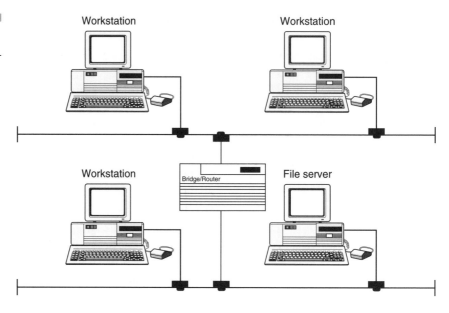

Ware Core Protocol (NCP) that services requests from the users' workstations and returns responses to these requests.

Version History

For those who are new to NetWare, a brief description of the history of NetWare is necessary. When NetWare was first introduced, it was a small operating system primarily offering networked file and print services.

Advanced NetWare 86.   The 8086-based operating system required a keycard for copy protection. This was a hardware-based security adapter that was installed in the PC-based server. This version shipped with an adaptation of Xerox's networking-layer software known as Internet Packet Exchange (IPX).

Advanced NetWare 286 version 2.0a.   Designed for the 80286 microprocessor, this version also required a keycard, and the installation procedure was extremely complex (supported IPX only).

Advanced NetWare 286 version 2.11.   With this release, Novell released its version of Xerox's transport-layer software known as Sequence Packet Exchange, which guaranteed reliable packet delivery. The complex installation procedure was simplified, and the keycard procedure was now locally administered.

Advanced NetWare 286 version 2.12.   With this release, Novell removed all keycard copy protection.

Advanced NetWare 286 version 2.15.   This release supported the Apple Filing Protocol (AFP) for support of the Apple Macintosh workstation and LocalTalk cabling.

Advance NetWare 286 version 2.15c.   This was the last release before 2.2 and the most frequently installed NetWare to date.

Advanced NetWare version 2.2.   As of this writing, this is the latest of the Novell 2.x releases. This version is basically the same as earlier releases but with a new installation program that combines all versions of 2.15 into one package. This includes SFT NetWare and all user versions.

Advanced NetWare 386 version 3.0.   This was the first release supporting the 80386 microprocessor chip. The operating system was completely redesigned. It did allow coexistance with previous versions of software. The first release supported only file and print services. Hard disks are now deemed reliable, and COMPSURF (a proprietary disk-scanning utility that detected errors on a hard disk before the installation of the operating system) is now an option before the installation.

Advanced NetWare 386 version 3.1.   With any new major release of software, major bugs are sure to be found. This release cleaned up most of the bugs in version 3.0. This version also supported more LAN adapters and provided better performance.

Advanced NetWare 3.11.   As of this writing, this is the latest of the NetWare releases. This version added support for TCP/IP. It allowed NetWare packets to be encapsulated in an IP header for transmission on an IP network. The installation of the server was made very easy. Copy a few files to a DOS-formatted hard disk, type the executable file name of "server," and basically feed diskettes for the rest of the installation. DOS can then be removed from the hard disk after the installation. This is done for speed purposes. The server may be brought down with the DOWN command, and control is returned to DOS (if DOS was not removed from the hard disk).

NetWare 4.x provides us with many new features. The most notable was NetWare Directory Services, or NDS. NDS is fully integrated into NetWare and IntraNetWare and is becoming the standard directory services for many other operating platforms as well. There is even a version of NDS for Microsoft NT. The NDS database can be distributed through partitioning and replication to give you access to all network resources, regardless of their global location.

Other features include flexible resource accounting and licensing management, easy installation and migration, high-performance Web server capabilities, multiprotocol routing, integrated Internet Protocol support, unsurpassed file and print services, symmetric multiprocessing, and network security and reliability.

NDS replaces the database used prior to 4.0. This was called the bindery and it contained all the information that NDS can distribute to multiple servers in a single server. The binder contained informational rights based on the user's access to network services. The binder had to be manually replicated on each server throughout the internet. Although NDS replaces the bindery, it is completely compatible with it.

Concepts

Novell NetWare is a LAN workgroup network operating system that permits workstations and their servers to communicate. The entities of NetWare that allow this are:

Access protocols (Ethernet, Token Ring, ARCnet, ProNET-10, FDDI)

Internet Packet Exchange (IPX)

Routing information protocol (RIP)

Service advertising protocol (SAP)

NetWare Core PrOtocol (NCP)—run in the server and the workstation; in the workstation, known as the shell

Access protocols were covered previously and will not be further discussed in this chapter.

Figure 6-2 shows the relationship between the OSI model and these NetWare processes. The ability to send NetWare commands and user

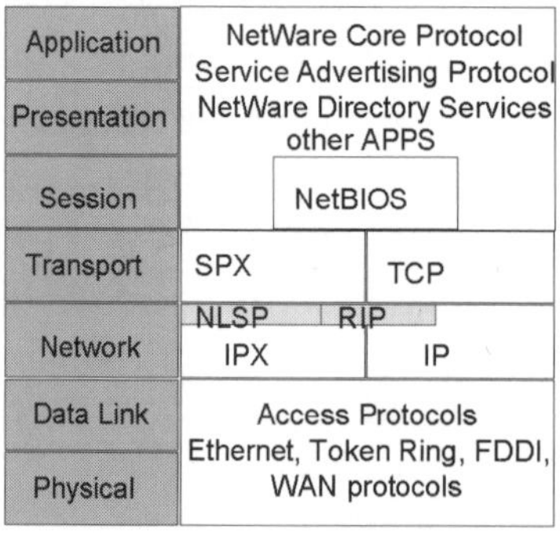

Figure 6-2
The OSI model and Novell NetWare.

Application	NetWare Core Protocol
	Service Advertising Protocol
Presentation	NetWare Directory Services other APPS
Session	NetBIOS
Transport	SPX / TCP
Network	NLSP / RIP / IPX / IP
Data Link	Access Protocols Ethernet, Token Ring, FDDI, WAN protocols
Physical	

NCP

data across the network relies on an XNS network layer Internet Datagram Protocol (IDP) derivative protocol known as IPX. IPX is a network-layer protocol implementation that allows data to be transferred across a local or a wide area network (through the use of routers in the WAN). IPX was derived using the IDP protocol of XNS.

The protocols that allow users access to their file servers is known as the NetWare Core Protocol (NCP) and the workstation shell program. NCP contains the ability to communicate directly with IPX, SPX, or NetBIOS. Novell's small protocol stack (on the workstation) and ability to communicate directly with the network layer gave NetWare its speed. Therefore, the remainder of this chapter will be divided into two parts: IPX and NCP (including the shell program). IPX will be discussed first.

Internet Packet Exchange

Before networks and the architecture of distributed computing, files that resided on a personal computer remained on the personal computer and were transferred to another computer by copying the file to a diskette and physically transporting that diskette to the other computer. This delivery system was either by human intervention or by addressing the diskette and having a mail service deliver the disk. Initial network protocols did allow for files to be exchanged between computers, but they were limited to that. They did not allow the files to be down-line loaded to the requesting station and to be seamlessly executed (like the files that were located on a local hard disk) once they were loaded to the requesting station.

With the advent of a network, data are still transported to another computer but the data to be transferred need to be formatted so that the network will understand what to do with them. There are network delivery commands that need to be transferred between the users' workstations and their file servers so that the data are delivered to the proper place. Whether the data are user data or network commands, they need to be formatted so that they may be transferred to the LAN. The Internet Packet Exchange (IPX) software is the network software used to allow this. IPX provides the delivery service of data.

IPX is the interface between the file server operating software Network Control Program (NCP)/workstation shell (the Novell program that runs on the workstation) and the access protocols (Ethernet, Token Ring, or ARCnet). IPX accepts data from the workstation's shell or NCP and formats the information for transmission onto the network. It also accepts data from the LAN and formats it so it can be understood by the shell or NCP. This protocol follows Xerox's XNS network-layer protocol of Internet Datagram Protocol (XNS IDP). This protocol was implemented for use on local and remote networks. Novell followed the XNS architecture and adapted it for use in its environment. Simply stated, IPX formats (provides addressing) data for network transmission and provides the delivery system for the data. NCP is the program that determines, on a DOS workstation, whether the data are destined for a network device or for a local device (local device on the workstation).

There are two purposes for using IPX as the network-level interface. First, since it follows Xerox's Xerox Network System (XNS) protocol, and this architecture was specifically built to run on top of Ethernet, it was designed to run on LANs. It was the only networking protocol designed to run on top of Ethernet. Other protocols such as TCP/IP, AppleTalk, etc., were adapted to run on top of Ethernet. Since the architecture was already written, and it was an open specification, Novell simply had to implement software to the architecture.

Second, it not only carries data but can easily route the data. This has many advantages. It allowed NetWare to support multiple network architectures easily. NetWare provided a routing stack that allowed ARCnet, Ethernet, Token Ring, and StarLAN networks (AT&T supported network based on 1 Mbps transmission on UTP, eventually upgraded to 10 Mbps but too late to have any impact on network installations) to exchange data between the different LANs transparently. It is true that bridges did not enter the commercial marketplace until 1985, but this was well after Novell had established itself. With the release of NetWare 3.x, Novell has changed the term and now properly refers to this ability as *routing*.

Finally, by implementing only the network-layer stack and a user interface (the shell), the amount of RAM consumed in a user's workstation was very small (sometimes as small as 20K of RAM). This was very important, for the first generation of PCs was limited to 640K of RAM to run application software. Applications could not be loaded into upper memory (later versions of the PC operating system allowed for this). As application software grew larger, the NetWare software remained the same and allowed larger applications to run.

Specifically, IPX is a full-implementation XNS IDP protocol with NetWare-adapted features. The original maximum packet size was 576 bytes (although Novell does not strictly adhere to this when data are exchanged between two devices on a local LAN). Today the data size is negotiable, up to the maximum transmission unit of the media (e.g.,

Ethernet). There are two entry points and two exit points for IPX. Data and commands are entered to and from either an application or the LAN. Data are transmitted from IPX from the same two points. It all depends on the direction of the data flow. This size does not include packet headers by the data-link layer.

Since data are transmitted from the network-layer software, data are delivered on a best-effort basis (remember that the transport layer is the protocol that provides reliable packet delivery). By implementing proprietary transport-layer software in NCP and simple transport software in the shell (the user interface to the NetWare operating system, discussed later), NetWare did not have to implement full transport-layer software in the protocol stack of the client workstation. This saved valuable RAM at a time when DOS workstations had a limitation of memory in which to run applications. By implementing a small, reliable protocol stack such as this, the speed at which stations communicated with each other increased, especially when the medium was upgraded from ARCnet to Ethernet. Sequence Packet Exchange (SPX) was developed later, and it is implemented into the shell. It is used primarily for peer-to-peer communications and utility programs such as RCONSOLE, SNA gateways, etc.

Many functions are provided by IPX, and they can be grouped into two categories: packet formatting and data delivery. Packet formatting is discussed first.

IPX Routing Architecture: The Data Delivery System

Packet Structure. Before developing a complete understanding of how data are transmitted and received on a network, some intermediate functions need to be addressed. The following text will address this.

Any data that are to be transmitted on a LAN need to be formatted for transmission to the network. All data handed down to IPX from an upper-layer protocol are encapsulated into an entity known as *datagram*. When the data-link layer applies its headers, it is called a packet. This process is similar to writing a letter. You write the letter and then put the letter into an envelope and address it to the receiver (the destination). On the envelope, you put the destination address and your return address so that the post office knows where to deliver the letter and the receiver knows where to send a response (if needed) to the letter. This is similar to the process IPX uses to format the data to deliver them to the network.

Figure 6-3
Novell proprietary
(non-LLC) IPX packet
format. *(Courtesy of
Novell, Inc.)*

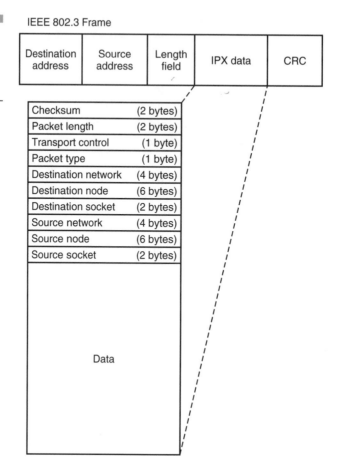

Figure 6-3
Novell proprietary
(non-LLC) IPX packet
format. *(Courtesy of
Novell, Inc.)*

Let's take a look at the packet structure of IPX. Figure 6-3 shows a proprietary IPX packet, and you will see that the IPX packet contains the following fields:

Checksum. This field contains the checksum (16 bits) for the IPX packet. The checksum can be thought of as a fancy parity check. Its objective is to ensure that the bits transmitted are the same bits that are received—in other words, that no bits in the packet were transposed during the transmission. The sending station performs the checksum algorithm on the packet and puts the result of the checksum in this field. The receiving station also performs a checksum on the IPX portion of the packet and generates a checksum. That checksum is checked with the checksum in the packet. If there is a match, the packet is said to be good. If the two do not match, that packet is said to contain an error and the packet is discarded. Since this algorithm is performed at the data-link layer also (a CRC of 32 bits provides better accuracy than a 16-bit CRC), this is a configurable feature for IPX. Originally, it was considered to be redundant and time-consuming—causing unnecessary overhead because of the high reliability of the network it was transmitted over (a LAN). With IPX, if this feature is disabled it is set to FFFF to indicate that checksumming is turned off. (Some IPX routers use this field to check whether the packet is a Novell encapsulated IPX packet or of another type. This is explained in more detail later.)

Length.   This field is used to indicate the total length of the IPX packet, including the IPX header checksum field, meaning the length of the IPX header and data fields. The minimum length allowed is 30 bytes (the size of the IPX header fields), and its maximum number is 576 (indicating a maximum of 546 bytes total data field). For communications on a LAN, the number may be as high as the tranmission medium allows: 1500 bytes for Ethernet, 1496 for IEEE 802.3 (including IEEE 802.2 headers), and 4472 for 4-Mbps Token Ring. Under Novell's packet burst mode, large packets may be transferred between two stations residing on different LANs. This is available under NetWare 4.x and is discussed later in this chapter.

Transport control.   This field is initially set to 0 by the sending station. This field counts the number of hops (the number of routers) the packet encountered along the way. Since the maximum number of routers a packet is allowed to traverse is 15 (a network 16 hops away is considered unreachable), the first four bits are not used. This is also used by routers that support Service Access Protocol (SAP) reporting and other file servers to indicate how far away a server (providing certain services) is from the recipient of the packet. The SAP process will be discussed later.

When a packet is transmitted onto the network, the sending station sets this field to 0. As the packet traverses each router (if needed) on its way to the destination, each router increments this by 1. The router that sets it to 16 discards the packet.

Packet type.   This field is used to indicate the type of data in the data field. This is the Xerox registration number for Novell NetWare. It identifies the XNS packet as a netware packet. Since IPX is a derivative of XNS's IDP protocol, it follows the assigned types given by Xerox as shown in Table 6-1.

Destination network.   This 32-bit field contains the address of the destination network on which the destination host resides. An analogy is that

TABLE 6-1

Xerox-Assigned Packet Types in Hex

Protocol	Packet Type, Hex
Unknown	0
Routing information	1
Echo	2
Error	3
Packet exchange protocol (PEP) used for SAP (service advertisement protocol)	4
Sequence packet exchanged (SPX)	5
Experimental	10-1F
NetWare core protocol	11
NetBIOS	14

NOTE: Packets are set to a 11 (17 decimal).

a network number is like the area code of the phone system. It is used by IPX in routers and workstations to deliver the packet to the local network or to use routers to deliver the packet to another network on the internet (the destination network number is not on the same LAN as the transmitter).

Destination host. This 48-bit field contains the physical address of the final (not any intermediate hosts, i.e., routers, that it may traverse on the way to the destination) destination network station. An analogy to this is the address displayed on the letter. Another analogy is the seven-digit number (not including the area code) on the phone system. If the physical addressing scheme does not use all 6 bytes (ARCnet and ProNET -10), then the address should be filled in using the least significant portion of the field first and the most significant portion should be set to 0s. For Ethernet and Token Ring, it is the 48-bit physical address of the NIC (network interface card). Remember that this address indicates the address of the ultimate (final) destination. It does not indicate any physical address of any intermediate stops along the way.

Destination socket. This 16-bit field is an indicator of the process to be accessed on the destination station. Remember from our previous general discussion on XNS that a socket number is an integer number assigned to a specific process running on a network station (for example, the file service that runs on a file server). Each and every service that runs on a file server will be assigned a socket number. For example, the file service is assigned a socket number of 0451 (hex). Any workstation requesting this service must set this field to 0451 for it to be properly serviced by the file server. Novell Static sockets will be shown in a moment. Since IPX follows the XNS standard, the socket numbers shown in Table 6-2 are reserved. Other socket numbers that are reserved and that may not be used without the permission of Xerox Corporation are in the range of 1-BB8. A number other than this may be used dynamically. This is covered in more detail in the source socket description.

Source network.   This 32-bit field contains the network number of the source network. It indicates the network number from which the packet originated. A network number of 0 indicates that the physical network where the source resides is unknown. Any packet of this type received by a router will have the network number set by the router. When IPX is initialized on a workstation, it may obtain the network number by watching packets on the LAN and derive its number from there. It may also find its network number from the router. Network numbers are not assigned to the workstation.

Source host.   This 48-bit field contains the physical address of the source host (the network station that submitted the packet). It represents the host number from which the packet originated. Like the destination host field, if the physical address is less than 6 bytes long, the least significant portion of this field is set to the address and the most significant portion of the field is set to 0s (user for ARCnet and ProNET - 10). Otherwise, it is set to the 48-bit address of the LAN interface card.

Source socket.   This 16-bit field contains the socket number of the process that submitted the packet. It is usually set to the number in the dynamic range (user-definable range).

Source, destination host and source, and destination network are pretty much self-explanatory. Sockets are a little more elusive.

Multiple processes may be running on a workstation (OS/2, UNIX—usually not DOS), and definitely multiple processes will be running on

TABLE 6-2

Assigned Socket
Numbers in Hex

Registered with Xerox	0001-0BB8
User definable	0BB9 and higher

a file server. Sockets are the addresses that indicate the endpoint for communication. A unique socket number indicates which process running on the network station should receive the data. Sockets represent an application process running on a network workstation. There are two types of sockets: static and dynamic. Static sockets are reserved sockets that are assigned by the network protocol or application implementor (in this case, Novell) and cannot be used by any other process on the network. Dynamic sockets are assigned randomly and can be used by any process on the network.

For example, to access the file services of a server, IPX would fill in the destination and source network, the destination host number of the file server, and the source host number of its workstation. The destination socket number would be set to 0451 (hex). This is known as a well-known socket number, for it is static (it will never change). It has been defined by Novell. The source socket (assigned by IPX at the source workstation) will be a dynamic number and IPX will pick it randomly from the range of 4000 to 6000 hex. The source socket is used by the destination as a socket number to reply to. It indicates the socket number that made the request. In this way, when the packet arrives at the server, the server knows that the packet is destined for it (the host number) and also knows the transmitting station is requesting something from the server (socket 0451). Deeper into the packet is a control code to indicate exactly what the transmitter of the packet wants (create a file, delete a file, directory listing, print a file, etc.).

Once the command is interpreted and processed, the server returns data to the transmitter of the packet. But it needs to know which endpoint of the workstation will receive the data (which process submitted the request). This is the purpose of the source socket number. The file server formats a packet, reverses the IPX header fields (source and destination headers), sets the destination socket number to the number indicated in the received packet of source socket number, and transmits the packet. (See Table 6-3.)

Finally, the socket number (source or destination network number, source or destination host number, and source or destination port number) is the absolute address of any process on the network. With the combination of these fields, any process on any network and network station can be found. IPX controls all socket numbering and processing.

That was one of the functions provided by IPX: the formatting of data into a packet so that they may be transferred across the network. How network numbers are found and how host numbers are found will be discussed later in this chapter.

TABLE 6-3

Netware Assigned
Port (Socket) Values

File server	
0451h	Netware core protocol
Router static sockets	
0452h	Service advertising protocol
0453h	Routing information protocol
Workstation sockets	
4000h–6000h	Dynamically assigned sockets used for workstation interaction with file servers and other network communications
0455h	NetBIOS
0456h	Diagnostic packet

The next (not in any order of precedence) function of the IPX protocol is the ability to route packets directly to a workstation on the same LAN or to a network station on a remote LAN. The following is a description of the Novell implementation called *routing* that allows this.

Encapsulation at the Data-link Layer.   Although this was covered in a previous chapter, Novell provides for support for a few more encapsulation methods not discussed there. (Given its long history, it had to do this to support so many different vendors' network interface cards.)

The methods of data-link encapsulation that Novell supports are:

1. Novell proprietary
2. IEEE 802.3 with IEEE 802.2
3. Ethernet V2.0
4. IEEE 802.3 with SNAP
5. Token Ring, native or with a SNAP header
6. ARCnet
7. FDDI with IEEE 802.2 on a SNAP header

Some of these packets are shown in Fig. 6-4. When installing a Novell network, the installer must choose between the encapsulation methods. Selection of one is needed. The software will not try to figure out the format. Once set, a transmitting station will format the packet and the receiver will read the packet according to the setting during the installa-

Figure 6-4

Encapsulation methods (data-link frame encapsulation).

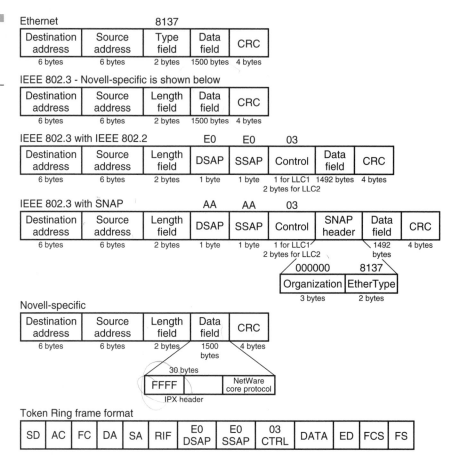

tion. Two communicating stations must use the same encapsulation type.

Some network installations prefer the Novell proprietary. This is basically the IEEE 802.3 MAC header encapsulation. Immediately following the length field is the beginning of the IPX header. This is set to FFFF, indicating that checksumming is turned off. Some router vendors use this field to indicate that it is a Novell packet. IEEE 802.2 is not included in the packet.

The reason for all the packet formats is compatibility. During the ramp-up of Ethernet, different vendors supported different encapsulation techniques. The Ethernet packet header was the first encapsulation technique used with Ethernet networks. When IEEE 802.3 formally adopted CSMA/CD (Ethernet), the packet format was changed. Novell again changed the packet format to include this new type.

With the IEEE 802.3 packet format, Novell supports both the IEEE 802.2 and the SNAP protocols. Furthermore, Novell decided to support its own packet format.

The installer of a Novell network may choose any of the preceding. It all depends on the type of network interface cards that are installed in the network.

Of course, NetWare supports Token Ring encapsulation methods. The two methods supported are IEEE 802.5 with IEEE 802.2 and IEEE 802.5 with IEEE 802.2 SNAP. Figure 6-4 has some of the fields filled in. On the Ethernet frame, the Ethertype of 8137 identifies a Novell frame. On the IEEE 802.3 with IEEE 802.2, EO is the SAP address assigned by the IEEE to Novell. On the SNAP packet, the organization of 000000 identifies an encapsulated Ethernet packet and 8137 is the Ethertype for Novell.

IPX Routing Functions: Novell Routers

The routing function allows packets to be forwarded to different networks through the use of a device known as a router. An analogy would be when a phone number such as 749-4232 is dialed in Virginia, the local switching office knows by the first three digits (the local exchange number) that the number is local and should be routed to a destination on the local phone system. The call is switched between exchange offices (if necessary) in the local area to its final destination. But when a number such as 212-337-4096 is dialed, the local switching office knows the call is to be routed to a distant location. In this case, the call is to be routed to an exchange in the state of New York. This is a crude analogy, but shows how network IDs and routers work in networks.

There are two available types of routers on a Novell network. First, Novell implements a routing function in its operating system. Previous to Novell 3.x, Novell documention called its routers "bridges." But in reality they are routers. As of this writing, Novell has officially changed the names to *internal* and *external router.*

The internal router is one that is usually performing some other tasks, as well as the routing function. These tasks may be file and print services or a gateway service to SNA. The external router is a workstation (for example, a personal computer) consisting of multiple network interface cards, and its sole function is to route packets. No other functions will be provided by this external router.

Second, independent router manufacturers (Bay Networks, 3Com, Cisco Systems, etc.) can also participate in a Novell network, providing

IPX routing and service advertisement protocol (SAP) functions. These independent router manufacturers are fully compliant with the Novell routing and SAP scheme. The independent router manufacturers usually provide multiprotocol routers that will route other types of packets (TCP, AppleTalk, DECnet, etc.) as well as NetWare packets. The protocols are routed simultaneously in the same router. These types are also high-performance routers—specifically built as routers and not as personal computers acting as routers. Novell continues to sell its Multiple Protocol Router (MPR), which enables simplex routing to occur. This is further enhanced through Novell's IntraNetWare platform.

To route packets, routers accept only packets directly addressed to them and determine the best path on which to forward packets they accept. This process involves multiple processes, which are explained next. First, we discuss routing tables.

IPX Routing Tables   Routers need to know of all other available routers and therefore all other active networks on its internet. The IPX router keeps a complete listing of the networks listed by their network numbers. This is known as a *routing table*. Each router in a NetWare internetwork contains a table similar to Table 6-4. The entries in the routing table let the router determine the path to forward a packet to.

Figure 6-5 shows a network depicting the aforementioned routing table. To help us discuss this figure briefly, network numbers are 32 bits in length. As shown in Fig. 6-5, the networks are separated by routers. The combination of all the networks together is called an *internet* (or *internetwork*). Each of the routers needs to know of each network on the internet, and the routing information protocol (IPX RIP) process is the method for exchanging this information. Among other things, Novell's IPX protocol changes XNS IDP RIP implementation slightly to add a timer. This is known as *ticks*. It provides the ability of a distance vector algorithm with true cost attributes, not just a hop count. Ticks are the amount of time that is required to reach that path. Most IPX routing implementations do not perform tick counts. It is also used to set timers on a NetWare workstation. Ticks will be explained in detail later. This routing table is the way it would look in the PC file server routing table shown in Fig. 6-5. This PC is an example of an internal router and has networks 20, 30, and 40 directly connected.

The first entry of the router table, the network number, contains the network numbers that are in place on the internet. A router will exchange its routing table with other routers on the network. The

TABLE 6-4 IPX Routing Table

Number Network	Number of Hops to Network	Ticks to Network	NIC	Intermediate Address of Forwarding Router	Net Status	Aging Timer
00000020	1	2	A	Local		0
00000030	1	2	B	Local		0
00000040	1	2	C	Local	R	0
00000050	2	3	B	02608C010203		1
00000060	2	3	A	02608C040506		2
00000070	3	4	A	02608C010304		2

entries in the table that are exchanged with other routers are network number, hops, and ticks. This information is transferred via a RIP packet (explained later), shown in Fig. 6-6. The other entries pertain to the local router and are not distributed by each router. Every router builds its own entries. A router receives these updates (routing tables) from other routers on the internet through a process known as RIP, which will be explained in a moment. From this received information, a router builds its own table and thus a picture of the internet.

The second entry shows the number of routers that must be traversed to reach this network. Anytime a packet must traverse a router to reach a destination, the process of traversing the router is known as a *hop*. Therefore, if a packet must cross over four routers to reach the final network, the network is said to be four hops away. The term *hops* is also called a *metric*. Four hops is the same as a 4-metric count.

The next entry is the tick counts. This number indicates an estimated time necessary to deliver a packet to a destination on that network. This time is based on the segment type. A tick is about 118 seconds. This number is derived from an IBM-type personal computer clock being ticked at 18 times a second. In actuality there are 18.21 ticks in a personal computer clock for every second elapsed. At a minimum, this field is set to 1; 18 would indicate 1 second.

For locally attached segments (excluding serial lines) with more than 1-Mbps transmission speed (Ethernet and Token Ring), the NIC driver

Figure 6-5
Network number
assignments with
routers.

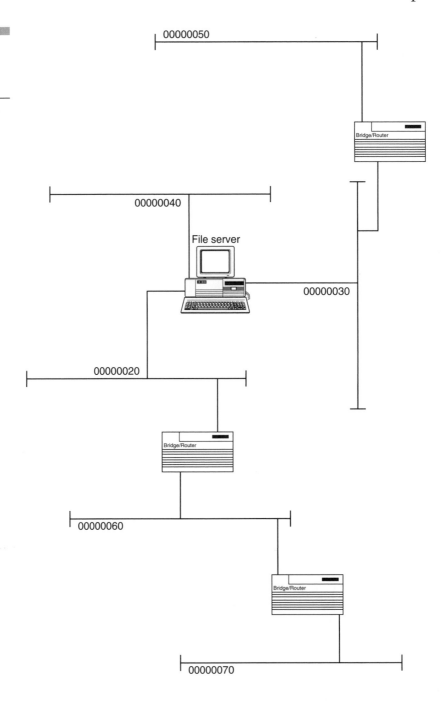

Figure 6-5
Network number
assignments with
routers.

Figure 6-6

IPX RIP packet
format. (*Courtesy
Novell, Inc.*)

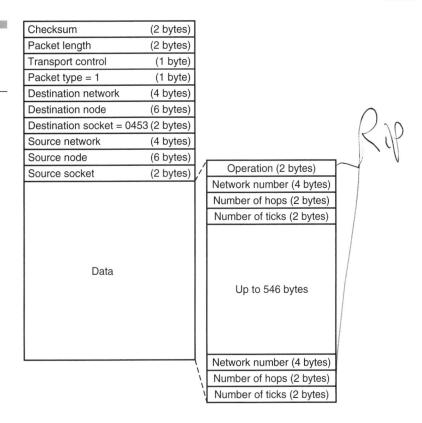

Checksum	(2 bytes)
Packet length	(2 bytes)
Transport control	(1 byte)
Packet type = 1	(1 byte)
Destination network	(4 bytes)
Destination node	(6 bytes)
Destination socket = 0453	(2 bytes)
Source network	(4 bytes)
Source node	(6 bytes)
Source socket	(2 bytes)

Data

Operation (2 bytes)
Network number (4 bytes)
Number of hops (2 bytes)
Number of ticks (2 bytes)
Up to 546 bytes
Network number (4 bytes)
Number of hops (2 bytes)
Number of ticks (2 bytes)

RIP

assumes a tick of 1. For serial network segments (X.25, synchronous line of T1 and 64 kbps, and asynchronous), the driver periodically polls to determine the time delay. For a T1 circuit, the tick counter is usually 6 to 7 ticks per segment. Any changes in this time will be communicated to the router and propagated to other routers on the network. These numbers in the tables are cummulative. This means that as each router broadcasts its routing table, this number is not reset. It is the sum of all the paths' tick counts to reach a destination network.

The network interface card (NIC) entry field records the NIC number from which the network can be reached. It indicates the controller card from which the router received this reachability information. A Novell file server can hold four NICs. It is the same as a physical port number in a standalone router (not a personal computer file server acting as a router). The intermediate address entry contains the physical node address of the router that can forward packets to each segment. If the network is directly attached, the entry is empty. If the network to be reached requires the use of another router, this entry contains the physical address of the next router to send the packet to. This physical address is extracted from RIP updates (broadcast routing tables) sent by those routers. An entry in the NIC field is valid only if the router is located in a PC. Otherwise, this field indicates the router physical port number.

The net status entry indicates whether the network is considered reliable.

The age entry is used to indicate how long it has been since a routing update has been made. This field is used to age-out (delete) entries for networks that have not been heard from for a certain amount of time. These timers follow the XNS specifications. This number can be in seconds or in minutes, depending on the manufacturer of the router.

In short, a routing table contains a listing of network numbers and associated paths (whether direct or indirect) for delivering packets to their final destination networks.

One last note is that, with the exception of the next hop router address, the entries in the routing table do not contain the physical addresses of the network stations that reside on the internet (this is true of most routing algorithms). The only physical address in the table is that of another router to which a packet, destined for a remote network, may be addressed. Routers do not know which other end stations are on the networks they connect to. The final destination (physical address of the final destination) is embedded in the IPX header (the destination host). Once the router determines that the final destination network number is directly attached to the router, it extracts the destination host number from the IPX header and addresses the packet and delivers it to the directly attached network segment. This will be discussed in detail later.

IPX Routing Information Protocol

For routers to exchange their tables with other routers on the internet, IPX uses an algorithm known as *routing information protocol,* or RIP. The RIP algorithm is the most widely used routing algorithm today. Variations of this protocol exist on TCP/IP, AppleTalk, IPX, XNS, and a host of other proprietary XNS vendor implementations. IPX RIP is not compatible with other protocol versions of RIP. This is why a multiprotocol router will force the network administrator to configure RIP for each protocol that is implemented on the router. The origins of RIP date back to the XNS protocol.

The functions of the RIP protocol are:

1. To allow a workstation to attain the fastest route to a network by broadcasting a route request packet that will be answered by the routing software on the Novell file server or by a router supporting IPX RIP

2. To allow routers to exchange information or update their internal routing tables

3. To allow routers to respond to RIP requests from workstations

4. To allow routers to become aware that a route path has changed

There are multiple ways for a workstation (not a server) to determine its network number. IPX on a workstation extracts information from the network by watching the RIP responses on the network to dynamically determine a workstation network number. It also may retrieve from the router during a RIP request or during a SAP request (discussed in a moment). The point here is that a workstation is not predefined (during installation) with a network number. Server and router network numbers, however, are implemented during the installation of that server or router. If the network number is obtained during a SAP request during workstation startup, the network number will not be known. The workstation will send out the SAP request with the IPX header network number entry set to all 0s. The destination node field will be set up to all FFs.

IPX will format an IPX RIP packet, as is shown in Fig. 6-6. Notice that the destination socket field is set to 0453(h) to indicate the destination process. The packet type is set to a 1. The static socket for RIP is 0453 (hex). It should be noted that RIP data (the routing table) is enveloped by an IPX header (the IPX packet will be encapsulated by the data-link header), and the packet is then transmitted onto the network. All network requests will use IPX, as shown in later figures. The RIP request is placed in the data area of an IPX packet. Not all packets contain user data. Some network packets will contain information known as *control information,* also known as *overhead.* It is necessary data for the network, but it is still overhead. This is one of those packets.

Refer to Fig. 6-6. A RIP request has an operation field of 1, and a RIP response has an operation field of 2. Following the information field is one or more (depending on the number of known routes) sets (network number, distance, and tick count) of information. The sets of information contain a network number followed by the number of hops to that network and the associated delay cost (ticks) to that network. Due to the

576-byte limitation of an IPX packet, a single RIP packet can contain 68 sets of known route information. If more routes (a larger table) are known, multiple packets will be sent by a router.

XNS's IDP protocol does not define a ticks field in an XNS IDP RIP packet. NetWare incorporated this so that a shell could determine the delay time in receiving a response from a file server (the receive time-out). If there are multiple paths to the same destination of equal hop count, the routers will use the route with the shortest tick time. These fields (ticks) are only valid for an IPX RIP response packet.

Router Operation

An individual router, like any other network attachment, attaches directly to a LAN. A router will have at least two attachments. The router separates two cable segments. The cable segments that it attaches to are known as the *directly connected cable segments*. That term will be used throughout this book.

When an IPX router first initializes, it places the network numbers of the directly connected routes into the routing table. These network numbers are manually entered during the configuration of the router during installation. Once they are integrated, the router sends a broadcast packet to the network (on each of its directly connected cable segments) containing these routes (the network numbers of the directly attached cable segments), which the router will now make available. Other routers on those cable segments read this information and update their tables. The RIP cost that is entered into the table at initialization is usually set to a 1. This is configurable in some routers.

The router then transmits another RIP packet *requesting* information from other routers on its directly attached network segments (a RIP request). This request is responded to by any other active routers on the directly connected segments. The term *directly connected segments* is used here because request and response packets are sent in broadcast mode. This means that all stations on the local network receive and process this packet. These broadcast packets are not forwarded to other networks by the routers. Routers update their tables and, in turn, broadcast their updated tables to their directly connected cable segments. Figure 6-7 shows router A sending (split horizon is not used here) out its routing table to network 2. Router B picks up this packet and adds 1 to each hop count entry in the received table. The router then compares the received table to its own table.

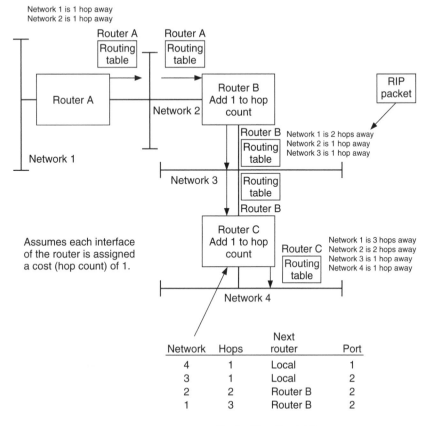

Figure 6-7
A routing table
update (RIP) not
implementing split
horizon.

Network 1 is 1 hop away
Network 2 is 1 hop away

Router A
Routing
table

Router A
Routing
table

Router A

Network 2

Router B
Add 1 to hop
count

RIP
packet

Router B   Network 1 is 2 hops away
Routing   Network 2 is 1 hop away
table   Network 3 is 1 hop away

Network 1

Network 3   Routing
              table

Router B

Assumes each interface
of the router is assigned
a cost (hop count) of 1.

Router C
Add 1 to hop
count

Router C   Network 1 is 3 hops away
            Network 2 is 2 hops away
Routing   Network 3 is 1 hop away
table   Network 4 is 1 hop away

Network 4

Network	Hops	Next router	Port
4	1	Local	1
3	1	Local	2
2	2	Router B	2
1	3	Router B	2

Router C routing table

If a router finds a network number in the received table that is not in its table, it adds this entry into its table. If a network number matches a network number in its table, the router checks the hop count. If the received table's hop count is lower than the one in the router's table, the router changes the entry in its table for the entry from the received table. If the hop count of the received table is larger, the entry is discarded.

Once its table is checked, the router broadcasts its table (if the 60-second RIP update timer expired) to network 3. Router C performs a similar comparison. (In actuality, it broadcasts table entries out both of its ports, but because of the algorithm of split horizon, only network numbers of its table that it did not learn from a port will be broadcasted out that port. Until split horizon is discussed, this example will do.)

Once these events have taken place, the IPX router places itself in the operation of receiving information (processing RIP requests, routing packets, and maintaining its routing table). In addition to these updating tasks, all routers broadcast their routing tables every 60 seconds. This is different from other implementations of RIP. For example, TCP/IP RIP broadcasts every 30 seconds and AppleTalk broadcasts every 10 seconds. These other routing tables are received by other routers so that they know about other networks on the internetwork. In this way, all routers remain synchronized. Every router may transmit its table at different times. Each router must broadcast it every 60 seconds. In other words, the routers do not broadcast their tables at the same time.

Periodically, routers will be brought up and will go down. If the router is operating on a file server, and the file server is brought down by the DOWN command at the file server console, that file server broadcasts a packet to its locally attached network segments indicating that it is being brought down and will no longer be able to route. All the other routers on the internetwork receive this information and update their tables (delete the entry) accordingly. In this way, the router tables may provide new information as to how to get to a network. If a file server is powered off or crashes, there is no way to notify the other routers on the network. Therefore, routers have the ability to age-out an entry in a table. Commercial IPX routers (Bay Networks, Cisco, 3Com) usually do not implement the feature of telling other routers that they are being shut down. They will time out an entry.

Figure 6-7 shows an update in one direction only. This is shown here for simplicity. In actuality, routing tables are broadcast out all active ports of a router every 60 seconds based on the router's own periodic timer. Most RIP-type routers employ the split horizon algorithm, which is explained subsequently.

Split horizon instructs the router not to broadcast a learned route back through a port from which it received the learned information.

Refer to Fig. 6-7, in particular router B and router C. With split horizon set, router B should broadcast out (in the direction of router C) information about network 1, 2, and 3 only. It will not broadcast information about network 4. Network 4 was learned from the port that it is sending its table out on. Furthermore, router A would not send any information out (in the direction of router B) about networks 3 and 4. It learned about these networks from router B on that port.

Upon receipt of this information, router C updates its routing table. When router C transmits its table in the direction of router B, it includes information only on networks 3 and 4. It learned about networks 1 and 2 from router B (actually it received information about those networks from that particular port), and it will therefore not broadcast the same information about those networks out that port.

Router C will still broadcast information about networks 1, 2, 3, and 4 out the router port connected to network 4.

Aging of the Table Entries.   Every time routing information is received about a certain route, the route's entry is set to 0. If this entry is incremented to four minutes, the router assumes that the path to that network is down and broadcasts this information to other routers on its local segments. This information is propagated on to other routers on the network to update their routing tables.

Determining a Local or Remote Path.   When a router is fully operational, it makes network numbers known on the internet, and other network stations may then use that router to forward a packet to those remote networks. Anytime a network station wishes to send information to a destination station, it must have the network address as well as the physical (data-link or MAC) address of the destination station. If the two stations are communicating on the same network (they have the same network number), the transmitting station can send the packet directly to the destination without invoking the use of a router.

However, if the destination station lies on a different network than the transmitting station, the transmitting station must find a router to submit the packet to for final delivery. In order to find a router, the network workstation must transmit a RIP request packet. Inside this packet is the destination address (network number) of the final destination. This request is answered by routers only on the immediate (same network as the requesting station) network. Routers that are not directly connected to the same network do not see this request because it is transmitted using data-link broadcast (all FFs in the data link address fields) destination address. Local routers can respond to this type of packet but do not forward this or any direct-broadcast packet. Any router that has a path to that destination will respond, and the network workstation will choose the router to forward the packet to. Usually, this will be based on the lowest tick or hop count.

In the response packet from a router is the router's physical address (in the data-link header). The requesting network station uses this address to physically address its packet to the router. In the case of multiple responses to a route request (there are multiple routers on the requesting station's network), the workstation chooses which router to submit its packet to.

When the originating station receives the router response, it extracts the router's MAC address out of the response packet and prepares its data for transmission in the following way (Fig. 6-8*a* shows the dialogue between network station A communicating to file server C through the intermediate router B):

1. It places the final destination's, not the router's, full internetwork address (a socket address consisting of network address, host address, port address) in the IPX destination header information fields (refer to Fig. 6-8*a*).

2. It places its own full internetwork address in the source header information fields of the IPX header (network number, host number, and a dynamically assigned socket number).

3. The router's physical address that was extracted from RIP response packet is then placed in the data-link destination header field (Fig. 6-8*a*) and its own physical address in the source address data-link header field.

Once all the fields have been filled out, the network station transmits the packet. When the router receives this packet, two possibilities exist for the router:

1. If the network number of the end-station destination (indicated by the network address in the IPX header) is directly connected to the router, the router extracts the data-link destination address (the host address) from the IPX destination host address field, and places it into the data-link destination address header. Next, it places its own physical host address in the data-link source address field and sends the packet to the directly attached network. Throughout the course of a packet's being forwarded, the IPX header is read to make route determinations. It is never written into. It is never changed, with the exception of transport control and the checksum field (if used).

2. If the destination network is not directly connected to the router, the router must send the packet to the next router in the path to that network. To do this, the router places the data-link MAC address of the next router in the path in the data-link destination address field, places its own address in the data-link source address field, and submits the packet to the network that has the next router attached. It knows the next router's data-link MAC address, for it is one of the fields in the routing table. Refer to Table 6-4. The router does not touch the original data in the IPX header field (except for the transport control field, which will be incre-

Figure 6-8

(a) Abbreviated version of data transfer using IPX headers

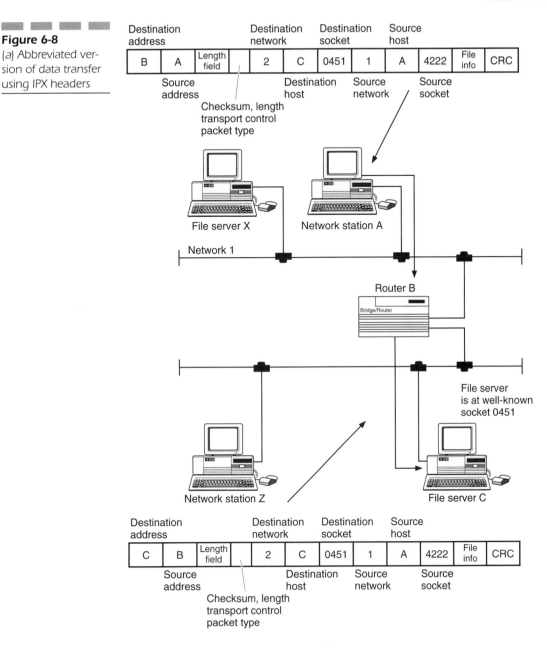

(a)

Figure 6-8
(b) Summary of
router functions.

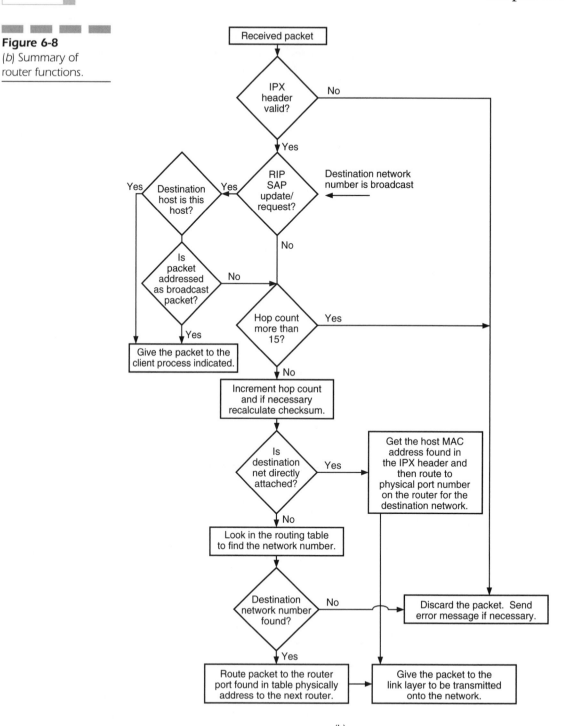

(b)

mented by 1). When the packet reaches the next router, it starts this decision process over again starting from the preceding number 1.

   If the transport control field is set to 16, indicating that 15 routers have been traversed, the router discards the packet. Upper-layer protocols on the originating station will time-out, and the packet will be retransmitted.

This is how routers know where to deliver a packet, and this is the purpose of the destination host field in the IPX header. Again, when a packet reaches the destination router (the router that is directly connected to the destination network as indicated by the destination network number field in the IPX header), the destination router determines if it is connected to the final network and then extracts the end-station host number (destination host), puts that number in the destination address of the data-link header, and transmits the packet to the end network workstation.

The sending station of a packet finds out about the final destination host number through a protocol known as *service access protocol.* This is discussed in a moment.

Routing Supplementals.   With NetWare version 2.15c and later versions, if more than one route (multiple routers to the same destination network) to a network exists, only the routing information that is equal (same number of hops and ticks) will be kept in the routing table; for a path will be selected only if it is the fastest route. Prior versions used to keep tables of all available routes, regardless of hops and ticks in the routing table.

The routing table shown in Table 6-4 is exclusive to Novell and its routing scheme. It uses a process similar to other routing schemes (TCP/IP, XNS, and AppleTalk) in that it uses the functions of a table and RIP to distribute that table. You will hear that if you know XNS you know IPX. This is somewhat true; but the differences are:

- Router table (tick entry)
- Format of the IPX RIP packet
- Not broadcasting complete tables across slower-speed serial lines (telephone lines)
- Broadcasting information that a server is going down (routing functions will cease and may be a path to a network)

■ The workstation can obtain its network address by watching the network cable and "gleaning" the address from other packets on the network. Upon start-up, if IPX has not determined the network number for the network station (not the server), and NETx.com has started (loading the shell program is discussed later), IPX will determine the network address from a RIP or SAP response packet that it uses during loading of the shell and the login program.

This completes the NetWare data delivery system known as IPX. Remember, until NetWare 5.0 any process that has to transmit or receive over the Novell NetWare environment must use IPX to deliver and receive the data.

Finally, Fig. 6-8*b* presents a summary of the router's functions in a flowchart.

The Service Advertising Protocol

Workstations and servers initiate communication with each other through the use of a name. Naming entities in a network is an easier way for users to access services on the network. Requiring users to remember network numbers, host numbers, and socket numbers will prohibit efficient use of a network. The Service Advertising Protocol (SAP) provides a directory service of sorts for services that are running on NetWare servers. It does not provide any information about users on the network and so on. It is dynamic and provides immediate information to the network. This is exclusive of the new directory system, NetWare Directory Services (NDS, discussed later).

In order for the workstation to find a server's name, log in to the network, or use printing or email services, it must be able to locate a server and the services running on that server. Routers are intricately involved in this process and keep tables of server names, their full internet address, the services they provide, and how far away they are. This process is known as the *Service Advertising Protocol* (SAP).

Many services are provided by the Novell NetWare server, as shown in Table 6-5. To provide this information to any network station on any network (an internet), Novell NetWare implements the protocol known as SAP. A NetWare internetwork cannot operate without this protocol. It allows services of a server to be announced to local and remote networks, no matter how large the NetWare internet may be. This service

TABLE 6-5

NetWare Service and Type Identi- fication

Description	Object Type	Description	Object Type
User	1	Time Synchronization Server	2D
User Group	2	Archive Server SAP	2E
Print Queue	3	Advertising Print Server	47
File Server	4	BTrieve VAP 5.0	4B
Job Server	5	SQL VAP	4C
Gateway	6	Xtree Network Version	4D
Print Server	7	Btrieve VAP 4.1	50
Archive Queue	8	Print Queue User	53
Archive Server	9	WANcopy Utility	72
Job Queue	A	TES-NetWare for VMS	7A
Administration	B	NetWare Access Server	98
NAS SNA Gateway	21	Portable NetWare	9E
NACS	23	NetWare 386.x1	07
Remote Bridge Server	24	Communications Executive	130
Bridge Server	26	NNS Domain1	33
TCP/IP Gateway	27	NetWare 386 Print Queue	137
Gateway	29	Wildcard	FFFFFFFF

not only provides the capability of providing information to a network station about the network, but also allows for a distributed server name service. In other types of networks, there are sometimes distributed name servers that keep a centralized database of names and their full internetwork addresses. This is true for the XNS and TCP/IP protocols. In order to find a particular station on the internet, the network workstation must transmit a packet to the name server to which the name server will respond with the full internet address of that remote network workstation. Once the originator of that packet receives a response from the name server, it can then attempt a connection.

While a centralized name service is good, it does have its deficiencies. The name server's full internet address must be manually entered in every network station on the network; otherwise, the network station will not know where the name server is. (This entry is entered into a static table and does not have to be entered in every time a name service is needed.) If the centralized name service is down, users must manually find and enter the full internet address of the remote network station. While all these problems are easily overcome, NetWare provides dynamic name service capabilities, plus much more, through its SAP protocol. This protocol is used for server names only.

NetWare name services operate just a little differently than standard name services such as the Clearinghouse Name Service of the XNS protocol. There is not a central name server on the network. The NetWare operating system provides a function that allows any server on the internet to broadcast to the network some of the server services that it can provide by name and type. These services are shown in Table 6-5. This information is propagated throughout a NetWare internet by the routers on the internet (including any bridges or routers) through the use of the following types of packets: broadcast SAP information packets and SAP query packets (also known as service identity packet).

A service identity packet is broadcast from each active server to inform the routers or other servers on the internet of the services that particular server offers. A service query packet is used the find any other active server on the network. The SAP is a protocol that uses IPX encapsulation to transmit and receive these packets. Routers and other file servers, not user workstations, maintain tables that contain the address of the server and the service that it provides to the network.

The SAP makes the process of adding and removing servers and their services on the internetwork dynamic. When file servers are initialized, they broadcast their available services (again, the services are shown in Table 6-5) using the SAP. When the file servers are brought down, they broadcast that their services will no longer be available using the SAP process. Routers, in turn, then delete this information from their tables.

The SAP process allows workstations on the network to find a service and to connect to that service. This is accomplished completely transparently by the user. All the user wants to know is where a server is in order to connect to it. Using the SAP, this is easily accomplished.

The propagation of this information is very similar to the way routing tables are updated using the RIP protocol. Using the SAP, network workstations can determine which services are available on the network. Inside a SAP broadcast packet is up to seven entries. Each entry contains a server name, the service number, and the full internetwork address of the server that contains a service. Obtaining the full internet address of the server is important for a workstation and server to communicate with this address. They do not communicate using names.

SAP tables are maintained in the routers. The SAP is not part of the routing services, but each router will update its SAP table, just as in the RIP protocol. Each router on the IPX internet contains a table of SAP entries. Each router maintains a server information table—all services known on the entire internetwork are maintained in each router. Any service that a client wishes to connect to will be provided by the client's local router.

Figure 6-9 shows servers and workstations on an internet. The file server broadcasts information about the services that it is providing to

Figure 6-9
SAP and RIP processes on the network.

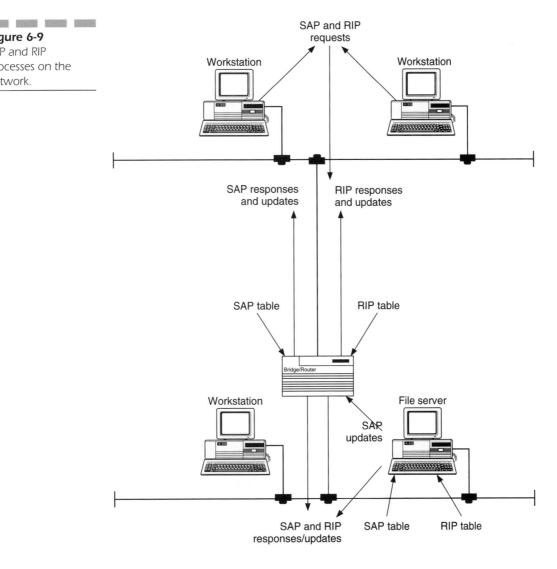

the network. This information is received by the router and maintained in the SAP table of the router. When the workstations at the top of the figure need to communicate with a file server, they broadcast SAP query packets to the network. The router performs a SAP table lookup for the workstation and, if a service name is found, returns the full internet address of the server so that the workstation may connect to it. At the bottom of the figure, the workstation ignores any SAP updates that the file server broadcasts. It performs only SAP queries, just like the workstations at the top of the figure. File servers have the routing function built in so they can act as both requestors and responders for RIP or SAP. The file server and router will update each other.

Like all other NetWare services, the SAP process uses IPX to broadcast its services. Figure 6-10 shows this packet format. The first entry in this packet identifies the operation being requested. It can perform five different operations as follows:

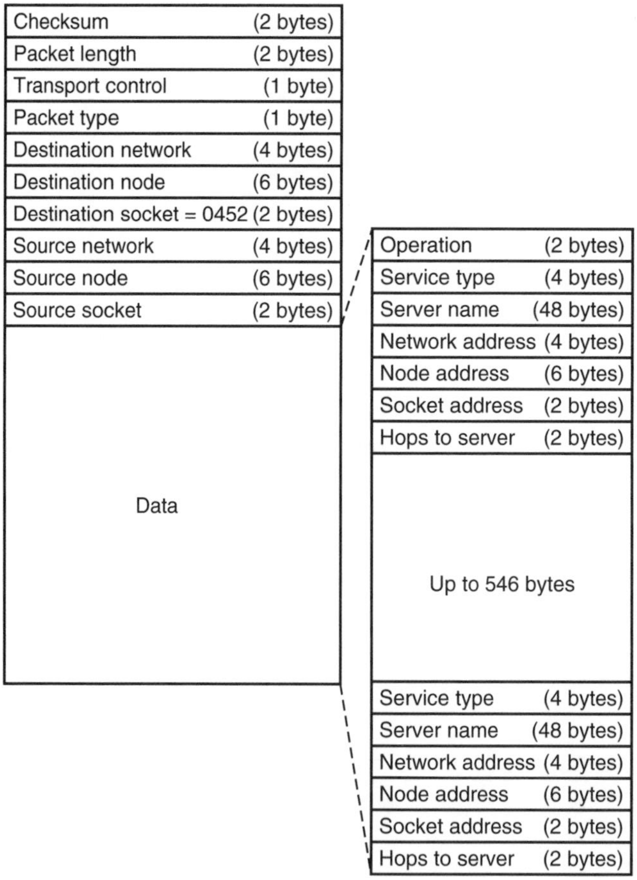

Figure 6-10

SAP packet format. *(Courtesy Novell, Inc.)*

1. A network workstation request for the name and address of the nearest server of a certain type
2. A general request, by a router, for the names and addresses of either all the servers or all the servers of a certain type on the internetwork
3. Respond to Get Nearest Server request (from a workstation initializing onto the network, discussed later in this chapter) or a general request
4. Periodic updates by servers or routers
5. An information change broadcast

Any server that is implementing this protocol will contain a small database (a SAP table) to store accumulated broadcasted server names, service types, and their associated network addresses. This table is shown in Table 6-6. These tables are kept in the servers and the routers that are present on the network. With these tables, a NetWare workstation may then broadcast a packet to the network to query information from the servers or routers. The purpose of keeping these tables in the routers is that doing so allows file servers to be located on a different LAN than the workstation. Routers will respond to workstation queries.

Each broadcast SAP update packet transmitted by IPX may contain information on a maximum of seven file servers. Multiple packets will be transmitted if the SAP table is larger than this.

TABLE 6-6 A Sample SAP Table

Name	Server's Full Internet Address*	Server Type†	Hops to Server‡	Time, Since Changed, s	NIC Number§
Server 1	02608c01020: 00000002 0451	4	1	20	3
Print server	02608c025678 00000003 0451	7	2	50	4

*MAC address, network number, socket number
†As listed in Table 6-5.
‡Number in the transport control field.
§If in a PC server; otherwise, port number on the router.

SAP Operation

When the file server is first started, its SAP operation places the appropriate entries (the services that users may connect to) in its SAP table in the format shown in Table 6-6. The SAP process then broadcasts its table out all its directly connected ports (network interface cards, serial lines, etc.). These SAP broadcasts are transmitted with the physical address set to all FFs. This is a local network broadcast in that the routers will not forward this onto other networks. Once a file server/router has updated its table (including adding one to the hop count for that router), that router broadcasts its updated SAP table to its directly connected networks. In this way, all information is eventually distributed (propagated) to all file servers and routers on the Novell internet. This will update all other servers and routers on those directly connected networks on the fact that a new server has become available, and all other servers will know which services are available on that server.

After this initial broadcast the server then requests SAP tables from other servers on its directly connected network. Other file servers and routers respond to this request, and the information returned allows the new server to find other servers on the network and build a SAP table for its use.

Once fully operational, the server broadcasts its SAP table to its directly connected network every 60 seconds. This is known as a *periodic broadcast*. The initiation procedure for those devices that are not file servers (i.e., routers only) is the same, but it does not accomplish the first initiation broadcast. This is because routers only route packets and do not provide server services. Routers initialize their SAP table and transmit an initial request for SAP information from other routers or file servers on the network and then start a timer that will broadcast its SAP table every 60 seconds.

When a router or file server is gracefully brought down (i.e., it has not crashed or been abruptly powered off), it notifies the network of this by broadcasting to the network that it will no longer broadcast its SAP table of servers. This is transmitted on the directly connected networks of that file server or router. This information is eventually propagated to the rest of the network by the remaining file servers or routers on the internet.

SAP requests to find a server, using the server's name, are usually made by a workstation to find a server or some type of service. The SAP process in the router answers this request. With this response, the workstation extracts the full internet address of the server out of the response packet and attempts a connection using that information. This internet address returned by the router may be a server on the local network or a server on a remote network. If the server is across a router or many routers, the network workstation issues a Get Local Target packet (RIP request) to the local network, and that is how it will find a router to use. It addresses the packet accordingly and transmits the packet to the router.

Now that the networks providing services have been defined, the final part of this chapter explains how NetWare uses these entities to provide file and print services on a network. All the functions previously discussed must be operational for NetWare to operate. The file/print service architecture (workstation-to-server communication) operates over these functions. The discussion on the NetWare interface will be accomplished in two parts: the workstation and the file server. This will be discussed in a moment.

Data Integrity.   Novell NetWare does not use a separate transport layer (for example, SPP). Transport-layer services provide reliable packet delivery. It uses connection IDs, sequence numbers, and acknowledgment packets to number or acknowledge the incoming packets or outgoing packets. In this way, the transport-layer software can determine if a packet has been lost during transmission. The file server knows what the last sequence number was, and if the next packet (Novell only, in this case) is not one above that previous sequence number, a packet was lost somewhere.

Sequence numbers are intialized to 1 following the connection sequence packet exchange between a workstation and its connected file server, and are incremented by 1 as each packet is received. Novell (and XNS for that matter) use packet-level sequencing as opposed to byte-level sequencing. Byte-level sequencing, like the one used in TCP/IP, is accomplished by assigning a sequence number for each data *byte* in the packet. Packet-level sequencing is incremented by 1 for each *packet*, not bytes, submitted. Packet-level sequencing provides for and is accurate for high-performance LANs (Ethernet and Token Ring), for they use a highly reliable access method scheme. TCP/IP is a byte-level sequencing scheme since its initial use was over noisy copper synchronous lines. TCP/IP was not architected for LANs. Therefore, its sequencing had to be more regimental in order to guarantee the reliability of the data.

NetWare has embedded the transport-layer function into its network operating system of NCP. Novell uses a stop-and-wait mechanism (or a window size of 1) packet sequencing. This means that each packet transmitted by the workstation must have an acknowledgment by the file server before the next transmission from the workstation is allowed to occur. This type of sequencing is used with the current release of NetWare—every packet sent must have an acknowledgment. There is a new version of the shell and NCP currently in beta testing (as of this writing), known as *packet burst mode*. This product allows for multiple packets to be transmitted without an acknowledgment for each packet. This means that the server will acknowledge a sequence of packets, not just one at a time. This speeds up some NetWare networks 400 to 500 percent. As of this writing, this software has not yet been released by Novell. The following text will refer to the older method of sequencing. The new methods are fully discussed at the end of this chapter. This is available under Novell 4.0.

Right after a connection has been established, the connected *file server* tells the *shell* what unique (not duplicated) connection number has been assigned to it. This connection number allows the file server to differentiate between users that are attached to this file server.

The shell (on the workstation) starts the packet transfer by setting the initial data packet with a sequence number of 1. The server acknowledges this packet and returns the same sequence number. The server places this sequence number into an entry in its connection table (shown later). The workstation waits for a response from the file server with that same transmitted sequence number. Once the workstation receives that response packet from the file server, the next packet (for that connection to that file server) is incremented by 1.

If the shell does not receive a response packet, it submits the request again until it receives a response from its server or it reaches a certain upper limit.

When the server receives the next packet from a particular connection, it extracts the sequence number from the received packet and compares it with the sequence number entry as its connection table. If the sequence number equals the sequence number in its connection table, the server responds to the request again (duplicates the response). If the sequence number is one more than the entry in the connection table, this represents a new request and the file server responds to the new packet. This discussion will be continued in more detail later in the section "File Server Concepts."

This concludes the discussion on the LAN operating interface for NetWare.

An Introduction to Workgroup Client-Server Computing

Throughout the following text, refer to Fig. 6-11. This figure shows a network that has file and print services available from a server. The user's workstation accesses the server for these services over the network.

The 1980s brought on the importance of file servers and the ability of users to communicate to file servers through a LAN. The applications that originally ran on a user's PC were now located on the file server, located somewhere on the network. This file server would service user requests (file and print) from their PCs. The majority of requests made to a server are file and print.

Figure 6-11
A network with virtual file and print services.

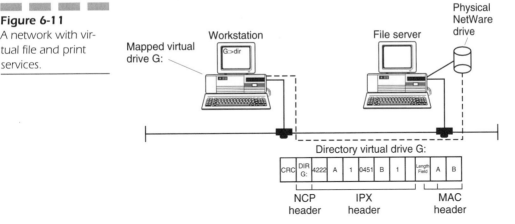

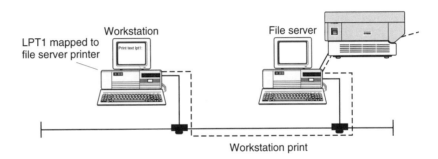

Physical and Virtual Drives

A connection to a file server is accomplished through *virtual drives*. Drives that are physically located on a user's PC are known as physical drives. DOS is the operating system that allows access to these physical drives (usually denoted A:, B:, C:, D:, and E:). These are the physical drives that are located on the local workstation. The network operating system (NetWare Control Protocol, or NCP) grants access to the file server's file system through the use of these virtual drives. In other words, the user sends a command to the file server, requesting not only a connection to the file server but access to the file server's file and directory system. If a workstation has a local drive assignment of C:, that letter should not be used for a virtual drive (a drive assigned to the file server connection).

On the user's workstation, the virtual drive identifier goes right along with DOS. The virtual drive to the workstation may have a drive identifier of F: or G:. The drive identifier is requested when the user asks for a connection to a server. With Novell, the command is the MAP command. The MAP command, used from the DOS workstation command prompt, allows the user to use the file server as an extension of its own file service.

For example, if an application resides on the file server and the user logged in successfully, the user (at the command prompt on the workstation) might use the MAP command to link to a directory on the file server using the following command:

```
MAP G: NOVA/sys:public
```

Using this command would give the user access to the public directory on the file server named NOVA. If access rights to this directory have been granted for the user, the user will switch to that virtual drive and begin to use that directory of files as if the files were located on the user's physical drives. The shell on the workstation will know whether any commands typed on the user's workstation are meant for the server or whether the command should be passed to DOS on the user's local workstation. Figure 6-12 shows how the shell intercepts all DOS commands to see if the commands are meant for the network. If a command is not, the shell passes the command back to DOS for local processing. Otherwise, the shell formats a network command and gives this information to IPX to be passed over the network. The shell is part of the NCP.

Figure 6-12
Shell command inter-
ception.

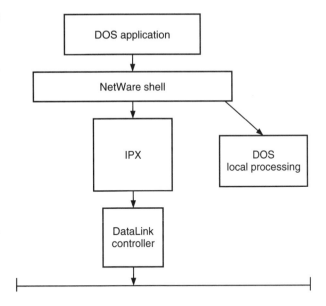

Running an Application from the Virtual Drive. When an application is requested from a virtual drive, the application does not run on the file server. Instead, it is as if the user had the application stored locally. The file server downloads the file to the requesting workstation, and then the application is run locally on that workstation. The file server remains idle (except for housekeeping chores) until the next request is received. When the user is done with the application, the user exits the application. The application file is exited as a normal DOS application exit. The connection to the file server is still active. Only the application was exited. The file server does nothing here.

The application is not uploaded to the file server. If the user wants to store data files from the application on the file server, the data files are uploaded from the user's PC to the file server and then the application exits.

Printing Using a Server

File servers also contain the ability to accept print requests from users on the network. This is accomplished using queues. Using the MAP command, the user redirects the local printer port (LPT1:, for example) to a named printer queue on the file server. The shell program intercepts all calls made to the local LPTx: port from application requests. The shell maps such a request to a network call and gives the request to IPX to deliver to the file server. The file server decodes the request and sends the information to a print queue on the file server. A print queue is nothing more than an area that contains print requests. Usually, the entries in the file server's print queue are stored as files on the file server's local disk. When an entry in a print queue is ready to be printed, the print service on the server reads the information from its disk and prints it to its printer. By redirecting (or mapping, in NetWare terms) the workstation's local printer port to a file server queue, any information that was printed on the local printer is redirected to the file server, and it prints the information out on the file server printer. The shell intercepts all printer activity to see if the information needs to be redirected, or mapped, to a server service.

There are many advantages to client-server computing:

1. *Software updates.*   An application program needs to be updated only on the file server and not on all the workstations.

2. *Lower printer costs.*   Only one laser printer needs to be purchased, instead of one for each user.

3. *Central respository for files.*   Only one copy of an application program needs to be loaded (depending on the number of users).

4. *Relaxed PC requirements.*   Only the file server has to be the most powerful computer (depending on the user's application).

5. *Electronic messaging (electronic mail).*   Files and messages can be moved between users through the file server. This eliminates the "sneakernet" and reduces the amount of paper associated with corporate interoffice mail. This concept will be covered in greater detail in the following section. Users are granted a connection by the file server, and then the users can access the files on the file server (given the correct permission requirements).

The NetWare Interface

There are two NetWare interface methods. The first (and oldest) functions are described here, and the newer NetWare interface is described at the end of the chapter. Another entity on a Novell network is the NetWare interface, more commonly known as the user interface. Actually, there are many interfaces that a workstation or server may use to transmit information to a network. These interfaces include a datagram interface, a virtual connection interface, a session interface, and a workstation shell interface. The subject of the following is to explain the workstation shell interface and the file server interface. It should be noted that any application may use any of these interfaces to transmit data on the network. In other words, an application running on a network workstation need not use the NetWare shell to talk on the network. Any application may call the IPX interface directly, therefore bypassing the shell, or may choose to call a session interface such as NetBIOS (email programs, SNA programs).

DOS. All DOS function calls in a NetWare environment are passed to DOS from the shell. The shell intercepts all calls from the workstation and determines whether the information is for the network or if it should pass the information to DOS. DOS does not know that the call was passed to it from the shell program and will process the call as if the call were made directly from the application program.

This method of intercepting the DOS calls before DOS gets a chance to read them started from the initial release of NetWare. In those days, most system calls (disk calls, screen calls, serial port calls, and parallel port calls) were made directly to the hardware, sometimes through a PROM known as ROMBIOS. Therefore, the shell interface intercepted these calls and then made the decision: Is it a network call or is it a local DOS function call? With these earlier releases of DOS, DOS had no idea that a network was installed and could not pass network calls to NetWare. So NetWare passed the calls to DOS.

Today, DOS is network-aware, and Microsoft provides for a simple way to make system calls. Most system calls are made through a software interrupt known as INT21. All DOS system calls are made using this interrupt. The shell still intercepts these calls just as it intercepted the previous hardware calls.

IPX. This software interface handles the transmission and reception of network data to and from the shell program. It envelopes the data handed to it from the shell and transmits them to the network. Upon reception of a packet from the network, it strips off the IPX envelope and passes the data to the NCP shell program or the server NCP program.

Network Interface. This interface may be in the form of a network interface card (Ethernet, Token Ring, or ARCnet). Data passed to it from IPX are enveloped with the particular network interface header information (network addressing and CRC error checking) and transmitted on the physical cable plant. This network interface also receives packets from the cable plant, strips off their headers, and passes the rest of the information to IPX for processing.

Application.   An application that is run on the workstation operates as a normal DOS application. Normally, the application is not network-aware (i.e., it does not know that it is operating on a network). An application is simply started on the workstation as if it is to run locally. Every call that is made from the application is intercepted by the NetWare shell to see if the call is made to a network resource. Special application programs such as a mail package or a 3270 SNA terminal emulator may also run on this workstation. These types of applications are usually written to a special interface called NetBIOS. All NetBIOS calls are translated into IPX calls for transmission on the network. NetBIOS over IPX has a packet type 20 (decimal).

A PC application on the workstation makes system (DOS) calls through an interrupt sequence known as INT21. These are DOS system calls to ask the workstation to provide some type of processing function (read a file, get a directory listing, delete a file). All INT21 calls are intercepted by the shell program to see if these calls are meant for a network device. If they are, the DOS INT21 function call is mapped to an NCP call and passed to IPX for transmission on the network. If the call is for a local device, the shell passes the call to DOS and waits for the next INT21 function call.

IPX data contain the data submitted by the application program and maybe control information from NCP.

The Workstation

NetWare clients operate on many different types of operating systems (OS/2, UNIX, DOS, and even SNA). The following text conforms to the DOS interface. A summary of the other operating environments is given at the end of the chapter.

There are two files that are executed in a DOS workstation that allow a NetWare system to communicate: NETx.com and IPX.com. These are not device drivers but are executable programs started from the command line interface (usually started from a start-up file like autoexec.bat on DOS PCs). For a workstation on a Novell network, the NetWare interface is contained in an executable program commonly named NETx.com (x signifies the major version of the current active DOS being used on the workstation). There is a version of the shell called

NETx.com (where x is literally part of the file name) that will automatically determine the operating system version and start up the correct shell for that version of DOS. This is available under NetWire on CompuServe.

Any data destined for a local device on the workstation (the disk drive or printer) are transferred to DOS by the shell for local processing. The shell maps the DOS request to a network-type request and passes this mapped information to IPX for delivery to the network.

In order for data to pass between the workstation and the file server, there must be some mechanism in the workstation to capture the data and format them for use on the network. This mechanism is known as the NetWare *shell.* Refer to Fig. 6-12. NETx.com is the shell program, and it is a terminate-and-stay resident (TSR) program in the workstation that intercepts all input (from the user at the workstation) as if it were the actual DOS command interpreter. In other words, the shell intercepts all information on the workstation to see if the data are for the network or if the data should be passed to DOS for local processing. Also, the NetWare shell is the interface between an application (WordPerfect, Lotus 1-2-3, etc.), running on a workstation and its associated file server. For the purposes of this chapter, input to the shell comes from two places: from an application or directly from the user.

As stated before, it is the NetWare MAP command that redirects local device identifiers (D:, E:, LPT1:, etc.) to the server, whether this device is a virtual disk driver or a printer queue. The MAP command also establishes a table of mappings. In this table is a listing of all the MAP redirect requests that have been issued successfully. This table contains a listing of the drive identifier and the mapped name of the server that it belongs to. For the printer, the table contains the local physical printer port identifier (LPT1:, etc.) and the name of the server and printer queue to which the mapped printer port was redirected. In this way, the shell interface knows whether the imputed DOS function call should go to the network or be handed to DOS for processing. If the call does not map to a virtual drive or printer queue in the table, it is handed to DOS.

For example, if you map the E: drive to the file server NOVA and type after the command prompt "DIR E:", the shell will intercept this and determine that you want a directory listing of the E: drive. The shell does a lookup in its table and determines that the E: drive is a mapped virtual drive to the file server NOVA. The shell formats a packet with the NCP command for the DIR command and submits this to IPX for delivery. Once the file server receives this packet, it processes the packet (does a directory command on the mapped portion of its hard disk) and returns the results to the workstation that requested the DIR (using IPX for delivery). The shell then prints the directory listing to the workstation's screen.

This action is taken whether an application submits a DOS function call or whether the user types in the function call directly to the command prompt.

These operations of the shell include establishing, maintaining, and disconnecting a connection from a file server, as well as file-related operations such as creating, reading, or deleting a file. The shell provides any file operations that you can perform on your workstation without a network.

Connection Establishment

Before any communications take place between a workstation and its associated file server, a connection must be established between the two. Table 6-7 shows how this is accomplished. The numbers under the call heading show in the sequence of events. The shell uses a combination of the processes of NCP, RIP, and SAP to establish a connection to a file server. There are two methods that the shell can use to connect to a file server: the shell program and the preferred server shell.

The Shell Program. When IPX.com and NETx.com (IPX is loaded first) are loaded, the workstation automatically attaches to the first server that responds to it. This provides for fast login service to the network.

TABLE 6-7

Initial Connection Sequence of the NetWare Shell

Call	Source	Destination	Protocol
1. Get nearest server	Client	Broadcast	SAP
2. Give nearest server	Router	Client	SAP
3. Get local target	Client	Broadcast	RIP
4. Give local target	Router	Client	RIP
5. Create connection	Client	File server	NCP
6. Request processed and connection number assigned	File server	Client	NCP
7. Propose packet size	Client	File server	NCP
8. Return maximum packet size	File server	Client	NCP

During this process, the workstation is able to identify its network number, if it does not already have it.

The response is an automatic connection to virtual drive seen as F: on the user's PC screen. This gives the user access to two other utilities: LOGIN.exe and ATTACH.exe. These applications allow the user to log in to a file server and attach to another server (not the login server). Login.exe provides user authentication. Without this, the user has just logged into the network but has yet to log into any server. The user may then use the ATTACH commands to reach a desired network server. Table 6-8 shows this sequence of events.

To accomplish this, the shell must find the file server's address and the nearest route to that server. To find the server's address, the shell invokes the services of SAP. This is called the Get Nearest Server request. All routers on the local network (the network that the workstation is attached to) should respond with the SAP information. The contents of this information are the nearest server's name, its full internetwork address, and the number of hops required to reach the server (how many routers are between the shell and its requested server).

Once the shell has received this information, it tries to find the best route to a server. The shell invokes the routing services of IPX for this request. This process is known as the Get Local Target request. When the response packet is received, IPX compares the network number returned through the Get Nearest Server request with its own network

TABLE 6-8   LOGIN.exe Sequence

1. Query bindery for preferred shell	Client	File server X	NCP
2. Address of preferred server	File server X	Client	NCP
3. Get local target	Client	Broadcast	RIP
4. Give local target	File server X router	Client	RIP
5. Create connection	Client	File server Y	NCP
6. Request processed and connection number assigned	File server Y	Client	NCP
7. Propose packet size	Client	File server Y	NCP
8. Return maximum packet size	File server Y	Client	NCP
9. Destroy service connection	Client	File server X	NCP

number. If the two numbers match, IPX informs the shell that the server is located on this local network and that it should send requests directly to the server on the local network.

But if the network numbers do not match, IPX submits a broadcast routing information request to the network through RIP request packet. The routers located on the local network respond with the known routes to this server. IPX finds the shortest route, discards all others, and returns the address of that router to the shell. The shell then uses the address and submits a Create Connection Request packet to the router (for a connection to the server, not to the router). These requests are illustrated in Table 6-8.

The IPX header contains the actual addresses of the file server that it wishes to communicate with. The packet is then transmitted on the network and accepted by the router. The router looks at the final destination address (destination network number) and routes the packet to that network.

Preferred Server Shell. A new shell, known as the *preferred server shell*, contains new features that enable the user to tell the shell which server it wishes to connect to. This may be either entered at the command line or located in a file called SHELL.CFG. As shown in Table 6-9, the first eight steps are still the same. The ninth step causes a lookup in the nearest server's database table to acquire the preferred server's address. Steps 11 and 12 might not be used if the preferred server is on the same local network (not separated by a router). The shell skips this and submits a connection request directly to the server.

One other difference between the old shell and the new shell is the process of Give Nearest Server responses. Previous shells accept the first response they receive and discard all other responses. The problem that arose with this is that a server will respond even if it has no free connections available for the shell. Shells are then not be able to establish a connection to that server. The preferred server shell accepts the first response, but it saves up to the next four responses it receives. They are used in case a connection cannot be made to the first response. If the first server response cannot be connected to, the shell uses the next response and tries to establish a connection to that server.

Logging In to a Server. Anytime after the initial connection is made, users may then log in to a file server. The utility LOGIN transmits the user's name and password to the file server for validation. LOGIN can also create a new connection to a file server, if prompted to on the com-

TABLE 6-9 Preferred Server Shell Connection Sequence

Call	Source	Destination	Protocol
1. Get nearest server	Client	Broadcast	SAP
2. Give nearest server	Router	Client	SAP
3. Get local target	Client	Broadcast	RIP
4. Give local target	Router	Client	RIP
5. Create connection	Client	File server X	NCP
6. Request processed and connection number assigned	File server X	Client	NCP
7. Propose packet size	Client	File server X	NCP
8. Return maximum packet size	File server X	Client	NCP
9. Query bindery for preferred shell	Client	File server X	NCP
10. Address of preferred server	File server X	Client	NCP
11. Get local target	Client	Broadcast	RIP
12. Give local target	File server X router	Client	RIP
13. Create connection	Client	File server Y	NCP
14. Request processed and connection number assigned	File server Y	Client	NCP
15. Propose packet size	Client	File server Y	NCP
16. Return maximum packet size	File server Y	Client	NCP
17. Destroy service connection	Client	File server X	NCP

mand line. Table 6-8 shows this. Note that steps 3 and 4 are not needed if the file server is located on the same local network.

During this connection process, the shell and the file server exchange a few packets. (This is commonly called *handshaking.*) The workstation requests of a file server that a connection number be assigned to the shell, and the two need to negotiate a maximum packet size that each will accept.

A workstation may attach to eight different servers at one time, no matter where the servers are located on the NetWare internet. This means that a workstation does not have a one-on-one relationship with just one server. A shell maintains a local table that has entries in it for all the connections that it has on the network. Each entry in the shell connection table (Fig. 6-13) contains the name and full internetwork address

Figure 6-13
Shell's connection
table.

Server's name	Full Internet address	Intermediate router's node address	Packet sequence number	Connection number	Receive time-out	Maximum time-out

of the server it is connected to. If the shell and its connected server are separated by a router, the address of that router is placed in the router's node address field. Next, the shell's connection number to that server is stored, along with the current sequence number. Finally, two time-outs are stored: receive time-out and max time-out.

The receive time-out is dynamically set and is the amount of time that a shell will wait for a response from a server. If this timer expires, the shell retransmits a packet to the server. Once set, the timer will change due to changing network conditions. Changing conditions may be a router going down, the file server becoming congested (experienced by the number of retransmissions count), or the file server response speeds up.

The max time-out value is preset at operating system initialization and represents the maximum value for timing out a connection.

Once the preceding connection sequence and a login have been established, the shell places all needed entries into a table. Using this table, the shell knows which packet is which and where a packet should be sent.

Again, note that the full internet address consists of the MAC address, the network number, and the socket on which the service is available. The combination of these numbers is called the *full internet address* or the *socket address*.

File Server Concepts

The Server. When a workstation requests a service of a file server, it builds a packet and submits the packet to the network for delivery to the server. Once the server receives the packet, it needs to know what operation to perform for the workstation. It could be a connection request, a data request, or maybe a print request. In any case, the server needs to know the operation.

In the packet received by the server is an NCP function code that the server will use to determine which operation is to be performed by the requesting packet.

Novell file servers provide many services as detailed below:

Accounting services keeps track of transactions that occur on the server and provides a detailed output of service usage.

AppleTalk filing protocol (AFP) provides the capability to allow both DOS and Apple files to reside on the server. These files can be accessed by both DOS and Apple personal computers.

Bindery services: A bindery is simply a database and is used in the server as a lookup table to provide information on resources to clients.

Communication services has many features, including providing an interface for application programs to extension services of asynchronous communication services.

Connection services have to do with the connection between a client workstation and its file server.

Directory and file service allow users to manage files on the file server.

Message service allows broadcasting of messages (up to 55 bytes in length) to other servers, clients, or both.

Print service allows a client to access the server for use of the printer services through the use of queues. Network workstations do not attach directly to a server's printer. They attach to a queue name, which, in turn, is used by the print service to access the server physical printer.

Value added processes (VAPs): For NetWare 2.1x, these are applications that are not Novell-native. They are external programs that run in a Novell server such as a database program or an electronic mail (email) system. In other words, they are applications that run external to the server's NCP, but act like a direct server service.

NetWare loadable modules (NLM): For NetWare 3.x, these are applications similar to VAPs. The applications access the NetWare operating system and act as a NetWare service. These applications are developed from third-party application developers. They are, in a sense, VAPs for NetWare 3.x.

A file server may accept many connections from remote workstations and applications, and the process of managing these connections is assisted by the use of a server connection table. As shown in Fig. 6-14, the file server maintains a table of connections and associated specifics to each connection.

Figure 6.14

File server connection table.

Connection number	Node address	Network address	Socket number	Sequence number	Watchdog count	Watchdog timer	NIC number	Intermediate router's address

Address Entries.   Address entries are used for supplying the MAC header addresses to response packets. They are also used for security. When the server is presented with a service request, the request contains the connection number of the requestor. The server matches the packet's full IPX internetwork address (see Table 6-10) with this connection and associated address in the table. If there is not a match, the server discards the request.

Sequence Number.   The sequence number is used for packet-level error checking. The data-link layer provides bit-level data integrity, and if there was any error in the packet during reception, the data link discards the packet. The problem here is that the data link does not inform the upper-layer software (the shell or IPX) that an error has occurred. A method is needed to ensure that all packets sent are received correctly and in the same order they were sent. This is the purpose of the sequence number, and the method is part of Novell's implementation of transport-layer software. Each single packet sent must be responded to by the file server. This is commonly called a packet window with a length of one (known as the *stop and wait*). There is a new protocol design called *packet burst mode*, which allows multiple packets to be sent by the shell, which requires only one acknowledgment. The workstation initiates this number and uses it to ensure that it has received a response to a particular request.

In NetWare, the sequence number is the responsibility of the shell. Each service request that a shell submits to the server contains a sequence number (an integer number). NCP for the server responds to this request to the workstation with the same sequence number. This request packet (received by the server) must contain the number that is one more than the previous packet; otherwise, the file server will think

TABLE 6-10

NCP Types

System Calls		
1111	Create service connection	
2222	Request (to the file server)	
3333	Reply (from the file server)	
5555	Destroy service connection	
7777	Request burst mode	
9999	Request being processed response (ACK with wait)	

that it is a duplicated packet. If the sequence number is 2 or higher, the server informs the workstation of the last known sequence number.

Error Detection and Correction.   When a client is communicating with a file server, there will be conditions when a file server may not respond to a client's requests. These conditions are depicted in Figs. 6-15 and 6-16. Three variations of this follow.

First, the packet could be lost either going to the server or coming from the server. In either case, the shell does not receive a response. (See Fig. 6-15.)

Figure 6-15
Packet timeouts and duplicate requests.

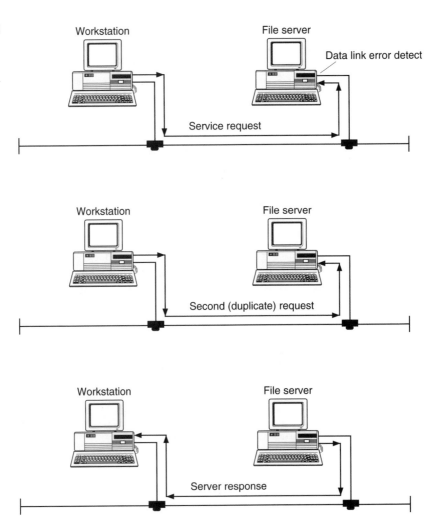

Figure 6-16
Queued packets on a
busy file server.

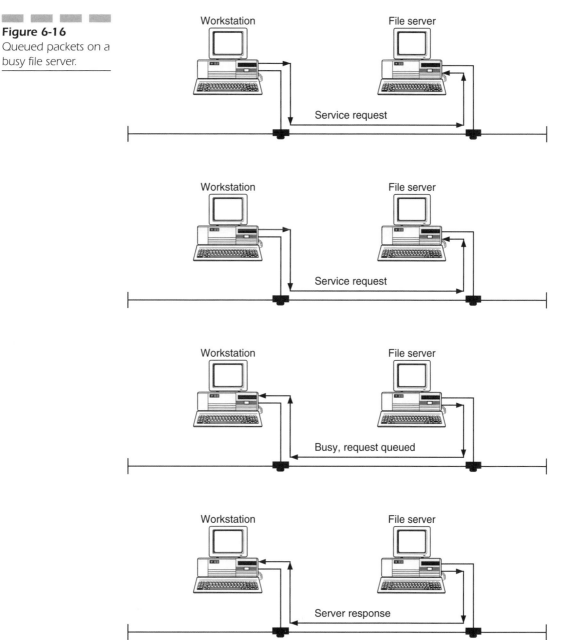

Second, the file server missed the packet (it was busy doing something else) or it has not yet finished processing the request. The common cause here is a database lookup on a large file. In any case, the shell will time out and retransmit the request. (See Fig. 6-16.)

Figure 6-15 shows a request being made of a file server where the data link (Ethernet card) finds an error in the packet. The packet is discarded, and the data link does not inform the upper-layer software of this error. The shell times out, adjusts its receive time-out value, and retransmits the request. This time, the packet is accepted and responded to by the server.

Figure 6-16 shows a file server that is too busy to respond to a request. The shell times out and retransmits the request. This time the file server was able to respond, but only with a busy response packet stating that it received the original packet and will respond later. The file server does this to stop any more retransmissions from the shell, enabling itself to complete the task at hand; when it is free, it will respond to the service request.

One other serious error may occur: a *complete failure*. What if the server is powered off, or the server's disk crashes? What happens when the same conditions occur on the workstation? These conditions are known and are easily handled by NetWare. If the shell does not receive a response to a request, it will time out and retransmit the request up to the max time-out value or IPX retry count.

If there is no response in this time period, the shell will display a message to the user. The user can ask the shell to retransmit the request or to abort the connection. If the connection is aborted, the shell will remove all entries for that connection in the shell's connection table. If there are no other connections in this table, the shell will automatically try to connect to the nearest server using the sequence explained previously. If no servers respond, the shell will display the following message to the user: *You are not connected to any file servers. Current drive no longer valid.*

Since unused connections require additional RAM and processing power, it is beneficial to clear all unused connections on the server. The watchdog timer is a process that provides this. If the workstation fails, the server will still have a connection for that station in its table. If, in a certain period of time, the server does not hear from that workstation, the server invokes a process known as the *watchdog process*. There are two entities in this process: the watchdog timer and the watchdog counter. There is one timer for each of the file server's connections. This process is active all the time the server is active.

If the server has not heard from client connection for a period of five minutes (the initial watchdog timer has expired), the server sends a poll packet to that client. A poll packet is simply a packet asking the client shell to respond to the request. (For those who know AppleTalk, this is known as a *tickle packet*.) For the TCP/IP protocol, a client or server can send a packet with a known bad sequence number to see if the other side responds. The other side of the connection, if alive, should respond with the next sequence number expected.

If the client responds to that packet, the watchdog timer is reset for that connection. If the poll is not responded to, the watchdog process resets the timer to 0 and increments the watchdog counter from 0 to 1. It then submits a poll every minute. If the client shell does not respond, the watchdog process increments the counter by 1 until it has reached the count of 10 (indicating that it has not heard from the client shell for 15 minutes—the original 5 minutes plus the poll count). At this point, that connection is cleared from the server's connection table. If the client responds, all watchdog timers and counters are reset and the process starts from step 1. Current versions of Novell 3.1x support adjustable watchdog timers.

Server System Calls

The protocol that a workstation and a server use to communicate with one another is called the NetWare Core Protocol, or NCP. This primarily runs at the session through the application layers of the OSI model. It is a language that the workstation and the server use to communicate.

This communication enables the workstation and the server to understand what each side of the connection needs. NCP is a proprietary protocol, and the specification has not been released by Novell.

The two main messages that NCP provides are request and reply messages. A request is usually made by the workstation and the reply comes from the server. NCP builds its own header into the packet, as is shown in Fig. 6-17a and b.

There are literally hundreds of requests that a workstation may ask of the server, but the basic requests are usually file reads and writes and printer queue requests. The request-field functions are shown in Table 6-10.

When a workstation requests a connection to a server, it uses the Create Service Connection request, which is type 1111. When the worksta-

Figure 6-17
(a) Client NCP request packet format. (*Courtesy of Novell, Inc.*)

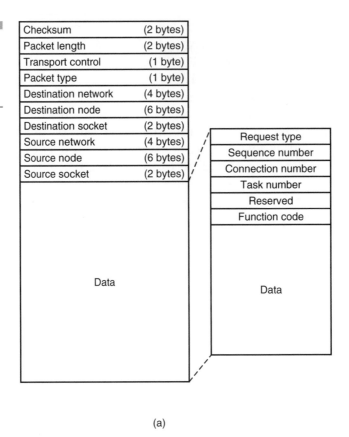

Checksum	(2 bytes)
Packet length	(2 bytes)
Transport control	(1 byte)
Packet type	(1 byte)
Destination network	(4 bytes)
Destination node	(6 bytes)
Destination socket	(2 bytes)
Source network	(4 bytes)
Source node	(6 bytes)
Source socket	(2 bytes)

Data

Request type
Sequence number
Connection number
Task number
Reserved
Function code

Data

(a)

tion no longer needs this connection to the server, it issues a Destroy Service Connection request (5555).

Refer to Fig. 6-17a. As stated before, once a connection has been established between the server and the workstation (known as the *client*), there are literally hundreds of requests that the workstation may ask of the server. Instead of providing a separate request function for each request, Novell provides a general request function (2222) for the workstation and then further down in the header places the specific request function, called the *function code*. The function code could be a 76 to tell the file server to open a file. The file to open would be found further inside the packet. The function code could be a 66 to tell the file server to close a file. The workstation shell program (NCP) creates an NCP header based on the information that it intercepted at the workstation. For example, if the user is in a WordPerfect file and requests WordPerfect to open a file that is located on a server, NCP (the shell) intercepts this call and builds an NCP header with function 2222 and function

Figure 6-17
(b) Server NCP reply
packet format. (Courtesy of Novell, Inc.)

Checksum	(2 bytes)
Packet length	(2 bytes)
Transport control	(1 byte)
Packet type	(1 byte)
Destination network	(4 bytes)
Destination node	(6 bytes)
Destination socket	(2 bytes)
Source network	(4 bytes)
Source node	(6 bytes)
Source socket	(2 bytes)

Request type
Sequence number
Connection number
Task number
Reserved
Completion code
Connection status

Data

Data

(b)

code of 76. The name of the file would be further down in the packet. The workstation then requests IPX to send this packet to the server.

The connection number contains the service connection number that was assigned by the server and given to the workstation during the connection login attempt between the workstation and the file server.

The task number indicates to the server which client task is making a request. The server uses the task number to deallocate resources when a task is completed.

The NCP reply packet is shown in Fig. 6-17 b. The reply type can be:

1. 3333—Service reply

2. 7777—Burst mode connection

3. 9999—Request being processed

The only difference between the reply header and the request header is the addition of the completion code and the connection status fields. This allows the server to inform the workstation as to whether the previous request was successfully completed. This is indicated by the field-of-completion code. A completion code of 0 indicates a successful completion. Any other value indicates it was not completed.

The connection status flag is used by the workstation to indicate the status of its connection to the file server. Every incoming packet received by the workstation has this field. This packet is used by the server when the command DOWN is given to the file server. The DOWN command is used to gracefully bring the file server down. A workstation that receives this packet will know that the file server is no longer up.

When burst mode is being used to transfer files between the workstation and the server, the reply code of 7777 is used.

NCP transfers information back and forth between the workstation and its associated server using these headers. The connection is established, maintained, and destroyed by means of the NCP header.

Finally, Fig. 6-18 shows a fully integrated Novell NetWare internetwork.

NetWare Supplementals

Multiple NetWare Protocols

IPX is not the only protocol over which NetWare currently runs. With the advent NetWare 3.1x, Novell has allowed access to its system using the protocol suite of TCP/IP. At first, access was allowed with the Network File System (NFS). NFS was written by SUN Microsystems to allow UNIX machines to share disks on a network. This allows NetWare workstations not only to have access to NetWare servers but also to mount UNIX disk drives. Next, Novell allowed IPX to be encapsulated in a TCP/IP packet so that it may be routed on TCP/IP network. However, Novell is recommending that sites upgrade to NetWare 4.x which allows IntraNetWare to operate. IntraNetWare is a more robust implementation of TCP/IP for a NetWare network. However, if sites remain on 3.1x, then running dual protocols stacks in the workstations must do.

Finally, Novell has announced direct support for TCP/IP. Currently in beta testing, this will allow NCP or NetWare to use TCP/IP directly as a transport protocol, and not use IPX.

UNIX machines may be accessed through their TCP/IP user interface, known as LAN Workplace for DOS. This allows users to run TCP/IP

Figure 6-18
Fully integrated Novell NetWare internetwork.

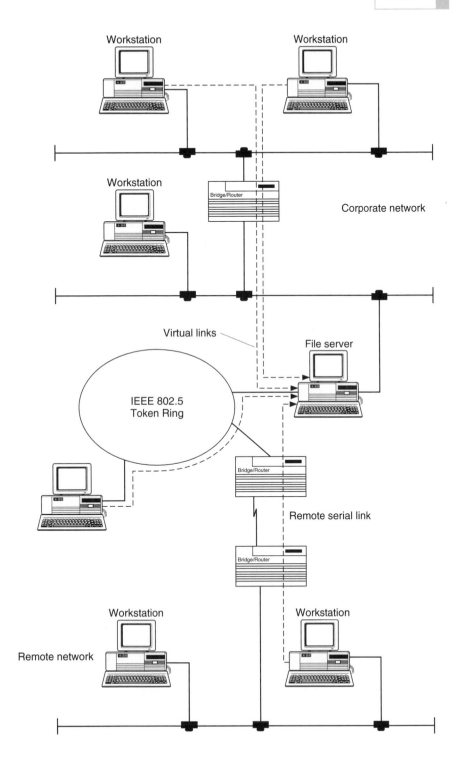

applications on a NetWare workstation, giving access to UNIX machines. Finally, there is NetWare NFS, explained previously.

LAN Workplace for DOS 4.0 uses a data-link library interface known as Open Data Link Interface. This interface is a library of routines that allows multiple protocols to reside on a workstation. In essence, these protocols are terminate-and-stay resident protocols that can be brought up and down at the user's discretion. This means that IPX can load as the transport for NetWare commands and data. At the same time, TCP/IP can be loaded so that connections may be made to both a Net-Ware file server and a TCP/IP host. ODI can multiplex both protocols on the same data link.

IntraNetWare

IntraNetWare is Novell's answer to providing a full range of TCP/IP services to corporate intranets. Novell is trying to head off a massive conversion from NetWare to applications based on TCP/IP, which includes not only Microsoft NT but also Java-based applications that are being used directly with Web browsers. The IntraNetWare intranet platform enables you to integrate your existing network with intranet and Internet technologies. You can integrate IntraNetWare with almost any other Novell product to enhance its functionality. For example, NetWare Web Server runs on NetWare 4 and on IntraNetWare server.

IntraNetWare features NetWare Web Server, which is a set of NLMs that enable you to create a Web server on your Novell server 4.x. This is a full-function Web server that supports PERL, CGI, and Novell's scripting language, which is based on Visual Basic. It also includes Netscape Navigator Internet browser, and File Transfer Protocol services for NetWare. The FTP services run on the server and enable users to use the FTP client protocol to access files on a IntraNetWare server (access rights are enabled).

Novell's IPX/IP gateway is an application that allows for your NetWare network to remain intact, and there is one node that acts as a NetWare to TCP/IP gateway. Instead of all the NetWare workstations upgrading to TCP/IP, one node on the network can run the TCP/IP protocol and allow for one IP address to be used. The users connect to this gateway, which performs the IP/IPX translation allowing users to gain access to the Internet or intranet-based Web servers. This function translates between IP and IPX so that your current NetWare network can stay intact and the browser on the user's workstation can communicate with the gateway to provide for Internet access.

Also, included is the IntraNetWare Print Services for UNIX. In addition, IntraNetWare offers security through authentication and auditing capabilities, and symmetric multiprocessing. It also includes Novell Directory Services (NDS) and provides for directory, security, multiprotocol routing, messaging, management, Web publishing, and file and print services.

IntraNetWare supports clients on multiple platforms, including Windows NT, Windows 95, Windows 3.1, UNIX, OS/2, Mac OS, and DOS clients.

IntraNetWare is used on both servers and clients. However, the functions are vast, and not all clients and servers have to be upgraded in order for IntraNetWare to work.

Packet Burst Mode Technology

Currently, each packet transmitted on a NetWare connection must be acknowledged. The next packet may not be transmitted without the previous packet being acknowledged. This type of system is still full-duplex connection, but the NetWare implementation is requested with a response type of protocol. This makes NetWare a very slow protocol when large file transfers are to take place.

The NetWare operating system of NCP has now been modified to allow a requestor to transmit a single request for up to a 64K segment of data. The file server places the data into a packet (according the maximum data size for the medium of transmission) and then replies with all packets to the requestor without a single reply packet being issued until the final packet has been transmitted. In some instances, this has improved response time by 300 to 400 percent.

This process is known as *providing a data window.* The window is how many packets may be transmitted without an acknowledgment. Packet burst mode takes this a couple of steps further. NCP provides a sliding window with a theoretical maximum of 125 (512-byte) packets before a reply packet needs to be generated.

The second process enacted by this protocol is the between-packet delay timer. This allows a window of packets to be transmitted with a settable delay between the packets. This allows for the receiver to free up buffer space to hold more information, and also prevents a station from hogging the network while it transmits packets back-to-back.

Determination of the window size and the packet delay timer is autonomous to each client on the network. Each workstation's shell configures these parameters based on the clock speed of the workstation, the amount of free RAM associated with the receiver's communication buffers, and the latency time between two communication workstations. Certain conditions—such as received bad packets and missed packets—are monitored while two network workstations are communicating. As these conditions improve, the parameters may change. This means that the values are dynamic in that they can change automatically depending on the conditions on the network.

One last feature of this new shell is that only the packets that are missed are retransmitted. This means that if you transfer a small file and it takes 200 packets to transmit it and all of the packets but 40 and 85 were received, the shell can request the server to resend packets 40 and 85 and not have the server retransmit the whole transmission.

In effect, all packets are still assigned a sequence number. These sequence numbers start at the connection negotiation between the client and the file server (NCP 7777 request). Each packet sent is given a sequence number, and the receiver monitors this. When a packet is received out of sequence, the shell flags this and asks for a retransmission of that packet indicated by the missing sequence number.

There is also less traffic on the internetwork. Since multiple packets may be sent without any acknowledgment (as indicated by the window size), there are that many fewer packets on the network.

Speed is also affected, especially when the packets have to traverse multiple routers. With the previous shell, each packet transmitted had to be acknowledged. This means that if the connection is across multiple routers, the acknowledgment must traverse the same routers. This can be a slow process.

The new shell (version 3.26) is available for NetWare 3.11 servers and their workstations. NetWare 3.12 and 4.x and VLMs enable this automatically. This new mode of operation still allows the shell to communicate with nonpacket burst mode servers while at the same time communicating with packet burst mode servers.

Why invent a new protocol? Why not use the SPX protocol? It has a technology similar to burst mode built in, and it is an existing NetWare protocol. The problem with SPX is that it achieves its efficiencies at the expense of NetWare's LAN speed. SPX is a windowing protocol that achieves efficiencies when a WAN is involved. SPX is a tried window protocol. A window protocol allows $n+x$ packets to be sent, where n is the next packet in line to be sent, and x is the window size, meaning the number of packets (in a stream of frames to be sent) that are allowed to be sent with the next transmission. In other words, if 100 packets are awaiting transmission and the window size is 12, then 12 packets may be sent in a single burst without an acknowledgment.

With SPX, once an acknowledgment returns from the server saying that it received one more than the number of packets sent (an ACK is always one more than the number received; it indicates the sequence number of the next packet to be received), the window can slide over the next 12 packets to be sent. The problem with this type of protocol is that, if the window size is 100, 100 packets are sent, and if frame 52 is damaged, then packets 52 through 100 must be resent. This is inefficient. Also, the protocol does not contain a sliding window. The size of the window is preset and cannot change as congestion on the LAN changes.

With burst mode only the damaged packet is resent, and then when it is received it is reinserted into the original stream at the destination end. From there it gets handed to the application. Also, with burst mode, the window size increases with the number of good transmissions and decreases with a number of bad transmissions.

Burst mode uses the concepts provided in other windowing protocols, but it takes them a couple of steps further. Burst mode operates at the application level, allowing it to know the transaction's contents. Burst mode bypasses the SPX protocol altogether. It talks directly to the IPX protocol. Burst mode does require a new packet format, because it places its headers into the packet.

Burst mode allows a client to issue a single read or write request for blocks of data up to 64K. Therefore, burst mode provides a window with a theoretical maximum of up to 128 packets per window (64K divided by 512-byte packets, the minimum frame size) before a reply packet needs to be generated. Remember that that is only the theory! The window is much smaller than this. The window is simply the number of packets to be transmitted at a time, requiring only one acknowledgment.

Besides the window size, another process enacted by this protocol is the between-packet delay timer. This allows a window of packets to be transmitted, but there is a settable delay between the transmission of packets. This allows for the receiving station to free up buffer space to hold more information, and also prevents a station from hogging the network while it transmits packets back-to-back.

Furthermore, a throttling technique is involved as well. As each window is transmitted, the burst mode protocol can dynamically increase the window size. With each successful transmission of a window, the protocol increases the size by 100 bytes. Conversely, the protocol reduces the window size upon any failure (any event that requires retransmission of the packet).

Inside the burst mode header are the details that enable the receiver to determine where each packet resides in the stream of packets transmitted. In this way, when a packet has to be transmitted and the sender retransmits it, the receiver can properly place the retransmitted packet into the stream. In this way, what was sent will be what is received.

Burst mode does require more memory in the client. It usually needs between four and five kilobytes just for implementing burst mode, and then you need to add more memory depending on how many buffers you have set aside for burst mode. The size of the buffers may vary, because they depend on the medium's transmission size (Ethernet versus Token Ring, for example). With the BNETX (burst mode enable shell), the maximum number of buffers allowed is 10. Using a VLM, there is no limit, although the default is set to 16 packets per window for burst reads and 10 packets per window for burst writes.

Operation. Upon connection setup, the workstation queries the server to see if the server can provide burst mode service. If the server does not reply with the expected response, the client assumes that the server cannot provide burst mode operations and assumes the normal method of NCP operations (non-burst mode). Servers can have connections with clients that use burst mode and with other clients that do not use burst mode.

After a connection has been set up, the packet size has been negotiated between the client and the server, and the client has determined that the number of packet buffers specified in the NET.CFG file on the client can be satisfied, transactions will occur between the client and the server. For example, let's assume that the client placed a write request to the file server. The packet size negotiated between the server and the client is set to 1024 bytes. The number of bytes to write is 64K bytes. Therefore, assuming defaults (10 is the window size), the client will set up six windows with 10 packets of 1K each and one window with 4K packets. The client will transfer each window and then wait for a reply. Here the client transferred 64K of data and only needed four replies for the complete transfer. If the client wanted to read 64K from the server, it would set up four windows (16 is the default) with 16 packets each. Again, this can change depending on the ability to use larger packets. For example, fewer windows would be used with Token Ring if the negotiated packet size is 4K.

NetWare Link Services Protocol

No, this is not NetWare Link State Protocol. Prior to reading this section, you should read the routing protocol section, especially if you read the OSPF section in the TCP/IP chapter. NetWare Link Services Protocol (NLSP) was designed to overcome the limitations of the RIP and SAP protocols. NLSP is classified as a link state protocol, whereas RIP and SAP are considered distance vector protocols. The advantages of link state protocols are many, and they are described at the end of this section.

Link state protocols have been around for a while, but the horsepower to run them has not been. Link state protocols are CPU intensive, albeit for short amounts of time. NLSP could have been employed much earlier, but the CPU power to run the protocol was simply not ready. Today, there are CPUs that make this technology viable.

NLSP is designed to function only with routers. NLSP does not require any change in operation from the workstation level. It is a routing protocol change. Basically, the workstation does not know what routing protocol the router is running. Therefore, NLSP can be installed without notification to the workstations.

NLSP allows for large NetWare internetwork to be designed and implemented, and it is not constrained by the limitations of the RIP protocol. It is designed as a hierarchical topology. In doing this, NLSP divides a NetWare internetwork into three times:

1. Local network segments within an area

2. Areas, which are collections of routers within the internetwork

3. Domains, which are parts of the routing hierarchy

In order to accommodate this, NLSP provides for three levels of routing:

1. Level 1 routers, which route data within an area

2. Level 2 routers, which route data between areas

3. Level 3 routers, which route data between domains

Figure 6-19 depicts this.

A level 2 router also assumes the role of the level 1 router. Likewise, a level 3 router assumes the role of both a level 2 and a level 1 router. There are many local network numbers within an area, and there can be many areas within a domain.

Dividing the internetwork up like this has many advantages. For example, a level 1 router must store link information about every link in

Figure 6-19
NLSP hierarchy.

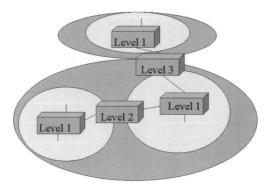

its area. But it does not store link information about links in other areas. To route data to another area, it simply has to know about its nearest level 2 router. For data packets destined for another area, a level 2 router simply hands the packet off to the level 2 router for further forwarding.

Likewise, level 2 routers only exchange level 2 (area) information between them. They do not exchange level 1 information between them. Similarly, level 3 routers only exchange domain information between them.

NLSP Technical Details. NLSP is known as a link state protocol. Link state protocols were developed to address the demands and complications of large internetworks. As shown in previous sections on the RIP protocol, information about networks travels slowly, and every router is dependent on every other router for correct information. If any router makes a mistake in the computation of its routing table, this mistake is propagated throughout the internet. The RIP protocol is also based on point-to-point links with other routers. Simply, it is a two-way conversation.

NLSP is called a link state protocol because the router maintains the status of very link in its area. A link can be a router port (LAN or WAN port). Link state protocols scale better than distance vector protocols (larger networks can be built reliably) and adapt easily to topology changes (e.g., losing a link and having to update the other routers). In general, upon initialization, a router running the link state protocol forms adjacencies with other routers on their LAN segment. The routers exchange information about their directly connected links. Furthermore, the routers exchange information about the other adjacencies that they know about. A router floods information about itself and its adjacent neighbors to every other router in an area (an area is simply a subnet of routers in an internetwork). Once the information is obtained, an algorithm in run on each router to determine the best route to any other network on the internet.

Each router determines its own routing table based on information it gathers from the network. With NLSP, router updates are made only when there is a change. If no changes occur on a network, the router remains stable with the information in its table. If it receives information about a change in the status of a link, it alters its table. RIP and SAP update on a periodic basis, typically every 60 seconds, even when there is not a change on the network. This requires each router to accomplish validity checks on its table every 60 seconds. If the network grows, the routing and SAP tables grow and require 60-second checks as long as the router is active. This is very inefficient for the router, for the bandwidth on the LAN, and for the serial links as well. When there are no changes, NLSP requires that a complete link state update occur every 2 hours.

NLSP does not depend on a hop count to determine entries in the routing table. Each link is assigned a cost, represented by a positive integer number. This 16-bit number allows costs on the link to range from 1 to 65,535. These costs are very important. They are used in the decision algorithm to determine the best path to a destination. The costs are assigned when the router is configured. The sum of the costs to a destination network will determine the shortest path to the final network. The lower the cost, the better the path. Higher-speed paths are assigned to lower costs, and lower-speed paths have higher costs. NLSP is based on an algorithm known as the Dijkstra algorithm (after Edsger W. Dijkstra, who devised the algorithm used in the forwarding method).

Link State Databases. NLSP forms a series of databases. Information about the state of each link of each router in each area is exchanged between routers using the link state packet.

The adjacency database keeps track of the router's immediate neighbors (routers on the same segment) and the information about its own link. The routers exchange messages called "hello" messages. These maintain the database. Routers send hello messages to find other routers on directly attached segments. Sending these hello packets allows for routers to discover the identity of other level 1 routers on directly attached segments. Sending these hello packets allows for routers on the same LAN to discover the identity of other level 1 routers on the same LAN. The router also listens for hello messages from other routers to build the adjacency database. Any changes in the state of the hello messages are reflected in this database. The adjacency database is simply a subset of the link state database. Routers transfer information about their links to other routers that are known from in the adjacency database.

When a router or a link of a router is initialized, the network numbers configured on its ports are put into the link state database. Using information in the link state packet, each router in an area that receives this information is able to build its own map of that area. This map is the link state database. The information stored in this map records the area's routers and server, the links that connect them, and other information. The map contains information about all the other routers and their links in the area. The link state database does not contain any information on end nodes. Every router in an area should contain the same information in its link state database that its neighbor does. NLSP provides for this synchronization. Flooding is the means by which the link state database in all routers synchronize. Each router sends information from its adjacency database to each of its neighbors.

This information is sent via the link state. When a new link state packet arrives at a router, two events happen. First, that link state packet is retransmitted on all links of that router except those on which the packet was received. Second, the packet is merged into the router's link state database. Therefore, the link state database is nothing more than a collection of link state packets obtained from the routers' own adjacency database and received from other routers through the flooding procedure. When a link state changes, each router detecting this event floods a link state packet indicating that change. The receiving routers mark the link as down but do not remove it.

Another method of determining that a link state has changed is the router's own timer for a link state packet. As each entry is made in the table, a countdown timer is applied to that entry. If a new link state packet arrives to update that entry, the timer resets and starts to count down. If no link state packet arrives to update that entry, the timer expires, that entry is purged, and that router floods a link state packet to the area to indicate this. All other routers pick up this change and recalculate their tables.

The final database is called the forwarding database or forwarding table. This is where a network number entry and it associated cost are kept. Unlike the forwarding database, each router contains it own unique forwarding table. A forwarding table is simply a table that maps a network number into the next hop. This is the same as the routing table in a RIP network. The decision process builds or rebuilds the forwarding table. The decision process is based on the Dijkstra algorithm. The algorithm is beyond the scope of this book. Suffice it to say that this algorithm determines the shortest path to every other network in an area. It is very complex and requires large resources (CPU and memory) to compute the forwarding table. A router uses this table to determine the next hop for a received packet.

One of the benefits of using this algorithm is that a change in one area will not cause the algorithm in another area. In other words, only in the area where the link state change occurred will cause the algorithm to run. This outcome requires all routers to flood information throughout the area in order to synchronize the link state databases in all the routers there. Other areas are not affected, and merely receive the computed changes through level 2 routers. In a stable environment, this update should not occur often, and when it does its very short in duration (taking less than 2 seconds on average for an area to converge). Compare this with up to 15 minutes for a RIP network to reconverge following a change. This is the worst-case scenario for RIP, but it could occur.

Novell's NLSP protocol can coexist with a network that is currently using RIP and SAP protocols. Therefore, a router that is running the NLSP protocol can also process RIP packets that it might receive, and in return it can broadcast RIP packets. There are three modes of operation for a router:

1. NLSP only

2. RIP only

3. Auto

Auto means that the router should run with the NLSP protocol, but when a RIP packet is received, the algorithms used to process a RIP packet are started and maintained. AN NLSP router will be able to respond to RIP requests and responses on a network.

NLSP is available for servers running NetWare 3.11 and above. NetWare 2.x servers must be upgraded to at least NetWare 3.11 before they can implement NLSP. NLSP cannot be run on the older Novell router software known as ROUTEBEN. These devices must upgrade to Novell's multiprotocol router software before implementing NLSP.

NLSP Addressing.   NLSP still uses the 32-bit network number as it pertains to IPX. This was done to maintain packet format compatibility for the vast installations of NetWare. However, the 32-bit address is used differently than before; the concept of masking has been introduced. The 32-bit network number is now split to allow for an area ID and a local network number to be assigned.

A typical 32-bit address could resemble this one: 2222C098. What is different is that a 32-bit mask must be applied to this network number. The mask for this address could be FFFFFF00. Applying this type of mask yields an area address and a local network address. In this case, the network address is 2222C0. The last two digits represent the local network address. A mask shows how much of the network number should be used for the area, and how much of the address should be used for the local network number. The above mask allows for 254 local network numbers to be assigned to this area (all 0s or all 1s are not allowed in the network number). The router maintains a listing of network IDs and area IDs in its routing table.

Packets received with a destination area ID that matches the router's area ID can be routed through L1 (level 1) routers. When a packet is received that contains a destination real ID different than the routers, the router must find a level 2 router to route the packet to. The level 2 router simply hands the packet off to the level 2 router without any network database lookup. The receiving level 2 router routes the packet to the appropriate router containing the area ID, and then the packet is routed to its local network ID using L1 services.

Assigning your own network numbers is still allowed. However, Novell is requesting that NetWare installations register their networks with the Novell network registry. Although this is not required, the purpose of this program is to ensure that no two sites have the same network number range assignment. Why is this important? This requirement is important when two Novell NetWare networks merge into one network (company mergers, etc.). Sites having similar network numbers cannot interoperate. This would be confusing to the routers and the servers. Similarly, sites cannot have server names and other objects named the same.

There is no hard and fast rule, but Novell recommends 400 networks to an area.

NLSP Advantages.   Link state protocols offer many advantages over distance vector protocols. Besides the obvious advantages of using MAC multicast addresses for updates and less CPU utilization because updates are only made during a change, there are many other advantages to using the NLSP protocol. NLSP maintains the original IPX packet format for workstation and servers to communicate over a LAN. NLSP does not affect the normal client/server operation of NetWare. Workstations and servers do not register themselves with the routers. Only the routers will perform new functions, and they only talk to other routers (including servers).

Unlike the RIP protocol, which simply stores the next hop address for a forwarded packet based on information that was previously computed by another router, NLSP mains an entire map of the network based on unmodified information given to it by another router. NLSP allows each router to build its own map of the network. This map allows it to make a more intelligent decision about which path may be the best. Each router determines its own entries in the routing table, and is not dependent on other routers tables to build its tables. Best paths are calculated based on a true cost number, not just a simple router hop. After the initial exchange of information between neighbor routers, routers further communicate with other routers only if there is a links state change in their area. This is unlike the RIP protocol, which periodically transmits its routing table whether there is a change or not. NLSP uses a special multicast MAC address when initializing or updating. This is important, because only those stations that are assigned the same MAC multicast address will process the received packet. All other stations will discard it. In other words, it is filtered at the NIC before even reaching the IPX stack in the computer. The RIP/SAP protocol uses broadcast addresses. In the best case, NLSP must do a complete topology update every 2 hours even when there are no changes at all.

NLSP not only works on routing tables, it also works with SAP broadcasts. The same rules apply to the SAP protocol. NLSP only advertises a service change when there is one. Otherwise, the SAP protocol does not broadcast its table periodically. NLSP also allows for more than seven services to be advertised within one update. Where all this has significance is over WAN links. These are generally lower-speed serial links. Conserving bandwidth on these links is very beneficial, especially when the WAN link is X.25 or Frame Relay.

Along with this, NLSP also uses IPX header compression when used across a WAN link. This type of compression technique allows a packet header to be compressed. It uses a modified version of the Van Jacobson version (RFC 1144). IPX compression is specified in RFC 1533: Compressing IPX headers of WAN media. This technique can allow for up to 30 bytes (the standard IPX header length) to be compressed into one byte in the best case or seven bytes in the worst case. The IPX compression protocol can also compress NCP request and reply headers of 37 bytes to between 2 and 8 bytes. The variance is due to whether the packet is using the CRC checksum and whether the data length can be determined at the MAC level.

NLSP supports load splitting. If there are parallel paths to a destination and the cost of these paths are the same, NLSP will split the traffic load over these parallel paths (load balancing). This is an option, and you can tell NLSP to not use it.

NLSP is better suited for scaling up to larger internetworks. Using RIP, the diameter of the network is limited to 15 routers (between two communicating stations). In other words, a packet cannot traverse more than 15 routers before being discarded. NLSP supports network diameter of up to 127 routers (an unrealistic number).

Finally, one of the biggest advantages of NLSP is cost assignment. Each link of a router is assigned a cost. Although IPX RIP allowed for this as well, it was linked to the 15-hop diameter. Assigning a higher cost, using RIP, to a link effectively reduced the diameter of the network.

For example, in Fig. 6-20 we have a simple network topology. The link between router A and router B has a WAN link speed of T1. Router B and router C are linked together also using a T1. The link between router C and router A has a WAN link speed of 56 kbps. Assuming a link cost of 1 for each link, a network station attached to router C that is conversing with a network station on router A will automatically take the 56-kbps link, because it has the shorter cost (hop count). Otherwise, it would take router C to router B to router A. RIP would not allow for this because it is one more hop than the link between router C and router A. Here is where you could artificially assign a RIP cost on the link between router C and router A. We could bump this up to 3 on each side. This would force RIP to take the path of the T1s and not the 56 kbps. But again, this would reduce the diameter for the rest of the network.

NLSP assigns a true cost to every link, but not one based on a simple hop count value. Using the same network design as before, the router C to router A link could be assigned a cost of 500, while the two T1 links could be assigned a cost of 100 each. In each router's table would list the paths of the two T1 links as the shortest paths for node A and node B networks.

Figure 6-20
NLSP cost.

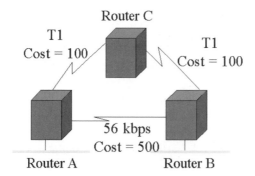

The DOS Requester and Associated VLMs

With the introduction of NetWare 4.x, Novell introduced a new client system called the DOS Requester. It is essentially a group of individual modules that perform the same functions as NETx.com but provide a much richer set of functions as well. It works with DOS version 3.1 and higher. Figure 6-21 shows the old and new methods of client software. VLM stands for *virtual loadable modules*. The shell was simply broken up into individual modules that can be loaded if those functions provided by that module are needed in the client. This is all accomplished through the file NET.CFG, which contains a listing of the modules to load.

The older type of client software was called the "shell hell" because it came in a variety of different versions: NETx.com, to allow it to be loaded into the upper memory block, and XMSNETX.exe and EMSNETX.exe, which allow it to run in expanded or extended memory (memory managed above the 1 MB DOS barrier). All of the shells had one thing in common: they all ran in front of DOS. That is, they intercepted all calls made by an application to see if the call was being made for service provided by the network. If it was not, the shell would hand the call of to DOS for local processing.

With all the new features that were being thought about and released with NetWare, the shell program had to grow along with them. For example, with the new directory services provided in NetWare 4, the shell had to be modified in order to function with it. To provide for a new client software interface and to limit the amount of memory these new features would require, Novell decided to make the client interface

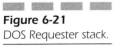

Figure 6-21
DOS Requester stack.

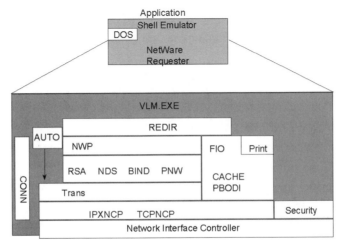

more modular. This allowed for it to be easily maintained and expanded, and to utilize DOS memory efficiently. What it did was wrap all the features of the previous shells up into one new interface, called the DOS Requester VLM. The VLMs are described below:

CONN.VLM. This is the Connection Table Manager. It spans all three layers of the requester, and provides for the maintenance of the connection table (client to server connections). It allows a client to connect up to 50 different servers, whereas the shell provided for up to eight different server connections.

IPXNCP.VLM. This VLM provides for the building of a packet with an NCP header and giving this information to the IPX protocol for transmission. It is not a replacement for the IPX protocol; it merely assists that protocol.

TRAN.VLM. This is the Transport Protocol Multiplexor. IT provides for the demultiplexing of protocol types. NetWare is able to run over several types of protocols (IPX/NCP and TCP/IP/NCP). It is this VLM that deciphers which type is in use and processes it accordingly.

SECURITY.VLM. This VLM provides security for NCP session protection.

AUTO.VLM. This is the auto-reconnect VLM. It provides for reconnection to a server after a session loss. This may happen due to a server being DOWNed by the LAN management, or crashing for various reasons. This module reconnects to the server and rebuilds the user environment with that server, including the connections status, drive mappings, and printer connections. Open files may be restored as well.

NDS.VLM and BIND.VLM. These provide NetWare's directory services and bindery services. These VLMs provide the requester with the ability to work the NetWare 4 directory services or the older (but more used) bindery.

FIO.VLM. The file input/output VLM is used for accessing files on the network. This VLM includes file cache, packet burst, and large internet packet (LIP or the sending of packets larger than 512 bytes over routers) functions.

PRINT.VLM. This VLM handles the print redirection services for the requester for NDS or the bindery.

NWP.VLM. This is the NetWare Protocol VLM; it coordinates requests to the correct module. It handles logins and logouts, and allows for connections to the correct services.

GENERAL.VLM. This module performs general housekeeping types of duties. These include creating and deleting search drive mappings, getting connection information, print and server queue information, and machine names.

REDIR.VLM. This provides the DOS redirect portion of the DOS requester. It provides the redirector interface to DOS through DOS's INT 2F redirection capabilities, and it works with DOS internal tables. It performs services for applications that DOS cannot perform (such as retrieving information from a network server). It works with DOS to eliminate the redundancy provided by previous versions of DOS and NetWare (i.e., providing two similar drive mappings).

VLM.EXE. This provides for VLM management. It manages all requests and replies made to the requester. It oversees where the VLMs are loaded in memory, and the order they are initially loaded in.

NETX.VLM. This VLM provides for the emulation of the NETX shell offered in previous versions of NetWare. It is used with the bindery services to ensure compatibility with the utilities that come with NetWare 3.x and earlier versions.

RSA.VLM. This VLM provides for Novell's RSA encryption algorithm.

The most noticeable feature of this interface is the ability to load or not load any of these modules. For example, if the servers on your network consists of only 4.x servers, or there is no application that was specifically written for the shell program, NETX.VLM does not have to be loaded. IF you do not require encryption service, RSA.VLM does not need to be loaded.

Unlike the shell program, the DOS Requester operates behind DOS and takes advantage of the DOS Redirector Interface. This is the DOS interface that allows it to recognize foreign file systems such as CD-ROMs and networks. It is REDIR.VLM that interacts with DOS.

The DOS Requester essentially replaces any form of the shell. In fact, the DOS requester VLM and the older NETX shell cannot be used at the same time in the same workstation. To provide for the shell, the DOS Requester has a shell emulator called NETX.VLM that can be loaded with the Requester. This is provided for those applications that made calls directly to the shell using a certain set of software interrupts. The shell emulator only looks for that specific set of software interrupts. This provides for backward compatibility. All of the modules put together compose the new client interface for NetWare. These modules are only intended for the client and are not loaded to the server.

EXAMPLE   Assuming that a previous connection has been made between a file
server and a client and that a particular file has been previously opened,
we can look at an example of this new client interface. When a request
is made by an application running on the workstation, NETX.VLM (the
shell emulator) intercepts the call to see if it should handle the call.
Because the request is not within its range of acceptable interrupts,
NETX.VLM passes the call to DOS. DOS will determine that it cannot
handle the request, for it was made to a network service. DOS then
hands the request to a redirector (REDIR.VLM) for it to handle the
request. The redirector checks DOS's system file table and determines
that the request is for a network device. REDIR.VLM passes this request
to FIO.VLM for further processing.

FIO.VLM receives this call from REDIR via the VLM manager. The
FIO gathers previously set up information about the file to be read
from, such as whether it is read-only or sharable, etc. The FIO also determines whether the Requester can cache reads and if it can use burst
mode. If the request can be fulfilled by simply reading the workstations
cache, it will perform the read and send the data back to the application
(through REDIR and DOS) without having to send any request packets
across the network. If the request can only be partially fulfilled by reading the cache, it will send the cache information to the application and
transmit a request to the file server through IPXNCP.VLM for the rest
of the information. Remember, the workstation's cache is used for read-
aheads. In other words, when information is read from the file server,
the request always reads more than it was requested to. The additional
information that was not requested is placed in the workstation's cache
under the assumption that the application will request it sometime in
the near future. If the amount of data request warrants the use of packet burst mode, then the FIO will fulfill the request using it.

When transmitting information to the server, the FIO transmits the
request to the IPX NCP module for NCP header information, out
through TRAN, VLM, and finally to IPX for delivery onto the network.
Throughout the handing of information to the different modules, each
module must invoke the use of the VLM.EXE program to ensure proper
handling between VLMs.

Once the file server returns the requested read information to the
client, it is received by the DOS Requester, which then forwards the
information to the REDIR.VLM module. This module updates DOS's
file table, and the data and return codes are given to DOS and finally
back to the application.

NetWare Directory Services

New for NetWare 4.0 is the availability of a new name service. Netware Directory Services (NDS) is based on X.500 directory service as published by the CCITT. It does not strictly follow this standard; this looseness allows it room for future expansion based on NetWare releases. NetWare 4.x contains many new graphical-based objects and utilities. Everything that is on a network, including users, file servers, printers, Web servers, etc., is contained through items known as objects in NDS. You can easily create, move, or modify NDS objects. For example, you can merge or rename entire NDS trees, move subtrees, or rename NDS containers.

NDS replaces the database used prior to 4.0. This was called the bindery, and it contained all the information that NDS can distribute to multiple servers in a single server. The bindery contained informational rights based on the user's access to network services. The bindery had to be manually replicated on each server throughout the internet. Although NDS replaces the bindery, it is completely compatible with it.

The name service is unlike any other naming service available with other protocols. This is a directory service. NDS is a true globally distributed database that can be replicated throughout the internetwork. It maintains information on network resources such as users, groups, servers, volumes, and printers. The database is structured in a tree topology, and can be in place over multiple servers. NDS provides this service completely transparently to the user.

With bindery service, if a user was not registered on a server, the user could not log in. NDS allows for a network login instead of a server login. Users no longer have to know which servers they have access rights to. NDS knows which services the user has access rights to, and will transparently allow or forbid interaction. NDS handles all address resolution in the background.

The directory is formed by placing objects together to from a tree. The structure of the NDS database is: object, property, value. There are three types of objects:

- Physical
- Logical
- Other

A physical object is used to hold information about a user or a printer. A logical object is used to hold information about user groups and print queues. Other objects are used to help organize and manage the previous two groups by providing more definition of the object.

Objects are further defined by categorizing the information in the object. This categorization is called a *property*. There may be many properties per single object. Each of the properties contains a value. A value contains the actual information that is needed. Properties for a user object could be:

- Login name
- Telephone number
- Email address
- Password restrictions
- Group membership

The values for these properties could be:

- Gerald Fisher
- 919.588.7784
- gfisher@doitright.com
- none
- Admins

Figure 6-22 shows an NDS tree. This figure shows three types of objects:

- Root object
- Container object
- Leaf object

Figure 6-22
NDS tree structure.

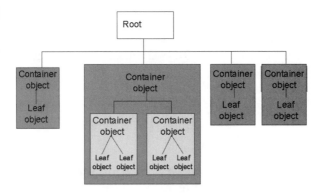

The root object is placed at the beginning of the tree during NetWare installation time. Each branch of the tree will be a container object. Container objects are nothing more than buckets that hold information or other container objects. Leaf objects are at the end of the branch and do not contain any other objects.

The root object cannot be manually created. Once created it cannot be deleted or renamed. It is created by the server installation program and is always placed at the top of the tree. The root object can have trustees, like the Admin user. The Admin user is assigned supervisory rights to the directory tree. This allows for the Admin to gain access to all the rights to all of the objects, starting at the root. This is needed simply to allow the Admin to set up the rest of the directory tree. After the tree is established, the Admin trustee can modify any object.

Container objects are simply a place-holder for other objects. This allows for a grouping method to be applied. You can think of this as being exactly the way the DOS directory structure is set up. A directory in DOS simply holds files or other directories. It establishes a parent-child relationship. In Fig. 6-22, the container directly under the root is the parent container for any directly branched containers it holds.

To allow for some structure in this environment, containers are grouped into two classes:

- An organization
- An organizational unit

An organization allows for structuring other objects in the tree. An organization can be a division of the company, like the engineering division.

An organizational unit (OU) provides for structure for the leaf objects in the tree. This allows for setting up a default login script or setting up a user template (something that allows for defaults to be set with every user created) in that organizational unit. Using our previous example, an organization could be engineering, and an organizational unit could be development engineering, systems engineering, sustaining engineering, etc. Organizational units further define an organization. An OU must be placed below an organization or below another OU.

A leaf object represents such things as users, servers, printers, etc., and cannot contain any other objects. The types of leaf objects are:

- User-related
- Server-related
- Printer-related

- Informational
- Miscellaneous

User-related objects contain information on the user. One object must be created for every person who logs in and uses the services of the network. When a user object is created, a home directory is also created for that user. Users working with 4.x client software simply have to know their directory name. This includes any organization of organizational unit names as well. This can get rather lengthy, but it can be put into the NET.CFG file for ease of use. These types of users can be placed anywhere in the NDS tree.

Users who are not using the 4.x client software can still log in to a server that is running NDS. Their user objects must have been previously created in a container where the bindery was being emulated. These types of users will log in to one server, and the bindery emulation for NDS must be on that server.

A user profile can also be set to a group instead of individually. This allows for rights to be set to a group instead of a single user. The rights that are assigned to the group object are also the rights assigned to the users that are members of that group.

The leaf objects that are associated with the server are NetWare server, volume, and directory map. The server leaf represents a server running NetWare on the network. Information included here is the server's location and what services the server provides. A volume object represents a physical volume of the network. The properties in this object are for information about the NetWare server, the volume on which it is located, and the name of the volume. Other properties deal with information for mapping drives (the ability to share a drive for use by the network). The last leaf object for servers is the directory map. It contains information about a particular directory path or a file in the file system for any server. It is the MAP command that uses this object. This object is useful in login scripts, because they are set to point to directories that contain commonly used applications. The useful feature about this object is its ability to change a property, and the effect is automatic to any login script that was using this object. There is no need to individually change login scripts.

The printer-related leaf objects contain information about the print queue, the print server, and the printer. A print queue object contains information about a print queue, and there must be one object created per print queue. Likewise, the print server object contains information about a network print server. The printer object contains information about a network print server. The printer object contains information about a physical printing device on the network.

Information leaf objects simply contain information about network resources, and have no effect on the operation of the network. The objects included in here are the AFP (Apple File Protocol) server and computer. The AFP object contains information about an Apple server such as its network address and the users for that server. The computer object contains information about such devices that are not server-related. This can be an external router or a user's workstation. Properties such as the serial number, the user's name assigned to that serial number, and the corresponding network address are all kept here. This may seem petty, but this information may be retrieved remotely by a network administrator's management system.

Miscellaneous leaf objects contain the alias, bindery, bindery queue, and unknown objects. The alias object is a point to another object. It can be placed in the tree to point to any other object in the tree. The bindery and bindery queue objects are used for backward compatibility. The unknown object represents an invalidated object that cannot be identified or properly located.

Object Review

Root:

- Automatically placed at the top of the tree
- Cannot be deleted or renamed
- Can be assigned trustees rights, of which the most notable is supervisory rights

Container object:

- Holds other objects
- Allows for logical organization of other objects in the tree
- Does not have a common name assignment
- Type organization, organizational unit

Leaf objects:

- Located at the end of the branch and do not contain any other objects
- Represent actual resources
- Defined by a common name
- Contain information on server, volumes, users, groups, printers, and other resources

Figure 6-23
Object names in a
tree

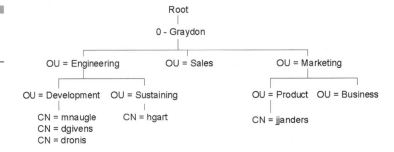

In Fig. 6-23, you will see a realistic directory structure setup. The full
name for reaching Matt Naugle in this structure is

CN = MNAUGLE.OU = DEVELOPMENT.OU = ENGINEERING.O = GRAYDON

The above example shows name types. This is signified by the CN, OU,
or O in front of the object. These explicitly state where we are going in
the tree. It could have been written as

MNAUGLE.DEVELOPMENT.ENGINEERING.GRAYDON

The above example shows user-related objects. It could show any other
objects, such as server-related ones. The structure is the same.

NDS should be replicated across many different servers for redundancy. This is not to say the database on one server should be completely replicated on every other server on the internetwork, but simply that
parts of the database should be replicated on other servers on the internetwork so that the whole NDS database is duplicated.

The last thing to consider is the ability of NDS to work in coordination with SAP. Because NDS provides for authentication of users to
access services of the servers, the services themselves are now contained
in the NDS database. Therefore, NDS is a replacement for SAP. However,
NDS servers still broadcast their existence using the SAP process. Workstations find out about NDS servers by using the SAP process. Once this
has been accomplished, the SAP process is not used, and the workstation
and the NDS server communicate directly. This means the workstation
must be using the 4.x client software. NDS server-to-server advertisements also use the SAP protocol.

Time Services

NetWare 4.x allows for time synchronization between servers. This is necessary because NDS requires a time stamp for each event that occurs. There are four types of time servers:

- Secondary time server
- Primary time server
- Reference time server
- Single-reference time server

All of the timer servers provide time references to the network, with the exception of the secondary time server. A secondary time server obtains the time from one of the other time servers and uses this to provide the time to workstations and other applications. It does not have the ability to determine the correct network time.

A primary time server is primarily used on larger networks. It synchronizes network time with at least one other primary or reference time server. It provides time for up to 150 secondary time server and clients. It polls other primary or reference time servers, and votes with them to synchronize the time.

A reference time server provides a time to which all other servers and clients synchronize when it is important to have a central point configured for time. It gets its time from an external source, such as the Navel Observatory radio clock. This time server polls other primary or reference time servers, and then votes with them to synchronize the time. This type of time clock does not change its time.

A single reference time server is used on smaller networks, and provides time to secondary time servers and clients. When this time server is used, you cannot have primary or reference time servers on the network. Its time is set by the network administrator. This is also the default time server installed with the server.

Time servers find out about each other by two methods:

- SAP
- Custom configuration

By default, all single-reference, primary, and reference time servers use SAP to announce their presence to the internetwork. The secondary time servers will use this information to find and choose a time server from which to get their network time. Primary and reference time servers use this information to find other servers to poll in order to determine the network time.

Using the SAP method has its advantages. It dynamically and quickly allows the timer servers to find out about each other. This is useful when time servers are added or deleted frequently. However, all this comes at a cost. The SAP process uses bandwidth (RIP method). The SAP protocol transmits its table every 60 seconds.

The custom configuration method allows you to assign specific time source servers that another time server should contact for time information or polling. What this means is that a static table can be built into every server. This list contains a listing of other servers to use. This requires careful planning and can create problems when a known time server is taken off line or placed on line; each table must be updated. For networks that have multiple time servers, this may create problems. Using this method, the time server does not use the SAP broadcast in any way.

An Introduction to TCP/IP

Introduction

TCP/IP is actually a suite of protocols under one common name. It will be the only protocol that most commercial entities will use in their companies. There are many other protocols out there, but the trend is to use TCP/IP. Transmission Control Protocol/Internet Protocol, or TCP/IP, is becoming the world's most widely implemented network protocol. You will not find too many customers that have not implemented the TCP/IP protocol, although most of them do not even know that it is TCP/IP that they are using. All they know is that they have a connection to the World Wide Web. Many people confuse the Web with the Internet; the difference is simply that the Web is an application of the Internet that allows us to easily access information available on the Web. The Web uses the communications facilities of the Internet to provide data flow between clients and servers. The Internet is not the Web and the Web is not the Internet.

Those who are reading about the TCP/IP protocol for the first time and would like to connect to the Internet or have access to the World Wide Web (WWW) have thought about the TCP/IP protocol—maybe not directly, but they will be using it to access the Internet.

TCP/IP is slowly overtaking the current reigning champ of all client server network protocols, the Novell NetWare protocol, and its datagram delivery mechanism of Internet Packet Exchange (IPX) and Sequenced Packet Exchange (SPX). In the 1970s everyone had some type of Wang machine in their office. In the 1980s and early 1990s Novell's NetWare applications started taking over. Today, NetWare continues to dominate the network arena with its installed base of client/server network applications, but the TCP/IP protocol, combined with Internet browsers such as NetScape's Navigator and Microsoft's Internet Explorer and Web programming languages such as JAVA and Visual Basic Script with Active/X, are producing powerful corporate networks known as *intranets*, which mimic the facilities of the Internet on a corporate scale. Intranets from different companies or simply different sites can communicate with each other through the Internet. Consumers can access corporate intranets through something called an extranet, which is simply the part of a corporation's intranet that is available to the public. A great example of this is electronic commerce, which you use when you purchase something via the Internet. TCP/IP allows for print services through an application known as lpr (line printer). Directory services are provided through Domain Name Service (DNS). File services are provided through the browser software, and database services are provided through all major database applicatons including Oracle, Sybase, Informix, and even Microsoft Access. Browsers are getting "smarter" with protocols such as ActiveX (from Microsoft) and JAVA (from NetScape and Sun Microsystems). Finally, the ultimate in full connectivity is the Internet, which allows the corporate intranets to interconnect (within the same corporation or between different corporations), providing global connectivity unmatched by any network application today. Within a short time (possibly by 1998), very powerful applications will be built, utilizing the TCP/IP software suite, that will eventually rival NetWare.

Another key factor of TCP/IP is portability. How many people can you name that use NetWare out of their house for corporate or commercial connectivity? Programs such as remote node and remote control allow NetWare clients to be accessed remotely, but not as seamlessly as TCP/IP does. TCP/IP allows you to move your workstation to any part of the network by dialing in from any part of the world to gain access to any accessible network. This brings us to another point: How many networks interact using NetWare? Theoretically, with TCP/IP you can access (excluding security mechanisms for now) any TCP/IP network in the world from any other point in the world. Addressing in TCP/IP is handled on a global scale to ensure uniqueness. Novell attempted global addressing but failed. Novell addresses are unique to each private installation, such as a single company, but are probably massively duplicated globally. For example, there are many installations with the Novell address of 1A somewhere in their network. Not everyone is going to renumber their network for uniqueness, but one method of doing so is to match the 32-bit address of TCP/IP subnets to your Novell network. Then convert each octet of the 32-bit address of TCP/IP into hex, and use that as your NetWare address.

Novell has entered the TCP/IP fray with their IntranetWare and support for native IP. IntraNetWare allows NetWare workstations to access TCP/IP resources. NetWare 4.x is supposed to allow for NetWare to run directly on top of TCP/IP (this is known as native TCP/IP support).

Microsoft and their emerging NT platform can also use TCP/IP as a network protocol. Two versions are available:

- Native TCP/IP and its applications (Telnet, FTP, etc.)
- RFC-compliant (RFC 1001 and 1002) TCP, which allows file and print service

These provide the ability to Telnet into a NT server or workstation and transfer files to that workstation or server, as well using native TCP/IP. For file and print services in a TCP/IP environment, NT can be configured to use NetBIOS over TCP/IP. This enables NT to be involved in a routed network. There are many other protocols that NT can run, as well, but they are beyond the scope of this book.

This all does not mean that protocols other than TCP/IP are being disbanded. Novell NetWare continues to run with the IPX protocol. As of 1997, NetWare was still the best-constructed client/server platform available. Tens of thousands of programs have been written directly for the NetWare interface, and it is used in corporate networks, schools, and state, local, and federal government. They are not simply going to disconnect their NetWare networks and move to TCP/IP overnight, so NetWare using the IPX protocol will be around for a long time, albeit in a diminishing role.

Most Fortune 1000 companies still depend on large mainframes for their day-to-day processing. In the early 1990s and late 1980s many corporations were convinced that smaller UNIX platforms using a distributed architecture could replace their "antiquated" Systems Network Architecture (SNA) networks. They were wrong? Although some networks have been converted to this architecture, many have not. There are many factors involved here. Time and money play an important role, but the rule continues to be, "if it ain't broke, don't fix it." Huge applications such as the airline reservation system and the banking system are built using SNA, and even if a perfect solution were found, it would take years to convert these programs over to a new system. Therefore, SNA is still being used, and there are even some sites that have converted from distributed systems running on UNIX platforms back to SNA mainframes because the mainframes were best suited to their particular situations.

Today there are also Web servers that front IBM mainframes. However, IBM fully supports TCP/IP, and the program known as TN3270 allows for 3270 terminal emulation over the TCP/IP protocol. The details of this are beyond the scope of this book, but it is important to note that although TCP/IP is very popular, other protocol schemes still provide many benefits, and will continue to be used for years to come.

One would tend to think that TCP/IP was developed by a large-scale R&D center such as that of IBM or DEC. It wasn't. It was developed by a team of researchers, but they were college professors, graduate students, and undergraduate students from major universities. This should not be hard to believe. These are people who not only enjoy R&D work but who also believe that when problems occur, the fun starts.

Many years from now we will look back on TCP/IP as the protocol that provided the building blocks of future data communications. However, take notice: TCP/IP is a never-finished project. It is fully functional today, but work on the project continues. There are over 75 working groups of the Internet Engineering Task Force (IETF), and as new needs continue to arise for the Internet, new working groups will be formed and new protocols will emerge. In fact, the IP version of the existing protocol (known as IPv4, or IP version 4) may be replaced. IP version 6 (IPv6) is currently being implemented experimentally around the Internet. It will be a few years before a complete switchover takes place, but this is a good example of the never-ending story.

While the ARPANET and the Internet were being built, other protocols such as System Network Architecture (SNA) and protocols based on XNS were developed. Client/server applications that allowed for file and print services or use on personal computers were built using protocols based on XNS, such as Novell NetWare (using IPX) and Banyan VINES. SNA was alive and well in the mainframe, and DECnet controlled the minicomputer marketplace. DEC also supported LAT (local area transport) for terminal servers, which also supported printers. DECnet started out before commercial Ethernet, and DEC's mini-computers were connected via local interfaces. Later, around 1982, DEC started to support Ethernet, but still with the DECnet protocol.

All of these protocols could run over Ethernet, Token Ring, or FDDI (Fiber Distributed Data Interface). In this respect, they did openly support the LAN protocols. However, aside from the LAN protocol, these protocols were proprietary, i.e., *vendor-dependent*. Protocols other than TCP/IP are proprietary, and the internals of those systems are known only to their respective company owners. Users and network administrators were held to proprietary network environments and proprietary network applications, which deterred network development and enhancement in all corporate environments. For example, just because a vendor supported XNS did not mean that its product would interoperate with other vendors' products running XNS. Running XNS on one system did not guarantee compatibility of communication with any other system except one from the same vendor. This was good for the vendor, but it tended to lock users into one vendor's products.

The only public wide area network (WAN) access protocol was X.25, and not everyone supported all features 100 percent, which again lead to compatibility problems. All of us remember X.25 as a slow (primarily 9.6 kbps or 19.2 kbps) WAN access protocol.[1] Alternatively, leased lines based

---

[1]This is not to bash the X.25 protocol—there were many valid reasons for running it at slower network speeds, including error correction and control, and faster speeds such as T1 were not available for data connection transfers.

on proprietary protocols of the network vendors were an option, but one that only allowed the corporate's networks to be interconnected. Ethernet was another option, but host interfaces and standardized network protocols were not readily available.

The Internet started as a tool (ARPANET) to link the government and research facilities. The ARPANET was an entirely separate network from the Internet, and it was shut down in the late 1980s. At that time, only a handful of people knew about the Internet. The Internet didn't really have anything to offer the commercial world, but engineers and scientists loved it. No one knew of the advantages of the TCP/IP protocol. It was not until the GUI interface was developed that the Internet "took off" and TCP/IP came with it. Therefore other protocols such as SNA and Novell NetWare sprouted in corporate America. Basically, there was no other choice.

One of the better protocols was AppleTalk. Yes, AppleTalk. AppleTalk was free, and it provided a remote dial-in feature that allowed remote users to appear to be local on the AppleTalk network. AppleTalk incorporated a naming scheme, a routing scheme, and many other great features at a time when all other vendors were forcing their customers to buy network equipment and software in order to connect their PCs together. This is not to say that AppleTalk did not have its problems—it had many, especially when building large Appletalk networks.

AppleTalk is actually software; LocalTalk is the hardware. It was Apple's version of networking Mac computers. The protocol was simple to use and install. It was built into every Mac. Cables were all that was needed to connect Apple computers in a simple network, and file and print services were built in, too. It was known as a true peer-to-peer protocol because each workstation could see every other workstation, and each workstation could be a server and share any of its resources. Each node had its own name service. Each node picked its own physical address. Even dialing in to an AppleTalk network was easy using the AppleTalk Remote Access (ARA) protocol. AppleTalk soon became a very popular method of hooking Macs together into a network. However, it was not envisioned as a protocol to handle large internets of Apple computers, and the inefficiencies of the protocol soon became apparent. It was about as close as you could come to a network operating system that allowed for simplicity and ingenuity.

TCP/IP eliminated proprietary network operating systems. TCP's beginnings were rough (because of interoperability issues), but the protocol stabilized, and interoperability between different computers and operating systems became a reality. For example, a DEC system running the VMS operating system combined with TCP/IP running as the network operating system can communicate with a Sun Microsystems UNIX workstation running TCP/IP. The two systems can communicate by taking advantage of the protocol and the specific applications written for the protocol, primarily by being able to log on to one another and by being able to transfer files between each other across a network.

When computers and their operating systems are interconnected with TCP/IP, it does not matter what the hardware architectures or the operating systems of the computers are. The protocol will allow any computer implementing it to communicate with another. What follows are the methods used to accomplish this.

TCP/IP allows public access to network protocols and seamless integration between all computing environments that wish to operate in a network environment. The commercial success of this protocol was unexpected; TCP/IP was never envisioned as a commercial network system and was not envisioned to gain the widespread use it has today. The idea of open-architecture networking was first introduced by Robert Kahn shortly after he arrived at DARPA in 1972. Had the open protocol of TCP/IP not flourished, the network environment would possibly be in the same situation as was the operating system in the 1970s and early 1980s, when everyone had their own proprietary operating system and users were stuck with one operating system. Users can choose TCP/IP without having to choose a particular vendor along with it. Suffice it to say that TCP/IP is the protocol of choice for future network installations.

History

A TCP/IP network is generally a heterogeneous network, meaning there are many different types of network computing devices attached to it. The suite of protocols that encompass TCP/IP was originally designed to allow different types of computer systems to communicate as if they were the same system. It was developed by a project underwritten by an agency of the Department of Defense known as the Advanced Research Projects Agency (DARPA).

There are many reasons the early TCP/IP became popular, three of which are paramount. First, DARPA provided a grant to allow the protocol suite to become part of Berkeley's UNIX system. When TCP/IP was introduced to the commercial marketplace, UNIX was mentioned in every story about it. Berkeley UNIX and TCP/IP became the standard operating system and protocol of choice for a lot of major universities, where it was used with workstations in engineering and research environments. Second, in 1983, all U.S. government proposals that included networks mandated the TCP/IP protocol. (This was also the year that the ARPANET was converted to the TCP/IP protocol.) Finally, a graphical user interface was developed to allow easy access to the system. TCP/IP and its applications were difficult to use if you had not had experience with them. Finding information on the Internet was a formidable task. Before the browser, TCP/IP applications were accessed from a command line interface with a few basic applications that allowed you to call a remote system and act as a remote terminal, transfer files, and send and receive mail. Some companies built graphical interfaces to their applications, but they were still rough and would not have gained commercial success. The browser hid all the complexities of the TCP/IP protocol and its applications and enabled the display of graphics, as well as text. By clicking on either the graphics or the text, users could go anywhere on the Internet (within security limitations). The browser also made it easier to find information on the Internet.

It was not very long before the capability of TCP/IP to allow dissimilar systems to communicate through the network became widely known. It could be used without a "forklift upgrade" to mainframe, minis, and personal computers; it simply ran on existing computers. TCP/IP became, and remains, a very popular network operating system.

Refer to Fig. 7-1. TCP/IP originated when DARPA was tasked to bring about a solution to a difficult problem: allowing different computers to

Figure 7-1
Internet *time line.*

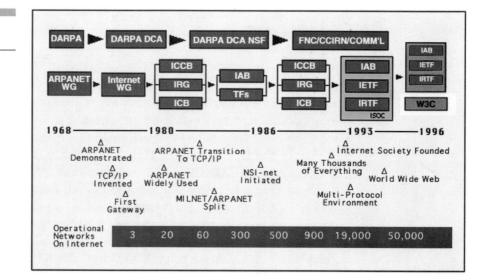

communicate with one another as if they were the same computer. This was a difficult task, considering the fact that all computer architectures in those days (the early 1970s) were highly guarded secrets. Computer manufacturers would not disclose either their hardware or software architectures to anyone. This is known as a *closed,* or *proprietary,* system.

The architecture behind TCP/IP takes an alternative approach. TCP/IP developed into an architecture that would allow computers to communicate without grossly modifying the operating systems or the hardware architecture of the machines. TCP/IP runs as an application on those systems.

Before it came to be called TCP/IP, the original protocol was known as the network control program (NCP). It was developed to run on multiple hosts in geographically dispersed areas through a packet-switching internet known as the Advanced Research Project Agency Network, or ARPANET.[2] This protocol was primarily used to support application-oriented functions and process-to-process communications between two hosts. Specific applications such as file transfer were written to this network operating system.

In order to achieve the goal of enabling dissimilar government computers to communicate, DARPA gave research grants to UCLA, the University of California at San Bernardino, the Stanford Research Institute, and the University of Utah. A company called BBN provided the Honeywell 316 interface message processors (IMPs, which have evolved into today's routers) to provide the internet communications links. In 1971, the ARPANET networking group dissolved and DARPA took over all the research work. The first few years of the design process proved to be an effective test and revealed some serious design flaws. A research project was developed to overcome these problems. The outcome of this project was a recommendation to replace the original program, known as the NCP program, with the Transmission Control Program. Between 1975 and 1979, DARPA had begun the work on the Internet technology that resulted in the TCP/IP protocols as we know them today. The protocol responsible for routing packets through an internet was called the Internet Protocol. Today, the common term for this standard is TCP/IP.

With TCP/IP replacing NCP, the NCP application-specific programs were converted to run over the new protocol. The protocol was mandated in 1983, when ARPA demanded that all computers attached to the ARPANET use the TCP/IP protocol. The formal switchover for the

---

[2]The ARPANET was taken down in 1993. The Internet that we run today was built during the ARPANET period as a parallel network.

ARPANET was in January 1983. Imagine being able to simply "flip a switch," and routers and hosts are running a new version of the protocol. In 1983, the ARPANET was split into two networks:

1. The Defense Data Network (DDN), also known as the MILNET (military network)

2. The DARPA Internet, a new name for the old ARPANET network.

Outside the ARPANET, many networks were being formed, such as the CSNET (Computer Science Network); BITNET (Because It's Time Network), used between IBM systems; UUCP (User to User Copy), which became the protocol used on USENET (a network used for distributing news); and many others. All of these networks were based on the TCP/IP protocol. All were interconnected using the ARPANET as a backbone. Many other advances were also taking place, with local area networks using Ethernet and other companies making equipment that enabled any host or terminal to attach to the Ethernet. The original route messengers known as IMPs (interface message processors) were now being made commercially and were called routers. These routers were smaller, cheaper, and faster than the ARPANET's IMPs, and they were more easily maintained. With these devices, regional networks were built that could now hook up to the Internet. However, commercial access to the Internet was still very limited.

One experiment that was successful, the CSNET (Computer Science Network), provided the foundation for the National Science Foundation (NSF) to build another network that interconnected five supercomputer sites. The five sites were connected via 56-kbps lines. This was known as the NSFnet. The NSF stated that if an academic institution built a community network, the NSF would give it access to the NSFnet. Not only would this allow regional access to the NSFnet but it would allow regional networks (based on the TCP/IP protocol) to communicate with one another. The NSFnet was formally established in 1986. It built a large backbone network using 56-kbps links, which were upgraded to T1 links in July 1988. Anyone that could establish a physical link to the NSFnet backbone could gain access to it. In 1990, the NSFnet was upgraded to 45-Mbps links.

Once the word of NSFnet spread, many regional networks sprang up, for example, NYSERnet (the New York State Educational Research Network) and CERFnet (the California Educational Research Network). The regional networks were supported at their level, not by the NSF.

The NSFnet was very useful beyond its conception of linking supercomputers to academic institutions. In 1987, the NSF awarded a contract to MERIT Network (along with IBM and MCI) to upgrade the NSFnet to T1 links and to link six regional networks, the existing five supercomputer centers, MERIT, and the National Center for Atmospheric Research into one backbone. This was completed in July 1988. In 1989, a nonprofit organization known as ANS (Advanced Network and Services, Inc.) was spun off from the MERIT team. Its goal was to upgrade the NSFnet to a 45-Mbps backbone and to link together 16 regional sites. This was completed in November 1991.

More commercial entities were building regional networks via TCP/IP, as well. To allow these entities access to the backbone, a concept known as the Commercial Internet Exchange (CIX) was developed. This was a point on the backbone that allowed commercial regional networks access to the academic NSFnet backbone.

The original ARPANET was expensive to run, and the interest inside DARPA began to wane. Major promoters of the ARPANET had left DARPA to take positions elsewhere. The ARPANET was taken completely out of service in 1989, and what emerged in its place is called the Internet. The term *Internet* was coined as an abbreviation of *Internet Protocol* (IP).

The NSFnet was basically a mirror image of the ARPANET. NSFnet and ARPANET were running in parallel. Regional networks based on the TCP/IP protocol were interconnected via NSFnet. NSFnet has connections to the ARPANET. More connections were being made through NSFnet because it was higher-speed, easier to hook into, and cheaper.

It was determined that the original network, the ARPANET, should be shut down. Sites on the ARPANET found new homes within the regional networks or as regional networks. NSFnet provided the backbone for interconnection of these regional networks.

Word quickly spread about the Internet, and around 1993, the NSF decided it could not continue supporting the rapid expansion directly. It produced contracts for outsourcing the operation of the Internet. The functional responsibility of running the Internet was given to many different companies. A concept called Network Access Points would take the place of the NSFnet. These are points located throughout the United States at which private companies who built their own backbones could interconnect and exchange route paths. Also with this comes the concept of peering; by allowing its backbone to be used by another provider to move their customers' traffic over, a provider is peering with another provider. There is a lot of controversy over this concept. Who should a backbone provider peer with or not peer with? Why should a provider let another provider use its backbone as a transit for its customers for free? Because NSF said so. The rising demand for Internet access put this issue on to the back burner temporarily, but as the Internet slows down, the concept is raising its ugly head again.

NAPs are basically the highest points in the Internet. They would enable many backbones that were privately built to be interconnected through the NAPs. Initially there were four official NAPs, but this number has grown to 13 as of this writing. Even with the commercialization of the Internet, no one company owns any part of the Internet. Everyone associated with the Internet must abide by the rules in place. External companies simply provide a specific service required to run the Internet. For example, Network Solutions, Inc., was granted the right to control domain name registration. However, they do not own this capability—they are still under the authority of the Internet Assigned Numbers Authority, which is run by Jon Postel (as of this writing) at the University of Southern California. AT&T was granted the right to host many document databases required by the Internet user community. All the functions of running the Internet were contracted out by the NSF. Any company (with lots of money) can build a backbone. To provide access to others, their backbone must be connected to others at an NAP. Individual backbone providers then interconnect multiple connections known as points of presence, or POPs, which are where individual users or businesses connect to the Internet. In April of 1995, the NSFnet backbone was shut down, and the Internet as we know it today took its place.

One last distinction of TCP/IP: To run the protocol on any network does not require a connection to the Internet. TCP/IP can be installed on as few as two network stations or as many as can be addressed (possibly millions). When a network requires access to the Internet, the network administrator must call their local registry or their internet service provider (ISP) to place a request for access and to be assigned an official IP address.

The World Wide Web

Great application programs have been available on the Internet for dozens of years, so why all the hype since 1994? The World Wide Web commercially came to us in 1994. This is when corporate money became involved. However, the idea started way back in 1981 with a program called Enquire written by Tim Berners-Lee. A program known as Mosaic became available as freeware in 1993. It was written by the co-founder of NetScape, Marc Andresson, at the U.S. National Center for

Supercomputer Applications (NCSA). Mosaic allowed text and graphics on the same Web page and was the basis for the NetScape Navigator browser and the Microsoft browser Internet Explorer, a screen from which is shown in Fig. 7-2.

First and foremost, the Web allows anyone, including nontechnical people, instant access to an infinite amount of information. You can get stock reports, access information from the library, order a book, reserve airline tickets, page someone, find that long-lost friend through the yellow pages, order a data line for your house, check your credit card statement, provide computer-based training, or attend a private (video and audio) meeting. And yes, you can send email.

But there is more! You don't have to be content with what is offered by online services such as CompuServe, Prodigy, and America Online—you can create your own Web page, as well. It's not too hard to do—the language required to create a Web page is pretty much English. Millions of ideas are available. There is a pull-down menu on the browser that allows you to see the source code (the basic instructions that tell the Web server how to format a page) of any Web page. By 1995, companies known as internet service providers (ISPs) were advertising their ability to put you on the Web for a low price of $19.95. Today, most professional

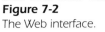

Figure 7-2
The Web interface.

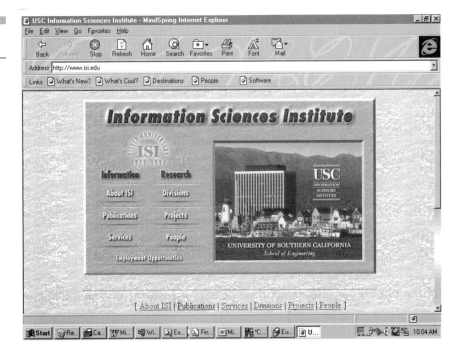

ISPs give you space on their server (a small amount but sufficient to get started) for you to create your Web page at no charge.

The Web provides point-and-click access to any type of information that you would like. You do not have to know an operating system to move around the Web. No other "cyberspace" provider has the rich simplicity of the browser. One click and you could be on a server in Japan, videoconference to California, send an email to your friend in England, or simply plan a vacation in Breckenridge, Colorado. The Internet had always provided information, but it was the simplicity and combination of text and still pictures on the same page that catapulted the World Wide Web into American homes.

Virtually anything that you want to check on, you can do on the Internet, and you do not have to remember IP addresses, directory commands for DOS and UNIX, file compression, how to execute the TAR command, or print to a postscript printer, etc. The Web is also global due to the fact that it uses the Internet as its transport.

On the application front, more and more applications are being written geared toward the most common Internet interface, known as a browser. This is an application that allows the Internet to be accessed graphically using icons and pictures and a special text language known as Hypertext Markup Language, or HTML. For platform independence in writing applications for the Web, the JAVA language was created.

What is the downfall of the Internet? Connectivity is generally not the problem. ISPs can be a problem, but even they are manageable. The biggest problem with the Internet is its biggest asset: information.

The Internet can often be confusing. With anyone allowed to create content and then post it, there is a lot of old information on the Internet. Web pages are not kept up. Web pages are not written correctly and contain too many slow-loading graphics. Many links are embedded to other Web pages that no longer exist. Information is posted without having validity checks. Remember, no one entity owns the Internet or the Web.

Some companies that once created Web pages no longer exist. All Web pages are not created equal; some take an eternity to write to your browser, while others take a minimal amount of time. Also, all ISPs are not created equal. An ISP is your connection to the Internet. Test out your ISP for service and connectivity. I recently switched from a major name ISP to a local ISP and found a 4× improvement in speed. However, the local ISP does not provide national service (local phone numbers around the United States), so when I started traveling, I switched to another ISP, one with national coverage *and* speed.

Be careful when scrutinizing the Internet. Make sure the data is reputable (i.e., referenceable). There are many charlatans on the Internet posting fiction as fact.

The Internet introduced the concept of "try something for free." We old timers expected this. Postings to the Internet were always free and commercialism was a no-no. Years ago, when I was developing software, it was the Internet that many times came to my rescue with a posting of source code that assisted in my development projects. This source code was available for free, and often the person who posted it did not mind an occasional email with a question or two. Another concept that developed on the Internet is shareware, wherein free samples of applications range from severely crippled ones (lacking many of the full-version features such as printing abilities) to the full-blown version of the software. The Web combined the two concepts, and the marketing concept really took hold when the Internet came into the business world. Every business sponsoring a Web page will give you something if you purchase something. The concept is actually very old but has been brought to life again via the Internet.

Most of us try a free sample before purchasing. This is what shareware allows, but payment is expected eventually. This leads to another big problem for the Internet: How and when do you charge for something? Most Internet users expect to surf the Internet, picking up what they want for free and then signing off. But we don't live in a free world, and eventually you should pay if you use a product. Unfortunately, there are those out there that continue to download software and not pay for it. If this continues, shareware will no longer be available, and we will end up with pay first, try later.

Another problem on the Internet is the spread of viruses. Workstations should be protected with some type of antiviral software before being used to download anything from the Internet. Most protection schemes are dynamic in that they are constantly checking for viruses, even during an email download or a file transfer. Here is where some online providers do provide an advantage. Private online providers such as America Online and CompuServe make every effort to test uploaded software and generally do not allow for content to be written to their servers. You will find those services more protected than the Internet.

The Internet is still the best thing going. Applications from all types of businesses are available on the Internet, and many experiments are being conducted on the Web, as well, including audio/visual applications such as movies, radio, and even telephone access.

Intranets and Extranets

Basically, an *intranet* is a TCP/IP-based internet used for a business's internal network. Intranets can communicate with each other via connections to the Internet, which provides the communication backbone. An intranet does not need an outside connection (to the Internet) in order to operate. It simply uses all the TCP/IP protocols to give you a "private" internet.

When a business makes part of its internal network available to the outside community, this is known as an *extranet*. You may have used an extranet when browsing through a Web page at General Electric or ordering some diskettes via a reseller's Web page. Users do not have complete access to a corporation's network but merely the part of it that it wants you to have access to. It can set filters on its routers and put firewalls (pieces of software or hardware that allow access to resources based on a variety of parameters such as IP addresses, port numbers, domain names, etc.) in place that prevent users from accessing anything more than a subset of their intranet.

Governing Bodies of the Internet

Who governs the Internet and the Web? No company or person owns the Internet. Some say that it is a miracle that the Internet continues to function as well as it does. In order to function it requires the complete cooperation of thousands of companies known as internet service providers (ISPs), telecommunications companies, standards bodies such as the IANA, application developers, and a host of other entities. Their main goal is to provide ubiquitous information access. Those who try to divert the Internet to their own advantage are usually chastised. However, this is becoming less true now that ISPs are competing for traffic patterns. Furthermore, all those who participate in the Internet, including all companies that have IP connections to the Internet, must abide by the rules, as well.

Refer to Fig. 7-1. The TCP/IP protocol suite is governed by an organization known as the Internet Activities Board (IAB). In the late 1970s, recognizing that the growth of the Internet was accompanied by a growth in the size of the interested research community and therefore an increased need for coordination mechanisms, Vint Cerf, then manager of the internet program at DARPA, formed several coordination bodies—the International Cooperation Board (ICB), to coordinate activities with some cooperating European countries focused on packet satellite research; an internet research group that was an inclusive group providing an environment for general exchange of information; and the Internet Configuration Control Board (ICCB). The ICCB was an invitational body to assist Cerf in managing burgeoning Internet activity.

In 1983, continuing growth of the Internet community demanded a restructuring of coordination mechanisms. The ICCB was disbanded, and in its place a group of task forces was formed, each focused on a particular area of the technology (e.g., routers, end-to-end protocols, etc.). The Internet Activities Board (IAB) was formed by the chairs of the task forces.

By 1985 there was a tremendous growth in the more practical, engineer-oriented side of the Internet. This growth resulted in an explosion in attendance at Internet Engineering Task Force (IETF) meetings and was complemented by a major expansion in the Internet community. No longer was DARPA the only major player in the funding of the Internet. In addition to NSFnet and the various U.S. and international government-funded activities, interest in the commercial sector was beginning to grow. Also in 1985, there was a significant decrease in Internet activity at DARPA. As a result, the IAB was left without a primary sponsor and increasingly assumed the mantle of leadership.

The growth continued, resulting in even further increase in substructure within both the IAB and IETF. The IETF combined working groups into areas and designated area directors. The Internet Engineering Steering Group (IESG) was formed of the area directors. The IAB recognized the increasing importance of the IETF and restructured the standards process to explicitly recognize the IESG as the major review body for standards. The IAB also restructured itself so that the rest of the task forces (other than the IETF) were combined into the Internet Research Task Force (IRTF), with the old task forces renamed as research groups. The growth in the commercial sector brought with it increased concern regarding the standard-setting process itself. Starting in the early 1980s and continuing to this day, the Internet has grown beyond its primary research roots to include both a broad user community and increased commercial activity. Increased attention was paid to making the process open and fair. This, coupled with a recognized need for community support of the Internet, eventually led to the formation of the Internet Society in 1991, under the auspices of the Corporation for National Research Initiatives (CNRI).

In 1992, the Internet Activities Board was reorganized and renamed the Internet Architecture Board, operating under the auspices of the

Internet Society. More of a "peer" relationship was defined between the new IAB and the IESG, with the IETF and the IESG taking a larger responsibility for the approval of standards. Ultimately, a cooperative and mutually supportive relationship was formed between the IAB, the IETF, and the Internet Society, with the Internet Society taking on as a goal the provision of service and other measures that would facilitate the work of the IETF.

This community spirit has a long history beginning with the early ARPANET. The early ARPANET researchers worked as a close-knit community to accomplish the initial demonstrations of packet-switching technology described earlier. Likewise, the packet satellite, packet radio, and several other DARPA computer science research programs were multicontractor collaborative activities that heavily used whatever available mechanisms there were to coordinate their efforts, starting with electronic mail and adding file sharing, remote access, and eventually World Wide Web capabilities.

An Overall View of the Internet

Figure 7-3 shows a "picture" of the Internet. It looks complicated, but it really is not. The figure shows (not to scale nor geographically correct) the pieces required to make up the Internet. There is a lot more going on than what this picture shows, but for the most part this is what makes up the Internet. As was briefly mentioned earlier, it is a miracle that the Internet actually runs. It requires the complete cooperation of the telephone companies, ISPs (which could also be telephone companies), individual users, etc. The Internet is run by everyone who uses it.

TCP/IP Technical Documents

If TCP/IP is such an open protocol, where does one find information on it? There are documents known as Requests for Comments (RFCs) that define the processing functions of TCP/IP. These documents are available online or may be purchased. They can be found online on any of these three registries: InterNIC (United States), RIPE (Europe), and APNIC (Asia Pacific). This information is best described in RFC 1534.

For example, *http://ds.internic.net/rfc/rfc-index.txt* will lead you the latest index (updated almost daily) of RFCs. It is a good idea to save this as a file in your local computer. You will return many times to this docu-

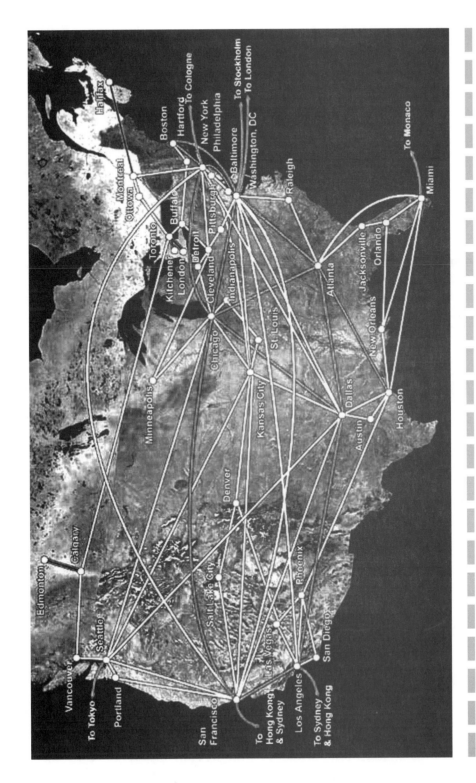

Figure 7-3 The Internet.

ment to find more information about a particular aspect of a protocol. Use the Find tool under the Edit pull-down menu to perform a search. Note that just because you type in a word, the search engine may not find specifically what you are looking for, so you may have to know a few things before venturing in to find something. For the most part, however, this is the best method of weeding through the RFCs. After finding an RFC, change *rfc-index* in the URL to *rfcxxxx.txt*, where *x* is the RFC number, and you now have the RFC online. If you save the RFCs that you will return to the most on your local directory, it will save you time later, because they can take some time to download.

The RFC is the definitive document for the TCP/IP protocol suite. I asked some systems engineers two things:

- When was the last time you reviewed a question by reading an RFC?
- Have you read RFCs 1812, 1122, and 1123?

The answer to the first question was generally "I don't know," and the answer to the second question was "What is in those RFCs?" How can any systems engineer claim to know the TCP/IP protocol without having read these three RFCs? The Web makes it so easy to review an RFC. Simply point your browser to *ds.internic.net/rfc/rfcxxxx.txt* or, for an index, *ds.internic.net/rfc/rfc-index.txt*. Get the RFC electronically, save it, and then use the search commands to find what you are looking for.

RFC Details

This section of the book has been expanded because everyone seems to be getting away from RFCs. Also, there are still many people just getting into the TCP/IP protocol who may have never seen an RFC.

The Requests for Comments are documents that define the TCP/IP protocol suite. They are the Internet's technical (mostly) documents. I mean mostly, for some are intellectually humorous (e.g., "A view from the 21st century by Vint Cerf," RFC 1607). An RFC can be written and submitted by anyone. However, all documents submitted do not automatically become RFCs. A text document becomes a draft RFC first. A draft RFC is a public document that undergoes a peer review process, during which comments are continually made on the draft. It is then decided whether or not it becomes an RFC.

Steve Crocker wrote the first RFC in 1969. These early memos were intended to be an informal, fast way to share ideas with other network researchers. At first, the RFCs were printed on paper and distributed via "snail mail." As the File Transfer Protocol (FTP) came into use, the RFCs were prepared as online files and accessed via FTP. The number of RFCs (as of this writing) numbers over 2000, and they contain information on all aspects of any Internet protocol. Development engineers study these documents and produce applications based on them.

Systems engineers do not need to study most of the RFCs, but for a basic understanding there are three RFCs that must be read. Therefore, in the spirit of the RFC action words, you MUST read RFCs 1122, 1123, and 1812 before being able to state that you understand the TCP/IP protocol suite. The majority of RFCs can be summed up in those three. The reading is not difficult, and many things are explained there.

RFC Submission

Memos proposed as draft RFCs may be submitted by anyone. One large source of such memos is the Internet Engineering Task Force (IETF). The IETF working groups (WGs) evolve their working memos (known as internet drafts, or I-Ds) until they feel they are ready for publication. Then the memos are reviewed by the Internet Engineering Steering Group (IESG) and, if approved, sent by the IESG to the RFC editor. The primary RFC must be written in ASCII text, including all pictures. Remember, simplicity and availability for all is the overall tone of the Internet. In a digital world, it is guaranteed that everyone at least has ASCII terminal functions either through a computer terminal or on a PC. The RFC may be replicated as a secondary document in PostScript (this must be approved by the author and the RFC editor), which allows for an easy-to-read RFC.

The format of an RFC is outlined by RFC 1543, "Instructions to Authors," and is shown in Fig. 7-4. Each RFC is assigned a number in ascending sequence, and the numbers are never reassigned. Once issued, RFCs do not change—revisions are issued as new RFCs. Some of the newer RFCs, however, only replace *part* of the older RFC, for example, by replacing an appendix or updating a function. They may also simply add something to the older RFC, which is indicated by an "updated-by:" statement on the first page. If a new RFC completely replaces another RFC, the new RFC reads "Obsoletes: RFC *NNNN*" in the upper left corner. The index of RFCs at the URL given earlier, does contain the information about updates.

RFCs are continuing to evolve as the technology demands. For example, the wide area network connection facility known as the Frame

Figure 7-4
RFC format.

Network Working Group
Request for Comment <place number here>
Obsoletes/Updates: <Place RFC number here>
Category:

Author Name (first initial and last name)
Author Organization
Submission Date

Running header

Title

Status of this memo

Abstract

Table of Contents

Running Footer

Relay specification is becoming very popular, and there are RFCs to define how to interface TCP to the frame relay protocol. RFCs also allow refinements to enhance better interoperability. As long as the technology is changing, the RFCs must be updated to allow connection to the protocol suite. IPv6 is well documented by many RFCs.

As of this writing, the IETF has more than 75 working groups, each working on a different aspect of Internet engineering. Each of these working groups has a mailing list to discuss one or more draft documents under development. When consensus is reached on a draft document, it can be distributed as an RFC.

RFC Updates

RFC announcements are distributed via two mailing lists: the IETF-Announce list, and the RFC-DIST list. You don't want to be on both lists. To join (or quit) the IETF-Announce list, send a message to *IETF-Request@cnri.reston.va.us*. To join (or quit) the RFC-DIST list, send a message to *RFC-Request@NIC.DDN.MIL*.

Figure 7-5
OSI model and
TCP/IP.

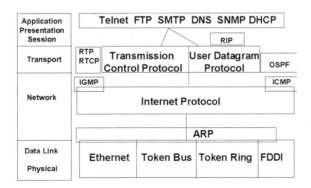

RFC Format

Refer to Fig. 7-5, which shows the format for the front page on an RFC. The details of the RFC (protocol coverage) make up the bulk of the document and are included after the cover page. RFCs have changed dramatically over the years and now include topics for discussion and action words, discussed later, which indicate which parts of the protocol MUST be implemented, which parts are optional, and others that should not be implemented. Following is a list of the sections of an RFC, and requirements for their format:

Network Working Group. The traditional heading for the group that founded the RFC series. This appears on the first line on the left-hand side.

Request for Comments NNNN. Identifies the document as a request for comments and specifies the number. Indicated on the second line on the left side. The actual number is filled in at the last moment before publication by the RFC editor.

Author. The author's name (first initial and last name only), indicated on the first line on the right side.

Organization. The author's organizational affiliation (company name, college division, etc.), indicated on the second line on the right side.

Date. The month and year of the RFC. Indicated on the third line on the right side.

Updates or Obsoletes. If the RFC updates or makes obsolete another RFC, this is indicated on the third line on the left side.

Category. The category of the RFC: standards track, informational, or experimental. Indicated on the third (if there is no Updates or Obsoletes indication) or fourth line on the left side.

Title. The title appears, centered, below the rest of the heading. If there are multiple authors and if the authors are from multiple organizations, the right-side heading may have additional lines to accommodate their names and organizational affiliations.

Running Header. The running header in one line (on page 2 and all subsequent pages) lists the RFC number on the left (RFC *NNNN*), the title (possibly a shortened form) centered, and the date (*month year*) on the right.

Running Footer. The running footer in one line (on all pages) lists the author's last name on the left and the page number on the right (Page *N*).

Status Section. Each RFC must include on its first page a "Status of this memo" section that contains a paragraph describing the type of the RFC. The content of this section will be one of the three following statements:

1. *Standards Track.* This document specifies an Internet standards track protocol for the Internet community, and requests discussion and suggestions for improvements. Please refer to the current edition of the "Internet Official Protocol Standards" (STD 1) for the standardization state and status of this protocol. Distribution of this memo is unlimited.

2. *Experimental.* This memo defines an Experimental Protocol for the Internet community. This memo does not specify an Internet standard of any kind. Discussion and suggestions for improvement are requested. Distribution of this memo is unlimited.

3. *Informational.* This memo provides information for the Internet community. This memo does not specify an Internet standard of any kind. Distribution of this memo is unlimited.

Introduction Section. Each RFC should have an introduction section that, among other things, explains the motivation for the RFC and, if appropriate, describes the applicability of the protocol described.

Discussion. Contains the following statement: The purpose of this RFC is to focus discussion on particular problems in the Internet and possible methods of solution. No proposed solutions in this document are intended as standards for the Internet. Rather, it is hoped that a general consensus will emerge as to the appropriate solution to such problems, leading eventually to the adoption of standards.

Interest.   Contains the following statement: This RFC is being distributed to members of the Internet community in order to solicit their reactions to the proposals contained in it. While the issues discussed may not be directly relevant to the research problems of the Internet, they may be interesting to a number of researchers and implementers.

Status Report.   Contains the following statement: In response to the need for maintenance of current information about the status and progress of various projects in the Internet community, this RFC is issued for the benefit of community members. The information contained in this document is accurate as of the date of publication, but is subject to change. Subsequent RFCs will reflect such changes. These paragraphs need not be followed word for word, but the general intent of the RFC must be made clear.

References Section.   Nearly all RFCs contain citations of other documents, and these are listed in a references section near the end of the RFC. There are many styles for references, and the RFCs have one of their own.

Security Considerations Section.   All RFCs must contain a section near the end of the document that discusses the security considerations of the protocol or procedures that are the main topic of the RFC.

Author's Address Section.   Each RFC must have at the very end a section giving the author's address, including name and postal address, telephone number, and, optionally, a FAX number), and email address.

RFC Action Words

The first RFCs led to ambiguity in the protocol. Not everyone reads and interprets alike. Therefore, most RFCs include the following action words to indicate precisely what should be implemented and what is optional.

MUST.   This word or the adjective *REQUIRED* means that the item is an absolute requirement of this specification.

MUST NOT.   This phrase means the item is an absolute prohibition of this specification.

SHOULD.   This word or the adjective *RECOMMENDED* means that there may exist valid reasons in particular circumstances to ignore this item, but the full implications should be understood and the case carefully weighed before choosing a different course.

SHOULD NOT.   This phrase means that there may exist valid reasons in particular circumstances that the listed behavior is acceptable or even useful, but the full implications should be understood and the case carefully weighed before implementing any behavior described with this label.

MAY.   This word or the adjective *OPTIONAL* means that this item is truly optional. One vendor may choose to include the item because a particular market requires it or because it enhances the product, for example, whereas another vendor may omit the same item.

TCP/IP

Most of the pieces that make up the Internet have been explained, but what about the actual protocol—what is TCP/IP? Let's start by studying the placement of the functions and protocols in the Open Systems Interconnect (OSI) model. In Fig. 7-5, we can see that there are distinct protocols that run at each layer from the network layer to the application layer of the OSI model. The heart of the TCP/IP network protocol is in layers 3 and 4. The applications for this protocol (file transfer, mail, and terminal emulation) run at the Session-through-Application layer.

As you can see, this protocol runs independently of the Data Link and Physical Layer. At these layers the TCP/IP protocol can run on Ethernet, Token Ring, FDDI, serial lines, X.25, etc. It has been adapted to run over these protocols. TCP/IP was first used to interconnect computer systems through synchronous lines, not high-speed local area networks. Today it is used on many types of media, including serial lines (asynchronous and synchronous) and high-speed networks such as FDDI, Ethernet, Token Ring, and Asynchronous Transfer Mode (ATM).

Figure 7-5 shows the bulk of the TCP/IP protocols that will be covered in this book. There are many, many more protocols than the ones shown here, but these are the ones required for the basic operation of TCP/IP. TCP/IP is actually a family of protocols working together to provide a path for Internet data communication. In this section, the TCP/IP protocol suite will be discussed in three phases:

1. The Internet Protocol (IPv4 and IPv6): RIP, OSPF, ICMP, IGMP, RSVP, and ARP

2. The Transmission Control Protocol and the User Datagram Protocol (TCP and UDP)

3. The suite of applications that were specifically developed for TCP/IP:

   Telnet

   File Transfer Protocol (FTP)

   Trivial File Transfer Protocol (TFTP)

   Simple Mail Transfer Protocol (SMTP)

   Domain Name Service (DNS)

   Real-Time Protocol (RTP)

   Real-Time Control Protocol (RTCP)

   Boot Protocol (BOOTP)

   Dynamic Host Configuration Protocol (DHCP)

This section gives a brief overview of the protocols as an introduction to the protocol suite of TCP/IP. Much more in-depth information is provided in the rest of this book. This section gives you a chance to read about each protocols briefly and then, if you wish, flip to the corresponding chapter in the book to learn more.

The Internet Protocol

The Internet Protocol (IP) is designed to interconnect packet-switched communication networks to form an internet. It transmits blocks of data called datagrams, received from IP's upper-layer software, to and from source and destination hosts. It provides a best-effort or connectionless delivery service between the source and destination—connectionless in that it does not establish a session between the source and destination before it transmits its data. This layer is also responsible for IP protocol addressing.

In order for multiple IP networks to interoperate, whether they are autonomous networks (single networks under control of one administrative authority) or multiple subnets inside one autonomous system, there must be a mechanism to provide flow between these differently addressed systems. The device that routes data between differently IP-addressed networks is called a router. The router is basically a traffic cop—you tell it where you want to go and it points you in the right direction. Routers contain ports that are physical connections to networks. Each of these ports must be assigned a local address. With more than one router, each router must know the others configured information. We could configure all the IP addresses and their associated ports on a router statically, but this is a very time-consuming and inefficient process. Therefore, we have protocols that distribute the IP address information to each router. These are called routing protocols.

The two main types of routing protocol for IP networks are RIP (Routing Information Protocol, version 1 or 2) and OSPF (Open Shortest Path First). Both are known as Interior Gateway Protocols (IGPs). These are protocols that run within a single autonomous system. An autonomous system is a collection of networks and routers that is under one administrative domain. For example, if a company known as Timbuktu Company has seven regional offices in the United States, all communications between those offices is accomplished via routers running RIP under one domain, known as *Timbuktu.com*. Connection to the outside world allows communication via the Internet (which is another domain) or to another company that is under another administrative domain.

There are two versions of IP: IPv4 (version 4, the current IP) and IPv6 (version 6, the experimental IP). IPv4 continues to operate admirably, but it has become burdened with "patches" to keep it working with all the new applications that are continually being placed on it such as Voice over IP and a slew of multimedia applications. The latest patch is the address scheme, and IPv6 was partially motivated by the inability of IPv4 to deal with running out of IP Class B addresses. IPv6 is a natural evolution of IP and not only extends the address space to 128 bits but also cleaned out a lot of unused functions.

However, IPv4 is becoming more entrenched, as newer schemes to use the old 32-bit address are being devised, such as using private addressing on your network and using address translators and other mechanisms to spoof your address to the Internet.

RIP. The origins of RIP are based in the origins of the Internet, but historically it came from the Xerox Network System (XNS) protocol. It was freely distributed in the UNIX operating system, and because of its simplicity it gained wide-spread acceptance. Unfortunately, there are many deficiencies associated with this RIP, and there have been many patches applied to it to make it work more reliably in large networks. For smaller networks though, the protocol works well.

Because IP is a *routable protocol*, it needs a *routing protocol* to enable it to route between networks. RIP is known as a *distance*-vector protocol. Its database (the routing table) contains two fields necessary for routing: a vector (a known IP address) and the distance (how many routers away) to the destination. The table contains more fields, but they will be discussed later in the book.

RIP builds a table in memory that contains all the routes that it knows about and the distance to a given network. When the protocol initializes it simply places the IP addresses of its local interfaces in the table. It associates a cost with each interface, which is usually set to 1 (explained in a moment). The router then solicits information from other routers on its locally attached subnets, or it may wait for information to be supplied to it. As other routers report (send their tables) to other routers, eventually the original routers will have all the information needed about all routes on its subnets or internetwork.

Any IP datagram that must traverse a router in the path to its destination is said to have traversed 1 hop. When a router receives a packet and examines the destination address in the datagram, it then performs a table lookup based on that destination address. The router also finds the port associated with the destination address in the database, and it forwards the datagram through that port to the final destination.

In RIP, all routers compute their IP network address tables and costs and send them to each other. A Router receiving these tables adds the cost assigned to the incoming interface (received port) to each of them. The router then decides whether or not to keep any of the information in the received table, in which case it stores the information in its own routing table. This information is then passed on to other routers.

OSPF. OSPF (Open Shortest Path First) is also a routing protocol, but it resembles RIP only in that it, too, is an IGP. When the Internet was created, the processors were no where near as powerful as what we have today. In fact, a Honeywell 516 minicomputer was used as the first router (then called an Internet Message Processor, or IMP). The only micro-CPU in those days was the Z80 from Zilog. RIP worked very well on the routers of that time. It also had very low overhead, in terms of computation. OSPF is a better protocol, but there was no machine that could feasibly run it at the time.

With today's faster processors and plentiful memory, OSPF is the routing protocol of choice (for open protocols, that is). It is very efficient when it comes to networks, although it is a complicated protocol, and it is very CPU-intensive when it builds its routing table.

OSPF is an IGP protocol. It exchanges routing information within a single autonomous system. It can be used in small, medium, or large internetworks, but the most dramatic effects are noticed on large IP networks. As opposed to RIP (a distance–vector protocol), OSPF is a link-state protocol. It maintains the state of every link in the domain. This information is "flooded" to all routers in the domain. Flooding is the process of receiving information on one port and transmitting it to all other active ports on the router. In this way, all routers receive the same information. This information is stored in a database called the link-state database, which is identical on every router in the autonomous system (or every area, if the domain is split into multiple areas). Based on information in the link-state database, an algorithm known as the Dykstra algorithm runs and produces a shortest-path tree based on the metrics, using itself as the root of the tree. The information this produces is used to build the routing table

Other Protocols.   The Internet Control Message Protocol (ICMP) is really as an extension of the IP layer. This is the reason that it uses an IP header and not a UDP (User Datagram Protocol) header. The purpose of ICMP is to report or test certain conditions on the network. IP delivers data and has no other form of communication. ICMP provides an error-reporting mechanism for IP. Basically, it allows internet devices (hosts or routers) to transmit error or test messages. An error message may indicate that a network destination cannot be reached, or it may generate or reply to an echo request packet (ping).

The Internet Group Membership Protocol (IGMP) is an extension of the IP protocol that allows multicasting for IP. The multicast address already existed for IP, but there was not a control protocol to allow it to exist on a network. IGMP is a protocol that operates in workstations and routers and allows the routers to determine which multicast addresses exist on their segments. With this knowledge, routers can build multicast trees, allowing multicast data to be received and propagated to their multicast workstations. IGMP headers are used as the basis for all multicast routing protocols for IPv4.

The Resource Reservation Setup Protocol (RSVP) provides some semblance of quality of service to exist in IP. It used to be that one could increase the speed of a network, which allowed more bandwidth for "hungry" applications. With this capability, quality of service was essentially ignored. However, the bandwidth cannot continually expand. The Internet was not set up for quality of service, and RSVP is the first attempt to allow for it. Its benefits show up best in multicasting applications, but it can be used with unicast applications, as well. It allows stations on a network to reserve resources via the routers on the network.

The Address Resolution Protocol (ARP) is not really part of the network layer; it "resides" between the IP and data-link layers. It is a protocol that translates between 32-bit IP addresses and 48-bit local area network addresses. ARP is only used with IPv4; it is not needed with IPv6. Because IP was not designed to run on a LAN, an address scheme was implemented to allow each host and network on the Internet to identify itself. When TCP/IP was adapted to run over the LAN, the IP address had to be mapped to the 48-bit data-link or physical address that LANs use, and ARP is the protocol that accomplishes it.

The Transport Layer Protocols

Because IP provides a connectionless delivery service of TCP (Transmission Control Protocol) data, TCP provides application programs access to the network, using a reliable connection-oriented transport-layer service. This protocol is responsible for establishing sessions between user processes on the Internet and also ensures reliable communication between two or more processes. The functions that it provides are

- To listen for incoming session-establishment requests
- To request sessions to other network stations
- To send and receive data reliably using sequence numbers and acknowledgments
- To gracefully close sessions

The User Datagram Protocol (UDP) provides application programs access to the network using an unreliable connectionless transport-layer service. It allows the transfer of data between source and destination stations without having to establish a session before data is transferred. This protocol does not use the end-to-end error checking and correction that TCP uses. With UDP, transport-layer functionality is there, but the overhead is low. It is primarily used for those applications that do not require the robustness of the TCP protocol, for example, email, broadcast messages, naming service, and network management.

TCP/IP Applications

Telnet is an application-level protocol that allows terminal emulation to pass through a network to a remote network station. (For users new to the TCP/IP protocol, this is not the same thing as Telenet, a packet-switching technology using the CCITT standard X.25.) Telnet runs on top of the TCP protocol and allows a network workstation to appear to a remote device (i.e., a host) as if it were a local device.

The File Transfer Protocol (FTP) is similar to Telnet in terms of control, but this protocol allows for data files to be reliably transferred on the Internet. FTP resides on top of TCP and uses it as its transport mechanism. The Trivial File Transfer Protocol (TFTP) is a simplex file transfer protocol (based on an unreliable transport layer called UDP). It is primarily used for boot-loading of configuration files across an internet.

The Simple Mail Transfer Protocol (SMTP) is an electronic mail system that is robust enough to run on the entire Internet system. This protocol allows the exchange of electronic mail between two or more systems on an internet. Using SMTP along with a system known as Post Office Protocol, individual users can retrieve their mail from centralized mail repositories.

The Domain Name Service (DNS) is a centralized name service that allows users to establish connections to network stations using humanly readable names instead of cryptic network addresses. It provides a name-to-network address translation service. There are many other functions of DNS, including translation from mail server name to IP address. Mail service would not exist if not for the DNS.

The Real-Time Protocol (RTP) and the Real-Time Control Protocol (RTCP) allow for real-time applications to exist on an IP network. RTP resides at the transport layer and works alongside the TCP protocol. It is a replacement for the TCP protocol for real-time applications. RTCP provides feedback to the RTP application. It lets the application know how things are going on the network. The protocols are actually frameworks rather than protocols; they are usually included in the application itself rather than residing as separate protocols with their own interfaces.

Multimedia applications such as voice and video are moving from experimental to emerging status. However, voice and video cannot simply be placed on a connectionless, packet-switched network. They need some help, and RTP and RTCP provide this help. RTP and RTCP, in conjunction with RSVP, are paving the way for real-time applications on the Internet.

The Boot Protocol (BOOTP) and Dynamic Host Configuration Protocol (DHCP) allow for management of IP parameters on a network. These protocols do not provide router configurations but end-station configurations. BOOTP was the original protocol that provided not only a workstation's IP address but possibly its operating image as well.

DHCP is best known for its management allocation scheme of IP addresses. However, DHCP is a superset of BOOTP, providing extended functions of IP, as well as IP address management.

Now that a brief introduction to the protocols has been made, let's take a deeper look into the protocol suite of TCP/IP, starting with the Internet Protocol (IP).

The Internet Protocol

Introduction

The main goal of IP is to provide interconnection of subnetworks (the interconnection of networks, explained later) to form an internet in order to transmit data. The IP protocol provides four main functions:

1. Basic data transfer
2. Addressing
3. Routing
4. Fragmentation of datagrams

The Internet Protocol—Connectionless, Best-Effort Delivery Service

The IP layer provides entry into the delivery system used to transport data across the Internet. Usually, when one hears the term *IP,* one thinks of networks connected together through devices commonly known as *routers,* which connect multiple subnetworks together. It is true that IP performs these tasks, but it also performs many others. The IP protocol runs on all the participating network stations that are attached to subnetworks so that they may submit their packets to routers or directly to other devices on the same network. It resides between the data-link layer and the transport layer. IP provides connectionless data delivery between nodes on an IP network.

The primary purpose of IP is to provide the basic algorithm for transfer of data to and from a network. In order to achieve this it implements two functions: *addressing* and *fragmentation.* It provides a connectionless delivery service for the upper-layer protocols. This means that IP does not set up a session (a virtual link) between the transmitting station and the receiving station prior to submitting the data to the receiving station. It encapsulates the data handed to it and delivers it on a *best-effort basis.* IP does not inform the sender or receiver of the status of the packet. It merely attempts to deliver the packet and will not make up for the faults encountered in this attempt. This means that if the data link fails or incurs a recoverable error, the IP layer will not inform anyone. It tried to deliver (addressed) a message and failed. It is up to the upper-layer protocols (TCP or even the application itself) to perform error recovery. For example, if your application were using TCP as its transport layer protocol, TCP would time out when an error occurred and then resend the data. If the application were using UDP as its transport, it then would be up to the application to perform error-recovery procedures.

IP submits a properly formatted data packet to the destination station and does not expect a status response. Because IP is a connectionless protocol, it might receive and deliver data (data sent to the transport layer in the receiving station) in a different order from the order in which it was sent, or it might duplicate the data. Again, it is up to the higher-layer protocols (layer 4 and above) to provide error-recovery procedures. IP is part of the network delivery system. It accepts data and formats then for transmission to the data-link layer. (Remember, the data-link layer provides the access methods to transmit and receive data from the attached cable plant.) IP also retrieves data from the data link and presents them to the requesting upper layer.

Refer to Fig. 8-1. IP adds its control information (in the form of headers), specific to the IP layer only, to the data received by the upper layer (the transport layer). Once this is accomplished, it informs the data link (layer 2) that it has a message to send to the network. At the network layer, encapsulated data are known as *datagrams* (Rumor has it that this term was coined referring to a similar message delivery system known as the telegram.) When the data-link layer adds its headers and trailers to a message, it is called a packet. When a message is transmitted onto the cable, the physical layer frames the information it has received from the data-link layer (basically with signaling information such as the preamble for Ethernet or the flag field for Frame Relay and X.25) and therefore it is called a *frame*. But for most of us, the terms *frame* and *packet* are

Figure 8-1
Data encapsulation.

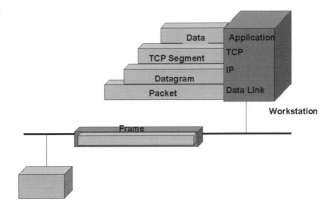

interchangeable. For the sake of simplicity, considering the fact that the primary focus of the book is network protocols over high-speed networks, *packets* and *frames* will be considered synonymous. Frames will not be mentioned unless the original specification mandates that term. It is important to remember that IP presents datagrams to its lower layer (the data-link layer). We will use the term *datagram* when specifically discussing the IP layer. We will use the term *packet* when specifically discussing the access layer (data link and physical).

The IP protocol does not care what kind of data are in a datagram. All it knows is that it must apply some control information, called an IP header, to the data received from the upper-layer protocol (presumably TCP or UDP) and try to deliver it to some station on the network or the Internet.

The IP protocol is not completely without merit. It does provide mechanisms on how hosts and routers should process transmitted or received datagrams, when an error should be generated, and when an IP datagram may be discarded. To understand the IP's functionality, a brief discussion at the control information it adds to the packet, the IP header, will be presented.

IP Header Fields

Figure 8-2 shows the IP header encapsulated in an Ethernet packet. It indicates the position of the header in the packet and shows the standard datagram header for an IP datagram. The following sections explain the header information in an IP datagram.

Version

This field defines the current version of IP implemented by the network station. Version 4 (IPv4) is the latest version, although IPv6 is being experimentally deployed. Table 8-1 lists the different Internet versions and their assigned numbers.

Header Length, Service Type, and Total Length

The Header Length field (HLEN in Fig. 8-2) defines the length of the IP header excluding the IP data field. Not all the fields in the IP header must be used. The Header Length field contains the number of 32-bit

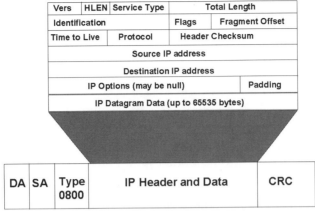

Figure 8-2
IPv4 header.

Vers	HLEN	Service Type	Total Length	
Identification			Flags	Fragment Offset
Time to Live		Protocol	Header Checksum	
Source IP address				
Destination IP address				
IP Options (may be null)				Padding
IP Datagram Data (up to 65535 bytes)				

DA	SA	Type 0800	IP Header and Data	CRC

Ethernet Data Field

TABLE 8-1

Assigned Internet
Version Numbers

Decimal	Keyword	Version	References
0	Reserved		
1–3	Unassigned		
4	IP	Internet Protocol	RFC791
5	ST	ST Datagram Mode	
6	IPv6		RFC 1883
7	TP/IX	TP/IX: The Next Internet	
8	PIP	The P Internet Protocol	
9	TUBA	TUBA	
10–14	Unassigned		
15	Reserved		

words in the header. The shortest possible IP header is 20 bytes. Therefore, this field would contain a minimum entry of 5 (20 bytes = 160 bits; 160 bits/32 bits = 5). This field is necessary, because the header can be variable in length, depending on the IP Options field.

The Service Type field is rarely used, so it is usually set to 0. This is an entry that allows applications to choose which type of routing path they would like. For example, a real-time protocol would choose low delay, high throughput, and high reliability, whereas a file transfer would not need this. A Telnet (TCP/IP remote terminal program) session could choose low delay with normal throughput and reliability. But there is another side to this story. The router must support this feature as well, which usually means building and maintaining multiple routing tables for each TOS entry and each routing protocol. The Service Type field is made up of the Precedence and Type of Service subfields.

The Precedence field can contain an entry from zero (normal precedence) to 7 (network control), which allows the transmitting station's application to indicate to the IP layer the priority of sending the datagram. This is combined with Type of Service (TOS) identifier bits: D (delay), T (throughput), and R (reliability) bits. These bits indicate to a router which route to take:

D bit—request low delay when set to a 1

T bit—request high throughput when set to a 1

R bit—request high reliability when set to a 1

For example, if there is more than one route to a destination, the router could read this field to pick a route. This becomes important in the OSPF routing protocol, which is the first IP routing protocol to take advantage of this system. If the transaction is a file transfer, you may want to set the bits to 0 0 1 to indicate that you do not need low delay or high throughput, but you would like high reliability. TOS fields are set by applications (i.e., Telnet or FTP), not routers. Routers only read this field. Based on the information read, routers select the optimal path for the datagram. It is up to the TCP/IP application running on a host to set these bits before transmitting the packet on the network. It does require a router to maintain multiple routing tables—one for each type of service. Not all routers support this field.

The Total Length field contains the length of the datagram (not packet), measured in bytes. This field allows for 16 bits, meaning the data area of the IP datagram can be up to 65,535 bytes in length.

Fragmentation

Although fragmentation should be avoided if possible, there are times when a packet transmitted from one network may to be too large to transmit on another network. The default datagram size[1] (this size

---

[1]This is known as the path MTU, or Maximum Transmission Unit. It is defined as the size of the largest packet that can be transmitted or received through a logical interface. This size includes the IP header but does not include the size of any link layer headers or framing.

includes the data and IP headers but not the Ethernet packet headers of the physical frame headers or trailers) is set to 576 bytes when the datagram is to be sent remotely (off the local subnet).[2] But why not utilize networks that support large packets? If a TCP connection path is from FDDI to Token Ring, why should the datagram size be automatically set to 576 bytes, when these media types support much larger packet sizes? The answer is that we cannot guarantee that any intermediate media types between the Token Ring and the FDDI support those large sizes. For example, suppose the source is a Token Ring station and the destination is a FDDI station, but in between the two stations are two Ethernet networks that support only 1518-byte packets. There are no tables in the routers or workstations that indicate media MTU (maximum transmission unit). There is a protocol (on path MTU discovery, see RFCs 1981 for IPv6 and 1191 for IPv4) that allows for this, but under IPv4 whether the router and workstations implement it is optional. Therefore, in order to be safe, instead of implementing RFC 1191, a transmitting station will send a 576-byte or smaller datagram when it knows the destination is not local.

Another example is a host initialized on an Ethernet that can send a request for a host server to boot it. Let's say the bootstrap host is on an FDDI network. The host sends back a 4,472-byte message, and this is received by the bridge. Normally, the bridge would discard the packet because bridges are capable of fragmenting IP datagrams. However, some bridge vendors have placed the IP fragmentation algorithm in their bridges to avoid this problem.

Although a router will fragment a datagram, it will not reassemble it. It is up to the receiving host to reassemble the datagram. Considering the implication of CPU and memory required to reassemble every datagram that is fragmented, this would be an overwhelming task of the router. If 2000 stations were communicating, and all stations were using fragmentation, a router could easily become overwhelmed, especially in those days.

A fragmented IP datagram contains the following fields:

- *Identification:* identifies a group to which datagrams belong so that they do not get mismatched. The receiving IP layer uses this field and the source IP address to identify which fragments belong together.

---

[2]Reference RFC 1812.

■ *Flags:* indicate

1. Whether more fragments are to arrive or whether no more data are to be sent for that datagram (there are no more fragments).

2. Whether or not to fragment a datagram (a don't-fragment bit). If a router receives a packet to be forwarded that it must fragment and the don't-fragment bit is set, it will discard the packet and send an error message (through a protocol known as ICMP, discussed later) to the source station.

■ *Offset.* The IP headers from each of the fragmented datagrams are almost identical. This field indicates the offset (in bytes) from the previous datagram that continues the complete datagram. In other words, if the first fragment has 512 bytes, this offset would indicate that the second datagram starts the 513th byte of the fragmented datagram. It is used by the receiver to reassemble the fragmented datagram.

Using the Total Length and the Fragment Offset fields, IP can reconstruct a fragmented datagram and deliver the reconstructed datagram to the upper-layer software. The Total Length field indicates the total length of the original packet, and the Offset field indicates to the node that is reassembling the packet the offset from the beginning of the packet. It is at this point that the data are placed in the data segment to reconstruct the packet.

Time to Live

There are two functions for the Time to Live (TTL) field: to limit the lifetime transmitted data and to end routing loops.

The initial TTL entry is set by the originator of the packet. This field (currently defined to be set in seconds) is set at the transmitting station and, as the datagram passes through each router, is decremented. With the speed of today's routers, the usual decrement is 1. One algorithm is that the receiving router notices the time a packet arrived and, when it is forwarded, decrements the field by the number of seconds the datagram sat in a queue waiting for forwarding. Not all algorithms work this way. A minimum decrement will always be 1. The router that decrements this field to 0 will discard the packet and inform the originator of the datagram (through the ICMP protocol) that the TTL field expired and the datagram did not reach its destination.

The Time to Live field may also be set to a certain time (i.e., initialized to a low number like 64) to ensure that a packet stays on the network for only a set time. Some routers allow the network administrator to set a manual entry to decrement. This field may contain any number from 0 to 255 (an 8-bit field).

Protocol and Checksum

The Protocol field is used to indicate which higher-level protocol should receive the data of the datagram (i.e., TCP, UDP or possibly other protocol like IGMP for multicasting). This field allows multiplexing of many protocols over the same IP interface. There are many protocols that may reside on top of IP. IP is not specific as to the protocol that runs on top of it. Currently, the most common transport implementations are TCP and UDP. The purpose of this field is to tell the IP how to correctly deliver the packet to the correct entity above it. If the Protocol field is set to a number that identifies TCP, the data will be handed to the TCP process for further processing. The same is true if the frame is set to UDP or any other upper-layer protocol.

The Header Checksum field is a cyclic redundancy check (CRC) of 16 bits. The idea behind it is to ensure the integrity of the header. A CRC number is generated from the data in the IP data field and placed into this field by the transmitting station. When the receiving station reads the data, it computes a CRC number. If the two CRC numbers do not match, there is an error in the header and the packet will be discarded. Each router that receives the datagram recomputes the checksum because the TTL field is changed by each router that the datagram traverses.

IP Options

This field is found on IPv4 packet headers. It contains information on source routing (nothing to do with Token Ring), tracing a route, time-stamping the packet as it traverses routers, and security entries. (For more information on these entries, please refer to the TCP/IP references at the end of this book.) These fields may or may not be in the header (which allows for its variable length). It was found that most of these features were not used or were better implemented in other protocols, so IPv6 does not implement them as a function of the header.

Source routing is the ability of the originating station to place route information in a datagram to be interpreted by routers. Routers forward the datagram based on information in the Source Route fields, which in some cases is blind. The originator indicates the path it wishes to take, and the routers must obey, even if there is a better route.

There are two types of source routes: loose and strict. The difference between the two is relatively simple. Routes (IP addresses) are placed in a field of the IP header. This IP address indicates the route the datagram would like to take to the destination. Loose source route allows a router to forward the datagram to any router it feels is correct to service the next route indicated in the Source Route field. A complete list of IP addresses from the source to the destination is probably not in the IP header, but some points in the Internet should be used to forward the datagram. For example, IP multicast uses LSR for tunneling its IP multicast datagrams over the nonmulticast-enabled IPv4 Internet. Strict source route forces a router to forward a datagram to its destination completely based on the routes indicated in the Source Route field.

The Traceroute is a very useful utility. It allows the echoing of the forwarding path of a datagram. With this option set, the points to which the datagram is routed are echoed back to the sender. This allows you to follow a datagram along its path. It is often used in troubleshooting IP networks. If you have Windows 95, you have this utility. Type in (at the DOS prompt) `tracert <IP address>` and watch the echo points on your screen.

IPv6 eliminated this field, along with those functions that were not used or were better implemented by other protocols.

Source and Destination Address (IPv4)

The addressing scheme for IPv4 has been altered very little during the last 17 years, but the protocols that use it have changed drastically. Therefore, this portion of the book has been expanded to fully explain IPv4 addressing. IPv6 addressing is covered later. Users will be most aware of this when starting their workstations or trying to access other stations without the use of a domain name server or an up-to-date host file. These fields indicate the *originator* of the datagram and the *final* destination IP address that the packet should be delivered to and the IP address of the station that originally transmitted the packet. All hosts on an IP internet will be identified by these addresses. IP addressing is extremely important, and a full discussion follows.

Currently, these address are set to 32 bits, which allows for o
lion addresses. This may sound like a lot, but, unfortunately, many mistakes were made in assigning IP addresses to corporations and individuals. (This topic is fully discussed later in the chapter.)

IPv6, the next version of IP (currently being implemented as autonomous islands in the sea of IPv4) accommodates 128 bits of address, which basically allows thousands of billions of hosts to be numbered. With IPv6, an efficient allocation scheme was developed to hand out IPv6 addresses, as well.

There are two types of network addressing schemes used with IPv4:

■ *Classless:* The full address range can be used without regard to bit reservation for classes. This type of addressing scheme is not primarily used in direct host assignment. The scheme is directly applied to the routing tables of the Internet and ISPs.

■ *Classful:* The original (RFC 791) segmentation of the 32-bit address into specific classes denoting networks and hosts.

Classful addressing is used on corporate networks. Classless addressing is used on the backbone of the Internet but not usually on a corporate networks. The range of addresses (32 bits for IPv4) available is used for both classless and classful addressing. We will first discuss classful addressing, because it started first and continues to be used on many networks. In fact, most of us will never deal with classless addressing. It is primarily a concern for internet service providers and their allocations from the Internet Registry and assignment to their customers.

There are protocols that understand each type. Why not do one or the other? Classless routing is performed primarily by the routers of ISPs. Customers will not run this type of addressing scheme. They could, but not everyone will have the latest version of IP, and most customer sites will not be reconfigured to support classless IP addressing.

A later section of this chapter will deal with classless addressing and the concepts of CIDR (classless interdomain routing), and variable-length subnet masks (VLSMs), and supernetting will be presented in the following section.

Classful Addressing

RFC 760 introduced IP. The beginnings of the IP addressing scheme were very simple and flat. The RFC didn't include a concept of classes (not to be confused with the term *classless* as it is used today); addressing was an 8-bit prefix that allowed as many as 200 + networks and many hosts per network. RFC 791 obsoletes RFC 760 and included the concept of classes. RFC 950 introduced us to subnetting, and RFC 1518 introduced the CIDR (classless) protocol. There have been many enhancements to the IP addressing scheme, but they continue to operate on the bases of Class and Classless.

The ideas and concepts from which TCP/IP evolved were devised before any data-link protocols of Ethernet and Token Ring were developed. Hosts were not attached to a local high-speed network (such as Ethernet or Token Ring). Hosts communicated with each other through low-speed point-to-point serial lines (telephone lines). Therefore, an addressing scheme to identify TCP/IP hosts and where they were located was implemented. The addressing scheme used to identify these hosts is called 32-bit IP addressing. This is also known as a protocol address.

Addressing's purpose was to allow IP to communicate between hosts on a network or on an internet. Classful IP addresses identify both a particular node and the network number at which that node resides on an internet. IP addresses are 32 bits long, separated into four fields of 1 byte each. This address can be expressed in decimal, octal, hexadecimal, and binary. The most common IP address form is written in decimal and is known as the *dotted decimal notation* system.

There are two ways that an IP address is assigned. It all depends on your connection. If you have a connection to the Internet, the network portion of the address is assigned through your internet service provider, which only provides the network range (a continuous IP network address segment) that you can work with. The ISP does not assign host numbers, nor does it assign the network numbers to any part of your network. If your network does not have a connection to the Internet, the IP addresses for your internet can be locally assigned by your network administrator. However, these addresses will require translation before an attachment to the Internet can be made through an ISP.

IP Address Format

Each host on a TCP/IP network is uniquely identified at the IP layer with an address that takes the form of *<netid, hostid>*. The address is not really two separate sections—it is read as a whole. The whole address is always used to fully identify a host. In fact, when an IP address is written, it is hard to tell the distinction between the two fields without knowing how to separate them, because there are no apparent delimiter. The following shows the generalized format of an IP address:

<Network Number, Host Number>

In decimal, the address range is 0.0.0.0 through 255.255.255.255. An example of an IP address is 128.4.70.9. When looking at this address, it is hard to tell which is the network number and which is the host number, let alone a subnet number (discussed below). Any of bytes 2 through 4 can indicate either a network number or host number. The first byte always indicates a network number. In order to understand how this is accomplished, let's look first at the how IP addresses are divided. The structure of an IP address takes the form shown in Table 8-2. There are 32 bits separated into 4 bytes that represent an IP address. The network number can shift from the first byte to the second byte to the third byte. The same can happen to the host portion of the address. The letters *xxx* represent a decimal number from 0 to 255 in decimal.

IP addresses are divided into five classes: A, B, C, D, and E. (RFC 791, which classified these types, did so before the development of subnets.) The classes allow various numbers of networks and hosts to be assigned. Classes A, B, and C are used to represent host and network addresses. Class D is a special type of address used for multicasting (for example, OSPF routing updates use this type of address, as well as IP multicast). Class E is reserved for experimental use.

In order to figure out this addressing scheme, it is best to know the binary numbering system and be able to convert between decimal and binary. IP addresses are sometimes expressed in hexadecimal, so it is helpful to know that system, too. IPv6 uses only hexadecimal, but the most common form for IPv4 is decimal. This book shows most addresses in binary and decimal.

TABLE 8-2

Structure of an IP Address

Byte 1	Byte 2	Byte 3	Byte 4
xxx	*xxx*	*xxx*	*xxx* Decimal, HEX, or octal (usually decimal)
<Network	Number	AND Host	Number>

Figure 8-3

IPv4 address classes.

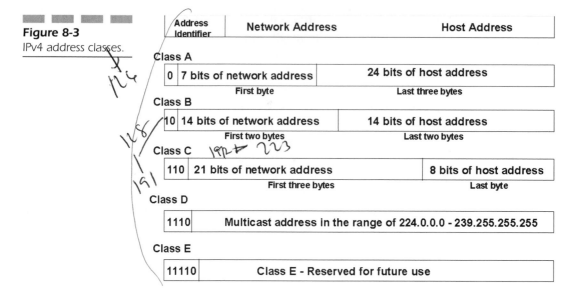

IP Class Identification

Classes A, B, and C are the most commonly used classes. Figure 8-3 shows how the classes are actually defined. How does a host or internet device determine which address is of which class? Because the length of the network ID is variable and dependent on its class, a simple method was devised to allow the software to determine the class of address and, therefore, the length of the network number.

The IP software determines the class of the network ID by reading the first bit(s) in the first field (the first byte) of every packet. Hosts and routers read in binary. Figure 8-3 breaks the IP addresses down into their binary equivalents. If the first bit of the first byte is a 0, it is a class A address. If the first bit is a 1, the protocol mandates reading the next bit. If the next bit is a 0, it is a class B address. If the first and second bits are 1 and the third bit is a 0, it is a class C address. If the first, second, and third bits are 1, the address is a class D address and is reserved for multi-cast addresses. Class E addresses are reserved for experimental use.

Class A. Class A addresses take the 4-byte form *<network number.host.host.host>*, bytes 0, 1, 2, and 3. Class A addresses use only the first of the four bytes for the network number. Class A is identified by the first bit in the first byte of the address. If this first bit is a 0, it is a class A address. The last three bytes are used for the host portion of the address.

Class A addressing allows for 126 networks (using only the first byte) with up to 16,777,214 million hosts per network number. The range for class A is 1–126. With 24 bits in the host fields (last three bytes), there can be 16,277,214 hosts per network (disregarding subnets). This is actually $2^{24}-2$. We subtract 2 because no host can be assigned all 0s (reserved to indicate a default route or all 1s. For example, 10.255.255.255 is not allowed as a unique address.

If all seven bits are set to 1 (starting from the right), this represents 127 in decimal, and 127.*x.x.x* is reserved as an internal loopback address and cannot be assigned to any host as a unique address. This can be used to indicate whether or not your local TCP/IP stack (software) is up and running. The address is never seen on the network. If you look at your machine IP addresses (usually by typing `netstat -r` at the command line), you will notice that every machine has 127.0.0.1 assigned to it. The software uses this as an internal loopback address. You should not see this address cross over the LAN (via a protocol analyzer such as a Sniffer™.) In fact, 127.*anything* is proposed as the loopback, so 127.1.1.1 delivers the same results as 127.0.0.1. It is an incredible waste of address space to have an entire IP address number assigned for one task.

About half of the Class A addresses were originally assigned. Today, Class A addresses are being handed out but through a different method involving internet service providers and the classless interdomain routing protocol (CIDR).

Class B. Class B addresses take the form <*network number.network number.host.host*>, for bytes 0, 1, 2, and 3. This has been the most requested class of address. It is the easiest to assign subnets to. Class B addresses use the first two of the four bytes for the network number and the last two for the host number. The class is identified by the first two bits of the first byte. If the first bit is a 1, the algorithm checks the second bit. If the second bit is a 0, this identifies a class B address.

Class B addressing allows for 16,384 network numbers (10111111.11111111.*host.host*, or 2^{14}, with each network number capable of supporting 65,534 ($2^{16}-2$)hosts (*net.net*.11111111.11111110). There are 16 bits in the first two fields, which should allow for 65,535 networks, but class B reserves the first two bits to identify the class type (in binary, a 10*xxxxxx* in the first field), so there are limited address numbers that can be used in the first field (the valid range becomes 2^{14}). This translates to 128 to 191 (in decimal) as the allowable network number range in the first field. Because the first field identifies the class, the second field is free to use all eight bits and can range from 0 to 255. The total range for network numbers for class B addresses is 128 to 191 in the first field, 0 to

255 in the second field, and *xxx.xxx* (representing the host ID) in the third and fourth fields.

Class B provides the largest range of addressing possibilities. However, new class B addresses are not available, because the class is exhausted and they are no longer given out.

Class C. Class C takes the form of *<network number.network number.network number.host>*, bytes 0, 1, 2, and 3. Class C addresses use the first three out of four bytes of the address for the network number and the last field for the host number. This allows lots of networks with fewer hosts per network. A class C address is identified by the first three bits of the first field. If the first and second bits are 1 and the third bit is a 0, this identifies a class C address (110*xxxxx*). Because the first three bits in the first field will always be 110*xxxxx*, the allowable network range is 192–223 in the first field. This allows for 2,097,152 (2^{21}) possible network addresses. All the bits in the second and third fields are allowed to be used (including all 0s and 1s). Therefore, the whole allowable range for class C network addresses is 192 to 223 in the first field, 0 to 255 in the second field, and 0 to 255 in the third field. The last field will range from 1 to 254 for host assignment. This allows 2,097,152 network numbers, each capable of supporting 254 hosts (all 0s and all 1s are still reserved, no matter what type of routing and addressing you are using). Notice that the largest possible number in the first field is 223. Any number over 223 in the first field indicates a class D address. Class D addresses are reserved as multicast addresses and are not used as unique addresses.

Class C addresses are the most commonly assigned addresses by the NIC. Class B addresses have been exhausted, so ISPs and regional Internet Registries are only assigning class C and class A addresses.

Class D. Class D addresses are special addresses and are known as multicast addresses. This address type is assigned to a group of network workstations and is not assigned to represent a unique address. In other words, multiple stations can have the same address. They are used to send IP datagrams to a group of but not all the hosts on a network. Multicasting has many uses including addressing router update messages and delivering data, video (conference), and voice (conference) over IP. It is more efficient than "broadcasting" because the upper-layer software is not interrupted every time a broadcast packet arrives. With broadcasting, every station that receives the broadcast packet automatically passes it to the upper-layer software without regard to the address. Every station that receives a broadcast packet must process it.

With a multicast address, each individual IP station must be willing to accept the multicast IP address before the transport-layer software is interrupted. Each NIC card registers a MAC-layer multicast address on its adapter card, just like a unicast address. In this way, the NIC card can discard a packet without interrupting the upper-layer software (there is some duplication of multicast and MAC addresses). The NIC is already set up to receive a broadcast packet. This is one address known as *FF-FF-FF-FF-FF-FF.*

As of this writing, RFC 1700 (on assigned numbers) fully explains the mapping of class D addresses to MAC addresses, and it also indicates assigned and registered multicast addresses. Multicasting is covered more completely in another section of this book.

Classes A–D Review.   For anyone new to this protocol, the easiest way to remember IP addresses and their associated class is this: the *first byte* always identifies the *class address. A* is the *first letter* in the alphabet, and a class A network address is only the *first byte,* leaving the last three fields for host addressing. *B* is the *second letter* in the alphabet, and the network portion of the address is the first *two bytes* of the address, leaving the last two fields for the host address. *C* is the *third letter* in the alphabet, and the network portion takes up the first *three bytes* of the address and leaves one field for host addresses. As for remembering which number is associated with which class, the only field that is important is the first field. Memorize the starting network number for each class.

Subnetting

You should review RFC 950 for more details on subnetting. Another name for a subnet mask is extended network prefix, but this book uses the term *subnetting.*

Implementing classes in network numbers gave some hierarchical structure to the Internet. Using class assignment, a network number could be selected based on the number of hosts that are on or would be on a network. However, the range was very limited. Class A allowed a lot of hosts but just a few networks. Class B allowed a balance of hosts and networks, and class C allowed many networks and a few hosts. Not much choice—either you had a lot of networks or a lot of hosts. The most requested network address type was class B, but many class B assignments were not fully exploited—it is rare to have 65,535 hosts on a single network. Too many class C addresses fill up routing tables, and most did not fully use all 254 available host addresses. Some sites even requested multiple addresses to fulfill their needs.

Not many class A addresses were assigned. In fact, after about 63 assignments, class A assignments were not handed out at all. Class B addresses were popular and were the most frequently asked for address class. Class C addresses could be and were assigned, but with only 254 hosts available, many Class C addresses were already assigned. Again, using class assignment, the routing tables started to fill up and most of the bits were wasted when implemented. It was like being given a five-passenger car, but you never had anyone in the other seats.

Subnetting allows tremendous efficiency not only in Internet routing tables but also on customer networks. It allows reassignment of some of the bits normally used by the host portion of the address to the network portion of the address. Figure 8-4 shows a subnetted network topology connected to the Internet. It has a class B address and uses an 8-bit subnet mask. The Internet recognizes the IP address 130.1.0.0, but it does not know about the subnets involved. This allows the Internet address (routing) tables to remain smaller while also allowing efficiency on the customer network by subnetting.

Reasons for Subnetting

As network numbers were assigned, many sites implemented routing on their local sites. This had many benefits, but using only RFC 791 and class assignment, a site had to be assigned multiple network numbers in order to accomplish this. This started to fill up the ARPANET routing tables, and created other problems, as well.

Many networks that accessed the Internet were creating and implementing their own "home-grown" subnetted environments. Before all networks ceased communicating because of incompatibilities, RFC 950 was released, defining a standard method for subnetting IP addresses. A network mask that covers only the network portion of the address (no

Figure 8-4
Subnetting.

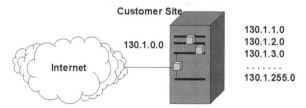

portion of the address is subnetted) is known as a natural mask. Subnet masks are used in routers and network stations.

Subnetting Examples (Classes A, B, and C)

Any of the classes A, B, and C can be subnetted, although this is easier with some than with others. Figure 8-5 shows the three classes of networks, each with an address. Each of the addresses has been assigned a subnet mask. A mask is the portion of an address that is subtracted from the original address. It indicates how many bits are masked out of the original address to use as a subnet address. Subnet masks are variable in length and move from the first host bit to almost the last. In other words, they move to the right of the address. Moving a mask to the left of the network address, beyond its natural mask, is known as supernetting, which will be discussed later.

In this example of class A and B addresses, all available bits following the network ID portion of the address are shown to indicate a subnet. The class C address uses the first six bits of the host portion of the address for the subnet. With any of the addresses, any of the host bits (except for two bits at the end of the address—there must be at least one host on a network) may be used for subnetting. For example, a class B address may use all of the third octet and two bits of the fourth octet for subnetting. This would give 1024 possible subnetwork numbers. In order to have 1024 subnet addresses, we must use all 0s and all 1s in the subnet field as valid subnet addresses. This may seem contrary to host and network ID assignment, but it is not. All 0s and all 1s can be used in the subnet portion of any address (they still cannot be used in the host

Figure 8-5
Subnet bit ranges.

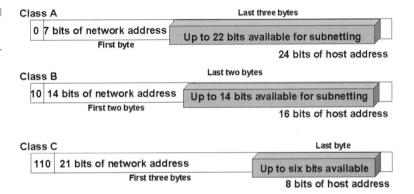

or network portion of the address as unique addresses). (Refer to RFC 1812.) This causes problems with subnet broadcasts, which will explained later. Using the above example (10-bit subnet on a class B address), each subnet can support up 62 hosts (63 would indicate 11 hosts broadcast).

Some considerations about subnets include the following:

Hosts and routers must implement subnetting (there is a way around this that is discussed in the section "Proxy ARP") and locally must have the same mask.

The router must be able to distinguish between all 1s as a subnet address and a subnet broadcast.

The routing update protocol must support subnetting.

More Subnet Examples

Subnetting confiscates unused bits, allowing more efficient use of address space. They are being taken away from host assignment and given back to identify a subnet of a network address. Figure 8-6 depicts two examples of real networks. They are subnets under the network numbers. As shown in the example, when subnetting a class B address, we can take any number of the bits of the third or six bits of the fourth byte (they should be contiguous, starting from the left) of the IP address and make them part of the network number (a subnet under the network number). The format of the IP address would now be: *<network number, subnet number, host number>*. For example, if the address assigned

Figure 8-6
More subnet examples.

Original Network and Host		Subnet Mask	Network Subnet Host		
128.1	1.1	255.255.255.0	128.1	1	1

• We recovered some of the host bits and made them into a subnet
• This reduces the amount of hosts we can have but who cares!

xxxxxxxx.xxxxxxx.11111000.xxxxx
Subnet mask 255.255.248.0

Subnet starts at 0 and increases in multiple of 8 for a range of 2^5.
Subnet mask range is 0 through 248, in increments of 8.

to a particular host is <u>128.1.1.1</u>, the network portion would be 128.1 and the host portion would be <u>1.1</u>. With subnetting (assuming all eight bits of the third field are used for a subnet address), the address would be defined as network number 128.1 and subnet 1, with a host ID of 1.

To illustrate further, if the class B network address of 130.1.0.0 is assigned to a site, it can be subnetted into the following: 130.1.*xxx*.0, *x* here indicating the decimal-formatted field that can be used for subnetting. The field can be subnetted using any of the bits in the field. (Translating binary into decimal, you can have one subnet using the most significant bit of the third field or use all the bits in the third field, which would give you 256 subnets with 254 hosts per subnet.)

The bottom of Fig. 8-6 shows that you do not have to take all the bits to make a subnet. It really is a case-by-case decision based on needs and expansion plans. Suppose the first five bits (starting from the left; they should start from the left and remain contiguous going to the right) are reserved in the third field for assigning subnet numbers. Convert those first five bits of that octet to binary. All five of these bits are now assigned to the subnet number and cannot be used for host IDs. Five bits yields 32 subnet numbers. Now, what are those numbers?

If we start from the left and go five bits to the right, we get *x.x.*11111000.*x* as a subnetwork number. The binary numbers are taken literally and will yield subnets in multiples of 8 (8 is the first binary bit set to 1). This gives us 0 ,8 , 16, 24, 32, 40, 48, 56, 64, 72, 80, 88, 96, 104, 112, 120, 128, 136, 144, 152, 160, 168, 176, 184, 192, 200, 208, 216, 224, 232, 240, and 248.

Physical and Logical Addresses

As shown in Fig. 8-7, the address is still read as if subnetting has not been turned on—it is only written differently. The address is 130.1.9.1, but, as shown, this is network 130.1, subnet 8, and host 257. Logically,

Figure 8-7
Physical and logical addresses.

10000010	00000001	00001001	00000001
130	1	9	1

Subnet Mask 255.255.248.0
Logical Address is:
Network: 130.1.0.0
Subnet: 8
Host: 257

Physically the host is number 257 on this subnet.

each field can never read beyond 255, but by using subnetting, you can have a host number *physically* higher than 255.

Writing a Subnet Mask

Let's break down the preceding example. Identifying the subnets is a little tricky. Figure 8-7 shows the binary equivalent of the conversion. If you are not familiar with binary, this may be tough, but you should be able to see the decimal equivalents of the subnet mask and add them up.

In our example, with each of those subnetwork numbers, we could have up to 2046 hosts per subnetwork number. This is a little more realistic than not subnetting, which would allow 65,534 hosts. We were assigned one IP address, and subnetting allows us to make better use of the address without having to reserve more addresses, or network numbers. Also, with subnetting only one IP address is in the Internet routing tables, even though we have 32 subnets on our network. The Internet routing tables do not care about subnets. We used one class B network number and have 32 subnets available to us from that one network. Without subnetting, we would have one network number and up to 65,534 hosts assigned to it.

How did we get 32 possibilities? Using five bits for the subnet mask gives us 32 (2^5) possible combinations (0 to 31). Remember, we can move the mask anywhere in the 14 available bits. The subnet mask could have used all eight bits in the third octet, which would give us 256 subnet numbers (all 0s and all 1s being allowed).

How do we write a subnet mask? It is always written in decimal and shows the number that will be used to mask the bits. For example, let's use the IP address 130.40.132.3. Using the first five bits of the first host field (the third octet) yields 248 (convert the first five bits to binary: 11111000). The byte is read as a whole eight bits even though part of it is used for the subnet and part for host assignment. This means the subnet mask for that IP address will be 255.255.248.0 in decimal. This is the mask that we have assigned to the network address of 130.40.132.3. We will always use 255 in the network portion of the subnet mask. The number 248 tells a network station to use the first five bits (five bits binary is 248 decimal) of the network address, not for a host ID but for a subnet. It tells the network station which bits to use for a subnet mask. The remaining eleven bits (the remaining three bits of the third octet and eight bits of the fourth octet) should be used for the host ID. This allows for 32 subnets with 2046 hosts on each subnet. Therefore, the IP address of 130.40.132.3, with a subnet mask of 255.255.248.0, yields network number 130.40, subnet number 128, and host ID 1027.

An Example Conversion. (*Hint:* Convert the address to binary, apply the mask in binary, and then convert it back to decimal as shown in Fig. 8-8a. When a bitwise AND operation is performed on an IP address, the IP address is ANDed with the subnet mask to allow the network station to determine the subnet mask.

Figure 8-8b shows the mask operation. In the middle of the figure is the IP address in binary. This address is logically ANDed with the mask. The bits that drop out of this operation indicate the network address to any TCP/IP station. This process masks out the host address and leaves the network address.

Figure 8-8
Masking. (a) Writing a subnet mask. (b) Using a nonaligned mask.

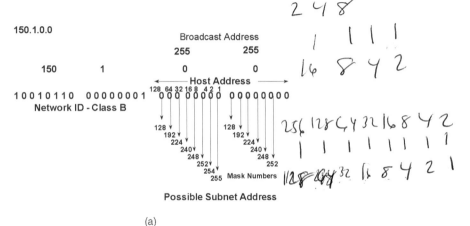

(a)

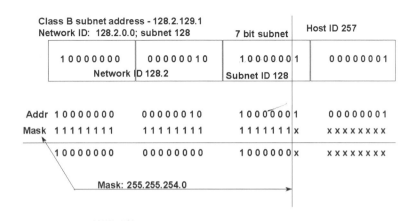

CIDR Prefix /23

(b)

Class A addresses can use the second, third, or fourth field (but not the whole fourth field) for subnets, and Class B addresses can use the third or fourth field (but not the whole fourth field). Class C is tricky—the only field left is the single host field. Subnetting this is allowed, but you can only use up to six of the bits in the fourth field, because you need to have address space for hosts.

Figuring out How Much. You have been assigned an address of 130.1.0.0. You need 50 subnets with at least 140 hosts per subnet. What is the acceptable subnet mask? You should realize that one mask does not fit all and that each customer must find the mask that best fits its site and allows for expansion. Later, we will discuss how to get around assigning one mask per network number through the concept of variable-length subnet masks, but for now, let's try a conversion.

The first step is to find out how many bits are needed for 50 subnets. In binary, six bits represents 64 (2^6) possible subnets. Therefore, we need to subnet off at least six bits for the subnet. This leaves 14 possible expansion subnets as well. Masking off six bits allows for 10 bits for the host address. This allows a possible 1022 hosts per subnet, well above the requirement of 100. We assign subnets from the left and work to the right. We assign the hosts from the right and work to the left. You must also define the broadcast address for a subnet. For the example above, the all-hosts broadcast is 130.1.3.255 (all the host bits are set to 1). You must fill in the submit number. This is just an example showing setting all the host bits.

Another example would be to define the mask for a network to support 40 hosts per subnet using the class address 195.1.10.0. First off, we determine that this address is a class C address and that only the last octet can be used for subnetting. The number 2^6 represents 40 hosts. This may seem like a lot, but the nearest mask would be 2^5, which would give us 30 possible host IDs, and this is not enough. The number 40 converted to binary is 101000. However, in the conversion we must remain contiguous, and we cannot interleave host and subnet bits. Therefore, we move to the left six bits, and we can use all six bits to the right. However, this only leaves two bits for a subnet. We can have four subnets with 62 (2^6-2) hosts per subnet. If the site needed more subnets, we would have to assign more class C addresses to the site.

A Possible Restriction. A routable protocol such as IP needs a routing protocol that allows the propagation of network IDs and subnets to other routers for the purpose of universal routing. For an autonomous system, these are known as Interior Gateway Routing Protocols. These are discussed later, but one related concept should be discussed here.

Depending on the routing protocol used (RIP version 1, RIP version 2, or OSPF), the network may be restricted as to variable subnet masks. When using RIP version 1, when a single network ID is subnetted, the whole network (all stations assigned to that network number) must use the same mask for that network address. When the network number changes (not the subnetwork number), the subnet mask may change. You cannot assign one network number with different subnet masks.

For example, when the RIPv1 routing protocol (explained later) is used, the subnet mask must remain the same throughout a single class B assignment. For example, if the network assignment is 130.1.0.0 and the subnet mask assigned is 255.255.255.0, this subnet mask must remain the same throughout the 130.1.0.0 network. If the network address changes—for example, to 131.1.0.0—the subnet mask may also change for this new network number.

RIP version 2 and OSPF do not have this restriction, because they broadcast their subnet mask in the table with the network IDs.

Subnet Mask Decisions. Let's say you are assigned one network number and you are using RIPv1. The subnet mask must be the same throughout your network. You decide how big the subnet mask should be, how many hosts per subnet will there be, and whether to allow for expansion. These are considerations you must make when assigning a subnet mask. With RIPv1 it is a trade-off (between more subnets than you need or more hosts). OSPF and RIPv2 do not have this trade-off, but care must still be taken when assigning network masks to a network number. This is shown completely in a later section on advanced IP addressing concepts.

Since RPv1 requires one subnet mask per network number, the serial line is caught in the problem with only two host needing asssignment. This restriction of one subnet mask becomes readily noticeable when assigning an IP address to a serial line (two routers using a leased phone line to connect). Some router vendors have come up with methods that accommodate the restriction of IP address assignment for a serial line. However, if the serial link needs an address assignment and you are not using RIP version 2 or OSPF, a whole subnet number is wasted on this point-to-point link. This is because the serial line (as far as IP is con-cerned) looks like a LAN and can be assigned network numbers just like stations attached to an Ethernet network. A serial link consumes a net-work number and associated host IDs, so a unique network number will be assigned and, instead of being able to use all available hosts IDs, it will be possible to use only two hosts IDs (there will be only two addressable points on that network).

The rest of the host IDs will be lost for that network number. They will be assigned and used for that serial link and will not be assignable to any other links. If you have a large site that encompasses many serial links and you do not have the ability to assign a large number of network numbers, use subnet addressing and the OSPF routing protocol. OSPF supports variable-length subnet masks, which collapse that serial link into two hosts within a network number, and, therefore, no host numbers are wasted on serial links. Variable-length subnet masks allow a single network number to use multiple masks. (Unlike RIP version 1, version 2 allows VLSM.) This allows more bits to be assigned back to the network, which results in a more efficient use of the address.

Another consideration is, if the network station moves to a new network, does the IP address for that station change? Like the current telephone system, IP addresses must change when the network station is moved to a new network that employs a different network number. If the network station is moved on the same logical network, the IP address may remain the same. For example, if a network station is moved to a different part of the same subnet, the whole IP address can stay the same. If the network station is moved to a different subnet (with a different subnet number), the IP address of the network station must change.

Multinetting

Multinetting is known by many different names. Basically, it is the assignment of multiple network numbers (regardless of the host number) to a single interface. This can be accomplished on routers to extend a multiple class C address on the same cable segments. This is not a function of TCP/IP but more of a trick that can be implemented by routers or workstations supporting routing protocols (see Fig. 8-9). This means that one network can employ more than one network number on the same physical cable plant. In order to accomplish this, a router must be used. Network stations continue to believe they are communicating with a remote network station, but the router is simply providing the address translation. The packet goes in one port and then right back out the same port. The two nodes actually reside on the same network segment. A router takes the steps necessary to allow network stations to converse on the network. Implementations differ in this regard, so the number of network numbers that can be assigned to the same cable plant varies.

Figure 8-9
Multinet.

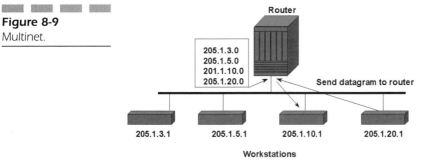

For example, as shown in Fig. 8-9, multiple class C network numbers can be assigned to the same cable plant. Class C addresses allow only for 254 host IDs per network number. This is a rather low number, and some sites have more than 254 network stations attached to a cable plant. This means that multiple stations on the same cable plant may have different network addresses. A router must be used to translate between two stations that are located on the same cable plant with different network addresses. This is called multinetting an interface.

Routers and Subnet Masks

Routers and routing update protocols are discussed fully in a later section. However, some readers may be far enough along to understand the following. How does a router that uses a simple routing update protocol such as RIP understand subnet masks? Very simply, it assumes (see Fig. 8-10).

The routing protocol RIP version 1, requires that a subnet mask be uniform across an entire network ID. The address 150.1.0.0 must contain one network mask. The fault here is the inability of RIPv1 to supply a

Figure 8-10
Routers and subnet masks, RIP version 1.

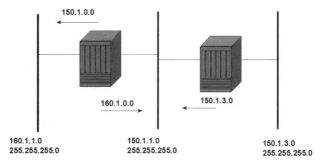

subnet mask entry in its routing updates to be used by other routers. RIPv1 is forced to make assumptions. It assumes that the mask is the same for learned subnets of the same network ID as its configured ports. This means that if a subnet route is learned on a port that has the same network ID as the port, RIP applies the assigned mask to that learned route as the port. If the learned subnet route has a different network ID than the port from which it learned the subnet route, it assumes the learned subnet route is not subnetted and applies the natural mask for that class.

Let's look at an example, a router that has two ports (see Fig. 8—10). The first port is assigned an address of 150.1.1.1, subnet mask 255.255.255.0. Port 2 has an address of 160.1.1.1 with a subnet mask of 255.255.255.0. If the router learns of a route 150.1.3.0, it applies the 24-bit subnet mask because it has the same network ID as its port. However, if the router learns a subnet route of 175.1.6.0, this network ID is not on either one of its ports and it applies a natural subnet mask of 255.255.0.0 to that address before updating its table for learned routes.

When does a router apply the subnet mask to a route and then include it in the routing update? The same rule applies. Taking the network numbers from the example above, when the router would like to broadcast its table, it applies the subnet mask of 255.255.255.0 to the learned route of 150.1.3.0 when it sends it update out port 1. However, it sends the address of 150.1.0.0 when sending the update out port 2. Port 2 has a different network ID associated with it, and therefore the natural mask is applied before sending out the table. These are the reasons that RIPv1 supports only one subnet mask for each network id.

Classful IP Address Review

All IPv4 addresses are 32 bits in length and are groupings of four bytes that represent both a network number and a host number. This address is usually represented in decimal. With the first bit reserved (set to 0$xxxxxxx$) in a class A address, the network numbers can range from 1 to 126. Number 127 is reserved as a local loopback IP address and must not be assigned to a network number and transmitted onto the network. With the first two bits reserved (10$xxxxxx$) in a class B address or three bits (110$xxxxx$) in a class C address, the network numbers for class B range from 128.1.0.0 to 191.255.0.0, and for class C they range from 192.1.1.0 to 223.255.255.0. Here are some examples:

192.1.1.1	node assigned with a host ID of 1, located on a class C net of network 192.1.1.0
200.6.5.4	node assigned with a host ID of 4, located on a class C network of 200.6.5.0
150.150.5.6	node assigned with a host ID of 5.6, located on a class B network of 150.150.0.0
9.6.7.8	node assigned with a host ID of 6.7.8, located on a class A net work of 9.0.0.0
128.1.0.1	node assigned with a host ID of 0.1, located on a class B network of 128.1.0.0

Notice that to represent only a network number, only the network number is written. The host field is set to 0. This type of network number display becomes apparent when you look at routing tables.

Those not familiar with binary should memorize the starting and stopping points of the first byte of an IP address:

Class A	1 to 126 in the first field	0
Class B	128 to 191 in the first field	10
Class C	192 to 223 in the first field	110

Subnetting provides the ability to move a mask over the bits normally associated with a host address and reclaim these bits as a subnet number. A mask can use 22 bits for a class A address, 14 bits for a class B address, and 6 bits for a class C address.

Restrictions on IP addresses include the following:

- Addresses cannot have the first four highest bits (in the first field) set to 1111. This is reserved for class E networks (a reserved network classification).

- The class A address 127.x.x.x is for a special function known as the loopback function. It should never be visible on the network. It is installed as an address on every active IP node on a network. Check your local table to see it. It can be used to test whether the IP stack is functional on your workstation.

- The bits that define the host portion of the address cannot be set to all 1s or all 0s to indicate an individual address. These are reserved addresses. All 1s indicates a local subnet all hosts broadcast and all 0s indicates a network number.

- All 0s and all 1s are allowed in the subnet portion of an address. However, you must be careful when assigning all 1s to the subnet portion of the address. This is allowed (according to RFC 1812), but it can reek havoc on those networks that use all subnets broadcast. If the subnet portion of the address is set to all 1s, this can be used as a *directed broadcast*. Routers forward this type of datagram if told to do so (they have to be configured).

- Any address with all 0s in the network portion of the address is meant to represent "this" network. For example, 0.0.0.120 is meant as host number 120 on "this" network (the network from which it originated).

- There is an old form of broadcasting known as the *all-0s broadcast* that takes the form of 0.0.0.0 or network ID, or the host bits are set to 0. The first form should not be used because 0.0.0.0 indicates a default router.

- You can assign your own IP network numbers if you will *never* have access to the Internet or you plan to use something like a network address translator (NAT, RFC 1631). RFC 1918 allows three IP addresses to be used for private networks.

- Addresses cannot exceed the 255 (decimal) range for any for the four bytes. Therefore, an address of 128.6.200.655 is not a valid address. Likewise, an address of 420.6.7.900 is not a valid address assignment.

Address Allocation (The Internet Registry)

Refer to Fig. 8-11. RFC 2050 describes the registry system for the distribution of globally unique Internet address space and registry options. This RFC is different from most others. The upper left corner contains the category "Best Current Practice." It gives an accurate representation of the current practice of IP address registries.

In order to achieve address uniqueness, distribution of global Internet addresses, and, most of all, conservation of IPv4 Internet addressees, the Internet Registry hierarchy was established. It consists of IANA, regional IRs, and local IRs.

The IANA is the Internet Assigned Numbers Authority, and it has overall authority for the number space used in the Internet. This number space includes port number, address, IP version numbers, and many other significant number assignments. Read RFC 1700 for a full description of the IANA.

The regional IRs operate under the authority of IANA. They operate in large geographical areas such as continents. Currently, there are three defined:

- InterNIC, which serves North America
- RIPE, which serves Europe
- APNIC, which serves the Asian Pacific regions

These three IRs do not cover all areas. It is expected that each IR cover its immediately adjacent area not specified. Local IRs are established under the authority of the regional IR and IANA. They cover national dimensions.

Addresses are allocated by regional registries to ISPs, which in turn assign them to their customer bases. ISPs that exchange routing information directly with other ISPs get their address allocation from their geographic IR. Other ISPs are referred to these ISPs for address assignment. In other words, if your address block has a reasonable chance of being propagated through the global Internet routing tables, your address allocation will come from the IR. Otherwise, you will get your address assignment from your upstream ISP. Customers (commercial corporations) need not worry about this. They get their address assignment from the ISP they sign up with.

Figure 8-11
Internet registry (IR).

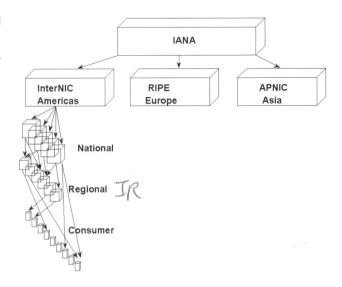

Address Resolution Protocol

The Address Resolution Protocol (ARP) protocol is contained in RFC 826. The Internet (not the ARPAnet) developed with high-speed local networks such as Ethernet, Token Ring, and FDDI. The TCP/IP protocol did not. Ethernet was commercially available in 1980 and started to gain more recognition when version 2.0 was released in 1982. TCP/IP was initially installed on serial lines through route message processors known as IMPs (Interface Message Processors). The hosts connected to the IMP, and the IMP connected to the phone lines that interconnected all ARPAnet (not today's Internet) sites. The IP address identified the host (and later the network and subnetwork). There was not a need for physically identifying a host because their was only one host per physical connection to the IMP. Multiple hosts could connect to an IMP, but each had an IP address that the IMP forwarded directly to the host.

Because multiple stations were meant to connect to a network (single-cable segment) such as Ethernet, they had to be physically identified on the Ethernet. The designers of local area networks (LANs) allotted 48 bits to identify a network attachment. This is known as their *physical address,* or *MAC address.* Physical addresses identify stations at their data-link level. IP is an addressing scheme used at the network level. On a local area network, two communicating stations can set up a session only if they know each other's physical address.

Because the LAN address is 48 bits and IP is 32 bits, a problem existed which was resolved in an RFC. The resolution was simple and did not affect the already established IP addressing scheme. It is known as the *Address Resolution Protocol,* or *ARP.* The format of ARP headers is shown in Fig. 8-12. This is an IP-address-to-physical-station-address resolution and is explained as follows.

When you are trying to communicate to a host on the same network number as the one on which you currently reside, the TCP/IP protocol uses ARP to find the physical address of the destination station. If the network number of the destination station is remote, a router must be used to forward the datagram to the destination. The ARP process is used here, as well, but only to find the physical address of the router.

There have been individual enhancements to this protocol, although not through an RFC. Some stations listen to all ARP packets because the originator sends them in broadcast mode. All stations receive such packets and "glean" the information that they need. These packets contain the senders' hardware and IP address mapping, which is used by other stations to build their ARP caches. Many stations flush their ARP tables

Figure 8-12
ARP header encapsu-
lation.

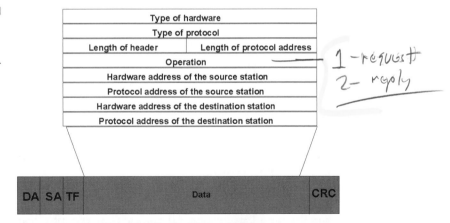

Type of hardware
Type of protocol
Length of header | Length of protocol address
Operation
Hardware address of the source station
Protocol address of the source station
Hardware address of the destination station
Protocol address of the destination station

1 – request
2 – reply

DA | SA | TF | Data | CRC

periodically. This is to reduce the cycles needed to refresh the cache and to conserve memory. It is also used to keep the table up to date. If a station moves from one subnet to another and stations on the old subnet do not empty their tables, they will continue to have an entry for that old hardware address in their tables.

ARP Operation.   As shown in Fig. 8-13, in order to connect to another station on a TCP/IP network the source station must know the designation station's IP address. For example, station 129.1.1.1 wants a connection with 129.1.1.4 (no subnet addressing is used here). Therefore, the class B network address of this station is 129.1.0.0 and the personal computer's host address is 1.1; hence the address 129.1.1.1.

With ARP, it is assumed that the IP address of the destination station is already known, perhaps through a name service (a central service or file on a network station that maps IP addresses to host names, explained in more detail later). To reduce overhead on the network, most TCP network stations maintain a LAN physical-address-to-IP-address table on their host machine. The ARP table is nothing more than a section of

Figure 8-13
Address resolution
protocol.

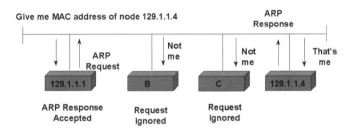

Give me MAC address of node 129.1.1.4

ARP
Response

ARP
Request

Not
me

Not
me

That's
me

129.1.1.1 | B | C | 129.1.1.4

ARP Response
Accepted

Request
Ignored

Request
Ignored

RAM that contains data-link physical-address-to-IP-address mappings that it has learned from the network.

Once the IP address is known for the destination station, IP on the source station looks into its ARP table to find the physical address for that destination IP address. If a mapping is found, no ARP request packet is transmitted onto the network. IP can bind the IP address with the physical address (place the physical address on the data-link header of the packet) and send the IP datagram to the data link for transmission to the network.

If the address is not located in the ARP table, the ARP protocol builds an ARP request packet and sends it physically addressed in broadcast mode (destination address *FF-FF-FF-FF-FF-FF*). Because the packet is sent out in broadcast mode, all stations on the physical network receive the packet, but only the host with that IP address will reply. For example, host 129.1.1.4 will reply to the request packet with an ARP response packet that is physically addressed to station 129.1.1.1.

The host whose IP address is in the request packet responds with an ARP reply packet that is not addressed to destination broadcast but has its source address set to its own address (physically and inside the ARP reply packet), and the destination address is the originator. Once the originator of the request receives the response, it extracts the physical address from the source address in the packet and updates its ARP table. Now that it has the mapping, it will try to submit its IP datagram to the destination station using the proper addresses (IP and physical address).

This process is completed as an involuntary act by the user. The user is typically using one of TCP's applications (Telnet for terminal service, SMTP for mail service, or FTP for file transfer service) attempting a connection. This ARP request and reply happen automatically in the connection, but most TCP vendors supply a utility program that allows a user see the entries in the ARP table.

To improve the efficiency of the protocol, any station on the physical network that receives an ARP packet (request packet) can update the ARP cache. An example of an ARP cache (table) is shown in Table 8-3. The sender's physical and IP addresses are in the packet, so all stations can update their ARP tables at the same time.

Figure 8-12 shows the ARP packet format. It is encapsulated in an Ethernet packet as shown. (See also Table 8-4.) This ARP process works for stations communicating with each other on the same LAN (with the same network number). If they are not on the same LAN, the ARP process still works, but the address of a router will be found. This is fully explained later.

TABLE 8-3

ARP Table for Station 129.1.1.1

Physical Address	IP Address
02-60-8C-01-02-03	129.1.1.1
FF-FF-FF-FF-FF-FF	129.1.1.255
FF-FF-FF-FF-FF-FF	255.255.255.255
00-00-A2-05-09-89	129.1.1.4
08-00-20-67-92-89	129.1.1.2
08-00-02-90-90-90	129.1.1.5

TABLE 8-4

Definition of the ARP Packet

Type of hardware	Normally indicates IEEE 802 network for local area networks. It could also indicate other types of networks.
Type of protocol	Would indicate IP for TCP/IP networks. It could also indicate AppleTalk.
Length of header	Indicates the length of the ARP header.
Length of protocol address	Since this header is used for other types of networks (AppleTalk), this field indicates the length of the protocol address (IP or AppleTalk address, not the physical address).
Operation	Indicates the operation of the header: ARP request or response.
Address of the	Physical address of the source station. This would be filled in source station by the requester.
Protocol address of the source station	IP address of the source station.
Hardware address of the destination station	Physical address of the destination station. This field is usually, but not always, set to 0s if it is a request of packet. This field would be set to the physical address of the destination station if it is an ARP reply. This field is filled in by the responding destination station.
Protocol address of the destination station	Set by the source station (ARP requester). This will contain the IP address of the wanted destination station. Only a station whose IP address matches this will respond to the ARP request.

Rules for ARP

- ARP is not a part of the IP protocol and therefore does not contain IP headers. ARP works directly on top of the data-link layer.

- ARP requests and responses are transmitted with a destination physical broadcast address (all *F*s) and therefore never leave their logical subnet.

- Because ARP is not part of the IP protocol, new EtherTypes were assigned to identify this type of packet: 0806 is an ARP request and 0806 is an ARP reply. Some ARP implementations can be assigned the 0800 EtherType, because IP is able identify the packet as an ARP request or ARP reply packet. Not all implementers of IP use these types. Some still use the EtherTypes of 0800 for ARP.

- Some implementations have an ARP aging capability. This allows ARP to delete entries that have not been used for a period of time. This reduces the ARP lookup time and saves memory.

- If a machine submits an ARP request for itself, it must reply to the request.

Reverse Address Resolution Protocol

The Reverse Address Resolution Protocol (RARP) is used when a network station knows its MAC address but does not know its IP address. When would this happen? Diskless workstations are a good example. In Fig. 8-14, the requesting client machine on the left sends out a RARP

Figure 8-14
RARP.

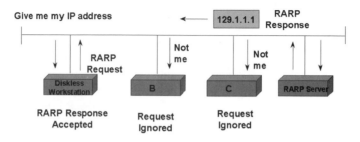

- Same packet type used as ARP
- Only works on local subnets
- Used for diskless workstations

request to a RARP server located somewhere on the physical network. This RARP server responds to the request with that particular station's IP address.

The format for a RARP packet is the same as that for an ARP packet. The only difference it that the field that is filled in is the sender's physical address. The IP address fields are empty. A RARP server receives this packet, fills in the IP address fields, and replies to the sender. It is the opposite of the ARP process.

Other protocols similar to this are BOOTP and the Dynamic Host Configuration Protocol (DHCP). DHCP is more powerful than RARP, but it does supply one of the same functions: resolving an IP address. Aside from being less functional than DHCP, RARP only works on single subnets. RARP works at the data-link layer and therefore cannot span subnets. DHCP can span subnets.

Proxy ARP

After subnetting was introduced, it was implemented over time. Virtually all networks and hosts now use and understand subnets. By the time IP subnet addressing was adopted, there were already a tremendous number of hosts established that used TCP/IP as their networking protocol and so could not using subnetting. Although proxy ARP is not used very much anymore, it is still worthy of mention. Proxy ARP is the capability of a router to respond to an end-station (host) ARP request for a host that it thinks is on the local LAN. If a host did not support subnet addressing, it could incorrectly mistake an IP subnet number for a host number. The router tricks the transmitting station into believing that the source station is on the local LAN.

Figure 8-15 depicts this situation. Host 130.1.2.1 thinks host 130.1.1.1 is on the local LAN. Host 130.1.1.1 supports subnet addressing and end station 130.1.2.1 does not. By deciphering the IP address, the first two fields (containing the network ID) are the same. Therefore, end station 130.1.2.1 sends out a local ARP request packet when it should be submitting the packet to the router so that the router can deliver the packet to the end host. If the router has proxy ARP enabled, the router answers for host B. The router, which supports subnetting, looks up the ARP request and then notices that the subnetwork address is in its routing table. The router responds for host 130.1.1.1. Host 130.1.2.1 receives this response and thinks it is from host 130.1.1.1. There is nothing in the physical address of a packet to indicate where it came from. The host then submits all

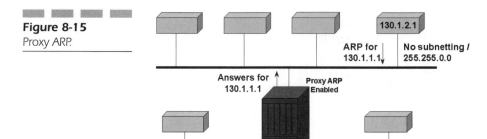

packets to the router, and the router delivers them to end station A. This communication continues until one end terminates the session.

Proxy ARP is a very useful protocol for those networks that have been using bridges to implement their IP network and are migrating to a routed environment. There are other useful situations for proxy ARP, but its use is waning. Today, most hosts on TCP/IP internets support subnet masking, and most IP networks are using routers.

A potential problem in using proxy ARP is for those networks that implement the mechanism to ensure that single IP addresses are on each network. Most TCP/IP implementations allow users easy access to their network number (that is, they can change it with a text editor). This allows any hacker to change his or her number to another in order to receive datagrams destined for another host. Some implementations of TCP/IP will detect this. Routers that implement proxy ARP will get caught, because they will answer for any station on a different network, thereby giving the impression that there is one physical address for multiple IP addresses. There is a trust on any IP network that IP addresses will not be arbitrarily assigned. There should be one IP address for each physical address on a internet.

Circuit and Packet Switching

The future of circuit and packet switching is a topic of debate because of voice over IP. Currently, 99 percent of all voice traffic runs over circuit-switched networks. However, it is estimated that in 10 years 5–10 percent of voice traffic will be running over packet-switched networks such as the Internet. This topic is not debated here, but some simple definitions are discussed.

TCP/IP allowed open communication and the proliferation of LAN-to-LAN and LAN-to-WAN connectivity between multiple operating environments. However, its topology and architecture were not based on the methods employed by the phone company, i.e., circuit switching.

The phone companies (AT&T, before the breakup) basically laughed at the idea of a packet-switched networks and publicly stated that it could never work. They thought that information that could find its own way around the network was impossible, as was a network where every transmitted packet of information has the same chance for forwarding. They stood by their stance that circuit switching was the only method that should be used for voice, video, or data. Circuit switching by definition provided guaranteed bandwidth and therefore quality of service. At that time, they were correct, but only for voice. Voice and video cannot withstand delay beyond a small time frame (about 150 ms or 0.150 s), but data can! In packet switching, the path is found in real time. Each time the path should be the same, but it may not be. Even so, the information will get from point A to point B.

There are many differences between circuit and packet switching. One main difference is that in circuit switching, a path is prebuilt before information is sent, whereas packet switching does not predefine or prebuild a path before sending information. For example, when you make a phone call, the phone company physically builds a circuit for that call. You cannot speak (transmit information) until that circuit is built. This circuit is built via hardware. It is a physical circuit through the telephone network system. (This is no longer completely true, because the phone company is currently employing methods of "virtual circuit switching" through technologies such as Asynchronous Transfer Mode, or ATM (beyond the scope of this book), but for our purposes of comparison, a voice path is prebuilt on hardware before information is passed). No information is contained in the digitized voice signal to indicate to the switches where the destination is located. Each transmitting node has the same chance in getting its information to the receiver.

In packet switching, the information needed to get the message to the destination station is contained in its header. Stations, known as routers, in the network read this information and forward the information along its path. Thousands of different packets of information may take the exact same path to different destinations.

Today, we are proving that not only is packet switching viable, it can be used for voice, video, and data. Newer, faster stations on the network and faster transmission transports have been invented. Along with this are new quality-of-service protocols that establish priorities on the network. This allows certain packets of information to "leapfrog" over other packets to achieve faster transmission.

Supernatural Acts for IP Addressing

Let's get back to the topic at hand, addressing. In this section we discuss some advanced topics for IPv4 addressing.

We have 32 bits of address space. Ignoring the rules of class addressing, this 2^{32} allows for 4,294,967,296 unicast addresses to be assigned (in some formation of networks and hosts). This seems like a lot of addresses. Now, we hear about the exhaustion of IP address space. Can this be true, with over 4 billion available addresses? Yes! The original problem was that there were three types of classful addresses and address allocation without a plan. It used to be that anyone who wanted an address was given one arbitrarily and addresses were allocated without knowledge of their location or a full understanding of their network requirements, which would have led to the proper assignment of an address. In 1992, a study concluding that not only was the address space near depletion (classes A and B) but also that assigning the remaining 2 million class C addresses would cause the Internet's router array to melt down. The Internet backbone routers were already congested and slow with the current routing tables of fewer than 30,000 routes.

With the vast explosion of connectivity to the Internet starting in 1994, the Internet was soon running out of IPv4 addresses. Class As in the range of 64–126 were not assigned, Class Bs were at the point of exhaustion, and Class C, although plentiful, only allowed for 254 host addresses per network number assignment. Class C subnetting is not exactly easy. Most sites were given multiple class C addresses, and this was quickly filling up the Internet routing tables (some predictions had it as high as 85,000 routes on the global routing tables, those tables held by national Internet Service Providers such as Sprint and MCI). Yet the computing power of the router and availability of RAM to hold those tables in the router were not ready yet. The doubling of the Internet was occurring every 9 months, yet the computing power of the routers was doubling every 18 months. But instead of producing faster and more powerful routers (as we did with mainframes in the 1970s and 1980s), we became smart and invented a hold-over solution using the existing equipment and current Ipv4 addressing scheme.

Some organizations and network providers had multiple contiguous networks assigned. Yet, as we learned in the previous section on addressing, each address is a network and holds one record slot in the routing database. The idea of supernetting was introduced in RFC 1338 as a means of summarizing multiple network numbers (one entry details multiple network IDs), further reducing the number of routes reported. This was a 1992 RFC intended as a three-year fix. It matured into CIDR.

Extending the IPv4 Address Space

The following was taken from RFC 760:

> Addresses are fixed length of four octets (32 bits). An address begins with a one octet network number, followed by a three octet local address. This three octet field is called the "rest" field.

This was taken from RFC 791, p. 6:

> Addresses are fixed length of four octets (32 bits). An address begins with a network number, followed by local address (called the "rest" field). There are three formats or classes of Internet addresses: in class a, the high order bit is zero, the next 7 bits are the network, and the last 24 bits are the local address; in class b, the high order two bits are one-zero, the next 14 bits are the network and the last 16 bits are the local address; in class c, the high order three bits are one-one-zero, the next 21 bits are the network and the last 8 bits are the local address.

RFC 950 introduced us to subnetting:

> While this view has proved simple and powerful (two level model, assigning a network number per network), a number of organizations have found it inadequate, and have added a third level to the interpretation of Internet addresses. In this view, a given Internet network is divided into a collection of subnets.

RFC 1517-1520 introduced us to classless interdomain routing (CIDR):

> It has become clear that the first two of these problems (routing information overload, and class B exhaustion) are likely to become critical in the near term. Classless Inter-Domain Routing (CIDR) attempts to deal with these problems by defining a mechanism to slow the growth of routing tables and reduce the need to allocate new IP network numbers.

Back to eliminating classes of addresses and subnets altogether and replacing it with a 32-bit address and a variable 32-bit prefix (mask). This section deals primarily with the IPv4 address extensions. Included in this are subnetting (a review, variable-length subnet masks, route aggregation, and CIDR). IPv6 should be included in this as well with the 128-bit address. However, discussion of IPv6 is held off until after the IPv4 discussion. The CIDR discussion fully reveals the address problem and what was done about it.

IP Address Assignment (The Old Method)

Originally, using RFC 791 without subnetting, an organization that has a complex (more than one) network topology has three choices for assigning Internet addresses:

1. Acquire a distinct Internet network number for each cable; subnets are not used at all.

2. Use a single network number for the entire organization, but assign host numbers in coordination with their communication requirements (flat networks segmented using bridges).

3. Use a single network number, and partition the host address space by assigning subnet numbers to the LANs ("explicit subnets"). Independent implementors of TCP/IP were doing this on their own.

Employing number 1 above caused routing tables to grow. RFC 950 allowed for subnet addressing to take place within an autonomous system, which allowed for a site to continue to subnet its AS, but the subnets were never propagated to the Internet routing tables. (See Table 8-5.) Subnetting and VLSM allowed for the global routing tables to stop growing exponentially and allowed sites to control their own networks as

TABLE 8-5

Masks and Prefixes*

IP Network Address	Prefix	Subnet Mask
150.1.0.0	/16	255.255.0.0
150.1.8.0	/21	255.255.248.0
205.16.16.128	/25	255.255.255.128

*The address 205.10.16.0/16 and 205.10.16.0/255.255.255.0 mean the exact same thing.

well. However, network numbers were plentiful and subnets slowed the expansion of the Internet routing tables. But this was before the commercialization of the Internet, starting in 1994.

The adverse effects of bridges in complex networks are well known. Since the bridge revolution, routers have become the mainstay of the corporate backbone. This worked well for shared environments, but by the late 1980s technology was changing in that the network attachments were becoming faster and more powerful. The bridging revolution came back as switches in that each desktop could now have its own 10-Mbps pipe. The switches build a small flat network and should be used to front-end routers, thus allowing for *microsegmenting* but not *microsubnetting.*

Subnetting one network number slowed the growth of Internet routing tables. This works well with class B addresses. Class C networks forced the Internet routing tables to grow, and Class A addresses were not handed out. Also, with over 50 percent of businesses today being to small and medium-sized businesses, class C addresses were needed. Again we were in a predicament. We needed a solution.

Address Efficiencies—The Terms

There are four terms used in this section:

Variable-length subnet masks (VLSMs): the ability to place variable-length subnet masks on an single IP network address. Refer to RFC 1817. VLSMs were explained in detail in the OSPF section.

Supernetting: a mask that is shorter than the IP network address natural mask.

Classless interdomain routing (CIDR): an advertisement mechanism that allows for advertising routes without regard to class assignment. The route could be identified by a supernet or by an extended subnet mask.

Address aggregation: the ability to summarize contiguous blocks of IP addresses as one advertisement.

The ability to manipulate IP addresses is affected not only on customer sites but within the global Internet as well. Class-oriented IP addresses are still used in the customer environment, whereas classless IP addressing is used in the Internet itself. Customers are free to use any mechanism that efficiently uses the address that is assigned to them. No longer are they restricted to use only one subnet mask for their assigned network number. OSPF and RIP2 give us more flexibility when using the subnet mask. These routing update protocols distribute the subnet mask for each entry in its table. They allow us great flexibility in mask assignment and allow for more efficiency of the network address. For a single network ID, we can move the mask around to various masks for the single network ID. A given site can make very efficient use of its assigned network ID using VLSM. We could move the mask down to 255.255.255.252 for serial lines, allowing two bits for the host, and then move the mask around again for various number of hosts. OSPF also allows for summaries in the routing updates, which enable routers to send out one network number with a single mask as an update indicating all networks included in the mask were handled by that router. This is very efficient.

The rapid expansion for connectivity to the Internet and the exploding corporate infrastructure of corporate environments initially caused problems on the Internet. IP addresses were assigned sequentially to requesting organizations without regard to the requestor's location or method of Internet connection. What this means is that a requesting company simply called in for an IP address assignment and was assigned an IP address from a list of sequentially listed numbers. A company in California could be assigned 150.1.0.0, and then a company in Virginia would be assigned 151.1.0.0 and maybe 40 class C addresses. Then a company in Texas could apply for 160.1.0.0 and 50 class C addresses. They could then sign up for any ISP they desired with their newly assigned IP addresses. This had negative effects on the Internet routing system in that the routing system filled up with smaller IP addresses across multiple, long hops of routers, instead of large contiguous addresses. Supernetting, CIDR, and address aggregation provided address flexibility and efficiency to the ISP and the Internet. CIDR is very similar to VLSM. Today, blocks of addresses (as indicated in Table 8-6) are handed out to Internet service providers (ISPs) in blocks (or ranges) through the Internet Registry (RFC 2050 fully explains this). For example, an ISP may be assigned the address block of 205.24.0.0/16. First, this assignment allows the ISP to hand out addresses in the range of 205.24.0.0 through 205.24.255.0. In this way, the global routing tables only know that addresses 205.24.0.0 though 205.24.255.255 go in one direction to an ISP. Second, all of these addresses are summarized into the single global routing table entry 205.24.0.0/16 for what would have been 255 entries using the old method. The global routing tables do not care about the individual network assignments.

TABLE 8-6

Current IANA
Address Block
Assignments

Address Block	Registry—Purpose	Date
000–063/8	IANA	Sept. 81
064–095/8	IANA—reserved	Sept. 81
096–126/8	IANA—reserved	Sept. 81
127/8	IANA	Sept. 81
128–191/8	Various registries	May 93
192–193/8	Various registries—multiregional	May 93
194–195/8	RIPE NCC—Europe	May 93
196–197/8	Internic—others	May 93
198–199/8	Internic—North America	May 93
200–201/8	Internic—Central and South America	May 93
202–203/8	APNIC—Pacific Rim	May 93
204–205/8	Internic—North America	March 94
206/8	Internic—North America	April 95
207/8	Internic—North America	Nov. 95
208/8	Internic—North America	April 96
209/8	Internic—North America	June 96
210/8	APNIC—Pacific Rim	June 96
211/8	APNIC-Pacific Rim	June 96
212-223/8	IANA—reserved	Sept. 81
224-239/8	IANA—multicast (class D)	Sept. 81
240-255/8	IANA—reserved (class E)	Sept. 81

The ISP subdivides this block to hand out individual address to its customers as classful addresses. But how an ISP cuts up the address and assigns these blocks follows the above-mentioned protocols. This whole block is assigned not to one company but to multiple companies. A company requiring Internet connection through that ISP requests address space from the ISP and explains its topology in detail. The ISP (knowing it has to assign network numbers sparingly) then assigns the correct number and network range to its downstream customers. The range is then entered into its (the ISPs) routing table perhaps as one address, even though multiple classes were given to the customer.

Masks and Prefixes. RFC 1820 claims that subnet masks are of historical value only and that the term *network prefix* should be used. However, this is not how millions of network professionals see it, and they continue to use the term *subnet mask*. Throughout this text I will use both *decimal subnet mask* and *prefix*. A mask and a prefix are essentially the same thing. For example, a subnet of 255.255.255.0 and a prefix of /24 are the same thing. Let's take a few subnet examples, starting with address assignment at a company site.

Determining the Subnet: An Example. A customer has the base network address 150.1.0.0 with a subnet mask of 255.255.0.0, or a /16 prefix. A network is to be divided in such a way that it can support 80 hosts on each subnet. To support 80 hosts, 7 bits are needed, which allows for 126 addresses (2^7-2), which will allow for future growth. (The next lowest mask yields only 2^6-2, or 62 addresses.)

Next we must determine the subnet mask for the network number. Since we are reserving 7 bits for host assignment, 25 bits are left for the network mask (32 bits-7 bits = 25 bits). This gives a subnet mask of 255.255.255.128, or a /25 prefix. The natural mask for class B is 255.255.0.0. This mask is 255.255.255.128, which allows for 9 bits to be assigned to the subnet mask, allowing for 512 subnets to be defined. The subnet numbers range from 0 to 521. This gives the range of subnets of 150.1.0.0 (providing for the zero subnet) through 150.1.255.128 (using all 9 bits including the all-1s subnet). Now that we have separated the subnets from the hosts, we should list them. See Table 8-7.

Variable-Length Subnet Masks (VLSM). The above examples show how to split a network up for subnets assuming one mask per network ID. This issue was discussed extensively previously in the book, and the foregoing example was placed there for review. A concept known as variable-length subnet mask (VLSM) enables us to assign variable masks per

TABLE 8-7

Subnet and Host Range

Subnets	Host Range
150.1.0.0 through 150.1.255.128	1 through 125 (2^7-2)
150.1.1.0 (x = host-reserved bits)	Host 1 (x = network/subnet-reserved bits)
10010110 . 00000001 . 00000001 . 0xxxxxxx	xxxxxxxx.xxxxxxxx.xxxxxxxx.x0000001
150.1.1.0	Host 127
10010110 . 00000001 . 00000001 . 0xxxxxxx	xxxxxxxx.xxxxxxxx.xxxxxxxx.x1111111

network ID. We can move the mask around the single network ID. VLSM is used when routing protocols such as OSPF and RIP2 are used. These protocols transmit the subnet mask along with the network ID in the routing update message (this is detailed in the RIP and OSPF sections of the book).

VLSM can get very, very confusing. One rule you should follow is, Do not make it overly complicated. Efficiency is important, but you must sit down with your team or customer and determine the network topology. For example, if you use that address 150.1.0.0 with /16 prefix (255.255.0.0), a very effective method of using VLSM is to use a /24 mask for subnets with lots of networks, a /27 mask for subnets with fewer hosts or higher-powered, network-hogging applications, and a /30 mask for the serial lines. This is shown next. You can try to develop a mask for every subnet, but a few bits left over is fine. Also, this method is not efficient, because you will be spreading different subnets through the network in a noncontiguous fashion, which can become burdensome on the route tables and does explain the variable-length subnet feature.

As shown in Fig. 8-16, your base address is 150.1.0.0/16. This goes at the top of the chart. From here we create 256 subnets using the /24 subnet mask, as shown one level down from the top level. No hosts have been assigned yet. After this we have decided that we have 50 serial (point-to-point) lines to work with and our future growth for 100 more remote sites over the next two years. Therefore we need 150 subnets for the serial lines, and only two host addresses are needed per serial line. From the previous subnetted addresses, we have reserved the 150.1.56.0, 150.1.57.0 (not shown) and the 150.1.58.0 (not shown) subnets for serial lines. The 150.1.56.0 network is further subdivided (sub-subnetted) using the first six bits of the fourth octet (255.255.255.252 or /30), yielding 64 subnets for

Figure 8-16
Variable-length subnet masks.

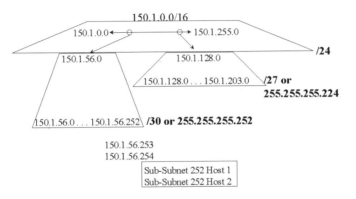

serial lines. With each subnet (56, 57, and 58) supporting 64 subnets, we now have 192 subnets allotted for serial lines. We leave two bits, which allows for two host addresses to be assigned (all zeros and all ones are not allowed as host addresses). Seventy-five of the subnets will be assigned a another mask (/27) to allow for sub-subnets (subnets of subnets) with a smaller number of hosts per subnet.

Longest Match Rule.   As you can see, playing with the address leads to a lot of ambiguity. It can become very complicated. Usually, company network managers do not have to overly concern themselves with this schema.

One rule that must be understood before any of this can work is the longest match rule. This is also discussed in the OSPF section of the book. When a network ID is encountered that matches prefixes of different lengths, the router always takes the path indicated by the longest mask. For example, if a router received an IP datagram with the destination address 200.40.1.1 and a route table lookup found 200.40.1.0/24 and 200.40.0.0/16, the router would forward the datagram out the path indicated by the longest mask, 200.40.1.0. Therefore, you must make sure there are no hosts assigned to 200.40.0.0/16.

The longest mask rule is implemented because the longer the mask is, the better granularity the router has in exactly defining the correct route. Therefore you must be aware that the router will route to the route determined by the longest mask match. If there are two entries for the same route, the longest mask wins.

Very Tight Address Assignment: An Example.   Refer to Fig. 8-17. Let's look at another example. The ISP block is 200.24.0.0/16. A customer of the ISP needs three subnets, with each supporting 60 hosts. Remem-

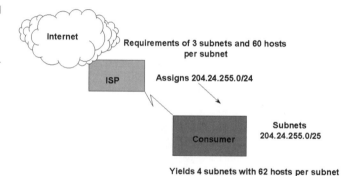

Figure 8-17
Too stringent address assignment.

ber that we assign the mask contiguous starting from the left. Since subnets are divided evenly (because of the binary nature of the address), we cannot have three subnets without dividing the address to provide for four subnets. The address assigned to the customer is 200.24.255.0/24.

1. How many bits are needed in the subnet mask to support three subnets?

2. Since $2^2 = 4$, two bits are required in the subnet mask.

3. This leaves six bits left for host assignment. Since 2^6 leaves 62 ($2^6 = 64$, and we subtract 2 for we cannot have all 0s or all 1s in the host portion of the address) address assignments for hosts, and we can use this single network address assignment for our company.

Refer to Fig. 8-18. Although we were able to be very stringent with the address assignment, this is not a good way of assigning or masking the address. It does not leave much room for growth on the host side as well as the network side. For example, what if the company expands to 100 hosts per subnet and requires two more subnets? It could call its ISP back and request another address assignment. But by now the ISP has handed out a few more addresses and the next address for the customer is 200.24.64.0/24. This is not contiguous with the customer's original assignment, and the ISP has to add another entry in the ISP's table. But this could have been avoided. To anticipate for this expansion, the customer could have been assigned four class Cs. The ISP block assigned to the customer could have been 200.1.252.0/22 (one entry in the ISP routing table), which yields the class C addresses of 200.1.252.0 and 200.1.253.0, 200.1.254.0, and 200.1.255.0. The customer would have been free to assign any subnet mask it wished to the addresses without notifying the ISP.

From here the customer could assign 1 bit of subnet mask on the address of 200.1.252.0, which allows for 7 bits of address space, yielding

Figure 8-18
A better way of address assignment.

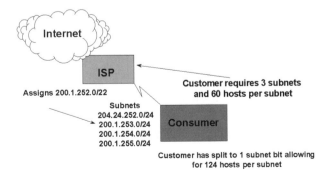

Internet

ISP

Assigns 200.1.252.0/22

Customer requires 3 subnets and 60 hosts per subnet

Subnets
204.24.252.0/24
200.1.253.0/24
200.1.254.0/24
200.1.255.0/24

Consumer

Customer has split to 1 subnet bit allowing for 124 hosts per subnet

125 (2^7-2) hosts per subnet. The other address could remain intact, or it, too, could be split with 1 bit of subnet mask. The customer could also have simply used all the bits in the fourth octet, using no subnet mask. One bit of subnet mask is allowed on a class C address: a 0 subnet and a 1 subnet. Review RF1812. This will lead to problems only if the site is using all subnets broadcast. However, some other router vendors, for example, Cisco, do not support one-bit subnet addressing.

A better method is to use a routing algorithm that supports variable-length subnet marks. Identify those network segments that require fewer hosts and assign them a mask of more than one bit. Do not assign a subnet mask for cable segments with lots of hosts. Then use switching products to adjust the data flow on the network.

This is a simple example of how you must think about your network design before calling an ISP. You can call the ISP before this, but the ISP does not know your network as well as you do. You know how many hosts and what traffic patterns are on the network. IP addresses are in short supply, and ISPs do not hand them out haphazardly. They must take into consideration their routing tables as well.

Supernetting Exposed.   The previous example shows the ability of an ISP to assign a block of addresses to a customer site. This was small (four class C addresses), but it shows up as one entry in the routing table 200.1.252.0/22. Notice that the mask at the ISP is pushed back beyond a natural class C mask. This is known as *supernetting*. Supernetting is the ability to push the mask back to the left beyond the natural subnet mask of a class address.

The current approach (lacking IPv6) is to provide large contiguous blocks of class C addresses (and possibly other classes) and provide them by more local levels in a hierarchical fashion. For example, a national backbone provider (call it ISP1) with connections to other national backbone providers through network access points (NAPs) is assigned a large block (one that will last two years) of class C addresses. In turn other regional service providers (call them ISP2) that utilize ISP1 are assigned a block of addresses from ISP1's address block assignment. In turn, ISP2 provides address assignment to its customers from the block it was assigned. See Fig. 8-19. This allows for very efficient and manageable global routing tables (those routing tables on the top-level providers).

Route Aggregation.   In the previous example, you were introduced to a concept known as *route aggregation*. It allows a router to summarize a

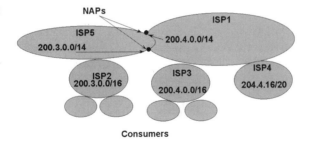

Figure 8-19
Supernetting
exposed.

group of routes as one advertisement. Imagine having one entry in the routing table to represent a large group of addresses. The router simply needs to know the prefix. This is completely possible with route aggregation. However, it is only useful when the routes are contiguous. Reducing the continuity of the routes reduces the efficiency of this concept.

To show this benefit clearly, I have chosen a class A example as shown in Fig. 8-20. The network address is 20.0.0.0. The natural mask for this is /8, or 255.0.0.0. We first subnet the address using a /16 prefix, or 255.255.0.0. This allows for addresses in the range of 20.0.0.0 through 20.255.0.0. We take the 20.127.0.0 subnet and further subnet it with a prefix of /24 (255.255.255.0). Finally we apply a /27 prefix to the 20.127.1.0 subnet.

Route aggregation is based on the concept of a common prefix assigned to a group of IP addresses. For example, the 20.127.1.0 was subnetted to /27. However, all the subnets that are created by this can be advertised as one route: 20.127.1.0/24. This is detailed in the following example. All of the addresses in this range have the same prefix. This would indicate to all other routers that any network in the range of

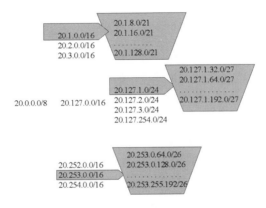

Figure 8-20
Route aggregation.

20.127.1.0 should be forwarded to that router. The other routers do not care about any of the particular subnets beyond that address. The router that receives the datagram to be forwarded to any subnet below 20.127.1.0 will forward it to the correct network.

The rules are simple here:

1. Write down the addresses in the range.
2. Convert each address to binary, one below the other.
3. Check for a contiguous, common prefix.
4. Move the prefix to the last bit of the contiguous binary digit.
5. Write the address starting at the first address and apply the step 4 prefix.

Remember, do not make this complicated. It is confusing enough. Three variable subnet masks are enough to work with for most networks (business networks and ISPs are different).

Determining a Common Prefix: An Example.   The /27 prefix allows for the following address range:

20.127.1.32

20.127.1.64

20.127.1.96

20.127.1.128

20.127.1.160

20.127.1.192

20.127.1.224

We convert these to binary:

20.127.1.32	000010100.01111111.00000001.00100000
20.127.1.64	000010100.01111111.00000001.01000000
20.127.1.96	000010100.01111111.00000001.01100000
20.127.1.128	000010100.01111111.00000001.10000000
20.127.1.160	000010100.01111111.00000001.10100000
20.127.1.192	000010100.01111111.00000001.11000000
20.127.1.224	000010100.01111111.00000001.11100000
Common prefix	000010100.01111111.00000001.00000000

Therefore, applying rule 4 and rule 5 we have that 20.127.1.0/24 represents all of the addresses.

Another Look at Route Aggregation. I want to give an example where aggregation is somewhat less efficient but you would not know that from the address. The following addresses appear to be contiguous:

155.1.140.0

155.1.141.0

155.1.142.0

155.1.143.0

155.1.144.0

But when we translate them to binary to find the common prefix to all of the addresses, we find a noncontiguous bit pattern:

155.1.140.0/24	10011011.00000001.***10001100***.00000000
155.1.141.0/24	10011011.00000001.***10001101***.00000000
155.1.142.0/24	10011011.00000001.***10001110***.00000000
155.1.143.0/24	10011011.00000001.***10001111***.00000000
155.1.144.0/24	10011011.00000001.***10010000***.00000000
Common prefix	10011011.00000001.***100011*xx**.00000000 -

The common prefix is 100011xx in the third octet, because we do not know where 145 or higher is. We have to see which ones have the same prefix and then use that. Any other numbers must be separate entries in the table. This would give us a route aggregation of 155.1.140.0/22. But this leaves out the 155.1.144.0 subnet. Depending on the range that this address is in, it could be listed in another route aggregation prefix. But since this is all the information we were given, 155.1.144.0 must be listed as a separate route: 155.1.144.0/24 (subnet mask of 255.255.255.0). This is so because this address not within the range of the common prefix of the other addresses even though the decimal address is contiguous. Networks do not calculate routes in decimal! Humans do, and this is why we make mistakes.

You should also notice that this allows us to have one route entry instead of four. This may not seem like much of an advantage, but when this concept is applied to a larger range of addresses (such as those on the Internet routing tables), one route entry is used to aggregate thousands of individual addresses. In the example the common prefix is 100011, which allows us to aggregate those routes to 155.1.140.0/14.

Classless Interdomain Routing

With Classless Interdomain Routing (CIDR), network numbers and classes of networks are no longer valid for routing purposes. This is where the network IP address format changes to <IP Address, prefix length>. This is for the Internet routing tables (ISPs). Class addressing is continuing to be used in customer environments. Classless addressing could operate in a customer environment, but most hosts would not understand this type of implementation. The millions and millions of hosts that are attached to the Internet are still operating in a class environment. Therefore, we have simply created a hierarchical routing environment that does not affect the customer environment whatsoever. Let's start out this discussion by assigning a prefix to the well-known class addresses. CIDR could operate in a customer environment, but that would require upgrading all routers and hosts to understand CIDR. This is not going to happen. CIDR is primarily used on the Internet routers.

Class A networks have a /8 prefix.

Class B networks have a /16 prefix.

Class C networks have a /24 prefix.

What we have changed to is the network prefix. A network number is basically a network prefix. Nodes on a classless network simply determine the address by finding the prefix value. This value indicates the number of bits, starting from the left, that will be used for the network. The remaining bits are left for host assignment. The prefix can range anywhere from /0 to /32, allowing us to move the network portion of the address anywhere on the 32-bit number.

Imagine an address of 198.1.192.0/20. This looks like a class C address, but the natural mask for a class C address is 24 bits, or a /24 prefix. This one allows for only 20 bits as the network assignment. But this prefix could be assigned to any address regardless of class. It could be assigned to 15.1.192.0 or 128.1.128.0. The prefix does not care about class. This is the capability of CIDR. The following section assumes that you can convert binary to decimal and vice versa.

This leads up to the next step in understanding IP addresses and Internet routing. It is called CIDR (pronounced *cider*). CIDR is explained in RFC 1517 - 1520. I am not detailing the CIDR specification here, just the concept. The concept is simple: implement a generalization of variable-length subnet masks and move from the traditional class A, B, C address toward the idea of a 32-bit IP address and a prefix (without the concept of a class). In CIDR there are 32 bits and a prefix. To understand CIDR, you must place the concept not on your local network but on the Internet routers. You can employ CIDR on your network, but there is really no reason too (since your hosts would have to be configured to understand supernets). The Internet routing tables were expanding at a exponential rate (without CIDR, they would have passed well over the 80,000 routes today). The Internet routers are simply those devices that move data toward a destination indicated by IP address and therefore do not have large subnets off of them to support hosts. CIDR works on the notion that we are routing an arbitrarily sized network address space (a range) instead of routing on class A, B, and C. CIDR route tables are built based on the prefix number. CIDR does not care about individual network IDs. It routes data on the range. For example, the address of 200.15.0.0/16 could be an entry in the Internet routing table. One entry indicates a range of addresses. Any IP datagrams received by that router with the first 16 bits indicating 200.15 would be forwarded out the port indicated in the routing table. This prefix could be assigned to any range of addresses for CIDR does not associate a prefix with a class. The prefix assignments are given in Table 8-8.

We must look at this concept through the ISP networks. ISPs give us the ability to communicate over the Internet. You cannot attach to the Internet unless you connect with an ISP. There are different kinds of ISPs: some are large and provide access to other ISPs and individuals, and others are small and provide Internet connectivity only to individuals and businesses. ISPs are allocated contiguous blocks of addresses. CIDR first used class C addresses since class B addresses were exhausted and class A addresses were not handed out (they are being handed out today). The basic idea of the plan is to allocate blocks of class C network numbers (at first; other class A and B addresses are to follow) to each network service provider. (It would be very helpful here to read RFC 2050 before continuing this section.) The customers of the provider are then allocated bitmask-oriented subnets of the service provider's address. Table 8-6 indicates the assignment of blocks to the Internet Registry.

Another Example: ISP Splice

Refer to Fig. 8-23. An ISP has been assigned the following block from the InterNIC: 209.16.0.0/16. At first glance this looks like a class C address, but the prefix does not match a class C natural subnet mask address. It is a class B prefix. Again, this is known as supernetting and shows that CIDR

TABLE 8-8

Prefix Assignments

Number of Prefix	Dotted-Decimal Address	Number of Addresses	Number of Class
/13	255.248.0.0	512k	8 class B or 2048 class C
/14	255.252.0.0	256k	4 class B or 1024 class C
/15	255.254.0.0.	128k	2 class B or 512 class C
/16	255.255.0.0	64k	1 class B or 256 class C
/17	255.255.128.0	32k	128 class C
/18	255.255.192.0	16k	64 class C
/19	255.255.224.0	8k	32 class C
/20	255.255.240.0	4k	16 class C
/21	255.255.248.0	2k	8 class C
/22	255.255.252.0	1k	4 class C
/23	255.255.254.0	512	2 class C
/24	255.255.255.0	256	1 class C
/25	255.255.255.128	128	$\frac{1}{2}$ class C
/26	255.255.255.192	64	$\frac{1}{4}$ class C
/27	255.255.255.224	32	$\frac{1}{8}$ class C

does not care about classes. With a prefix of /16, this would represent 256 class C addresses. However, in CIDR the ISP is free to choose any method of segmenting this address up and handing it out to its customers. The ISP also knows that IANA and the InterNIC do not just hand out lots of address. Therefore the ISP is very careful about carving up the addresses.

The ISP pulls off a portion of the address space using a /20 prefix: 209.16.16.0/20. This represents small portion of the addresses, namely 16 class C addresses. The ISP leaves the upper four bits of the address reserved for future use. We will work with the address space 209.16.16.0/20. Based on some surveys with its customers, the ISP cuts the address into two pieces, yielding 209.16.16.0/21 and 209.16.24.0/21 (For those not familiar with binary, shifting right one bit divides the number by two. Shifting left one bit multiplies the number by 2.) One half of the address, 209.16.16.0/21 (eight class C addresses), is assigned to a single customer. The other half of the address, 209.16.24.0, is cut up again into three pieces:

209.16.24.0/22, representing one-fourth of the address (four class C addresses)

209.16.28/23, representing one-eighth of the address (two class C addresses)

209.16.30.0/23 representing one-eighth of the address (two class C addresses)

How is this done? Refer to Table 8-9. As we see in the table:

Customer A gets the class C address range of 209.16.16.0 through 209.16.23.0.

Customer B gets the class C address range of 209.16.24.0 through 209.16.27.0.

Customer C gets the class C address range of 209.16.28.0 through 209.16.29.0.

Customer D gets the class C address range of 209.16.30.0 through 209.16.31.0.

See Fig. 8-21. If you use the addresses given above and count up in binary using the table, you will get a better picture of how this operates.

So CIDR is at the ISP, and class addressing is at the customer site. What does this buy us? It does not necessarily buy the customer anything more (except for a faster network with the ISP), but it does great things for the ISP's routing tables and therefore the Internet routing

TABLE 8-9

IP Address Division

Action	Address Space	Binary Equivalent
ISP segments off 16 addresses of the original address.	209.16.0.0/16 becomes 209.16.16.0/20	11010001.00010000.00000000.00000000 11010001.00010000.0001 | 0000.00000000
ISP splits this new address in half, yielding two address ranges.	209.16.16.0/21 209.16.24.0/21	11010001.00010000.00010 | 000.00000000 11010001.00010000.00011 | 000.00000000
Based on a customer survey, 209.16.16.0/21 is given to a single customer.	Action yields 8 class C addresses.	
209.16.24.0/21 is split up again.	209.16.24.0/22	11010001.00010000.000110 | 00.00000000
	209.16.28.0/23	11010001.00010000.0001110 | 0.00000000
	209.16.30.0/23	11010001.00010000.0001111 | 0.00000000

Figure 8-21
CIDR example.

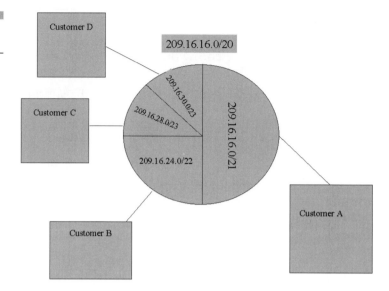

tables. Whereas the ISP would have had 16 entries in the routing table, it now has four. Whereas the Internet routing tables would have had 256 entries in the global routing table, they now have 1. Now multiply this by the number of ISPs worldwide, and I think you will begin to see the efficiencies of this protocol and without it the explosion of the Internet routing tables.

Comparison between CIDR and VLSM

CIDR and VLSM seem similar. In essence, they are. Why not use VLSM instead of CIDR? The difference is that CIDR allows for the efficient routing mechanism to take place by the ability to of the recursive allocation of an address block. Routing is then based on this address block allocation and not on an individual class address. This block is handed down by the IANA to the IR, to the upper-level ISP down through the ranks of downstream ISPs, and finally to the customer.

VLSM permits recursion as well but more so on an individual address space in use by the customer. A customer division of an address space is not visible to the Internet. VLSM still operates with class addresses.

Variable-length masks allow for variable-length subnets for a network ID based on an address assignment by an ISP. This allows one network number to contain different masks and allows a better use of an IP address. With VLSM, a lot of the bits in an address space are wasted. An example is the assignment of an IP address to a point-to-point WAN link. This wastes 252 address bits.

Variable-length masks allow for greater flexibility when dividing up a network ID as to subnets and hosts. Without VLSM, you have to choose between having enough networks, with close to the right amount of hosts, or having the right amount of hosts with close to the right amount of networks.

Zero and Ones Subnet

RFC 950 stated that we should preserve all bits equal to 0 and all 1s in the subnet field because they had special meaning in certain fields indicated by IANA-assigned RFC 1700 numbers. For example, the address 130.1.255.255 could be interpreted as meaning all hosts on the network 130.1, or the address 0.0.0.1 could be interpreted as meaning host 1 on this network. According to RFC 950, p. 5, "It is useful to preserve and extend the interpretation of these special addresses in subnetted networks. This means the values of all zeros and all ones in the subnet field should not be assigned to actual (physical) subnets."

But due to increasing demand to make full use of all of the bits in the 32-bit wide address, subnet 0 and all 1s subnet are allowed. However, you must use caution when doing so. RFC 1812 (Requirements for IPv4 Routers) states, "All-subnets broadcasts (called multi-subnet broadcasts) have been deprecated." Also it states, "In a CIDR routing domain, wherein classical IP network numbers are meaningless, the concept of an all-subnets-directed-broadcast is also meaningless." Basically, there are no subnets in CIDR.

The above discussion about the CIDR router domain could be misread as referring to any routed domain. Many router vendors interpret RFCs in different ways. For example, 3Com has the ability to turn ASB (all subnets broadcastrouting) on or off, thereby allowing the all 1s subnetwork number free to be assigned.

Why would you want to place an ASB? This can be useful when multicasting. As of this writing, the multicast protocols are not being used on customer networks, mainly due to the inexperience and nervousness of the router support staff and their management. Routed networks are tricky enough without thoroughly understanding multicasting. Therefore, multicast application software vendors support ASB to route their information in a nonmulticast network. Unruly, but it works.

This thinking may be propagated down to the lowest levels of routing in the Internet, the customer AS. If the customer AS has "deprecated" ASB, then you can implement all 0s and all 1s subnets. However, if a customer network has implemented it all 1s subnets, then a packet addressed to an ASB will be routed to the subnet represented by the all 1s.

Internet Assigned Numbers Authority

This topic was discussed previously, but I want to mention it here again. Addressing in the Internet would be unmanageable if it were not controlled by one authority. A great example of this is the Novell IPX network number assignment. Every site is free to assign any IPX network number it wishes. This leads to complications in the mergers of two IPX networks, by way of a company merger, for example. No two network numbers may be assigned to the same internetwork.

As defined by the Internet Engineering Task Force (IETF) and its steering group (the IESG), the Internet protocol suite contains numerous parameters, such as internet addresses, domain names, autonomous system numbers (used in some routing protocols), protocol numbers, port numbers, and management information base (MIB) object identifiers, including private enterprise numbers for management, and many others. Certain fields within IP and TCP are required to be unique. Imagine a port number that is arbitrarily assigned for FTP or an IP address that is allowed to be assigned by any site and then connected to the Internet. It is the task of the IANA to make those unique assignments as requested and to maintain a registry of the currently assigned values.

The Internet Assigned Numbers Authority (IANA) is the central coordinator for the assignment of unique parameter values for Internet protocols. The IANA is chartered by the Internet Society (ISOC) and the Federal Network Council (FNC) to act as the clearinghouse to assign and coordinate the use of numerous Internet protocol parameters.

As of this writing, RFC 1700 contains the compilation of assigned numbers. An up-to-date ftp site is available at ftp://ftp.isi.edu/in-notes/iana/assignments.

Requests for parameter assignments (protocols, ports, etc.) should be sent to <iana@isi.edu>.

Requests for SNMP network management private enterprise number assignments should be sent to <iana-mib@isi.edu>.

The IANA is located at and operated by the Information Sciences Institute of the University of Southern California. If you are developing a protocol or application that will require the use of a link, socket, port, protocol, etc., contact the IANA to receive a number assignment (refer to RFC 1700).

Internet Control Message Protocol

Because IP is a connectionless, unreliable delivery service, allowing routers and hosts on an internet to operate independently, there are certain instances when errors will occur on the Internet. Some examples of these errors could be when a packet is not routed to the destination network, the router is too congested to handle any more packets, or a host may not be found on the Internet. There is no provision in IP to generate error messages or control messages. Internet Control Message Protocol (ICMP) is the protocol that handles these instances for IP. The purpose of these control messages is to provide feedback about problems in the communication environment, not to make IP reliable.

ICMP controls many entities on an internet. ICMP datagrams are routable because they use IP to deliver their messages. IP resides on top of IP and does not use TCP or UDP for its transport. ICMP is a separate protocol from IP, but it is an integral part of IP's operation. ICMP does not use a transport layer and runs directly on top of IP. Therefore, ICMP is an "unreliable" function, and no error message is sent for an ICMP message.

For example, when a transmitting station transmits a packet using indirect routing to a remote destination, what would happen if a final router could not find the end station (the end station was not in the router's ARP cache and it did not respond to the router's ARP request)? This is one of the reasons for the ICMP service. The router would send an ICMP message back to the originator of the datagram, end station A, that the destination node could not be found. This message is transmitted to the user as an error message on the user's screen.

Notice that each ICMP message has a Type and Code field. The Type field identifies the ICMP datagram and the Code field provides further granularity. For example, the Type code 3 indicates that the destination was unreachable, but with a Code field of 1, this gives us a further clue that the destination host (and not the port or network) was not reachable. This could mean the network was found but no station responded to an ARP request when transmitted by the sending router. The receiver of that ICMP datagram would then post a message either to a screen or to a log file interpreting the ICMP message. To try this on your own, try pinging a device that you know does not exist.

ICMP will also copy the first 64 bits (IPv6 increases this to 512 bytes) of data from the original datagram into its datagram. This provides some information about the offending datagram and can be used in troubleshooting. ICMPv6 provides more of the original data.

ICMP Ping

One of the most common uses for ICMP is the ping program. Ping (not originally named but commonly called the packet internet groper) is an ICMP message that tries to locate other stations on the Internet to see if they are active or to see if a path is up. Ping is an echo program. As you can see from the argument list with Ping, you can also test round-trip delay using Ping. The originator of a datagram sends a ping request and the destination station should echo this request. Information can be contained in the ping datagram that the destination station should echo, as well. Ping has the following format:

```
ping [-t] [-a] [-n count] [-l size] [-f] [-i TTL] [-v TOS]
[-r count] [-s count] [[-j host-list] | [-k host-list]]
[-w timeout] destination-list
```

The options include the following:

-t	Ping the specified host until interrupted.
-a	Resolve addresses to hostnames.
-n count	Number of echo requests to send.
-l size	Send buffer size.
-f	Set Don't Fragment flag in packet.
-i TTL	Time To Live.
-v TOS	Type of Service.
-r count	Record route for count hops.
-s count	Time-stamp for count hops.
-j host-list	Loose source route along host list.
-k host-list	Strict source route along host list.
-w timeout	Timeout in milliseconds to wait for each reply.

Notice that you can test many things using the ping cor
ing, source route, route recording, and data can be used in tlus
mand to test many things along the path to a destination. Another use
of the ping command is to check for network delays along a path. The
response to a ping request can indicate the response delay. This delay is
usually measured in milliseconds.

A lot of network management software uses this command to determine the status of a given station. Network management software builds
maps to show the topology and placement of network stations on the
map. Using colors (green for active, yellow for possible errors, and red for
not responding) a network manager can trace problems on the network.
A lot of the work is done through the use of the ping utility. A simple
example is shown in Fig. 8-22.

More ICMP Functions

ICMP has added functions over the years, as indicated by RFCs 1256
(Router Discovery) and 1393 (ICMP Traceroute). Another use is to find
the address mask of the local network. ICMP running in a router can
respond to a host's request to find the subnet address mask for its network. A host, upon startup, can request of a router the subnet mask
assigned to the network.

Although not really used any more (there are better methods for controlling traffic, such as the Slow Start algorithm discussed elsewhere),
source quench is the end station's ability to indicate to the originator of a
message that the host cannot accept the rate at which the sender is submitting the packets. A source quench packet is continually generated to
the originator until the rate of data flow slows down. The intended

Figure 8-22
Ping.

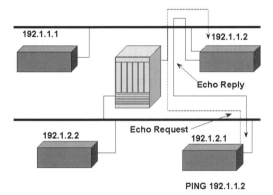

recipient of a source quench continues to slow down its data rate until it receives no more source quench packets. The station that was asked to slow down then starts to increase the data rate again. This is similar to a flow control, except that it is more like throttle control. The data are not stopped, merely slowed down and then increased again. It is generated by any network station on the Internet to indicate that the node cannot handle the rate of incoming data. This ICMP type was not included in ICMPv6. It was found that other protocols handled congestion better than forcing the routers to handle it.

There are many other uses of the ICMP protocol. When a router receives a datagram, it may determine a better router that can provide a shorter route to the destination network. This is an ICMP redirect, and this message informs the sender of a better route. If the TTL field is set to 0, a router will inform the originator of this through an ICMP message (time exceeded). A user's workstation can request a time stamp from a router, asking it to repeat the time when it received a packet. This is used for measuring delay to a destination.

A summary of message types follows:

0	Echo reply
3	Destination unreachable
4	Source quench
5	Redirect
8	Echo
11	Time exceeded
12	Parameter problem
13	Time stamp
14	Time-stamp reply
15	Information request
16	Information reply

IP Routing

Introduction

Now that IP addressing has been completely covered, except for IPv6 addressing, which will be covered later, it is time to jump directly into another function provided by IP: routing. One usually thinks of IP routing in terms of large, complex networks combined through a series of complicated devices known as routers. However, IP performs direct routing as well. Its function is to deliver datagrams, whether local or remote.

A packet-switched network (in contrast to a circuit-switched network) is based on a unit of information, known as a datagram, and its ability to make its way through the network to its destination. The datagram may be routed locally (i.e., the destination is on the same subnet as the originator), which is known as *direct routing*, or it may invoke the use of a forwarding device such as a router if the destination is remote (i.e., on a different subnet than the originator). The latter is known as *indirect routing*, which implies hierarchical routing. A datagram that is sent may invoke both direct and indirect routing.

Why not just have one large, flat network and place everyone on the same network? Eliminate indirect routing completely. Switched networks tried to do this. Flat networks do have their place—in small networks or WAN protocols or extended subnets through switches or bridges. With the current suite of network protocols, a large, flat network is inefficient (it does not scale well), especially in view of the millions of addressable stations attached to it. And the protocols that currently run on networks are broadcast oriented. This means that multiple stations can be attached and grouped to a single network and these stations will see all data no matter who sent it and who it is for. The protocols were built for shared environments. These networks were invented before the advent of switches. Second, when stations need to communicate, the initial communication could be sent in broadcast mode. Communication between certain devices is always done in broadcast or multicast. This is a special type of packet that enables all stations to receive the packet and hand it to their upper-layer software to filter or process. As you scale for growth, a network cannot remain flat. There must be some sort of hierarchy to allow for efficiency.

Not all stations need to see each other. As a network scales, it must maintain its manageability. To make a network more manageable, it is split into many networks called subnets (virtually any network today, whether split or not, is called a subnet). To make these subnets manageable, they are in turn split into sub-subnets. The interconnection of these subnets is accomplished using forwarding devices known as routers. Routers enable data to be forwarded to other networks in a very efficient manner. It is always easier to manage many smaller networks than it is to manage one large network. Also, broadcast data stay on their network or subnet; broadcast data are not forwarded by routers (although exceptions occur, such as DHCP or a directed subnet broadcast).

In order for routers to forward data to other networks, they use special protocols (known as routing protocols) to enable them to draw an internal map of the entire internet. To accomplish this, there are two types of protocols used: *interior gateway protocols* (*IGPs*) and *exterior gateway protocols* (*EGPs*). The exterior gateway protocol that is used with IP is known as the Border Gateway Protocol (BGP). The IGPs that will be explained here are known as the Routing Information Protocol (RIP and RIP2) and Open Shortest Path First (OSPF).

Routing

This section gives a brief introduction to direct routing. Throughout this section, different network numbers will be used. The examples will not employ the use of subnets. Subnets effectively act like network numbers, and they are also separated by a router. For example, in Fig. 9-1 the network numbers could be 140.1.1.1 on the network with end station B and 140.1.2.1 on the network containing host A. Using a subnet mask of 255.255.255.0 would yield two different networks: 140.1.1.0 and 140.1.2.0. For simplicity in explaining routers, I have chosen to use completely different network numbers.

How does a network station know whether the packet has to be directly (local) or indirectly (remote) routed? For the network station, it is a relatively simple process. The whole basis for routing is in the IP network number assigned to the network station.

Remember that an IP address contains the network number as well as the host number. With the first 1, 2, 3, or 4 bits of the 32-bit IP network address identifying the class of the address, this allows any network station (workstation or router) to quickly extract the network portion out of the class of the IP address. In other words, by reading up to the first 4 bits of the IP address, a network station can quickly determine how much of the IP address to read to determine the network number of the address. The sending station will compare the packet's destination network number to its own network number. If the network number portion of the destination IP address matches its own, the packet can be

Figure 9-1
Routing.

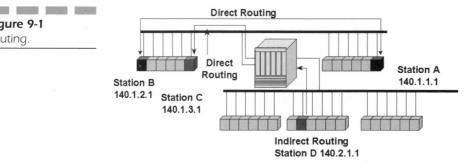

routed directly on the local LAN, without the use of a router. The packet is simply transmitted to the station (using ARP if necessary).

Once this determination is made and the packet is destined for a local route, the network station checks its ARP table to find the IP-to-physical address mapping. If one is found, the packet is physically addressed and transmitted onto the network. The physical destination address (located in the data-link header) will be that of the receiving station. If the station's address is not in the ARP cache, the ARP request process is invoked.

Referring to Fig. 9-1, end station B and host A are located on the same network. They communicate through direct routing. However, station C and Station D are on separate networks. They communicate through the use of indirect routing. However, notice that when the router sends the datagram to station C, this IP function uses direct routing.

A distinction should be made here. There is a difference between a *routing protocol* and a *routable protocol*. A routable protocol is one that allows for routing, such as NetWare (IPX) and TCP/IP. NetBIOS and LAT (a DEC terminal/printer protocol) are not routable protocols. Examples of a *routing protocol* are RIP, OSPF, etc. These are protocols that enable the routing functions to work properly.

Indirect Routing

If the host resides on a network with a different network number (not on the local subnet), the transmitting station will have to use the services of a router. The transmitting station will assign the physical destination address of the packet to that of the router (using ARP, if necessary, to find the physical address of the router) and submit the packet to the router. Each workstation may be able to determine the address of its closest router or is preconfigured with the address of its default router.

The router will, in turn, deliver the packet either to its locally attached network (ARPing for the destination's physical address and submitting the datagram directly to that network station) or to another router for delivery of the data. Notice here that the destination physical address is that of the router and not the final destination station. This type of routing is indirect routing. The final destination IP address is embedded in the IP header.

Sending a packet to its final destination might involve both direct and indirect routing. For example, when a packet is to be delivered across an internet, the originating station will address it to the router for delivery to its final network. This is indirect routing. No matter whether the final destination network ID is directly connected to that router or whether the packet must traverse a few routers to reach its final destination, the last router in the path must use direct routing to deliver the packet to its destination host.

Depending on the options field settings, none of the IP routing protocols will alter the original IP datagram, with two primary exceptions: the TTL (time to live) field and the cyclic redundancy check field. If an IP datagram is received by a router and it has not arrived at its final destination, the router will decrement the TTL field. If TTL>0, it will forward the packet based on routing table information. Otherwise, the IP datagram's header contents will remain the same (with the exception of an error detection field known as the cyclic redundancy check, or CRC). Since the TTL field has changed, the CRC must be recalculated throughout all the networks and routers that the datagram traverses. Otherwise, the only alterations that are made are to the data-link headers and trailers. The IP addresses in the IP header will remain the same as the datagram traverses any routers in the path to its destination.

IP routers forward datagrams on a connectionless basis and therefore do not guarantee delivery of any packet. They operate at the network layer, which provides best-effort or connectionless data transfer. Routers do not establish sessions with other routers on the internet. In fact, IP routers do not know of any workstations (nonrouters) on their subnets.

These routers forward packets based on the network address of the packet (in the IP header) and *not* on the physical address (the 48-bit address for broadcast networks) of the final destination (the receiver) of the packet. When the router receives the packet, it will look at the destination network address (embedded in the IP header of the packet) and determine how to route the packet. Routers only route packets that are directly addressed to them. They do not operate in promiscuous mode (watching all LAN traffic) for forwarding datagrams.

Figure 9-2
IP routing flowchart.

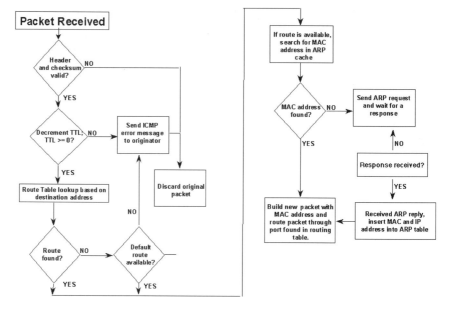

Figure 9-2 shows the flowchart of the routing process.

Some features need to be explained about the IP layer that allow the internet to operate. When a router receives a packet, how does it know where and how to send the packet? The router must know which path to use to deliver the packet. This is accomplished through IP routing algorithms, which involve the maintenance of a table of known routes (network numbers) and a means of learning new routes (network numbers) when they become available.

Distance-Vector. Information is kept in the router that allows to it to know all the networks or subnets in its domain and the paths to get to those networks. This information is grouped in a table. There are two standard methods for building this table: distance-vector and link state. Link state will be covered later. In the distance-vector method the information sent from router to router is based on an entry in a table consisting of the parameters *<vector, distance>*. *Vector* means the network number, and *distance* means what it costs to get there. The routers exchange this network reachability information by broadcasting their routing table information consisting of these distance-vector entries. This broadcast is local, and each router depends on other routers for the correct calculation of the distance.

Each entry in the table consists of a network number (the vector) and the amount of routers (distance) between the router and the final network (indicated by the network number). This distance is sometimes referred to as a *metric*. For example, if the source station wants to transmit a packet to a destination station that is four hops away, there are probably four routers separating the two networks.

Each time a datagram must traverse a router (thereby passing through a new network number) is considered a hop (metric). For RIP, the maximum diameter of the internet is 15 routers (hops). A distance of 16 is an indication that the network is not reachable. Remember, RIP is an IGP that is under one domain. The Internet itself encompasses many domains, and the diameter of the Internet is much larger than 15 hops.

As shown in Fig. 9-3, each router contains a table with starting entries of those networks that are directly attached to it. There are actually more header entries in a routing table, but the significant portions are shown in the figure. From this table, we know that networks 134.4.0.0 and 134.3.0.0 are directly connected to this router. Network 134.4.0.0 is assigned to port 1 of the router. It is running the RIP protocol, and *xxx* indicates how long the route has before it is deleted from the table.

RIP Updates. This figure shows the updating process of RIP. Parts of a router's table (the network number and the hop count) are broadcast to the local networks to which the router is directly attached. There are a few exceptions, which will be explained shortly. Any router that is located on the same network will receive the packet, read the routing table data, update its table if needed, and then discard the packet. Routers do not forward any update packets they receive. In other words, all routers for-

Figure 9-3
Routing distance-vector table.

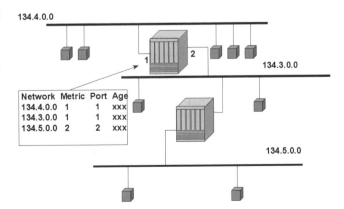

ward their tables out each active port. As each table is received, the routers build a picture of the network. As each broadcast is transmitted, more and more information is propagated throughout the network. Eventually, all routers know of all networks on their internet.

There are three events that can cause a router to update its existing table based on newly received information:

1. If the received table contains an entry to a network with a lower hop count, the router will *replace its entry* with the new entry with the lower hop count.

2. If a network exists in the received table that does not exist in the router's own table, it will *add the new entry.*

3. If the router forwards packets to a particular network through a specified router (indicated by the next-hop router address) and that router's hop count to a network destination changes, it will *change its entry.* In other words, if router A normally routes data for network X through router B, and router B's hop count entry to that network changes, router A *changes its entry.*

Router Table Updates.   Figure 9-4 shows what happens when router A submits its routing table out of its port connected to network 2. (For simplicity, the figure shows the updating through one port only. In reality, routing tables are submitted out all ports of a router, with a few restrictions on which entries of the table get transmitted.)

Figure 9-4
Router table updates.

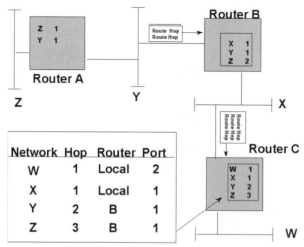

Router A transmits a table containing two networks: Z and Y. Each of these networks is one hop away (each is directly connected). Router B will receive this packet and add 1 to each hop count entry in the received table. (This is accomplished assuming that the RIP cost assigned to that port of router B is 1. It could be configured to be something else.)

Router B will examine its table and notice that it does not have an entry for network Z. It will add this entry to its table in the following format: network Z, available through port 1, two hops away. It will then check the next entry. Network Y will not be added, for router B already has network Y in its table with a cost of 1. Since the incoming table reports that network Y has a cost of 2, router B will ignore this entry. (There are rules that will prevent router A from sending out information about network 2, and these rules will be discussed later.)

Once its table is updated, router B will transmit its table every 30 seconds out of its ports (again, for simplicity only one port is shown). Router C will receive this table from router B and perform the same steps as router B. Eventually, all information about all networks on the internet is propagated to all routers.

IP Routing Tables. As illustrated in Fig. 9-5, the significant entries in a routing table consist of three elements:

1. Network number
2. Hops to that network number
3. The next router in the path to that network

Routing table fields vary according on the update mechanism used. The table in Fig. 9-5 is an example of a routing table used by the routing

Figure 9-5
IP routing tables.

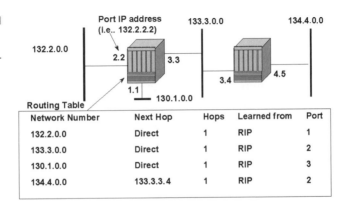

Network Number	Next Hop	Hops	Learned from	Port
132.2.0.0	Direct	1	RIP	1
133.3.0.0	Direct	1	RIP	2
130.1.0.0	Direct	1	RIP	3
134.4.0.0	133.3.3.4	1	RIP	2

information protocol (RIP) for the IP protocol. The table values are defined as follows:

Network number.   A known network ID.

Next router to deliver to.   The next router that the packet should be delivered to if the destination network is not directly connected. A directly connected network is one that is physically connected to the router. Most routers today have more than two connected networks.

Hops.   This is the metric count of how many routers the packet must traverse before reaching the final destination. An entry of 1 indicates a local route.

Learned from.   Since many routing algorithms may exist in a router (i.e., RIP, OSPF, and BGP may exist in the same router), there is usually an entry in the table to explain how the route was acquired.

Time left to delete.   The amount of time left before the route will be deleted from the table.

Port.   The physical port on the router from which the router received information about this network.

The Routing Information Protocol (Version 1)

Dynamic updating is the process by which routers update each other with reachability information. Before the advent of dynamic updating, most commercial vendors supported manual updates for their router tables. This meant entering network numbers, their associated distances, and the port numbers manually into the router table. The Internet was then known as the ARPANET, and it employed a routing update scheme known as the Gateway Information Protocol and later the Gateway to Gateway Protocol (GGP). This is beyond the scope of this book and is not used anymore. Independent router vendors did not have that many routers and subnets to update, so placing a manual entry in the routers was not all that complicated. As networks grew larger, this became a cumbersome way of building tables. RIP is the protocol that enables automatic updates of router tables.

RIP is based on the distance-vector algorithm just described. Implementations of this protocol first appeared on the ARPAnet in 1969 using the Gateway Information Protocol. However, it was devised by Xerox Corporation as the routing algorithm used by the internet datagram protocol of XNS.

RFC 1058 defined RIP for TCP/IP, and it was formally adopted by the IAB in 1988. Although it was not primarily intended as *the* routing algorithm for IP, it gained widespread acceptance when it became embedded in the Berkeley UNIX operating system through a service known as *routed* (pronounced "route d"—*d* is for the daemon process that runs the protocol in UNIX). The protocol was actually in widespread use long before 1988; it was distributed in Berkeley 4BSD UNIX and gained widespread acceptance because of the vast number of installations of this operating system. Incorporating the functions of RIP in an RFC allowed interoperability and certain detailed functions to exist.

With RIP information, any router knows the length of the shortest path (not necessarily the best) from each of its neighbor routers (routers located on the same network) to any other destination. There are many deficiencies to this protocol, and they are identified at the end of this section.

Figure 9-6 shows the RIP header and data encapsulated in an Ethernet packet.

RIP Operational Types

Two types of RIP packets traverse a network (indicated by the command field). One type of RIP packet requests information and the other gives information (a response packet). Naturally, a response packet is generated for a request packet and is used for periodic RIP updates. Most RIP packets that traverse a local network are the periodic RIP table updates.

Both routers and individual hosts can implement the RIP protocol since RIP has two modes: *active* and *passive*. In active mode, RIP both lis-

Figure 9-6
RIP packet encapsulation.

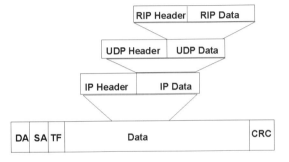

tens to RIP broadcasts from other network stations (and builds it own internal tables) and transmits its own broadcasts to respond to requests from other stations.

In passive mode, RIP listens only for RIP updates. (It may build its own tables or it may not. If it does build a table, it will not broadcast the table. It will build a table so that it will not have to request information from other routers on the network.) Passive mode can be used for non-routing network stations. These devices have no reason to broadcast updates, but have every reason to listen for updates. Today, most DOS PC computers use a concept of a default gateway, explained later. Even Windows 95 uses a default gateway if prompted. It can build a routing table, but Windows 95 is not RIP enabled.

Using the RIP passive protocol allows the host to maintain a table of the shortest routes to a network and designate which router to send the packets to. This does consume a considerable amount of RAM for both the table and the algorithm. Without it, TCP/IP requires the use of a default gateway or static entries indicating that when a packet is destined for a remote network, the host must submit the packet to a specified gateway for it to process, even if this gateway does not have the shortest path to that network. Passive implementations add no overhead to the network, for they listen only to routing table updates that are on the network. Without passive RIP, these devices have to maintain their own tables or implement a default or static routes.

For simplicity, most workstations do not invoke active versions of the RIP protocol They do not build tables and keep track of networks. To communicate with a router, workstations generally use their default gateway parameter.

Remember that RIP packets do not leave their local network. All participants in the RIP protocol (for example, routers) receive the packet, update their tables if necessary, and then discard the update packet. They compute the reachability of networks based on adding a cost (usually 1) to the just-received table's count entry and then broadcast their tables out their ports (usually being mindful of a protocol named split horizon, which is explained a little later).

RIP Field Descriptions

The RIP header and data fields are shown in Fig. 9-7. The fields in the RIP packets are identified as follows:

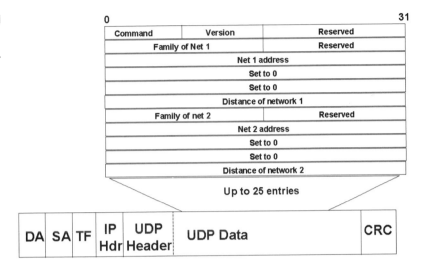

Figure 9-7
RIP field descriptions.

This field consists of a number corresponding to a command, as follows:

1   Request for partial or full routing table information

2   Response packet containing a routing table

3,4   Turn on (3) or off (4) trace mode (obsolete)

5   Sun Microsystems internal use

Version. Used to indicate the version of RIP. The field is currently set to 1 for RIP version 1.

Family of net x. Used to show the diversity of the RIP protocol. This field is used to indicate the protocol that owns the packet. It is set to 2 for IP. Since XNS could possibly run on the same network as IP, the RIP frames would be similar. This indicates that the same RIP frame can be used for multiple protocol suites. AppleTalk, Novell NetWare's IPX, XNS, and TCP/IP all use the RIP packet. The packet is changed a little for each protocol.

IP address. Indicates the IP address of a specific destination network. This field is filled in by the requesting station. An address of 0.0.0.0 indicates the default route, explained later. The address field needs only 4 of the available 14 bytes, so all other bytes must be set to 0. If this is a request packet and there is only one entry, with the address family ID of 0 and a metric of 1, then this is a request for the entire routing table.

Distance to network. Only the integers of 1 to 16 are allowed. An entry of 16 in this field indicates that the network is unreachable.

The next entry would start with the IP address field through the metric field. This would be repeated for each table entry of the router to be broadcast. The maximum size of this packet is 512 bytes.

Although not mentioned until later, the RIP protocol relies on the transport-layer protocol of the user datagram protocol (UDP), discussed in the section on transport-layer protocols. In this is the specification for the length of the RIP packet. Also, for those interested, RIP operates on UDP port number 520 (port numbers are discussed in the UDP section).

Default Routes

On a TCP/IP network, there is a concept known as the *default route.* This is not a standard feature of any other network protocol (XNS, AppleTalk, IPX, etc.). The default route can be maintained in two places: the router and the end station. Default routes are similar to static routes.

For an end station that does not support the active or passive function of the RIP protocol, thereby allowing it to find a route dynamically, the default router (commonly called a default gateway) is assigned to it. This is the 32-bit address of the router the workstation should route to if remote routing is necessary. The IP layer in the end station would determine that the destination network is not local and that the services of a router must be used. Instead of implementing the RIP protocol, the end station may submit the packet to the default router as assigned by the default route number. The router will ensure that the packet reaches its final destination. If that router does not have the best route, it sends a message (using the ICMP protocol) to the end station to inform it of a better route. This will be explained later.

A router may also be assigned a default route, indicated by a 0.0.0.0 in its routing table. This is implemented for the condition when a router receives a packet and does not have the network number in its table. The router will forward the packet to another router for which it has as an assigned default route. This means that when a router has received a packet to route, and its table does not contain the network number indicated in the received packet, it will forward the packet to its default router hoping that the default router has the network number in its table and will be able to properly forward the packet. The default router receives the packet and, if the network number is in table, it will forward the packet. If the network number is not in its table, it, too, may have a default router—and it will forward the packet to that router. If there is no route and there is not another default route, the last router sends a control message (through ICMP) back to the originating station.

The problem with default routes is that a workstation's default router may go down and the workstation will not know if there is another router on the network. The network number may change, or there may be a better path for the workstation to take. The default gateway does allow for the elimination of routing tables in the network station and routers, and reduces the routing tables by allowing groups of networks to become available through the default route.

Figure 9-8 shows how default routes are implemented.

Disadvantages of the RIPv1 Protocol

As noted before, the acceptance of RIP in the Internet community was based on its implementation in the popular Berkeley 4BSD UNIX operating system through a process known as *routed* (pronounced "route-d," two words). Before RIP was implemented, router tables often had to be constructed manually. RIP allowed these tables to be updated dynamically, which was a real advantage at that time. Unfortunately, it was implemented before the rapid growth of the TCP/IP. It has many disadvantages that were not considered limiting at the time it became accepted.

RIP understands only the shortest route to a destination, which may not be the fastest. RIP understands only hop counts. For example, there may be two paths to a destination—one that traverses two T1 lines (three hops) and another that has two hops but uses a 9600-baud serial line. RIP

Figure 9-8
Default routes.

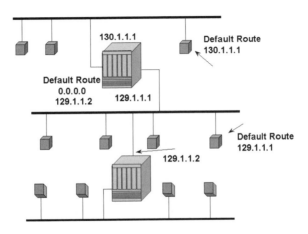

would pick the 9600-baud line, since it is shorter (two hops). Variations of RIP allow the network administrator to assign an arbitrary RIP hop count or cost to a route to avoid this problem. However, this creates another problem. This incremented RIP number adds to the upper limit of a 15-hop diameter in RIP. The limit for the number of hops that a network may be from any network station is 15; a hop count of 16 is considered unreachable. If you artificially add hops to a path, you decrease the total number of routers allowed in a path.

With RIP, received routing table updates are only as accurate as the router that submitted them. If any router made a computational error in updating its routing table, this error is received and processed by all other routers.

What may also be apparent is the fact that the routing tables could get very large. If the network consisted of 300 different networks (which is not uncommon in larger corporations), each routing table of every router would have 300 entries. Since RIP works with UDP (connection-less transport-layer service), the maximum datagram size of a RIP packet is 512 bytes (576 including all media headers). This allows for a maximum of 25 *network number, distance* combinations in each packet. Therefore, it would take 13 packets for each router to broadcast its routing table to all other routers on all the local networks in the internet. This information would be broadcast every 30 seconds by each of the 300 routers, and there is a possibility that nothing had changed from the previous update! This is an unnecessary consumption of bandwidth and other resources, especially over slow serial lines.

This leads to the second disadvantage. RIPv1 broadcasts (using a data-link physical address of all FFs) to the network, normally every 30 seconds, even across slower serial links. This makes the data link pass the packet up to the upper-layer protocols on all stations on the network, even if the stations do not support RIP. This becomes even worse for those installations that have AppleTalk and IPX on their networks running RIPv1. Not only will RIP for TCP/IP broadcast every 30 seconds, but AppleTalk broadcasts every 10 seconds and NetWare using IPX (RIP and SAP) every 60 seconds.

Scaling with RIP. RIP was designed for small, stable networks. This is stated in the Xerox documentation. RIP does not handle growth very well. This problem has two consequences. The first limitation is that a destination network may be no more than 15 hops away (a distance of 16 in any routing table indicates that the network is unreachable). RIP is not recommended for large networks to be based on RIP. Careful planning is needed to implement large-scale networks based on the RIP protocol.

The other scaling problem involves the propagation of routing information. Four terms need to be understood here, for they are used quite frequently: *split horizon, hold-down timer, poison reverse,* and *triggered updates.*

Refer to Fig. 9-9. With router A directly attached to network Z, initially it advertises that route through *all* its ports as a distance of 1 (whatever the RIP-assigned cost of the port that attaches to that network is). Router B receives this and updates its table as network Z with a distance of 2. Router B then broadcasts its table (at the 30-second update timer), and router C receives this and update its table as network Z with a distance of 3. Notice that all routers broadcast all the information in their tables through all ports (even the ports from which they received the update).

Why would router B broadcast a reachability of network Z when router A already has a direct attachment to it? Wouldn't this confuse router A? Normally it would, but remember that the only time a router makes changes to its tables is when the hop count distance is lower, when there is a new entry, or when the router path taken to a network changes its hop count. Since that hop count is higher, router A will simply ignore that particular entry in the update table.

Using the original algorithm, a serious problem occurs when router A loses its reachability to network Z. It will update its table entry for that network with a distance of 16 (indicating not reachable) but will wait to broadcast this information until the next scheduled RIP update. So far so good, but if router B broadcasts its routing table before router A (notice that not all routers broadcast their tables at the same time), router A will then see that router B has a shorter path to network Z than it does (a distance of 2 for router B versus a distance of 16 for router A). Router A will change its entry for network Z. Now, router A, on its next RIP update broadcast, will announce that it has a path to network Z with a distance of 3 (2 from the table entry received from router B plus 1

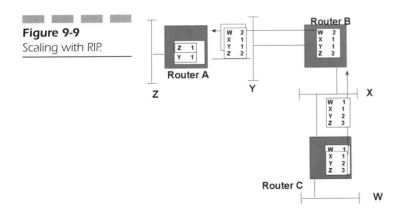

Figure 9-9
Scaling with RIP.

to reach router B). There is now a loop between routers A and B. When router B receives a packet destined for network Z, it will forward the packet to router A, router A will forward it back to router B, and this will continue until the TTL field reaches 0. The RIP protocol works extremely well in a stable environment (an environment where routers and their networks rarely change). The process of clearing out dead routes and providing alternative paths is known as *convergence*.

Even future RIP updates will not quickly fix the convergence in this case. Each update (every 30-second default) will add 1 to the table entry, and it will take a few updates to outdate the entry in these routers. This is known as *slow convergence*, and it causes errors in routing tables and routing loops to occur. What if you have a network diameter of 15 routers, with successive routers exactly opposite on the timer in broadcasting their updates? In other words, when one router broadcasts its table, the receiving router has just finished its broadcast. The lost route could take many minutes to update those routers at the end of the network.

To overcome the limitations, a few rules were added to the IP RIP algorithm:

1. *Split horizon.* Implemented by every protocol that uses a variation of RIP (AppleTalk, IPX, XNS, and IP), this rule states that a router will not broadcast a learned route back through the port from which it was received. Therefore, router B would not broadcast the entry of network Z back to router A. This keeps router B from broadcasting back to router A the reachability of network Z, thereby eliminating the possibility of a lower hop count being introduced when network Z becomes disabled. Figure 9-10 shows how the routers would send their tables using split horizon. Notice which routes are not included in their updates.

Figure 9-10

Split horizon demonstrated.

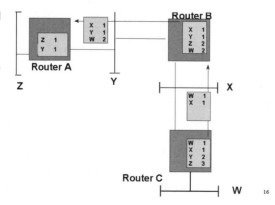

2. *Hold-down timer.* This rule states that once a router receives information claiming a known network is not reachable, it must ignore all future updates that include an entry (a path) to that network, typically for 60 seconds. Not all vendors support this in their routers. If one vendor does support it and another does not, routing loops may occur.

3. *Poison reverse* and *triggered updates.* These are two additional rules that help to eliminate the slow convergence problem. They state that once the router detects that a network connection is disabled, the router should keep the present entry in its routing table and then broadcast "network unreachable" (metric of 16) in its updates. These rules become efficient when all routers in the internet use triggered updates. Triggered updates allow a router to broadcast its routing table immediately following receipt of this "network down" information.

The two most commonly instituted rules are split horizon and poison reverse.

Split Horizon Demonstrated. In Fig. 9-10, there are three routers, labeled A–C, and four subnets, labeled W–Z. Upon startup, the routers learn of their immediate subnets. Ignore the routing tables shown for the time being. Router A learns about subnets Y and Z. Router B learns about subnets X and Y. Router C learns about subnets W and X. These are directly connected ports, and those networks appear first in the routing table. The routers may or may not automatically broadcast their tables out after initialization (this is vendor dependent).

Next, router C transmits its table containing subnets W and X. It transmits this information out of the ports connecting to subnets W and X. Router B transmits its table containing subnets X and Y out both of its ports, and router A transmits its table containing subnets Y and Z out both of its ports. Each of the costs in these tables is set to 1. Usually, these table broadcasts are not synchronized, but it is shown here that way for simplicity.

Router B picks up the information transmitted by router C and makes some decisions. It adds to each entry the cost associated with the port that it received the information on. In this case, that port was assigned a cost of 1. Therefore, it now has two entries in the received table each with a cost of 2. It then compares this information to its table. It already has a entry for subnet X and it has a cost of 1, so it discards the new information. The next entry is for subnet W with a cost of 2. It does not have that entry, so it adds this entry to its table with a cost of 2. Router B figures it is now complete.

Router C will update its table with the entry for subnet Y (propagated by router B after router B received this information). Router C now has the entries in its table of subnet W cost 1, subnet X, cost 1, subnet Y cost 2, and subnet Z cost 3.

The periodic timer has expired (after 30 seconds), and router C is ready to broadcast its table. Out of the port associated with subnet W it lists the entries for subnets W through Z. However on the port associated with subnet X, it will only include those entries for subnet X (some routers do not include this entry if they know of another router on this segment) and subnet W.

It will not include the entries for subnet Y and subnet Z. The rule for split horizon is not to rebroadcast a known route back over the port that the router learned it on. If router C did not learn about subnet Y and subnet Z through the port associated with subnet X, then it would have included these entries in the table when it broadcast its table on subnet X. This eliminates the problem mentioned before about a router losing its connection to a network and then learning about another path to that network and updating its table.

RIP Version 2

In November 1994, RIP underwent a change. It was modified with some additions (extensions) to overcome some of the shortcomings of the protocol. RIP version 1 continues to exist on many routers and as of this writing, it continues to outnumber OSPF networks. However, there is no reason not to implement version 2 of the protocol. Version 2 is completely backward compatible with version 1 and contains all of the capabilities of the version 1 protocol.

RIP version 2 implemented the following features:

- Authentication—simple text password
- Subnet masking—allowing variable-length subnet masks to be implemented
- Next host
- Multicast
- Route tag—used to provide a method of separating RIP routes from externally learned routes
- Compatibility switch—to allow for interoperability with version 1 routers

Authentication

There really is no room in the RIP update datagram for authentication. But since this has become commonplace (OSPF), room was made for it. Refer to Fig. 9-11. The address family identifier (AFI) is used for authentication. If the AFI contains a 0xFFFF, then the first entry in the route entry list is the password to be used for authentication. The header of the RIP datagram changes as shown in the figure. The authentication type is type 2 (simple password), and the next 16 bytes contain this password (any amount of characters up to 16 bytes). RIPv1 will ignore this entry (the first entry) since AFI is not set to an address family of IP.

If a RIPv2 router is configured with no authentication, it will accept and process both RIPv1 and v2 unauthenticated messages and discard authenticated messages. If the RIPv2 router is configured for authentication, it will accept RIPv1 and v2 messages that pass authentication. Remember, not all v1 implementations follow the RFC. They may play with the fields and still be able to be processed by RIPv1 routers, although this is not recommended. Unauthenticated RIPv2 messages will be discarded.

Subnet Mask Field

As we learned in other sections of this book, one of the biggest problems with RIP is its inability to support a subnet mask in the routing update. This led to the shortcoming of one subnet mask per network ID. Subnet masking really extends the life of RIP. RIPv1 does not indicate a subnet mask in a route entry. See Fig. 9-12. This can create many problems, two of which are learning and updating. How does RIPv1 know

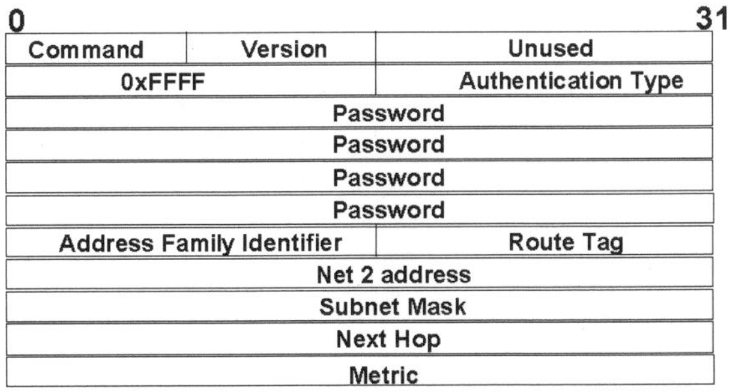

Figure 9-11
Authentication.

Command	Version	Unused
0xFFFF		Authentication Type
Password		
Password		
Password		
Password		
Address Family Identifier		Route Tag
Net 2 address		
Subnet Mask		
Next Hop		
Metric		

0   31

how to apply a subnet mask for a learned IP address? How does RIP provide a mask for its updates? Good questions! The answer is not so good, though. RIP assumes that the IP address uses the same subnet mask as it does, providing the IP network ID portion of the address is the same as its own and there is a subnet mask applied to its interface.

For example, suppose a router has two interfaces. On interface 1, the IP address is 130.1.1.1 and has a subnet mask of 255.255.255.0. Interface 2 has the address 205.1.1.1 (subnet mask of 255.255.255.0). When interface 1 receives a routing update with any entry that has the same network ID as its own (130.1.x.x), it will apply the subnet mask that is configured to its port to those entries. You will see an entry in the routing table for that learned address. So if a RIPv1 update was received on interface 1 and the update contained the entry 130.1.4.0, then the interface will record 130.1.4.0 in its routing table. However, if interface 1 received 155.1.1.0, it would only place 155.1.0.0 into its table for its does not know the subnet mask for the address of 155.1.0.0.

When the router must transmit its table, how does it know to apply a mask to any of the entries in the table? It depends on the interface that is transmitted out of. On interface 1, the router will transmit two entries (150.1.1.0 and 200.1.1.0). This was demonstrated earlier in Fig. 9-10.

So, RIPv1 and subnet masks did not understand each other. RIPv2 fixes that. Notice in Fig. 9-11 that the format of the RIP datagram is preserved. The two fields in RIPv1 labeled "set to 0" are not used for the subnet mask and next hop entries. Each route entry in the datagram will have a subnet mask associated with it. If the field contains a 0, then

Figure 9-12

RIP version 2.

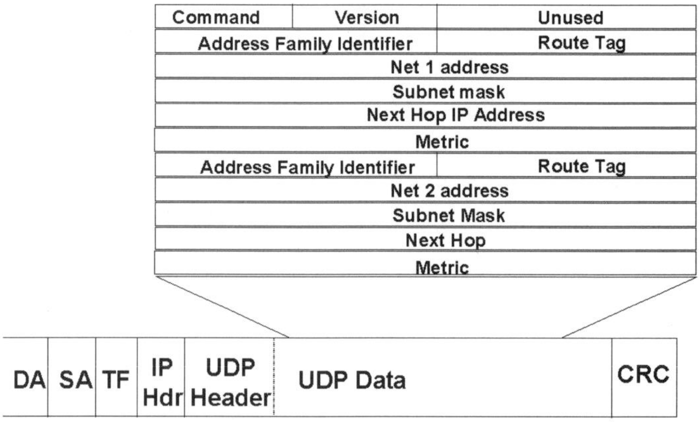

Command	Version	Unused	
Address Family Identifier		Route Tag	
Net 1 address			
Subnet mask			
Next Hop IP Address			
Metric			
Address Family Identifier		Route Tag	
Net 2 address			
Subnet Mask			
Next Hop			
Metric			

DA	SA	TF	IP Hdr	UDP Header	UDP Data	CRC

there is not a subnet mask associated with the route entry. Also, in coordination with RIPv1 routers, a mask that is shorter than the class's natural mask should never be advertised.

Route Tag and Next Hop Fields

The route tag entry is used to advertise routes that were learned externally (not in this IGP). OSPF has this capability and allows the OSPF IGP to learn about routes external to the IGP. For example, if the routes were learned via BGP-4, the route tag entry could be used for setting the autonomous system where the routes were learned from.

The next hop field allows the router to learn where the next hop is for the specific route entry. If the entry is 0.0.0.0, then the source address of the update should be used for the route. Over a point-to-point link (to routers connected by a serial line) there is not much use for this entry (the next hop could be extracted from the source IP address in the IP header of the packet). This field does have considerable utility in instances where there are multiple routers on a single LAN segment using different IGPs to communicate to multiple LANs.

Multicast Support

A key improvement for the RIP protocol is the ability to use a multicast address for its packets and for its datagram IP header. The multicast address for RIPv2 is 224.0.0.9 with a MAC address of 01-00-5E-00-00-09. Of course, this must be mapped to an Ethernet multicast address (for more information on this, refer to the section on multicasting in this book or RFC 1700).

If you read the section on multicast, you know that the benefits of multicast are great. RIPv1 uses a broadcast address that interrupts not only the NIC but also the IP service layer, even if the packet is not destined for that host. Why interrupt the host when the packet/datagram is destined for some other host? All broadcast packets must be received and processed. This was not a problem when RIPv1 was introduced to the IP community (there were not many hosts to contend with). Multicast allows only those hosts that have designated their NICs to receive multicast packets to receive and process them. All other multicast packets will be ignored.

Even though multicast is used, IGMP is not to be used, for the address is a local multicast address. This means that the packet will never leave the network it was transmitted on (it will not be forwarded by routers).

Compatibility with RIPv1

So if all of these new "features" are used, how do we communicate with version 1 RIP routers? There are two sides to this story. If a RIPv2 router receives a RIPv1 update, it will process it as a v1 update and will not try to convert any of the information received to RIP features. If a RIPv1 request is received by a RIPv2, the RIPv2 router should respond with a version 1 response.

Now, there are many changes (multicast, broadcast, etc.) in the way in which a v2 router could respond. Therefore, during the configuration of a v2 router, there are various parameters that allow the v2 router to act in many different ways.

- RIP-1—Only version 1 messages will be sent.
- RIP-1 compatibility—RIP-2 messages are sent with broadcast addresses (IP header and MAC).
- RIP-2—Messages are multicast.
- None—No RIP messages are sent.

Although not required, some routers have implemented a receive parameter listing that allows for

- RIP-1 only
- RIP-2 only
- Both

Also for compatibility, RFC 1058 states that the version field should be used according to the following format:

- Any version field of 0—Discard the entire packet.
- Any version field of 1 and MBZ (must be zero) fields nonzero—The packet is discarded.
- Any version greater than 1—The packet should not be discarded simply because the MBZ fields contain a value other than zero.

Therefore, routers that strictly adhere to RFC 1058 may be able to process RIP version 2 updates and build routing tables based on that information. RIP version 1 routers will ignore the subnet mask and next hop fields. Also, they will ignore the route tag field (it is a reserved field in RIP version 1). Also, RIPv1 will ignore any AFI field that is set to FFFF (for RIPv2 authentication) and the route that applies to the AFI. (For RIPv2, it will be the first entry of a RIPv2 datagram. All other entries will be valid RIP route entries.)

Static versus Dynamic Routing

The last topic of discussion is the capability of routing protocols to accept information for their tables from two sources: the network and a user.

Although the RIP protocol allows automatic updates for routing tables, manual entries are still allowed and are known as *static entries*. A default route is a static entry. An end station that is configured with a default router is said to have a static route entry. A static route can be configured to be included or not included in a dynamic update. Static routes refer to the process of manually placing an entry in the table. For any given port on the router, the network administrator may update that port table with a static route. Static routes usually override dynamic routes.

Static tables have many disadvantages. First, as discussed before, static tables are not meant for large networks that incur many changes such as growth. As the topology changes, all the tables must be manually reconfigured. Second, in the case of router failure, the tables have no way of updating themselves. Dynamic tables overcome these disadvantages.

The primary advantage of a static entry is security, for static tables can be configured *not* to broadcast their routes to other routers. In this way, users can customize their routers to become participants on the network without their network identities being broadcast to other routers on the network. Static routes also allow the user to update a routing table with a network entry that will be used in end stations with the dynamic function turned off. This allows the user to maintain the routing table.

Static entries are also used in various IP topologies. For example, in a hub-and-spoke topology, where a business has a centralized corporate office and many remote offices (such as a bank), there really is no reason to fully enable RIP at the branch offices. Why not turn RIP supply (the ability to broadcast routes but not listen for any) at the remote branch and add a default route, in the remote branch router, pointing to the upstream router located at the corporate office? In this way, the upstream router dynamically learns about all the remote offices (and learns when they go away), and the branch office has one simple entry in its table (besides its attached subnets), a default route to the upstream router. Since there is no other path besides that one link to the upstream router, the router simply passes on any packets it receives from its attached workstation on the network. This reduces the amount of memory and processor power needed at the remote branch, also enabling a cheaper router to be placed out there.

This is an example of why OSPF need not be turned on for a complete network. There is no reason whatsoever to run OSPF out at the branch offices. There is plenty of reason to run it at the corporate offices. OSPF will simply pull in the RIP networks as external networks (but this could possibly build large routing tables).

Remote Networks

There are times when geographically separated networks must be connected. This means that the networks cannot be connected by the conventional means of a LAN interconnect. Imagine trying to cable a network together with one subnet in California and another in Virginia. Ethernet cannot stretch that far. The only feasible way of doing this is by using some type of WAN service from the telephone company. AT&T, MCI, and Sprint all provide WAN services for data networks. The choices available are Frame Relay, Switched Multimegabit Data Service (SMDS, primarily used in metropolitan areas and not cross-country), Integrated Service Digital Network (ISDN), and leased lines. For simplicity, only leased lines using the point-to-point protocol will be explained here. Figure 9-13 shows a simple example of a remote network topology using serial lines.

Just as the router has physical connectors for connection to a LAN, the router has a connection that enables a remote network connection. Instead of the LAN interface on the router, there is a serial line interface. The connector for this is usually a V.35, EIA-232 (formerly RS-232-D), or RS-449 connector. The connection will be made to a device known as a Data Service Unit/Customer Service Unit (DSU/CSU). This is a box that receives the serial signal from the router and repeats it to the telephone switching office.

Figure 9-13
Remote networks.

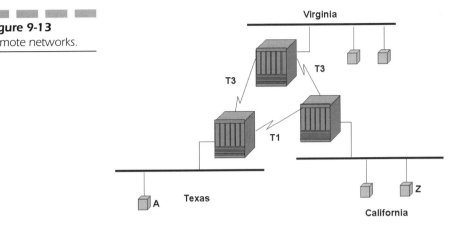

The leased line is a specially conditioned line that is provided by the phone company. This line has been conditioned to handle high-speed digital traffic. It is not the normal line that is used for voice switching. This line is permanently switched to provide a connection between two points. Therefore, it is sometimes called a point-to-point link. It is analogous to dialing a number, receiving a connection, and never hanging up.

The router at the remote end is also attached to a DSU/CSU and can receive the signals generated at the remote end. The typical speed at which these lines run varies. The most common speeds are 56 kbps and 1.544 Mbps (for T1). These lines are called leased lines because the customer does not own the line. It is leased from the phone company, and the rate varies depending on the length of the line. Rates are usually cheaper for short runs (where the other point of the network is a few miles away) and higher for longer runs. Rates also vary depending on the speed of the line.

The serial line provides a simple interconnect between two routers that cannot be connected directly by a LAN. A real problem in using them in an IP internet is that they consume a full network number or a subnet number. There have been methods to overcome this using variable-length subnet masking, which is available with the routing algorithms of OSPF and RIPv2. Otherwise, they generally act as a full network even when there are only two points connected.

Open Shortest Path First (RFC 2178)

The following represent the major shortcomings of the RIP protocol:

- The maximum distance between two stations (the metric, measured in router hops) is 15 hops. A destination (network ID) whose hop count is 16 is considered unreachable

- The cost to a destination network is measured in hops. RIP determines a route based on a hop count, which does not take into consideration any measurements except the number of routers between the source and destination network. A two-hop high-speed network will be bypassed for a one-hop low-speed link. A path can be tricked into taking a better path by adjusting the hop count metric on the router port, but this reduces the available network diameter.

- RIP updates its entire table on a periodic basis, consuming bandwidth using the broadcast address.[1]

- RIP sends its update in a 576-byte datagram. If there are more entries than 512 bytes, multiple datagrams must be sent. For example, 300 entries require 12 back-to-back 512-byte datagrams

- Slow convergence: In the worse case, a RIP update can take over 15 minutes end to end. This can cause blackholes, loops, etc.

- RIPv1 does not support VLSM.

The first shortest-path-first routing protocol was developed and used in the ARPANET packet-switching network way back in 1978. This research work was developed and used in many other routing protocol types and prototypes. One of those is OSPF.

OSPF Features

- Shortest-path routes based on true metrics, not just a hop count

- Updates the routing tables only when needed or every 30 minutes using a multicast address

---

[1]This applies to RIP version 1; RIP version 2 uses multicast or broadcast.

- Pairs a network address entry with a subnet mask
- Allows for routing across equal paths, performing load balancing
- Supports type of service (TOS) routing
- Permits the injection of external routes (routes from other autonomous systems)
- Authenticates route exchanges
- Quick convergence
- Direct support for multicast

An OSPF Network

Refer to Fig. 9-14. OSPF is composed of a large (usually) topology that is generally separated into areas. An area is a grouping of networks into a single common place. Areas allow for smaller, more manageable groups of networks to work together as a single entity. Areas communicate with each other through the use of a special area known as the backbone area. When OSPF is set up, and there are no plans to separate the topology into multiple areas, then one area is created, known as the backbone area, and all networks are combined into this one area. External networks (IGPs or EGPs) are allowed to work with an OSPF network as well, through the use of special routers.

Figure 9-14
An OSPF network.

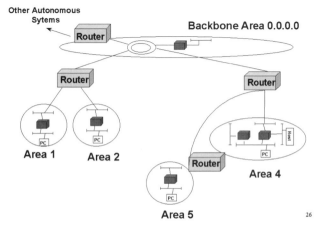

A Routing Protocol Comparison

When reading the following discussion on the OSPF protocol, keep the main goal in mind: network design. OSPF allows us to build very efficient networks through segmenting an autonomous systems into small groups called areas, variable-length subnet masks, type of service routing, and a host of other betterments compared with the RIP protocol. Table 9-1 shows the comparison between RIP and OSPF.

OSPF Overview

There are two types of standardized IGPs, RIP (version 1 or 2) and OSPF. Like RIP, OSPF is an IGP, which means that it is designed to run internal

TABLE 9-1 OSPF Compared with RIP Versions 1 and 2

Function/Feature	RIPv1	RIPv2	OSPF
Standard number	RFC 1058	RFC 1723	RFC 2178
Link-state protocol	No	No	Yes
Large range of metrics	Hop cout (16 = infinity)	Hop count (16 = infinity)	Yes, based on 1–65,535
Update policy	Routing table every 30 seconds	Routing table every 30 seconds	When link state change or every 30 minutes
Update address	Broadcast	Broadcast, multicast	Multicast
Dead interval	300 seconds total	300 seconds total	Variable settings
Supports authentication	No	Yes	Yes
Convergence time	Media delay + dead interval	Variable (based on number of routers × dead interval)	Variable (based on number of routers × dead interval)
Variable-length subnets	No	Yes	Yes
Supports supernetting	No	Yes	Yes
Type of service (TOS)	No	No	Yes
Multipath routing	No	No	Yes
Network diameter	15 hops	15 hops	65,355 possible
Easy to use	Yes	Yes	Complex Setup

to a single autonomous system (AS). It exchanges routing information within a single AS.[2] It can be used in small, medium, or large internetworks, but the most dramatic effects are readily noticed on large IP networks. As opposed to RIP (a distance-vector protocol), OSPF is a link-state protocol. It maintains the state of every link in the domain. This information is "flooded" to all routers in the domain, but within a single area. Flooding is the process of receiving the information on one port and transmitting it out all other active ports on the router. In this way, all routers receive the same information and can compute their own routes. This information is stored in a database called the *link-state database*, which is identical on very router in the AS (or every area, if the domain in split into multiple areas). Based on information in the link-state database, an algorithm known as the Dykstra algorithm runs and produces a shortest-path tree based on the metrics, using itself as the root of the tree. The information this produces is used to build the routing table.

OSPF Network Support

OSPF supports broadcast and nonbroadcast multi-access (NMBA) and point-to-point networks. Broadcast networks, such as Ethernet, Token Ring, and FDDI, can support one or more network attachments together with the ability to address a single message to all those attachments—a broadcast network. Alternatively, nonbroadcast networks, such as X.25, ATM, or Frame Relay, support one or many hosts but do not possess a method for broadcasting. Point-to-point is exactly that—a link that has two connection points. Two routers connected through a serial line is an example of a point-to-point link. There can be no other connections between these two points.

Router Types and Routing Methods

There are three types of routing in an OSPF network:

- Intra-area routing—Routing within a single area
- Inter-area routing—Routing between two areas
- Inter-AS routing—Routing between autonomous systems

---

[2]An AS is described as those networks and routers grouped into a single domain under one authority.

When there is only one area, there is basically only one specialized type of router (besides the DR and BDR, which are explained shortly). That is the one that deals with external routes. When an OSPF environment is split into multiple areas, there are multiple router types. There are four types of routers in an OSPF environment:

■ Backbone router (BR)—A router that has an interface to the backbone

■ Area border router (ABR)—A router that has interfaces to multiple areas

■ Autonomous system boundary router (ASBR)—A router that exchanges routing information with routers attached to different autonomous systems

■ Internal router (IR)—A router whose attachments all belong to the same area

Some of these router types have overlapping roles. For example, an ABR can also be a BR. Two other kinds of routers are not necessarily included with the ones above but are mentioned here to provide a single point of reference for the reader. One router on a subnet is selected as the designated router (DR). All other routers on the subnet form an adjacency (a logical point-to-point connection on a subnet) to this router. Information about networks to and from the subnet is transferred over the DR. The DR generates network link-state advertisements (LSAs) on behalf of its subnet and floods this information throughout its area. This advertisement identifies all routers adjacent to the DR and records the link states of all the routers currently attached to the network. The backup designated router (BDR) backs up the DR in case the DR fails.

Figure 9-15 shows the placement of these different types of routers. Remember, not all of them are in place all the time.

Message Types

OSPF routers pass messages to each other in the form of link-state advertisements (LSAs). Each LSA describes a piece of the OSPF routing domain. All LSAs are then flooded throughout the OSPF routing domain but within a single area. (A single area can be an entire OSPF domain.) The flooding algorithm is reliable, ensuring that all routers have the same collection of link-state advertisements.

Figure 9-15
OSPF routers.

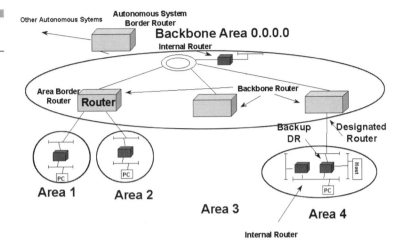

LSAs are categorized as follows:

Type 1: Router links advertisement.   This message is flooded within an area and contains information about a neighbor's router links (basically the IP address of an interface and the cost associated with that interface). Every router originates a router links advertisement.

Type 2: Network links advertisement.   This message is flooded within an area. It is generated by the designated router and includes information on all routers on this multi-access network. Whenever the router is specified as a designated router, it originates a network links advertisement.

Type 3: Summary links advertisement.   This message is flooded into an area by an area border router and describes reachable networks from outside the area (in other areas of the OSPF domain).

Type 4: AS boundary router summary link advertisement.   This message is flooded into an area by an ABR. It describes the cost from this router to an AS boundary router.

Type 5: AS external link advertisement.   This message is flooded to all areas except stub areas and describes an external network reachable via the AS boundary router that generated it.

Type 6: Multicast group membership LSAs.

Type 7: Multicast OSPF.

LSAs contain 32-bit sequence numbers. This number is used to detect old LSA packets and duplicate LSA packets. Each new LSA uses an incremented sequence number. Therefore, OSPF routers maintain their LSA databases with fresh information by updating it with an LSA of a higher sequence number. This also allows the OSPF router to flush out old entries.

Another method employed by OSPF to maintain its LSA database is the age field. Each LSA entry has an expiration timer that allows the database to get rid of old entries.

Metrics (Cost)

A cost is associated with the output side of each router interface. This cost is a configurable parameter on the router. When LSAs are transferred between routers, the cost of the individual links is added. This information is added up in a router (receiving LSAs) before the Dykstra algorithm runs. Multiple paths can be found to a destination, and the path with the lowest cost will be placed in the routing table. Simply stated, the lower the cost of a router port, the more likely the interface is to be used to forward data traffic.

According to RFC 1253 (OSPF version 2 MIB), a default value can be used for costing a link:

$$\text{Metric} = 10^8/\text{speed}$$

It is only a recommendation, and any number can be used. For example, if you are using a higher-speed link (such as those available with the ATM protocol). This yields a number having the typical values listed in Table 9-2.

Generic Packet Format

As of this writing, there are seven types of advertisements (listed previously). All OSPF packets have the same header, but the body of the packet varies. This is shown in Fig. 9-16.

The Hello Protocol

Routers periodically transmit hello packets not only to find other OSPF routers on their subnet but also to transmit and make sure that certain

TABLE 9-2

Costs Associated with Different Network Speeds.

Network Type/Bit Rate (Speed)	Metric (Cost)
≥ = 100 Mbps	1
Ethernet/802.3	10
E1 (2.048 Mbps)	48
T1 (ESF or 1.544 Mbps)	65
64 kbps	1562
56 kbps	1785
19.2 kbps	5208
9.6 kbps	10,416

Figure 9-16
OSPF packet header (Ethernet header).

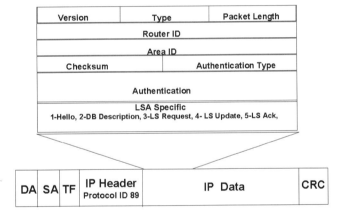

parameters are set to the same values in all the routers on that subnet. The hello packet format is shown in Fig. 9-17. The hello packet stays on the local subnet; it is not forwarded by the router. The hello packet contains

- The router's selection of DR and BDR
- The router's priority used to determine the DR and BDR
- Configurable timers that include the hello interval (time in which a router expects to hear hellos) and the router DeadInterval (the time period before a router is declared down)
- A list of neighbor routers that this router has received hellos from

Figure 9-17
Hello protocol.

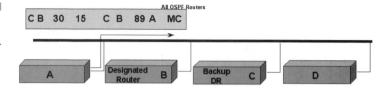

The most basic exchange between routers is called the hello protocol. This protocol allows OSPF routers to discover one another (within a single area) and allows the building of relationships between routers. This is the protocol that allows the DR and BDR to be selected. Once the DR is selected, adjacencies are formed.

For multi-access networks, when a router transmits a hello packet it is sent using the ALL-SPF-Routers multicast address 224.0.0.5.

OSPF routers build and maintain their relationships through periodic exchanges of hello packets. Included in the transmitted hello packets is a list of all the routers a router has heard from (has received hello packets from). When a router sees its address in a received hello packet, it knows that the router that transmitted that packet has seen it. Once this is accomplished, the DR and the BDR are selected. Any DR with a priority of 0 counts itself out of the selection. There is one DR and DBR per subnet or LAN segment.

These packets are continually sent every hello period specified in the packet. This is how a router can detect that another router is down (the DeadInterval, which it uses to wait and build a new database with the Dysktra algorithm).

Adjacency

After the hello discovery process has enabled the DR and BDR to be selected, routers on a single LAN segment determine whether to form an adjacency with one another. An adjacency allows two routers to exchange routing information through link-state advertisements. The following are the requirements for establishing adjacency:

- The link is a point-to-point link or a virtual link (discussed later).
- The router is the DR or BDR.
- The neighbor is the DR or BDR.

You can see that if the router is the DR or BDR, an adjacency is formed between the DR/BDR and a router. If these conditions are not met, an adjacency is not formed. That is, not all routers form adjacencies with each other.

As the adjacency is formed, the "adjacent" routers' databases must become "synchronized." That is, the exact same information must be contained in each router. A series of steps are required before full adjacency. The reason for this is to synchronize the link-state database. The adjacent routers transmit to adjacent neighbors a summary list of LSAs using the *database description packet*. The router takes this information, compares it to its own LSA database, and then builds a request list of

- LSAs that are in the received summary list but not in its LSA database
- LSAs that are in the database but not in the received information from its adjacent neighbor

This newly built request list in then transmitted to its neighbor using the link-state request packet. Each router that receives this request list responds to each requested record in the list. The router that received the request packet responds with a link-state update packet. Neighbors are considered to be *fully adjacent* when they have received all responses to the requests. They become fully adjacent on a one-on-one basis with each router that has formed an adjacency. After the routers become fully adjacent, each runs the SPF algorithm using the information supplied in the database. Execution of the algorithm results in OSPF routes, which are added to the routing table.

The procedure for adjacency is shown in Fig. 9-18.

Maintaining the Database

After the Dykstra algorithm (which builds a routing table) is run, the databases are continually checked for synchronization between adjacencies using LSAs and the flooding procedure. The flooding procedure is simple: receive a LSA, check for the information in the database, and determine whether or not to forward it to another adjacency using a LSA. To ensure reliability, flooding uses an acknowledgment procedure.

Reliability is also built into the protocol. When a LSA is transmitted, it is acknowledged. An unacknowledged packet is retransmitted by the router until it is acknowledged.

Figure 9-18

OSPF adjacencies.

OSPF Adjacency

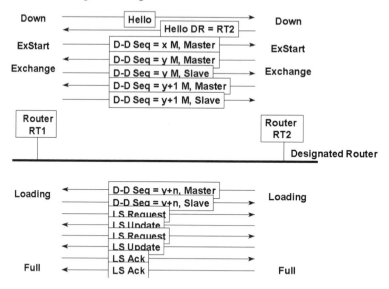

Every LSA contains an age field. This field is used to age out old entries in the database. If an entry is aged out, then this information is flooded throughout the domain (a single area), and the Dykstra algorithm is run again to build a new router table.

Sequence numbers are generated for all LSAs. When a router transmits a LSA, it applies a sequence number to it. In this way the receiving router will know if it is receiving the most "recent" information from another router. The sequence number is 32 bits long and is assigned to a LSAs in ascending order.

Changes in the LSA database require a rerunning of the SPF algorithm and an update of the routing table, depending on the outcome of the algorithm.

OSPF Areas

There may be as a little as one area for the entire OSPF AS, or there may be many areas. Refer to Fig. 9-19. An area is a grouping of contiguous networks and the associated routers that have interfaces belonging to those networks and hosts. Each area runs its own copy of the link-state routing algorithm, which allows each area to build its own topological database. It is important to note that an area limits the flooding of an LSA. LSAs do

Figure 9-19
OSPF areas.

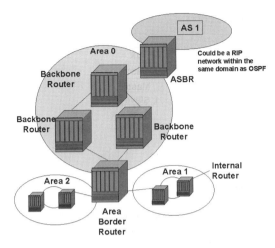

not leave the area in which they were originated. Furthermore, splitting a domain into areas allows for routing traffic bandwidth savings.

Each area is identified with a 32-bit number known as the *area ID*. This number is formatted in the same manner as an IP address. However, it does not have anything to do with the IP addressing scheme of your network; it simply identifies an area. Common area IDs are 0.0.0.0, which is reserved for the backbone area (a single area must be configured with the area ID of 0.0.0.0, or a multiple area must have one of its areas labeled 0.0.0.0, known as area 0), and 1.1.1.1 or 0.0.0.1 to identify area 1. There is no strict guideline for area ID numbering except for area 0.0.0.0.

Each router in an area is assigned a router ID regardless of its area ID assignment. This number is a 32-bit number that uniquely identifies that router in the autonomous system. Typically, the router ID is chosen from the IP address assignments of one of the router interfaces.

The topology of an area in not known to any other area. This means that the internal routers of an area do not contain any information about the OSPF topology outside their area, giving the benefit of reduced routing overhead. A single area that is spread over sparse environments (WAN links) must contain the same topology database for the entire area, no matter how large it has become. So how does an area determine routes that are not within its area? This is accomplished via the backbone area and the summary links advertisement.

ABRs play an important role in an OSPF network. Since areas do not know the topology in areas other than their own, some mechanism must be provided to allow network reachability information to traverse different areas. After all, it wouldn't do much good to be able to dynamically route in your own area and then have to statically point to networks in other areas. ABRs compact the topological information for an area and transmit it to the backbone area. Routers in the backbone area make sure that it is forwarded to the areas that are attached to it. In order to accomplish this, ABRs run multiple copies of the OSPF algorithm, one for each area (including the backbone area). Areas also provide the advantages of hierarchical topologies.

The Backbone Area

One of the areas is a specialized area known as the backbone area and labeled 0.0.0.0 or area 0. When a domain is split into areas, the areas communicate with one another through the backbone area. This area contains those routers and networks not contained in any other area and routers that connect to multiple areas (ABRs). Its primary responsibility is to distribute routing information between areas. The backbone area contains all the standard properties of an area; its topology is not known by any other area, and the backbone area does not know the topology of any other area. What causes the information to be in the backbone area?—the ABR.

The Area Border Router

There is a special router type known as the area border router (ABR). Its job is to connect an area to the backbone and to summarize its area topology information (for all areas that it connects to) to the backbone area, where it is received by other ABRs to be included in their tables. ABRs also receive other area summaries from the backbone and propagate this information to their area. ABRs also are part of the backbone area. Therefore, an ABR belongs to a minimum of two areas: its own area and the backbone area. If there is only one area in the AS (the backbone area), there are no ABRs.

Since an ABR belongs to two or more areas, it has a separate database for each area that it belongs to. It also executes a single copy of the routing algorithm for each area it belongs to. A typical ABR maintains connections to its area and to the backbone area. For its area, it receives flooded LSAs within its area and maintains a synchronized database for the area. The other copy of the algorithm runs for the attachment to the backbone. An ABR does not flood learned information about its area to the backbone. It summarizes this information using summary link advertisements. These advertisements are pushed to other ABRs on the backbone, allowing those areas to learn about each other without directly participating in the backbone's routing advertisements (remember, the backbone is a real area too).

Since area reachability information is propagated over the backbone area, every area must touch the backbone through the use of an ABR. An area is not allowed to be segmented from the backbone. A special condition does exist, however allowing an area to be extended off an area that is not the backbone through a concept known as a *virtual link.*

Virtual Link

A virtual link is shown in Fig. 9-20. The backbone must be contiguous. This means that an ABR must connect directly to the backbone and not to another area. Area-to-area communication is not allowed directly, only through the backbone. However, some designs force the creation of an area that does not have direct connectivity to the backbone so that the backbone is no longer contiguous. Backbone connectivity is restored through a virtual link.

Virtual links can be configured between any two ABRs that have an common interface to a non-backbone area. The virtual link is configured on each ABR and acts like a point to point link. Virtual links belong to the backbone. The routing protocol traffic that flows along the virtual link uses intra-area routing only.

As shown in Fig. 9-20, the two endpoints of the link are ABRs, and the link is configured in both routers. The two ABRs used in the virtual link also must belong to a common (not the backbone) area.

Figure 9-20
Virtual link.

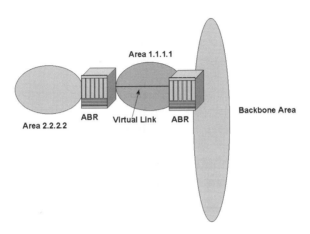

Simply stated, a virtual link is like a point-to-point link between two fully adjacent ABRs that allows an area to receive and transmit summary information (learn routes within the AS) when it is not directly connected to the backbone area.

Inter-Area Routing

Routing a packet between areas involves transmission of the packet from its source through its internal area to the ABR. The ABR transmits the packet over the backbone area to another ABR, where is transmitted internally on the area to the destination host. Areas cannot route directly to other areas!

Again, the backbone area is a special area. Its main function is to distribute information from areas to other areas. It consists of networks that are not contained in any other defined area, and routers that are attached to an area or areas.

Why use areas? Breaking the AS into routable areas greatly reduces the amount of overhead in the form of routing information updates that need to be distributed throughout the AS. While this may not seem like much, remember that each area can be unique. One area can have a majority of WAN links, others may be mostly networks, and others may be a combination of network types. Why make the update process very complex, and why bother other areas with your information? Remember, when the routing algorithm runs, every router in an area must run it. If one router in an area runs the algorithm, the routers in other areas may not have to run it. Dysktra runs in an area only. ABR will have a minimum of two copies of the Dykstra algorithm, one for each area it connects to.

Figure 9-21 shows the communication path of area 1 to area 2. Notice that the path must at least touch a router that is connected to the backbone. It does not necessarily loop around the backbone routers and then forward to area 1 or area 2.

Information from Other Autonomous Systems

What about communicating with other autonomous systems (outside the OSPF domain)? Through the use of a special router type, the autonomous system boundary router (ASBR), an OSPF network can communicate with another AS. This adds another level of hierarchy to

Figure 9-21
Interarea routing.

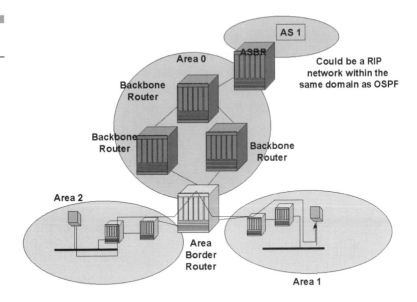

the OSPF routing. The first is the intra-area routing. The second level is area-to-area routing through the backbone. The third level is routing to external autonomous systems.

ASBRs run the OSPF protocol and some type of Exterior Gateway Protocol (such as Border Gateway Protocol, or BGP, defined in RFC 1403, BGP, or even RIP). RIP is seen as an external network and its routes are imported into a link-state database as such. An external AS need not be another AS in the sense of a BGP. OSPF treats any routing protocol unlike itself as an external AS. This type of protocol allows information to be exchanged between ASs. The EGP type of protocol only runs on the interfaces that are between the ASs. OSPF runs on the interfaces internal to the AS. An ASBR does not have to directly attach to the backbone.

To allow for this, another type of advertisement is used, known as the external links advertisement. Each ASBR in the AS generates one of the advertisements. This is the only advertisement that is flooded into every one of the areas in the AS. These advertisements describe routes that are external to the AS. There is one entry for every external route. As you can see, this information could quickly fill up a routing table.

The external route uses one of two types of metric: type 1 or type 2. Type 1 metrics are the preferred route and are used when considering the internal cost of the AS. Type 1 metrics include the link-state metric as well as the metric assigned to it. Therefore, any router that receives this type of update for an external route must use the internal (AS) metrics to reach the ASBR advertising that external route. So the computation of the cost to reach that route uses metrics that are internal to the AS and the AS that was supplied in the advertisement.

Type 2 metrics use the same metric that was advertised by the ASBR. Internal AS metrics are not added to the ASBR metric for the route in computing a path (based on cost) for that external route.

Stub Areas

In some instances a majority of the entries in the routing table will simply be routes external the OSPF domain. There is one entry for every external route. As you can see, this could quickly fill up a routing table with external routes.

Stub areas were created to reduce these entries. If an area has a single entry or exit point that is used for all external traffic, it can be configured as a stub area. A stub area blocks the import of the AS external link advertisements into the area. This reduces the number of entries in the stub area's database.

RFCs Related to OSPF

2178 DS   J. Moy, "OSPF Version 2," 07/22/1997. (Pages = 211) (Format = .txt) (Supersedes RFC 1583)

2154 E   S. Murphy, M. Badger, B. Wellington, "OSPF with Digital Signatures," 06/16/1997. (Pages = 29) (Format = .txt)

1850 DS   F. Baker, R. Coltun, "OSPF Version 2 Management Information Base," 11/03/1995. (Pages = 80) (Format = .txt) (Supersedes RFC 1253)

1793 PS   J. Moy, "Extending OSPF to Support Demand Circuits," 04/19/1995. (Pages = 31) (Format = .txt)

1765 E   J. Moy, "OSPF Database Overflow," 03/02/1995. (Pages = 9) (Format = .txt)

1745 PS   K. Varadhan, S. Hares, Y. Rekhter, "BGP4/IDRP for IP—OSPF Interaction," 12/27/1994. (Pages = 19) (Format = .txt)

1587 PS   R. Coltun, V. Fuller, "The OSPF NSSA Option," 03/24/1994. (Pages = 17) (Format = .txt)

1586 I   O. deSouza, M. Rodrigues, "Guidelines for Running OSPF Over Frame Relay Networks," 03/24/1994. (Pages = 6) (Format = .txt)

1585   I   J. Moy, "MOSPF: Analysis and Experience," 03/24/1994. (Pages = 13)
              (Format = .txt)

1584   PS   J. Moy, "Multicast Extensions to OSPF," 03/24/1994. (Pages = 102) (For-
              mat = .txt, .ps)

1403   PS   K. Varadhan, "BGP OSPF Interaction," 01/14/1993. (Pages = 17) (Format
              = .txt) (Supersedes RFC1364)

1370   PS   Internet Architecture Board, "Applicability Statement for OSPF,"
              10/23/1992. (Pages = 2) (Format = .txt)

Datagram Routing

Now that routing fundamentals, the RIP and OSPF protocols, and routing tables have been discussed, this section shows a routed packet using
direct and indirect routing. This is independent of the routing protocol.
As shown in Fig. 9-22, a PC, host 129.2.1.2 is trying to pass a datagram to a
host machine, host 129.1.1.2. The host machine is one hop (one router)
away. The IP layer of the PC knows that it must use a router (the source
and destination network addresses are different) and will use RIP (network request from the router) or the default gateway to determine the
IP address of the router to use. Upon determining the router's physical
address, it will physically (MAC layer) address the packet to router A. The
source and destination IP addresses in the IP header of this datagram
will be the PC as the source and the destination IP address as the host.
The source (PC) and final destination (the host) IP addresses will be
embedded into the IP header and will not change throughout the routing of this datagram.

Router A will receive this packet and extract the network number
from the final destination IP address in the received IP header. The
physical address headers will be stripped. The extracted network number will be compared to the router's internal routing table.

Router A will determine that the destination network can be reached
directly through one of its ports (the destination network is directly
attached). The router will determine the destination station's physical
address through its ARP table (or it may request it through the ARP
process). Router A will then build a packet with the original datagram
sent by the PC to submit to the host. The physical source address will be
the router's; the physical destination address will be the host's. The
packet is then transmitted to the host.

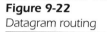

Figure 9-22
Datagram routing

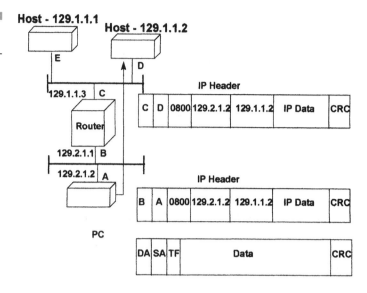

Another example is a multihop destination. This is not shown here, but the end station is still trying to reach host B, only this time it is two routers away. Router A will determine that the destination can be reached through router B. It will physically address the packet to send to router B, with its physical address as the source address and the physical destination address as that of router B. This is for physical addressing only. It will then submit the packet to the network that has router B attached.

Router B will receive this packet, extract the destination IP network address from the packet, and compare it to its routing table. From its table, it will determine that the final network is directly connected to port 2 of its router. Router B will perform an ARP lookup in its ARP table to find the physical address of the host. If an address is there, it will physically assign its own physical address to the source address of the packet, and the destination address will be that of the final destination station—in this case, the host. This is the only time that the packet will actually carry the physical address for the destination host.

To return the packet, the host will start over again, will notice that the destination network is remote, and will submit the packet to router A.

IPv6

Introduction

After 20 years, IPv4 still operates in the same manner as its designers intended it. There are extensions and enhancements, but much continues in this manner. So let me state emphatically that IPv6 is not revolutionary. It is the next step in the datagram delivery protocol known as IP. It is not a replacement (per se), and there are many new and some revised functions of the protocol that improve on it. Currently, there are enough fixes and extensions to the IPv4 protocol (not that there are many problems with IPv4) to make it last well past the year 2000. I have heard people ask over the years, "Why implement a new version of IP when this one is working just fine?" IPv4 patches are simply Band-Aids™ in a time of need to further enhance the Internet to reach more people and business requirements. IPv4 cannot handle its own continued growth, nor the expansive needs and requirements of the Internet, on into the twenty-first century. The Internet will be moving much more than textual data in the next century.

"The next IP." "Version 6." "Completely redone IPv4." If you hear statements like this, ignore or correct them. IPv6 is not a new network layer protocol. Remember this, if anything, about IPv6: It is an evolutionary step for IP. IPv6 has become an efficient IPv4 that is extensible.

As you read through this section, you should start to understand that the timing of this upgrade to IP is about right. The capabilities of IPv4 compared with IPv6 require a much more sophisticated computer than was required with IPv4. Generally, IPv4 could run on low-powered routers and end stations. The requirements of IPv6 will make use of the higher-powered routers and workstations. When IP was changed, no other protocol was dramatically changed. Again, this is the advantage of modular protocols. TCP and UDP stayed the same. The calls to the API are different—the socket interface known as Berkeley sockets (UNIX) or, for PCs, the Winsock interface—but the basic functions of TCP/UDP and the applications that use them are the same. The other protocols that must change are things that directly interface with IP. These are Domain Name Server, DHCP, OSPF, RIP, ICMP, etc. Most of these changes concern the 128-bit address change.

You will hear a lot about IPv6 over the next couple of years. Even with the release of this book in 1998, IPv6 implementations will continue to remain islands in the IPv4 Internet. This is the correct approach for IPv6. You cannot "flip the switch" as was done in January 1983 with IPv4. The Internet is extremely large today, and it is very commercial.

There are still quite a few studies in progress to determine IPv6 addressing allocation, effects of IPv6 on IPv4 networks, tunneling, etc. Implement slowly but surely. Test before implementing. Apply applications that have a need in the marketplace to IPv6. Work out the kinks before commercialization.

What ever happened to IPv5? Well, it exists and is known as the Internet Stream Protocol, or ST2, and it is defined in RFC 1819. ST2 is an experimental resource reservation protocol intended to provide end-to-end real-time guarantees over an internet. It allows applications to build multidestination simplex data streams with a desired quality of service. The revised version of ST2 specified in RFC 1819 is called ST2+.

ST2 operates at the same layer as connectionless IP. It has been developed to support the efficient delivery of data streams to single or multiple destinations in applications that require guaranteed quality of service. ST2 is part of the IP protocol family and serves as an adjunct to, not a replacement for, IP. The main application areas of the protocol are the real-time transport of multimedia data—e.g., digital audio and video packet streams and distributed simulation and gaming across internets. ST2 can be used to reserve bandwidth for real-time streams across network routes.

IPv6 builds on IPv4. Like most great things in life, you build on a foundation, something that you know works. Cars, over the years, are still built in the same fashion and still have tires, transmissions, engines, and bodies. But after many years, the extensions of those basics have contributed to more than just basic transportation. Many efficiencies and add-ons have been applied to the basic car to make it more safe, better for the environment, etc.

A word that is stressed throughout this text is the word *dynamic*. Multicasting is used extensively for automation and other dynamic functions. There is no broadcast address in IPv6! Routers and hosts discover each other dynamically, and hosts can configure themselves dynamically. There is even a replacement for the DHCP protocol that enforces (and efficiently utilizes) IP addressing, and then, of course, there is the biggest change of all for IP—the address! Placing IPv6-capable nodes on a network with other IPv6 nodes and routers will enable an IPv6 network to be established immediately via dynamics. Neighbor-discovery protocol will initiate and find the nodes on the network and nodes will autoconfigure their addresses. Routers simply need to have their interfaces configured and enabled, and off we go! But IPv4 networks prevail; probably about 99.99 percent of all networks are IPv4. Therefore, we must make IPv6 work within the bounds of the existing IPv4 network.

IPv6 Main Points

There will be a phased-in approach for the next couple of years, and today IPv6 is up and running through a series of islands that run autonomously and also use part of the current IPv4 Internet. It is known as the 6Bone, and complete information can be found on *http://www.6bone.net.*

IPv6 can be grouped into the following categories:

- *Expanded addressing capabilities:* IPv6 increases the IP address size from 32 to 128 bits to support more levels of addressing hierarchy, a much greater number of addressable nodes, and simpler auto-configuration of addresses. There are three types of addresses: unicast, anycast, and multicast. The scalability of multicast routing is improved by adding a "scope" field to multicast addresses.

- *Header format simplification:* To make IPv6 more efficient, some of the header fields have been dropped, and the header is a static 40 bytes.

- *Improved support for extensions and options:* Because the IP header is a static 40 bytes, changes cannot be made in the header, so the concept of header extensions is in. This provides greater flexibility for introducing new options in the future.

- *Flow-labeling capability:* A new capability has been added to enable the labeling of packets belonging to particular traffic "flows" for which the sender requests special handling, such as nondefault quality of service or real-time service.

- *Authentication and privacy capabilities:* There is added support for authentication, data integrity and, optionally, data confidentiality through extensions.

IPv6 Proposals

IPv6 was not developed overnight, and, in the spirit of the Internet, many people came up with many ideas. These ideas were studied over a 2-year period, and the proposals can still be reviewed. Many proposals were presented, including some with wonderful names:

ISO CLNP (Connectionless Protocol) which was demonstrated as TUBA (TCP and UDP over Bigger Addresses) (RFCs 1247, 1526, and 1561).

IP version 7 (also known as TP/IX, RFC 1475) included many changes to the TCP and IP layers and eventually evolved into the CATNIP (RFC 1707).

IP in IP evolved into IPAE (IP Address Encapsulation). It proposed running two layers of the IP protocol, one for the worldwide backbone and one for the regional IP networks. This eventually evolved into Simple IP, or SIP. It moved the address to 64 bits and did away with some of the unused features of ICMP.

During 1992 and 1993, the Pip internet protocol, developed at Bleacher, was one of the candidate replacements for IP. It had many improvements in routing strategies. In mid-1993, Pip was merged with the Simple Internet Protocol (SIP), resulting in SIPP (SIP Plus).

SIPP (RFC 1710) is a new version of IP that was designed to be an evolutionary step from IPv4. It is a natural increment to IPv4. It can be installed as a normal software upgrade in internet devices and is interoperable with the current IPv4.

While it is true that IPv6 solves the addressing problem, as you can see from the above list it has a few other properties that improved upon the IPv4 protocol. Table 10-1 identifies the IP version numbers. There are probably more than you thought!

	Decimal	Keyword	Version	References
TABLE 10-1	0	Reserved		
IP-Assigned Version Numbers	1–3	Unassigned		
	4	IP	Internet Protocol	RFC 791
	5 JWF	ST	ST Datagram Mode	RFC 1190,
	6	IPv6		RFC 1883
	7	TP/IX	TP/IX: The Next Internet	
	8	PIP	The P Internet Protocol	
	9	TUBA	TCP and UDP over Bigger Addresses	
	10–14	Unassigned		
	15	Reserved		

Figure 10-1
IPv6 header format.

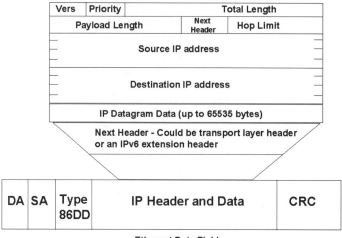

IPv6 Header Format

The format of the IPv4 header is shown in Fig. 10-1. Notice the changes between the IPv4 header and IPv6 (shown in Fig. 10-1). IPv6 seems to be missing a few things, but they are there—they just cannot be seen (yet). In fact, the only field that seems not to have changed or moved positions is the Version field. This field was originally going to be used as the delineating factor to determine whether a received IP packet was based on IPv4 or IPv6. In other words, the EtherType field of an Ethernet packet would remain 0800(h), and the version field of the header would determine the processing of a received IP datagram. This changed, though, and IPv6 now has its own Type field, 86DD(h) (and SAP in IEEE 802 networks).

IPv4 uses four key mechanisms in providing its service: the Type of Service, Time to Live, Options, and Header Checksum fields in the header.

The Options fields provide control functions needed or useful in some situations but unnecessary for the most common communications. The options include provisions for time stamps, security, and special routing (strict and loose source route—nothing to do with Token Ring). However, over the years, it was noticed that these Options fields were not being used by the majority of Internet hosts. IP datagrams that contained an option cannot be simply forwarded. They require special attention. They are placed in their own queue, and the router operates on this queue separately from the received datagram queue. Second, if they were not used very often, many implementers of routers did not optimize their software to operate on datagrams that included special options. This generally results in a performance penalty on the router.

So why have them? Well, their functions *were* used in some cases. IP multicasting, for example, uses the loose source route option when incorporating the tunneling mechanism for DVMRP (refer to the multicast section in this chapter to understand more). It was decided that IPv6 should allow for it to be extensible, so IPv6 implements the concept of an *extension header.* Before we discuss this, let's review the IPv4 Options fields and see how some of these functions were moved.

IPv4 Options Review

With the IP header becoming fixed, all the Options fields were "eliminated" from the header. They were not eliminated completely, however—they merely changed forms. Basically, IPv4 options are now IP header extensions. The following are the IPv4 options:

- *Security:* Used to carry security, compartmentation, user group (TCC), and handling restriction codes compatible with DOD requirements.

- *Loose source routing:* Variable in length and used to route the internet datagram based on information supplied by the source.

- *Strict source routing:* Variable in length and used to route the internet datagram based on information supplied by the source.

- *Record route:* Variable in length and used to trace the route an internet datagram takes.

- *Stream ID:* Used to carry the stream identifier.

- *Internet time stamp:* A right-justified, 32-bit field that indicates a time in milliseconds since midnight UT (Universal Time). There are place holders for multiple time stamps and a Flags fields to indicate time stamps only. A time stamp is preceded with the internet address of the registering entity. The internet address fields are prespecified, or an IP module only registers its time stamp if it matches its own address with the next specified internet address. This can be used for measuring the transport layer protocols and other utilities.

IPv6 Extension Headers

The extension headers of IPv6 allow it to become extensible beyond a specified (and limited) Options field. It can be modified at later dates to include other options. The current IPv6 specification calls for the following headers (in the order they should appear in the datagram):

- ▪ *IPv6 Header*   (not directly part of the extensions but shown here to show header order).

- ▪ *Hop-by-Hop Options* (RFC 1883):   Used to carry optional information that must be examined by every node along a packet's delivery path. One of the options is the jumbo datagram option

- ▪ *Destination Options* (RFC 1883):   Used to carry optional information that need be examined only by a packet's destination node(s).

- ▪ *Routing (Type 0)* (RFC 1883):   Used by an IPv6 source to list one or more intermediate nodes to be "visited" on the way to a packet's destination. This function is very similar to IPv4's Source Route options.

- ▪ *Fragment* (RFC 1883):   Used by an IPv6 source to send packets larger than would fit in the path MTU to their destinations.

- ▪ *Authentication*   (RFC 1826).

- ▪ *Encapsulating Security Payload*   (RFC 1827).

- ▪ *Upper-layer header*   (not part of the extension header, but shown here to show order).

In end-to-end communication, these fields should be ignored by all stations that may receive it. They are generally built and used by the source and destination stations only. The exception is the Hop-by-Hop options field, which may be reviewed by routers in the path to the destination.

An example of an extension header is shown in Fig. 10-2. Notice that the numbering by IANA is still preserved. UDP is the next header, and it has number 17 (the same port number whether it is IPv4 or IPv6).

Sizes and Fragments

In IPv4, fragmentation was discouraged. In IPv6, it continues to be discouraged, but fragmentation responsibility is passed to the source node, instead of to the router as in IPv4. In order to send a packet that is too

Figure 10-2
IPv6 extension head-
er example.

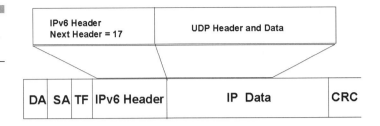

large to fit in the MTU of the path to its destination, a source node can divide the packet into fragments and send each fragment as a separate packet, to be reassembled at the destination.

Fragmentation caused problems mainly due to efficiency of the routers and end stations. Any missing fragment caused the whole TCP segment to be retransmitted (RFC 1122, p. 58). This creates bandwidth problems, memory problems, and uses CPU cycles. In IPv4 it consumes considerable resources on the router (fragmentation) and host (reassembly). IPv6 encourages implementing RFC 1191. This is the specification for *dynamic path MTU discovery,* or having the host dynamically find out maximum packet sizes for the path to the destination (that is, determine the networks that the datagram will transit). MTU is an acronym for maximum transmit unit, i.e., how large a datagram can be transmitted to the destination station. By enabling this, we eliminate the packet identification, the control flags, and the fragment offset. If any fragment is lost, all the fragments in the segment must be retransmitted (RFC 1122, p. 58).

Most hosts simply avoid the problem of discovering the maximum size of a packet on the destination path (if not local) and set the packet size to 576 bytes (the accepted minimum packet size for most IP networks). They still do this even after RRC 1191 presented a simple way to determine the maximum size. Some implementations transmit a large packet and wait to see if an ICMP "datagram too big" message is returned. The originating host then returns to using 576-byte packets.

Picking the right size is a very complex matter, and most hosts stick to 576 bytes for nonlocal destination hosts. This eliminates fragmentation and associated problems, but it also creates an inefficiency in that some networks allow for large packet sizes. Imagine transferring information between Token Ring networks separated by an FDDI backbone—not all networks are Ethernet. Why move the size down to 576 bytes just because the data traverses routers? If the file is 100 Mbytes in length, this can have considerable consequences. Therefore, MTU discovery is more efficient than fragmentation because it spreads the responsibility around to multiple entities.

Path MTU discovery is discussed in RFC 1191. It changed part of the ICMP RFC 793 in that it recommends using one of the previously unused header fields as the next hop MTU size indicator. This is an interesting aspect of ICMP for IPv6—it really does *improve* upon previous experience and knowledge without intending to *replace* it.

As shown in Fig. 10-3, the fragmentation process now uses extension headers. Basically, the same information is kept in the header as the IP fragmentation of IPv4, including the ID, offset, etc.

Header Differences

The first thing to notice about the IPv6 header is that it is a static 40 bytes in length—the length of the packet header is not variable. Also, the checksum was removed. IP first ran over copper serial lines that tended to be noisy (which shows up as static in audio). Checksums abound in the TCP/IP protocol suite and the access methods of FDDI, Ethernet, and Token Ring. The removal of the checksum allows all systems that forward IP datagrams to speed up, because they do not have to compute the checksum at every hop.

IPv4's Total Length field is replaced by a payload length. This is not significant except that IPv6 is a static 40 bytes, so the payload length is truly a measurement of the payload, not the IPv6 header as part of the sum. This field is 16 bits in length, which allows a maximum 65,355-byte payload. However, IPv6 includes a new concept known as jumbo datagrams (or jumbograms). These allow various network attachments such as I/O connections between high-speed computers that can process data segments larger than 64 k (see RFC 2146).

One of the more interesting changes to IP with version 6 is the concept of concatenated headers. Concatenation is accomplished by using

Figure 10-3
Fragmentation.

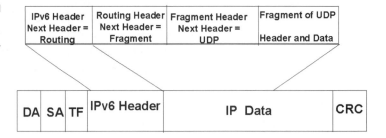

the Next Header field on the IPv6 header. In IPv6, the Protocol Type field is set, and that header would immediately follow. For example, if the payload is UDP, the protocol type is set to 17 (decimal) and the UDP header would immediately follow.

The Time to Live (TTL) field is one of the more versatile fields in IP. It is used to prevent datagrams from constantly looping, keep packets on a local network, and, in multicast datagrams, to indicate scope (hearing range). It probably has many other private uses, as well. In IPv6, this field is renamed to Hop Limit. Basically, what it is really used for is to count down by 1 counter. The original intention of the field was to indicate a time (in seconds) that could be used, for example, by a router. If the router could not forward the packet within the amount of time indicated by the TTL field, it would discard the datagram and generate an ICMP message. However, over time, most router delays were being measured in milliseconds, not seconds. The accepted decrement of the field was set to 1, and therefore a hop count became an indication of time.

Priority and Flow Label

The two new fields in the IPv6 header are for flow label and priority. These fields are still under study, but a brief explanation is provided here. More information is available in the RFC.

The Priority field distinguishes the datagram from other datagrams on the network. It provides priorities for two types of traffic: congestion- and non–congestion-controlled traffic. The following list gives the seven specific priorities for congestion-controlled traffic:

0   No specific priority

1   Background traffic (news)

2   Unattended data transfer (mail)

3   Reserved

4   Attended bulk transfer (file transfer)

5   Reserved

6   Interactive traffic

7   Control traffic (routing protocols and network management)

The second type of flow, non–congestion-controlled, doesn't care about congestion on the network. This type can include delay-sensitive applications such as real time audio and video. Further clarifications based on case studies will soon come out.

As experimental as the priority field is, so is the Flow Label field. A flow distinguishes traffic. Flows are traffic streams from one sender to one or more receivers.

IPv6 Addressing

Now for the fun part: Addressing is probably the best-known feature of IPv6. Ask anyone to talk about the differences between IPv4 and IPv6, and the change from 32-bit to 128-bit addressing will be the first thing out of their mouth. There are also many other things about the address space and its allocation that were carefully crafted. The first expansion of the address space was to 64 bits. It was later increased to 128 bits. IPv6 addresses provide the same function as in IPv4, as identifiers for interfaces and sets of interfaces. Even though 128 bits are written for use, only 15 percent of the available address space is allocated for use. There are three types of addresses:

- *Unicast:* An identifier for a single interface. A unique address delivered to a single destination.

- *Anycast:* New for IP (version 6), an anycast address is an identifier for a set of interfaces (typically belonging to different nodes). This is similar to a multicast, but a packet sent to an anycast address is delivered to one of the interfaces identified by that address (the "nearest" one, according to the routing protocol's measure of distance).

- *Multicast:* An identifier for a set of interfaces (typically belonging to different nodes). A packet sent to a multicast address is delivered to all interfaces identified by that address.

In IPv6, broadcast addresses are not defined. They were superseded by multicast addresses.

In IPv4, we identified addresses by their 32-bit value. Normally, an address was written in a form known as dotted-decimal notation. For example, 132.1.8.10. An IPv6 address is written in hexadecimal (a change from dotted-decimal notation) and consists of 8 groupings, containing 4 hexadecimal digits, or 8 groups of 16 bits each:

XXXX:XXXX:XXXX:XXXX:XXXX:XXXX:XXXX:XXXX

For example,

FEDC:BA98:7654:3210:FEDC:BA98:7654:3210

is an IPv6 address.

Another example is the following unicast address:

1080:0000:0000:0000:0008:0800:200C:417A

Writing an IPv6 address has become unwieldy (DNS becomes very important here), so there are provisions to condense the address into its smallest available form. For example, the above unicast address can be compressed into

1080:0:0:0:8:800:200C:417A

or even

1080::8:800:200C:417A

The double colon has special significance: It is a demarcation point for compressing leading 0s. Notice here that only leading zeros can be compressed, and the :: symbol can be used only once during the compression. Therefore, if you had the address

1080:0:0:5698:0:0:9887:1234

you cannot write it as

1080::5698::9887:1234

The algorithm that runs the expansion for the address would get confused. How many 0s go in each of the colon-compressed slots? The correct way to compress this address would be to compress one of the two groups of leading 0s:

1080::5698:0:0:9887:1234   or   1080:0:0:5698::9887:1234

As with IPv4, the first few bits of the IPv6 address tell something about the address (this has nothing to do with class addressing). Table 10-2 shows this. The format prefix does not deal with classes of address, but rather with where the address has been assigned or an address type.

To determine the fraction of the address space that is used is simple—the formula is $(1/2)^x$, where x is the number of bits used. For example, if the first eight bits are 0000 0000, the fraction is $(1/2)^8$, or $1/256$.

TABLE 10-2

IPv6 Addressing Prefixes

Allocation	Prefix (binary)	Fraction of Address Space Used
Reserved	0000 0000	1/256
Unassigned	0000 0001	1/256
Reserved for NSAP Allocation	0000 001	1/128
Reserved for IPX Allocation	0000 010	1/128
Unassigned	0000 011	1/128
Unassigned	0000 1	1/32
Unassigned	0001	1/16
Unassigned	001	1/8
Provider-based Unicast address	010	1/8
Unassigned	011	1/8
Reserved for geographically-based unicast addresses	100	1/8
Unassigned	101	1/8
Unassigned	110	1/8
Unassigned	1110	1/16
Unassigned	1111 0	1/32
Unassigned	1111 10	1/64
Unassigned	1111 110	1/128
Unassigned	1111 1110 0	1/512
Link local-use addresses	1111 1110 10	1/1024
Site local-use addresses	1111 1110 11	1/1024
Multicast addresses	1111 1111	1/256

Prefixes just like in the CIDR environment are also used in this environment. A /30 indicates that the first 30 bits are used for routing. Also, notice that fields in certain types of addresses are given names to further identify subaddress portions.

There are three address types that are assigned from the 0000 0000 format prefix space. These are "unspecified addresses," loopback addresses, and IPv6 addresses with embedded IPv4 addresses. This allocation supports the direct allocation of provider addresses, local-use addresses, and multicast addresses. Space is reserved for NSAP addresses, IPX addresses, and geographic addresses. The remainder of the address space is unassigned and reserved for future use. This can be used for expansion of existing functions (e.g., additional provider addresses) or the addition of new ones (e.g., separate locators and identifiers).

A value of *FF* (11111111) identifies an address as a multicast address. Multicast addresses are used extensively in autoconfiguration of addresses and neighbor discovery. Anycast addresses are taken from the unicast address space and are not syntactically distinguishable from unicast addresses.

Methods of Deploying IPv6

IPv4 will continue to dominate the Internet and intranets for years to come. Therefore, we must use a hybrid of IPv4 and IPv6 implementations. Following are two examples:

- *Dual IP layer:* IPv4 and IPv6 running at the network layer in both hosts and/or routers. These nodes have the ability to send and receive both IPv4 and IPv6 datagrams. Requires a node to be configured with both an IPv4 and IPv6 an address, which may or may not be related. In this way IPv4-only hosts can access services that exist on a IPv6 host.

- *IPv6 over IPv4 tunneling:* The process of taking an IPv6 datagram and wrapping an IPv4 header on it, for transit across IPv4 hosts or routers. Two methods available: *configured* and *automatic tunnels*. These nodes are configured with an IPv4-compatible IPv6 address. This is that special unicast address that has a 96-bit prefix of all zeros. The next 32 bits contain an IPv4 address.

   Configured tunneling: The IPV4 tunnel endpoint address is determined by the encapsulating node.

   Automatic tunneling: The IPv4 tunnel endpoint is determined from the IPv4 address of the IPv6 packet.

A node may be capable of automatic and/or configured tunneling. Automatic tunneling uses the IPv4-compatible address scheme. A node does not have to support dual-stack IP to support tunneling.

The transitional implementation will also have special nodes:

- *IPv4-only:* a node that understands only IPv4
- *IPv4/IPv6:* a node that is running dual IP stacks and understands both IPv4 and IPv6
- *IPv6-only:* a node that understands only IPv6

Unicast Addresses

The unicast address space is a contiguous bit-wise, maskable address that is similar to the addressing scheme used in IPv4 CIDR. The address types for unicast addressing are divided into two groups: provider based and special use. Provider-based addresses include:

- Global provider based
- Geographically based
- NSAP
- IPX hierarchical

Special-use addresses include:

- Unspecified: 0::0 (used when an address is not known)
- Site local-use
- Link local-use
- IPv4-capable host
- Loopback: 0::1—same function as in IPv4

A type of address that is expected to be very common is the IEEE 802.*x* (or Ethernet) LAN MAC address, which is shown in Fig. 10-4. The IEEE 802.*x* MAC address is 48 bits in length, and because of its registry, every card has a unique number assigned to it. However, where these addresses are not available, a method of using E.164 (telephone) addresses could be implemented, as well. An interesting point is that by using the IEEE 802.*x* MAC address, an IPv6 node could simply listen to the cable plant for router advertisements, which would yield the subnet ID for itself. Putting the two together would give it a unique address to use. This is autoconfiguration.

Global communication using IPv6 is provided by the unicast addressing scheme of globally based providers (see Fig. 10-5). The first three bits identify the address as a provider-oriented unicast address. The *registry ID* identifies the internet address registry (currently IANA, RIPE, APNIC, or Internic) that assigns provider identifiers, indicated in the *Provider ID*

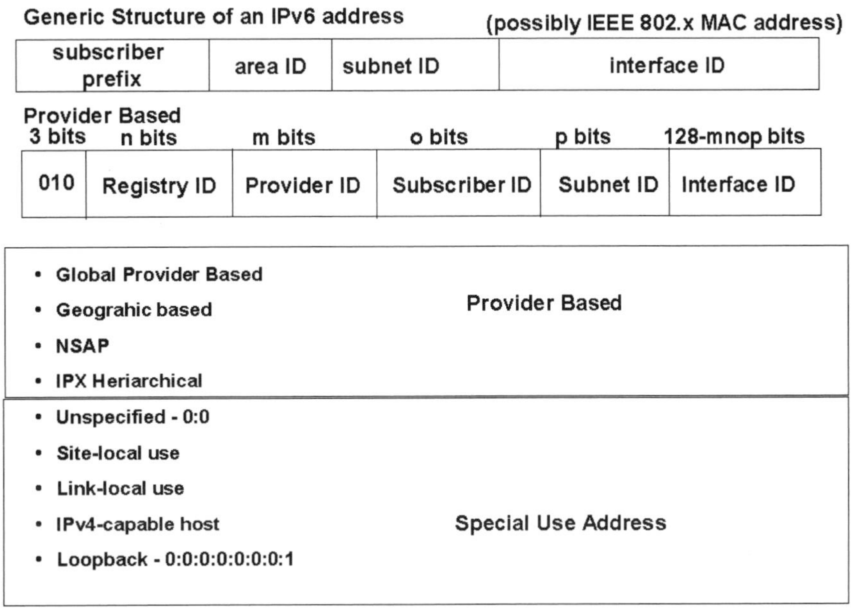

Figure 10-4
Unicast addresses.

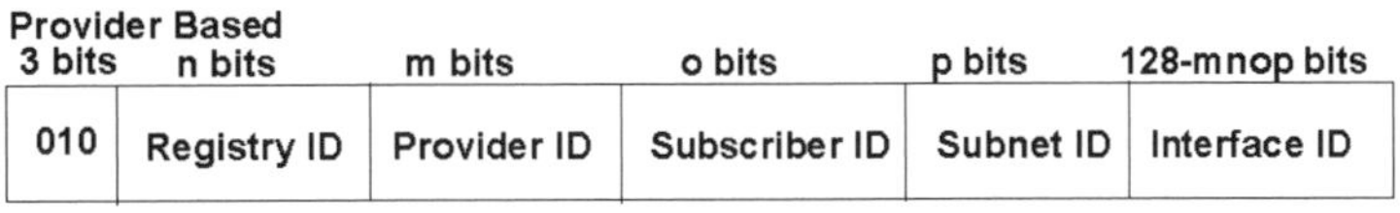

Generic Structure of an IPv6 address   (possibly IEEE 802.x MAC address)

subscriber prefix	area ID	subnet ID	interface ID

Provider Based

3 bits	n bits	m bits	o bits	p bits	128-mnop bits
010	Registry ID	Provider ID	Subscriber ID	Subnet ID	Interface ID

- Global Provider Based
- Geograhic based   **Provider Based**
- NSAP
- IPX Heriarchical

- Unspecified - 0:0
- Site-local use
- Link-local use
- IPv4-capable host   **Special Use Address**
- Loopback - 0:0:0:0:0:0:0:1

Figure 10-5
Provider-based IPv6 addressing.

Provider Based

3 bits	n bits	m bits	o bits	p bits	128-mnop bits
010	Registry ID	Provider ID	Subscriber ID	Subnet ID	Interface ID

field, to internet service providers, which then assign portions of the address space to subscribers. This is a process similar to the address-assignment policy used with CIDR and described in RFC 2050. The *subscriber ID* distinguishes among multiple subscribers attached to the internet service provider identified by the provider ID. This is like a customer number. The *subnet ID* identifies a specific physical link. There can be multiple subnets on the same physical link; however, a specific subnet can not span multiple physical links. The *interface ID* identifies a single interface among the group of interfaces identified by the subnet prefix.

Local-Use IPv6 Addressing

Local-use addressing is used exactly as the name implies, locally. This addressing can be subnet local or subscriber local, which gives the two names: *link local* and *site local*. Figure 10-6 shows the format of these two

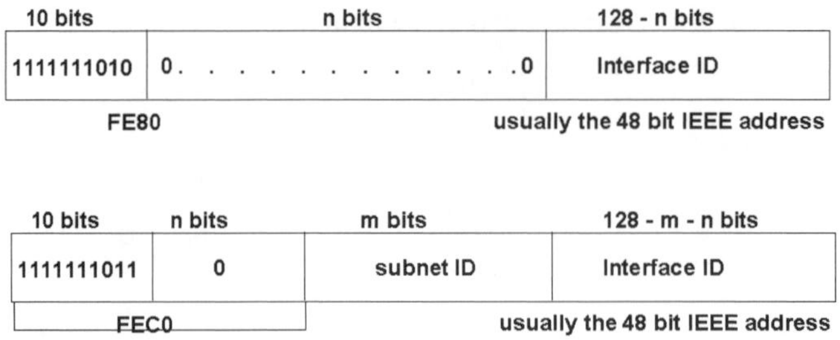

Figure 10-6
Link and site local
address formats

10 bits	n bits	128 - n bits
1111111010	0. 0	Interface ID
FE80		usually the 48 bit IEEE address

10 bits	n bits	m bits	128 - m - n bits
1111111011	0	subnet ID	Interface ID
	FEC0		usually the 48 bit IEEE address

addressing types. Notice that both addressing types use the reserved multicast prefix format of *FF.*

Stations that are not configured using a provider-based address or a site local address use the link local address. This is an address that can be used between two stations on a single link or a network. This type of address will not be processed by a router, so it cannot span subnets. It can be used by a station that is starting up and does not know its location on the network. Basically, it is the concatenation of its 48-bit MAC address and the well-known link local prefix of FE80:: <MAC address>. Where the MAC address cannot be used, a serial number or some other unique identification of the card can be used.

A site local address allows a site to configure its network without being connected to the Internet. In IPv6, a network can devise and implement a completely addressed internet network that will allow that site to communicate with all interfaces at the site (it may span globally). However, none of these stations can communicate over the Internet. The reasons for this may be many. Some companies may not want a connection to the Internet until some specified time in the future. An example of this is a bank that set up its complete internet based on private addressing (in IPv4, that is). The problem was that it did not use the RFC 1918 private address space allocation to accomplish this. Its site was up and operational for 2 years without a hitch (at least, not too many). When it wanted to connect to the Internet, it had a choice of either using network address translation (NAT, RFC 1631) or readdressing their network. Based on many factors, which included scalability and peer-to-peer communication, the bank readdressed its site.

With IPv6 site local addressing, this would not have happened. Any entity can pick addresses within the range of site-local scope and configure its site (basically, all a company's locations). If at some later time it was assigned a global provider address prefix, it would simply add a provider number and subscriber ID to the address.

A site local address cannot be routed over the Internet, because it has a site-local prefix assigned.

IPv6 Addresses with Embedded IPv4 Addresses

The transition from IPv4 to IPv6 will be key for the successful implementation of IPv6. IPv6 would never be accepted if it were necessary to perform a one-time complete cutover. IPv4 is working—stumbling with addressing and routing table explosion, but working. It is embedded. Therefore, installations must be allowed to "try before they buy" into IPv6. This is one reason why other protocols that have tried to replace TCP/IP have failed (e.g., OSI). You must have a compatible procedure in place, as did, for example, Microsoft when they brought out Windows 95™: It allowed for most Windows 3.*x* programs to run on it. Another example is OS/2: It did not run the very popular Windows 3.1 programs very well and basically required a major cutover to make it work. We now see where Windows 95 is and OS/2 is not, so this shows how important it is for IPv6 to be backwards compatible with IPv4. A transition scheme is provided courtesy of RFC 1933, which is a fairly easy-to-read document.

IPv6 hosts can use the IPv4 network as a virtual interface that enables these hosts to reach other hosts through tunnels. The address that is used for this is a special type of link local address called the IPv4-compatible address, shown in Fig. 10-7. Also necessary for the transition is that hosts become dual stack (supporting both IPv4 and IPv6 IP stacks) and tunneling. A mechanism is provided in the IPv6 addressing structure to institute this.

There was another addressing method, known as the IPv4-mapped address, which allowed for translation. But this method is out of favor, and the method most used today is the one just described.

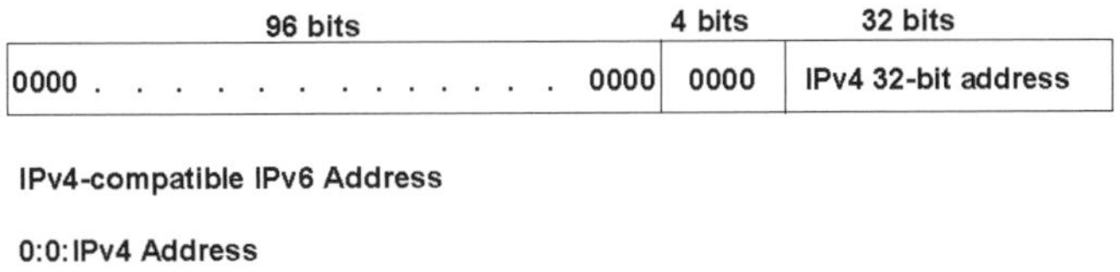

Figure 10-7
IPv6 address with embedded IPv4 address.

96 bits	4 bits	32 bits
0000 0000	0000	IPv4 32-bit address

IPv4-compatible IPv6 Address

0:0:IPv4 Address

Autoconfiguration

One improvement of IPv6 over IPv4 is the ability of a node to start up and dynamically attain its node and network address. This saves a lot of time installing new hosts in a network. It is called Autoconfiguration. There are two types of autoconfiguration: stateful and stateless.

In stateful autoconfiguration, some external device assists the node at startup to determine its network address (prefix) and node address, and, maybe, some router addresses. A possible use for this is for DHCP to enable the configuring of an initializing node.

Stateless autoconfiguration means that a node will configure itself and find resources on the network through the use of multicast addresses. This allows the node to start up and send out request messages to which other nodes will respond, and from these response the node can determine its network address and prefix and node address. IPv6 nodes start this behavior by joining the all-nodes multicast group upon startup. This is accomplished by initializing the interface to the all-nodes multicast address of *FF02::1*. These nodes can solicit information from the router by using the all-routers multicast address of *FF02::2* as the destination and their own link local address as the source.

Stateless autoconfiguration has its advantages in that it is really automatic and very simple to use. However, this type of configuration is susceptible to hackers in that anyone could simply place their network station on the subnet and immediately gain access to the resources on that subnet. Stateful autoconfiguration was designed to combat such a threat.

Neighbor Discovery

One of the most important RFCs on IPv6 is 1970, which covers neighbor discovery. Although it uses the Internet Control Message Protocol (ICMP), do not expect to find its listing in the ICMPv6 RFC (RFC 1885). The Address Resolution Protocol is not used with IPv6. Goodbye ARP! Nodes (hosts and routers) use neighbor discovery to determine the link-layer addresses for neighbors known to reside on attached links and to quickly purge cached values that become invalid. Hosts also use neighbor discovery to find neighboring routers that are willing to forward packets on their behalf. Finally, nodes use the protocol to actively keep track of which neighbors are reachable and which are not and to detect changed link-layer addresses. When a router or the path to a router fails, a host actively searches for functioning alternatives.

This sounds like a happy medium between ARP for IPv4 and the methods employed by ES-IS procedures of the CLNP (Connectionless Network Protocol) from the OSI suite. In ES-IS (the routing update protocol for the OSI protocol suite), the active end stations send hello packets to which the active routers on a network listen and which they use to build a database. In this database is a listing of all the end stations that the OSI router has heard from. The OSI router transmits a packet to make itself known on the network, as well. Workstations record the router's address so that they can send packets to it, for either the first packet transmitted locally or for all off-network forwarding. The OSI router informs the node about the location of the destination station.

CLNP was one of the first protocols recommended to replace IPv4, but it is actually a clone of IP and was basically outdated by the time the IPng (IP next generation) group formed in 1992 and it was pushed aside. The IPv6 neighbor discovery protocol corresponds to a combination of the IPv4 protocols ARP [RFC 826], ICMP router discovery [RFC 1256], and ICMP redirect (RFC 791).

A question that may be asked here is, With all the dependency on dynamically discovering link-layer addresses between hosts and routers, how can an ICMP message be sent if the media (link-layer) address is not yet known (i.e., the neighbor discovery procedures have not yet determined the link-layer addresses for all dependencies on a node local link). This is easily solved by using a well-known IPv6 multicast address. ICMP cannot work without a known media address.

A special multicast address at the MAC layer has been invented, as well. All stations should be listening to their special MAC multicast address. This is formed by using *3333* and the last 32 bits of the IPv6 address as one of the addresses to listen for on the NIC card. If an address of 3333 is received by the NIC, it will also process the last 32 bits. If this matches its own address, it will pass it on to the IPv6 IP layer of its upper-layer software.

With the exception of nonbroadcast multi-access (NBMA) networks or a link-layer interaction that is specified in another document, RFC 1970 applies to all link layer types. However, because neighbor discovery uses link-layer multicast for some of its services, it is possible that on some link types (e.g., NBMA links) alternative protocols or mechanisms to implement those services will be specified in the appropriate document covering the operation of IP over a particular link type. The services described in this document that are not directly dependent on multicast, such as redirects, next-hop determination, neighbor unreachability detection, etc., are expected to be provided as specified in this document. The details of how one uses neighbor discovery on NBMA links is an area for further study.

Types of Neighbor Discovery   What types of nodes can be discovered using neighbor discovery? The neighbor discovery protocol solves a set of problems related to the interaction between nodes attached to the same link. It defines mechanisms for solving each of the following problems:

Router discovery: How hosts locate and identify hosts on their local link.

Prefix discovery: How hosts discover the set of address prefixes that define which destinations are on-link for an attached link. (Nodes use prefixes to distinguish destinations that reside on-link from those only reachable through a router.)

Parameter discovery: How nodes learn such link parameters as the link MTU or such Internet parameters as the hop-limit value to place in outgoing packets.

Address autoconfiguration: How nodes automatically configure an address for an interface.

Address resolution: How nodes determine the link-layer address of an on-link destination (e.g., a neighbor), given only the destination's IP address.

Next-hop determination: The algorithm for mapping an IP destination address into the IP address of the neighbor to which traffic for the destination should be sent. The next hop can be a router or the destination itself.

Neighbor unreachability detection: How nodes determine that a neighbor is no longer reachable. For neighbors used as routers, alternative default routers can be tried. For both routers and hosts, address resolution can be performed again.

Duplicate address detection: How nodes determine that an address they wish to use is not already in use by another node.

Redirect: How routers inform a host of a better first-hop node to reach a particular destination.

Also contained in this RFC 1970 is the original ICMP redirect message (RFC 792), in which a router sends to a host stating, "I will forward the packet that you sent to me to my next hop port, but there is a better path to the destination that you indicated, and it is through another router *x*. Here is the address of that router."

Neighbor Discovery (ND) and IPv4 In IPv4 there is no generally agreed-upon protocol or mechanism for neighbor unreachability detection, although Hosts Requirements (RFCs 1122 and 1123) does specify some possible algorithms for dead gateway detection (a subset of the problems neighbor unreachability detection tackles).

Router discovery is part of the base protocol set; there is no need for hosts to "snoop" the routing protocols. Router advertisements, which enable address autoconfiguration, carry link-layer addresses; no additional packet exchange is needed to resolve the router's link-layer address. Router advertisements also carry prefixes for a link so there is no need to have a separate mechanism to configure the "netmask." Routers can advertise an MTU for hosts to use on the link, ensuring that all nodes use the same MTU value on links lacking a well-defined MTU.

Address resolution multicasts are "spread" over 4 billion (2^{32}) multicast addresses, greatly reducing address resolution–related interrupts on nodes other than the target. Moreover, non-IPv6 machines should not be interrupted at all. Because redirects contain the link-layer address of the new first hop separate address resolution is not necessary upon receiving a redirect.

Multiple prefixes can be associated with the same link. By default, hosts learn all on-link prefixes from router advertisements. However, routers can be configured to omit some or all prefixes from router advertisements. In such cases, hosts assume that destinations are off-link and send traffic to routers. A router can then issue redirects as appropriate.

The recipient of an IPv6 redirect assumes that the new next hop is on-link (i.e., on its own subnet). In IPv4, a host ignores redirects specifying a next hop that is not on-link according to the link's network mask. The IPv6 redirect mechanism is analogous to the Xredirect facility. It is expected to be useful on nonbroadcast and shared media links in which it is undesirable or not possible for nodes to know all prefixes for on-link destinations. Neighbor unreachability detection is part of the base, significantly improving the robustness of packet delivery in the presence of failing routers or partially failing or partitioned links and nodes that change their link-layer addresses. For instance, mobile nodes can move off-link without losing any connectivity due to stale ARP caches. Unlike ARP, neighbor discovery detects half-link failures (using neighbor unreachability detection) and avoids sending traffic to neighbors with which two-way connectivity is absent.

The router advertisement messages do not contain a Preference field, as in IPv4 router discovery. The Preference field is not needed to handle routers of different "stability"; the neighbor unreachability detection will detect dead routers and switch to a working one.

The use of link local addresses to uniquely identify routers (for router advertisement and redirect messages) makes it possible for hosts to maintain router associations in the event of a site's being renumbered to use new global prefixes.

Using a hop limit equal to 255, neighbor discovery is immune to off-link senders that accidentally or intentionally send ND messages. In IPv4, off-link senders can send both ICMP redirects and router advertisement messages.

Placing address resolution at the ICMP layer makes the protocol more media-independent than ARP and makes it possible to use standard IP authentication and security mechanisms as appropriate.

IPv6 Tunneling

Tunneling is, very simply, the method of transporting IPv6 packets over IPv4 routing topologies. It is being used today with the 6Bone (*http://www/6Bone.com*). The following two tunneling methods are used with a router endpoint:

- *Host to router:* Dual IP stack hosts can tunnel IPv6 packets to an intermediary dual-stack IP router that is reachable via a IPv4 infrastructure.

- *Router to router:* Routers that are running the dual-stack IP interconnected by an IPv4 infrastructure can tunnel IPv6 datagrams between themselves.

With these types of tunnels, the tunnel endpoint is an intermediary router that must decapsulate the IPv6 packet and forward it on to its final destination. The endpoint of the tunnel is different from the destination of the packet being tunneled. Therefore, the address in the IPv6 packet being tunneled does not provide the IPv4 address of the tunnel endpoint. The tunnel endpoint address must be determined from information that is configured on the node performing the tunneling. This is the configured tunnel approach. The endpoint is explicitly configured.

Tunnels are characterized by two endpoint IPv4 addresses. The IPv4 protocol identifier is 41, the assigned payload type number for IPv6.

For host endpoints, we have the following two possibilities:

- *Router to host:* Dual-stack IP routers can tunnel IPv6 packets to their final-destination dual-stack IP host.

■ *Host to host:* Dual-stack IP hosts can tunnel IPv6 packets between themselves over an IPv6 infrastructure (without the use of a router).

These two types provide tunneling all the way to the final destination. In these cases, the tunnel endpoint is the node to which the IPv6 packet is addressed. This is automatic tunneling, which allows IPv6 packets that are to be sent to IPv6 destinations using the IPv4-compatible address and that are located remotely (off-link) to be encapsulated in IPv4 headers and sent through the IPv4 infrastructure. The four tunnel types are depicted in Fig. 10-8.

IPv6 Flowcharts   Which type of tunneling is used where, and how does it work? This depends on the destination address. If the address is an IPv6 address and the destination is local (on-link), the packet is sent. If the address of the end node is an IPv4 address and it resides on a different subnet, an IPv4 router must be used.

Figure 10-8
IPv6 tunnels.

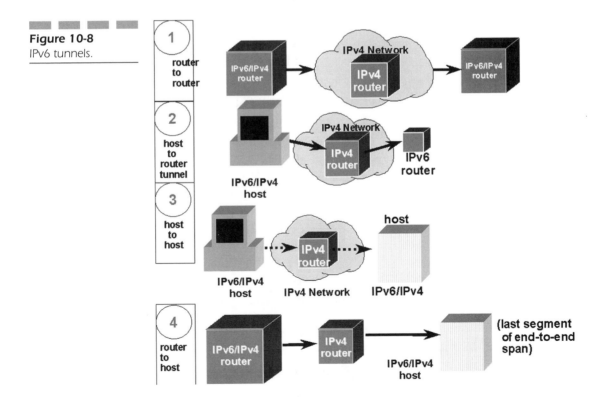

The key to all of this is the special address *0:0:0:0:0:0:<IPv4 32-bit address>*. Dual-IP-stack hosts recognize the special address and immediately encapsulate the packet with an IPv4 header. This is called an end-to-end tunnel. The receiving station will unencapsulate the datagram (strip off the IPV4 header) and read it as an IPv6 datagram.

IPv6-only hosts can also use the IPv4 Internet through dual-stack IP routers. The IPv6-only host transmits the IP datagram as an IPv6 datagram. The dual-stack IP router recognizes the special address and wraps it in an IPv4 header (using the last 32 bits of the special address in the IPv4 destination IP address).

Finally, if the address is an IPv6 address but not the special address, a configured tunnel can be used instead of an automatic tunnel (which recognizes the special address). This requires configuration of the IPv6 host. More or less, this is a manual configuration, and it tells the IPv6 host where to send the packet.

IPv6 and IPv4 Dual-Stack Strategy   Figure 10-9 shows the IPv6 and IPv4 dual-stack capability of a host. Based on the EtherType field of the received packet, the NIC software can hand off the unencapsulated packet to either of the TCP/IP software stacks.

Figure 10-10 shows an IPv6 tunneling strategy. IPv6 and IPv4 modes can coexist harmoniously on a network using tunneling. The IPv4/IPv6 router in the figure can attach to an IPv4 router. Note that the IPv4 and IPv6 hosts can exist with the IPv6/IPv4 routers. However, an IPv6-only host cannot use the IPv4 router. The flowcharts shown in Figs. 10-11, 10-12, and 10-13 show some routing examples for IPv6.

Figure 10-9
Dual-stack strategy.

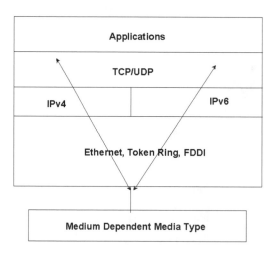

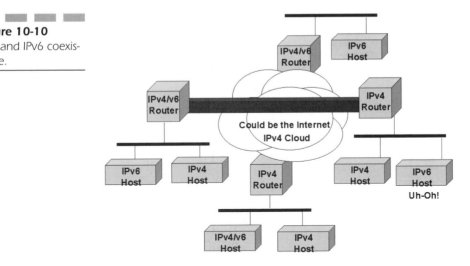

Figure 10-10
IPv4 and IPv6 coexistance.

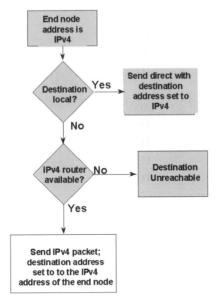

Figure 10-11
Dual-stack host tunneling, example 1.

Anycast Addressing

The next form of IPv6 addressing is the anycast address. This is new for IP and brokers on the functions of nearest hosts implemented by Novell NetWare. That is stretching it, but it may give some readers a better understanding of what the address really is. An anycast address is similar to a multicast address. The exception here is that a packet sent to an any-

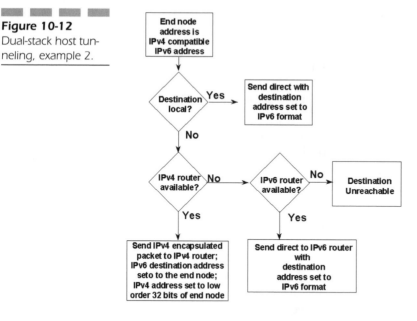

Figure 10-12
Dual-stack host tunneling, example 2.

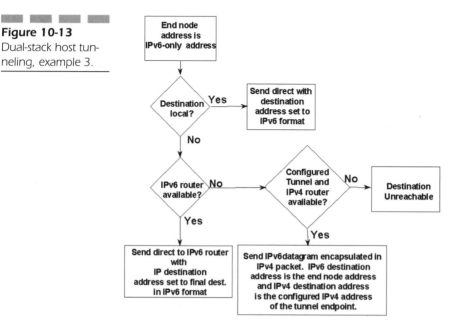

Figure 10-13
Dual-stack host tunneling, example 3.

cast address is routed to the "nearest" interface having that address, using distance as a factor.

A source node sends a datagram addressed as an anycast address, which will be recognized by all destinations of a given type. The routing system is key, here; it is the routing system that delivers the datagram to the nearest server. This helps to find servers of type file/print, timer, name, DHCP, etc.

This concept may seem familiar to those who know the Novell NetWare protocol. Functionally, it is implemented differently, but the concept is the same.

Multicasting

Multicasting for IP started in 1988 with IGMP. IANA also assigned a new class of addressing known as class D addressing. Multicasting is carried over to IPv6, and its addressing allows for more granularity. Multicasting is used extensively with IPv6. The format of the address is shown in Fig. 10-14.

- The first part of the address is the multicast reserved bits FF

- The scope is included in the overall reserved address

  - All name servers within a site local scope

  - All name servers within a link local scope

  - Same multicast function but different address

  - Same function as the TTL in MBONE

The first eight bits must be set to *FF.* The next four bits are called the flag bits, and only one is defined. The T bit is the transient bit. Setting this to 1 indicates that the multicast address is not permanently assigned by the IANA; A 0 indicates that it is permanently assigned.

The scope of the address is four bits in length and controls the "hearing range" of the multicast address. It performs the same function as the

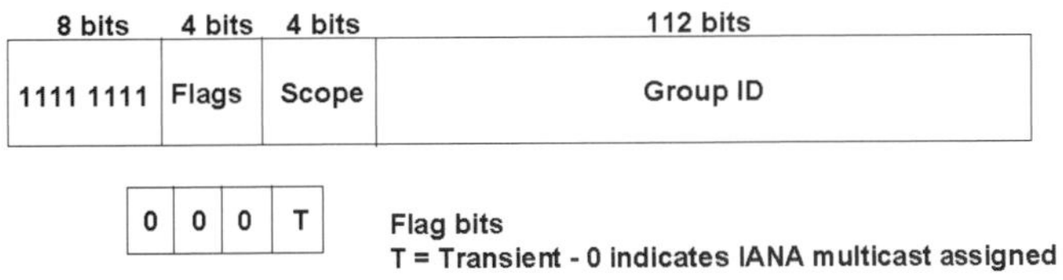

Figure 10-14
Multicast address format.

8 bits	4 bits	4 bits	112 bits
1111 1111	Flags	Scope	Group ID

0	0	0	T

Flag bits
T = Transient - 0 indicates IANA multicast assigned

TTL field in an IPv4 multicast packet. The following list indicates which scopes are currently assigned:

Scope	Range
0	reserved
1	node local scope
2	link local scope
3	unassigned
4	unassigned
5	site local scope
6	unassigned
7	unassigned
8	organization local scope
9	unassigned
A	unassigned
B	unassigned
C	unassigned
D	unassigned
E	global scope
F	reserved

Notice that in IPv6 multicast addresses, weaving the scope in as part of the address makes it possible to have multiple multicast addresses for the same function. The first part of the address is the multicast address identifier, but the scope is included in the overall address. For example, there is one multicast address looking for all DHCP servers in a radius of 3 hops. Another would allow for a radius of 10 hops, but it is still the same multicast function.

IPv6 Routing

Routing in IPv6 is almost identical to IPv4 routing under CIDR except that the addresses are 128-bit IPv6 addresses instead of 32-bit IPv4 addresses. With very straightforward extensions, all of IPv4's routing algorithms (OSPF, RIP, IDRP, ISIS, etc.) can be used to route IPv6.

IPng also includes simple routing extensions that support powerful new routing functionality. These capabilities include provider selection (based on policy, performance, cost, etc.), host mobility (route to current location), and auto-readdressing (route to new address).

The new routing functionality is obtained by creating sequences of IPv6 addresses using the IPv6 routing option. The routing option is used by a IPv6 source to list one or more intermediate nodes (or topological groups) to be "visited" on the way to a packet's destination. This option is very similar in function to IPv4's loose source and record route options.

OSPFv6 for IPv6, like IPv4, will run directly on top of IPv6. OSPFv6 runs as a separate protocol just like any other "ships-in-the-night" type of protocol in a multiprotocol router. It will have a separate link-state database from OSPFv4's. In short, nothing is shared between OSPFv4 and OSPFv6 (in the router, that is). Neither will know the other exists.

In order to make IPv6 operate with OSPFv6, some changes are necessary—most notably the 128-bit address. Router IDs, links, and areas will be associated with a 128-bit number.

RIPv6

How could we forget good old RIP? It is still a good, decent protocol for small networks, it is still very easy to implement, and it is still dominant.

The packet format is shown in Fig. 10-15 and is represented by RFC 2080. (It is variable in length and therefore number of entries per packet.) Notice that the same amount of space is taken up for the route table entries as in IPv4 (160 bits per entry). One new feature is that the packet size was extended to beyond the limit of 576 bytes, as in RIPv1 and v2. It was noted that these update packets would never traverse a router, and

Figure 10-15
RIPv6 packet format.

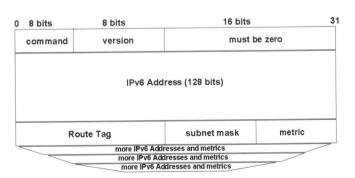

Variable in length and therefore number of entries per packet.

therefore the limit on the route table entries (RTE) is simply limited by the MTU of the medium over which the protocol is being used.

The formula is as follows:

```
Number of RTE = MTU−size of IPv6 headers)−UDP (header length)−RIPng
(header length)\RTE size
```

The eight-bit subnet mask indicates the number of bits in the prefix. Because there are eight bits, this gives us the capability of a 256-bit prefix, which is more than enough for the 128 bits of IPv6.

ICMP for IPv6

The Internet Control Message Protocol (ICMP) for IPv6 is found in RFCs 1885 and 1970. It is explained in RFC 1885, but its individual functions (for example, path MTU discovery) are detailed in existing RFCs such as RFC 1191. This may seem confusing, but IPv6 and its extension protocols used previous RFCs if they were found relevant to the protocol.

ICMPv6 is an ICMPv4 extension. It was originally documented in RFC 792 and is an integral part of the IP. Over the years, other functions that utilize ICMP, such as router discovery (RFC 1256), have been added. ICMPv6 is a version of ICMP for IPv6. There are currently two RFCs that define all the ICMP functions for IPv6: RFCs 1885 and 1970. RFC 1885 contains information on new functions and names the older functions that made it through the review process. RFC 1970 includes the discovery protocols of RFC 1256 and a few others. It also includes the redirect message.

ICMPv6, as defined in RFCs 1885 and 1970, is currently using control and information messages previously defined in RFCs 791, 1112, and 1191. Therefore, the procedures for certain ICMP functions continue to be defined in their respective RFCs. You must read the original RFCs to find full explanations of the procedures.

As was previously explained, the Internet Protocol is an "unreliable" protocol. ICMP is an add-on protocol. It does not make IP reliable; it is a control message protocol, and the purpose of these control messages is to provide feedback about problems in the communication environment, not to make IPv6 reliable. There are still no guarantees that a datagram will be delivered or a control message will be returned. Some datagrams may be undelivered, without any report of their loss. The higher-level protocols that use IPv6 must still implement their own reliability procedures if reliable communication is required.

ICMP messages typically report errors in the processing of datagrams. To avoid the problem of ICMP messages about ICMP messages etc., no such messages are sent.

ICMPv6 is used by IPv6 nodes to report errors encountered in processing packets and to perform other internet-layer functions such as diagnostics (ICMPv6 "ping") and multicast membership reporting. ICMPv6 is an integral part of IPv6 and *must* be fully implemented by every IPv6 node.

ICMPv6 and ICMPv4

ICMPv6 cleaned up ICMPv4, mainly by eliminating control messages that were not used. No longer specifically identified as ICMP messages are time stamp, time-stamp reply, source quench, information request, and reply. Most of these procedures are incorporated into other protocols. For example, source quench is not used any more because other mechanisms such as TCP "slow start and congestion control" were found to be more useful. Why put the onus on the routers when a mechanism is better defined and more useful elsewhere? Also, certain codes for a specific type have been moved or eliminated (for example, source route failed code for the destination unreachable type). The address fields obviously had to be extended to handle ICMPv6 128-bit addressing. The multicast control functions of ICMP were also included in ICMPv6 to allow for group membership query, report, and termination. With all of the above changes, ICMPv6 is not backwards-compatible with ICMPv4. It uses the next header function of IP and the next header type of 56. One last change is that ICMPv4 messages copied the original IP header and 64 bits of data in the returned message. The exception to this is echo request of echo reply. ICMPv6 allows as much data to be copied from the offending datagram so that the ICMPv6 datagram does not exceed 576 bytes.

The format of the ICMPv6 header is shown in Fig. 10-16. It is the same format as in ICMPv4. The Type field indicates the type of the message. Its value determines the format of the remaining data. Error messages are identified as such by having a zero in the high-order bit of their message Type field values. Thus, error messages have message types from 0 to 127; informational messages have message types from 128 to 255. The Code field depends on the message type. It further identifies

Figure 10-16
ICMPv6 header.

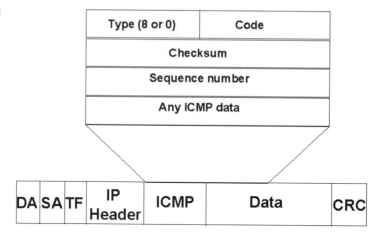

the ICMP message. The Checksum field is used to detect data corruption in the ICMPv6 message and parts of the IPv6 header. For the most part, the destination address is set to the source address of the previously received offending packet.

ICMPv6 Error Messages Error message type 1, destination unreachable, includes the following codes:

0: No route to destination. There was no corresponding route in the router's forwarding table (only for routers that do not possess a default route entry).

1: Communication with destination administratively prohibited—for example, by a firewall or other restrictive administrative command.

2: Not a neighbor. There was an attempt to deliver the datagram to a neighbor that was indicated in the strict source routing entries, but the next hop indicated is not a neighbor of this router.

3: Address unreachable. The router could not resolve the link-layer address for the indicated network address.

4: Port unreachable. The destination does not have a service port available. For example, the datagram is intended for TCP, but all available resources are taken for TCP (there are no listener ports available), or the datagram was sent to a service port that the destination does not support, for example, whois, finger, or the route daemon.

The first 32 bits after the ICMP header are unused, and they must be initialized to zero by the sender and ignored by the receiver. A destination unreachable message generally originates from a router but can also be generated by the originating node. It can be generated for any reason except congestion. There are no ICMP messages to indicate congestion (other protocols monitor and report this condition).

Error message type 2, packet too big, contains only code 0. The first 32 bits after the ICMP header indicate the maximum transmission unit (MTU) for the selected forwarding (next-hop) port. This error message is important because datagrams do not necessarily have to be 576 bytes in size. FDDI-to–Token Ring topologies, for example, do very well with streaming large packets. This is an indicator of how large a packet may be to be forwarded along the path from source to destination. This is the path MTU discovery procedure, which is outlined in RFC 1191. This error message also provides an exception to the rules: It can be sent in response to a packet received with an IPv6 multicast destination address, or a link-layer multicast or link-layer broadcast address.

Error message type 3, time exceeded, contains the following codes:

0: Hop limit exceeded in transit

1: Fragment reassembly time exceeded

The first 32 bits of the ICMP message are specified as unused for all code values, and they must be initialized to zero by the sender and ignored by the receiver. This can be sent if a router receives a packet with a hop limit of zero or a router decrements a packet's hop limit to zero. It *must* discard the packet and send an ICMPv6 time exceeded message with Code 0 to the source of the packet. This indicates either a routing loop or too small an initial hop-limit value.

Error message type 4, parameter problem, contains the following codes:

0: Erroneous header field encountered

1: Unrecognized Next Header type encountered

2: Unrecognized IPv6 option encountered

The first 32 bits of the ICMP message are a pointer that identifies the octet offset within the invoking packet in which the error was detected. It points beyond the end of the ICMPv6 packet if the field in error is beyond what can fit in the 576-byte limit of an ICMPv6 error message.

If an IPv6 node cannot process the datagram due to some error in the headers, it must discard the packet and should send an ICMPv6 parameter problem message to the packet's source, indicating the type and location of the problem. To indicate the type of problem, a pointer field is provided. The pointer field indicates the point in the originating header where the error was detected. For example, an ICMPv6 message with Type field 4, Code field 1, and Pointer field 40 would indicate that the IPv6 extension header following the IPv6 header of the original packet holds an unrecognized Next Header field value.

ICMP Informational Messages   Informational message type 128, echo request, contains only code 0. The first 16 bits of the ICMP message act as an identifier to aid in constructing an echo reply to an echo request. It may contain a zero. The next 16 bits are a sequence number to aid in specifically matching multiple echo requests from the same source. It, too, may contain a zero.

The remaining message is option data that may have been typed in on the request and is echoed on the reply. This is the ping command that you may have typed in. For example, PING 192.1.1.1 is an echo request to an echo server residing on host 192.1.1.1

Message type 129, echo reply, can contain only code 0. The first 16 bits (after the ICMP header) are an identifier field from the previously received echo request message. It is used to match replies with requests. The next 16 bits is a sequence number from the node that sent the echo request number message. This is useful to match multiple requests from the same host. The rest of the message contains echoed data copied from the received echo request.

ICMP and Neighbor Discovery   Address resolution is accomplished via neighbor discovery messages. Neighbor discovery messages are generated and processed by ICMP. Following are the neighbor discovery messages:

Router solicitation message format: Hosts send router solicitations in order to prompt routers to generate router advertisements quickly.

Router advertisement message: Routers send out router advertisement messages periodically, or in response to router solicitations. They contain information related to the local prefixes and whether the router can act as a default router.

Neighbor solicitation messages: Nodes send neighbor solicitations to request the link-layer address of a target node while also providing their own link-layer address to the target. Neighbor solicitations are multicast when the node needs to resolve an address and unicast when the node seeks to verify the reachability of a neighbor.

Neighbor advertisement message: Nodes send neighbor advertisements in response to Neighbor Solicitations and send unsolicited neighbor advertisements in order to (unreliably) propagate new information quickly. For example, if a node has determined some change, such a link-level address change, it can quickly update this information to its neighbors.

Redirect message: Routers send redirect packets to inform a host of a better first-hop node on the path to a destination. Hosts can be redirected to a better first-hop router but can also be informed by a redirect that the destination is, in fact, a neighbor. The latter is accomplished by setting the ICMP target address equal to the ICMP destination address.

ICMPv6 and Multicast   The functions of IGMP were moved into ICMPv6. The ICMPv6 group membership messages are used to convey information about multicast group membership from nodes to their neighboring routers. This function of ICMPv6 allows IGMP messages to be sent. These are group membership for query, reports, and reduction (or leaving a group with a termination message). Due to the dynamic nature of the IPv6 and its neighbor discovery protocols (routers and hosts), IGMP functions were moved into the ICMP protocol suite. For example, when a node initializes (in an IPv6 environment), it must immediately join the all-nodes multicast address on that interface, as well as the solicited-node multicast address corresponding to each of the IP addresses assigned to the interface (more information is contained in the IPv6 addressing section of this chapter).

In the IPv6 header, the destination address is set as follows: in a group membership query message, the multicast address of the group being queried, or the link local all-nodes multicast address; in a group membership report or reduction message, the multicast address of the group being reported or terminated.

The hop limit is set to 1 to ensure that this message does not leave the local subnetwork. The ICMPv6 fields are set to code 0 for the following:

Type 130:   Group membership query

Type 131:   Group membership report

Type 132:   Group membership reduction

The first 16 bits after the ICMP header are used for the maximum response delay. In query messages, it is the maximum time that responding report messages may be delayed, in milliseconds. In report and reduction messages, this field is initialized to zero by the sender and ignored by receivers. The next 16 bits are unused, and they are initialized to zero by the sender, ignored by receivers. The rest of the message is filled with the multicast address, which is the address of the multicast group to which the message is being sent. In query messages, the multicast address field may be zero, implying a query for all groups.

IPv6 Cache Entries

All of the following caches are built, in part, by the neighbor discovery process. Instead of the simplex ARP cache used with IPv4, IPv6 maintains four caches.[1] The destination cache contains information about destinations to which traffic has been sent recently. It includes both local and remote destinations and associates an IPv6 address of a destination with that of the neighbor toward which the packets were sent. This cache is updated with information learned from ICMP redirect messages. Other information such as the path MTU (PMTU) and round-trip timers maintained by transport protocols can be maintained in this cache. Entries are created by the next-hop determination procedure.

The neighbor cache is a record that contains information about individual neighbors (hosts or routers may be entries) to which traffic has been sent recently. It contains such information as the neighbor's link-layer address, an indication of whether the neighbor is a host or a router, a pointer to any queued packets waiting for address resolution to complete, etc. This information is also used by the neighbor unreachability protocol.

The prefix list cache is created from information received in router advertisements. It is a listing of the local prefixes and an individual expiration timer that indicate that a set of addresses is on-link. Nodes receive and store this information when it is transmitted from a router in this cache. This enables a node to determine a remote destination. A special "infinity" timer value specifies that a prefix remains valid forever, unless a new (finite) value is received in a subsequent advertisement. For example, the prefix of the local link that a node is attached to is considered to be on the prefix list with an infinite invalidation timer, regardless of

---

[1]Actually, four caches may not be maintained. Implementers can integrate this information any way they wish, including simply using one large table or four linked tables in one database, but all the required information must be gathered and maintained. The entries are shown here separately for the sake of simplicity.

whether routers are advertising a prefix for it, and received router advertisements cannot change this value.

The router list cache is built from received router advertisements. This list contains information about those routers to which packets may be sent. Router list entries point to entries in the neighbor cache; the algorithm for selecting a default router favors routers known to be reachable over those whose reachability is suspect. Each entry is paired with an associated expiration timer value (extracted from router advertisements). This timer is used to delete entries from which the node has not received advertisements.

IPv6 Algorithm

If you are looking for more information on how IPv6 routes datagrams, you must read RFC 1970. This is a very important RFC for understanding the IPv6 routing algorithm. It contains the neighbor discovery mechanism, and this includes everything on IPv6 subnets, such as hosts and routers. IPv6 does not use ARP, it uses neighbor discovery.

To route a datagram in IPv6, we consult the destination cache, the prefix list, and the default router list to determine the IP address of the appropriate next hop, an operation known as "next-hop determination." Next-hop determination is invoked to create a destination cache entry. The results of next-hop determination computations are saved in the destination cache (which also contains updates learned from Redirect messages). Therefore, a sending node first looks in the destination cache for a matching entry to the destination IP address. If one is not found in this cache, the prefix list cache is consulted. The sending node compares the destination prefix mask with the entries in the prefix list cache. If a match is found here, the node determines whether the destination is local or remote. If the destination is local, the next-hop address is simply the destination address of the datagram. If the destination is remote, the node must select a router from the default router list. If there are no entries in the default router list, then the destination is assumed to be local.

After the next-hop determination has been added to the destination cache, the neighbor cache is used to determine the media address of that next-hop neighbor. Once the IP address of the next-hop node is known, the sender examines the neighbor cache for link-layer information about that neighbor. If no entry exists, the sender creates one and then starts the address resolution procedure to complete the entry. The datagram to be transmitted must wait for this to complete. Once the neighbor entry is complete, it is used for subsequent transfers to that destination station. None of the other procedures are needed.

Address Resolution

The purpose of address resolution is to determine the link-level address of a destination, given only its IP address. This is performed only for those IP addresses that are local (hop count set to 1). When a multicast-able interface starts it must join both the all-nodes multicast group and the solicited-node multicast group. This enables the node to receive and process packets without having all of its addressing established. In fact, a node must keep the above multicast addresses until all addressing has been resolved. Address resolution consists of sending a neighbor solicitation message and waiting for a neighbor advertisement using multicast addressing. The solicitation is sent to the solicited-node multicast address corresponding to the target address. The solicited-node multicast address is a link local scope multicast address that is computed as a function of the solicited target's address. The solicited-node multicast address is formed by taking the low-order 32 bits of the target IP address and appending those bits to the 96-bit prefix *FF02:0:0:0:0:1* to produce a multicast address within the range *FF02::1:0:0* to *FF02::1:FFFF:FFFF.* For example, the solicited-node multicast address corresponding to the IP address *4037::01:800:200E:8C6C* is *FF02::1:200E:8C6C.* IP addresses that differ only in the high-order bits (e.g., due to multiple high-order prefixes associated with different providers), map to the same solicited-node address, thereby reducing the number of multicast addresses a node must join. In response to this request (which the sender may send several times if no response is found within a certain period of time), a neighbor advertisement should be generated by the remote node. The originating node should receive this packet and update its neighbor cache with the information in the received neighbor advertisement (the link-layer information). The MAC address is set as previously indicated by taking the low-order 32 bits of the target IPv6 address and prepending *3333* to that address, which is the IPv6 all-nodes MAC multicast address.

One more check is accomplished each time a neighbor cache (link-layer information) entry is accessed while transmitting a unicast packet: the sender checks neighbor unreachability detection–related information according to the neighbor unreachability detection algorithm. This is not so much a protocol as keeping an eye on the progression of the upper-layer protocols with this address. This unreachability check might result in the sender transmitting a unicast neighbor solicitation to verify that the neighbor is still reachable.

If at some point communication ceases to proceed, as determined by the neighbor unreachability detection algorithm, next-hop determination may need to be performed again. For example, traffic through a failed router should be switched to a working router. Likewise, it may be possible to reroute traffic destined for a mobile node to a "mobility agent."

Note that when a node redoes next-hop determination, there is no need to discard the complete destination cache entry. In fact, it is generally beneficial to retain such cached information, as well as the PMTU and round-trip timer values that may also be kept in the destination cache entry. Next-hop determination is done the first time traffic is sent to a destination. As long as subsequent communication to that destination proceeds successfully, the destination cache entry continues to be used.

RFCs Related to IPv6

The following is a list of RFCs pertaining to Internet Protocol version 6:

- RFC 1883 describes the IPv6 protocol. RFC 2147 updates (does not replace) the RFC.
- 2147, PS. D. Borman, "TCP and UDP over IPv6 Jumbograms," 05/23/1997 (pages = 3, format = .txt, updates RFC1883).
- 2133, I. R. Gilligan, S. Thomson, J. Bound, W. Stevens, "Basic Socket Interface Extensions for IPv6," 04/21/1997 (pages = 32).
- 2080, PS. G. Malkin, R. Minnear, "RIPng for IPv6," 01/10/1997 (pages = 19).
- 2073, PS. Y. Rekhter, P. Lothberg, R. Hinden, S. Deering, J. Postel, "An IPv6 Provider-Based Unicast Address Format," 01/08/1997. (pages = 7).
- 2030, I. D. Mills, "Simple Network Time Protocol (SNTP) Version 4 for IPv4, IPv6 and OSI," 10/30/1996 (pages = 18).
- 2019, PS. M. Crawford, "Transmission of IPv6 Packets over FDDI," 10/17/1996 (pages = 6).

- 1972, PS. M. Crawford, "A Method for the Transmission of IPv6 Packets over Ethernet Networks," 08/16/1996 (pages = 4).

- 1971, PS. S. Thomson, T. Narten, "IPv6 Stateless Address Autoconfiguration," 08/16/1996 (pages = 23).

- 1970, PS. T. Narten, E. Nordmark, W. Simpson, "Neighbor Discovery for IP Version 6 (IPv6)," 08/16/1996 (pages = 82).

- 1933, PS. R. Gilligan, E. Nordmark, "Transition Mechanisms for IPv6 Hosts and Routers," 04/08/1996 (pages = 22).

- 1924, I. R. Elz, "A Compact Representation of IPv6 Addresses," 04/01/1996 (pages = 6).

- 1897, E. R. Hinden, J. Postel, "IPv6 Testing Address Allocation," 01/25/1996 (pages = 4).

- 1888, E. J. Bound, B. Carpenter, D. Harrington, J. Houldsworth, A. Lloyd, "OSI NSAPs and IPv6," 08/16/1996 (pages = 16).

- 1887, I. Y. Rekhter, T. Li, "An Architecture for IPv6 Unicast Address Allocation," 01/04/1996 (pages = 25).

- 1885, PS. A. Conta, S. Deering, "Internet Control Message Protocol (ICMPv6) for the Internet Protocol Version 6 (IPv6)," 01/04/1996 (pages = 20).

- 1884, PS. R. Hinden, S. Deering, "IP Version 6 Addressing Architecture," 01/04/1996 (pages = 18, format = .txt).

- 1883, PS. S. Deering, R. Hinden, "Internet Protocol, Version 6 (IPv6) Specification," 01/04/1996 (pages = 37, updated by RFC 2147).

- 1881, I. IESG, "IPv6 Address Allocation Management," 12/26/1995 (pages = 2).

- 1809, I. C. Partridge, "Using the Flow Label Field in IPv6," 06/14/1995 (pages = 6).

User Datagram Protocol

Introduction

A transport layer allows communication to occur between network stations. Data are handed down to this layer from an upper-level application. The transport layer then envelopes the data with its headers and gives them to the IP layer for transmission onto the network. In TCP/IP there are two transport-layer protocols: UDP (user Data Protocol) and TCP (Transmission Control Protocol).

The functionality of UDP should be familiar. It is a connectionless, unreliable transport service. It does not provide an acknowledgment to the sender upon the receipt of data. It does not order the incoming packets, and it may lose packets or duplicate them without issuing an error message to the sender. This should sound like the IP protocol. The only advantage that UDP has is the assignment and management of port numbers to uniquely identify the individual applications that run on a network station and a checksum for simplex error detection. UDP tends to run faster than TCP, because it has low overhead (8 bytes in its header, compared to TCP's typical 40 bytes). It is used for applications that do not need a reliable transport. Some examples are network management, name servers, and applications that have built-in reliability.

Any application program that incorporates UDP as its transport-level service must provide an acknowledgment and sequence system to ensure that packets arrive and arrive in the same order as they were sent.

As shown in Fig. 11-1, applications data are encapsulated in a UDP header. The transport layer has its own header, independent of all other layers, that it prefaces to the data handed to it from its upper-layer protocol. The UDP header and its data are then encapsulated in an IP header. The IP protocol then sends the datagram to the data-link layer, which then encapsulates the datagram with its headers (and/or trailers) and sends the data to the physical layer for transmission.

Figure 11-1
UDP header encapsulation.

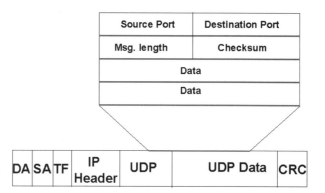

Upon receipt of the packet, the data link interprets the address as its own, strips off its header (and trailers), and submits the packet to the IP layer. IP accepts the packet on the basis of the correct IP address in the IP header, strips off its header, and submits the packet to the UDP-layer software. The UDP layer accepts the packet and now has to demultiplex the packet on the basis of the port number in the UDP header.

In Fig. 11-1 the packet header for UDP (8 bytes) is small, but functional. The message length indicates the size of the UDP header and its data in bytes. The minimum packet size for UDP is 8 bytes (the size of the header). The checksum is used to check for the validity of the UDP header and data. It does not have to be implemented and would be set to 0 if not implemented.

UDP Multiplexing and Demultiplexing

UDP accepts data from the application layer, formats them (UDP header) with its information, and presents them to the IP layer for network delivery. UDP also accepts data from the IP layer and, depending on the port value, present the data to the appropriate application. As shown in Fig. 11-2, UDP is responsible for directing the rest of the packet (after stripping off its headers) to the correct process according to the port number assigned in the UDP header. This process is called *demultiplexing*. There are many different types of port numbers to indicate any application program running on the network station. UDP reads the destination port field of the UDP header (demultiplex) and gives the data to the application. When the application program (identified by the port number) initializes, the station's operating system works in conjunction with it and provides a buffer area for information that may be

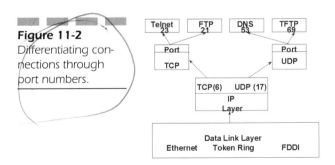

Figure 11-2
Differentiating connections through port numbers.

stored. UDP will place the data in this area for retrieval by the application program. UDP does provide one error mechanism for ports that are not valid: it can generate an "ICMP port unreachable" message to be sent to the originator of the packet.

Since the TCP/IP protocol suite includes applications that are specifically written for it (TFTP, domain name service, etc.), there are statically assigned port numbers that identify these applications. Certain port numbers may not be used by any unknown application program because they are reserved by the applications that are defined in the RFCs. The reserved port numbers are specified in RFC 1700, which contains the FTP site for up-to-date information on the protocol numbers.

Port Numbers

Review RFC 814. Since many network applications may be running on the same machine, a method is needed to allow access to these applications even though they reside on the same machine and the machine contains one IP address. One IP address and many applications? How do we decide which datagram belongs to which application?

It would not be advantageous to assign each process an IP address, nor would it be advantageous to change the IP addressing scheme to include a marker to identify a unique application in the machine. Instead, both the TCP and UDP protocols provide a concept known as *ports* (sometimes incorrectly called *sockets*). Ports, along with an IP address, allow any application in any machine on an internet to be uniquely identified.

There are three different types of port numbers: assigned, registered, and dynamic. The assigned numbers RFC (RFC 1700 at the time of this writing) contains assigned and registered numbers. The first 1024 ports are assigned and in specific use and cannot be reused by any application. The remaining addresses can be dynamic and registered (16 bits allow for 65,535 ports) and can be used freely, although IANA does request vendors to register their application port numbers with it.

When a station wishes to communicate with a remote application, it must identify that application in the datagram. For example, if a station needed to use a simple file transfer protocol known as *trivial file transfer program* (TFTP) on the station 130.1.1.1, it would address the datagram to station 130.1.1.1 and insert *destination port* number 69 in the UDP header. The *source port* number would identify the application on the local station that requested the file transfer, and all response packets generated by the destination station would be addressed to that port number on the source station. Generally, the source port is randomly generated by the source station. If the source port is not used (as with broadcast RIP update tables), it should be set to 0. When the IP layer demultiplexes the packet and hands it to UDP, UDP passes the data to the locally assigned port number for it to process the data.

Assigned, Registered, and Dynamic Port Numbers

Refer to RFC 1700 and the FTP site *ftp://ftp.isi.edu/in-notes/iana/assignments*. (Point your URL to this address on your browser.) As was mentioned before, in the TCP/IP protocol, UDP port numbers are of three kinds: assigned, registered, and dynamic. Assigned numbers are in the range 0–1023 and are fully controlled by the Internet Assigned Numbers Authority (refer to RFC 1700). These numbers deal with protocols such as Telnet, FTP, Network Time Protocol (NTP), and so on. No matter which implementation of TCP/IP (i.e., which vendor's TCP) is in use, those applications listed beside the port number will always be the same. They are known as *well-known port numbers* and are assigned by IANA. In this case, RFC 1700 spells out which processes are assigned to which port numbers. Any port number beyond 1023 but less than 65535 may be used by any application. Some companies have registered their port numbers with IANA, and other companies respect this by not using the same port number.

Here is an example of the use of an assigned port number. If station A wanted to access the TFTP process on station B, it would call station B in the UDP header with a destination port number of 69 (decimal). The source station requesting TFTP services also has a port number, which is dynamically assigned (discussed next) by its TCP/IP stack. The RFC suggests methods for assisting in assigning a dynamic port number. In this way, the server and client can communicate with one another using the port numbers to identify the service uniquely for that datagram.

Dynamic Port Numbers

TCP/IP also implements port numbers for dynamic use. Since the port number field in the UDP header is 16 bits long, 65,535 ports (minus the 1023 assigned ports) are available for individual use as registered and dynamic ports. One use for a dynamic port is for a source station that is requesting the services of TFTP on a remote station. The source station

Figure 11-3
Dynamic port numbers.

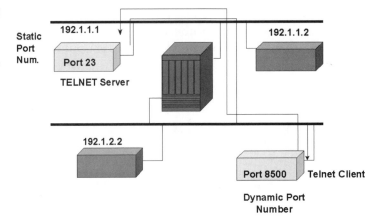

would dynamically assign itself an available port number (usually above size) to use so that the remote station would know what port to access when it transfers the file. Refer to Fig. 11-3. If a user initiates a Telnet (remote terminal program, to be discussed later), the Telnet request packet sent to the Telnet server will include in its UDP header a dynamic port number for the requesting network station that wants the Telnet, called the *source port*; let's say it was assigned port 8500. The destination port number would be 23, the assigned number for Telnet. In this way, the server will accept the packet, give it to the Telnet process in the host, and know how to address the port number in the response packet when the host responds. In the response packet, the server will fill out the UDP header with a destination port of 8500 and a source port of 23, and send the packet back to the requesting station.

Dynamic ports are also used when network vendors implement proprietary schemes on their devices—for example, a proprietary scheme for a network station to boot or a proprietary scheme to allow network management statistics to be gathered. All these applications are valid and may run on any TCP environment using a dynamic port assignment.

The disadvantage to dynamic ports occurs when a broadcast IP datagram is transmitted to the network using a dynamic port. That port could be in use by another vendor on the network, and another network station may invoke a process to accommodate that request. Such an event is rare, but has been known to happen.

Dynamic port numbers are assigned by the TCP/IP software at the local workstation. They can be duplicated from workstation to workstation without respect to the application, because an application on any network station is uniquely identified by the combination of the IP address (network number, host number) and the port number. When taken as a whole number, that combination is called the *socket number* and cannot be duplicated on an IP network except by negligence.

As a final note, some people confuse the terms *port number* and *socket number*. In proper IP terminology, a port number and a socket number are not the same thing, as just indicated.

Transmission Control Protocol

A communication facility needs the capability to transfer data reliably between two points. Imagine setting up a communication system that allowed only for unreliable data transfer. The U.S. Postal Service transfers most mail in this manner; when you mail a letter, you have no idea whether it really arrived unless you make the effort to check. That analogy should make you a little nervous about sending critical data over such a network. The problem is further exemplified by a packet switch network, in which the same communication channel is used by multiple entities all vying for the same path and each header contains its own directional information.

Originally, TCP/IP hosts were connected via telephone lines (commonly known as serial lines). This communication facility was not the same in the early 1970s as it is today. The lines were extremely noisy and were not conditioned to handle high-speed data. Therefore, the TCP protocol has strict built-in error detection algorithms. The following paragraphs explain the TCP protocol and show the strictness with which it is structured to ensure the integrity of the data.

TCP, like UDP, is also a transport-layer protocol. In contrast to UDP, the purpose of the TCP transport-layer software is to allow data to be exchanged reliably with another station on the network. Like UDP, TCP provides demultiplexing of port numbers to identify an application in the host, but TCP also provides reliable transport of data, including many different options that may or may not be sent by the originating station.

TCP Details

Not all networks use separate transport layer software to converse on a network. The best example of this is Novell, with its LAN workgroup operating system NetWare. Novell relies on the network layer software to transport its data and on the NetWare Control Protocol (as an application) to provide the sequence numbering of the packets. There is nothing wrong with this practice; it generally speeds up the communication process between two stations on the network, and the overhead of one role of the transport layer is diminished. However, that type of protocol was developed on high-speed, low-error-rate media such as Ethernet. TCP was not, and it is much more robust in its transport layer protocol. (TCP is actually a protocol and not a separate piece of software.)

The protocol of TCP uses sequence numbers and acknowledgments to converse reliably with another station on the network. Sequence numbers are used to determine the ordering of the data in the packets and to find missing packets. Since packets on an internet may not arrive in the same sequence as they were sent (for example, a single packet in a series of packets being transmitted may be discarded by a router or a receiving station may receive two copies of the same packet), the sequence number with acknowledgments ensures that the packets are read in the same order in which they were sent. This process is called full duplex, because each side of a connection will maintain its own sequence number for the other side.

TCP is a byte-oriented sequencing protocol, whereas Novell NetWare is a packet-oriented sequencing protocol, which applies a sequence number to each packet transmitted and not to each data byte in the packet. Byte-oriented means that every byte in a transmission is assigned a sequence number. This does not mean that TCP transmits packets containing only one byte. Rather, TCP will transmit data (many bytes) and assign the packet the sequence number of its first byte. Assigning one sequence number per byte in the transmission may sound repetitious, but remember that TCP/IP was first implemented over noisy serial lines, not over reliable, high-speed LANs.

Refer to Fig. 11-4. In this figure two datagrams are transmitted. Each datagram is assigned one sequence number according to the number of bytes in the TCP data field. Notice how the sequence number jumps by the amount of bytes that are in each packet. The receiver of these datagrams will count the amount of bytes received and increment its sequence number of received packets. The first packet received has a sequence number of 40 and contains 4 bytes. The receiver expects the next sequence number to be 44. It is, and that packet contains 7 bytes of data. The receiver expects the next packet to have a sequence number of 51. It does. This is how the byte sequencing of TCP works.

The sliding window scheme will be discussed later. First, the TCP header definitions will be presented.

Figure 11-4
TCP details.

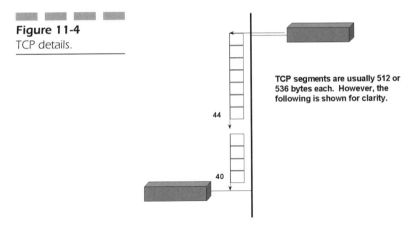

TCP segments are usually 512 or
536 bytes each. However, the
following is shown for clarity.

44

40

Figure 11-5
TCP header details.

Source Port	Destination Port
Sequence Number	
Acknowledgment Number	

Data Offset	Reserved	U A P R S F R C S S Y I G K H T N N	Window

Checksum	Urgent Pointer
Options	Padding

TCP Data	

TCP Field Definitions

Figure 11-5 shows the TCP header fields as encapsulated in an IP datagram.

Source Port. The port number of the originating station.

Destination Port. The port number for the receiving station.

Sequence Number. A number assigned to a TCP datagram to indicate the beginning byte number of a packet, unless the SYN bit is set. If this bit is set, the sequence number is the initial sequence number (ISN) and the first data byte is ISN + 1.

Acknowledgment Number. A number sent by the destination station to the source station, indicating an acknowledgment of a previously received packet or packets. This number indicates the next sequence number the destination station expects to receive. Once a connection is established, this field is always set.

Data offset. Indicates how long the TCP header is (i.e., the number of 32-bit words in the TCP header). It indicates where the data begin and the TCP header stops.

Reserved. Reserved for future use. Must be set to 0.

Control bits.

URG	Urgent. This is used to inform the destination that urgent data is waiting to be sent to it. For example, a destination station may have closed the receive window to the sender, but the receiver will still accept packets with this bit set.
ACK	If set, this packet contains an acknowledgment to a previously sent datagram(s).
PSH	Push function. This immediately sends data when read the segment.
RST	Reset the connection. One function for this is to not accept a connection request.
SYN	Used at startup and to establish sequence number.
FIN	No more data is coming from the sender of the connection.

Window. The number of data octets, beginning with the one indicated in the acknowledgment field, that the sender of this segment is willing to accept. It indicates the available buffers (memory) on the receiver.

Checksum. An error detection number.

Urgent Pointer. The urgent pointer points to the sequence number of the byte following the urgent data. This field is interpreted only in segments with the URG bit set.

Options. Variable in length, this field allows for TCP options to be presented. These are:

- End of option list
- No operation
- Maximum segment size (MSS)

TCP Services

As already noted, the primary purpose of TCP is to provide reliable, securable logical circuit or connection service between pairs of processes. To provide this service on top of a less reliable internet communication system requires facilities in the following areas:

- Basic data transfer
- Reliability
- Flow control
- Multiplexing
- Connections
- Precedence and security

All of these functions are fully discussed in the following paragraphs.

TCP Connection Establishment

Each function will be explained separately. The process will all be tied together at the end of this section.

Refer to Fig. 11-6. In contrast to the UDP protocol in TCP a connection must be established between two stations on a network before any data are allowed to pass between the two. Applications such as Telnet and FTP communicate using TCP through a series of function calls. These calls include OPEN and CLOSE a connection, SEND and RECEIVE (information) to that connection, and to receive STATUS for a connection.

When a connection to a remote station is needed, an application will request TCP to place an OPEN call. There are two types of OPEN calls: *passive* and *active.* A passive OPEN is a call to allow connections to be accepted from a remote station. This usually occurs when an application (such as Telnet or FTP), starts on a network station and it will indicate to

Figure 11-6
TCP connection
establishment.

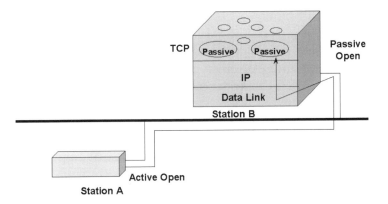

TCP that it is willing to accept connections from other stations on the network. TCP will note the application through its port assignment and will allow connections to come in. The number of connections allowed depends on the number of passive OPEN issued. This passive end of the TCP actions is known as the *responder* TCP. It will open up connection slots to accept any incoming connection request. This may be thought of as the server end of TCP. This passive OPEN call does not wait for any particular station to attach to it.

An active OPEN is made when a connection to a remote network station is needed. Referring to Fig. 11-6, station A wishes to connect to station B. Station A issues an active open call to station B. In order for the connection to be made, station B must already have issued a passive OPEN request to allow incoming connections to be established. In the connection attempt packet is the port number on station B that station A wishes to use. Station B's operating system will spawn a separate process on its system to maintain that connection. This process will act as if it is running locally on that station. TCP will then await another incoming connection request. This process is similar to the way a multitasking operating system handles multiple applications.

The Three-Way Handshake

A connection will be active only after the sender and receiver exchange a few control packets to establish the connection. This is known as the *three-way handshake*. Its purpose to synchronize each endpoint at the start of a TCP connection with a sequence number and acknowledgment number.

Refer to Fig. 11-7. Station A will place an active OPEN call to TCP to request connection to a remote network station's application. Station A will build a TCP header with the SYN bit (the sync bit, shown previously in the TCP Header Fields) set, assign an initial sequence number (it does not always start at 0 and can start at any number; I have chosen 100), and place it in the sequence number field. Other fields will be set in the TCP header (not pertinent to us at this time), and the packet will be given to IP for transmission to station B.

Station B will receive this packet and notice that it is a connection attempt. If station B can accept a new connection, it will acknowledge station A by building a new packet. Station B will set the SYN and the ACK bits in the TCP header (shown in Fig. 11-5), place its own initial

Figure 11-7
The three-way hand-shake.

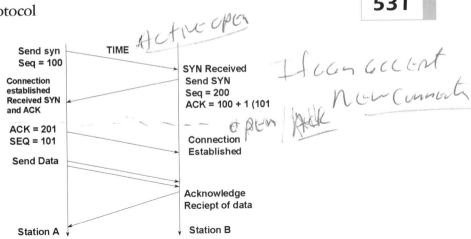

sequence number (200) in the sequence field of the packet, and set the acknowledgment field to 101 (the station A sequence number plus 1, indicating the next expected sequence number).

Station A will receive this response packet and notice that it is an acknowledgment to its connection request. Station A will build a new packet, set the ACK bit, fill in the sequence number to 101, fill in the acknowledgment number to 200 + 1, and send the packet to station B. Once this has been established, the connection is active, and data and commands from the application (such as Telnet) may pass over the connection. As data and commands pass over the connection, each side of the connection will maintain its own sequence number tables for data being sent and received across the connection. They will always be in ascending order.

Sequence numbers do not have to, and probably will not, start at 0. But it is fundamentally important to note that they will wrap to 0.

TCP Segment

The unit of transmission between two stations speaking TCP is called the *segment*. Segments are used to establish a connection, send and receive data and acknowledgments, advertise window sizes, and close a connection. A TCP segment will contain the TCP header (shown in Fig. 11-8) and its data. The data handed to TCP for transmission are known as a *stream*—more specifically, an *unstructured stream*. A stream is a flow of bytes of data, and an unstructured stream is a flow of data bytes of an unknown type. This means that TCP has no way of marking the data to indicate the ending of a record or the type of data that is in the

Figure 11-8
TCP segment.

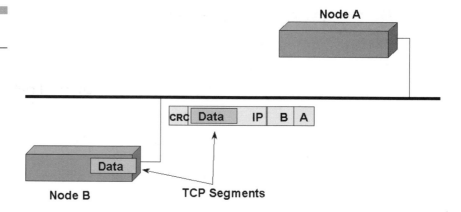

stream. When TCP receives a data stream from the application, it will divide the data into segments for transmission to the remote network station. A segment can have control or data information. It is simply an unstructured stream of data bytes sent to a destination.

A TCP segment may be as long as 65535 bytes (or longer in IPv6), but it is usually much shorter than that. Ethernet can handle only 1500 bytes of data in the data field of the Ethernet packet (Ethernet V2.0; 1496 bytes for IEEE 802.3 using IEEE 802.2). On the other hand, FDDI can handle a maximum of 4472 bytes of data in a packet, and Token Ring packet size varies depending on the speed—for 4 Mbps the maximum size is 4472 bytes, and for 16 Mbps the maximum size of the packet is 17800 bytes—but is usually set to 4472 bytes. To negotiate a segment size, TCP will use one of the Options fields in the TCP header to indicate the largest segment size it can receive (the MSS option), and submit this packet to the remote network station.

TCP does not care what the data are. Data in a TCP segment are considered a stream, which is constructed at the sender and sent to the receiver. The receiver reconstructs this stream from the variable segments that it receives. Once the connection is established, TCP's main job is to maintain the connection (or multiple connections if there are more than one). This is accomplished through the sequence numbers, acknowledgments and retransmissions, flow control, and window management.

Once the connection between stations A and B is established (by way of a successful three-way handshake), TCP must manage the connection. The first of the management techniques to be discussed is sequence numbers.

Sequence Numbers and Acknowledgments

A point should be made up front. Acknowledgments do not simply refer to a datagram or a TCP segment. TCP's job is to reconstruct a piece of data that was transmitted by the sender. Therefore, the acknowledgment number actually refers to the position in the stream of data being sent. Why? Because, IP is connectionless, and retransmissions may contain a different size than the original. The receiver collects information and reconstructs an exact copy of the data being sent.

Segments may also arrive out of order. It is TCP's job to place them back in the order in which they were sent. However, error may occur during this process, and TCP will only ACK the longest contiguous prefix of the stream that has been received correctly.

TCP calculates a sequence number for each byte of data in the segment taken as a sum. For each *byte* of data that is to be transmitted, the sequence number *increments by 1*. Let's say a connection was made between stations A and B as shown in Fig. 11-6. Station A sends a segment to station B with a sequence number of 40 and knows that the segment contained 4 bytes, so it will increment its sequence number to 44. On acknowledgment from station B (containing the ACK number of 44), station A will then transmit the second segment to station B, knowing that 7 bytes are being transmitted. Station A's sequence number increments to 44 + 7 = 51, and Station A will wait for an acknowledgment from station B.

Each datagram will contain as many bytes as will fit; the number is indicated by the transmission window sent by the destination (windows will be discussed shortly). The sequence number is set to the number of the first byte in the datagram being sent, which is the last received ACK number from the destination. The TCP segment (the data) is then given to IP for delivery to the network.

Each segment transmitted must be acknowledged. Multiple datagrams may be sent with one acknowledgment to all received good segments. This is called an *inclusive* or *cumulative* ACK. TCP accomplishes this bidirectionally across the same connection. Each datagram transmitted will have the TCP header ACK bit set. With the ACK bit set, TCP will read the acknowledgment field to find the next byte number of the segment that the other end of the connection expects. In other words, the number in the ACK field equals the sequence number of the original segment transmitted plus the number of the bytes successfully received in that segment plus 1. To make TCP more efficient, the ACK number is stuffed into a data datagram. Usually, there is no separate datagram on the network used just for ACK packets. All data bytes up to but not including the ACK number are considered good and accepted by the receiver.

Since TCP is a byte-oriented transport protocol, sequencing and acknowledgments are accomplished for each byte of TCP data. To ensure the integrity of the data, TCP had to become a robust protocol that took every byte and ensured that it would make it across to the destination. Local Area Network protocols such as Novell NetWare and Xerox XNS were developed to work on high-reliability mediums (shielded copper cable in controlled environments). Their sequence numbers are based not on bytes in their data segment but on the number of packets. TCP ACKs bytes of data, not packets, datagrams, or segments.

EXAMPLE

As shown in Fig. 11-9, connection has been established using an initial sequence number from the sender and an initial sequence number supplied by the receiver (the destination). Each side will maintain its own sequence number, which may be in the range of 0 to 2,147,483,647. Each side of a TCP connection knows the upper and lower limits of the sequence numbers, and once the limit has been hit, it will roll over to 0 (each side knows to include 0). The initialization sequence numbers are selected at random. Each side must ACK each other's received datagrams.

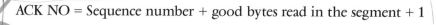

ACK NO = Sequence number + good bytes read in the segment + 1

This is a clean, fast, efficient way of determining which bytes were successfully received and which ones were not. The sender of data must retain a copy of transmitted data until it receives an acknowledgment for those bytes from the remote network station of a connection.

Acknowledgment packets are not necessarily separate packets with only the acknowledgment number in the packet; that would be inefficient. For example, if station A opened a connection to station B and station A and station B were sending data to each other, the ACK datagram can be combined with the response data packet. In other words, one

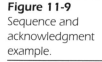

Figure 11-9
Sequence and
acknowledgment
example.

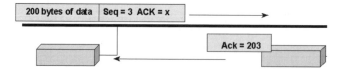

| 200 bytes of data | Seq = 3 ACK = x |

Ack = 203

datagram transmitted contains three things: the data from station B to station A, the acknowledgment from station B of the data previously sent from station A, and the sequence number for the data B is sending to A.

If the sender does not receive an acknowledgment within a specified time, it will resend the data starting from the first unacknowledged byte. TCP will time-out after a number of unsuccessful number of retransmissions. The retransmission of a datagram is accomplished using the Go-back-to-N routine. Any number of outstanding bytes may be not acknowledged. When the destination station does acknowledge the receipt of a series of bytes, the source will look at the ACK number. All bytes with sequence numbers up to but not including the ACK number will be considered received in good condition. This means that if a source station starts the sequence number with 3 and then sends two datagrams containing 100 TCP segment bytes each, then, when it receives an ACK from the destination of 203 (3 to 102 and then 102 to 202, leaving the ACK at 203 as the next expected byte), it will know that the data in both of the datagrams it sent previously are considered received in good condition.

The number of outstanding packets allowed is the next topic of discussion.

TCP Flow and Window Management

Two functions are required of TCP in order for the protocol to manage the data over a connection: *flow control* and *transmission control*. Do not confuse these functions with the ICMP source quench mechanism. Source quench is used by a host to inform the source of transmissions and that the host buffers are full and that the host would like the sender to slow its rate of transmission.

For those readers who do not understand flow control, it is a mechanism used to control the flow of data. For example, if data are being received at a destination station faster than that station can accept them, it needs to tell the source station to slow down or to stop completely until it can clear out some space (replenish buffers) to receive the data.

How many segments may be outstanding at any one time? Data management is accomplished using a "window," as shown by the following. Data for TCP to transmit to the remote network station will be accepted by TCP from an application. These data will be placed sequentially in memory, where it will wait to be sent across a connection by IP to a remote station. TCP places a "window" over these data, in which the data are classified as data sent and acknowledged, data sent but not acknowledged, and data waiting to be sent. This is called a *sliding window,* because the window will slide up the data segment as each data packet is sent and that packet acknowledged.

Refer to Fig. 11-10. Sequence numbers 100 to 104 have been transmitted to the destination station and the destination station has acknowledged receipt of these segments. Packets containing sequence numbers 105 to 108 have been transmitted by the source station, but it has not received an acknowledgment for those data. Segments containing sequence numbers 109 to 114 are still in the source station and are waiting to be sent. Packets containing 115 to 117 are not yet in the window. (The actual segment size is usually 512 or 536 bytes each, but for clarity a much smaller size is shown in Fig. 11-10.)

The important thing to notice in the figure is the black box covering the segments. This represents the window. It will constantly move in ascending sequence order upon receipt of acknowledgments from the destination station.

When the receiving station (the destination station) is running low on buffer space (an area of memory to store incoming data), the receiver of this data can tell the sender to slow its transmission rate by reducing the window size, indicated in the Window field in the TCP header of an acknowledgment. This field will contain the number of bytes (by indicating the range of sequence numbers) that the destination station is willing to accept. Figure 8.18 shows the TCP header, specifically the window field.

When the remote network station cannot accept any more data, it may set this window field to 0. That is, the sender can send data to a host, and if the host cannot accept the data, it should respond with the ACK set to the previous ACK sent and a window set to 0. It will continue to submit these zero packets until it can accept data again. When buffer space is freed up, it can again submit a packet with the window

Figure 11-10
TCP flow and window management.

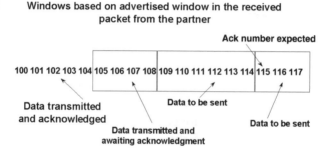

Windows based on advertised window in the received packet from the partner

Ack number expected

100 101 102 103 104 | 105 106 107 108 | 109 110 111 112 113 114 | 115 116 117

Data transmitted and acknowledged

Data to be sent

Data transmitted and awaiting acknowledgment

Data to be sent

size set to a nonzero number to indicate that it can again accept data. However, if the URG bit is set, this indicates to the receiver that the sender has urgent data waiting to send.

This connection management technique allows TCP to maintain control of data transfer over a connection by informing TCP on the sending side how much data the receiver is willing to accept. This enables a sender and receiver of data on a connection to maintain consistent data flow over the connection.

TCP Retransmission

One last function to discuss is TCP's capability to know when to send a retransmission of a packet. This is a fairly complex subject that will be discussed only briefly here. Since data runs on an internet that has varying delays caused by routers and low-speed or high-speed networks, it is nearly impossible to determine an exact timer delay for an acknowledgment. The acknowledge could show up one time in a matter of milliseconds and at another time in a matter of seconds, because of the heterogeneous nature of an internet. TCP accommodates this varying delay by using an adaptive retransmission algorithm. This allotted time is dynamic (not held to one number) and is accomplished as follows: When TCP submits a packet to be sent, it records the time of the transmission and the sequence number of the segment. When TCP receives an acknowledgment to that segment, it will again record the time. Using the difference between the two times, TCP builds a sample round-trip delay time. TCP uses this time to build an average time for a packet to be sent and an acknowledgment to be received. When it calculates a new value from another sample, it will change its timer delay for the waiting of the ACK packet. It does not change this timer abruptly but slowly.

Slow Start and Congestion Avoidance

Another feature of TCP is the slow start and congestion avoidance mechanisms. Original versions of TCP would start a connection with the sender transmitting multiple segments into the network, up to the window size advertised by the receiver (set up during the three-way handshake). Local subnets were not affected as much as communication between subnets. If there were routers and slower links between the sender and the receiver, problems could arise; a huge amount of data could possibly be sent at startup. The method to avoid this problem is slow start. Secondly, when congestion occurs, TCP must slow down its transmission rate of packets into the network, and then invoke slow start, congestion control, or both to get things going again. In practice, they are implemented together.

Slow start operates by observing that the rate at which new packets should be transmitted on the network is the rate at which the acknowledgments are returned by the other end. Slow start adds another window to the sender's TCP: the *congestion window*, which is not advertised in the TCP header, but assumed (i.e., kept local). When a new connection is established with a local or remote host, the congestion window size is initialized to one segment. Segment sizes are variable depending on the computer or LAN type used, but the default, typically 536 (yes, 536 for it is the segment size [TCP layer] that we are working with) or 512 could be used. Each time an ACK is received, the congestion window is increased by one segment. The sender can transmit up to the minimum of the congestion window and the advertised window. The distinction here is that the congestion window is flow control imposed *by the sender*, while the advertised window is flow control imposed *by the receiver*.

The sender starts by transmitting one segment and waiting for its ACK. When that ACK is received, the congestion window is incremented from one to two, and two segments can be sent. When both of those two segments are acknowledged, the congestion window is increased to four. This rate continues to increase exponentially until the TCP advertised window size is met. Once the congestion window size equals the advertised window size, segments are continually transferred between stations using the window size for congestion control on the workstations (just as if slow start had never been invoked).

However, upon congestion (as indicated by duplicate ACKs or timeouts), the slow start algorithm kicks back in, but also starts up another algorithm known as *congestion control*. When congestion occurs, a comparison is made between the congestion window size and the TCP advertised window size. The smaller number is then halved and saved in a variable known as the *slow start threshold*. This value must be at least 2 segments, unless the congestion was a timeout, in which case the congestion window is set to 1 (slow start). The TCP sender then can either start up slow start or congestion avoidance. Once ACKs are received, the congestion window is increased. Once the congestion window matches the value saved in the slow start threshold, the slow start algorithm stops and the congestion avoidance algorithm starts. This algorithm allows for a more linear growth in transmission, because it multiples the segment size by itself (by 2) and divides this by the congestion window. The rate is continually increased based on this algorithm each time an ACK is received. This allows for growth on the TCP connection but at a more controlled rate (linear rather than exponential rate as in the slow start algorithm).

The effects of introducing these algorithms on the Internet were dramatic. All versions of TCP/IP software now include these algorithms, and their effects are not only based on remote connections; these algorithms are placed into action between two stations on a local subnet as well.

Differentiating Connections through Port Numbers

As shown in Fig. 11-11, TCP, like the UDP transport protocol, uses ports to enable access to an application program. The port numbers have the same functions as the UDP port numbers: They are used to identify an application. These assigned and registered port numbers are contained in the same RFC that references all TCP/IP numbers. As of this writing it is RFC 1700, although the FTP site is kept up to date. When a connection is established between two network stations, the originating station must be able to tell the receiving station which application it would like to use. Likewise, the originating station must indicate to the receiving station where to send data back. As shown in the figure, ports are assigned to allow this communication.

Think of two personal computers with connections established to a host computer. They each indicate to the host station that they want to

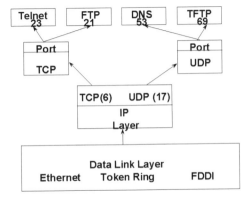

Figure 11-11
Differentiating connections through port numbers.

24

establish Telnet communications with it. The TCP/IP software understands this request because the incoming connection packet states that the port is number 23, the well-known port assigned to the Telnet application. The personal computer's port assignments (source port indicated in the TCP header) were chosen at random by the TCP software running in the personal computer. As with UDP, any random port assignment above 1023 can be issued by the personal computer's TCP software. The field for port number in the TCP header is 16 bits wide, up to 65,535 ports (minus the first 1024 assigned ports) may be dynamically assigned.

As stated before, the combination of an internet address and a port is known as a *socket*. Port numbers indicate only the end connection point (application program identifier) of the connection. When a port number is used with the internet address of a network station, it becomes the socket number.

Another thing to notice in this scenario is that the host Telnet server has accepted two Telnet connections. Both personal computers gained access to the host computer using destination port 23. What makes each connection unique in this case is the dynamic port assignment on the personal computer. Each Telnet connection initialized on the personal computer uses a different source port number.

Each *active* call on a TCP/IP internet will assign its own port numbers above 1024. The host has two Telnet connections connected to port 23 on the host computer. The host computer will spawn each connection as a process to the operating system. Each connection will be uniquely identified by the socket address. The host computer will then return to waiting for any incoming connections, depending on how many *passive* OPEN calls it has allowed. (See Table 11-1.)

TABLE 11-1 TCP Connections	**Connection**	**Connection State**	**Local Address**	**Local Port**	**Remote Address**	**Remote Port**
	1	Connected	192.1.1.1	69	192.1.1.2	33000
	2					
	3					
	x					

Connection State. This describes the current status of the connection. This could be in the listen mode, in the process of termination, closed, etc.

Local Address. This contains the local IP address for the TCP connection. If this connection is in the listen state, it should be filled in with 0.0.0.0.

Local Port. This contains the local port number for the connection.

Remote Address. This contains the remote IP address for the connection.

Remote Port. This contains the remote port number for the connection.

Termination

Finally, TCP must be able to terminate a connection gracefully. This is accomplished using the FIN bit in the TCP header. TCP offers a full-duplex connection, so each side of the connection must close the connection. Refer to Fig. 11-12 for two communicating devices, end station A and host station B. The application running on end station A indicates to host B that it wishes to close a connection by sending a packet to host station B with the FIN bit set. Host station B will acknowledge that packet and will now no longer accept data from end station A. Host station B will accept data from its application to send to end station A, though, and end station A will continue to accept data from host station B. This way, station A can, at a minimum, accept a FIN packet from host station B to close the connection completely. To finalize the closing of this connection, host station B will send a packet to end station A with the FIN bit set. End station A will ACK this packet, and the connection will be closed. If no ACK is received, FINs will be retransmitted and will eventually time out if there is no response.

Figure 11-12
Termination.

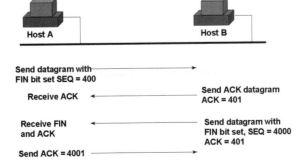

Host A   Host B

Send datagram with FIN bit set SEQ = 400

Send ACK datagram ACK = 401

Receive ACK

Receive FIN and ACK

Send datagram with FIN bit set, SEQ = 4000 ACK = 401

Send ACK = 4001

Real-Time Protocol and the Real-Time Control Protocol

Another protocol that is placed at the transport level (IPv6) and as an application of UDP (IPv4) is Real-Time Protocol (RTP). It used to be that data were moved over the Internet in order for people simply to communicate with one another. However, moving data such as a file or an email is relatively simple act and it is very forgiving. Other applications, such as viewing a video clip as it is being downloaded to a network station, require special attention. This type of data movement is not very forgiving. Dropped packets cause faded pictures and jerky or intermittent audio.

Examples of multimedia are the transmission of corporate messages to employees, video and audio conferencing for remote meetings and telecommuting, live transmission of multimedia training, communication of stock quotes to brokers, updates on the latest election results, collaborative computing with times such as electronic whiteboarding, transmission of live TV or radio news and entertainment programs over networks, and many others.

Data transfer, whether it is text or voice and video, can be classified as real-time and non-real-time. Multimedia can include both real-time and non-real-time data. Real time is the ability to see and hear the event while it happens as the data representing it are transferred over a network. For example, viewing a video clip as it is downloaded to a network station is classified as real-time. A company executive's speech captured by a camera and distributed to thousands of desktops through IP video servers for immediate viewing consists of real-time data. Real-time applications have specific requirements, as you will see in a moment.

Non-real-time data transfer allows an event to be viewed later. Compared to real-time data transfer, time is not a consideration. For example, you can download a multimedia file and view it at your leisure. Other non-real-time examples are Web browsers. It may take a few seconds or minutes to receive a Web page, but once all the data are received, the page is accurate, not fuzzy or missing pieces.

The Real-Time Transport Protocol (RTP) provides end-to-end data delivery for real-time (time-sensitive) applications such as those required by transporting voice and video. It accomplishes this through payload type identification, sequence numbering, timestamping, and delivery monitoring. It does not provide any quality-of-service guarantees.

Because real-time data needs special treatment, protocols were developed to handle it. An early attempt to move real-time data across IP networks was adopted as the Streams 2 (ST-2) protocol (RFC 1819). Also known as IP version 5, it was the IP replacement for streaming data. IPv4 handled delivery of non-real-time data, whereas ST-2 was the IP protocol to handle real-time streaming. It included the ability to do multicast, transport, and quality of service in one protocol. However, the ability of ST-2 to scale was limited by its requirement of static (manual) binding to end node addresses. Besides, the user community wanted the ability to do both bursty data and streaming data over a common IP layer. Therefore, the IETF working groups came out with multiple protocols: RSVP, multicast support, and the new streaming protocol, RTP.

However, RTP is more of an architecture than a protocol, and, as stated in the RFC, "RTP is a protocol framework that is deliberately not complete." The RFC includes descriptions of those functions that should be common across applications that develop toward RTP. It provides a framework for a protocol in that it defines the roles, operations, and message formats. Applications written toward RTP usually incorporate the functions of RTP into their application, thereby adding to the RTP.

RTP follows the architecture known as application layer framing (ALF), which allows for a more cooperative relationship between protocol layers than a strict adherence to them. RTP is considered to be adaptable and is often implemented in the application rather than a separate protocol, such as TCP. RTP replaces the TCP layer for applications and in most cases, RTP works with the UDP layer for socket addresses (multiplexing) and checksum capability. RTP works in conjunction with a control protocol known as the Real-Time Control Protocol (RTCP). A port number of 5004 has been registered for the RTP and 5005 for RTCP. The source and destination port are the same for the sender and the receiver in RTP.

From the preceding text, your mind has probably wandered, and now you are thinking about multimedia and multicast. This is the right way of thinking, in that audio and video are generally used in group receivers. RTP is designed for multicast operation. This is obvious when one sender transmits to many receivers, but it also applies to the control protocol that is used for feedback of control information.

Feedback is not simply delivered to the sender. If feedback is transmitted in multicast, all participants in the multicast receive this feedback information. Feedback is sent by a receiver of the multicast, but the feedback information is sent with an IP multicast destination. This allows all those involved in the multicast group to receive the feedback information and process it. This also allows all receivers to determine whether they are the only ones having problems or whether other receivers are having problems as well.

Translators

The RTP protocol is open, allowing many different encoding schemes to be used with this protocol. This provides many advantages, because some protocol schemes are better used in different topologies than others. To compensate for differences in encoding schemes and in transmission and reception rates, the concept of mixers and translators are used. Translators and mixers reside in the middle of a multimedia network as network attachments, like any other attachment. Their application makes them reside logically between senders and recipients, and they process the RTP information as they receive it and then retransmit the information.

The translator functions are the easiest to explain, so we will start with the translator. As shown in Fig. 11-13, a translator simply translates from one payload format to another. Take the example of a network station whose user would like to participate in a stream but that is located beyond a WAN link that provides very little bandwidth. The high-speed workstations could simply reduce their capabilities to provide for the low-bandwidth link, but with the translator they do not have to. The translator can simply receive the high-bandwidth signal and translate it to a low-bandwidth signal for the remote network station. In this way, receivers with high-quality links can continue using them, while other receivers with low-bandwidth links may participate as well through the translator.

Mixers

Mixers perform a vital service as well. Instead of taking a source stream and translating it to another type of stream, mixers combine multiple source streams into one single stream and preserve the original format. For example, if you were having an audio conference between four network stations, and another user at a network station over a low-speed

Figure 11-13
Translators.

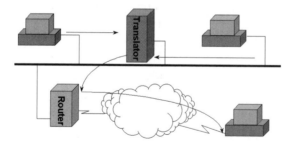

■■■■ ■■■■ ■■■■ ■

Figure 11-14
Mixer.

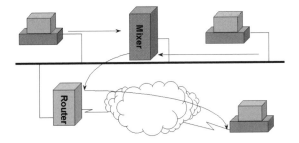

link would like to join, the mixer would simply pull the three network stations into one single stream instead of three for communication to the network station over the low-speed link. As shown in Fig. 11-14, the mixer accepts equal streams from all sources and combines them into one stream of the same speed as the ones it collected. If a digital audio conversation is being carried on between four workstations, each consuming 64 kbps for its own use, then when the fifth network station wants to join the conversation, but its link is only 64 kbps, the mixer combines all four higher-speed audio signals into one 64 kbps stream, allowing the network station over the low-speed link to join the conversation and allowing the other network stations to maintain their high-speed and probably high-quality audio links.

Basically, mixers and translators allow for variances to occur in an multimedia stream. Whether for translating streams from one format to another or for allowing mixing of signals to accommodate differences, these two types of devices are very much a part of the RTP protocol.

Support for Time-Sensitive Applications

RTP supports sequencing and time stamps, synchronizes different streams (audio or video), and also incorporates information describing the payload type. That information allows RTP to support multiple compression types, such as MPEG and H.261. In fact, an RTP receiver can receive information that is encoded by two different methods, and it can produce one single stream from this. This process, known as mixing, is explained later.

A digitized audio signal may be produced using a simplex encoding scheme known as pulse code modulation (PCM; see Fig. 11-15).

■ RTP has the ability to replay data as they were originally transmitted. This is important for real-time applications in that the video and audio must be replayed as it was sent.

Figure 11-15
Support for time-
sensitive apps.

**Transmitted with
out respect to time**

160 byte packets
every 20 ms

Retimed at
destination

- TCP/IP is not time sensitive when rebuilding the original data at the remote end.

- RTP supports sequencing, time stamps, etc., to synchronize different streams.

- The datagram includes the payload type which identifies the encoding, such as MPEG or H.261.

Say, for example, that PCM builds 160-byte packets every 20 milliseconds for a sampled voice stream. This information is transmitted through an internet using IP. The digitizing of the voice signal is very sensitive to time. The received stream of voice must be put back into the original timing as it was transmitted; otherwise there will be an uneven flow for voice at the receiver, and it will not be received as it was spoken at the originator. IP does not care about timing, sequencing, or delays—it only has to deliver the data. IP will probably deliver these packets at different times and may deliver them in varying order. Therefore, an RTP application must put the packets back in the original order and reapply the timing between the packets. RTP provides information on this but does not accomplish this directly. Time stamps, which mark the relative beginning of the event, are provided with the packet, and they provide enough information to the application to rebuild the original audio stream.

Payload Type

The payload type may be one of the following:

0   PCMU audio

1   1016 audio

2   G721 audio

3   GSM audio

4   Unassigned audio

5   DVI4 audio (4 kHz)

6	DBI4 audio (16 kHz)
7	LPC audio
8	PCMA audio
9	G722 audio
10	L16 audio (stereo)
11	L16 audio (mono)
12	TPS0 audio
13	VSC audio
14	MPA audio
15	G728 audio
16–22	Unassigned audio
23	RGB8 video
24	HDCC video
25	CelB video
26	JPEG video
27	CUSM video
28	nv video
29	PicW video
30	CPV video
31	H261 video
32	MPV video
33	MP2T video
34–71	Unassigned video
72–76	Reserved
77–95	Unassigned
96–127	Dynamic

Not that you would care about all of these payload types, but there are a few that you should recognize.

The RTP Header Fields

Refer to Fig. 11-16. The Vers field consists of two bits that indicate the version, which as of this writing is 2 (Version 1 was the draft spec, and 0 was used for the public domain Visual Audio Tool [VAT]). The P bit indi-

Figure 11-16
The RTP fields.

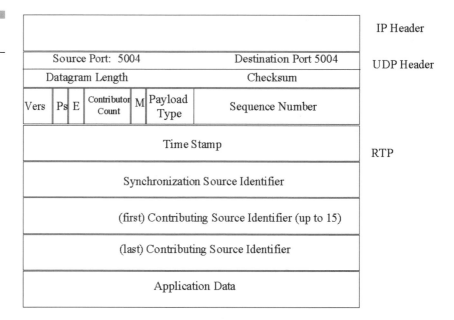

cates that padding is used at the end of the datagram and the last byte indicates how many bytes of padding. The E bit is the extension bit, which indicates that the RTP fixed header is followed by an extension header. The extension mechanism is provided to allow individual implementations to experiment with new payload-format-independent functions that require additional information to be carried in the RTP data packet header. No extensions have yet been defined. The next field, the contributor count and indicates the number of contributing source identifiers the message contains. The field is four bits wide, allowing for 15 contributors; it is used in the mixing operation. The next field (M) is the marker field and is used by applications to mark information in the data they are transmitting; for example, it could be set to 1 to indicate that this is the last message of a stream. The marker bit is followed by the seven-bit payload type, as described in the previous section. Next, the sequence number is 16 bits long and increments by 1 for each message sent. As in TCP, the sequence can be started anywhere within a 16-bit range. The time stamp indicates a number reflective of the instant of the transmission of the first byte in the RTP data packet. Time stamps are used to restore the exact timing as it was sent from the source. Several messages may have the same time stamp, which could indicate that they were sent at the same time and belong to the same video frame.

The Synchronization Source (SSRC) field is a 32-bit number that identifies the originator of the message that inserted the sequence number and the time stamp for the data, so as not to be dependent on the IP address. Suppose, as shown in Fig. 11-14, that there are two sources of audio data. Each source will have its SSRC identifier for the packets that it sends. The selection of the identifier is beyond the scope of this book, but a random number is generated for this field by the source, allowing each source to be unique; if two sources do select the same identifier, RTP has the mechanisms to detect and correct this. This field could indicate an alternate source if the received message was originated by a mixer. If two packets enter a mixer, the mixer inserts its 32-bit number as the source and push the previous SSRC numbers into the Contributing Source Identifier.

The Contributing Source Identifier fields indicates a source or sources (the original Synchronization Source identifiers) of a stream of RTP packets that were involved in the combined stream produced by a mixer.

Providing Control for RTP

RTCP (Real-Time Control Protocol) is the control mechanism for RTP; it provides information, in the form of feedback, on network conditions and reception quality. Using RTCP allows RTP applications to adapt to variances in the media. For example, a router carrying the stream could become overloaded and slow down the forwarding of packets. Another application on the network may be using considerable bandwidth so that the receivers of RTP cannot receive as many packets as fast as they want to. RTCP enables control information to be distributed not only to the server but also to the receivers. This allows receivers and senders to make their own decisions about the quality. Another feature of RTCP is the gathering of user information. RTCP reports on users that are attending a "session."

RTP uses the same format for all of its messages. RTP can work alone but usually does not. RTP relies on RTCP to carry control information. See Fig. 11-17.

- RTP carries only RTP data.
- Control is provided through a separate protocol known as real-time control protocol (RTCP).
- RTCP defines five message types
  1. Sender Report: transmission and reception statistics from participants that are active senders

Figure 11-17
Providing control for
RTP.

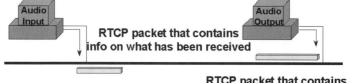

2. Receiver Report: reception statistics from participants that are not active senders
3. Source Description: Source description items, including CNAME
4. Bye: indicates end of participation
5. Application Specific: application specific issues

Sender and Receiver Reports

Reports under RTCP allow senders and receivers to communicate with each other for quality reasons. Each sender sends a report giving statistics for its transmissions. Receivers consume these data and also send out reports to indicate how well they are receiving data. The senders use these data to tune their transmissions.

RTP receivers may send two types of reports. These are the Sender and Receiver reports (SR and RR, respectively). The type of report sent depends on whether the receiver is also a sender. RTP receivers provide reception quality feedback using the RTCP receiver reports, whereas the sender report tells receivers what they should have received. In both types of report, the V field in the header is for the version number, which should be set to 2. The R Cnt is the receiver block count, which contains the number of receiver blocks in the message. The next field is the packet type. Next, the length field indicates the length of the packet in bytes.

As you will see, both types of reports are very similar. The only difference between the two reports, besides the packet type, is the 20-byte sender information section in the sender report. The Sender Report is issued to let the receiver know what it should have received. Both the SR and RR can include from zero to 31 blocks (notice the blocks are simply replicated). Each block is for each of the synchronization sources from which the receiver has received RTP data packets since its last report.

Refer to Fig. 11-18. For the sender report (packet type 200), the SSRC field ties this sender report to any RTP data packets the source may have sent. The NTP time stamp is the actual time of day. It is used as a control

Figure 11-18
RTP sender report.

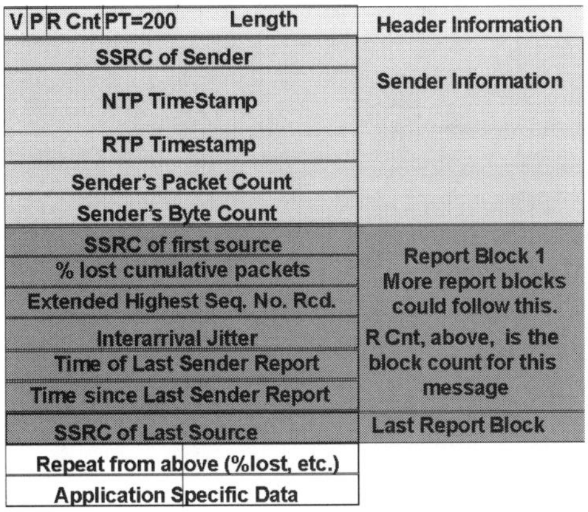

to measure the time difference delta for time stamps extracted from reception reports. This allows an estimate to be made as to the round-trip propagation delay to those receivers. The RTP time stamp allows receivers to put this message into order relative to the RTP packets. The last fields indicate the number of packets and bytes the sender has transmitted.

The next sections are receiver blocks. In the sender report, they allow the sender to report its transmitted data but also any RTP data that it has received. Each block represents one remote source. The block indicates the fraction of packets from that source that were lost since the last report and the cumulative total number of packets lost since inception. The extended highest sequence number received is the highest sequence number from that source. The interarrival jitter field allows the receiver to estimate the variance of the source's interarrival times. A high value indicates that this receiver is receiving a stream of packets irregularly. The last two fields indicate when the last report from this source arrived.

Receiver Reports

Refer to Fig. 11-19. The receiver report is basically the same as the sender report, except the sender information (NTP and RTP time stamps, sender packet count, and sender byte count) was stripped out. The packet type field is set to 201.

Source Description Packet

Refer to Fig. 11-20. The source description packet (SSDP) is used to provide more information about the source. The RTCP header is used, and

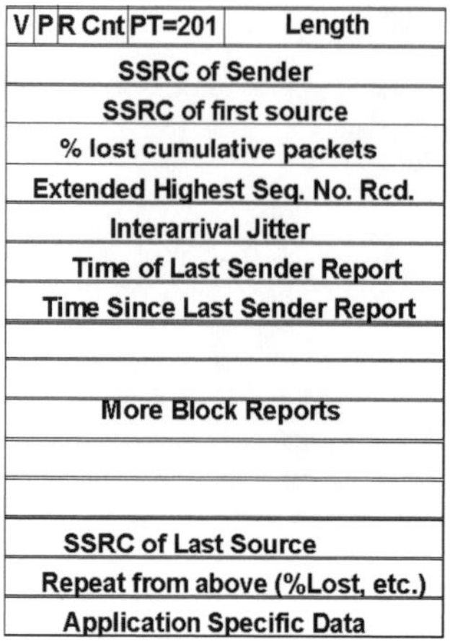

Figure 11-19

RTCP receiver report.

V	P	R Cnt	PT=201	Length
			SSRC of Sender	
			SSRC of first source	
			% lost cumulative packets	
			Extended Highest Seq. No. Rcd.	
			Interarrival Jitter	
			Time of Last Sender Report	
			Time Since Last Sender Report	
			More Block Reports	
			SSRC of Last Source	
			Repeat from above (%Lost, etc.)	
			Application Specific Data	

Figure 11-19

RTCP receiver report.

V	P	R Cnt	PT=202	Length
			SSRC or CSRC of first source	
			SDES Items	
			SSRC or CSRC of second source	
			SDES Items	
			SSRC or CSRC of last source	
			SDES Items	

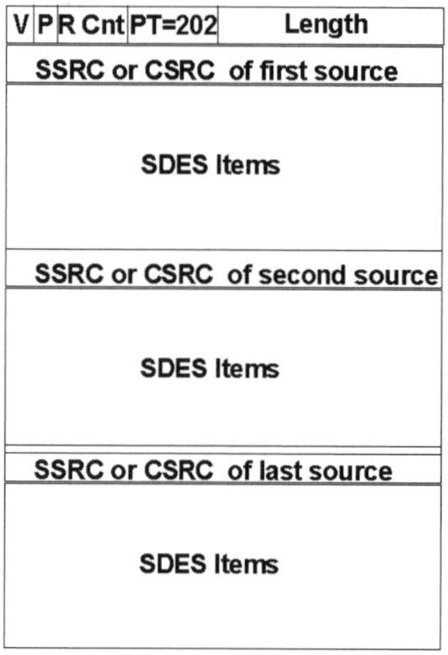

Figure 11-20

Source description
packet.

the packet type is set to 202. The first field contains the SSRC or CSRC identifier of the source. Applications are free to put their own items in as well.

Bye Packet

Refer to Fig. 11-21. For a source to leave a conference, it uses the Bye packet. Eventually, all participants in the conference would notice that the source is missing, but this message allows this to be quickly learned.

Figure 11-21
Bye message.

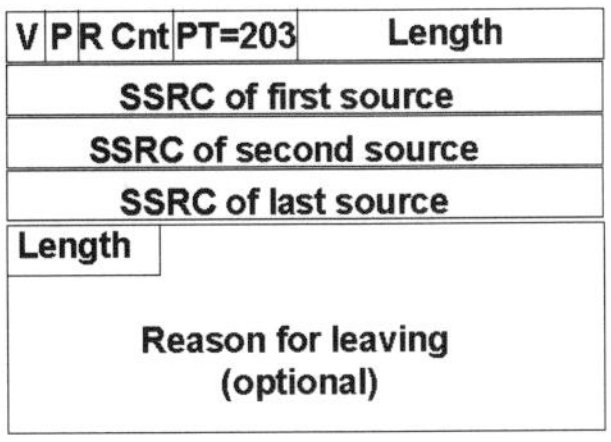

V	P	R Cnt	PT=203	Length

SSRC of first source
SSRC of second source
SSRC of last source
Length

Reason for leaving
(optional)

Included in the packet is a field that allows the source to identify the reason that it is leaving. This field is optional.

Application-Specific Messages

This type of message allows for experimentation of the RTCP protocol. It allows an application developer to place its own messages in a packet to be used by a receiver or sender of its application. These messages are another example of how this protocol is really not finished and never will be, allowing for uniqueness and extensibility.

Caveats

RTP does not provide any quality-of-service (QoS) parameters such as those that allow for timely delivery. RTP does not guarantee delivery of out-of-sequence packets. It provides sequencing, but only for the transport sequencing of placing the order of packets in operations like decoding; however, the packets can be decoded out of receiving sequence. Like most other TCP/IP protocols, a protocol provides a unique function and not two functions. RTP expects other TCP/IP protocols to provide QoS services.

RFCs and Web Sources

To see what is playing on the Mbone, use *http://www.precept.com/cgi-bin/iptv/iptvmain.pl* or *http://www.cilea.it/collabora/MBone/agenda.html*.
RTP and RTCP are contained in RFC 1889.

CHAPTER 12

Selected TCP/IP Applications

Introduction

This chapter gives an introduction to TELNET, FTP, TFTP, SMTP (including POP), and DNS. Many applications written for the TCP/IP environment exist today. There are word-processing systems, CAD/CAM systems, mail systems, and so on. The most common applications that run on the TCP/IP network system are

1. Telnet (remote terminal emulation)
2. File Transfer Protocol (FTP)
3. Trivial File Transfer Protocol (TFTP)
4. Simple Mail Transfer Protocol (SMTP)
5. Post Office Protocol (POP)
6. Domain Name Service (DNS)

These applications are fully documented in the RFCs and almost always are delivered with any TCP/IP protocol suite in the market today. This means that you can switch to almost any type of computer using TCP applications software, and the commands and functions of these programs will be the same.

The applications were written specifically for TCP/IP and basically provide almost all the applications that users need to access any network. Database programs, word-processing programs, and so forth are all viable programs but are not pertinent to the operation of a TCP/IP network. Using these applications, a user can find any other needed application on the Internet. The ones listed previously are the bare minimum needed to create a networked user environment in which all users can actively communicate and share data across the network.

NOTE: *One nice thing about these available network applications is that they run on TCP/IP no matter which operating system is being used. The commands, their connection techniques, the commands that control the application, and the interface to the user almost always will be the same. So if you normally work with UNIX and then switch for a day to DOS, the same FTP commands that worked on the UNIX machine will be there in the DOS machine. It is hard to say that for most applications running on any system today.*

Telnet

The Telnet connection simply allows a terminal service to operate over the TCP/IP network as if the terminal were directly connected. Remember that computers and terminals were connected by a cable, and the terminals were directly attached to the host computer. Telnet provides a terminal service for the network. It allows any terminal to attach to any computer over the network. It can emulate many different types of terminals, depending on the manufacturer of the Telnet program. There are Telnet programs that emulate DEC VTxxx series terminals, IBM 3270 and 5250 terminals, etc.

The advantage to the Telnet program is that, with access to this program, a user may log on to any host on the TCP/IP internet (provided that security options are allowed). Sessions are set up over the TCP/IP network. Figure 12-1 shows a typical Telnet connection on a TCP/IP network.

The Telnet protocol uses TCP as its transport protocol. The user starts the Telnet protocol at his or her workstation, usually by typing TELNET <*domain name or IP address*>. The Telnet application may be started with or without an argument. The argument invokes a simpler procedure in which the Telnet process automatically tries to connect to the host signified by the argument statement. The Telnet application attempts to establish a connection to the remote device by accessing the services of the domain name server or directly using the IP address. If an argument is not supplied, Telnet waits for the user to open a connection using the DNS or an IP address.

Figure 12-1
Telnet.

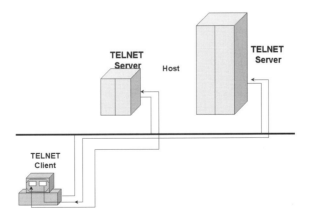

Telnet Options

The Telnet program is extensible through the use of options. Each side of the connection tells the remote end of the connection which of these options it can support and which ones the other end should support. This provides for symmetry.

The Telnet protocol was written so that it can work on a variety of operating systems. Therefore, before a connection is made to the remote device, the Telnet protocol must do some work to synchronize the connection with the remote device. For example, the DOS operating system for personal computers requires the use of a CR-LF (carriage return–line feed) to terminate a line of text. Systems such as UNIX require a line of text to be terminated with a LF. Another example concerns the echoing of characters. Upon the connection attempt, the Telnet protocol negotiates with the remote device as to which end will do the echoing of typed characters to the initiator of a connection. During the connection attempt between a source and destination station, the two stations will communicate their options, which indicate how each end of the connection will respond over the Telnet connection. These options include

1. The ability to change from seven-bit text to eight-bit binary

2. Allowing one side or the other to echo characters

3. Specifying a terminal type

4. Requesting the status of a Telnet option from the remote connection

5. Setting a timing mark to synchronize the two ends of a connection

6. The ability to terminate a record with an EOR code

7. Setting line mode so that strings of characters may be sent as opposed to character-at-a-time transmission

8. Stopping the go-ahead signal after data

The options are negotiated between the two network stations in the following manner:

Request	Response
WILL <*option*>	DO or DON'T <*option*>

For example, WILL ECHO from station A is a request that station A provide the echoing of characters. The response will either be DO ECHO, meaning that the remote end agrees, or DON'T ECHO, meaning that the remote end will not allow station A to echo.

Agreement between the two Telnet ends for a DO <*option*> statement will be communicated with WILL <*option*> or WON'T <*option.*>. For example, if the Telnet application is running on a DOS personal computer and it is set up for local echo, upon the connection setup the Telnet request from the PC would be WILL ECHO and the response should be DO ECHO. If the PC were set up without the local echo option and the user wishes the remote end to provide echo, the PC should negotiate DO ECHO and the response would be WILL ECHO.

Using WILL/WON'T and DO/DON'T provides symmetry. Either side of the connection can provide the request or the response. One side provides services in exactly the same manner as the other side.

File Transfer Protocol

Telnet provides users with the ability to act as a local terminal even though they are not directly attached to the host. One other TCP/IP application that provides network services for users is a file transfer protocol. With TCP/IP, there are three popular types of file access protocols in use: File Transfer Protocol (FTP), Trivial File Transfer Protocol (TFTP), and Network File System (NFS). The FTP protocol allows files to be transferred reliably across the network under the complete control of the user. FTP is transaction based. Every command is responded to using a number scheme similar to the SMTP protocol (discussed later).

FTP is very robust. Remember the previous discussion on how ports and sockets are established and used. Most connections between two network stations are made via one source port and one destination port. The FTP protocol actually uses two port assignments (and therefore two connections): 20 and 21. A network station wanting a connection to a remote network station must connect to two ports on the destination station in order for FTP to work. Port 20 is used for the initial setup of the connection and is used as the control connection. No data pass over this circuit except for control information. Port 21 is used to pass user data (the file to be transferred) over the connection.

To establish the connection, simply type *FTP* <*domain name or IP address*>. From this the command line should read <FTP> (this depends on your application). With the advent of Windows and Windows-like operating systems, FTP now has a GUI interface to take some of the harshness out of the protocol. After the connection is established, the server process awaits a command from the client. To transfer a file from the server to the client, the user types GET <name of a file>, which will be transmitted to the remote network station. With this, a second connection is established between the server and client FTP process known as the *data connection*. Now we have two connections, but only during the file transfer process. Once the file is transferred, the data connection port is closed.

This is the well-known (or assigned) FTP data port. A user could also type in *FTP* and wait for the FTP prompt. At the prompt, the user would use the OPEN command to establish the connection.

FTP Commands

There are a lot of FTP commands, but in reality only a few are used. They are

OPEN	Open a connection to a remote resource.
CLOSE	Close a connection to a remote resource.
BYE	End this FTP session.
BINARY	The file transfer will be a file of binary type (i.e., executable file, Lotus file, etc.).
GET	Get a file from the remote resource: GET <filename>, MGET <multiple files, wildcards included>.
PUT	Put a file to the remote resource: PUT <filename>, MPUT <multiple files, wildcards included>.
CD	Change the directory on the remote device (to change the directory on the local end, use LCD).
DIR	Get a directory listing on the remote device (to get a directory listing on the local end, use LDIR).
HASH	Display hash marks on the screen to indicate a file is being transferred.

FTP Data Transfer

Refer to Fig. 12-2. Commands are sent over the control port. Once the connection is established, file transfer actually occurs over the data port. If a user wanted to establish an FTP connection between itself and

Figure 12-2
FTP data transfer.

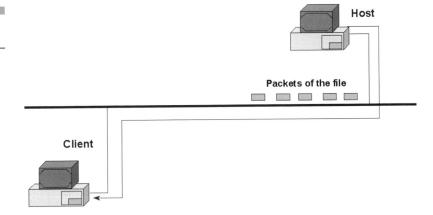

a remote host, the following sequence of events would take place on a DOS PC (for other operating systems, the prompt would change, but the commands are all the same in every FTP implementation).

If multiple files are needed, the user could use the command MGET or MPUT, which stands for *multiple get* and *multiple put.* If the file wanted is a binary file (a spreadsheet and an application are examples of binary files), the user would have to type in the keyword BINARY at the FTP prompt. This indicates to the FTP program that the file to be transferred is a binary file. Any of the commands may be entered at the FTP prompt. The protocol is transaction based, and the numbers preceding each line are for the node to interpret the nature of the next command. The text is for human consumption.

```
C:\WINDOWS>ftp
ftp> open mnauglepc
Connected to mnauglepc.
220 mnauglepc FTP service (NEWT v4.01) ready for new user.
User (mnauglepc:(none)): mnaugle
331 mnaugle, please enter your password.
Password:
230 User mnaugle logged in.
ftp> pwd
257 "c:\" is the current directory.
ftp> lcd
Local directory now C:\WINDOWS
ftp> get autoexec.bat autoexec.002
200 PORT command successful.
150 Opening ASCII mode data connection for autoexec.bat.
226 File transfer complete.
1911 bytes received in 0.00 seconds (1911000.00 kbps)
ftp> bye
221 Goodbye.
```

Trivial File Transfer Program

An alternative to the FTP program is the Trival File Transfer Program (TFTP). This is a simplex file transfer program and is primarily used to bootstrap diskless network workstations (the program is small enough to fit in a ROM chip on the diskless workstation to initiate a boot across the network) or even network components (bridges and routers). The FTP program is an extremely robust and complex program, and hence a larger file, and situations exist that require file transfer capabilities without complexity. TFTP is a small file and provides a more restrictive file transfer process (for example, no user authentication); it is also a smaller executable software program.

There are other differences between FTP and TFTP. TFTP does not provide a reliable service; therefore, it uses the transport services of UDP instead of TCP. It also restricts its datagram size to 512 bytes, and every datagram must be acknowledged (no multiple packet windowing). There are no windows for packets to be acknowledged. It could be said that it has a window of 1.

The protocol is very simple. The first packet transmitted from the client process to the server process is a control packet, which specifies the name of the file and whether it is to be read from or written to the remote workstation (GET or PUT command). Subsequent packets are data packets, and file transfer is accomplished with 512 bytes transferred at one time. The initial data packet is specially marked with the number 1. Each subsequent data packet number is incremented by 1. The receiving station will acknowledge a packet immediately upon receipt, using this number. Any packet less than 512 bytes in length signifies the end of the transfer. Error messages may be transmitted in place of the data in the data field, but an error message will terminate the transmission. Also notice that only one connection is made to the remote resource, whereas FTP has one for data and one for control information. The GET and PUT commands are used in the same way as in the FTP program.

The sequencing of the data is accomplished through the TFTP, not the transport-layer service of UDP. UDP provides only unreliable, connectionless service. TFTP keeps track of the sequencing of the blocks of data and the acknowledgments that should be received. For those readers familiar with NetWare, this is the same type of transaction accomplished between the NetWare Core Protocol (NCP) and its underlying delivery system, IPX.

Domain Name Service

Read RFCs 1034 and 1035. These contain the bulk of the Domain Name Service (DNS) information. These RFCs are supplemented by RFCs 1535–1537. DNS has many uses, but its main function continues to be the mapping of IP addresses to human-usable names.

There are millions of hosts on the Internet today containing even more millions of users. Most users have no idea what the underlying protocols are doing, nor do they care. But they would if they had to memorize IP addresses and determine other functions such as mail. Actually, most users would be frustrated by the numbering system, and the Internet would not be as popular today as it is. When the Internet was young, an early method of mapping the 32-bit address to a host name required the downloading of a file maintained (at the time) by the network information center (NIC). It was a single file, hosts.txt, that contained a simple mapping of Internet addresses to host names. This file was usually contained in the /etc subdirectory on a workstation, and various TCP/IP applications could access the information in this file. Not having this file meant that a user had to type in the 32-bit address for connectivity to a remote host.

As time passed, the population of the Internet became very diverse and more autonomous. In the 1980s prelude to the Internet was known as the ARPANET (now shut down), and the hosts were primarily time shared. More and more connections to the Internet were sites that had LANs installed, and connected to these LANs were mainframe and minicomputers or even personal computers. These sites administered their own names and addresses in the hosts.txt file but had to wait for the NIC to change hosts.txt to make changes visible to the Internet at large. Finally with the addition of more and more sites to the Internet, the applications on the Internet became increasingly sophisticated, creating a need for general-purpose name service. Refer to Fig. 12-3.

After many experimental RFCs, the global name system for the Internet became known as the Domain Name System, and the utility for implementing the system was called the Domain Name Service (DNS). DNS comprises three components: a name server, a database, and a name resolver. Name servers make information available to the resolvers. The information that name servers contain are IP addresses, aliases, mail information, etc. The resolvers usually reside on users' workstations and are embedded into applications of TCP such as Telnet and FTP. They are not separate programs. The name server is a separate program and

Figure 12-3
Domain Name
Service (DNS).

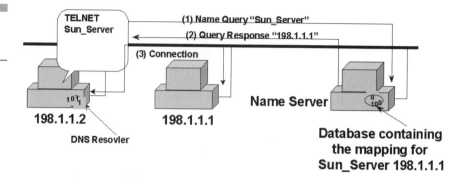

resides anywhere on a network, answering queries from the resolvers. The domain servers each maintain a portion of the hierarchical database under separate administrative authority and control. Redundancy is obtained by transferring data between cooperating servers (primary masters and secondary masters).

Your site may not require a DNS. You may have just a few hosts and can depend on another DNS to supply the information you need. For the Internet itself, a site must use the DNS system. A great example of the dependency on DNS occurred when a corrupted database file (containing "directions" to other hosts) was posted on the nine root servers. Millions of onliners were without the capability of attaching or communicating with other hosts on the network for hours. Without information (the IP address) concerning a remote system, two nodes cannot communicate. We can look up the information in the InterNIC database, but without prior knowledge of how to query the database manually, one is literally lost on the Internet. DNS provides information about hosts, not users, on the Internet.

DNS Structure

Refer to Fig. 12-4. DNS is very much like a file system on UNIX or DOS. It starts with a root, and branches attach from this root to give an endless array of paths. Each branch of a file system is given a directory name; in DNS it is called a label. Each label can be 63 characters in length, but most are far less than that. This means that each text word between the dots can be 63 characters in length, with the total domain name (all the labels) limited to 255 bytes in length.

The IP protocol mandates the use of IP addresses. Any user may use the IP address to connect to any service on the network. But for a user

Figure 12-4
DNS structure.

to remember the addresses of all the network servers on the network is an impossible task. Users are more likely to remember names than they are to remember numbers.

For those familiar with database environments, the domain name server is simply a database (consisting of information such as names and IP addresses, and much more) that any station on the network can query using the domain name resolver. The Domain Name System is not necessarily complex, but it is involved. It is based on a hierarchical structure as shown in Fig. 12-4. The assignment of names is relatively simple and is accomplished via the Internet registries, with the ultimate authority being the Internet Assigned Numbers Authority (IANA). The domain name is simply that—a name assigned to a domain. Domain names such as *isi.edu, cisco.com,* and *3Com.com* represent the top-level domain names at those companies or educational institutions. The naming within those domains (naming of the hosts) is left up to the individuals who are assigned those domain names; the InterNIC does not care. The hierarchical structure allows hosts to have the same name as long as they are in different branches of the structure or in different domains.

DNS Components

DNS includes much more than the name-to-address translation.

The Domain Name Space and Resource Records This is the database of grouped names and addresses that are strictly formatted using a tree-structured name space and data associated with the names. The database is divided up into sections called zones, which are distributed among the name servers. While name servers can have several optional functions and sources of data, the essential task of a name server is to answer queries using data in its zones. Conceptually, each node and leaf of the domain name space tree identifies a set of information, and query operations are attempts to extract specific types of information from a particular set. A query specifies the domain name of interest and describes the type of resource information that is desired.

Name Servers   These are workstations that contain a database of information about hosts in zones. This information can be about well-known services, mail exchangers, or hosts. A name server may cache information about any part of the domain tree, but in general a particular name server contains complete information about a subset of the domain space and pointers to other name servers that can lead to information from any other part of the domain tree. Name servers know the parts of the domain tree for which they have complete information; a name server is said to be an authority for these parts of the name space. Authoritative information is organized into units called zones, and these zones can be automatically distributed to the name servers to provide redundant service for the data in a zone. The name server must periodically refresh its zones from master copies in local files or foreign name servers.

Resolvers   These are programs that generally reside on users' workstations and that send requests over the network to servers on behalf of the user. Resolvers must be able to access at least one name server and use that name server's information to answer a query directly or pursue the query using referrals to other name servers.

When a DNS server responds to a resolver, the requester attempts a connection to the host using the IP address and not the name. Only part of a name could be used, such as *host*. This is known as a *relative name* and is part of a larger name known as the *absolute name*. The absolute name for the preceding example could be *host.research.Naugle.com*. This name would be in the domain name server. Most resolvers step through a preconfigured list of suffixes (in order of configured input), append it to the name, and attempt a lookup when the full DNS (absolute) name is not specified.

DNS Structure

DNS is hierarchical in structure, as shown previously. A domain is a subtree of the domain name space. The advantage of this structure is that, at the bottom, the network administrator can assign the names. Refer to Fig. 12-5.

From the root, the assigned top-level domains (TLDs) are as follows:

.gov	A government body
.edu	An educational body
.com	A commercial entity

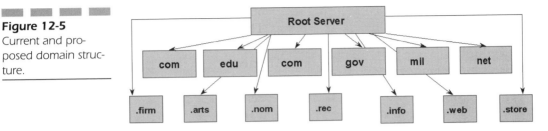

Figure 12-5
Current and proposed domain structure.

The bottom names here are prospected to be added to the Top Level Domain root name structure. They have not been formally approved.

.mil   Military

.org   Any other organization not previously listed

.con   Any country using ISO standard 3166 for names of countries As stated in RFC 1591, "the IANA is not in the business of deciding what is and what is not a country." Therefore, it is up to ISO to determine who is on that list.

An initial set of seven new generic TLDs was chosen by the International Ad Hoc Committee (IAHC), each with three or more letters. The proposed TLDs are

- *.firm* for businesses or firms
- *.store* for businesses offering goods to purchase
- *.web* for entities emphasizing activities related to the World Wide Web
- *.arts* for entities emphasizing cultural and entertainment activities
- *.rec* for entities emphasizing recreation/entertainment activities
- *.info* for entities providing information services
- *.nom* for those wishing individual or personal nomenclature

The above categories may not make it in, but are mentioned just to show you some of the ideas that are being passed around for expanding the TLD list. Broad categories were chosen, possible candidates were listed, and then the list was narrowed down to a set representing a synthesis of public comments, previous proposals, contributions (e.g., see *http://www.iahc.org/contrib/informal.html*), and discussions during the IAHC's activities. Seven was selected as the initial number to define initial changes using a relatively modest scale, with later evaluation and modification as appropriate. In addition, the trademark community has repeatedly urged that no new TLDs be created since it increases their responsibility to police domain name registrations for trademark violations. The number seven represents a compromise between different viewpoints within the IAHC and public comments concerning the number of TLDs to introduce in the first phase.

Going down the tree, we can pick out a domain name such as *research.Naugle.com*. This would signify the research department (which is a subdomain of domain *Naugle.com*) at Naugle Enterprises, which is defined as a commercial entity of the Internet. *Naugle.com* can be a node in the domain acting as a name server, or there may be different name servers for *Naugle.com*.

A user at workstation 148.1.1.2 types in the TELNET command and the domain name of *HOST1.research.Naugle.com*. This workstation must have the domain name resolver installed on it. This program will send out the translation request to a domain name server to convert the host name to an IP address. If the host name is found, the domain name server will return the IP address to the workstation. If the name is not found, the server may search for the name elsewhere and return the information to the requesting workstation, or return the address of a name server that the workstation can query to get more information. More on that in a moment. A domain contains all hosts with the corresponding domain names. A domain name is a sequence of labels separated by dots. A domain may be a subdomain of another domain. This relationship can be tested by seeing if the subdomain's name ends with the larger domain's name. For example, *research.Naugle.com* is a subdomain of *Naugle.com*. *Naugle.com* is a subdomain of *.com* and "" (root).

Special servers on the Internet provide guidance to all name servers. These are known as root name servers, and as of this writing there are nine of them. They do not contain all the information about every host on the Internet, but they do provide direction as to where domains are located (the IP address of the name server for the uppermost domain a server is requesting). The root name server is the starting point for finding any domain on the Internet. If access to the root servers ceases, transmission over the Internet will eventually come to a halt.

Name Servers

The programs that keep information about the domain name space are called name servers. The name resolvers do not usually store information, nor are they programmed with information as is a name server. All information is kept in the server.

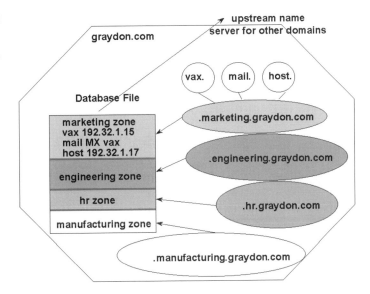

Figure 12-6
Name servers.

Refer to Fig. 12-6. Name servers keep information about a part of the name space called a zone. Name servers can be "authoritative" about one or more zones, meaning that a server is all-knowing about a zone. A server can be a primary name server for one zone and a secondary name server for another. However, these functions rarely cross; name servers are generally either primary or secondary for the zones they load.

There are two types of name servers: primary masters and secondary masters. The primary master builds its database from files that were pre-configured on its hosts, called zone or database files. The name server reads these files and builds a database for the zone it is authoritative for. Secondary masters can provide information to resolvers just like the primaries, but they get their information from the primary. Any updates to the database are provided by the primary. This system was set up primarily for ease of use. It is important to note that there should be more than one name server per zone or domain.

Let's use a simple example. Refer to Fig. 12-7. You are host on domain *Naugle.com*—specifically, *host1.research.Naugle.com*. You are looking for a host named *Labhost.bnr.ca.us*. You type in TELNET *labhost.bnr.ca.us*. The name server on your network is a primary and is not authoritative for the *.us* domain. Your name server then sends out a query to the root server that it knows about, and that root server refers you to the name server for the *.us* domain. Your name server sends out a request to that name server for *.ca*. The *.ca* name server refers you to another name

Figure 12-7
Name servers.

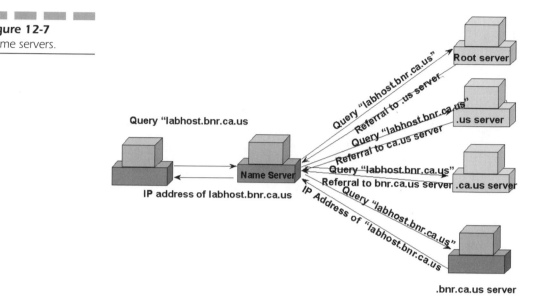

.bnr.ca.us server

server authoritative for the domain *bnr.ca*. Your server then sends one final request to that server for information on *labhost.bnr.ca*. That server responds with the IP address, which your server returns to your workstation. The Telnet protocol then uses that IP address to attempt to connect to your requested destination. A point to bring out here is that the information in the name server database is not dynamic in the sense that it knows the current status of any station. A station may be turned off, may not be accepting any new connections, etc. The name server simply responds to requests for information that is contained in its database.

Query Function Types

Two types of queries are issued: recursive and iterative. Recursive queries received by a server forces that server to find the information requested or to post a message back to the querier that the information cannot be found. Iterative queries allow the server to search for the information and pass back the best information it has. This is the type of query that is used between servers. Clients used the recursive query. This is shown in Fig. 12-8.

Generally (but not always), a server-to-server query is iterative and a client resolver-to-server query is recursive. Also, note that a server can be

Figure 12-8
Query function types.

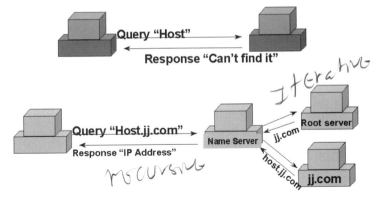

queried, or it can be the entity placing a query. Therefore, a server contains both the server and client functions.

A server can transmit either type of query. If it was handed a recursive query from a remote source, it must transmit other queries to find the specified name or send a message back to the originator of the query stating that the name could not be found.

Example DNS Database

A database is made up of records, and the DNS is a database. Common resource record types include

- A—host's IP address
- PTR—host's domain name, with the host identified by its IP address
- CNAME—host's canonical name, with the host identified by an alias domain name
- MX—host's or domain's mail exchanger
- NS—host's or domain's name server(s)
- SOA—indicates authority for the domain
- TXT—generic text record
- SRV—service location record
- RP—responsible person

When a resolver requests information from the server, included in the request will be one of the above record types. In this way, the server will know exactly what the resolver is requesting; this could be a mail server or an IP address translation, or simply a request for some generic information.

Some of the more useful records in the database are discussed next.

SOA Records The start of authority (SOA) record is the first entry in the database file. This record indicates that the name server is the best available source of information for this domain. The following is an example of a record using SOA.

```
Naugle.com.  IN    SOA  ns1.Naugle.com Matt.NT1Server.Naugle.com. (
1567                    ;Serial
18000                   ;Refresh after five hours
3600                    ;Retry after 1 hour
604800;                 Expire after 1 week
86400)                  ;Minimum TTL of 1 day
```

The numeric entries above are indicated in seconds. (Notice that anything following a semicolon is ignored.) The first entry on the first line indicates the domain this server is authoritative for: *Naugle.com*. The next field indicates that the class of data is Internet (other types are defined but not used today). The first name after SOA indicates the primary name server for this domain, and the field after this indicates the person to contact; replace the first dot with the @ symbol and send an email there for more information. The information between the parentheses concerns the secondary name server. For example, the serial number on the primary name server should be incremented when new information is placed in the database. In this way, the secondary servers will know that they have old information and should get updated by this primary.

Notice that some domain names are written with a dot at the end, and others are written without a dot. The ones with a dot on the end are known as absolute names, and they specify a domain name exactly as it lies in the hierarchy name space starting from the root. Those names that do not end with a dot are domain names that may trail from some other domain. This is best illustrated using the directory system. In DOS or UNIX to change directories, you use the CD (CHDIR) command. With this you can specify directly from the root which directory you would like to change to, or you can change directories relative to another directory. You do not have to type in the full directory path name each time you want to change directories. This is the same for DNS, where the dot is used to signify the full path name and its omission signifies a domain relative to another path name.

Name Server Records   The next entry in the server database is for name server resource records. If you have five name servers in your domain, you should list them here, as in the following example:

```
Naugle.com.    IN   NS   ns0.Naugle.com.
Naugle.com.    IN   NS   ns1.Naugle.com.
Naugle.com.    IN   NS   ns2.Naugle.com.
Naugle.com.    IN   NS   ns3.Naugle.com.
Naugle.com.    IN   NS   ns4.Naugle.com.
```

The above entries indicate that there are five name servers for domain *Naugle.com*. Name servers can be multihomed (one station connected to more than one subnet).

Address Records   The following list indicates the name–to–IP address mappings:

```
Localhost.Naugle.com.           IN  A  127.0.0.1
DatabaseServer.Naugle.com.      IN  A  128.1.1.1
HRServer.Naugle.com.            IN  A  128.1.15.1
EngServer.Naugle.com.           IN  A  128.1.59.150
NS0.Naugle.com.                 IN  A  128.1.1.2
NS1.Naugle.com.                 IN  A  128.1.15.2
NS2.Naugle.com.                 IN  A  128.1.16.190
NS3.Naugle.com.                 IN  A  128.1.59.100
NS4.Naugle.com.                 IN  A  128.1.59.101

;Aliases
NT1Server.Naugle.com            IN  CNAME DBServer.Naugle.com.
NT2.Naugle.com                  IN  CNAME HRServer.Naugle.com.
NT3.Naugle.com                  IN  CNAME EngServer.Naugle.com
```

This file has new types: A and CNAME. The A record type stands for Address (A for 32-bit addresses, and AAAA for IPv6 addresses). There can be more than one address for a name, as in the case of a multihomed host (a host with a connection to more than one subnet). This could be stated as

```
;multhomed hosts
NT5.Naugle.com    IN  A  128.1.60.5
                  IN  A  128.1.61.5
```

Name servers will return the closest address to the requester; this depends on the requester's address. If they are on the same network, the name server will place the closest address first. Since the DNS has no idea of routing tables, if the requester and its network address are different, it will return both addresses.

The aliases are just that—names for another name. When a request comes in and the server finds a CNAME record, it replaces the name with the CNAME. It will then do another lookup, find the address, and return this to the requester.

Mail Exchange (MX) Records   One of the largest uses of the Internet is for email, and DNS plays a major role here as well. It does not send or receive mail, but it does provide information on the mail servers for a given name. In the database is a resource record type of MX. DNS uses this single type of resource record for mail routing. It specifies a mail exchanger for a domain name. This mail exchanger is a host that will either process the mail or forward it. Processing is simply the task of providing for delivery to the addressee or providing a path to another mail transport. It may also forward the mail to its final destination or pass it on to another mail exchanger in closer proximity to the recipient using SMTP (explained later). An example record is

```
engineering.naugle.com.  IN  MX   5   mail.naugle.com.
```

This states that the mail exchanger for the domain *engineering.naugle.com* is *mail.naugle.com.* The number 5 is a precedence value. This number can range from 0 to 65,535. If there is only one MX for a domain, this field is useless. For example, suppose you are given the following record:

```
engineering.naugle.com.  IN  MX  5    mail1.naugle.com.
engineering.naugle.com.  IN  MX  10   mail2.naugle.com.
```

A mail program (such as SMTP) should use the mail exchanger with the lowest value (5) first. If this fails, then use the next one associated with that domain. If there are no records associated with a domain or host, then the mail program must be able to deliver the message directly, and some versions of mail senders have this capability. The most common mail transport service today is called *sendmail.*

Playing with the Database

There is a program available (not usually on Windows 95, but on all UNIX systems) called nslookup. This is a program that transmits queries to a specified name server. If you have that program, you can use it and another name server to supply information to you. An example of a command in this program is

```
nslookup <domain name such as starburstcom.com>) 198.49.25.10
```

With this command and these arguments you are asking a name server whose address is 198.49.25.10 (this is one of the InterNIC name servers) to supply information about a domain such as *starburstcom.com.*

If you have access to the Web, then head up to the InterNIC (in the United States; RIPE for Europe and APNIC for Asia Pacific), and point your browser to

http://ds.internic.net/cool/dns.html

All the records for the domains that it is authoritative for (which includes all the top-level domains—*.com, .net,* etc.) are there. Check "any information" plus the ANY radio dial button and see what comes back.

WHOIS **Command**

Another useful utility is WHOIS. This command is included with most UNIX systems, but for those others who do not have access to their UNIX systems or who are using Windows, the InterNIC provides this function at

http://rs.internic.net/cgi-bin/whois

```
> Whois ascend.com

Ascend Communications, Inc. ASCEND-DOM
  1275 Harbor Bay Pkwy
  Alameda, CA 94502

Domain Name: ASCEND.COM

Administrative Contact, Technical Contact, Zone Contact:
  Rochon, Lyle LR88 lrochon@ASCEND.COM
  (510) 769-6001

Record last updated on 10-Jul-97.
Record created on 05-Dec-90.
Database last updated on 3-Aug-97 04:39:20 EDT.

Domain servers in listed order:

DRAWBRIDGE.ASCEND.COM 198.4.92.1
NS.UU.NET
```

Say you want to find out who owned the 192.1 Class C block of addresses. At the WHOIS server you would type in WHOIS NET 192.1, and a listing would follow:

```
BBN Corporation NETBLK-BBN-CNETBLK BBN-CNETBLK
192.1.0.0 - 192.1.255.255
```

Actually, there was a lot more information listed, but it was too much for this page. Try it yourself and see what happens! You can use the InterNIC's whois server as shown above or use Telnet to get into their server by typing *Telnet whois.internic.net*. After you have a connection, type in the command WHOIS, and you should get the whois prompt. From there you can check on person (WHOIS PE ROBBINS), domain (WHOIS DOM BAYNETWORKS.COM), and network numbers (WHOIS NET 192.32). Pretty cool stuff and very, very open!

More DNS Information

DNS is available through a program known as BIND (Berkeley Internet Name Domain). It is available on most UNIX systems and Windows NT™. It is customizable through example files that are included. Most implementations simply set up for their hosts and point to an upstream root name server for reference to other sites on the Internet.

BIND (DNS) ships with most versions of UNIX. If for some reason it does not, you can download BIND from the address below. This site also contains information on DHCP and the Windows NT port of BIND (not supported).

http://www.isc.org

There are many sites around the Web to assist you (for a small charge) with DNS. DNS can be a daunting task, especially for large installations. You may want to consult help for your first installation. Once you get the hang of it and read a few books on DNS, you will see how simple it is. One of the best (and only) books about DNS is *DNS and BIND*, by Paul Albitz and Cricket Liu (ISBN 1-56592-236-0).

Simple Mail Transfer Protocol

Review RFCs 821, 822, and 974 for the Simple Mail Transfer Protocol (SMTP) protocol. RFC 822 defines the structure for the message, and RFC 821 specifies the protocol that is used to exchange the mail between two network stations.

We now have email to send to one another, completely bypassing the postal system. There are some who arrogantly call the postal system "snail mail." However, many people today still immensely enjoy receiving a handwritten letter from a family member, friend, or business correspondence through the postal system. Also, it is hard to send packages through the email system, and some that do get through (attachments) get banged up along the way. Yes, the postal system is old and cranky, but it works, in some cases better than email. Suffice it to say that the postal system will be here for many years to come.

But this is an electronic discussion, so I will stick to that. Email does have many, many advantages, and one of the top advantages is speed. Its biggest disadvantage: lack of emotion. Like everything else, email has its place, but it is not 100 percent of the pie; it is merely another form of communication.

To send and receive mail between users, there are actually two protocols (possibly three) that are used:

- SMTP—used for the actual transport of mail between two entities
- POP (Post Office Protocol)—a protocol that allows single users to collect their mail on one server
- DNS—used to identify the mail hosts for a domain or host name

Mail can be sent and received using only SMTP, but the involvement of the other protocols makes it much easier to use and more efficient. This section will concentrate on SMTP and POP. DNS was already explained in relation to the mail exchanger. SMTP was created before POP, so I will start with the SMTP protocol. This is a protocol that allows users to transmit messages (mail) to other users. It is one of the most widely used applications of the TCP/IP protocol.

SMTP Functions

The protocol is relatively simple. A message is created, properly addressed, and sent from a local application to the SMTP application, which stores the message. The server will then check (at periodic intervals) to see if there are any messages to deliver. If so, the mail server will try to deliver the message. If the intended recipient is not available at the time, the mail server will try again later. The mail server will try a few times to deliver the message. If it cannot, it will either delete the message or return it to the sender.

The address has the general format of *local-part@domain-name*. You should recognize the domain name format. For example, an address in the SMTP header could be *matt@engineering.naugle.com*. This would indicate that the message is addressed to a user named Matt in the domain of *engineering.naugle.com*. When DNS is used to look up the mail handler for Matt, it will have some sort of entry like

```
engineering.naugle.com   IN   MX   10   NT1mail_server.engineering.
                                         naugle.com.
```

Using this listing the name will be looked up and the mail will be delivered to that host.

There are two entities to this system: the sender SMTP and the receiver SMTP. To transport mail between two systems, the sender SMTP will establish communications with a receiver SMTP. Attachments are allowed with Internet email, but not directly with the protocol used in SMTP (sendmail). The Internet email mailer program SMTP (sendmail) can handle only text. Therefore, most email applications convert an attachment to text before sending it. A common type is MIME (Multipurpose Internet Mail Extensions, which is beyond the scope of this book). At the receiver end, the email application converts the attachment back to its original format.

SMTP Flow

The SMTP design is based on the following model of communication. Once you have filled out the header and body section of your mail message, the sender SMTP establishes two-way communication with a receiver SMTP. The receiver SMTP may be either the ultimate destination or a temporary stop on the way to the final destination. Commands are sent to the receiver by the sender SMTP, and replies are sent from the receiver SMTP to the sender SMTP in response to each of the commands.

Refer to Fig. 12-9. Once two-way communication has been established, a series of commands follows (which you can see in operation using some mail applications). The sender SMTP will send a HELLO command identifying itself to the receiver using its domain name. The receiver acknowledges this with a reply using its domain name. Next the server issues a MAIL command to the receiver. In this will be the identification

Figure 12-9
SMTP flow.

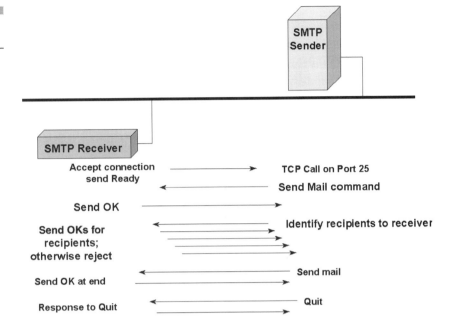

of the person (or place or thing) sending the mail. The receiver acknowledges this with an OK. The sender SMTP then sends a RCPT command to the receiver using the intended receiver name as an argument. Each recipient in the list is sent to the receiver one at a time, and the receiver acknowledges with an OK each recipient that it knows about. For those that it does not know about (different domain name), it will send back a reply stating that it is forwarding the message. For intended recipients that it does not have an account for, it will reply to the sender that no such user exists. After the intended recipients have been ACK'ed or NACK'ed, the SMTP sender sends the DATA command. The receiver SMTP will OK this and indicate what the end-of-message identifier should be. Once the ending identifier is received, the receiver SMTP will reply with an OK. Notice that all data are received, the ending identifier is received, and then a reply message is sent by the receiver.

If no problems occurred the sender ends the connection with a QUIT command. The SMTP receiver will reply indicating that the communication channel is closed. The minimum set of commands that a receiver must support are HELO, MAIL, RCPT, DATA, RSET, NOOP, and QUIT.

Depending on the mail program you use, the basic transaction between a recipient and sender of mail hasn't changed since RFC 821 was written. The interface allows you to complete the mail message by filling in the header (addresses and subject) and the body (text) of the letter. When you hit the SEND button, the transaction takes place. Some mail programs actually place the mail commands and state numbers on display while the transaction is taking place. It should be noted here that sending mail is immediate. It may get queued for a small length of time on different routers and transient mail servers, but this will not take long.

Most electronic mail today is sent via SMTP, and it will reside on your mail server host until you retrieve it using POP (discussed shortly). Today retrieving your mail does not mean that you have to run the SMTP protocol. A server host will accept mail messages directed to you on your behalf. Then you can sign on any time you want and retrieve your mail.

DNS Interaction for Mail

Refer to Fig. 12-10, which shows the interaction with DNS for the mail service. The MX record in DNS identifies a mail exchanger for the purpose of identifying hosts for recipients. A mail exchanger is a host that will either process or forward mail for the domain name. Processing means that the host will deliver it to the host it is addressed to or will hand it off to another transport service such as UUCP or BITNET. Forwarding means that the host will transfer the message to the final destination or to another mail exchanger closer to the destination.

There can be multiple entries for a mail exchanger in DNS. Each MX entry will have a precedence number indicating to the sender which mail host it should try first. If the precedence value is equal among MX records, then the sender can randomly pick one from the list. Once the sender has successfully delivered the mail to one of the MX hosts, its job is done. It is the job of the MX host to make sure that it is forwarded to its final destination. If there are no MX records for a domain name, it is up the mailer application as to what happens next. Some will try to deliver the mail to the IP address of the destination.

Post Office Protocol

The original mail program standard RFC 821 (which is the one in use today) was set up to send messages directly to a user logged onto a terminal and to store these messages to a mailbox. The commands allowed the

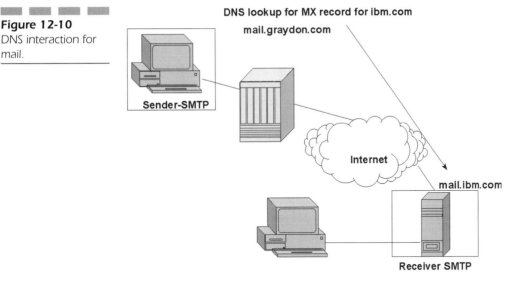

Figure 12-10
DNS interaction for mail.

receiver to determine if the user was logged on to a terminal (as opposed to a PC), if the user was accepting messages, and if not, if there was a mail box to deliver mail to. Messages were sent and received in seven-bit ASCII (the eighth bit was set to 0), which would not allow binary messages to be sent (i.e., no attachments). In fact, the original messages could not exceed 1000 characters.

To transfer mail, the host had to be operational (able to receive) all the time. Today, terminals still exist but personal computers have commonly taken their place. Therefore, the final recipient is the personal computer. Even though a personal computer retrieves its mail via Post Office Protocol (POP), it uses the SMTP functions to send its mail. Since SMTP expects to be able to deliver mail immediately, this means that users would have to have their personal computers on 100 percent of the time in order to accept mail. To receive mail, you would have to log on to a specific host and then read your mail.

To operate a mail server generally requires that the mail server be available for a majority of the time, have the ability to store many mail messages, and be able to fully run SMTP and accept mail from an SMTP sender. Although this may have been feasible for situations such as terminal-to-host connectivity, it is not feasible for the situation that we have today, namely, personal computers and mobile workers. SMTP is a very robust transaction-oriented protocol and requires use of the statements listed previously to operate fully.

What was needed was the ability for SMTP to operate (drop off the mail, like using a box at the post office) and the use of another protocol to download messages to our personal computers (similar to dropping by the post office and retrieving our mail from the post office box). POP is the protocol for this. Mail can be delivered to a drop-off point, and POP allows us to log in and retrieve our mail.

When you sign up with an Internet service provider, a POP account is assigned to you such as *mnaugle@POP3.ISP1.com*. You use this account when configuring your mail program. Also, when sending mail you must give the SMTP server name to the configuration program as well. The POP protocol is not used for sending mail.

POP Operation

POP3 (the latest version of POP) is used only to retrieve your mail from the drop-off point. Sending mail is a different story. Your personal computer has the ability to establish a TCP connection to a relay host (intermediate mail host) to send mail. Therefore, you should consider your host as having the ability to be an SMTP sender, and the SMTP protocol explained above applies. However, for retrieving your mail, POP3 comes into play.

Refer to Fig. 12-11. The client (your PC with a mail application such as Eudora™), once established with TCP/IP on its network, builds a connection to the POP3 server. The POP3 server configuration is built during the installation of your mail program on the PC. The connection between your PC and the POP3 server is a TCP connection on TCP port 110. Similar to SMTP, once the connection is established, the server will respond with a greeting such as "POP3 server ready."

The POP3 protocol then enters the authentication state. During this phase, you must identify yourself with a username and password. The RFC does not indicate which authentication mechanism you should use. The most common is the simple username-password combination. However, other options are available, such as Kerberos and APOP, which are beyond the scope of this book. Once you have been authenticated, the POP3 server puts an exclusive lock on your mailbox to ensure that no other transactions take place on the messages while you are retrieving your mail.

The server now enters the transaction state, in which each of the messages in your mailbox is assigned a number. This allows your client POP

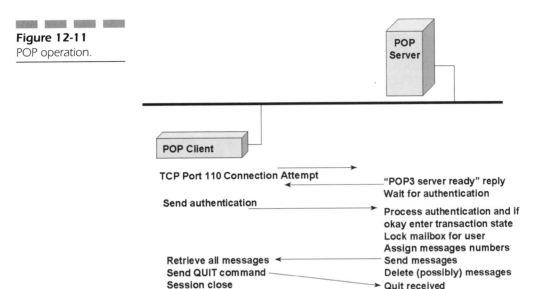

Figure 12-11
POP operation.

POP
Server

POP Client

TCP Port 110 Connection Attempt
"POP3 server ready" reply
Wait for authentication

Send authentication
Process authentication and if
okay enter transaction state
Lock mailbox for user
Assign messages numbers

Retrieve all messages
Send messages
Send QUIT command
Delete (possibly) messages
Session close
Quit received
Read messages locally
Perform update on mailbox

to indicate how many messages are in your mailbox. Each message can be retrieved one at a time, or all can be retrieved. Furthermore, you can instruct your client POP to delete messages as they are retrieved. This can be either good or bad. It would be nice to hold onto your messages as a backup on the server, but this requires disk space, which can be depleted quickly.

From here you retrieve your messages, and depending on how you configured your PC mail program, the messages are marked for deletion after your session. After you retrieve your messages, your mail program will send the QUIT command, which closes the POP3 session down. Next the update process begins on the server, which involves housekeeping work on the server (deleting messages, etc.).

At this point, you can read your messages locally on your PC. You are disconnected from the POP3 server, and you can manipulate the messages locally. There are many other options available for POP3, which may or may not be implemented. But from a user's point of view, they are transparent.

Figure 12-12 shows the relationship between SMTP, DNS, and POP.

Figure 12-12
SMTP, DNS, and POP
topology.

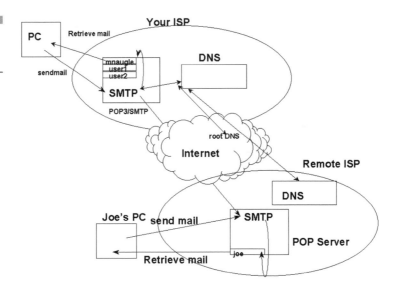

IP Multicast

Introduction

Three types of addressing are used with IPv4: point to point (unicast), point to multipoint (multicast), and point to all multipoints (broadcast). Multicast is the ability to push data from a single source and have it received by many receivers, even though it was sent as a single stream. Refer to Fig. 13-1.

You will often hear the pseudo-technical term "push or pull" technology. This means that you either have to go and get the information off of the Internet (pull) or the information comes to you (push). There is a vast array of information in the Internet, and finding it can be a daunting task. Furthermore, even when you supposedly know where the information is, you can have a hard time finding it. Push technology means that the information was pushed down to you. An example of this is an information news service that retrieves information on certain subjects throughout the day. A lot of readers have probably heard of PointCast™. As of this writing, the server uses an automated pull technology that will eventually become a full-blown push technology. Whenever information changes on that news server, stories related to your requests are pushed down to your workstation, as opposed to your having to go out and find the information and pull it down.

Another exciting development will be the ability to have voice and video run over an IP network. I know, it can be accomplished today, but I am not fond of viewing a 2×2 inch fuzzy screen that has incredible delay and non–lip-sync audio. It is like watching an old Japanese movie that has been translated to English. Frame reception at 5 fps is not great. It is fun to experiment with, but until it comes close to CATV (cable TV), users will not consider it as a serious requirement for the LAN infrastructure.

Figure 13-1
Multicast.

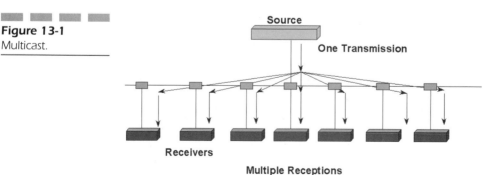

Transmitting voice and video over IP can cause problems on the Internet. To replicate a separate stream of data for every requested user to a single video source would easily overload the Internet. The ability to transmit one packet and have it replicated at certain points along the many paths to separate destinations is a much more efficient system of distributed data, voice, and video. Multicasting allows this.

Multicast Components

Refer to Fig. 13-2. There are four components to a multicast network:

- Multicast-enabled host NIC
- Multicast-enabled TCP/IP software
- Multicast-enabled infrastructure (routers, switches, etc.)
- Multicast-enabled applications

IP multicast usually does not operate alone. Other protocols are used in conjunction with it to provide for VVD (voice, video, and data) over IP. Such protocols as the Real Time Protocol (RTP) and the Real Time Control Protocol (RTCP), the Real Time Streaming Protocol (RTSP), and the Resource Reservation Protocol (RSVP) operate cooperatively to allow multicasting to work. Similar to this is IP. There are many components to the TCP/IP network, and IP is simply one of the components. It provides for datagram delivery. IP multicast is also one component of many. IP multicast is based on a few protocols, but all the other components are necessary to really make it work.

Figure 13-2
Multicast components.

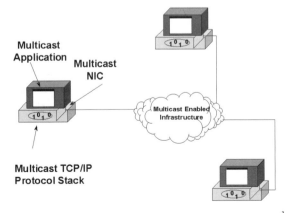

Berkeley Sockets for Multicast provides an API set that easily enables most UNIX hosts to become multicast ready, and Microsoft is supporting multicast in its sockets interface known as Winsock 2.0. With the APIs in place, multicast applications can be built.

Multicast Caveats

Multicast transmission and reception can take place anywhere multicast is enabled. There are obstructions to enabling multicast. Routers and switches must be multicast ready. The backbone of your internet must be multicast ready. The WAN must be multicast ready. Add to this the lack of general knowledge and ample studies on the effect of multicast, and you should be able to understand the slowness of its implementation. The Internet currently plays a part in multicast, but only as a simply transport mechanism between multicast-enabled networks. Tunnels can be built across the Internet using the loose source routing feature of IP. In this way, two multicast "islands" can be connected together. Otherwise, the Internet (as of 1997) is not multicast enabled and probably won't be for some years to come.

Many corporate networks have moved to frame relay as their WAN protocol. With this, there are two pieces: the customer device (called the CPE, for customer premise equipment—usually a router) and the frame relay provider's equipment (frame relay switches). However, it is not as easy as flipping a switch to make the frame relay cloud multicast ready. Many studies and tests have to be performed to see how multicast reacts to an existing network that has many customers who are already enabled and who expect 99.999 percent uptime. Maybe after this writing, the frame relay providers will multicast-enable their WAN networks, but it may not happen for a while. Some frame relay providers promise multicast readiness in the first part of calendar year 1998.

Corporate networks are not multicast enabled either. These environments have just redone their topologies to allow ATM and LAN switches to be employed and are now looking at multicast. However, most corporate environments do not even realize that their routers are multicast ready. Some think that new equipment must be purchased before multicast can be enabled. All routers support the multicast protocol of IGMP. Most routers support DMVRP and MOSPF, and a few support PIM. These will be discussed in detail later, but suffice it to say here that these are software features that can be turned on, but very carefully. Multicast may be an extension of the IP protocol, but routing multicast packets is a different story.

Figure 13-3
Unicast.

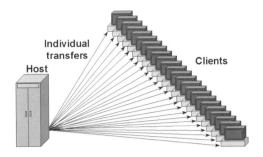

Unicast (versus Multicast)

Refer to Fig. 13-3. In unicast, the server must send information one transfer at a time to each of the recipients even though the data are the same for each transfer. For example, let's say that the end-of-the-month sales report is complete and needs to be transmitted to 200 file servers around the country. With unicast addressing, one could write a simple script to initiate a file transfer to each of the 200 file servers. If the file was 2 MB in length, you would have a 400-MB file transfer that unnecessarily consumes both time and bandwidth.

Enabling multicast would push that one file to the 200 file servers simultaneously. The 200 file transfers are reduced to one file transfer, and all 200 stations will receive the file just as if it had been unicast to them. Instead of a 400-MB transfer, you are back to a 2-MB file transfer.

Multicast file transfer is used extensively in software distribution. Microsoft and other application companies upgrade their software at least twice a year. In large environments, this is a daunting task. If the upgrade is 20 MB and it must be distributed to 50,000 desktops, it would require 10^{11}, or 1000 billion, bytes to be delivered. Assuming the transfer occurs at 512,000 bits per second, this will require 180 days to complete. Just when you have completed one upgrade, the next one is ready.

However, by using reliable multicasting, the file could be delivered to 50,000 desktops as a single stream. This means one transfer that looks like 50,000. With multicast, the same transmission would be completed in 5.2 h.[1] This example is theoretical, but if put into practice, I believe that it would be around the number indicated. Even if it took 10 hours to complete, look at the alternative.

---

[1] This number assumes no retransmissions. Retransmission costs are variable depending on the number of clients requesting, the status of the lines, etc. Even so, the worst cost is less than 25 percent of the original pass.

Figure 13-4
Multicast.

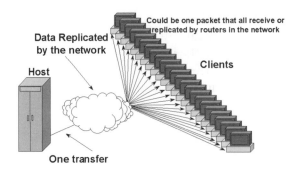

Thus, the advantages to multicast-enabled applications and networks are time savings and scalable bandwidth. As noted in Fig. 13-4, multicast involves one stream of packets that are seen by many receivers all using the same IP multicast address as their receiving IP interface. IP multicast places the burden of packet replication on the network via the IP multicast address and routers able to forward received multicast packets to more than one port.

Multicasting Types

There are two types of data transmission normally associated with multicasting: *real time* and *non–real time.*

Real-time transmission is the ability to work with the information while it is being transmitted. Examples of real-time applications are video and voice transmissions. Non–real-time transmission is the ability to transfer the information for use at a later time. That time may be directly after the transfer, or it could be days after the transfer. Examples include CBT files, kiosks, or any type of data file. Both types have their purposes in the multicast arena.

Real-time transfers such as video and voice are still in the experimental stage. Transferring information across a network that is packet switched without some capability for priority is experimental. In some environments, it works rather well (on a single subnet), but when the data are transported across an internet, the quality deteriorates, and the interest in the products does as well. There are protocols that are experimental as well (RSVP) that should assist real-time multicast protocols by providing bandwidth to the application instead of to the network.

Current real-time multicast transfers data using a simplex approach that makes it near–real time. They simply buffer the incoming data for about 30 s and start the playback. While the information is being used, the application continues to buffer the incoming data for playback. This is not perfect, but it is better than nothing.

Addressing Type Review

The example given above is a simple but very real example. However, there are many applications today that require one-to-many or many-to-many transmission and reception. Audio and video, stock tickers, workgroup applications, electronic whiteboards, and other applications can have one or more senders and one or more receivers. In IP, there are three types of addresses:

- Unicast—a unique address that allows one host to receive a datagram
- Broadcast—an address that allows every host to receive a datagram
- Multicast—an address that allows a specific group to receive a datagram

Unicast is the ability to uniquely identify a host on a subnet or internet. The transmission and reception are accomplished in a one-to-one relationship. A broadcast address is an address that is received by every host on the subnet. Routers (with exceptions such as DHCP and BOOTP) will not forward datagrams that have a broadcast address. A multicast address allows a specific group of hosts to receive a datagram while all others ignore it. For our purposes, we will stick to multicasting as it applies to the IP protocol.

Introduction to IP Multicast

Multicasting has its roots back in 1978, when Y. K. Dalal and R. M. Metcalfe (yes, the inventor of Ethernet) published an article called "Reverse Path Forwarding of Broadcast Packets" (*Communications of the ACM* 21(2):1040–1047, December 1978). For the Internet, in December 1985 an RFC was produced by Steve Deering (RFC 966, "Host Groups: A Multicast Extension to the Internet Protocol," which is now known as RFC 1112, "Internet Group Membership Protocol or IGMP"). There have been many iterations and flavors since then and the text of this book will cover only that information that is current. We will cover the algorithms of reverse path broadcasting, truncated reverse path broadcasting, reverse path forwarding, reverse path multicasting, and other techniques. Thirteen years later, we are still in the experimental stage, although many standards have emerged and are being implemented today. The main terms to learn are reverse path forwarding and the routing protocols of DVMRP, MOSPF, PIM (SM and DM), and IGMP. Otherwise, we will leave out the technology. With an experimental technology, there are many types of experiments and eventually a few developments will come out of it.

According to RFC 1112, "Host Extensions for IP Multicasting," the following is a description of IP multicasting:

IP multicasting is the transmission of an IP datagram to a "host group," a set of zero or more hosts identified by a single IP destination address. A multicast datagram is delivered to all members of its destination host group with the same "best-efforts" reliability as regular unicast IP datagrams, i.e., the datagram is not guaranteed to arrive intact at all members of the destination group or in the same order relative to other datagrams.

The membership of a host group is dynamic; that is, hosts may join and leave groups at any time. There is no restriction on the location or number of members in a host group. A host may be a member of more than one group at a time. A host need not be a member of a group to send datagrams to it.

As you learn more about IP multicasting, you will realize that this protocol is simply an extension of the IP protocol itself. It does not replace the IP protocol; in fact, it adds a few functions to the IP protocol to allow a host to send and receive multicast datagrams. This can be considered similar to the way ICMP works with IP.

There are three levels of IP associated with multicast:

Level 0—no support for multicast

Level 1—ability to send multicast but not receive, to allow for simplicity

Level 2—ability to both send and receive multicast packets

Extensions to the IP Service Interface

The normal IP transmission logic is as simple as the following:

```
if IP-destination is on the same local network,
        send the datagram local directly to the destination
```

```
else
        send datagram locally to a router
```

Since multicasting is nothing more than an extension of the IP protocol, the logic is simply revised as follows:

```
if IP-destination is on the same local network OR
        IP destination is a host group,
        send the datagram local directly to the destination
else
        send datagram locally to a router
```

Notice that the multicast host does not specifically look for a router, even though members of the host group may be multiple hops away. Multicast datagrams are not addressed to a router, but they can be sent through the Internet. They do not have to remain local. Multicast datagrams that span subnets require routers, and these routers must run a special multicasting protocol. When a host transmits a multicast packet, it simply transmits the packet out its interface using the normal IP datagram transmission logic. In this way, all the hosts that belong to the same group on the local network will receive and process this datagram. If the TTL field (known as the scope) is greater than 1, the multicast routers will receive and forward this packet through its interfaces to all other networks that belong to that group.[2] Therefore, the router is also a member of the host group. The receiving router decrements the TTL and forwards the packet as a local multicast on the networks participating in that group.

Receiving Multicast Datagrams

Multicast IP datagrams are received using the same "receive IP" operation as normal unicast datagrams. However, before any datagrams destined to a particular group can be received, an upper-layer protocol (an application) must ask the IP module to join that group. Thus, the IP service interface must be extended to provide two new operations:

JoinHostGroup (group-address, interface)

LeaveHostGroup (group-address, interface)

---

[2]How the router determines which interfaces belong to that group is discussed in the section on DVMRP.

The JoinHostGroup operation requests that this host become a member of the host group identified by "group-address" on the given network interface. The LeaveGroup operation requests that this host give up its membership in the host group identified by "group-address" on the given network interface. A unique interface is specified for those IP hosts having more than one interface. A host with more than one interface can join the same group on each of the interfaces. But this leads to reception of duplicate multicast datagrams. If more than one application requests to join the same group, the port numbers will differentiate the applications.

The possibility exists that an operation will not work. Since an application can join any group, multiple applications or a single application may join many groups, which can lead to resource problems. JoinHostGroup may fail due to lack of local resources. A multicast membership may persist even after an application has requested the LeaveHostGroup operation due to the fact that other applications may be using that host group.

Address Format

We know all about the address formats of classes A, B, and C. The format for multicast addressing is known as the class D address. Its range is reserved by IANA and is 224.0.0.0 to 239.255.255.255. The base address, 224.0.0.0, is reserved and cannot be assigned to any group. Furthermore, IANA has reserved the range of 224.0.0.1 through 224.0.0.255 for the use of routing protocols, topology discovery, and maintenance protocols. This is a rather large range (2^{28}). What is interesting is that no router that receives a datagram within this address range is allowed to forward it. It must either consume or discard (filter) the datagram. Other addresses of interest according to RFC 1700 are

224.0.0.0 base address (reserved)	[RFC1112]
224.0.0.1 All systems on this subnet	[RFC1112]
224.0.0.2 All routers on this subnet	
224.0.0.3 Unassigned	
224.0.0.4 DVMRP routers	[RFC1075]
224.0.0.5 OSPFIGP—all routers	[RFC1583]
224.0.0.6 OSPFIGP—designated routers	[RFC1583]
224.0.0.7 ST routers	[RFC1190]
224.0.0.8 ST hosts	[RFC1190]

224.0.0.9 RIP2 routers   [RFC1723]

224.0.0.10 IGRP routers   [Cisco]

224.0.0.11 Mobile-agents

224.0.0.12 DHCP server/relay agent   [RFC1884]

224.0.0.12–224.0.0.255 unassigned   [IANA]

Mapping to an Ethernet or IEEE 802 MAC Address

Refer to Fig. 13-5. Network interface cards are not interested in any type of layer 3 addressing. NICs receive and transmit data on the network using MAC (media access control) or hardware addresses. Therefore, some type of mapping must be used to map an IP multicast address to a MAC address. But the NIC plays an important role in receiving multicast packets.

Somehow, there has to be a MAC address for multicast, and a single one for all multicast packets is not efficient. Note that up to 32 different IP multicast groups may be converted to the same MAC address, because the upper five bits of the IP address are ignored. It may appear that the upper nine bits are ignored, but the first four bits of a class D address are always 1110 (which converts to 224 decimal, the starting number for class D addresses), and since nine bits are displaced in this procedure, only the next five bits are really ignored. If you read through RFC 1700, you will see that most of the assigned addresses are not affected by this procedure.

On an Ethernet or IEEE 802 network, the 23 low-order bits of the IP multicast address are placed in the low-order 23 bits of the Ethernet or IEEE 802 net multicast address. The IANA has been allocated a reserved

Figure 13-5
Mapping to an Ethernet or IEEE 802 MAC address.

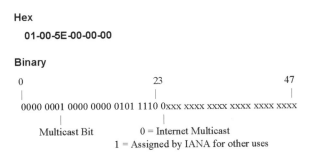

Hex

01-00-5E-00-00-00

Binary

block of MAC layer addresses. Therefore, a multicast MAC address always begins with 01-00-5E (hex).

For example, in Fig. 13-6 the IP multicast address 224.0.1.88 is mapped to a MAC address (converted to hex). First, the IP address must be converted to hex (it is usually written in dotted decimal notation as shown). The address 224 is E0 in hex, 0 is 00 in hex, 1 is 01 in hex, and 88 is 58 in hex. But only the low-order 23 bits are used. Therefore, the IP address of 224.0.1.88 converted to a MAC address is 01-00-5E-00-01-58.

In order for the NIC card to receive or transmit multicast packets, the following functions must be invoked to place the multicast address in the NIC card and to remove it.

JoinLocalGroup (mapped group address)—to allow the link layer to receive multicast packets for a particular host group

LeaveLocalGroup (mapped group address)—to allow the link layer to stop receiving multicast packets for a particular host group

The mapped group address is the MAC address that is mapped from the host group multicast address.

Protocols

Review RFC 1112 (or 2236, IGMP v2 proposed standard). Naturally, a few protocols are necessary to make multicasting work. We will start out with the most prevalent protocols. No matter which router-to-router protocol is used, one protocol is used for all of them. It is the Internet Group Management Protocol (IGMP). To support IGMP, a host must join the "all hosts" group (address 224.0.0.1) on each network interface at initialization time and must remain a member for as long as the host is

Figure 13-6
A converted IP multicast address.

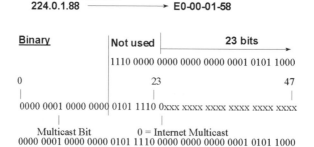

active. Therefore, at least one multicast address that every multicast host should be a member of.

IGMP is the protocol that runs between multicast hosts and their adjacent multicast routers.[3] IGMP is used to keep neighboring multicast routers informed of the host group memberships present on a particular local network. The IGMP header is used for all multicast communication, whether it is between hosts or routers. When you place a sniffer on a multicast LAN, it will show that an IGMP header is used for all multicast communication, whether it involves IGMP hosts or DVMRP, MOSPF, or PIM routers.

In order for an interface to receive a multicast datagram, it must have previously been set up to receive and process multicast datagrams. Since IGMP does not use a transport layer like TCP or UDP, the IP protocol field is set to 2 (as reserved by IANA RFC 1700) to identify the process (IGMP) using the IP service. Therefore, before any multicast packets are received, the upper-layer software must ensure that IP and the MAC layer interfaces are set up to receive multicast datagrams. For this to be accomplished, two simple functions (one for IP and one for the link-layer protocol) are added. First the NIC must be set up, and then the IP software must be made multicast capable.

A host may be a member of more than one group. In fact, there is no upper limit on the number of groups allowed (except for the upper limit of the IP multicast address). Since NICs have a very limited capability for receiving multicast packets, there a limited number of multicast packets that they can receive. In other words, when the user installs its version of IP for multicast, it must set up the NIC to receive multicast packets as well. Each host group will have a different multicast address and therefore will be mapped to a multicast MAC address as well. But the NIC card may only be able to hold a finite number of multicast addresses. Check with the manufacturer. Some have implemented the ability to receive all multicast packets. In this case, it will be up to the IP layer software to filter out unwanted packets.

IGMP Header

The IGMP router places queries to its subnets, and the hosts that belong to a group specified in the query respond with a report. The IGMP

---

[3]Router manufacturers can choose whether or not to implement multicast. Routers that participate in multicast must run a multicast protocol (beyond RIP, OSPF, etc.). Most major router manufacturers implement or are in the process of implementing these protocols.

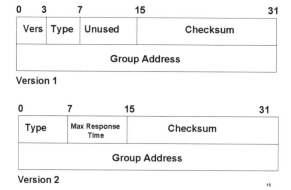

Figure 13-7
IGMP header.

header is shown in Fig. 13-7. This illustration shows headers for IGMP version 1 (RFC 1112) and version 2 (RFC 2236; you will find both types on your network). The Type field can be one of four types for host communication. The four types are

0x11—membership query

0x16—version 2 membership report

0x17—leave group, for IGMP version 1 compatibility

0x12—version 1 membership report, for backward compatibility

IGMP version 2 has a different header than version 1. The Version and Type fields of a version 1 header are combined into one field called the Type field. To allow a multicast router to determine the difference between the two, a new Type field was created for version 2: membership report message. IGMP version 1 and version 2 routers may coexist. An IGMP version 2 router must be able to act as a version 1 IGMP router. To determine this, the Max Response Time field is set to 0 in all queries (this maps to the unused field in version 1).

The membership query packet has two types:

General query—used to learn of members of any group on an attached network

Group-specific query—used to learn of members of a specific group

How do you tell the difference between the two? This is determined by the group address in the IGMP header. A general query uses all zeros in the group address field, and the group-specific query uses the exact group address. Both of these messages are sent using the IP header address of 224.0.0.1 and a mapped multicast MAC address of 01-00-5E-00-00-01.

The Leave Group message is new to IGMP version 2. It allows a router to immediately determine if there are any members of a group left on its interface that received the Leave Group message. This is explained later.

Router Functions of IGMP

A multicast router uses IGMP to learn which groups have members on each of its interfaces (see Fig. 13-8). This information will be used to build multicast trees for forwarding multicast data. The router will also have a timer for each of those group memberships. A router that runs IGMP is also a member of any host group that has members on one or more of its interfaces. The multicast router keeps a list of the memberships on its interfaces, and not of the individual hosts that belong to that group. There really is no need to keep track of the hosts. If only one host on a router interface wishes to join a group, the router has to forward multicast datagrams to that interface. It does not matter if there are 100 hosts or 1 on that interface; the router must be a member of that group as well and forward multicast datagrams out that interface. The multicast router will know if there are any members of a group left on its interface by sending a query packet out that interface.

For IGMP version 2, a multicast router may assume one of two roles: querier or non-querier. All multicast routers, upon initialization, assume they are querier routers. Since multicast routers periodically transmit a query to find hosts for groups, the new router will eventually receive a query if there is another router providing this function. If the router receives a query message from another router, and that router has a lower IP address, the new router will assume the role of a nonquerier. If the new router has a lower IP address, it will assume the role of querier and the other router will assume the role of nonquerier.

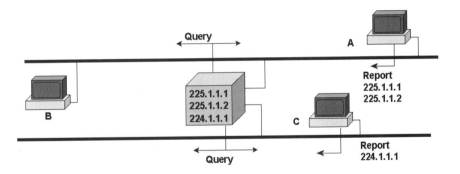

Figure 13-8
Router functions of IGMP.

When a host receives a query, it sets delay timers for each group that it belongs to (set between 0 and the time indicated in the Max Response Time field of the received query). For a host with more than one interface, each interface maintains it own timer. When the time is up, the host responds with a version 2 membership report. The TTL field of the IP header will be set to 1. This ensures that the packet will not be forwarded beyond the local network that it was transmitted on. If the host receives a report from another host in the same group, it will stop its timer and it will not send a report. This is to conserve bandwidth and processing time, and to avoid having duplicate reports on the network. All of this is accomplished using multicast addressing. 224.0.0.2 is the all-routers multicast address, and 224.0.0.1 is the all-hosts address (which can include routers).

If the router receives a report, it will add the group to its internal list, noting the interface that it received the report on. It then sets a timer for the next query message. If it receives more reports for a group, the timer will be reset to the Max Response Time value and started again. If no reports are received before this timer expires, the router assumes there are no members for that group, and it will not forward remotely received multicast datagrams on the interface.

Host Join

Figure 13-9 depicts the router functions of IGMP. When a host joins a multicast group, it immediately transmits a version 2 membership report two or three times (remember, IGMP does not use ICMP or TCP). This is done in case that host is the first member of the group, and it is repeated in case the first report gets lost or clobbered.

When a host leaves a group (IGMPv2), it transmits a Leave Group message. A host may or may not be able to determine if it was the last

Figure 13-9
Host join.

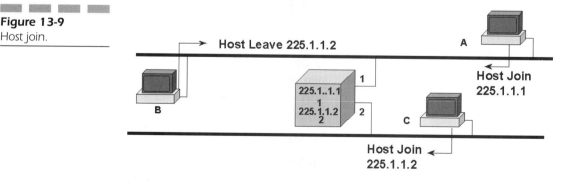

member of a group; this is a storage and processing decision on the part of the implementer. A host will transmit this message only if it can determine that it is the last host in the group. Implementations that cannot determine this may or may not send this message. The multicast router will determine if any hosts exist for a group using the query message in any event. When a multicast router receives this message, it will send group-specific queries to the group left. If no reports are received, then the multicast router will assume there are no members left in that group, and it will not forward any multicast datagrams for that group out of that interface.

IGMPv2 is a proposed standard (RFC 2236) and can be found on the Web as indicated at the end of this chapter. It mostly provides for the ability to conserve bandwidth by allowing a host to elect to receive traffic from specified sources (IP addresses) of a multicast group. Alternatively, it allows a host to specify which sources it does not want to receive information from. What is a source? It is simply a host that originated a multicast datagram. There may be many sources in any one group. With IGMP versions 1 and 2, a host is required to receive all information for a group of which it is a member, of no matter which source transmitted it. Also, the Leave Group message is enhanced to allow a host to specify which sources it no longer wishes to receive information from. The multicast router will receive this message and possibly stop sending information from that source to that group.

Now that we understand how the host operates with IP multicast and how a host interacts with a multicast router, we need to learn how multicast actually operates. First, we will study the algorithms and then we will take an in-depth look at one multicast protocol, the Distance Vector Multicast Routing Protocol, or DVMRP. As indicated at the end of this chapter, the draft RFC used is version 3 of DVMRP.

Multicast Algorithms

Many people do not even realize that they have already worked with multicasting. If you have worked with the spanning tree algorithm for bridging or the Open Shortest Path First (OSPF) protocol for IP routing updates, you have worked with a multicast algorithm.

IP multicasting for subnet routing uses the following protocols:

■ Distance Vector Multicast Routing Protocol (DVMRP)
■ Multicast Open Shortest Path First (MOSPF)

- Protocol-Independent Multicast (PIM)

  Sparse mode

  Dense mode

- Pretty Good Multicast (PGM), a Cisco Systems proposed algorithm, not discussed here

There are essentially three forwarding algorithms that can be used with IP multicasting:

- Flooding
- Spanning tree

  Simple spanning tree

  Reverse path broadcasting

  Truncated reverse path broadcasting

  Reverse path multicasting

- Core-based trees—used in sparse environments (environments that do not have a densely populated environment of hosts); CBT is a type of spanning tree algorithm, but different enough to merit its own category

The purpose of all the algorithms is to build a multicast tree for the forwarding of multicast datagrams. Some algorithms (CBT and sometimes PIM) build only one tree that all members of the group share, even if the tree does not supply the most efficient (shortest) path between all members of the group. Other protocols build shortest-path trees for all members in the group (DVMRP, OSPF). There may be multiple multicast trees built for each source/multicast destination on a network. These trees are built on demand. They are built dynamically when the first multicast datagram arrives from the source.

Multicast datagrams do not necessarily follow the unicast datagram path. The multicast tree is a dynamic logical tree that a router will build to forward multicast datagrams to their receivers.

Leaves, Branches, and the Root

Refer to Fig. 13-10. Throughout all the changes in IP multicast algorithms, one thing has remained constant. That is the topology view of a multicast tree. As was previously covered in the bridging section of this book, it is based on leaves, branches, and a root. A few terms should be explained here before we continue. Those familiar with the spanning

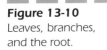

Figure 13-10
Leaves, branches, and the root.

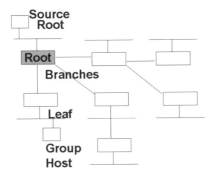

tree already understand leaves, branches, and the root. Those who don't should read on.

A spanning tree has leaves, branches, and a root. The leaves are simply the endpoints of the tree. If there are no forwarding paths beyond a router path, then the interface is considered a leaf interface. If there are more forwarding paths to a host group, then the interface is considered a branch. The root is the source of the multicast transmission. There can be many sources for a multicast network.

Think of a tree. A tree has a root, a trunk, branches, and leaves. Leaves are the outermost parts of the tree. In fact, they are endpoints. The branches contain the leaves, but leaves cannot contain branches. The root is the source of all life in the tree. Lose the root and the tree dies.

Spanning Tree and Flooding

The simplest and most inefficient algorithm for IP multicast forwarding is flooding. Essentially, when a multicast router receives a multicast datagram, it first checks to see if it has received this datagram before. If it has, it discards the datagram. If it has not, the datagram will be forwarded to all interfaces on the multicast router except for the one that the datagram was received on. You can see how simple this would be to implement. There is not much in the way of resources required to implement this algorithm. But the flooding algorithm does not scale well. In other words, as your network grows, the flooding algorithm will hog resources and become very inefficient. It will generate a large number of duplicate packets and it forwards them out all the interfaces that it has configured, even if there are no hosts downstream that belong to that multicast group. The downstream routers will have to process the datagrams as well. The router will have to maintain a table for each packet recently received, and a timer mechanism would have to be established to clean up the tables.

Therefore, the spanning tree algorithm looks more appealing. It requires more logic in the multicast routers, but the trade-off in efficiency is well worth the price. The first algorithm that was invoked was a simple spanning tree. It created one spanning tree out of the current Internet topology. See Fig. 13-11.

Once the spanning tree is built, if a multicast router receives a multicast datagram, it forwards the datagram out each of its spanning tree interfaces, except the one that it received the datagram on. This eliminates the loops provided in the simplex flooding algorithm, and the router is not taxed by having to maintain tables for recently forwarded packets (duplicates). Although the spanning tree adds more maintenance traffic to the network (to maintain the spanning tree topology) and involves more overhead processing in the router, the efficiencies provided make it superior to the flooding algorithm. But this method has drawbacks in that it may not provide the best paths to all destinations based on the group address and the source. It simply forwards multicast datagrams out its spanning tree interfaces without regard to the group address and the source and to whether there were any recipients on any part of the spanning tree. In other words, the complete spanning tree sees all multicasts, including non-members.

Reverse Path Broadcasting

Because the simple spanning tree was not efficient or scalable, another algorithm known as reverse path broadcasting (RPB) was devised. The

Figure 13-11
Spanning tree and flooding.

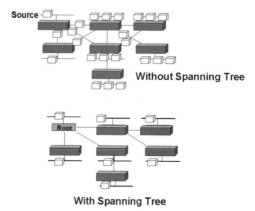

Source

Without Spanning Tree

Root

With Spanning Tree

function of this algorithm is to provide a group-specific spanning tree. For each group source (the host that originates the multicast datagram), a distinct spanning tree is built between that source and all the potential recipients. Refer to Fig. 3-12.

This algorithm may seem backward, and in a way it is. We know that multicasting is based on a (source, group) pair. If a multicast datagram is received on a router's interface, the router will then determine if the interface that the datagram was received on is the shortest path back to the source. Sounds like the opposite of RIP? It is. If the router determines that the interface does provide the shortest path back to the source, then it will forward the received datagram on every interface that is has active except the one that it received the datagram on. Otherwise, the router has determined that the interface does not provide the shortest path back to the source, and it will discard the datagram. The interface that the router determines is the shortest path back to the source is called the *parent link*. The interfaces that the router forwarded the multicast datagram on are called the *child links*.

Where does the router get the information to enable it to make the decision on what is the shortest path back the source? The routing tables of the router. If you are using a link-state routing update protocol such as OSPF, each router maintains a routing table for whole network. If you are using a distance-vector protocol such as RIP or RIP2, routing updates are done by either advertising the previous hop for the (source, group) pair or by relying on the poison reverse feature of a router. Don't worry about this now. We will explain it further in our discussion of DVMRP. DVMRP combines the simplex routing abilities of RIP and reverse path multicasting.

Figure 13-12
Reverse path broadcasting.

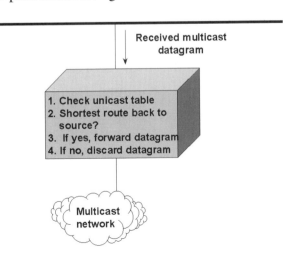

RPB involves many advantages over the previous mechanisms described. It is simple to implement. Multicast datagrams are forwarded over multiple links since RPB builds separate spanning trees for each (source, group) pair.

Truncated RPB and Reverse Path Multicasting

Once RPB was implemented, limitations soon arose. For example, there was no method of telling which network segments had multicast hosts for a group and which ones did not. Multicast datagrams were simply forwarded on all active paths of the spanning tree for each (source, group) pair.

To overcome these limitations, the *truncated RPB* (TRPB) algorithm was devised. The TRPB algorithm basically forwards the first multicast packet of every (source, group) pair to all participating routers in the (source, group) spanning tree. Using IGMP, a multicast router can determine the group memberships on each leaf subnetwork and thereby determine whether any of its segments have active host groups. If not, then it simply does not forward a multicast datagram out that interface. The tree is truncated at the branch. Refer to Fig. 13-13, which shows TRPB at the top. The router forwards multicast datagrams to the router below but truncates it on the left port.

That solved one problem, but to make the scheme still more efficient we want to be able to truncate at the branches as well. This is called *reverse path multicasting (RPM)*. Since the TRPB protocol sends the first packet to all interfaces on the (source, group) spanning tree, a router can determine whether hosts exist for the group on its interfaces. TRPB allowed the router to truncate only at the leaf level. But what if there are no hosts associated with that group on any of the router's interfaces? This is shown at the bottom of Fig. 13-13. RPM allows the router to transmit a *prune* message back through the interface that it received the multicast datagram on (its parent link), which will enable its upstream neighbor to basically shut off the interface to that downstream router. There is no need to forward multicast datagrams to the router if it is only going to throw them away. Prune messages are only sent once. If the upstream router does not have any leaf networks for a host group and other branch interfaces all sent back a prune message, then that upstream router may send a prune message to its upstream router (its

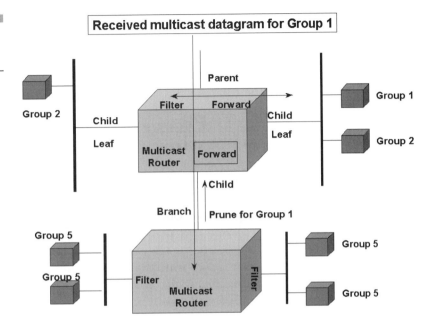

Figure 13-13
TRPB and reverse
path multicasting.

parent link) as well. The next upstream router would then shut off its interface to that downstream router. You can prune all the way back to the root.

This cascading of prune messages creates a true spanning tree topology that will forward multicast datagrams only to those interfaces that have active group hosts. How do we re-grow a branch or create leaves? Periodically, the prune interfaces are removed from the router's table, and the branches and leaves grow back. This allows the forwarding of multicast datagrams down those branches, resulting in a new stream of prunes message to create the true spanning tree.

This algorithm eliminated all problems except for one: scaling. It still does not allow for growing the network to thousands or tens of thousands of routers with hundreds or thousands of multicast groups. The first multicast packet is received by all routers, and then constant pruning messages are needed to keep the spanning tree efficient.

Distance Vector Multicast Routing Protocol (DVMRP) uses RPM today. IGMP is also used today. The other protocols have historical significance only and are presented here simply to show the progression of IP multicast algorithms.

Core-Based Trees

The previous algorithms built spanning trees based on a source host. Multiple trees could be built all from different sources. The source host is basically the root of the spanning tree, and the spanning tree branches out from the source. If there are many sources, there are many roots. If there are many multicasts, each has its own multicast tree. With the core-based tree (CBT), a single forwarding tree is shared by all members of a group. The core of this tree (the root) is based on the core router, not the source of the multicast datagram.

Refer to Fig. 3-14. CBT works by building a backbone consisting of at least one core router. Multicast messages for a group are transmitted in this direction. Any host that wishes to receive multicast information for a specific group transmits a join message toward the core backbone. Each source must be configured with at least one IP address of the core routers. The core consists of at least one router. There can be multiple core routers, and if so, the links that connect these core routers become the core backbone.

If there are multiple groups in the network, multiple trees may be built. This method is not based on the concept of one multicast tree for all groups. However, there is only one multicast tree for each group.

After the join message is issued, each intermediate router marks the interface and the multicast group and then forwards the message toward the core. In doing this, the router is able to forward multicast data toward the core for that group. When the core routers receive this data, they in turn multicast the data back out all ports. A core router will multicast the data back over all ports except the one that it received the data on.

Figure 13-14
Core-based trees.

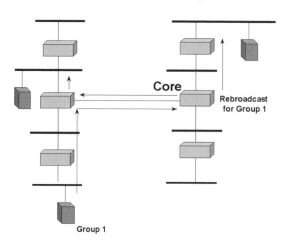

CBT may simply remain an algorithm for which other protocols will be developed. To become a protocol, issues of dynamic selection of the core backbone and management must be settled. The most notable protocol using it is the Protocol-Independent Multicast (PIM) sparse mode (covered later). There are advantages and disadvantages to the algorithm. Since each group is based on single tree rooted at the core, state information on the router is easier to maintain, which reduces the resources required on the router. Information that must be passed between routers to maintain the state is also reduced, resulting in better efficiency of the bandwidth. However, since all individual data and control messages travel toward their specific core, congestion may be inevitable.

Distance Vector Multicast Routing Protocol

The following text about the DVMRP routing protocol is based on the DVMRP version 3 specification, which as of this writing is still a draft RFC. DVMRP uses the reverse path multicasting (RPM) algorithm to dynamically build multicast delivery trees. DVMRP is used to determine the router's position in the multicast tree in reference to the source subnet of a multicast datagram. DMVRP is a "broadcast and prune" multicast routing protocol. It uses reverse path forwarding to see if a multicast datagram should be forwarded downstream. A multicast forwarding tree is built between a source and all members of the group (receivers). For those familiar with this protocol, it should be noted up front that the latest version of this protocol for UNIX, mrouted 3.5, is also based on RPM but is significantly different from previous versions in the areas of packet format, tunneling, etc.

DVMRP combines the simplicity of the RIP routing protocol with reverse path multicasting and IGMP to produce a routing protocol for multicast packets. In order for DVMRP to work, two tables must be built: a unicast routing table and a forwarding table. The unicast route table is used to determine if a multicast datagram was received on the correct port (the upstream interface). The forwarding table is used to determine which interface a router should forward a multicast datagram on (the downstream interface).

To build the unicast routing table, DVMRP passes route reports containing entries for source subnets. This table is processed as with RIP, and the shortest distance back to a source is computed and placed in the table. The forwarding table is built by broadcasting the first multicast datagram received and then waiting for other routers to send back prune and graft messages to indicate who wants and who does not want the datagram. Why not simply use the unicast routing table? The forwarding table allows multicast traffic to follow a different path than the unicast traffic and provides support for a tunnel interface, which unicast traffic does not understand.

DVMRP routers support two types of interfaces: router and tunnel. The multicast router interface is obvious, but the tunnel interface is not. To allow for non-multicast routers to exist in a multicast network, the concept of tunnels is used. Multicast datagrams are encapsulated in unicast IP packets (using IP in IP encapsulation), and these are sent over the unicast routers. Contained in the IP header is a route list that the unicast routers should use. The last entry in this list is the end of the tunnel and is a router that again supports IP multicast. The last router strips off the unicast information and send the datagram on.

DVMRP is a protocol that uses a distance-vector distributed routing algorithm allowing each router to determine the distance from itself to any IP multicast traffic source. When DVMRP determines this, it creates IP multicast delivery trees between a source and its distributed group hosts.

DVMRP and IGMP

Figure 13-15 shows a comparison between the multicast routing protocol of DVMRP and the host membership protocol IGMP. The protocol of IGMP runs between the hosts and the routers, and DVMRP runs between the routers.

It should be noted here that with IPv4, the encapsulation of DVMRP data is accomplished using IP encapsulation protocol type 2, which is IGMP. If you place a protocol analyzer on the network, you will see DVMRP communicating using IP protocol type 2 for IGMP encapsulation. This does not mean that IGMP is running between routers; it is simply using the IGMP encapsulation to send its data.

Neighbor Discovery

A DVMRP router can discover neighbor DVMRP routers through a process known as neighbor discovery using the *probe packet*. When a DVMRP router is initialized, it transmits these discovery packets to

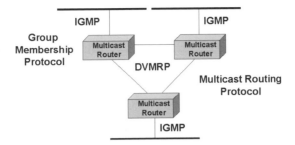

Figure 13-15
DVMRP and IGMP.

inform other DVMRP routers that it is operational. These messages are sent periodically (every 10 s) to the All DVMRP Routers multicast address. Each message should contain a list of neighbor DVMRP routers that it knows about on that interface. Other routers on other interfaces are not included in this listing; it is local only. A router should see its IP address in its neighbor's messages.

The probe packets enable other DVMRP routers to discover each other and to also detect when a neighbor router no longer exists. If a DVMRP router does not detect this message from a neighbor within 35 s, it considers that neighbor to be down. Contained in this message is a listing of all other DVMRP neighbor routers that the router knows about. If a router does not receive any probe messages, it considers the subnet to be a leaf network only.

Route Reports

Unicast routing information is sent between neighbors using a special packet called a *route report*. For those familiar with RIP, a route report is a RIP update. Contained in the route report is the unicast (source) subnet and its mask, and the metric (cost) associated with that subnet. A route learned through route reports should be refreshed within 140 s (2 × report interval + 20), after which it can be replaced with the next best route to the same source. If no update and no alternative route exists and 200 s have passed, the route is discarded from the routing table.

A route report is sent out every 60 s. Anytime during this interval any number of route reports can be sent. In this way, a router is not burdened by a periodic update like RIP that could consist of thousands of routes. At any time during this interval "flash updates" can be sent. These reports indicate changes in the network but contain only the source subnet that has changed. This reduces the chances of loops and other catastrophes that may result when paths for source networks change.

Receiving a route report is a different matter. There are many checks done on the received information. Route reports are processed only if they are sent by a known neighbor; otherwise they are discarded. Generally, two rules are followed. (1) If the route entry is new and the metric is less than infinity, the route is added. That is the simple rule. The second one is more complicated. (2) If the route entry exists, perform the following checks:

New metric < infinity AND new metric > existing metric	If the same neighbor is reporting it, update the entry; otherwise discard the entry.
New metric < infinity AND new metric < existing metric	Update the entry with the route, and if necessary update the reporting neighbor.
New metric < infinity AND new metric = existing metric	Refresh the route and if the new neighbor has a lower IP address, update that entry.
New metric = infinity AND new gateway = existing gateway	Route is now unreachable; update the entry
New metric = infinity AND new gateway ≠ existing gateway.	Ignore
New metric is between infinity and 2 × infinity	Neighbor considers the receiving router to be upstream for the route indicated, and that router is dependent on the receiving router for the route indicated. If the receiving neighbor router considers that router to be downstream, the receiving router marks that neighbor as dependent for that route; otherwise the packet is discarded, for a dependent router cannot be considered to be upstream.
New metric > 2 × infinity	Ignore.

DVMRP Tables

When an IP multicast datagram is received by a router running DVMRP, the router first looks up the source network in its DVMRP unicast routing table. To ensure that all DVMRP routers have a consistent view of the unicast path back to a source, a unicast routing table is propagated to all DVMRP routers as an integral part of the protocol. This is a separate unicast routing table for DVRMP. As stated previously, there are two tables used in multicast routing: a routing table and a forwarding table. What is contained in these tables? The DVMRP routing table contains source subnets and From gateways. It has the shortest-path source-rooted spanning tree to every participating source subnet in the Internet. Compare this with the entries from a typical IGP table such as RIP, which contains destinations and next-hop gateways. The forwarding table is created because the routing table is not aware of group memberships.

Table 13-1 shows a simple DVMRP routing table. The headings are defined as follows:

- Source subnet—the subnetwork that contains a host source
- Subnet mask—the subnet mask of the source subnet
- From gateway—the immediate upstream router that leads back to the source subnet
- TTL (time to live)—indicates how long the entry stays in the table before being removed

Table 13-2 shows a DVMRP forwarding table. If an interface is designated a leaf network, or the router has downstream routers for the source group, it will be included in the downstream ports. The upstream port is determined by the unicast routing table as follows: if this interface has the shortest route back to the source subnet of this group, it is registered as the upstream interface with that port designa-

TABLE 13-1

A Simple DVMRP Routing Table

Source Subnet	Subnet Mask	From Gateway	Metric	Status	TTL
150.1.0.0	255.255.0.0	150.1.1.1	5	Up	400
150.2.0.0	255.255.0.0	150.1.2.1	3	Up	350

TABLE 13-2

A DVMRP Forwarding Table

Source Subnet	Multicast Group	TTL	Upstream Port	Downstream Ports
150.1.0.0	224.0.1.1	430	1	2, 3
150.2.0.0	224.0.1.2	500	1	3
	224.0.2.5	300	3	5

tion. The forwarding table is created to represent the local router's understanding of the shortest-path source-rooted delivery tree for each (source, group) pair.

To build a unicast routing table, the upstream DVMRP router depends on downstream routers for information. The information that is sent to other DVMRP routers is called a route report. The metric is the most important field of the report. This is not only used to build the source subnet table (indicating the source subnet and their reachability), but it also enables the building of a forwarding table to indicate to the router who the downstream routers are that depend on that upstream router for the forwarding of multicast datagrams. Upstream routers send route reports to their downstream neighbors indicating source subnets and their metrics. As with RIP, the metric to a source subnet is the cumulative cost of all the incoming interfaces so far. The route reports will be sent to a DVMRP neighbor router. Contained in this report are source subnets and metrics (in the range of 1–31).

If a downstream router wishes to indicate to an upstream router that it is dependent on it for receiving multicast datagrams for a particular source subnet, that downstream router will echo the route back to the upstream neighbor with a metric higher than 32. Infinity for DVMRP is considered to be 32. Therefore, the downstream neighbor will add 32 to the incoming metric and echo this back to the upstream router. This relies on a technique known as *poison reverse*. When the upstream router receives this update and sees that the metric for the source subnetwork lies in the range between infinity and twice infinity, the upstream router will add the downstream router to its list of dependent routers for that source. The infinity value of 32 indicates that a source network is not reachable. The full range of metrics is 1–63, where 1–31 is the original metric of the source, 32 means not reachable, and 33–63 is the poison reverse metric of a downstream router telling its upstream router that it wants to be added to its table for multicast datagrams of a given source.

Why not just use the existing unicast routing table such as a RIP2 table? The reason is that not all routers will be running DVMRP, and multicast routers must be able to interact with non-multicast routers. In order to accomplish this, tunnels must be built across non-multicast routers. Tunnels effectively force the path that the multicast datagram will take. The tunnel may take one route and the regular unicast packet may take another route. In this way, a router's unicast table may not coincide with a DMVRP router's unicast table. Therefore, we use the unicast information in DVMRP exclusively to determine the shortest route back to the source subnet of a multicast datagram.

Pruning and Grafting

As shown in Fig. 13-16, DVMRP broadcasts (floods) the first packet that it receives and then waits for prune messages from the downstream interfaces. The prune messages come back from DVMRP routers in response to unwanted multicast traffic at the leafs of the multicast tree. Therefore, finding the leaf networks for any multicast tree is important. Those routers that connect to leaf networks start the pruning process, identifying which downstream interfaces do not belong to the multicast group. Routers that identify those interfaces prune those interfaces. If, after pruning its own interfaces, the router finds that none of its interfaces belong to that group, it sends a prune message upstream to its neighbor. If the router continues to receive multicast datagrams for that source, it will continue to send prune messages (increasing the delta between them) until the multicast traffic for that group stops. A pruned interface's lifetime is about 2 hours.

To join new receivers to the tree, the *graft message* is sent. This message is sent hop by hop to each multicast router. Each message is acknowledged between routers to ensure that it was received, thereby guaranteeing end-to-end delivery. Routers that receive graft messages can make a series of decisions.

If the receiving router has a prune state for the (source, group) pair, it acknowledges the graft message and sends a graft message of its own to

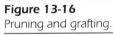

Figure 13-16
Pruning and grafting.

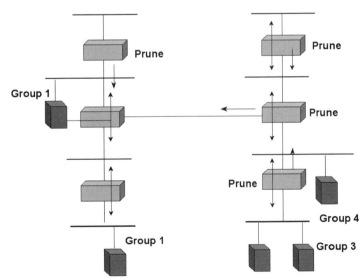

its upstream router. If the router has some pruned downstream interfaces but not a pruned upstream interface, it simply adds that interface to the list of downstream interfaces in its routing table. It also sends an acknowledgment to the source of the graft message. If the router has no state (pruned or otherwise) for the (source, group) pair, then any received datagrams for the (source, group) pair should be automatically flooded. A graft acknowledgment is sent to the source of the graft message as well.

Tunneling

RFC 2003 describes IP encapsulation or tunneling. RFC 1075 also describes tunneling for multicast. It is not necessary to have all routers running a multicast protocol. For example, most routers on the Internet do not run a multicast protocol. With tunneling, you build "islands" of multicast autonomous networks, and they communicate by tunneling the multicast datagram over the Internet. This is shown in Fig. 13-17.

DVMRP supports the ability to tunnel a multicast datagram through non-multicast routers. The multicast datagram is encapsulated in a unicast IP packet and addressed to the routers that support native multicast routing. In other words, we wrap the multicast packet in an IP header and tell it what path to take to a destination multicast router.

To encapsulate an IP datagram using IP in IP, as shown in Fig. 13-18, an IP header is inserted before the existing IP datagram header. The source and destination addresses of the outer IP header describe the input and output of the tunnel or the tunnel endpoints. The original IP

Figure 13-17
Tunnels.

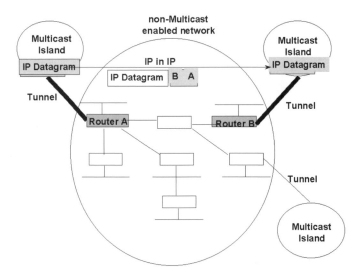

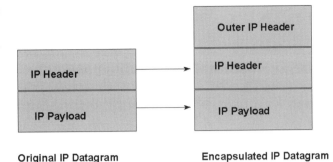

Figure 13-18
IP in IP packet format.

IP Header

IP Payload

Original IP Datagram

Outer IP Header

IP Header

IP Payload

Encapsulated IP Datagram

header contains the IP source and destination addresses of the originator and final destination of the datagram.

Protocol-Independent Multicast

DVMRP provides great mechanisms for multicasting. Using the latest version of IGMP and DVMRP, a fully functional multicast tree or trees can be dynamically built and utilized for voice, video, and data. But there are some disadvantages as well. It does not scale well. DVMRP is a distance-vector multicast routing update protocol. DVMRP broadcasts the first multicast packet it receives from the source and then prunes the tree based on feedback from other routers. DVMRP is known as a *dense-mode* multicast protocol. It is most efficient when the clients are densely located on the network. It does not scale well when the protocol is being used over WAN links or when there are simply a few clients scattered throughout the customer's network. This broadcast and prune mechanism, along with multicast routing updates, causes unnecessary overhead over low-bandwidth media types. Furthermore, DVMRP routing tables are based on a RIP-like update. DVMRP also requires the routers to keep state information. This includes group and source information as well as use of this information to calculate a tree.

If all the members of a multicast group are located in a bandwidth-rich region (supported by high-speed LANs and not low-speed WANs), then the group members should be supported by a dense-mode protocol such as DVMRP, MSOPF, or Protocol-Independent Multicast (PIM) dense mode (PIM-DM). This can be limiting in that the scope of the group cannot include any members beyond the scope of the domain without placing the unnecessary burden of broadcast and prune messages and possibly multicast routing updates over the links that include that remote receiver.

PIM Dense Mode

PIM offers two versions for multicast routing: dense mode and sparse mode. Dense mode is the easiest to explain, especially if you have read the previous section on DVMRP. It functions similar to DVMRP in that it uses RPM to build source-routed multicast trees. But PIM does not rely on an independent unicast routing protocol, whereas DVMRP does. It is protocol independent.

When a multicast packet arrives on a PIM-DM interface, it is checked against the unicast forwarding table (of any type: RIP, OSPF, etc.) to make sure it was received on the interface that provides the shortest route back to the source. If it was, it is forwarded to all interfaces, until the branches are specifically pruned. Unlike DVMRP, PIM-DM will continue to forward multicast packets until specific prune messages are received. DVMRP uses a routing table to determine if there are downstream routers that want to receive the multicast datagrams for a specific group. DVMRP, relying on a routing table that is sent to all multicast routers is more selective when it forwards messages during the construction of a source-rooted multicast tree. The reasoning behind this is that simplicity and protocol independence were selected as higher priorities than the additional overhead caused by packet duplication. Building a unicast routing table virtually eliminates duplicate packets. PIM-DM accepts duplicate packets as an alternative to becoming dependent on a unicast routing protocol and therefore avoids building yet another routing database. PIM-DM assumes that all downstream interfaces want to receive multicast datagrams. PIM was actually built for sparse-mode multicast networks, and DM was added for simple functionality. PIM-DM does not contain the concept of rendezvous points, and there are no periodic joins (however, the join message is still used). A draft RFC allows for "border routers" that enable PIM and DVMRP interoperability.[4] PIM-DM is less complex than DVMRP.

PIM-DM uses three mechanisms to build a multicast tree:

- Prune
- Graft
- Leaf network detection

---

[4]This can be found at *http://netweb.usc.edu/pim.*

Protocol Operation

The following is also presented as a flowchart in Fig. 13-19. When a multicast datagram is received, its incoming interface is looked up in the unicast routing table. Therefore, the router must be running some type of unicast routing protocol (remember the name *Protocol-Independent Multicast*). If the receiver interface is the one that the router would forward unicast datagrams back to that subnet on, the multicast datagram accepted and is forwarded to all ports except the incoming interface. If not, the datagram is simply discarded without error messages being sent (silently discarded). From here, the router checks for a forwarding state for the group address. If there is no entry for the group address, the router adds one. The router checks the outgoing interface list to see whether it should forward the datagram. The outgoing interface list contains a listing of interfaces for which the router has heard a group membership report or has heard PIM router messages. The PIM-DM router messages can be hello, prune, join, or graft. If there are active interfaces, the router forwards the datagram out those interfaces. If no interfaces are indicated, a prune message is sent.

The intended receiver router of that prune message is placed in the message (not in the IP header). The downstream router knows this address by doing an RPF lookup in the unicast routing table. When the receiver router receives this prune request, it schedules a deletion of that

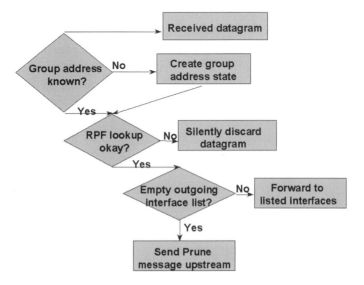

Figure 13-19
PIM dense-mode operation.

LAN interface for that group, meaning that it inserts a delay before deletion. It is waiting to see if any other routers respond. Other routers on the subnet also receive this prune message and in turn send a join message to that router, forcing that router to cancel the deletion of the LAN interface for the (source, group) pair. The fact that one router prunes does not mean that no other routers on that same LAN want to continue receiving information for the group address.

The absence of entries in the outgoing list could mean that there are no group members on the interface and the router has not received any PIM hello messages from other routers located on that subnet (this allows for leaf network detection, in that in absence of these messages, only multicast hosts reside on a subnet). A router not only keeps track of the leaf members (local group database built by IGMP) but also contains a listing of routers as well. When a router is not heard from in a specified amount of time, a router deletes that router's entry from the list.

Pruned states for any multicast entry are eventually timed out, forcing all multicast datagrams to be forwarded on all interfaces again until the multicast trees become pruned once again.

Adding Interfaces

A router can add an interface for which it received a graft message (rejoin a branch for a group) or received an IGMP membership report. If the router already has state information about a group (it has built an entry for the group), it simply adds or refreshes the interface entry on which the IGMP message or graft message was received. If the outgoing list entry is empty, the router will send a graft message upstream toward the source. Any router that receives this message will use that received interface as the outgoing interface for the existing (source, group) pair. If the router has no state information at all for the (source, group) pair, it will do nothing, for it knows that PIM-DM routers will deliver a multicast datagram to all interfaces when creating a state for the group.

PIM graft messages are positive acknowledged. A PIM graft message is unicast to upstream routers. The upstream-router changes the graft message into a graft acknowledgment and sends it back to the originating router.

PIM Sparse Mode

PIM-SM was designed to restrict multicast traffic to only those routers that have a need for the packet. In PIM-SM a specific router is known as the *rendezvous point (RP)*.[5] Senders and receivers join a multicast group by registering at the rendezvous router. Routers find out their RP and then send received multicast datagrams as *unicast datagrams* to the RP. The RP router redistributes multicast datagrams out the group trees that it has built. A rendezvous point is simply an IP address of a single router. These points are used by senders to announce themselves and for receivers to find out about new senders for a group.

PIM sparse mode is modeled after the core-based tree algorithm. However, the difference between the two is that CBT uses one tree, centered at a core router, instead of at the source of a multicast datagram. CBT builds a single tree for all members in a group.

All routers running PIM periodically (every 30 s by default) transmit hello messages to each other for the purpose of discovering other PIM routers. They use the 224.0.0.13 (ALL_PIM_ROUTERS) group address. This is local multicast that is not allowed to traverse a router. When a PIM router receives this message, it stores the IP address for that neighbor. Each PIM router entry has its own timer for repeat hello messages. This time is included in the received hello message, and the router notes this time in its table (set to 3.5 · hello period, or by default 105 s). If the router does not periodically hear from the neighbor, it times out and deletes that neighbor from the table. When the router known as the *DR* (designated router, usually selected by IGMPv2) receives a new entry (a new router), it unicasts its most recent RP address information to the new neighbor.

The DR is responsible for sending join/prune commands to the RP on behalf of its local receivers and sources. The choice of the DR is not based on the IGMP querier, nor is it based on the long-term, last-hop router for the group. The router with the highest IP address within all the received hello messages is elected DR. The last-hop router is the one that is the last router to receive multicast messages before they are delivered to the local receivers. If there is such a router, it will be the DR.

Types of Multicast Trees

Multicast trees are still built using PIM, and there are two types:

- A shared tree (rendezvous-point-rooted trees), indicated by (*,G) in the routing table, is a shared tree for the multicast group G.

---

[5]For redundancy and scalability, some PIM-SM implementations allow for more than one rendezvous point, but that is beyond the scope of this book.

■ A source-rooted tree (or SRT tree), indicated by (S,G) in the routing table, is a tree that has been built for the multicast group G and is sourced by the IP address S.

Like all other multicast routing protocols, PIM conveys its messages in IGMP header data packets. If a host (receiver) wants to join a group, it conveys its membership information through IGMP. When a PIM router receives this IGMP message, the DR looks up the associated RP. The DR creates a wildcard entry for the group, which is written as (*,G). The DR creates a join/prune message (both join and prune entries are included in the same message). This will be illustrated below. PIM works in conjunction with IGMP.

For a given (source, group) pair, a multicast tree is initially built around the RP router. This initial tree is called a shared tree because all members of the group converse using this single shared tree (although it may not be the shortest path between a source and a host). A shared tree is easy to construct, reduces the amount of overhead in the router (tables, state information, etc.), and is easy to implement. However, it may not be efficient. Shared trees are built based on the center point rendezvous router. A shared tree may not allow for the shortest path to be built between a source and some receiver hosts.

The PIM protocol can adapt here as well. Based on the data rate, after the shared tree is constructed (after meeting at the rendezvous point), a shortest-path tree can be built between a host receiver and a source. The router sends a join command directly to the source, and a multicast tree is built. The original path through the rendezvous router is torn down.

Joining a Group

Refer to Fig. 13-20. A router with directly connected neighbors must first join the shared tree. When the DR gets a membership notification from a host, it looks up the associated RP for that group (more information on the RP is coming up). The DR creates a wildcard multicast entry for the group in the form of (*,G). If there is no specific match for the group, the packet is forwarded according to this entry. The RP address is contained in a special field in the route entry and is included in periodic join/prune messages.

The DR sends a join message to the primary RP. The (*,G) entry indicates an (any source, group) pair. The intermediate router (B) forwards the unicast PIM join message. Router B also creates a forwarding cache entry

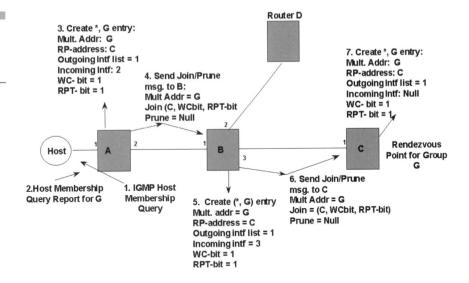

Figure 13-20
Joining a group.
(From PIM-SM draft
RFC)

for the (*,G) pair. It creates this so that it will know how to forward multicast datagrams for the group.

Sending Data Using PIM Sparse Mode

Refer to Fig. 13-21. When a host transmits a multicast packet to a specific group, the designated router (chosen by IGMPv2) forwards the multicast datagram as a unicast datagram to the RP. This unicast datagram is the multicast datagram encapsulated as a PIM-SM register packet. This type of packet informs the RP of a new source. The RP strips off the encapsulated (register) headers and redistributes the multicast datagram out the delivery tree. The active RP for that source transmits PIM join messages back to the source station's DR. The routers lying between the source's DR and the RP maintain the path information by the received PIM join messages. The routers do that so that when they receive non-register-encapsulated packets they will know what interfaces to forward them on. The RP will send the unicast datagram back out as a multicast datagram across the groups multicast tree. The source's DR will continue to encapsulate the multicast datagrams and send them to the RP. The RP sends register stop messages if the RP has no downstream receivers for the group or for that source. It also sends register stop messages if the RP has already joined the (S,G) tree and is receiving the data packets natively (unencapsulated). If a timer used by the DR expires, the DR

Figure 13-21
A host sending to a
group. (*From PIM-SM
draft RFC*)

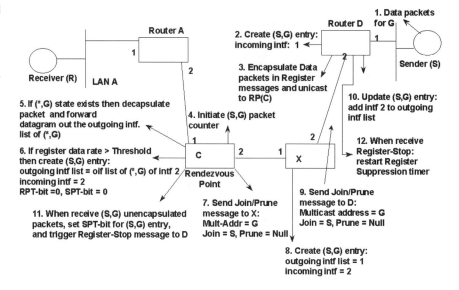

Figure 13-21
A host sending to a
group. (*From PIM-SM
draft RFC*)

starts to send the multicast datagrams encapsulated in register messages
again.

Refer to Fig. 13-22. The initial tree built is the shared RP tree. However,
based on data thresholds that are relative to time the tree can be converted to source-routed tree.

Rendezvous Points

PIM-SM uses the specific router known as the *rendezvous point (RP)* to
start out a shared tree. Senders and receivers join a multicast group by
registering at their rendezvous router. A rendezvous point is simply an
IP address of a single router. These points are used by senders to
announce themselves and for receivers to find out about new senders for
a group.

Where and how is the RP found? One router in a single PIM domain
(a contiguous set of routers that all implement PIM) is called the *bootstrap router (BSR)*. This router is responsible for sending out bootstrap
messages. The BSR is dynamically elected, and once elected it distributes
information about the RP. BSR information is sent to each router in the
PIM domain. To find out about RPs, all routers within a PIM domain
collect bootstrap messages and store the information contained in the
BSR messages.

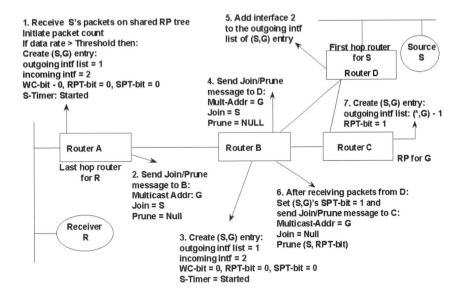

Figure 13-22

Converting to a source-rooted tree. *(From PIM-SM draft RFC)*

If a router wishes to work as an RP, it becomes a candidate RP (C-RP). C-RPs send out advertisement messages to the BSR for the domain. Inside the advertisements are the group address and the group mask (prefix) fields that it can become the RP for—in other words, the group address range it can support as an RP. This range can vary from one group to all groups. This allows the BSR to distribute RP information to other PIM routers in the domain using the all PIM routers message.

Comparison of Sparse-Mode and Dense-Mode Protocols

Table 13-3 compares the sparse-mode and dense-mode protocols.

Multicast Open Shortest Path First

Review RFC 1584. If you are not familiar with the OSPF (RFC 1583) protocol, review the chapter on that protocol. There are assumptions about that protocol that are made here.

Modifications have been made to the OSPF routing protocol that have enabled the protocol to route IP multicast datagrams. Three types of routing are provided:

TABLE 13-3

Sparse-Mode and
Dense-Mode
Protocols

Sparse Mode	Dense Mode
Requires explicit joining of senders and receivers.	Sends and stores explicit prune state information in response to unwanted packets.
Does not send packets where they have not been requested.	
Stores shared-tree join information in anticipation of data packets.	Stateless until data packets are sent.
Relies on an RP initially for senders and receivers to meet and build a shared tree.	No RP; the broadcast nature of the protocol builds the tree.
Is unicast protocol independent	Is unicast protocol dependent.
Requires periodic refreshing of explicit join/prune messages	No periodic updates on prune messages; event driven.

- Intra-area routing
- Inter-area routing
- Inter–autonomous system routing

Intra-area routing is the most basic routing algorithm provided. It runs in a single OSPF area and supports the forwarding of multicast datagrams within a single area. This could be a single area in multiple areas of an OSPF autonomous system or it could be a single autonomous system when there is only one area in the OSPF topology.

Inter-area routing is an OSPF topology that is split into several routing areas connected through a common area known as the *backbone area*. Decisions on forwarding multicast datagrams are still determined as in intra-area routing. The information contained in the forwarding cache is used. The difference between the two lies in the method of forwarding group membership information and the method of constructing the inter-area multicast tree. Selected area border routers (ABRs) are configured to perform a function known as *inter-area multicast* routing. These routers are responsible for the forwarding of group membership information, and they forward multicast datagrams between areas.

Inter–autonomous system routing involves a source path and a destination path that are outside at least one autonomous system (AS). Selected autonomous system boundary routers (ASBRs) are selected as inter-AS multicast forwarders. Multipath open shortest path first (MOSPF) makes the assumption that each inter-AS multicast forwarder is running a multicast routing protocol, such as PIM or DVMRP, that uses the RPF forwarding mechanism. This is the method used by MOSPF to leave its AS and route to another AS that could be running another routing protocol. This is the method that MSOPF uses to get across the unicast Internet since MSOPF does not support tunnels.

MOSPF Differences

MOPSF differs from DVMRP and PIM is many ways, but it should be noted up front that MOSPF does not broadcast the first multicast packet it receives. The protocol builds a source-rooted shortest path tree "on demand" when it receives the first multicast packet and then prunes the branches not associated with this group. In addition, MOSPF does not allow for tunneling as DVMRP does. Multicast datagrams are sent in native mode and are not encapsulated in any way (DVMRP uses IP-in-IP encapsulation for tunneling).

A difference between OSPF and MOSPF is that MOSPF does not allow for equal-cost multipath. Tie-breaking rules have been identified for paths that are found to be equal when the shortest-path tree is calculated. One of these rules is that given an equal-cost path to a destination, the router or LAN with the higher IP address will be chosen. The second tie-breaking rule is that a network is always chosen over routers (point-to-point links).

OSPF Caveats

OSPF allows for an autonomous system to be split into areas. When MOSPF is used in a multi-area OSPF, topology problems may occur because the ability of a router to have complete knowledge of the entire autonomous system is lost and incomplete shortest-path trees are built. Multicast datagrams will still get through, but they may take the most efficient paths.

Local Group Database and the Group-Membership LSA

MOSPF provides for the ability to forward multicast datagrams from one IP subnet to another. A new OSPF link state advertisement (group-membership LSA) has been added to accommodate this, allowing a source-rooted pruned shortest-path tree to be built. This new LSA augments the link-state database. Therefore, the MOSPF database is the link-state database of OSPF but with entries that pertain to multicast networks.

The new LSA places in the database the location of multicast destinations. Using this information in the database, MOSPF can build a shortest-path tree for a source group. MOSPF is an extension of OSPF, and MOSPF routers will interoperate in nonmulticast routers when forwarding unicast datagrams.

A MOSPF router bases its forwarding decision on the contents of a data cache known as the *forwarding cache*. Each entry in the cache represents a source/destination combination (and possibly TOS). This forwarding cache is built from two components:

- Local group database, built by IGMP
- Datagram's shortest-path tree

The local group database keeps track of the local membership for the routers directly attached networks. These entries are paired in the form (group, attached network). The attached network is the IP address of the network, and the group is the IP multicast address of the multicast group. As long as one host on that network indicates membership, the router will place an entry in the local group database. Similar to the other multicast routing protocols, this database allows the router to determine which port(s) it should forward a received multicast datagram. It is the IGMP protocol that assists in building this database.

To allow for multicast datagrams to be forwarded to all members of the group in an area, the local group database is flooded throughout the area (and is received by ABRs) using the group-membership LSA. There is a separate group-membership LSA for each multicast group in the router's group database. The router's group-membership LSA for a specified group lists those local router ports (i.e., the router itself and any directly connected transit networks) that should not be pruned from the group's datagram shared trees.

Role of the DR and BDR

It is the designated router (DR) that issues the host membership queries for the networks attached to it. An MOSPF router ignores reports for those networks for which it was not elected the DR. Any responses

(IGMP reports) received from the networks build entries based on groups in the database. A group address (member) in this database is deleted when the DR does not receive a report from that member.

Having the DR become the querier prevents unnecessary replication of packets, preventing multicast datagrams from being replicated as they are delivered to local group members. This allows for different entries in the local group database for the DRs in the autonomous system, meaning that each router in the autonomous system has a different local group database. However, the MOSPF link-state database and the datagram shortest-path trees are identical in each router belonging to the autonomous system.

The backup designated router (BDR) performs the same functions as the DR. It does not send out a query but processes the IGMP reports (host responses) so that it will contain a complete picture as the DR. In case the DR fails, the BDR can take over. One word of caution: you never want a non-MOSPF router to become the DR or BDR. To disable a non-MOSPF router from becoming the DR or BDR, set its priority to 0. If the DR or BDR is not an MOSPF router, MOSPF cannot operate properly.

When an IGMP report is received, the DR (all other routers except for the BDR discard this message) does some simple error checking (making sure the address is not in the local use range of 224.0.0.0–244.0.0.255). If there is no entry in the local group database, it creates one (using the format of IP group address, attached network number) and sets the age entry to 0. In the local group database, there is only one entry per multicast address, even if multiple hosts report membership. The DR may transmit a new group-membership LSA. Group-membership LSAs are flooded only to those neighbors that have indicated (through their database description packets) that they are multicast ready. They accomplish this by setting the MC (multicast capable) bit in the OSPF options field of all hello packets, database description packets, and link-state advertisements.

Local Group Database

The local group database consists of three components:

- Multicast group: a class D IP address
- Attached network: the IP address of the attached network that contains the multicast group
- Age: the number of seconds since an IGMP host membership report has been seen

Other routers find out about local group members through the group-membership LSA. There is one LSA for each multicast group that has one or more entries in the local group database. The link-state database indicates which routers or transit networks have attached group members.

Operation

The local group database enables the local delivery of multicast datagrams. A multicast datagram's shortest-path tree enables its delivery to distant (i.e., not directly attached) group members. Both the link state database and the local group database are used in the calculation of the forwarding cache.

The following are standard assumptions when using MOSPF:

- All MOSPF routers within the same area calculate the same shortest-path tree for a given multicast datagram. This is accomplished via synchronized link-state databases.

- Using the new group-membership LSA, link-state databases can be synchronized. With each router in an area having the same database, each router should be able to build a source-rooted shortest-path multicast tree without having to broadcast the first multicast packet.

- The shortest-path multicast tree is built on demand. This means that the tree is built when a router receives a multicast datagram for the as time for a given multicast group. An MOSPF router does not automatically forward the first multicast datagram, as DVMRP does. Since the synchronized link-state database contains group-membership LSAs, this allows an MOSPF router to perform "broadcast the first datagram" in its memory, and the tree is built on demand. The routers already know where the active group memberships are and can build the tree without forwarding the first datagram and then waiting for prune messages.

When the first multicast datagram arrives (at any router), the source subnet (IP address of the source) is located in the link-state database. This is sometimes called the MOSPF link-state database, but the single database (link-state database) contains entries for both unicast and multicast datagrams. A source-rooted multicast tree is calculated using the router LSAs and the network LSAs using the Dykstra algorithm (same as unicast

OSPF). Once the tree is built, it is pruned (to eliminate links that do not contain at least one member of a group) using the group-membership LSAs. The final result of the Dykstra algorithm is the pruned shortest-path tree that is rooted at the source (remember that OSPF calculates its shortest-path tree using itself as the root). This shortest-path tree is used to understand which ports should be used for the forwarding of multicast datagrams that are distant (i.e., there is no local group membership but there are members of the group downstream from this router) and which ports we should receive which multicast datagrams on (source).

Now we have two sources of information, the source-routed shortest-path tree and the local group database. Both of these will be used to determine the forwarding cache, which is the only place the router will look to determine the forwarding of a multicast datagram.

Forwarding Cache

The forwarding cache is used to determine how to forward a multicast datagram. A multicast datagram may be delivered locally or forwarded on a branch to another multicast router. I mentioned before that upon receipt of a multicast datagram, the Dykstra algorithm is run and the result is a pruned source-rooted tree. The forwarding cache is built using the shortest-path tree built by the Dykstra algorithm and the entries in the local group database. The router first finds its position in the shortest-path tree. Once the router discovers its position in the shortest-path tree, it creates an entry in the forwarding cache that contains the (source, group) pair, the upstream node, and the downstream interfaces.

The following entries are placed into the forwarding cache:

- Source network: the network number of the source
- Destination multicast group: a known destination group address to which multicast datagrams are currently being forwarded
- IP TOS
- Upstream node: the interface that datagrams addressed to (source, group) should be received on
- List of downstream interfaces: the interface(s) that a multicast datagram (indexed by source, group) should be forwarded on
- List of downstream neighbors, to assist in the forwarding of multicast datagrams in a hybrid (mixed OSPF and MOSPF) network

■ Time to live (TTL): the number of hops the datagram will travel to reach any outlying group members; this provides for efficiency in that the router can discard a multicast datagram if the received TTL is less than this TTL

TTLs are used by transmitting hosts to restrict the forwarding of a multicast datagram. This allows for efficiency in that a multicast datagram will only be forwarded (over routers) the number of hops indicated by the TTL of the received datagram. Notice that in the forwarding cache, each of the downstream neighbors is labeled with a TTL value. This is an optimizing feature because if a MOSPF router receives a multicast datagram whose TTL is lower than the entry in its routing table, the router will discard the datagram. The information contained in the cache remains stable until:

■ An OSPF topology change. This will force the cache to be flushed (all entries are deleted). Entries are not placed back into this cache until receipt of a multicast datagram, which will build a new entry.

■ A group-membership LSA is received that contains a change in the members of a group. A new tree will have to be constructed based on this information

Inter-Area MOSPF Routing

The basic algorithm for MOSPF works in a single area. Inter-area routing for multicast involves a source and one or more destination groups in different areas. The forwarding of multicast datagrams between areas is still decided by the information contained in the forwarding cache. The problem is that a complete shortest-path tree cannot be accurately built because detailed information about the other area's topology is not known (in OSPF it is summarized; in MOSPF it is not known). Furthermore, since LSAs are not flooded to different areas, the group LSA is not propagated to other areas. To compensate for this unknown information, estimates are made by using the wildcard feature of the area border router (new to MOSPF) and the summary link advertisements provided by the ABR. How MOSPF overcomes these limitations will be explained below after a few introductions.

In a multicast topology that is represented by multiple areas, a new function within the ABR called *inter-area multicast forwarders* is implemented. These forwarders pass group information and allow for the ability of multicast datagrams to cross areas. It is the ABR that is configured to perform this function. It runs as a separate function of the ABR and is used only for multicast datagrams.

An ABR implementing the multicast forwarder summarizes its area's group information (how is explained in a moment) and sends it to the backbone area through the use of group-membership LSAs. Group-membership LSAs are "injected" into the backbone area. The backbone routers receive this information and include it and the router it was received from in their link-state database by multicast group. The backbone routers process this information but do not forward any multicast information on to any other multicast ABR. Furthermore, no information regarding the backbone's group membership is forwarded to any area. It is asymmetrical. Information flows into the backbone, but the backbone does not flow the information into other areas. How does information flow between areas or from the backbone to an area? How does the multicast forwarder know of the groups in its area and know when to pass information from the backbone to an area? This involves the concept known as the *wildcard receiver.*

An MOSPF router may indicate that it wants to receive all multicast datagrams regardless of the destination. It can indicate this through its router LSA using a newly defined bit in the rtype field known as the *wildcard bit,* or W bit. This bit is used with inter-area multicast forwarders (ABRs), and permits a MOSPF router to receive all multicast traffic in an area regardless of the group. MOSPF routers that employ this function ensure that they remain on all pruned multicast trees and therefore able to receive all multicast datagrams regardless of group membership. By default, all multicast forwarder ABRs are wildcard receivers.

Inter-Area Multicast Example. Refer to Fig. 13-23. Having the multicast forwarders as wildcards enables them to receive multicast datagrams from the backbone and enables any multicast forwarder to be included in any multicast tree built in their area. When a multicast datagram is to be forwarded, it is received by the ABR (multicast forwarder) for forwarding to the backbone. The backbone routers know which groups are active in which areas, and since the ABR is part of the backbone, they can receive the information from the backbone to be forwarded to their area.

The backbone routers do not implement the wildcard function, because they inherently know about all multicast groups through the summary information by the multicast ABRs flowed into the backbone and received by the backbone routers.

Figure 13-23
Inter-area multicast
example.

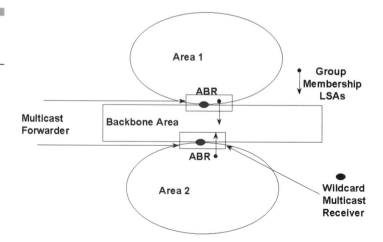

Figure 13-23
Inter-area multicast
example.

How is the forwarding cache built based on these assumptions? It depends on whether the source and the router building the tree are in the same area. The forwarding of multicast information is accomplished using the forwarding cache, but an accurate picture cannot be drawn. If the source and the router performing the calculation are in the same area, the wildcard feature of the ABR comes into play. It forces the router to be included in all multicast computations, allowing for the branches of the ABR to be included in the shortest-path tree. The ABR will not be pruned. If the source and the router performing the calculation are in different areas, then the summary link advertisements are used. This forces the inter-area multicast forwarders of the ABR to be included in the calculated tree.

A final note. Area border routers have separate link-state databases for each area they attach to—this is a normal OSPF process. But for multicast forwarding, this means that each multicast forwarder must build a separate forwarding tree for each area it attaches to. However, the forwarding information for all of the areas is contained in one forwarding cache, and as soon as this is built, the shortest-path trees for each area are dismantled. There is no need for them once the forwarding cache contains all the information needed for forwarding multicast datagrams.

Figure 13-24 shows the path between areas. Notice that the ABRs are wildcard receivers. This allows them to be included in all trees that are built.

Inter–Autonomous System Multicast

This type of multicast routing involves a source and a destination that reside in different autonomous systems. This type of routing includes routing between an MOSPF domain and a DVMRP domain but within the same AS. This is similar to the way OSPF treats RIP. Both OSPF and RIP can be used in the same AS, but they are treated as separate routing domains.

Just as in inter-area multicast routing, configured autonomous system boundary routers are configured to perform the inter-AS multicast forwarder function. This works under the assumption that each inter-AS multicast forwarder employs a multicast forwarding protocol that uses reverse-path forwarding such as DMVRP or PIM. The multicast ASBRs use the wildcard capability. With this, each ASBR acts as a multicast wildcard receiver for each of its attached areas. This ensures that the inter-AS multicast forwarders are included in all multicast trees. They are not pruned.

Most of the operation of this router is similar to that of the inter-area multicast forwarder. One case is different: that in which the source of the multicast datagram and the router making the calculation are in different ASs. Again, the details of the each AS topology will not be known. To compensate for this, information can be assumed by using the summary ASBR link and the AS external links that describe the source subnet. After the calculation is done, the multicast tree begins at the inter-AS multicast forwarder, with all branches stemming from this router.

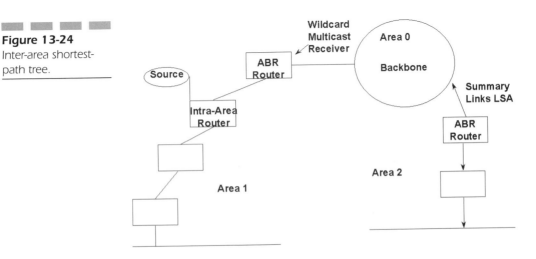

Figure 13-24
Inter-area shortest-path tree.

Multicast Conclusion

Multicasting is coming, and it is taking many forms on internets and the Internet. We have voice, video, and data multicasting. There is one common protocol message format, and that is IGMP. PIM and DVMRP use this framing for their messages as well. MOSPF uses its own LSA for communicating between routers. However, all multicast applications conform to IGMP to register themselves on the local subnet.

All of the protocols have their own advantages and disadvantages. MOSPF requires that you are running OSPF. A RIPv1 or RIPv2 network cannot simply install MOSPF. If the site chooses to use RIP, then it must use PIM or DVMRP. MOSPF converges instantly, whereas PIM and DVMRP may have routing loops during a slow convergence. However, installing MOSPF is not a simple task, because MOSPF is OSPF with multicast extensions. All of the rules regarding OSPF still apply. MOSPF does not support tunnels. It expects some other protocol, such as DVMRP, to be running on the multicast ASBR routers. DVMRP does not scale well, and neither does PIM without some tuning. DVMRP and MOSPF work better in densely populated environments, whereas PIM has the modes of operation, sparse and dense mode. PIM allows for duplicate packets in dense mode, in favor of a simpler protocol that is not dependent on a unicast routing mechanism. After all this, you must consider all the options before placing a multicast protocol on your network.

One of the things you must think about with multicast lies beyond simply receiving data. What you must think about is responding to that data. Multicasting does allow for one station to send one data stream to be received by literally tens of thousands of receivers. But what if there is a need for acknowledgement of that data? This poses a considerable problem. If two thousand receivers all need to send some type of receiver status message back to the sender, the sender must be able to handle this type of back-channel information flow, especially if the status is long. The inability to handle this, in which the originator is overrun by the back channel, is called *implosion*.

RFCs to Be Reviewed

- RFC 1700—assigned numbers
- *ftp://ftp.isi.edu/in-notes/iana/assignments/*—for current IANA number assignments

- RFC 1112—host extensions for IP multicasting

- *http://ds.internic.net/internet-drafts/RFC 2236, IGMP v2*

- http://ds.internic.net/internet-drafts/draft-ietf-idmr-dvmrp-v3-04—Distance Vector Multicast Routing Protocol draft, expires August 1997

- *http://ds.internic.net/internet-drafts/draft-ietf-dhc-mdhcp-00.txt*—multicast address allocation extensions to the Dynamic Host Configuration Protocol

- *http://ds.internic.net/internet-drafts/draft-ietf-mboned-mdh-00.txt*—Multicast Debugging Handbook, expires September 1, 1997

Web Sites

http://www.ipmulticast.com

http://netweb.usc.edu/pim/

The MBONE agenda of scheduled events can be found at *http://www.cilea.it/MBone/agenda.html,* and you can send email to rem-conf@es.net to request scheduling and to specify the bandwidth you expect to use. As always, it is best to read the FAQs for the MBONE to avoid problems. A FAQ URL is *http://www.research.att.com/mbone faq.html.* In addition, a good review article can be found at *http://www.cs.cul.ac.uk/mice/mbone_review.html.*

Other
TCP/IP Protocols

Boot Protocol

The Boot protocol has been around a long time. It was most often used with diskless workstations. It enabled these workstations to get their configuration and boot files to remotely boot over the network. The best example of this was Sun workstations and their Network File System (NFS). These diskless network stations could boot over the network (using a small bootstrap protocol found in a PROM). With this a machine could get its configuration parameters such as its IP address and subnet mask and then perform the boot sequence to boot up from a remote server.

The Bootstrap Protocol (BOOTP) is a UDP/IP-based client/server application originally promoted to allow diskless clients to boot remotely from a server on the same network or from a server on a different network for the purpose of obtaining the name of a file to be loaded into memory and executed, an IP address, and the address of its boot server. The RFC for BOOTP is RFC 951, but there have been a few supplemental RFCs added since then to clear up some "loosely defined" features of the protocol that can lead to misinterpretation and eventually incompatibilities between vendors supporting the protocol. The most recent one is RFC 1542, Clarifications and Extensions for the Bootstrap Protocol. Other configuration information such as the local subnet mask, the local time offset, the addresses of default routers, and the addresses of various Internet servers can also be communicated to a host using BOOTP.

BOOTP Operation

In BOOTP operation (Fig. 14-1):

- Two modes of operation: request and reply.
- Request contains the client's hardware address and IP address if known. It may also contain a requested server. Usually it contains a name of the file to use for booting this client.
- Response from the server will contain answers to the request.

The basic operation is as follows. There is a single packet type exchanged between the client and the server. One field in the packet is called the Opcode, which can have one of two values: BOOTREQUEST or BOOTREPLY. The client broadcasts a BOOTREQUEST packet that

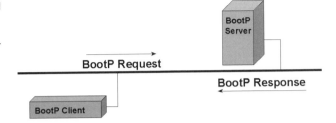

Figure 14-1
BOOTP operation.

contains its hardware address and its IP address, if it is known. The BOOTREQUEST may contain the name of the server the client wants to respond. This is to force the BOOTREQUEST packet to a particular server to obtain its information. This may occur if there are many servers on the network or if there are more than one image (older/newer) version that could be sent to the client. Inside the BOOTREQUEST may be a generic filename to be booted. Simple names like ipboot or unixboot are used. When the server replies with a BOOTREPLY, it replaces this entry with the full path name for the file on that server. Refer to the example BOOTP tab file later in this section for more information.

If the client does not know its IP address, the server must possess a database of MAC-to-IP address mappings. Once a match is found, this IP address is placed in a field in the BOOTREPLY packet.

The BOOTP headers and descriptions are shown in Fig. 14-2, and Table 14-1 lists the field definitions.

Client Side (BOOTREQUEST)

Following is the series of steps taken by the client, which are flowcharted in Fig. 14-3. The client builds a BOOTREQUEST packet as follows:

- The IP destination address (in the IP header, not the BOOTP fields) is set to 255.255.255.255. Optionally, it may be set to the server's IP address, if it is known.
- The IP source address is set to its IP address (if known), otherwise it is set to 0.
- The UDP port numbers are set to 67 for the UDP destination port (the server) and 68 for the UDP source port (the client).
- The Op field is set to 1 (BOOTREQUEST).

Figure 14-2
BOOTP (DHCP) field
definitions.

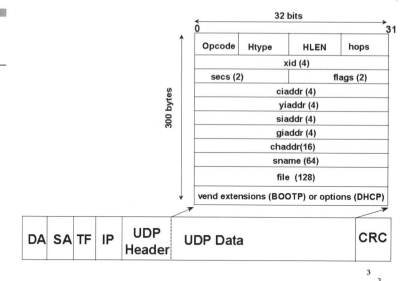

- The htype is set to the hardware address type and the Hlen is set to the length of the hardware address.

- The xid is set to a random number.

- The secs is set to the number of seconds that have elapsed since the client started booting.

- The ciaddr field is set to the IP address of the client, if known; otherwise it is set to a 0 and the chaddr is set to the client's hardware address.

- If the client wishes to restrict booting to a particular server name, it will set this name in the sname field; otherwise, the sname field is set to 0.

- The file field can be set to 0 to indicate to the server it wishes to boot from the default file for its machine. Setting this field to 0 could also indicate that the client is interested in finding out client/server/gateway IP addresses and does not care about a file to boot from. The field could be set to a simple generic name such as ipboot or unixboot, indicating that it wishes to boot the named program configured for the client. Finally, the field can be set to the full path name on the server where the boot file resides.

- The vend field is set to whatever the vendor wishes. However, it is recommended that the first four bytes be set to a "magic

TABLE 14-1

BOOTP Field Defin-
itions

Field	Bytes	Description
op	1	Packet operation code/message type. 1 = BOOTREQUEST, 2 = BOOTREPLY.
htype	1	Hardware address type, same as ARP section in "Assigned Numbers" RFC. For example, 1 = 10-Mb Ethernet.
hlen	1	Hardware address length. For example, 6 (bytes) for 10-Mb Ethernet.
hops	1	Client sets to 0. Optionally used by gateways in BOOTP relay.
xid	4	Transaction ID, a random number used to match this boot request with the responses it generates.
secs	2	Filled in by the client, indicating the number of seconds that have elapsed since the client started trying to boot.
ciaddr	4	Client IP address. Filled in by client in BOOTREQUEST, if known.
yiaddr	4	"Your" (client) IP address. Filled in by the server if the client doesn't know its own address (i.e., ciaddr was a 0).
siaddr	4	Server IP address, returned in the BOOTREPLY by the server.
giaddr	4	The gateway's IP address for the port that received the first BOOTREQUEST. It is used in the BOOTP relay function.
chaddr	16	The client's hardware (MAC) address. It is filled in by the client.
sname	64	Optional. The server host name being requested by the client. All other servers would then ignore this packet.
file	128	The boot filename. The "generic" name in the BOOTREQUEST. Refer to the BOOTP tab file example for more information.
vend	64	Optional vendor-specific field. Examples of its use could be a serial number, version number, etc. It is generally ignored by BOOTP.

number" (that is, you pick it). This allows the server to determine what kind of information it is seeing in this field.

If no reply is received by the client within a preset length of time, the client will retransmit the request up to a predetermined number of times. The retransmission is regulated in that it will randomly send on retransmission requests. This is to ensure that the network will not be flooded with requests should clients somehow sync their transmissions (purely by coincidence). Before retransmission, the secs field is updated.

Figure 14-3
Client Side (BOOTRE-
QUEST).

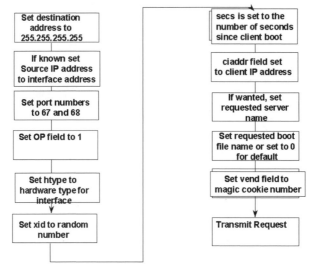

Figure 14-4
Server side.

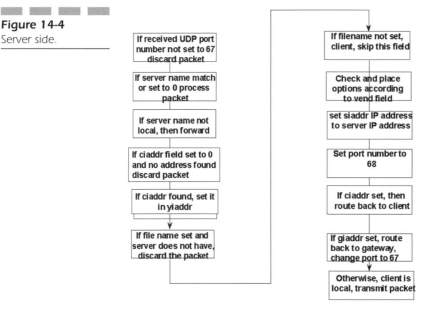

Server Side

The following steps are flowcharted in Fig. 14-4. When the server receives
the packet, it makes a series of decisions:

- If the UDP destination port is not set to 67, the server discards
  the packet.

- If the `server` field (`sname`) is set to 0 or matches our name, the server further processes the packet.

- If the server name does not match its own name but the name is local on the network, the server discards the packet.

- If the server name is not on the local network, the server may choose to forward the packet to that server. Usually, this is accomplished via the BOOTP relay service, explained in a moment.

- If the `ciaddr` field is set to zero, the client does not know its IP address, so the server looks it up in its database. If no match is found for this chaddr, the server discards the packet. Otherwise, the `yiaddr` field is filled in on the response packet.

- The `filename` field is then checked. If this field contains a 0, the client is not interested in a boot file or wishes to use the default boot file. If there is a filename specified, a default file is found, or the field contains a full-length path name that the server has, the `file` field is replaced with the fully specified path name of the selected boot file. If the field is not set to 0 and no match is found for this field on this server, the client is asking for a file that the server does not have, and the server discards the packet.

- Finally, the `vend` field is checked, and if a recognized type of data is provided, client-specific actions are taken and a response from these actions is placed in the `vend data` field of the reply packet. This field, for example, could contain configuration options that can be passed to the boot file that will be transmitted to the client after the BOOTP is finished.

- The `siaddr` field is set to the server's address, and the `op` field is set to a BOOTREPLY. The UDP destination port is set to 68

TABLE 14-2

BOOTP Field Settings

Ciaddr	Giaddr	B*	UDP destination	IP Destination	Link Destination
non-zero	X	X	BOOTP Client (68)	ciaddr	normal
0.0.0.0	non-zero	X	BOOTP Server (67)	giaddr	normal
0.0.0.0	0.0.0.0	0	BOOTP Client (68)	yiaddr	chaddr
0.0.0.0	0.0.0.0	1	BOOTP Client (68)	255.255.255.255	broadcast

*The broadcast bit that tells the router to use a broadcast and not a unique address.

(BOOTP client). Refer to Table 14-2 for the exact settings of this field.

■ If the `ciaddr` address of the BOOTREQUEST is set to a nonzero address, the packet is IP-routed back to the client. However, if the `giaddr` is set to a nonzero, the BOOTREPLY is sent directly to this router and the UDP destination port is set to BOOTP Server (67). Otherwise, the client is local and the BOOTREPLY is send back to the client on the local LAN.

Chicken or the Egg?

A question should come to mind here for those who have been following: If the client does not yet know its IP address, how is it going to respond to the server's ARP request when the server decides to respond? If the client knows its IP address (as indicated in the `ciaddr` field), the BOOTREPLY can be sent as a normal IP packet, because the client will respond to ARPs, but if the client does not yet know its IP address, it cannot respond to ARPs sent by the server. There are two options available. If the server has the capability to manually construct an ARP entry in its table for this client, it will do so using the `chaddr` and `yiaddr` fields that it is responding with in its BOOTREPLY (BSD UNIX has this capability). The server will then reply using IP services and skipping the ARP process. If the server does not have this capability, it simply send the BOOTREPLY with the IP address set to broadcast.

BOOTP Relay Agents (or BOOTP Gateway)

The BOOTP operation also allows for the possibility that clients and servers may not be on the same IP network or subnet (those network segments not separated by a router). Therefore, some kind of relay agent is needed to forward the BOOTP messages. This type of service is most commonly found as part of the router function. However, it is more than a router simply receiving a BOOTP message and forwarding it on to the next segment. Basically, the router receives the message, processes it as if the message were addressed to it, and then sends out a new BOOTP message to the appropriate forwarded port.

The forwarding (scope) of a BOOTP or DHCP packet can be limited by configuring the router to limit the scope. Each router increments the TTL field. If the received packet already has its limit set in the TTL field, the packet will be discarded. Also, the router must be configured as to which ports it should forward this packet on. It is not simply forwarded to one port. When the router is the first router to forward the packet, it places its address in the `giaddr` field of the packet. It knows that it is the first router if this field is set to 0.0.0.0 when it received it. When a router forwards the packet, it forwards it set to broadcast.

Dynamic Host Configuration Protocol

Which came first, BOOTP or Dynamic Host Configuration (DHCP)? Why do you hear about BOOTP every time something is written about DHCP? The protocols are based on a broadcast mechanism, so why and how do these protocols operate in a routed environment? In fact, why do we need another configuration mechanism, when we have DRARP (Dynamic Reverse Address Resolution Protocol) and ICMP (Internet Control Message Protocol)? DRARP addresses the problem of IP address assignment, and hosts can use ICMP to find out the subnet mask for a network and to dynamically discover routers. We consider a router to sometimes be a host—does DHCP provide configuration information for a router?

DHCP is gaining considerable attention because of tight address allocation restrictions,[1] which require efficient assignment or reassignment of IP addresses, and the emergence of Windows NT™ and the TCP transport in the corporate environment.

The DHCP protocol, for the purposes of this writing, can fully interoperate with BOOTP servers and clients.

Following are definitions of some terms we will use in our discussion:

- *DHCP client:* A host that is requesting configuration information
- *DHCP server:* A DHCP host that supplies configuration parameters to a requesting host
- *BOOTP relay agent:* The protocol that allows BOOTP and DHCP packets to traverse a router

---

[1]IP addresses are in short supply. They are handed out very carefully. DHCP offers the capability of handing them out statistically based on probability. Therefore, we can have many users and not use as many IP addresses.

■ *Binding:* Configuration parameters, including an IP address, that are "bound" to a host

DHCP provides a transport mechanism for passing configuration information (that is located on a server) to requesting hosts on a TCP/IP network. The parameter information is based on the host requirement RFCs (RFCs 1122, 1123, 1112, etc.). After being supplied with this configuration information, a host should be able to communicate with any other host on the Internet.

DHCP is based on BOOTP. It was designed to use some of the features of BOOTP (such as relay agent, for forwarding messages across routers) and to be interoperable with existing BOOTP clients (RFC 1534 describes the interoperability functions of BOOTP and DHCP). DHCP messages are in the exact same format as BOOTP (see Fig. 14-2). DHCP added the ability to support "leased" IP addresses and other functions. This allows requesting stations to get their IP addresses from a server and then return them when they are done. These added functions are described in RFC 2132.

DHCP consists of two parts: a protocol for delivering host-specific configuration parameters and the ability to allocate IP addresses. It is based on a client/server model where the host requests information from a server. A host can ask a specific server to supply information to it, or it may simply rely on any server to relay information to it. A server must be preconfigured to handle a specific client's request or the server will ignore the request.

The first service provided by DHCP is static storage of network parameters for requesting clients. This information is stored in a database (or table) on a host server. The entries are "keyed." This means that a unique identifier is used to single out the parameters of a requesting host. This identifier states the client identifier, or chaddr, and the assigned network address. This uniquely identifies the lease between the client and the server for DHCP.

IP Address Allocation

The next service that DHCP provides is IP address allocation. Three methods are supported:

■ Automatic allocation

■ Dynamic allocation

■ Manual allocation

Automatic allocation permanently assigns an IP address to a requesting host. Dynamic allocation gives an IP address to a requesting host for a specific amount of time. Manual allocation is the ability to reconfigure an IP address for a host; the server simply relays that information when the host requests its IP address. This is different than automatic allocation in that the IP address is preconfigured for the host by the system administrator, whereas automatic allocation gives an arbitrary address (from a pool of IP addresses) to a requesting host. The host is not preconfigured with the IP address for that host.

DHCP Messages

The following message types are used with client/server interaction:

- *DHCPDISCOVER:* A client broadcast that is used to locate available servers. It may be forwarded by routers to allow for server's segments.
- *DHCPOFFER:* A server-to-client response to a DHCPDISCOVER message with an offer of configuration parameters.
- *DHCPREQUEST:* A client message to servers for one of the following:
  - Requesting offered parameters from one server and implicitly declining offers from all others
  - Confirming correctness of previously allocated address after, e.g., system reboot
  - Extending the lease on a particular network address
- *DHCPACK:* A server-to-client message that contains configuration parameters, including a committed network address.
- *DHCPNAK:* A server-to-client message indicating that the client's network address is incorrect (e.g., client has moved to new subnet) or the client's lease has expired.
- *DHCPDECLINE:* A client-to-server message indicating that a network address is already in use.
- *DHCPRELEASE:* A client-to-server message relinquishing a network address and canceling the remaining lease.
- *DHCPINFORM:* New with RFC 2131, this is a client-to-server message asking only for local configuration parameters; the client already has an externally configured network address.

Figure 14-5
DHCP operation.

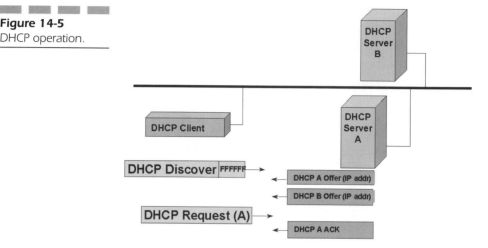

Refer to Fig. 14-5, which gives a graphical depiction of DHCP operation. First, a client transmits a DHCPDISCOVER message on its local physical subnet. The IP destination address is set to broadcast (0xFFFFFFFF), and the MAC destination address is set to broadcast, as well. The IP source address is set to 0x00000000, and the MAC source address is set to chaddr or the clients hardware address.

The client can place in this message some options that include an IP address and the lease duration. If the client has placed a "suggested" IP address in the option field, the client has been previously configured using DHCP and is now restarting and would like to use that address again. If the client was manually configured with an IP address, the client should use the DHCPINFORM message instead of the DHCPREQUEST message. The client can request a specific list of parameters by using the "parameter request list," which indicates to the server which parameters, by tag number, the client is specifically interested in. See RFC 2132 for more information on DHCP options.

A DHCP message is picked up by routers that implement the BOOTP relay and transmitted over the Internet. Again, the scope can be limited (how many routers it can traverse). The hops field (set to 0 by the client) is incremented (usually by 1) by each router, and the administrator of the router sets the maximum hop count. If the received packet has a hop count of 2, and the max hops parameter configured by the router is 3, the router sets the hops field to 3 and forwards the packet. If the received packet already has a 3 in the hops field, the router is not allowed to increment the field to 4, and it discards the packet. This is known as the scope (range) of the DHCP packet.

Each active server that receives this message may respond with a DHCPOFFER message that includes an IP address in the yiaddr field of the packet. Not all servers will respond. Some may be preconfigured to not respond to certain requests, and others may not have the binding for that client. It may also appear in various option fields, as well. The server does not have to take the offered IP address off the available list, but it does help when it removes this offered IP address from its availability pool. At this time, the server can check for current use of the offered IP address by sending an ICMP echo request using the offered IP address. This is configurable.

DHCP Responses

Refer to Fig. 14-6. Responses from the server can be addressed in a few different ways:

- If the giaddr field of the received client packet is set to 0 and the ciaddr field is set to a nonzero, responses are sent as unicast to the client, using the yiaddr as the destination IP address and the chaddr as the destination MAC address (giaddr field set to 0 indicates that the client is on the local subnet).

- If the giaddr field is set to a nonzero value, the response is sent to the BOOTP relay agent using the IP address indicated by giaddr field.

- If the broadcast bit is set, the server broadcasts (IP and MAC addresses) its responses to the client. Using the B bit allows the

Figure 14-6
DHCP response.

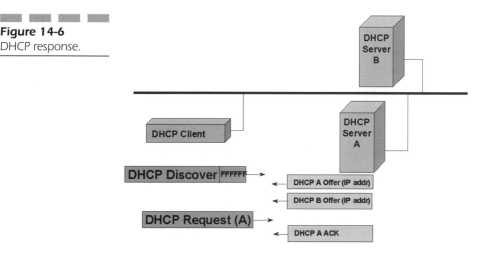

client to indicate to a potential server that it cannot receive uni-cast IP datagrams before its TCP/IP configuration has been set.

If the server receives a DHCPREQUEST message with an invalid requested IP address, the server should respond to the client with a DHCPNAK message and report this error in a log.

A client may receive one or more offers from different servers. The client selects a server from the responses that closely matches its original request parameters. If the client does not receive a response to its DHCPREQUEST, it times out and retransmits a DHCPREQUEST message.

To respond to a DHCPOFFER, the client transmits a DHCPRE-QUEST. In the options field of this message is the server identifier (the server's IP address), indicating which server the client has selected. All other servers will partially ignore this message, but they use that message to indicate that the client will not be using their services, and this releases the offered IP address back to the available pool.

The selected server commits this binding[2] to a place in memory, where it will be stored. The server responds to the client with a DHCPACK message containing all the client's configuration parameters. This binding indicates the client identifier, or the chaddr, and the assigned IP address.

When the client receives the DHCPACK, it performs some final checks such as ARPing for the newly assigned IP address to ensure that no one else is assigned to this address. If there are any inconsistencies, the client sends a DHCPDECLINE message to the server, and after wait-ing 10 s (at least 10 s), it restarts the configuration process. Also, if the server transmitted a DHCPNAK message to the client, the client restarts (after 10 s).

Refer to Fig. 14-7. To "gracefully" stop using an assigned IP address, the client should transmit a DHCPRELEASE message to the server (using the same keyed binding that it used to get the address from that server). This may occur before the lease duration is up.

DHCP Shortcuts

A simpler approach is used for clients that were previously serviced by DHCP. The client can skip the DHCP discover message and instead transmit a DHCPREQUEST message. A server will then respond with a

---

[2]A binding is a key that is used to look up information. For example, the binding could be an IP address to a client hardware (MAC) address.

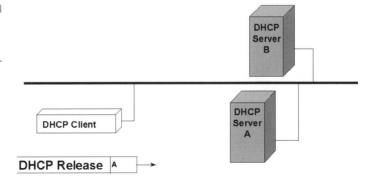

Figure 14-7
Releasing an IP
address.

DHCPACK message to fill in the client's configuration parameters. If any of the consistency checks come back as invalid, the server responds with a DHCPNAK message. This means that the client may not reuse the requested IP address that it thought it could use. The client then restarts the process, using the longer method (i.e., starting with DHCPDISCOVER).

The DHCPINFORM message can be used by a client to inform a server that it already has an IP address (through a manual configuration process) but it would like a download of some configuration parameters that may be set for it in the server's bindings. The reply from the server is unicast (a major difference from all of the other DHCP messages, which are broadcast) directly back to the client.

Lease Duration

Timers between the server and client may be different—they are not synchronized using DHCP. This may lead to a problem in that the erver assumes the lease is up before the client does (because of the possibility of the inaccuracy of the clocks). To compensate for this, the DHCP server can return a smaller lease than requested by the client. This is only in the response to the client; the server writes the original time into its binding table. This may or may not work—it depends on how the developer interpreted the RFC.

A client transmits a DHCPREQUEST for the use of an IP address for certain period of time. The server guarantees not to hand out an allocated IP address that is in use by another client. Furthermore, the server will try to reallocate the same IP address to a requesting client each time it requests an IP address. The length of time that a client uses an IP address is known as the *lease*. If a client needs to extend the lease, it can submit a request to the server. It is understood between the client and the server that if the server does not receive a message from the client indicating that the client would like to extend the lease, the server assumes that the client is no longer using the IP address, and the lease expires. However, a DHCP server does not automatically issue expired-lease IP addresses. It will continue down the list of IP addresses and assign those that have not been assigned before until its list is exhausted. Only when this occurs does the server reallocate IP addresses that were previously assigned but have expired. The server can probe the network to see if a lease-expired address is still being used by simply sending an ICMP echo request and waiting for a reply. As another consistency check, the host with the newly assigned address can issue an ARP to see if the address has already been assigned to another host. Note, however, that the RFC terms these checks as "SHOULD" which means that the implementer of the protocol should use this feature but it is not required to. You may want to ask about this before implementing a vendor's DHCP code.

DHCP implementations vary, but the lease time is a 32-bit unsigned integer. This number is expressed in seconds, which allows for a lease time in the range of 1 s to (approximately) 136 years. Longer than I plan to be in this business! According to RFC 2131, there is no minimum lease-time requirement.

Efficiencies

Clients usually do not need all the configuration parameters that are available with DHCP. To build some efficiency into this protocol, the protocol assumes defaults. The RFCs on host requirement name the defaults for the parameters. When a client receives the DHCPNAK packets that contain configuration information, a lot of information will not be included. The host assumes the default value for anything not contained in the DHCPNAK message.

Another question should have come into your mind by now (if you were paying attention). If a host is using DHCP for its initial configuration, how does it know to accept TCP/IP packets when TCP/IP has not been fully initialized in the host? To work around this problem, DHCP uses the flags field. The field is 16 bits in length, but only 1 bit is used. The other 15 must be set to 0. The bit that is used is called the broadcast bit, or B bit. Any station that cannot receive unicast datagrams (usually sent by BOOTP relay agents and servers on DHCPOFFER, DHCPACK, and DHCPNAK messages) must set this bit on DHCPDISCOVER and

TABLE 14-3 DHCP Operational Tables

Field	DHCPOFFER	DHCPACK	DHCPNAK
op	BOOTREPLY	BOOTREPLY	BOOTREPLY
htype	From "Assigned Numbers" RFC	From "Assigned Numbers" RFC	From "Assigned Numbers" RFC
hlen	Hardware address length in octets	Hardware address length in octets	Hardware address length in octets
hops	0	0	0
xid	Xid from client DHCPDISCOVER message	Xid from client DHCPREQUEST message	Xid from client DHCPREQUEST message
secs	0	0	0
ciaddr	0	ciaddr from DHCPREQUEST or 0	0
yiaddr	IP address offered to client	IP address assigned to client	0
siaddr	IP address of next bootstrap server	IP address of next bootstrap server	0
flags	flags from client DHCPDISCOVER message	Flags from client DHCPREQUEST message	flags from client DHCPREQUEST message
giaddr	Giaddr from client DHCPDISCOVER message	Giaddr from client DHCPDISCOVER message	Giaddr from client DHCPREQUEST message
chaddr	Chaddr from client DHCPDISCOVER message	Chaddr from client DHCPDISCOVER message	Chaddr from client DHCPREQUEST message
sname	Server host name or options	Server host name or options	Unused
file	Client boot filename or options	Client boot filename or options	Unused
option	Options	Options	Unused

DHCPREQUEST messages. DHCP servers processing the request mark this bit and transmit their responses as broadcast in both the IP header and the MAC header. If the broadcast bit is set to 0, the responses of the server are transmitted as unicast, with the IP address set to yiaddr and the MAC destination address set to chaddr.

Finally, a client can use the DHCPRELEASE message to gracefully shut down or to indicate to the DHCP server that the client no longer needs the IP address assigned by the server. The client is not required to use this message and may simply let the lease expire.

Tables 14-3 and 14-4 list the operational tables for DHCP. We recommend that you review the following RFCs on BOOTP and DHCP:

TABLE 14-4   Operational Tables

Field	DHCPDISCOVER, DHCPINFORM	DHCPREQUEST	DHCPDECLINE, DHCPRELEASE
op	BOOTREQUEST	BOOTREQUEST	BOOTREQUEST
htype	From "Assigned Numbers" RFC	From "Assigned Numbers" RFC	From "Assigned Numbers" RFC
hlen	Hardware address length in octets	Hardware address length in octets	Hardware address length in octets
hops	0	0	0
xid	Selected by client	Xid from server DHCPOFFER message	Selected by client
secs	0 or seconds since DHCP process has started	0 or seconds since DHCP process has started	0
ciaddr	O (DHCPDISCOVER) or client's network address (DHCPIN-FORM)	0 or client's network address (bound/renew/rebind)	0 (DHCPDECLINE) or client's network address (DHCPRELEASE)
yiaddr	0	0	0
siaddr	0	0	0
flags	Set broadcast bit if client requires a broadcast response	Set broadcast bit if client requires a broadcast response	0
giaddr	0	0	0
chaddr	Client's hardware address	Client's hardware address	Client's hardware address
sname	Options, if indicated in sname/file option; otherwise, unused	Options, if indicated in sname/file option; otherwise, unused	Unused
file	Options, if indicated in sname/file option; otherwise, unused	Options, if indicated in sname/file option; otherwise, unused	Unused
options	Options	Options	Unused

951: Bootstrap Protocol (BOOTP), including information on the relay agent

1534: Interoperation between DHCP and BOOTP

2131: Dynamic Host Configuration Protocol (DHCP) (obsoletes RFC 1541)

2132: DHCP Options and BOOTP Vendor Extensions (obsoletes RFC 1533)

Resource Reservation Protocol

Currently, QoS (quality of service) is limited to manual items such as filters, protocol prioritization (fancy filters), compression, network design, and fast pipes. Most of these techniques are applied to WAN ports. Although this works well, many applications such as voice and video are also running on LANs and WANs. There is another method of providing QoS with broadcast networks—it is RSVP. RSVP provides a general facility for creating and maintaining distributed reservation states across unicast or multicast environments. It is not supposed to be the QoS that rivals ATM's guaranteed QoS. It shows promise as the first of many attempts to build quality of service for existing-broadcast oriented networks without having to tear out those networks and replace them with ATM.

Quality of service has never been built into most protocols that are running on networks today. When Ethernet was invented in the late 1970s, 10 megabits seemed a huge enough pipe to give any bandwidth-hungry application more than enough room. However, bandwidth-hungry applications are not the culprit; it is millions of bandwidth-hungry users. Personal computers were not anticipated to have a great impact on the business world. Mainframes and minicomputers were expected to continue to be the computing hardware of choice. However, the personal computer rapidly became the dominant computer system of choice. After just a few years, the personal computer came to be able to handle sophisticated graphics, and many different options of voice and video soon became available. Connection to the Internet soon became a must-have, as well.

Shared Ethernet and Token Ring networks could not provide the bandwidth necessary to support not only bandwidth-intensive applications that are network-aware but the millions of personal computer users, as well. Ethernet has since scaled up to 100 megabits per second, and Gigabit Ethernet is making inroads, too. The virtually limitless scalability of the ATM protocol makes it the first commercial protocol that has scalable quality of service parameters built in. However, ATM still runs on fewer than 1 percent of all desktop installations, and many companies will not tear down their Ethernet or Token Ring networks and replace them with ATM just to get QoS and scalable bandwidth. Consumers want QoS, but they want it with their existing networks.

Alternatives

IP, since its inception, has had a field known as Type of Service (TOS). The TOS is for internet service quality selection. It is specified along the abstract parameters of precedence, delay, throughput, and reliability. These parameters are mapped into the actual service parameters of the particular networks a datagram traverses. File transfers can use a high-delay network, whereas terminal access can use one with low delay. In order to provide for this, router vendors would have to provide for TOS in their routers, and application vendors would have to build this into their applications. For routers, this can require the maintenance of multiple routing tables for each TOS. The application program is the program that sets these bits, and most application programs moved data, so basically there was no real demand for TOS. Over the years, we simply came up with faster networks to compensate for the millions of new users and bandwidth-hungry applications. The easy way to support QoS is to manipulate bandwidth.

Gigabit Ethernet now allows us three choices for Ethernet: 10, 100, and 1000 Mbps. This allows scaling but not for data QoS. Bandwidth is simply one factor in the equation. Also, what comes after Gigabit Ethernet? We are finally moving into the capabilities of ATM, but we still run into customer resistance to ATM conversion. Customers will place ATM on the backbone and possibly use it for the WAN but not yet use it on the desktop. We cannot keep producing more bandwidth without giving some consideration to taming the applications.

RSVP is the first widely known protocol to allow for some type of QoS on an existing broadcast-oriented network. As of this writing it is still an RFC draft, with the latest version being the functional specification of May 1997. It can be used with IPv6 or IPv4. RSVP covers the QoS portion of protocols optimized for real-time, streaming, and multimedia issues. It operates directly on top of IP, and it supports both unicast and multicast protocols. RSVP appears to have far greater advantages when used in a multicast network. It is not a transport protocol but a control protocol like ICMP or IGMP. With IPv4 it operates with UDP, but with IPv6 it will operate with on top of IP, using the extension header concept.

However, let's be up front. RSVP is not designed to provide QoS for the entire Internet. Its original design was intended to allow QoS for multimedia applications on smaller networks. It was designed to allow applications to choose among multiple, controlled levels of delivery service for their data packets. It reserves network resources along the transmission path of a data stream and communicates between sending and receiving hosts, but the creation of the reservation is accomplished by the receiving host, and only in one direction for data flows. Once the reservation is made, routers between the sender and receiver maintain the reservation. Finally, RSVP is not a replacement for the QoS offered in ATM. Many see it as a migratory step in moving to ATM.

Like TOS in the IP header (IPv4), applications must be RSVP-aware. The application is what makes use of RSVP. As of this writing there are just a few applications that do so. WinSock (the API for Windows applications), for example, is QoS-aware, starting with Winsock 2. Applications that are not RSVP-aware may be able to use RSVP toolkits or dialer programs, which are secondary applications that can make a request for you before starting your application. Applications that are making use of other Internet protocols such as the Real-Time Protocol (RTP), and the Real-Time Streaming Protocol (RTSP) are better suited to RSVP.

RSVP Operation

Reservation requests are made by the receiver, not the sender. Why? The receiver better understands its local parameters (such as LAN type) and is better able to make an intelligent request than a server. Otherwise, we would have to configure the server to know every aspect about its possible receivers. For example, if a server must make a bandwidth reservation, how would it know that a receiver is on Ethernet, Token Ring, FDDI, or ATM? How would a server know the type of computer making the reservation. Why does this matter? Speed. An ATM station should be able make a request larger than an Ethernet station simply because the bandwidth is available. An application on a host uses RSVP to request specific QoS from the network for particular data streams or flows from the application.

The operation of RSVP is based on two concepts: flows and reservation. RSVP reserves resources based on flow. Flows are traffic streams (data) from a sender to a receiver or possibly to multiple receivers. The flow is defined by the destination IP address and, optionally, a destination port. RSVP can also define the flow by using the Flow Label field in the IPv6 header in conjunction with the source IP address. In combination with the flow, RSVP determines the QoS that a flow requires. QoS determines the network resources for the flow. RSVP does not interpret the flow spec, but it does give that information to hosts and routers along the flow's path. Those systems can examine the flow spec to see if they have the resources to accept the reservation, and if they accept it they use the flow spec to reserve the required resources.

As stated before, the receivers using RSVP actually make the reservations. This is to relieve the burden of the server of being the overall administrator of all the possible receivers. Some receivers are located on Ethernet, others on Token Ring (speed differences). Some may want to leave the flow at any time. Receivers have better independent control over themselves, and this allows flexibility in RSVP.

The reservation is split into two functions; one is performed by the sender and one is performed by the receiver. Having the receiver make the reservation leads to a question. How does the receiver know the path by which the flow will be forwarded? The sender sends path messages that follow a path from the sender through routers. The path message describes the flow to any possible receivers. A path message notifies routers to get prepared for a possible flow. It identifies the flow to the routers and alerts the routers to the possibility of incoming reservation requests. For multicast, a path message is sent to a destination multicast address.

Path Messages   When a sender transmits a path message, it is received by routers along the path. A router inserts its own IP address as the message's last hop. As the path message is propagated through the network, each router notes the previous router's address and then inserts its own IP address before forwarding the path message. Having each router note the last router's IP address for a flow allows a router that receives a reservation request to know how to forward that request back in the direction of the sender. This ensures that the receivers will take the correct path for a particular flow. Why? Most network designs have more than one available path, and a receiver might make a reservation in a path that the sender did not specify.

Path messages can be sent at any time, and routers maintain the path state in what is known as a soft state. Routers maintain the path information only for a certain period of time, after which they delete the state. This allows for dynamic flexibility in the path. A new path (via topology changes) can be set up that renders the old path obsolete. A router might fail in the path when no alternative path is available, in which case the path information is obsolete and needs to be deleted.

RSVP and Routers   RSVP also runs in routers and works in conjunction with the requests being transmitted by a network application. RSVP is used in routers to forward QoS requests to all stations along the path or paths of a particular flow. It is also up to the routers to establish and maintain an RSVP state. In other words, if an application makes an RSVP request, each router must forward it to another router in the direction of the source—yes, the reverse path, receiver to sender. An RSVP process uses the local route table to obtain routes.

QoS is implemented by a collection of mechanisms known as *traffic control*. This includes mechanisms known as packet classifier, which determines the QoS class and possibly the router for each packet; admission control, which determines whether resources are available to accept or reject a request; and packet scheduler, which achieves the promised QoS for each outgoing interface.

Figure 14-8 shows the block diagram for RSVP. The application host interfaces are to an application API and to traffic control; the router host interfaces are to routing and traffic control. Two modules within RSVP, known as admission control and policy control, are utilized by an RSVP request. Admission control determines whether the node has the available resources to accept the request (similar to call admission control under ATM). Policy control determines the permission rights of the requestor. If either of these checks fail, the request is discarded and a message is sent back to the requestor (the application that made the request), indicating the type of failure. If both of these checks clear, parameters are set in the packet classifier and the packet scheduler in hopes of obtaining the resources required by the request.

RSVP Requests   The most basic RSVP request consists of a flow descriptor. A flow descriptor contains the following:

Flow spec: A reservation request that defines a desired QoS and is used to set parameters in a nodes packets scheduler

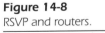
Figure 14-8
RSVP and routers.

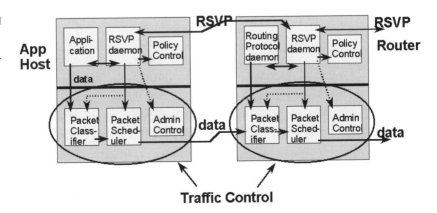

Traffic Control

Filter spec: Used to define the set of packets to receive QoS as defined in the flow spec and used to set parameters in the packet classifier

RSVP is based on sessions and defines a session as a data flow with a particular destination and transport-layer protocol. Each session is maintained independently. It is defined by a combination of three things:

Destination address: a multicast or unicast destination address.

Protocol ID: 46

Destination port: TCP or UDP port number or an application-specific port number. This can be omitted when the destination address is multicast.

There are two message types sent between senders and receivers for reservation of resources. These messages are not sent reliably because the program uses IP directly in path and reservation messages. Path messages sent downstream by the RSVP sender host. They are forwarded by routers using the unicast/multicast routing table. These messages store a path state in each forwarding node, which includes the unicast IP address of the previous-hop node. This is used to route the reservation messages (sent by a receiver in response to a path message) in the reverse path. In addition, the path message contains information on the format of data packets that the sender will generate and the traffic characteristics of the data flow, and it may carry advertising information known as one pass with advertising (OPWA). This is known as an *Adspec,* and it allows path messages to gather information en route to the receiver which the receiver, can use to predict end-to-end service.

Reservation messages are sent upstream by the receiver to the sender. They can either be distinct or shared, allowing unique reservations to occur for receivers or a single shared reservation that is shared among all packets of selected senders. A reservation message is sent upstream along the tree until it reaches a point where an existing reservation is equal to or greater than that being requested. At that point the reservation is already in place and does not need to be forwarded any further.

In a reservation request is a set of options that are collectively known as the reservation style. These options allow for *shared* or unique (*distinct*) reservations. Examples of shared reservations are those for invoking the use of multicast. Video and audio applications that make use of multicast are great examples of this. Why have multiple reservations for these receivers when one guaranteed pipe will do? Distinct reservations are for one-on-one applications such as a small desktop-to-desktop video conferencing or when some other type of high-priority, low-loss data stream is needed.

A receiver can request a confirmation with its reservation message by indicating this along with its address. One the reservation is confirmed as either unique or merged, a confirmation message is sent.

The basic reservation is completed in one pass. This means that the reservation message is sent from one router to another in the reverse path to the sender. Each router along the way has the right to reject a reservation request.

RSVP Control

Refer to Fig. 14-9, which shows the two options for the control of the selection of senders: *explicit,* which specifies a selected group of senders (each filter spec must match exactly one sender), or *wildcard,* which implicitly selects all the senders to the session (no filter spec is needed). These styles define how reservations from different senders are treated within the same session and whether the requests need to meet specific criteria or not. There are three types:

- *Wildcard filter type:* Implies both the shared reservation and the wildcard sender selection. This creates a single reservation shared by flows from all upstream neighbors. You can think of this as a big pipe, completely independent of the number of senders using it, which is shared by multiple inputs to the pipe. The size is simply the largest of the resource requests from all receivers. It automatically extends to new senders as they appear.

- *Fixed filter:* Implies distinct reservation and explicit sender selection. This very strict reservation allows a reservation to be set up for packets from a particular sender, which is not shared with other senders' packets, even from the same session. This style can quickly use up all available resources.

Figure 14-9
RSVP control.

Sender Selection	Reservations	
	Distinct	Shared
Explicit	Fixed-Filter (FF) Style	Shared-Explicit (SE) Style
Wildcard	(None Defined)	Wildcard-Filter (WF) Style

■ *Shared explicit:* Implies shared reservation and explicit sender. It creates a single reservation shared by selected, but not all, upstream neighbors.

Disabling Reservations and Handling Errors

Reservations are removed by *teardown* messages, which are not required, but are recommended. If a reservation is not removed by an application it will eventually be removed by the routers in that if a refresh message has not been received within a certain amount of time, the reservation must be removed. There are two types of teardown messages:

■ *Path Tear:* Received by all receivers downstream (from the point of initiation, not necessarily the sender). It deletes the path state (in routers, for example) and all dependent reservation states in each node that receives this information.

■ *Resv Tear:* Deletes reservation state and travels upstream toward all senders from its point of initiation (again, not necessarily the final receiver). This message specifies style and filters. Any flow spec is ignored.

The error messages that RSVP uses are PathErr and ResvErr. This is a simplex process in that an error message is sent upstream to the sender that created the error.

Merging Flow Specs

One last statement should be made here. It should have come up as a question about multiple reservations being made and the availability of resources to handle such requests. Given the state of today's routers, wouldn't we simply run out of resources within a short amount of time? The answer is hard to determine and depends on the manufacturer of the router. Some routers are high performance and possess multiple processors (some of them on the I/O card), lots of memory, high-speed interfaces, and so on. Some router vendors do not support this. Therefore, it is hard to predict how this control protocol (RSVP) is going to work on routers.

There are some efficiencies in the RSVP protocol itself. One of them is called *merging flow specs.* Multiple reservation requests from different next hops for the same session and the with same filter spec will have only one reservation on that interface. Contained in the reservation message forwarded to a previous hop is the "largest" of the flow specs requested by the next hops to which the data flow will be sent. In other words, flow specs can be cumulative or merged.

A Simple Example

Refer to Fig. 14-10a. A multicast RSVP example is given with the thought of real-time and non–real-time. Multicast tends to contain large data transfers, and with real time it will be a piece of the bandwidth that is continuously used. Also, with real-time multicast, frame loss is not acceptable in large amounts, nor is delay acceptable.

Before a session can be created, it must be identified by the triple (DestAddress, ProtocolID, and DestPort[optional]). This is must be propa-

Figure 14-10

RSVP example,

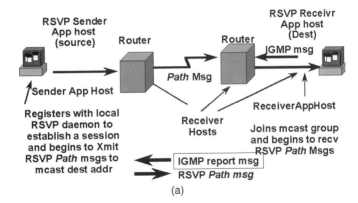

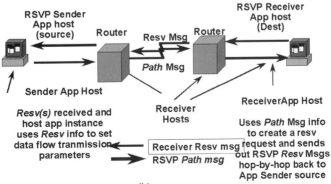

gated to all the senders and receivers. The following occurs during a session setup:

1. The receiver joins a multicast group using IGMP.
2. An RSVP-aware application starts to send path messages to the multicast destination address, which will be received by all receivers in the multicast group.
3. A receiver sends a reservation message, specifying the desired flow descriptors. These will be received by the sender, as shown in Fig. 14-10 *b*.
4. The sender starts sending the data packets.

Once the request has been accepted and processed, the resources are reserved, but they are in a "soft state." A soft state is one that has an entry but requires some maintenance to stay alive. If this maintenance is not applied, the entry will be deleted. The soft state maintains the reservation, and path and reservation messages are used to maintain the soft state. This means that the resources established can be modified as dynamically as changes occur. A soft state is maintained by RSVP's sending refresh messages along the path to indicate to the routers and nodes that they should keep the resources maintained. If these refresh messages are not received, an RSVP resource times out and is deleted.

RSVP Issues

Won't a router become overwhelmed by each receiver making RSVP requests? No. First, any router has the capability of rejecting a request. Second, RSVP is maintained in the router via a soft state. A reservation will be torn down when it is not needed. Third, RSVP allows for the concept of merging. This enables requests to be merged together (shared) when a reservation of equal size to that of the request is already in place.

Will RSVP work in areas that do not support it? Yes. The ability to simply "flip a switch" and have all routers on the Internet be RSVP-capable is not a reality. In 1983, we did "flip a switch," and all routers (IMPs) and hosts started running the TCP/IP protocol, but today we have millions of routers connected to the Internet, so moving slowly is the method of choice. Therefore, RSVP will be implemented slowly on the Internet.

We also need some RSVP-aware applications. RSVP works with non-RSVP environments, but non-RSVP environments cannot provide any reservation. RSVP path messages are forwarded without problems because they use their local unicast or multicast routing tables. In the path message is the IP address of the last RSVP-capable node before the message traversed a non-RSVP node. In this way a reservation message is forwarded directly to the next RSVP-capable router on the path back toward the source. Furthermore, there is a bit setting RSVP that sends to the local traffic control mechanism when it knows that there are non–RSVP-node hops in the path to a given sender. The router combines this with other sources of information to forward a message along the path to receivers using Adspecs.

RSVP Summary

- The Resource Reservation Protocol (RSVP) is used to reserve network resources along the transmission path(s) of a data stream. The goal is to obtain optimal QoS for that application instance.

- RSVP communicates between sending and receiving hosts, with the receiver creating the actual reservation for a session.

- Reservations are made by receivers *upstream* back toward the sender(s).

- The focus of reservations is on network-layer resources in interconnecting devices (i.e., routers or devices acting as routers).

- RSVP is designed for an integrated services framework to allow applications to choose among multiple, controlled levels of delivery service for their data packets.

- RSVP doesn't actually transport data; it is an Internet control protocol like ICMP, IGMP, and routing protocols. For instance, once a host joined a multicast group, it would use RSVP to reserve required resources along a path, but it would rely on routing protocols such as DVMRP, MOSPF, or PIM to determine the actual forwarded paths for the data streams.

- RSVP is an application's method of requesting and receiving the necessary QoS requirements for proper application performance and is supported in IPv4 and IPv6.

- RSVP operates on top of IP, occupying the place of a transport protocol, and works as an internet control protocol similar to ICMP.

- RSVP supports unicast and multicast protocols.

- RSVP is designed to accommodate large, heterogeneous groups of users with dynamic memberships and topology.

- RSVP is unidirectional, or only makes reservations in one direction for data flows.

- Once a reservation is made, it's maintained by using the "soft state" in the routers.

- Soft state provides graceful support for membership changes and adaptation to routing changes.

- RSVP makes resource reservations for both unicast and many-to-many multicast applications, adapting dynamically to changing group membership, as well as to changing routes.

- Because RSVP is simplex, or unidirectional, it treats a sender as logically distinct from a receiver. This applies even though an application process may act as a sender and receiver at the same time.

Conclusion

Individual user demands for better IP service are driving the need for some type of bandwidth reservation. The Internet continues to deliver all types of data on a first-come-first-serve basis. Internet routers still "drop" an extraordinary number of packets being sent over the Internet, causing retransmissions.

More applications are running over the Internet every day. Multimedia applications are the ones that require QoS, only because users demand it. We have come to expect it because of the standard set by the telephone and cable TV networks. RSVP will enable QoS, but it will remain only in pockets of networks throughout the Internet. It will place great demands on the routers. Today's (1998) routers have yet to prove they can handle anything more than simple data forwarding, and they are not doing that very well. Faster routers are coming onto the market, and they will help alleviate the problem.

The Internet is becoming channelized, which means that there will be streams of data running across the Internet that users can tune in to. The point is that QoS is made up of many factors, and RSVP is simply one of them. Do not think that by applying RSVP, all your troubles will disappear. You must continue to apply the other factors, as well, such as compression, filters, protocol prioritization, network design, OSPF, IP address summaries, etc. One more thing: Multimedia really requires (for best operation) that multicast be enabled. Only recently have ISPs started to multicast-enable their networks (even with the entire Internet being nonmulticast). Streaming real-time data across the Internet is not very efficient.

Lastly, you should be aware that RSVP is not an attempt to recover lost ground from ATM, as some Ethernet zealots would have you believe. ATM and other software and hardware technologies will continue to integrate. RSVP is the first attempt to provide some type of quality of service based on user-by-user need.

The RSVP Web page can be found at *www.isi.edu/rsvp.*

Simple Network Management Protocol

Network management can be broken down into five distinct categories according to Simple Management Protocol (SNMP):

- *Account management:* Gathers information on which users or departments are employing which network services
- *Fault management:* Includes troubleshooting, finding, and correcting failed or damaged components, monitoring equipment for early problem indicators, and tracking down distributed problems
- *Security:* includes authorization, access control, data encrypting, and management of encrypting keys
- *Configuration management:* Tracks hardware and software information, including administration tasks such as day-to-day monitoring and maintenance of the current physical and logical state of the network and recognition and registrations of applications and services on the network
- *Performance:* The monitoring of traffic on the network

SNMP comprises several elements that all must work together in order for SNMP to operate:.

- *Management server:* The network station that runs the management application to monitor or control the management clients.
- *Management clients:* The network stations contain the agent (a software component), which enables the management server to

control and monitor them The agent can be located in any directly attached network device such as a router, a PC, a switch, etc.

- ■ *SNMP:* A request/response protocol that allows the exchange of information between the server and an agent. This protocol does not define the items that can be managed.

- ■ *Management information base (MIB):* A collection of objects that contain information that is used by a network management server. It contains all of this information under an entity known as an object. Similar objects are placed together to form groups.

SNMP Manager

An SNMP manager is a software application that queries the agents for information. It can also set information on the client agent. The returned information is then stored in a database to be manipulated by other application software that is not defined by SNMP. The information gathered can be used to display graphs of how many bytes of information are transmitted out a port, how many errors have occurred, etc. SNMP simply sets or gathers information in a node.

The server comprises two things: management applications and databases. Management applications receive and process the information gathered by the SNMP manager. These applications also have some type of user interface to allow the network manager to manipulate the SNMP protocol. A network manager can set the SNMP node that it would like to talk to, send that node information, get information from that node, etc.

Databases store information from the configuration, performance, and audit data of the agents. There are multiple databases on the server:

- ■ MIB Database
- ■ Network element database
- ■ Management application databases:
  - ■ Topology database
  - ■ History log
  - ■ Monitor logs

All of this runs on top of SNMP. It is not necessary in order for SNMP to operate, but it does allow for the human factor.

Agents

Agents are simple elements that have access to a network element's (router, switch, PC, etc.) MIB. Agents are the interface from the network management server to the client MIB. They perform server-requested functions. When a server-requests information from a client, it will build its SNMP request (explained in a moment) and send it, unicast, to the client. The agent receives this request, processes it, retrieves or sets the information in the MIB of the client, and generates some type of response to the server.

Usually, agents only transmit information when asked to by a server. However, there is one instance in which an agent will transmit unsolicited information. It is known as a trap. There are certain things on a network station that force it to immediately notify the server. Some of these traps are defined by the SNMP RFC. Things such as cold/warm start and authentication failure are traps that are sent to the server. Most agent applications today permit the use of user-defined traps. This means that the network administrator of a router can configure the router to send traps to the server when certain conditions are met. For example, a router may send a trap to the server when its memory buffers constantly overflow or when too many ICMP redirects have been sent.

Another type of agent is known as a *proxy agent*. It allows one station to become an agent for another network station that does not support SNMP. Basically they serve as translators between servers and non–SNMP-capable clients. The reasons could be security, limited resources, etc.

Management Information Base

The Management Information Base (MIB) is a collection of objects that contain specific information, that together form a group. You can think of an MIB as a database that contains certain information that was either preset (during configuration of the node) or was gathered by the agent and placed into the MIB. Simply stated, the MIB is a database that contains information about the client that it is currently placed on.

The structure of management information is defined in RFC 1155, which defines the format of the MIB objects. This includes the following:

Syntax   Required. The abstract notation for the object type. This
                   defines the data type that models the object.

Access | Required. Defines the minimal level of support required for the object types. Must be one of read-only, read-write, write-only, or not accessible.

Status | Required. The status of the MIB entry. Can be mandatory, optional, obsolete, or deprecated (removed).

Description | Optional. A text description of the object type.

Index | Present only if the object type corresponds to a row in a table.

DefVal | Optional. Defines a default value that can be assigned to the object when a new instance is created by an agent.

Value notation | The name of the object, an object identifier.

Following is an example of an MIB entry:

```
OBJECT: ——-
            ifOperStatus { ifEntry 8 }
    Syntax:
            INTEGER {
                up(1), — ready to pass packets
                down(2),
                testing(3) — in some test mode
    }
```

Definition: The current operational state of the interface. The test-ing(3) state indicates that no operational packets can be passed. Access: read-only. Status: mandatory.

The Protocol of SNMP

SNMP is the protocol that is used between a manager and a client. SNMP uses a series of commands and protocol data units (PDUs) to send and receive management information. SNMP was intended to eventually be migrated to the OSI management scheme, but this never came about. Therefore, an encoding scheme known as Abstract Syntax Notation, or ASN.1, was used. Only the INTEGER, OCTET, STRING, OBJECT IDENTIFIER, NULL, SEQUENCE, and SEQUENCE OF codes are used. There are other, unused encodings. SNMP uses UDP as its transport layer.

Following is a list of PDU types:

GetRequest: Requests an agent to return attribute values for a list of managed objects.

GetNextRequest: Used to traverse a table of objects. Because the object attributes are stored in lexicographical order, the result of the previous GetNextRequest can be used as an argument in a subsequent Get-NextRequest. In this way a manger can go through a variable-length table until it has extracted all the information for the same type of object.

GetResponse: Returns attribute values for the selected objects or error indications for such conditions as invalid object name or nonexistent object.

SetRequest: Used to change the attribute values of selected objects.

Trap: Used by the agent to report certain error conditions and changes of state to the managing process. The conditions are cold-start, warm-start, link-up, link-down, EGP neighbor loss (though there is not much use for this one anymore), and authentication failure.

SNMP provides simple authentication process between the client and the server. This is known as the community string, and it must match between a client and the server. This string is embedded in the protocol packet, and if either side has a different entry, the received SNMP packet is discarded. The community string is configured manually on the server and the client. The problem is that it is not encrypted in any way when it is transmitted. Any protocol analyzer that is on the same link as this packet can see the community string name. It is plain text.

Figure 14-11 shows the encapsulation of an SNMP PDU between a SNMP client and a management station.

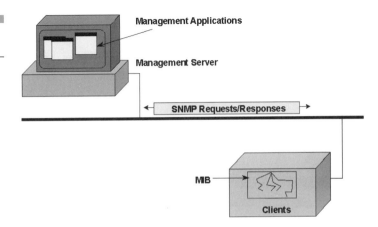

Figure 14-11
SNMP elements.

Management Applications

Management Server

SNMP Requests/Responses

MIB

Clients

AppleTalk

One of the protocols that I enjoyed researching is AppleTalk. The simplicity of this protocol is mirrored in the Macintosh computer system. AppleTalk is not a separate product for purchase; it comes with the computers. All that is needed is the wiring to connect the LocalTalk access hardware. With this hardware and the AppleTalk software together, you can build a network operating system for file and print sharing, accompanied by an easy-to-use name scheme and even the AppleTalk Remote Access Protocol (ARAP) for dialing into your network.

TCP/IP has taken over as the network operating system of choice, and AppleTalk has its disadvantages as well, as you will see. Suffice it to say, that AppleTalk will not be the protocol of choice for future Apple installations. TCP/IP works very well with this computer system and is extremely fast in comparison. However, many, many AppleTalk installations today are working quite well, and they will not be dismantled for the sake of TCP/IP. This chapter will explain the underlying AppleTalk architecture—the architecture that enables Apple computers and printers to be shared. It is assumed that the reader has some experience with an Apple Macintosh computer.

In 1983, Apple was about to introduce its new personal computer line known as the Macintosh computer. At this same time, networks were becoming popular in the business community, so Apple engineers decided to incorporate a networking scheme into the Macintosh. The major drawback of networks at that time was the complexity and cost associated with the implementation of a network. The AppleTalk networking scheme was designed to be innovative but to adhere to standards wherever possible. Apple's main goal was to implement a network in every Macintosh inexpensively and seamlessly. Their first implementation was to link a LaserWriter (Apple's laser printer) to the Macintosh. File sharing between Macintoshes came later.

The key goals of the AppleTalk networking system are

- Simplicity
- Plug and play
- Peer-to-peer
- Open architecture
- Seamless

In the design of this system, the standard Macintosh user interface was not to be disturbed. Users should be able to implement an Apple network and not know that they are running on one. A user should be able to look at the screen and notice the familiar icons and window interfaces as if the Macintosh were operating locally. All exterior actions on a Macintosh were implemented in the Chooser, and that is where the network extensions were also provided. The network menus should be as friendly to use as the operating system itself. The engineers overwhelmingly accomplished this.

Even the wiring scheme used fits into these goals. Users should need only to plug the network cable in for the Apple operating system to be able to detect the network and work accordingly. The beginnings of AppleTalk gave us two entities to work with: (1) LocalTalk, which is the access method (physical and data-link layers); and (2) AppleTalk (more commonly known as AppleShare), which is the network operating system (OSI network through application layers).

The following text is written to explain the underlying technology of the AppleTalk network operating system—the network system that cannot be seen. It is an uncomplicated system and was an inexpensive solution at a time when other networking solutions were expensive. Macintosh computers have two large components that have given Apple computers a clear advantage in the personal computer marketplace: built-in networking and an easy-to-use object-oriented operating system. The object-oriented operating system provides users with computing capabilities based on objects. When working with the Apple personal computer, components of the operating system or application programs are accessed with icons that appear on the screen. Users use the mouse to access the icons, which provide an entrance into the operating system or to an application program.

The focus of this chapter is the network portion of AppleTalk. The AppleTalk network operating system consists of two entities: network protocols and hardware. AppleTalk protocols are arranged in layers, with each layer providing a service to another layer or to an application. We will show how data flows on an AppleTalk network.

Figure 15-1a shows the layout of the AppleTalk protocols. Figure 15-1b shows the subset of the protocols to be fully described in the following chapters.

The network protocol is the software version of AppleTalk and the wiring scheme is the hardware component of AppleTalk. The hardware portion consists of the physical wiring, the connectors, and their physical interfaces. It also includes the network access methods used with AppleTalk. Currently, AppleTalk can run over LocalTalk (Apple cabling and access method scheme) and the two other most popular access methods: Ethernet and Token Ring.

The software portion of the AppleTalk stack consists of the network through application layers of the OSI model. The network layer has the ability to direct messages locally or through network-extending devices

Figure 15-1
(a) AppleTalk and the
OSI model. (*Courtesy
Apple Computer.*)

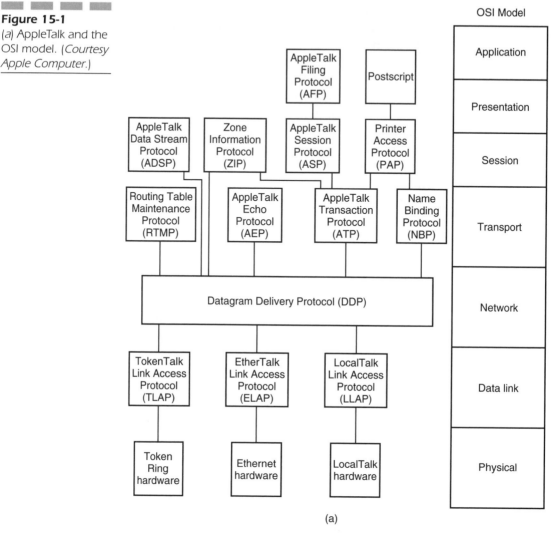

(a)

such as routers. The stack also contains transaction-oriented transport- and session-layer routines, which enable messages to be sent reliably through the network. Printing is also incorporated into the AppleTalk session layer. Finally, network workstations and servers may communicate through the network by the use of the Apple Filing Protocol (AFP), located at the application layer.

Since the hardware portion consists of the lowest two OSI layers, it will be discussed next. LocalTalk will be fully discussed and Ethernet

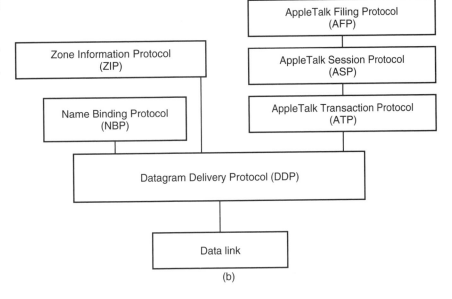

Figure 15-1
(*b*) AppleTalk stack.
(*Courtesy Apple Computer.*)

and Token Ring briefly discussed. One last note: There are two versions of AppleTalk—Phase I and Phase II. Throughout this text, all protocols are Phase II.

The Physical Layer—AppleTalk Hardware

AppleTalk can be used over many different types of media. This entity comprises the network controller and cabling systems used to allow users to communicate with each other. The current media capable of handling the AppleTalk system include LocalTalk (for Apple proprietary networking), EtherTalk (for use on Ethernet systems), TokenTalk (for use on Token Ring networks), and LANSTAR AppleTalk from Northern Telecom.

First, we'll discuss the hardware requirements of the access methods.

LocalTalk

Apple's low-cost network implementation of LocalTalk consists of the wiring (cable segments) and access methods to transmit data on the wiring. AppleTalk with LocalTalk is embedded into every Macintosh computer. The access method of LocalTalk is used only with Apple Macintosh computers and the related network devices such as printers and modems. It is not used with any other network protocol.

Just as Ethernet defines the access methods that govern the transmission and reception of data on a special type of cable plant, LocalTalk accomplishes the same functions and is an inexpensive way to connect workstations and their associated peripherals into a network. As shown in Fig. 15-2a, a LocalTalk network consists of the devices that connect AppleTalk network stations on a bus topology.

The devices used are

1. LocalTalk connector modules shown as the two device connectors
2. A 2-m LocalTalk cable with locking connector shown as the bus cable
3. A LocalTalk cable extender
4. The DIN connector shells

The device connector module is a small device with a cable attached to it that connects a network node to the LocalTalk cable. There are two types of connectors associated with it. An eight-pin round connector (DIN connector, shown as the upper-left device in Fig. 15-2a) is used to connect to the Apple IIe, Apple IIGS, Macintosh Plus, Macintosh SE, and

Figure 15-2
(a) LocalTalk devices.

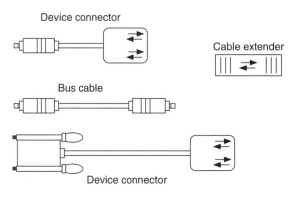

See basically two devices in LocalTalk networks
1. Device connector (2 types)
2. Twisted pair cable with DIN or RJ-connectors
Others include: cable extenders

(a)

AppleTalk

Figure 15-2
(*b*) A LocalTalk net-
work.

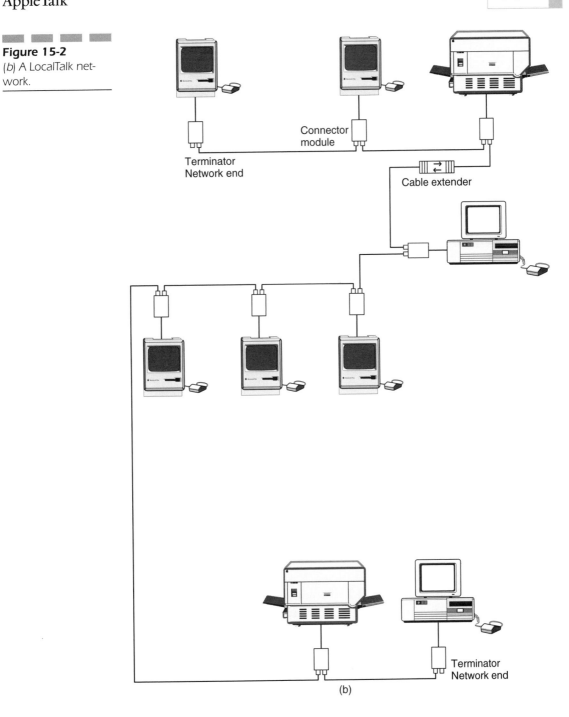

Connector
module

Terminator
Network end

Cable extender

Terminator
Network end

(b)

Macintosh II computers; the LaserWriter II NT and LaserWriter II NTX printers; and the ImageWriter II printer with the LocalTalk option installed. The nine-pin rectangular connector, shown as the lower device connector in Fig. 15-2a is used on the Macintosh 128 K, 512 K, and Macintosh 512 K enhanced computers and the LaserWriter and LaserWriter Plus printers.

LocalTalk cable is available in 10- and 25-m lengths. Apple also sells a kit, including the connectors, for custom-made lengths. To run a longer cable between two network devices, a cable extender adapter is used to connect these cables to make longer runs of cable. In order for a network station to transmit and receive on the cable, a special device known as a *transceiver* is used. The Apple engineers have built this device into every Macintosh computer, so all the user has to do is connect the cable directly into the printer port in the back of the Macintosh. Printers participate in the AppleTalk scheme as autonomous devices and need not be directly connected to the Macintosh.

The AppleTalk specifications state that the longest single cable length is 300 m. Other companies that have compatible wiring for AppleTalk have made modifications to this scheme and allow cable segments of up to 1000 m; the text in this chapter follows Apple Computer's recommendations. Cables are made available in different lengths, and the total length of the cable plant may run as high as 1000 m. This style of cabling is similar to the thin Ethernet cable scheme. Refer to Fig. 15-2b. In order to build the LocalTalk cable plant, the cable must be connected together. Devices are connected to the LocalTalk network by connecting the DIN (or rectangular) connector to the printer port of the Apple computer. The connector block at the other end of the cable has two RJ-11 or jacks. The bus cable is connected to these jacks, and network stations are more or less concatenated to each other.

As with Ethernet coaxial cable, the two end stations on the bus cable will have one connection to the bus, and the other jack will be self-terminating. Phonenet systems must have a terminator plug in the other jack. Figure 15-2b depicts this. Only the two end stations will have the terminator in their blocks. All other stations will have the bus cable plugged into both connector blocks.

The speed on LocalTalk is 230.4 kbps, much lower than the speed of Ethernet (10 megabits per second) or Token Ring (4 or 16 Mbps). The components used in the LocalTalk system were inexpensive, and the cabling system was built on inexpensive wire. The resulting limitations on the LocalTalk system forced a low speed.

Media Considerations for AppleTalk

With the capability of using different network media for AppleTalk, it becomes necessary to compare the three access methods. Table 15-1 shows the comparison.

Data-Link Functions

Functions in AppleTalk at the data-link layer include Control Panel software; the Link Access Protocol (LAP) manager; the AppleTalk Address Resolution Protocol (AARP); the Ethernet driver; and the Token Ring driver.

TABLE 15-1 LocalTalk and Ethernet Media Considerations*

	LocalTalk	Thick Ethernet	Thin Ethernet	LANSTAR AppleTalk
Medium	Twisted pair	Coaxial	Coaxial	Twisted pair
Link access protocol	LLAP	IEEE 802.3	IEEE 802.3	LLAP
Transmission rate	230.4 kbps	10 Mbps	10 Mbps	2.56 Mbps
Maximum length	1000 ft	Segment: 1640 ft	Segment: 656 ft	2000 ft to star
		Network: 8202 ft	Network: 3281 ft	
Minimum distance between nodes	No min.	8.2 ft	1.5 ft	No min.
Maximum number of nodes	32	Segment: 100	Segment: 30	1344
		Physical network: 1023	Physical network: 1024	
Maximum number of active AppleTalk nodes per physical cable segment	32	Unlimited	Unlimited	1344

*Cable segment is a piece of cable not separated by a repeater device. Network segment is the total number of devices on all cable segments not separated by a network-extending device such as a bridge or a router.

The Control Panel

AppleTalk makes extensive use of the Macintosh (Mac) user interface, and part of the window system in the Apple user interface is an icon called the Control Panel, which contains the network control device package. The Control Panel uses the System Folder to store all alternative AppleTalk connection files.

To allow connection to the LocalTalk bus, the user must use the Apple icon on the menu bar of the screen interface. Here, the operating system allows multiple choices into the Apple operating system. The user selects the Chooser menu, in which icons represent each possible action on a network. The user can select a network connection to use (LocalTalk, EtherTalk, or TokenTalk). Only one physical interface to the network may be in use at a time. That is, the network station may be attached to both an Ethernet network and a LocalTalk network, but it can send and receive data on only one or the other (not both). A user may attach to any network device on the *active* network.

The Link Access Protocol
Manager for LocalTalk

This entity (Link Acess Protocol or LAP Manager) is used to send and receive data over the selected media; it is the data-link layer for LocalTalk. For workstations that are connected to both LocalTalk and EtherTalk, the LAP Manager sends packets to the network connection that the user selected in the Control Panel. For AppleTalk networks operating over a LocalTalk medium, the access method is LocalTalk Link Access Protocol, or LLAP.

One of LLAP's responsibilities is to handle access to the cable plant. The method employed here is similar to the access method of Ethernet—with a small twist. The access method used by LocalTalk is Carrier Sense with Multiple Access and *Collision Avoidance* (CSMA/CA). Ethernet uses *Collision Detection* (CSMA/CD). Collision avoidance is explained in a moment.

LLAP Packet Types   The following types of packets are sent by LLAP:

lapENQ   Inquiry packet used by the station to assign itself an address

lapACK   A packet sent back in response to a lapENQ

lapRTS   Sent to a destination station to indicate that a station has data for it

lapCTS   Sent in response to a lapRTS to indicate to a source station that the destination station can accept data

The rightmost column of Fig. 15-3 shows a LocalTalk/AppleTalk packet structure. This figure will be used throughout the LAP section.

LLAP-Directed Transmissions (Station-to-Station Communication) The following text assumes that the source stations know the identity of the destination stations. How that happens will be discussed later.

Since there is only one cable plant and all stations need access to it, all active network stations compete for sole use of the cable plant for a limited amount of time. To gain access to the cable plant, one at a time, is the algorithm used in LLAP. LLAP uses Carrier Sense Multiple Access with Collision Avoidance (CSMA/CA) as follows. Refer to Fig. 15-4. First, it performs *carrier sense (CS)*: the process of checking the cable plant to ensure that no one is currently using the cable to transmit (by sensing electrical activity on the cable). If the cable is busy, the station that intends to transmit is requested to defer (to hold the data until the cable plant is clear).

Once the cable is quiet—and it has to have been idle for at least one *interdialog gap (IDG)*, which is 400 μs—it will wait an additional amount of random time that is a function of the number of times the node had to previously defer and the number of times it assumes a collision has occurred (explained in a moment). If the cable became busy at any time during this interval, the whole process will be started over.

If the cable remains quiet throughout this whole time period, the network station will send a Request-to-Send (RTS) packet to the destination station. The destination station must reply with a Clear-to-Send (CTS) packet to the originator within the *interframe gap (IFG)* of 200 μs. When the originator receives this packet (CTS packet), it knows the destination is active and willing to accept packets. The source is now able to send data to the destination station. It must send data within the IFG time period. If the destination needs to respond to the packet (an ACK or a data response), it must repeat the foregoing procedure, starting with an RTS packet. Once one station has transmitted any packet, the cable is free to anyone to try and gain control over it. This is one of the reasons why the network operates at a slow rate (LocalTalk operates at 230.4 kbps). This process is not used when AppleTalk runs over Ethernet or Token Ring; when run over these media, AppleTalk follows their access methods.

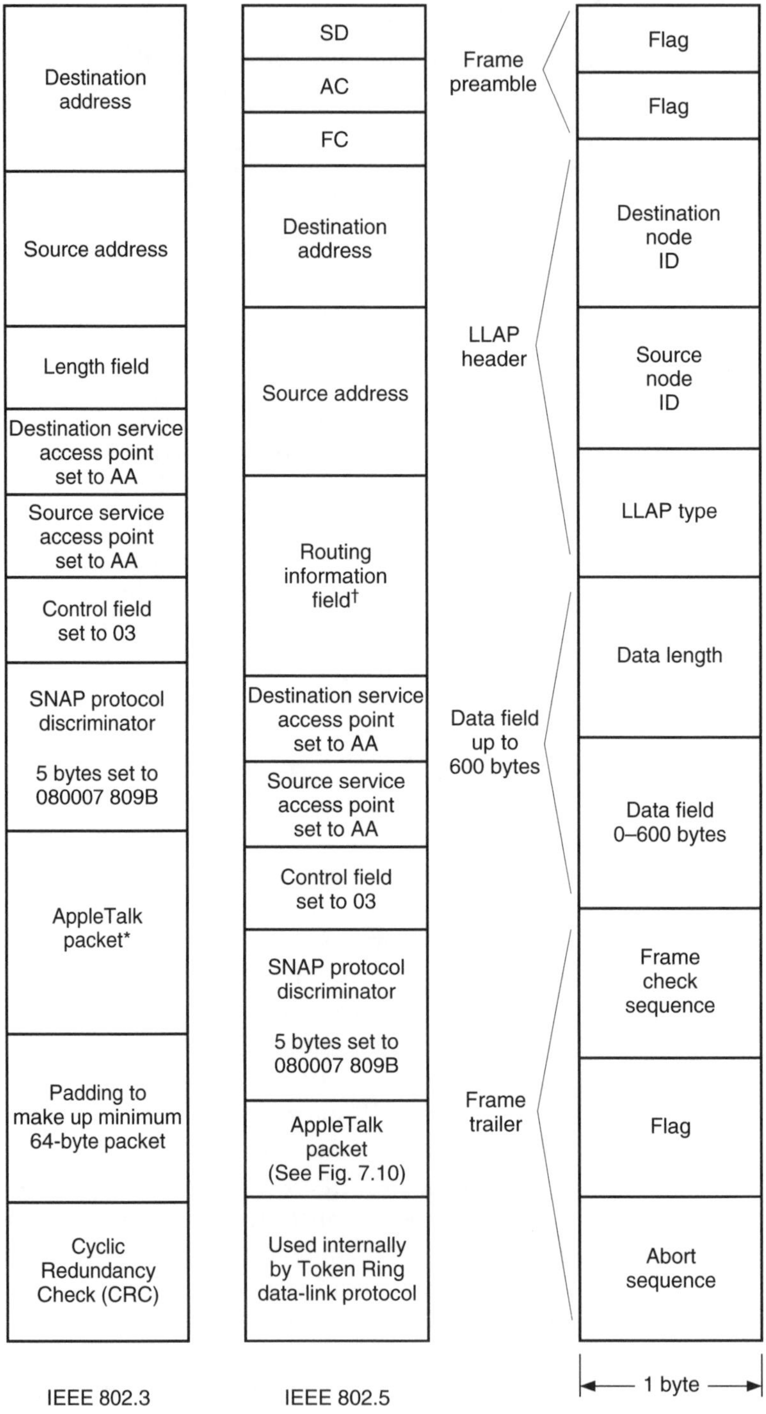

Figure 15-3
ELAP (IEEE 802.3),
TLAP (IEEE 802.5),
and LocalTalk packet
frames. *See Fig. 15-
10. †Refer to Chap. 2.
(*Courtesy Apple Com-
puter.*)

IEEE 802.3 | IEEE 802.5

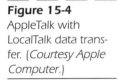

Figure 15-4
AppleTalk with
LocalTalk data transfer. (*Courtesy Apple Computer.*)

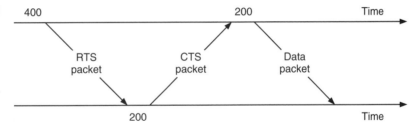

All time is in microseconds.

If a collision did occur (it should have occurred during the RTS-CTS handshake), the corresponding LLAP control packet (either the RTS or CTS) will be corrupted when received by one of the stations. The final result is that the CTS packet will never be received, and the originating station will *assume* that a collision has occurred. The originating station must then back off and retry. With this, the originating station is said to *assume* that a collision has occurred, because there is no circuitry defined in AppleTalk to determine that an actual collision has occurred. This reduces the cost of implementing AppleTalk with LocalTalk.

It will attempt 32 times to transmit a packet before notifying the upper-layer software of its inability to do so.

LLAP Broadcast Transmissions   Broadcast transmissions from different from directed transmissions. Broadcast transmissions are intended for all stations on the local network. If there are 20 stations on the network, all 20 stations should receive a broadcast packet.

Broadcast frames will still wait for the cable to be quiet for at least one IDG. They will then wait an additional amount of random time. If the link is still quiet, the broadcast frame will send an RTS packet with the destination address set to FF (hex) (255). If the line remains quiet for one IFG, the station will then transmit its data. It does not expect to receive any responses.

The broadcast packet has many functions, such as to send a message to all stations at one time or to find other stations on the network.

LLAP Packet Receptions   A network station will receive a packet if the packet's destination address in the packet received is the same as the receiving station's internal address and if the receiving station finds no errors in the packet's Frame Check Sequence (FCS).

A Frame Check Sequence is an algorithm to verify the data that was sent is the same data that was received. For simplicity, you may think of this algorithm as a complex parity checker. This data check guarantees the transmission of the packet to be 99.99 percent free of errors. The transmitting station computes the FCS while building the packet. When the receiving station receives the packet, it will compute its own FCS. It will then compare its FCS with the one received. If there is a discrepancy between the two, the receiving station will discard the packet.

The LLAP of the receiving station will also drop a packet for other reasons, such as one that is too large or too small or one of the wrong type. It will handle this without interrupting the upper-layer software.

The next protocol in the data-link layer of AppleTalk is the AppleTalk Address Resolution Protocol (AARP). This will be discussed by studying EtherTalk and TokenTalk.

AppleTalk Addressing   To communicate with another station requires more than sending RTS and CTS packets on the network. A packet must be addressed so that another station may receive this packet and decide whether it is meant for that station.

AppleTalk was designed to run on top of LocalTalk. With Ethernet and Token Ring, the physical address of the controller card, also known as the MAC (Media Access Control) address is usually assigned by the manufacturer of the controller card and is physically set in the hardware. During initialization, network software that runs on these controller cards will read the address of the controller card and assign this as the address on the network. In contrast, LocalTalk does not have an address "burned" into a chip in the LocalTalk hardware. AppleTalk has an addressing scheme, but the address of a network station is decided by the network station's AppleTalk software during the initialization of the network software. This type of address is also known as a *protocol address*. Originally, in AppleTalk with LocalTalk, there were no physical addresses—only AppleTalk protocol addresses.

As with any other protocol that operates over a network, any attachment on the network must be identified so that any other device may communicate with it. As previously discussed, each device on a network is assigned a unique node address and a group network address. Since LocalTalk was devised as the hardware complement to AppleTalk, the following text describes the AppleTalk addressing scheme, which will be used again when mapped to an Ethernet or Token Ring MAC-layer address. The AppleTalk addressing scheme was not changed when AppleTalk was migrated to Ethernet and Token Ring. The AppleTalk address is combined with the Ethernet or Token Ring physical address. This requires the use of a new protocol known as AppleTalk Address Resolution Protocol (AARP), which allows a translation of the AppleTalk software address to Ethernet and Token Ring physical addresses. AARP is discussed at the end of this section.

There are many addresses on every network: one for each attachment (workstation, router, server). On an AppleTalk network, each network station possesses a unique identity in the format of a numeric address. The assignment of this address to the network attachment is a *dynamic* process in AppleTalk. This creates two advantages over hard-coded addressing schemes (like the ones used in Token Ring and Ethernet). First, there is less hardware required because the address is not fixed (burned into a PROM). Second, there is no central administration of IDs from vendor to vendor (Ethernet and Token Ring vendor IDs are required from the IEEE standards committee). However, dynamic node assignment can create large problems when combined with network-extending devices such as a bridge.

The AppleTalk address will actually be two numbers: a 16-bit network address and an 8-bit node ID. A network address is similar to an area code in the phone system. The node ID is similar to the seven-digit phone number. The node ID ranges from 0 to 255, and the network number ranges from 0 to 65,535. Each network attachment on the network will be identified by these two numbers together. For example, a single network station can be found on any AppleTalk internet if the network number and node ID are known. Knowing the network number will track it down to a group of network stations, and knowing the node ID will single out a network station within that group. Again, AppleTalk addressing takes the form of network number, nodeID, with the network number being 16 bits in length and the node ID being 8 bits in length. This is the process and format for the AppleTalk address. This address will be mapped to an Ethernet MAC address through a process known as AARP, which will be discussed later.

For AppleTalk Phase II, there are two addressable types of networks: nonextended and extended. AppleTalk Phase I did not have the concept of extended networks. Nonextended networks are individual networks that contain one network number and one zone name; LocalTalk and AppleTalk Phase I/Ethernet are examples of nonextended networks. Likewise, extended networks are networks that can contain more than one network number and multiple zone names. Extended networks were devised to allow for more than 254 node IDs on a single physical network (i.e., for Ethernet and Token Ring).

Some restrictions are placed on the AppleTalk addresses. A network number of 0 indicates that the network number is unknown; it is meant to specify the local network to which the network station is attached. In other words, packets containing a network address of 0 are meant for a network station on their own local network segment. Network numbers in the range of FF00 (hex) through FFFE (hex) are reserved. They are used by network stations at start-up time to find their real network number from a router and at times when no router is available.

Node ID of 0 also has a special meaning. A packet with this address is destined for any router on the network specified by the network part of the address. Packets utilizing this address will be routed throughout an internet and will be received by a router whose network address is included in the network ID field. A protocol such as the Name Binding Protocol (NBP, discussed later) is one example of this type of addressing. It allows processes within routers to talk to each other without having to use up a node ID. Node ID 255 is reserved as a broadcast address. A packet containing this address is meant for all stations on the network. For AppleTalk Phase II, node ID 254 is reserved and may not be used.

With LocalTalk only, the available 254 node IDs are divided into two sections: those reserved for servers and those reserved for workstations, as shown in Table 15-2. This eliminates the chance that a station was too busy to answer an inquiry packet. It also prevents workstations from acquiring a server ID, which could be disastrous on an AppleTalk network. Because of the high speed and reliability of Ethernet and Token Ring, the EtherTalk and Token Talk protocols do not implement this separation of node IDs.

An AppleTalk node may acquire any address within the restrictions just stated. Remember, this is a dynamic process, meaning the number is randomly chosen by the AppleTalk protocol running on a network sta-

TABLE 15-2

Node ID
Definitions

Node ID Range	Assignment
0	Not allowed or unknown node
1–127	User node IDs
128–254	Server node IDs
255	Reserved for broadcast
Network number 0 and node ID	Networkwide or zone-specific
255	Broadcast

tion. All stations on the network will participate in the selection of a node ID for a network station.

Nonextended Node ID Address Selection   On a nonextended network, server and workstation node IDs range from 1 to 254 (0 and 255 are reserved). This type of network is assigned exactly one network ID and exactly one zone name (zone names will be discussed later). An example of this type of network is LocalTalk or Phase I AppleTalk with Ethernet framing. When the network station starts up, it will assign itself a node ID. The station will then send out a special packet, known as an inquiry control packet, containing this address to see whether any other station has already reserved this number for its use. If no response is received, the node will then use this number. If a response to this inquiry packet is received, the requesting node will randomly choose another number, and submit another inquiry control packet, and wait for a response. The node will continue this until it can acquire a unique ID.

The node will then send out a request to the router to find out the 16-bit network number that has been assigned to the particular network. If a response is not received, the node will assume that no router is currently available and will use network ID 0. If a router later becomes available, the node will switch to the new network number when it can. (It might not switch immediately, for previous connections may have been established.) If a response is received, the response packet will contain the network number assigned to that network, and the node will then use that.

Extended Network Node ID Selection   AppleTalk Phase II introduced a new network numbering scheme and the concept of an *extended network*. An extended network is a network cable segment that consists of multiple network numbers and, theoretically, may have up to 16 million network attachments on it. It can afford this expansion, for each network station is assigned a combination of a 16-bit network ID and an 8-bit node ID. Network IDs are like area codes in the phone system. They usually, but not always, identify groups of nodes with common network IDs assigned to them. Extended networks can also be assigned multiple zone names. (Zoning is discussed in a moment.)

With extended networks, there is one less node ID allowed. Node ID FE (hex) (decimal 254) is reserved; therefore, there are only 253 node IDs allowed per network ID. Implementation of AppleTalk network IDs is different from other network ID implementations in that multiple network IDs are allowed on the same cable segment. In order to allow

AppleTalk to run on Ethernet and Token Ring networks, this had to be taken into account, because the maximum number of network attachments on an extended cable segment of Ethernet is 1024. The maximum number of attachments on a Token Ring network is still 260.

For those network stations operating on an extended network, the network ID acquisition is a little different. First, the network station will assign a provisional node address to itself. This is assigned by the data link and its only purpose is to talk to a router. The start-up network ID is taken from the range of FF00 (hex) to FFEE (hex). This range is reserved for use with start-up stations and may not be permanently assigned to any network.

A unique twist to this acquisition process is that, if the network station was previously started on the network, it will have reserved its previous node ID and network ID on its disk. This is known as *hinting*. When the network station starts up, it will try this address first. If this address is no longer valid, it will start from the beginning.

If the network station had not been previously started, it will send out a special packet to the router. The router will respond with a list of the valid network IDs for that network segment. The node will then select a network number from that range. If no router is available, the reserved start-up network number is used and will be corrected later when the router becomes available.

One noteworthy point here for those familiar with the manual filtering capabilities of bridges or routers: Since node IDs and network IDs are dynamic, it is impossible to guess which network station is assigned to a network number. The manual filtering of bridges or routers for AppleTalk is generally reserved to zone names.

AppleTalk Phase I and Phase II   When AppleTalk Phase I was implemented, no more than 254 network stations could attach and be active on a single network cable segment. One network number and one zone name existed for each cable segment (not separated by a router). This protocol also supported the Ethernet framing format.

With AppleTalk Phase II, the network addressing was extended so that many network numbers could exist on a single cable plant. This network number is 16 bits wide, allowing for over 16 million network stations per network segment (in reality, an unrealistic number, but allowed). Support for the Ethernet framing format was also changed to the IEEE 802.3 with SNAP headers frame format (this format was shown in Chap. 2). AppleTalk Phase II allows 253 network station addresses per network number (one less than Phase I), but now you may have multiple network IDs on the same cable plant. AppleTalk Phase II also supports Token Ring networks using 802.2 SNAP frames. Most Apple network implementations today have switched over to AppleTalk Phase II.

The AppleTalk Address Resolution Protocol

The aforementioned AppleTalk addressing scheme is the one that Apple Computer developed to work with LocalTalk. Since AppleTalk's inception, Ethernet and Token Ring have taken over as the network implementation of choice (especially since the price has dropped considerably for the controller cards). LocalTalk is still used in smaller network environments and is by no means a dead access method.

In order to have AppleTalk run on an Ethernet or Token Ring network, the AppleTalk address previously described must be mapped internally to conform to the 48-bit MAC address (also known as the hardware, or physical, address) used in Ethernet and Token Ring. AppleTalk Address Resolution (AARP) is the protocol that accomplishes this.

There are three types of packets that AARP will use:

1. *Request packet.* This is used to find another node's AppleTalk address/MAC address. In order to send information to another station (on a local network), the station must know its MAC address and its AppleTalk address. This packet is sent out to find the address.

2. *Response packet.* This is used to respond to a node's request for an address mapping.

3. *Probe packet.* This is used to acquire an AppleTalk protocol address and to make sure that no one else on the local network is using this address. This packet is sent out up to 10 times, once every 200 ms (1/5 s), therefore 10 times in 2 s.

To identify all software and hardware protocol stacks operating within a network station, they are addressed with integer numbers. As previously discussed, when AppleTalk and LocalTalk were first devised, an 8-bit integer was selected for this purpose.

Ethernet uses 48-bit addresses for network attachment identification (6-byte source and 6-byte destination MAC-layer address). A software entity was devised to translate between the two. For those readers familiar with TCP, AARP is similar to IP's ARP protocol. AARP resides between the Link Access Protocol and the LAP Manager. AARP performs the functions described in subsequent paragraphs.

Functions Provided by AARP AARP provides the following services:

1. Selection of a unique address for a client
2. Mapping the protocol address to the specific physical hardware address
3. Determining which packets are destined for a specific protocol

These processes are used only when implementing AppleTalk on top of Ethernet or Token Ring. AARP is not implemented for LLAP (LocalTalk).

We have not replaced the AppleTalk software address. We have merely devised a scheme to work with it. This is similar to the process that TCP/IP uses to map the 32-bit TCP/IP addresses to the Ethernet or Token Ring MAC hardware address. If you know that protocol, called the Address Resolution Protocol (ARP), you will know how AARP functions.

AARP Address Mappings Each node that uses AARP maintains a table of address mappings. In the example of Ethernet, it maintains a table of the AppleTalk protocol addresses and their associated IEEE 802.3 physical addresses. This table contains the mappings for every network station on the network—not for every station on the Internet, just the local cable segment. This table is the Address Mapping Table (AMT). (See Table 15-3.)

Instead of requesting the mapping each time a station needs to talk to another station on the network, the AMT is a table that will have a listing of all known network stations on its local network. Each entry in the table consists of an AppleTalk address and the corresponding MAC hardware address. When a network station needs to talk to another

TABLE 15-3

An AMT Table

16.3	02608c010101
16.4	02608c014567
16.90	02608c958671
17.20	02608c987654

station on the network, it will first look up the mapping in the AMT. Only if it is not there will the station send out a request packet.

Once an entry in the AMT is made, AARP maintains this table. AARP will age out (delete after a certain time) old addresses and update the table with new ones.

AARP receives all AARP request packets, since they are sent out with a broadcast MAC destination address. If AARP does not need to respond to a packet, it will discard the packet, but it will check something first—the sender's hardware (48-bit address) and AppleTalk node address (8-bit address) embedded into the AARP request packet. It will extract this information, use it in its initialization table, and then discard the packet. Table 15-3 shows an example of an AMT for a given network station. This shows four entries. When a network station that contains this table would like to talk to another station, it must find the hardware address of the remote station. It will first consult this table.

In order to communicate with another station over Ethernet or Token Ring, a network station must know the destination's AppleTalk and MAC address. If the requesting station wants to talk to station 16.3, it looks in this table and finds the MAC address for 16.3. Therefore, it builds a packet and, in the data-link header for addressing, it puts the MAC address (02806c010101 from the AMT shown in Table 15-3). If the entry for 16.3 is not in this table, AARP builds a request packet and transmits the packet to the network. Upon receiving a response, it adds the contents of the response packet to the AMT and then builds a packet for 16.3.

AARP Node ID Assignment   When a network station initializes, AARP assigns a unique protocol address for each protocol stack running on this station. Either AARP can perform this assignment or the client protocol stack can assign the address and then inform AARP.

For AARP to assign the address, it must accomplish three things:

1. Assign a tentative random address that is not already in the Address Mapping Table (AMT)

2. Broadcast a probe packet to determine whether any other network station is using the newly assigned address

3. Permanently use this number for the workstation if the address is not already in use.

If AARP receives a response to its probe packet, it will then try another number and start the whole algorithm over again until it finds an unused number.

When a network operating system submits data to the data-link layer to be transmitted on the network, the protocol will supply the destination address. In AppleTalk, this will be supplied as a protocol address. This address will then be mapped to the corresponding MAC address for that particular destination station. This mapping is what AARP accomplishes.

Examining Received Packets   When AARP receives packets, it will operate only on AARP packets. All other packets are for the LAP. In other words, AARP does not provide any functionality other than AARP request, response, and probe packets.

Once AARP has provided the mapping of addresses, the data-link protocol may then accomplish its work. The access protocol of Ethernet remains unchanged to operate on an AppleTalk network. AARP does not interfere with the operation of Ethernet or Token Ring.

LAP Manager for EtherTalk and TokenTalk

Ethernet was developed at Xerox's Palo Alto Research Center (PARC). First known as the Experimental Ethernet, it was first utilized in 1976. In 1980, a cooperative effort by Digital, Intel, and Xerox led to a public document known as the "Blue Book" specification, whose formal title was Ethernet Version 1.0. In 1982, these three companies again converged and came out with Ethernet Version 2.0, the standard by which Ethernet operates today.

In 1982, the components that made up an Ethernet network were extremely expensive and were out of most companies' cost reach. Since Ethernet is an open publicly available architecture, not a proprietary one, many companies have jumped on the bandwagon and have developed Ethernet products of their own. Since Ethernet is an open specification, all Ethernet products are compatible with each other. You may buy some Ethernet products from one company and more Ethernet products from another company, and the two will work with each other (at least at the data-link layer). Changes in cabling strategies and mass production of the chip sets have also led to price reductions in Ethernet, and currently Ethernet is the most popular local area network—second in cost only to ARCnet networking scheme. For more information on Ethernet and related documents, refer to the References section of the book.

The origins of Token Ring date back to 1969. It was made popular when IBM selected it as its networking scheme. It was adopted by the

IEEE 802.5 committee in 1985, and it offered some advantages over the Ethernet scheme. Some of these advantages included a star-wired topology and embedded network management. The cost of Token Ring was extremely high at first, but the price reductions over the last few years, and its advantages over Ethernet have let this networking scheme become very popular.

AppleTalk was derived as an alternative to this high cost of networking. AppleTalk, with LocalTalk, operates at all layers of the OSI model. It is a true peer-to-peer network and is built into every Macintosh computer. It is a simplex "plug and play" type of networking that easily allows users to access the services of a network.

However, Ethernet and Token Ring are still the most popular networking schemes, with many advantages over the LocalTalk access method. Therefore, AppleTalk has been adapted to allow for this. To allow these networks compatibility with AppleTalk, the primary layer that is replaced is the LocalTalk data-link layer and the way LAP manager works with it.

When an AppleTalk network station wishes to communicate with another AppleTalk network station, it must provide its data link with an AppleTalk protocol address, which consists of a 16-bit network number and an 8-bit node ID.

When Ethernet and Token Ring are used as the medium, a major change is invoked here. The address format used with these access methods is 48 bits long. The AppleTalk protocol address must be translated into this format before a packet may be transmitted on an Ethernet or Token Ring network.

Extensions were made to AppleTalk's LAP in the form of EtherTalk Link Access Protocol (ELAP) and TokenTalk Link Access Protocol (TLAP) to accommodate Ethernet and Token Ring access methods. This LAP manager also uses the AppleTalk Address Resolution Protocol (AARP) to translate the 48-bit addresses to the AppleTalk addresses. ELAP and TLAP function according to the access methods of Ethernet (CSMA/CD) and Token Ring, respectively.

Theory   Simply enough, EtherTalk and TokenTalk are the Ethernet and Token Ring access methods that have AppleTalk running on top of them. The LocalTalk access method was stripped out and replaced with either Ethernet or Token Ring.

If you have an Ethernet network, EtherTalk will allow AppleTalk to run on that network. The same is true for TokenTalk on a Token Ring network. The major changes were made in the LAP Manager. The network layer and all layers above are still AppleTalk. Refer to Fig. 15-5,

Figure 15-5
AppleTalk on Ether-
net.

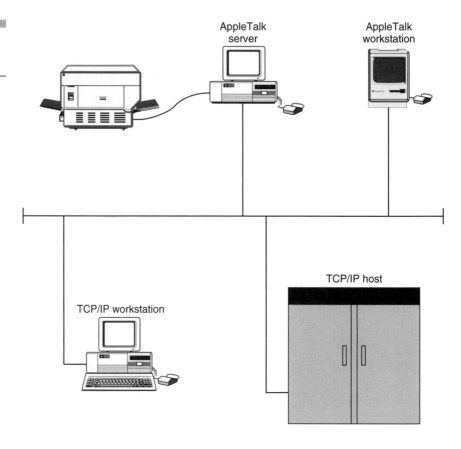

which shows the attachment for an Ethernet system). One noticeable change is the printers are now usually connected to an AppleTalk file server and not directly to the LocalTalk network. Ethernet and Token Ring interfaces are not in widespread use for direct attachment of Apple printers.

Compare the Token Ring packets to the AppleTalk with the LocalTalk packet headers shown in Fig. 15-3. As far as AppleTalk is concerned, it is running on top of LocalTalk. Only the LAP Manager is changed.

The easiest method to allow AppleTalk Phase II to operate over Ethernet and Token Ring was through the use of the Institute of Electrical and Electronics Engineers (IEEE) protocol specification called IEEE 802.2 Type 1. SNAP is discussed fully at the end of that chapter.

The IEEE 802.2 (pronounced "802 dot 2") data-link specification consists of two types. Type 1 is a connectionless protocol (meaning that an established link between a source and destination network station does not have to exist before data transmission occurs) and type 2 is connection-oriented (meaning the opposite of type 1). EtherTalk and TokenTalk use IEEE 802.2 type 1 packet formats for transmission over the network.

A subpart of that protocol is defined here, called SubNetwork Access Protocol or SNAP. Refer to Fig. 15-3. For those familiar with the Ethernet specification, the IEEE 802.2 specification replaced the concept of the type field with the concept of a Service Access Point, or SAP. SAPs allow distinction of a single protocol within a packet. When IEEE 802.2 was being written, it was recognized that many protocols may exist on a network station or on a network. When a packet is transmitted, the processing the protocol stack that submitted the packet and the process of the protocol stack to which the packet is destined should be known. For example, you may have AppleTalk, Xerox Network Systems (XNS), and TCP/IP all running on the same network station. All three may send and receive packets from the same network connection (in this case, Ethernet). To determine which protocol stack any received packet is for, a SAP number is used. SNAP allowed an easy port of those existing non-IEEE LAN protocols to run on the IEEE 802.*x* data link.

Packet Formats for Ethernet and Token Ring   The IEEE 802.2 protocol also allowed migration of existing packet types and network protocols to the IEEE packet type. The Type 1 protocol is the one most commonly used on Ethernet and Token Ring when the IEEE 802.2 protocol is used. To allow for this, a protocol known as SubNetwork Access Protocol (SNAP) was invented. As shown in Fig. 15-3, the Source Service Access Field (SSAP) is set to AA (hex), and the Destination Service Access Field (DSAP) is also set to AA (hex). AA is assigned by the IEEE to indicate that the packet is for the SNAP format and that all data-line drivers should read the packet as such. The control field is set to 03 (hex), to indicate an unnumbered information packet.

The next five bytes are the protocol discriminator, which describes the protocol family to which the packet belongs. First, three bytes of 0s in this field would define the packet as an encapsulated Ethernet framed packet. This would tell the data-link software that the next byte following the SNAP header is the type field of an Ethernet packet and to read the packet accordingly. However, the first three bytes could read 08–00–07 (hex), which would indicate that the frame is an encapsulated IEEE 802.3–framed packet. In that case the following two bytes are a protocol ID field. A value of 80–9B (hex) indicates that the packet is an AppleTalk packet. As explained in Chap. 2, the protocol discriminator allows proper translation when the frame is forwarded onto a network.

For example, if a frame traversed multiple media types (Token Ring, FDDI, and Ethernet), the frame format would change for each type of media traversed. So, if the frame were received on a Token Ring port of a router, and it needed to be forwarded to an Ethernet network, the router would need to know which type of frame and format to use on the Ethernet—Ethernet V2.0 or IEEE 802.3. This is the purpose of the protocol discriminator. If the first 3 bytes were 00–00–00, then it would use Ethernet V2.0 frame format. If this field were set to a number other than 0, the discriminator will use the 802.3 frame format.

The one exception to this occurs in bridging (not routing) an AARP frame. Bridges employ a translation table to indicate special occurrences, and will format this field correctly, according to the IEEE 802.1h specification. The SNAP address 00–00–00–80–F3 is used to identify AARP packets. Following this is the AppleTalk data field, which will contain the AppleTalk OSI network-layer protocol of Datagram Delivery Protocol (DDP) information.

TLAP packet formats are like ELAP packet formats, except for the internal fields for the Token Ring controller and the source routing fields for packet routing with bridges. All IEEE 802.2 SNAP fields are the same.

Referring to the middle column of Fig. 15-3, the first few bytes pertain only to the data-link layer of Token Ring. These bytes include the Access Control field and the Frame Control field. These fields indicate to the Token Ring data-link controller how to handle the packet and which type of packet it is. The next field, the destination address, like an Ethernet packet address, is the 48-bit physical address of the network station for which the packet is intended. The source address is the 48-bit physical address of the network station that transmitted the packet to the ring. The next fields are the routing information fields according to the source routing protocol defined by IBM and the IEEE 802.5 Committee, as was discussed in the beginning of the book. Following the routing fields are the IEEE 802.2 SNAP headers: DSAP, SSAP, and control. The DSAP and SSAP fields contain AA (hex), and the control field will contain 03 (hex). Following this is the protocol discriminator, which is set to 08–00–07–80–9B (hex) to indicate that AppleTalk is the protocol for this packet.

The ELAP packet format is shown in the leftmost column of Fig. 15-3. The first six bytes are the physical destination address of an Ethernet network station. The next six bytes are the physical address of the Ethernet station that transmitted the packet. The next field is the length field, which indicates to the data link the amount of data residing in the data field, excluding the pad characters.

What follows the length field is how AppleTalk easily resides on an Ethernet network. At byte 14 (starting from byte 0 at the top) is the IEEE-assigned SNAP DSAP header, set to AA (hex), and the next byte is the SNAP SSAP, also set to AA (hex).

Following the DSAP and SSAP bytes is the field to indicate the control. This byte is set to 03 (hex) to indicate to the data link that it is a type 1 (connectionless) IEEE 802.2 packet. Following this is the SNAP protocol discriminator. 08–00–07–80–9B indicates AppleTalk, and the data link should read the packet according to AppleTalk protocol specifications. If that particular node is not running the AppleTalk protocols, the network station software will simply discard the packet. If the node is running the AppleTalk protocols, it will accept and decipher the packet according the AppleTalk protocol specification.

Operation   When the AppleTalk protocol is started, ELAP or TLAP will ask AARP to assign a dynamic protocol address to it. All that is done here is to replace the link layer. AppleTalk above layer 2 remains the same; no changes are made to it. The data-link layer changes to accommodate the new access technique of Ethernet or Token Ring. Figure 15-6 shows how three LAP Managers may be installed, but only one may be used at a time. The upper-layer protocols may switch to any of the three protocols. There is one small change to the LAP Manager, and that comes in the form of AppleTalk Address Resolution protocol (AARP).

The foregoing are the data link and physical layers for AppleTalk. The software portion of AppleTalk begins here with the network-layer entity, the Datagram Delivery Protocol (DDP). The previous protocols can be intermixed and used throughout any networking scheme. The following discussion concerns AppleTalk.

The AppleTalk Network Layer: End-to-End Data Flow

Datagram Delivery Protocol

The Datagram Delivery Protocol (DDP) layer resides at the OSI network layer and allows data to flow between two or more communicating sta-

Figure 15-6
LAP Manager.

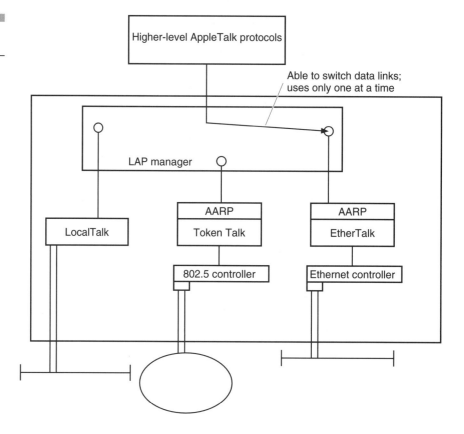

tions on the network. There are actually multiple entities that make up this layer:

- The Datagram Delivery Protocol (DDP)
- The Routing Table Maintenance Program (RTMP)
- The AppleTalk Echo Protocol (AEP)

Many functions are provided by the DDP, including data delivery, routing, and socket assignment. The first to be discussed is the concept of sockets.

The data-link layer can be thought of as nothing more than a "car" that carries "passengers." It accepts the passengers and will take those passengers to the destination they indicate. Whereas the data link delivers data (given to it by the network layer) based on a node-to-node relationship, DDP establishes the concept of *sockets*. When data is transmitted or received by a station, the packet must have some place in the network code to attach to. Because many processes may be running on a network workstation, the DDP must be able to identify which process the packet should be delivered to. This is the purpose of the socket.

With DDP, communication between two stations is now on a socket-to-socket basis for data delivery and reception. All the communication is accomplished on a connectionless service, meaning that once the data is delivered to a process (known as a socket), the receiving station does not acknowledge the originating station.

Sockets   To communicate with a process on another device, the initiator of a communication transfer will need an addressable software endpoint to indicate the final destination for these data. The socket number tells the network station software to deliver the incoming packet to a specific process or application in the network station. This is the only purpose of the socket.

Socket numbers are addressable endpoints in any network station that actually represent particular application programs and processes running in that network station. There will be one unique socket number for each process that is running in a single network station. Since many processes may be running in a network station at any one time, each process must be uniquely identified with a socket number. If you know the socket number, you will be able to communicate with the process or application that owns it. A socket number is also used to identify the process that submitted each packet. If a file server receives a data packet and must respond to it, the server needs to know the socket number in the source station that will receive the data.

All communications between a source and a destination station on the network will attach to each other through a socket. A socket is abstract to the user, not a physical device, and is used by the networking software as an end connection point for data delivery. There are source and destination sockets. The source socket identifies the sender of the connection, and the destination socket indicates the final connection point (the addressable endpoint on the destination station). When a source station initiates communication, a source socket will be identified in the DDP header of each packet so that the destination station will know where to attach when a response is generated.

AppleTalk implements sockets a little differently than other network protocols. In other protocols, sockets are assigned as static (well-known sockets) and dynamic sockets. AppleTalk well-known sockets are those sockets that are directly addressed to DDP only. There are no well-known sockets for applications that run on the Internet. The well-known sockets are listed in Table 15-4. When a packet that has a known

TABLE 15-4

Socket Values

DDP Socket Value, hex	Description
00h	Invalid
FFh	Invalid
01h	RTMP socket
02h	Names information socket (NIS)
04h	Echoer socket
06h	Zone information socket
80h–FEh	Dynamically assigned

socket is received by a network station, DDP will act upon the packet. These types of sockets are used by the DDP process and no other process. This is the difference between AppleTalk and other protocols, such as TCP/IP, NetWare, IPX/SPX that assign static sockets for every well-known process in the network station. A committee assigns these socket numbers, and once a process is assigned this socket, no other service may duplicate it. This well-known socket is universal. All applications that are written will be addressed to these socket numbers. AppleTalk allows its processes (file service, mail service, print service, etc.) to ask DDP for a socket number, and it could be different every time the service is started on the network. Like the node ID, socket numbers are assigned dynamically.

For example, the router table update process that runs in AppleTalk is statically assigned socket number 1. Any process that wishes to communicate to the router process must identify the packet with a destination socket number 1. A listing of the router sockets and their associated applications is shown in Fig. 15-7. All the well-known sockets used by DDP are shown in Table 15-4.

Dynamic sockets are assigned at process initiation. For example, when a process on a workstation initiates, it will request a socket number. It is the DDP layer that assigns the dynamic socket numbers. An application wishing to connect to a process that is not directed for DDP—for example, the Apple Transaction Protocol (ATP, the transport layer protocol for AppleTalk, discussed later)—would use a locally dynamically assigned socket to connect to the destination station's DDP process. DDP would accept this packet and look into the type field in the DDP header; for ATP, this field would contain a value of 3. DDP type fields are shown in Table 15-5. DDP would strip the DDP packet headers off of the packet

Figure 15-7
DDP. (*Courtesy Apple Computer.*)

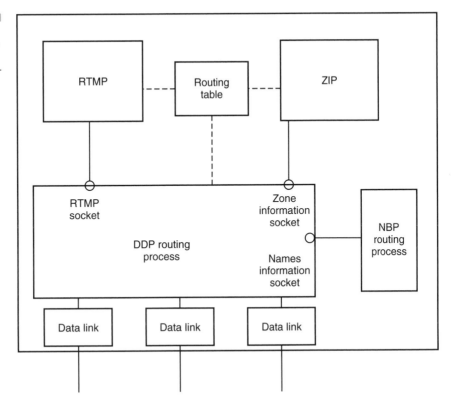

TABLE 15-5

DDP Type Fields

DDP Type Field Value	Description
00h	Invalid
01	RTMP response or data packet
02	NBP packet
03	ATP packet
04	AEP packet
05	RTMP request packet
06	ZIP packet
07	ADSP packet

and pass the rest of the packet to the ATP process. ATP would then act on the packet according to the information in the ATP headers.

Valid socket numbers are numbered 01 to FE (hex) and are grouped as follows:

- 01 to 7F (hex): Statically assigned sockets
- 3 and 05 (hex): Reserved for Apple Computer's use only
- 40 to 7F (hex): Experimental use only (not used in released products)
- 80 to FE (hex): Dynamically assigned for node-to-node communications

From previous discussions, we know that an AppleTalk node is assigned a 16-bit network number and an 8-bit node number. Sockets form the final part of the addressing scheme used by Apple. A network number is assigned to each network segment; each network station is dynamically assigned a unique node ID; now, with the socket number ID, any process running on an AppleTalk internet can be identified, no matter where the process is running. With this three-part addressing scheme, you can find the node, the network that the node lies on, and the exact process running on that node. This scheme, called the *internet socket address,* takes the form of network number—node number—socket number.

Now that sockets have been identified, DDP also has a type field in its packet to identify the process running on top of DDP that the packet is intended for. Table 15-5 shows these fields. DDP will accept the packet on the indicated socket number. It will, in turn, look at the type field to determine which process to hand the packet off to. For example, when DDP receives a packet from its data link, and the type field is 06, it will strip off the DDP headers and turn the rest of the packet over to the Zone Information Protocol (ZIP) for further processing. DDP's job is done, and it returns to listening for packets or for interrupts from the higher-level protocols (listed in Table 15-5) for packet delivery.

As mentioned before, when a process starts (it could be an application such as electronic mail or file server), it will ask the transport-layer protocol of Apple Transaction Protocol (ATP, discussed later) to assign it a socket number (a dynamic socket number). ATP will pass this call to DDP, which will find an unused dynamic socket number and pass it back to ATP. ATP will then pass this socket number back to the calling process. The calling process will then pass the socket number to the Name Binding Protocol (NBP, discussed later), which will bind this socket number to a name. DDP logs the number and name into a table to ensure that it will not be used again (socket numbers are not allowed to be duplicated in the same network station). All processes on an AppleTalk network are available to users through names and not socket numbers. NBP provides this service to the users, but uses socket numbers to find users and services on the AppleTalk internet. NBP and the use of sockets are discussed later.

Routers, Routing Tables, and Maintenance

The second function of the DDP is routing: the ability to forward packets that are destined to remote networks, allowing networks to form an internetwork. As stated before, an internet consists of a number of local LANs connected together into an internet through special devices known as routers (also incorrectly called gateways). These devices physically and logically link one, two, or more individual networks together. In doing this, a network is transformed into an internet. One of DDP processes is the process through which network stations may submit their packets to the router in order to route to a destination on a different network (LAN). DDP provides a dynamic routing protocol that is similar to the Routing Information Protocol (RIP) found on XNS, TCP/IP, and Novell NetWare networks. To accomplish this, DDP uses static sockets (socket 1) to deliver special messages known as routing table updates.

Router Description   A router is a special device that enables a packet destined for networks other than the local network they were transmitted on. Routers are usually separate boxes on the network that contain at least two or more physical ports, each connected to a cable plant. The router shown in Fig. 15-8 contains two physical ports (DB-15 Ethernet DIX connectors). They would be the physical connection point of the routers.

Figure 15-8
AppleTalk router.

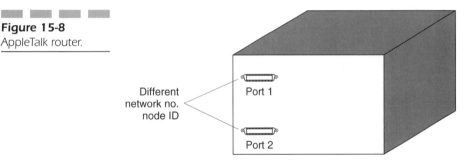

By Apple's definition, AppleTalk routers are available in three forms: local, half, and backbone. Local routers are attached to networks that are located geographically close together (on the same floor or between multiple floors in a building). Such networks are usually multiple network segments with *all* segments having station attachment. Local routers are connected directly to the networks; there is no intermediate device between the two networks and the routers. For example, an Ethernet-to-Ethernet router is a local router. See Fig. 15-9*a*. Each segment separated by a router will be assigned a network number or a range of

Figure 15-9
(*a*) AppleTalk internet.

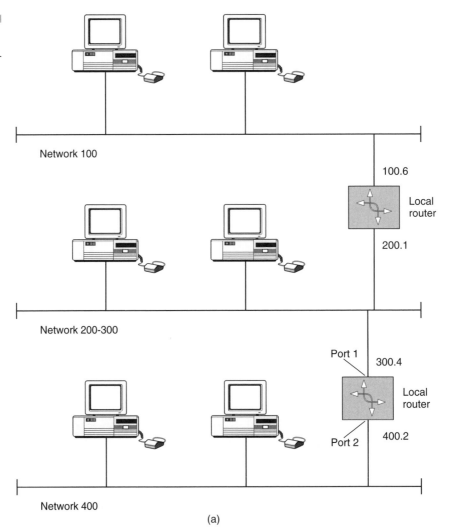

(a)

network numbers. This is shown by the middle network having the range of 200 to 300; network numbers from 200 through 300 are reserved for this network segment. Notice that different network numbers are given to the two routers connected to this network. Their network numbers are different, but they are assigned to the same cable segment. When assigning a range to a network, the network numbers should be contiguous.

Figure 15-9
(b) LocalTalk-to-back-bone connection.

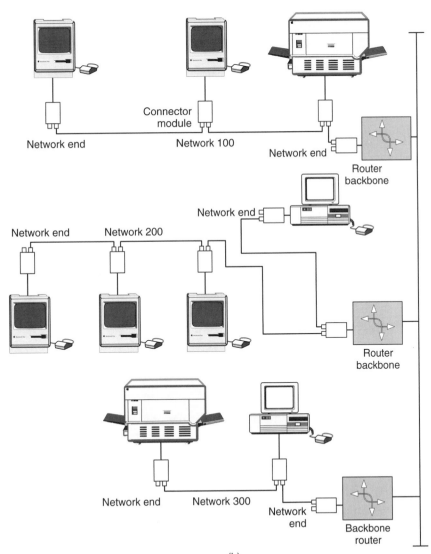

Connector module

Network end   Network 100
                                            Network end
                                                      Router backbone

Network end

Network end   Network 200

Router backbone

Network end   Network 300
                             Network end
                                        Backbone router

(b)

Half-routers are used to connect geographically distant networks together. This type of connection is usually done through the telephone system on what are known as *leased lines:* special lines that the phone company has "conditioned" to accept digital data. They are not standard voice lines, which are too noisy to carry most high-speed data. Typical data rates are 56 kbps and T1 (1.544 Mbps), although, now that most telephone companies are using fiber, T3 (45 Mbps) is becoming more available. Half-routers are important to note for their hop count. Since the two LANs are separated by a serial line, each router on each end of the line is considered a half-router, and the two taken together are considered one router. Therefore, no network number is assigned to the serial link, and a network separated by a serial line from another network is considered one *hop* away. Please note that not all router vendors support this method. Some router vendors will assign a network number to the serial line and the two networks separated by a signal line are considered two hops away. This will be discussed in a moment.

Backbone routers are routers that connect AppleTalk network segments to a backbone network segment: a backbone cable segment that is not AppleTalk. This type of network is shown in Fig. 15-9 *b*. Apple defines a backbone router as one that connects to a network having a higher throughput than the one to which it interconnects. Examples include FDDI backbones and 16-Mbps Token Ring.

Any time that a packet must traverse a router to get to another network, this action is known as a *hop*. Throughout an AppleTalk internetwork, a packet may traverse no more than 15 routers; therefore, a distance of 16 hops is considered not reachable.

Operation   Routing of AppleTalk packets reduces the maximum size of a DDP packet to 586 bytes, even though the frame capacity of Ethernet and Token Ring is much higher.

In an AppleTalk internet, there are two types of devices for data: routing and nonrouting nodes. For the purposes of this chapter a routing node will be a device that *only* routes data for the Internet and is usually located in a separate box known as a router, as shown in Fig. 15-8.

Routers are usually autonomous from the other devices on the network, and their main function is to receive packets from other nodes on the network and forward them to their appropriate network. Routers on AppleTalk internets accomplish more than this (zones), but that topic will be covered later.

The other type of node is called a nonrouting node, such as a user's workstation or even a file/print server. When a network station has data to transmit, it will determine whether the packet is to be transmitted locally (to another network station on the same LAN) or to a network station that is remote (separated by a router). The network station accomplishes this by comparing the destination network number to its known network numbers. Remember that with AppleTalk, there can be more than one network number assigned to a single network. Therefore, the destination network number is local if it matches any of the network numbers assigned to that network. If there is a match, the packet may be locally transmitted. If there is not a match, the network station must employ a router to get the packet to its final destination. In that case, the network station tries to find a router and addresses the packet to the router. The router will then forward the packet to its final destination. A router will discard a packet for an unknown destination network.

Simply stated, a router accepts packets directed to it, looks up the network address in a table, and forwards the packet to either a locally attached network or another router that will further route the packet to its destination.

The internal functions of a router are those of DDP, shown in Fig. 15-7. Each router will contain the following:

1. A data-link handler (ELAP, TLAP, or LLAP)
2. A DDP routing process
3. A routing table
4. A process to update the routing table (RTMP)
5. A physical hardware connector (known as a *router port*)

The router ports may be connected to any of the previously mentioned router types, but no two active router ports may be attached to the same network cable (this is done with other routers to allow for dynamic redundancy). You may connect two router ports to the same network as long as one of the router ports is disabled. When the active router port fails, the disabled router port can then be made active. This allows for manual redundancy.

Each router port contains a port descriptor, which contains the following four fields:

1. *A connection status flag.* This is an indicator to distinguish between an AppleTalk port and another type port (a serial link, a backbone network, etc.).
2. *The port number.* This is a number assigned to a physical port of a router and used to identify a port to forward packets.
3. *The port node ID.* This is a router node ID for that port.

4. *The port network number.* This is the particular network number of the LAN connected to the router port.

When a port is connected to a serial link (indicating a half router), the port node address and port network number are not used. When a port is connected to a backbone network, the port network number range is not used and the port node address is the address of the router on the backbone network.

It should be noted here, again, that not all router vendors support this method. Some router vendors treat AppleTalk routing like any other RIP routing protocol. In other words, they make no distinction between backbone, half router, and local router—a router is a router, and network numbers will be assigned to each and every port of the router no matter what the connection is.

This is necessary to point out because routing of AppleTalk packets may be different from one vendor to another. The concept of the backbone, half, and local routers is specific to the AppleTalk specifications. It is the method by which AppleTalk is recommended to be implemented.

The Router Table   To find other networks and their routers, AppleTalk routers use an algorithm similar to that of TCP/IP, XNS, and NetWare's IPX. It is known as a *distance-vector algorithm.* The router maintains a table of network numbers (the vector) and the distance to each network number (distance, hop, or metric number).

In order for the router to know where to forward the packet, it must maintain a table that consists of network numbers and the routes to take to get there. Each entry in an AppleTalk router table contains three things: the port number for the destination network, the node ID of the next router, and the distance in hops to that destination network. Refer to Table 15-6 and to the bottom router in Fig. 15-9*a.* The table consists of the following entries:

Network Range.   This is the vector. This is a known network number that exists on the Internet.

TABLE 15-6

AppleTalk Router Table

Network Range	Distance	Port	Status	Next Router
400	0	1	Good	N/A
200–300	0	2	Good	N/A
100	1	2	Good	200.1

Distance. This is the number of routers that must be traversed in order to reach the network number. By Apple's standards, this entry will have a 0 for locally attached networks.

Port. This is the physical port that corresponds to the network number on the router. If the hop count is 0, then this is the network number range assigned to that port. If the hop count is greater than 0, it is the port from which the network number was learned. Likewise, it indicates the port that the router will forward a packet to if the network number is so indicated in the packet.

Status. This indicates the status of the path to that network.

Next Router. If the network number is not locally attached to the router, this field indicates the next router in the path to the final destination network. If the network number is directly attached to the router, there will not be an entry in this field. The router will forward the packet directly to that cable plant.

A routing table like this one is maintained in each of the routers that are on the AppleTalk internet. The table simply tells the router the network number and how to get there. But where did the router get this information?

All router tables are constructed from information that comes from other routers on the network. In other words, each router will tell each other router about its routing table. A process must be invoked to allow the routers to exchange data (their routing tables) for periodic updates. This allows the router to find shorter paths to a destination and to know when a new router is turned on or when a router has been disabled. Possibly, a new path must be taken to get to a particular network. The protocol that enables this maintenance is called the Router Table Maintenance Program (RTMP).

Routing Table Maintenance Program (RTMP) The Routing Table Maintenance Program provides the logic to enable datagrams to be transmitted throughout an AppleTalk network through router ports. This protocol allows routes to be dynamically discovered throughout the AppleTalk internet. Devices that are not routers (workstations, for example), use part of this protocol, known as the RTMP stub, to find out their network numbers and the addresses of routers on their local network.

Some DDP packets are shown in Figs. 15-10*b* and *c*. Figure 15-10*a* shows the general DDP packet. The type field would be filled in appropriately for each packet type of RTMP, NBP, ATP, AEP, ADSP, and ZIP.

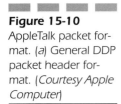

Figure 15-10
AppleTalk packet format. (*a*) General DDP packet header format. (*Courtesy Apple Computer*)

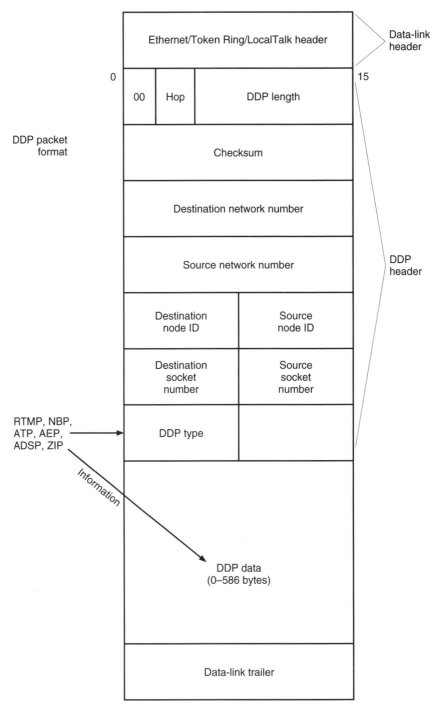

DDP packet format

RTMP, NBP, ATP, AEP, ADSP, ZIP

(a)

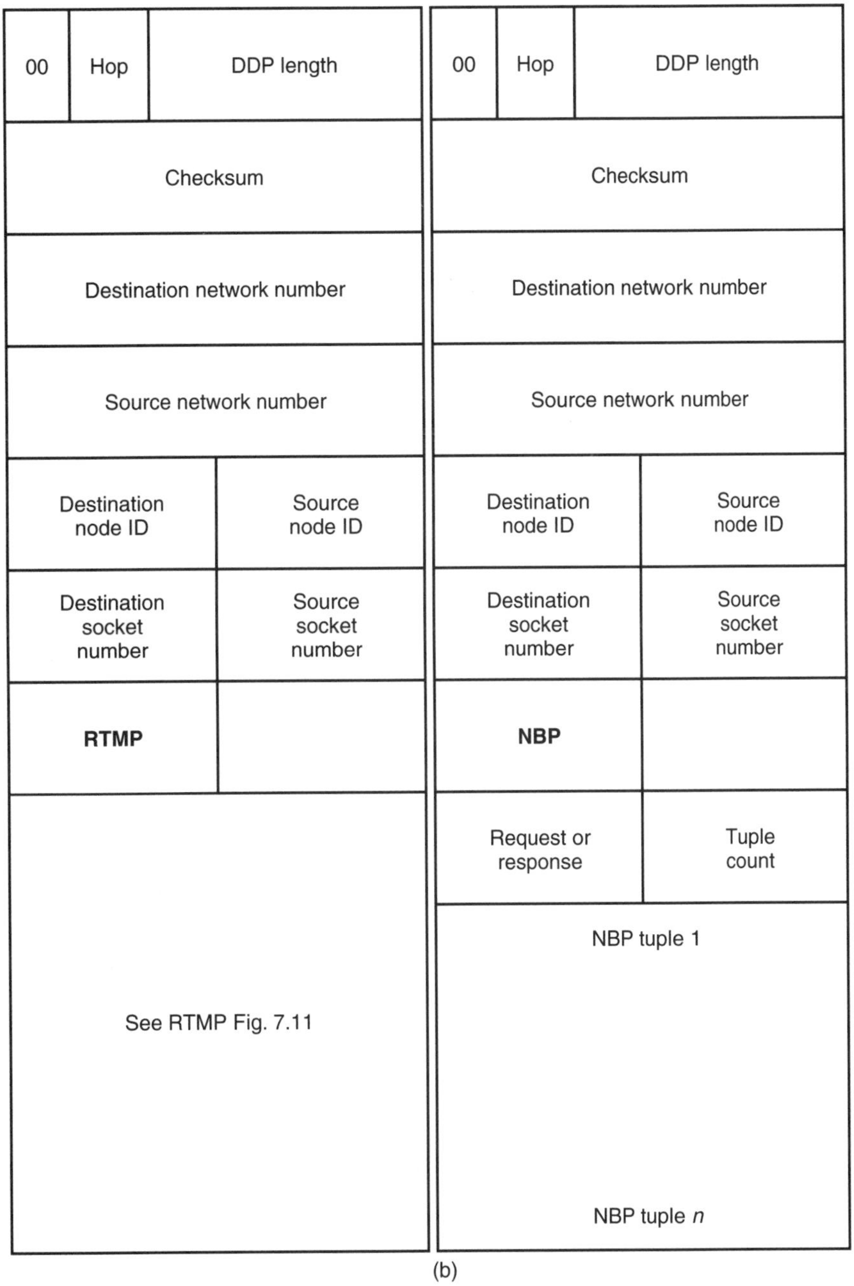

Figure 15-10
(*b*) RTMP and NBP
packet format. (*Courtesy Apple Computer*)

00	Hop	DDP length
Checksum		
Destination network number		
Source network number		
Destination node ID	Source node ID	
Destination socket number	Source socket number	
RTMP		
See RTMP Fig. 7.11		

00	Hop	DDP length
Checksum		
Destination network number		
Source network number		
Destination node ID	Source node ID	
Destination socket number	Source socket number	
NBP		
Request or response	Tuple count	
NBP tuple 1		
NBP tuple *n*		

(b)

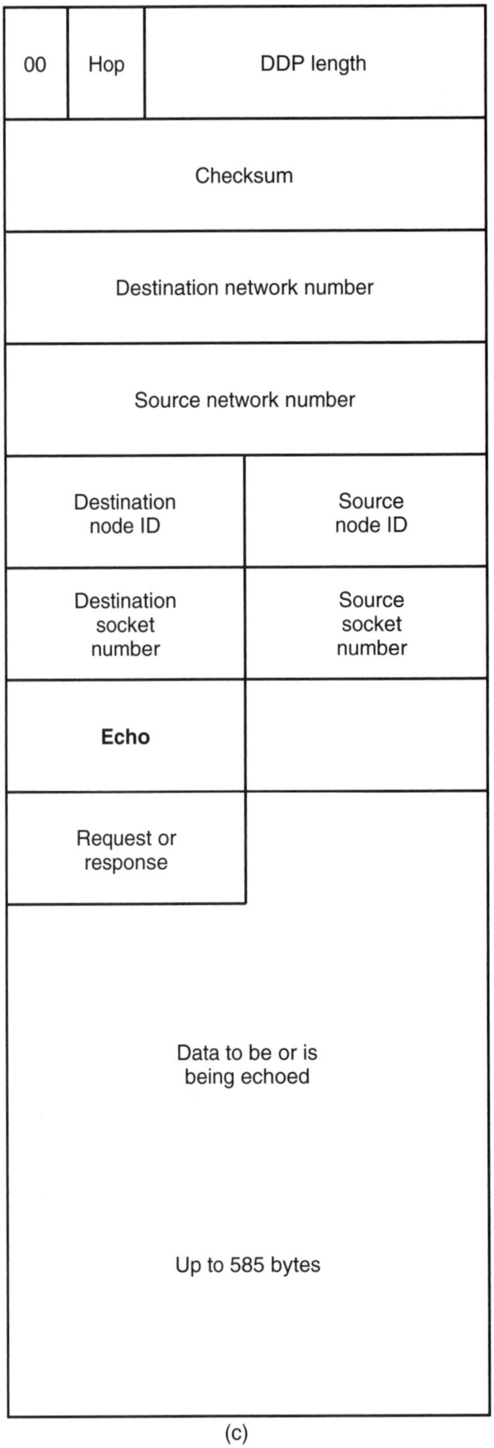

Figure 15-10
DDP packets. (c)
Echo packet format.
(Courtesy Apple
Computer)

00	Hop	DDP length
		Checksum

Destination network number

Source network number

Destination node ID	Source node ID
Destination socket number	Source socket number
Echo	
Request or response	

Data to be or is
being echoed

Up to 585 bytes

Figures 15-10*b* and *c* show the RTMP and NBP packet types. RTMP packets are further described in Fig. 15-11. The Hop field indicates how many routers a packet has traversed. AppleTalk is also one of the few routing protocols that use a hop count of 0. Any network number associated with a hop count of 0 is considered to be a local network number.

Routing tables in the routers are exchanged between routers through the RTMP socket. When a packet is addressed with this socket number, DDP, upon receipt of the packet, will know exactly what the packet is and will interpret it without passing it on to another process. In other words, the DDP socket number of 1 indicates a routing update or request packet.

Each RTMP (DDP socket number 1—source sockets may be any number, but the packet must be addressed to destination socket 1) packet includes a field called the routing tuple, consisting of network number (two bytes) and distance in hops (one byte), for nonextended networks. Extended networks contain the header consisting of network number range start, distance, network number range end, and an unused byte set to 82 (hex); then come routing tuples. When the router receives this packet, it will compare it to the entries already established in the table. New entries are added to the table and distances may be adjusted, depending on the information in the packet.

There are two ways a router may receive its network number. One way is to configure the router as a seed router. The network administrator must manually assign the network numbers to each port on the router. Each port must be configured as a seed router or a nonseed router port.

When a seed router starts up, it will enter the network numbers assigned to its local ports into the table. Other routers on the network will obtain their network numbers from the seed router. For a network that is served by multiple routers (each router has the same network numbers assigned to it), the seed router will distribute the network numbers to the rest of the routers on that network. There is at least one seed router per network, and there can be multiple seed routers per internet. Again, the purpose of the seed router is to inform the other routers of their network numbers.

Nonseed routers will learn their network numbers and zone names through the seed router. In an extreme case, all routers may be configured as seed routers. Configuring all routers as seed routers allows the network administrator to assign the network numbers on each of the router ports statically. The seed router will contain a list of network numbers on a per-router-port basis. The seed router can also contain a zone name listing on a per-port basis.

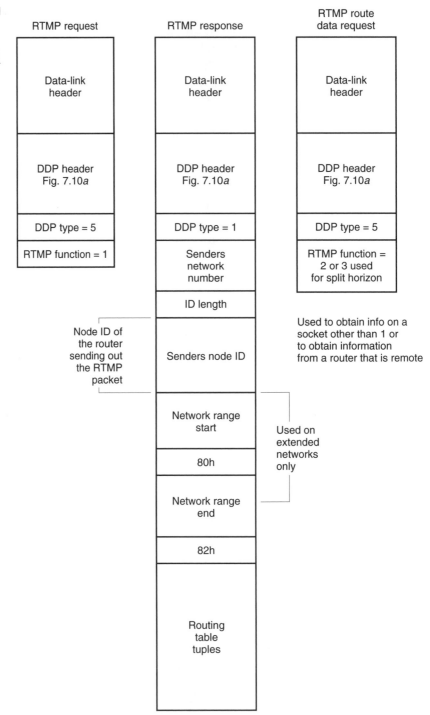

Figure 15-11
RTMP packets. (*Courtesy Apple Computer*)

RTMP request	RTMP response	RTMP route data request
Data-link header	Data-link header	Data-link header
DDP header Fig. 7.10*a*	DDP header Fig. 7.10*a*	DDP header Fig. 7.10*a*
DDP type = 5	DDP type = 1	DDP type = 5
RTMP function = 1	Senders network number	RTMP function = 2 or 3 used for split horizon
	ID length	
	Senders node ID	
	Network range start	
	80h	
	Network range end	
	82h	
	Routing table tuples	

Node ID of the router sending out the RTMP packet

Used to obtain info on a socket other than 1 or to obtain information from a router that is remote

Used on extended networks only

Once a router has found its local port network numbers and entered them in its table, it will submit its router table to its locally attached network segments, being conscious of split horizon (which will be discussed shortly). This will inform other routers on those segments of the networks that router knows about. Those routers will update their tables and then submit their tables to their locally attached cable segments. With this process, each router on the Internet will eventually know about all network numbers and their associated routers. When a router updates its routing table with a new or changed entry (network number), the hop count is incremented by 1. After this, the router will transmit its table to the directly attached cable segments. This does not occur for entries in the table that are not new or are not changed from the previous update.

Nonrouting nodes (i.e., user end stations) are not expected to maintain routing tables like a router. A nonrouting node may acquire its network number and a router address (a router to which to submit packets for forwarding data to another network) in one of two ways:

1. It may listen to the routing update messages sent out by routers on its local network. In these packets will be the local network number and the address of the router that sent the update.

2. It may also ask any router to respond by submitting a request for this information. With this, the requesting node should get back a response packet, which looks like a routing update packet but contains no routing update entries. Instead, embedded in this packet will be the network number and address of the router that sent the response packet. Also, this packet is a directed response packet. It is not sent in broadcast mode.

A nonrouting node that is on an extended networks will transmit a special packet known as the ZIP GetNetInfo request. This is a request to the router's Zone Information Protocol (ZIP). A router will respond to the packet with the network number range assigned to that zone. Normally, nodes on extended networks do not submit a routing request for network number information, but the AppleTalk specification does not disallow it. This request is studied in more detail later in the chapter.

The router that first responds to the request by an end node (a user's workstation, for example) on either an extended or a nonextended network is called the end node's A-router. This is the router to which the nonrouting node will send its packets that must be routed. The entry in a nonrouting node for its A-router may change if a different router responds to another routing request.

Aging Table Entries In actuality, the end station's (nonrouting node) A-router will change each time a routing update is transmitted by a router. Each entry in a router table must be updated periodically to ensure that the path to a destination is still available. Otherwise, a route could stay in the table but not be valid. To age out old entries, a timer is started on receipt of each routing tuple. After a certain amount of time (called the *validity timer,* set to approximately 20 s) has expired without notification of a particular route through RTMP (no routing tables received contained that network number), the router will change the status of the select entries from "good" to "suspect." After more time has expired, a suspect entry is changed from "suspect" to "bad," and then finally it is deleted from the routing table. If, at any time, the router receives an update pertaining to that entry in the routing table, it will place the entry status back to "good" and start the timer process over again. Aging of the A-router in a nonrouting node is set to 50 s.

RTMP packets are shown in Fig. 15-11. There are three types: a request packet, a response packet, and a router data request. The request packet is used to obtain information from a router about a network or networks. In response to a request, a router will send the response packet. This will contain information about a network number, such as how many hops away the requested network is. Another thing to notice in the response packet is the network range field. Any station on the network that reads this packet will know from this field what the network range for its network is.

The RTMP Router Data Request Packet is a special type of routing information inquiry packet that does not require network number updates. The router data request is used by a router to obtain information about another router, such as whether that router supports a protocol known as split horizon. Split horizon is the capability for a router not to announce information about a network on the same port from which it learned about the network. This means that if a router has learned about network 1 through port A, it does not announce this information back out port A.

RTMP routing update packets are transmitted by the router every 10 s. This means that a router will transmit its table to its locally backed cable segments every 10 s. This is the most frequent of any of the protocols that use RIP as their updating protocols. Novell's implementation of RIP is 60 s, and TCP/IP and XNS is 30 s.

Finally, Fig. 15-12 shows a flowchart for router information flow.

Figure 15-12
AppleTalk routing.
(*Courtesy Apple Computer*)

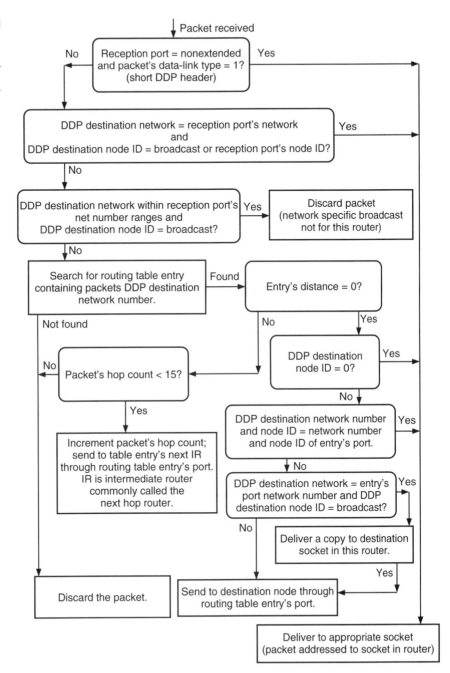

AppleTalk Echo Protocol

The AppleTalk Echo Protocol (AEP) resides as a transport-layer protocol and allows any station to send a packet to another station, which will return the same packet (echo it). This allows stations to find other active stations (even if you just wanted to test whether the station was active or not) on the network, but, more importantly, it allows a network station to determine packet round-trip delay times.

This protocol is used when a network workstation has found the destination station it wishes to talk to and submits an echo packet to "test" the path. It will submit an echo packet (DDP socket number 4) to the destination and will wait for a reply. When the reply does come, it notes the time and submits the packet to establish a connection to that station. This can be used to establish timers for packet time-outs. The packet format for AEP is shown in Fig. 15-10c.

That was the delivery system for AppleTalk. It included the data-link and network layers. All data are submitted to these layers for transmission on the network. The network and data-link protocols do not care what the data are. The only job that these protocols are tasked to do is to provide a transportation service for the upper-layer protocols.

The following protocols (transport- and session-layer protocols) allow for a session to be set up and maintained and ensure that the data is transmitted and received in a reliable fashion.

Names on AppleTalk

Throughout this section, refer to Fig. 15-10a. As discussed in the previous text, the delivery system depends on certain numbers in order to deliver data. There is one physical address for every network station on the Internet. There are network-layer addresses, so that the network layer is able to route the data over the Internet if needed. Socket numbers are needed, so that once data arrive at the final destination, they are forwarded by DDP to the appropriate software process. In other words, many addresses are used throughout the AppleTalk internet.

To eliminate the need for users to remember all the network addresses, node addresses, sockets, and so forth of the network and data-link layers, a naming scheme has been devised. Also, because network addresses may change frequently (dynamic node IDs, etc.), all services (file, print, mail, etc.) on a network are assigned user-definable (string) names. These string names may not change frequently, so AppleTalk uses a name process to identify network stations on an internet. This accommodates two concerns: (1) It is easier for a user on the Internet to remember names than numeric addresses, and (2) network stations can acquire different network addresses while retaining a static name. All of this is accomplished transparent to the users.

Names on an AppleTalk network are usually assigned by the network administrator. A user needing access to these services usually cannot change the names.

Names are used to request a service to be processed, such as attaching to a file server or sending a print job to a network-attached printer. All of these processes on the network are assigned numbers so that the network stations may communicate with each other; network stations do not communicate with each other using the string names we type into the network station. String names are only for users on the AppleTalk network. When network stations communicate with one another, they still use the full internet address of the network station, not the user-defined name of the network station.

Defining the Names on AppleTalk Within a network, AppleTalk has a concept known as *Network Visible Entities* (NVE). The actual physical devices (file and print servers, routers, etc.) can be seen by the user. The actual services within these physical devices cannot be seen by the user. These services are represented on the network as sockets, and each socket is assigned a numeric address. In this way, a network station may logically attach to the service, not to the actual device. For example, for those who use electronic mail, the user is not visible to the network but the electronic mailbox service is.

A Network Visible Entity (NVE) is any process on an AppleTalk internet that is accessible through a socket in DDP. In other words, the application processes to which users attach from their network stations are the NVEs.

An NVE can assign an *entity name* to itself. This name consists of three fields, and each field may contain a maximum of 32 characters. Any of the fields may be "user-defined," meaning that there is not a defined method for assigning the names. The form taken for this name is:

```
object:type@zone
```

An example of this name structure is Server1:Postoffice@Vienna. The naming of this entity name is case-insensitive. The object mentioned here is *Server1*, which appear to indicate the name of a server (although this is user-defined). The type is *Postoffice,* which could possibly indicate a mail server. This server is located in an internet zone named *Vienna.* This NVE form will become more apparent as the text proceeds.

Now that we have identified the names and what constitutes a name, these names must still be mapped to an internet address so that network stations on the network may attach and pass data. The protocols that handle names on the AppleTalk network are the Name Binding Protocol (NBP) and the Zone Information Protocol (ZIP). These protocols are described here briefly and in more detail later.

The name-binding process is the process by which a network station will acquire the internet socket address of a destination network station by use of its string name. This process is invoked by NBP. NBP maintains a table that translates between names and their corresponding full internet addresses. It converts the string name to the numeric address used internally by network stations. This process is accomplished completely transparently to the user. All of these name-to-address mappings are maintained in tables.

The Zone Information Protocol maps network numbers and zone names. A zone is a logical grouping of network stations, no matter where these network stations are located on the Internet; furthermore, zones do not reflect network addresses. Network stations grouped into a zone many have many different network addresses. The grouping of zones is called the AppleTalk internet. Each network number requires at least one zone name. Network station placement in a zone is fairly liberal. Network stations may participate in one zone or many zones. Zones may cross networks (across a router or routers). In other words, these zones may be spread through many different network numbers. Zone names originate in routers. The network administrator assigns the zone name and broadcast this to the network. This is what brings logic to this chaos. Routers maintain a listing of zone names and their associated network numbers through a table known as the Zone Information Table (ZIT). A simplex AppleTalk zone is shown in Fig. 15-13.

Before any NVE can be accessed, the address of that entity must be obtained through a process known as *name binding.* This is the process of mapping a network name to its internet socket address (the network number, the node ID, and the socket, or port, number). A Network Visible Entity (file, print, database, email, etc.) will start and ask ATP for a socket number. ATP will return a socket number (given to it by DDP), and the NVE process will then determine its network and node addresses (by methods previously discussed). All of these numbers combined (network ID, node ID, and socket number) form the internet socket address. The

Figure 15-13
Zone names

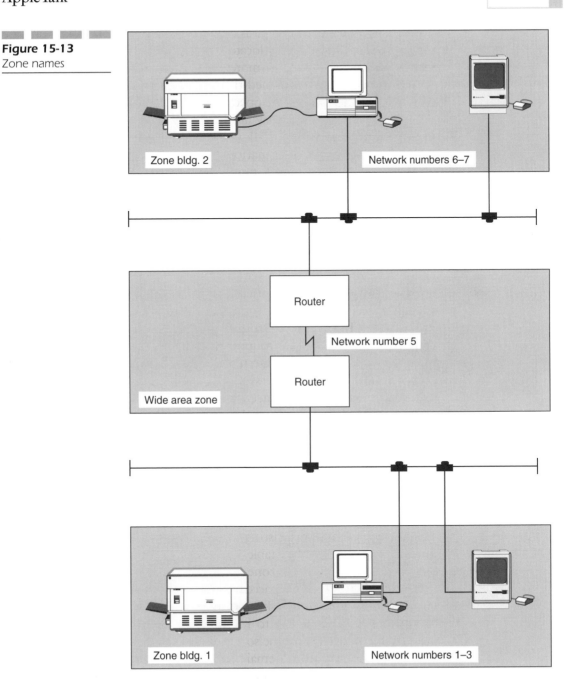

Zone bldg. 2

Network numbers 6–7

Router

Network number 5

Router

Wide area zone

Zone bldg. 1

Network numbers 1–3

process will then call NBP to bind the internet socket address to a name.

Each network station must maintain a *names table*, which contains an NVE name–to–internet address map for all entities (processes visible to the network, such as a print, file, or mail service) in that particular network station. NBP does not require the use of special nodes called name servers. NBP is distributed on all network stations on the AppleTalk internet; this means that each network station acts as a name server. Use of a centralized name service, a single station on which all names reside, is not usually found on AppleTalk networks. According to the AppleTalk specification, centralized name service is allowed to exist on an AppleTalk network, though. The process that maintains this table is available through the Names Information Socket. This socket (static DDP socket number 2) is also responsible for accepting and servicing requests to look up names from within the network station and from the network.

A separate table, called the *names directory*, contains a distributed database listing of all the names tables on the Internet. Whereas the names table on a network station contains only those internet socket address–to–name mappings of the NVEs on that local station, the names directory contains a listing for all NVEs on the Internet.

Simply put, NBP handles the names table and name lookup requests from client processes or requests from the network. Every network station on the AppleTalk internet has this process, whether it is providing services for the network or uses the services of the Internet.

The NBP process provides four types of services:

1. *Name registration.* Register an entity's name and socket number on its local node's name table.

2. *Name deletion.* Delete a name and its corresponding socket.

3. *Name lookup.* Respond to a name registration.

4. *Name confirmation.* Affirm a name registration.

Briefly, when any process starts on a network station, it will register its name and associated socket number in the names table. NBP will then place the NVE in the names directory. When this process removes itself from the network station, it will also ask NBP to remove its name and socket number from the names table. Name lookup occurs when a request for a name-to-internet address binding is required, and name confirmation occurs to check the validity of the current binding. Routers participate in this process through the use of zone mappings.

Name Registration Any NVE working on the Internet can place its name and corresponding internet socket address into its names table. This allows the entity to become visible on the network. First, the NBP process will check to see whether the name is already in use. If it is, the attempt is aborted and the process will be notified. Otherwise, the name is placed into the table. The NBP process then places this name-to-address mapping in the names directory. It should be noted that any entry in the names table is dynamic; it is reconstructed every time the network station is started. Using this process, every network user is registered on the Internet.

When the entity wants to delete itself from the names directory, it will do so by telling NBP to delete the entry.

On a single network, mapping a name involves three steps:

1. The requesting network station's NBP process broadcasts an NBP lookup packet addressed to the Name Information Socket number to the network. A lookup packet contains the name to be looked up.

2. Every active network station that has an NBP process will receive this packet and will perform a name lookup on its table not directory for a match to that name.

3. If a match is found, a reply packet is generated and transmitted to the requesting socket. Included in this packet is the name's address.

If there is no response to this packet, the requesting network station assumes that the named entity does not exist on the local network, and the name is added. This is a name confirmation.

On an internet (more than one network connected by a router), two extra steps are added to find an NVE. An NBP-directed broadcast request containing the NVE to be found is first sent out to all available routers on the station's local network. A router will pick up this request and will send out a broadcast lookup request for all networks in the requested zone. This process is done for zones and may stretch across multiple networks. The process then continues with the second step.

Zones A zone is a group of networks that form an AppleTalk internet. The Zone Information Protocol (ZIP) is the protocol that maintains an internetwide map (a table) of zone names-to-network numbers. NBP uses ZIP internet name mappings to determine which networks belong to a given name. It is easier to address network stations based on their zone.

An AppleTalk internet is a collection of network stations grouped together by zones.

An extended network can have up to 255 zones in its zone list. An AppleTalk internet can theoretically have millions of zone names. A nonextended network can have at most one zone name.

The ZIP process uses routers to maintain the mapping of the internet network numbers to the zone names. ZIP also contains maintenance commands so that network stations can request the current mappings of the zone names-to-network numbers. A router must participate in an AppleTalk internet zone protocol. Just as a routing table is maintained in a router, Zone Information Tables (ZITs) reside in each router on the AppleTalk internet. There is one ZIT for each physical router port on the router. Each table provides a list of the mappings between the zone name and internet address for every zone on the AppleTalk network. It looks like a routing table, but lookups are on the zone name instead of network numbers. This is similar to Novell NetWare's Service Advertisement Protocol. This table may consist of one zone name mapped to one network number, it may contain one zone name with multiple network numbers, or it may contain multiple zone names with multiple network numbers assigned to those zones. With the last, it should be noted that zones may overlap each other. An example of three logical zones is shown in Fig. 15-13. An example of a zone overlap would be that the printer in *Zone bld 2* could exist in *Zone bld 1*, except that it is still attached to the segment of network numbers 6 to 7. Not all routers support zone names on serial backbones, as shown in Fig. 15-13. Some routers do provide for support of half routers, which allow point-to-point links (such as the serial line) between the two routers, but there is no zone name on the point-to-point link.

To establish the zone table, the zone process in a router will update other routers through the ZIP socket. The requesting router will form a ZIP request packet, input into this packet a list of network numbers, and transmit it to the node's A-router. This router will reply with the zone names that it knows about (with the associated network numbers included). Since this process is accomplished over DDP (best-effort delivery service), the request contains a timer; on expiration of that timer, ZIP will retransmit a request.

To maintain a ZIT, ZIP monitors the router's routing table (not the ZIT) for changes in the entries in the table. If a new entry (meaning a new network number) has been found on the Internet, ZIP will send out request packets to other routers in an attempt to find a zone name associated for the new network number. Therefore, ZIP maintains its zone name table by monitoring the routing table.

When a network number is deleted from the routing table (eventually, the network number will be deleted from all routing tables on all routers on the Internet), the ZIPs on all routers will also delete the zone name entry for that network number.

Any network station on the Internet may request the mappings from ZIP by transmitting a ZIP request packet to the network. If the router does not know the zone name, it will not reply.

For a user's workstation, there are special ZIP packets that enable it to operate properly with the ZIP. These are:

GetZoneList	Requests a list for all zones on the Internet
GetMyZone	Gets the zone name for the local network (nonextended networks only)
GetLocalZones	Obtains a listing of all the zones on the requesters network

Just as a network station may keep its last known node ID stored (the hinting process), a network station, upon start-up, may have a previous zone name that it was using (the last time the station was active on the network). If so, the network station, upon start-up, will broadcast a packet called the GetNetInfo to the network. In this request packet will be the last zone name the station worked with, or if the zone name is unknown, the packet will not contain a zone name. Routers on this network will respond to this request with information on the network number range and a response to the zone name requested. If the zone name requested is okay, the network station will use this zone name. If not, the router will respond with the zone name used on that network.

To see the ZIP in action is as simple as signing onto the network and requesting connection to a service on the network. The Apple Mac Chooser will bring up a listing of zone names for logon or to access a service. This is an example of the ZIP protocol finding out all the zone names for the user.

Transport-Layer Services

Reliable Delivery of Data

The preceding documentation consisted of information for stations to identify themselves, gain access to a network, find zones, names, routers, and how to connect to an NVE via internet sockets. All of these features allow communication to exist with data being delivered on a best-effort, connectionless method. With the connectionless delivery system, there is no guarantee that the data were delivered or, if it were delivered, that the packets in a data segment transferred arrived to the destination in the same order in which they were sent.

Consider transferring a file that is 250 kbytes long. This is larger than any network protocol could transfer in one packet. Therefore, many packets are used to transfer the data. If we used only a connectionless protocol, the data might arrive in the wrong order, or one packet out of the thousand packets that were transferred might never make it. A protocol to enable information to flow reliably is called the AppleTalk Transaction Protocol (ATP). It is the next protocol to be examined.

Without transport-layer services, we would have to rely on the application itself to provide this service. To require every application to build this into its software is like reinventing the wheel. The network software should provide this service for every network application, and ATP is the protocol that does.

AppleTalk provides two methods for data delivery: transaction-based and data stream (the AppleTalk Data Stream Protocol, ADSP). Transaction-based protocols are based on the request-response method commonly found on client workstation–to–server communication. Data stream protocols provide a full-duplex reliable flow of data between network stations. The protocol to be studied here is the transaction-based transport layer. It offers something different from the other protocols shown in this book. Therefore, the ADSP protocol will not be discussed here.

The protocols that make up this OSI transport layer consist of the following:

1. AppleTalk Transaction Protocol (ATP)

2. Printer Access Protocol (PAP)

3. AppleTalk Session Protocol (ASP)

4. AppleTalk Data Stream Protocol (ADSP)—not discussed in this chapter

These protocols guarantee the delivery of data to their final destination. Two stations use the AppleTalk Transaction Protocol to submit a request, to which some type of response is expected. The response is usually a status report or a result from the destination to the source of the request.

AppleTalk Transaction Protocol

AppleTalk Transaction Protocol (ATP) is the protocol on AppleTalk that provides reliable delivery service for two communicating stations. There are three types of packets that are sent:

1. Transaction Request (TReq): a transaction request initiated by the requester
2. Transaction Response (TResp): the response to a transaction request
3. Transaction Release (TRel): a message that releases the request from the responding ATP transaction list

A Transaction Identifier (TID) assigns a connection number to each transaction. Each request and response between a source and destination is as simple as the three preceding commands taken as a whole.

Refer to Fig. 15-14. First, the requesting station submits a TReq. The destination station will respond with TResp. The requesting station will then acknowledge the transaction response by sending it a TRel. This

Figure 15-14
ATP transaction.
(*Courtesy Apple Computer*)

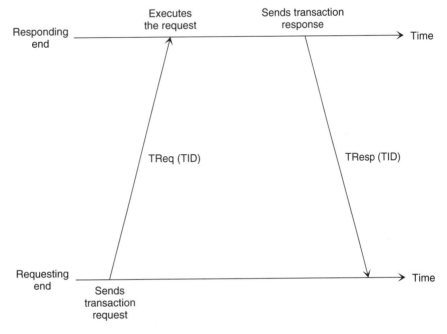

handshaking will continue until the session is released by the session layer.

ATP provides adequate services for most sessions. Because most network protocols restrict the size of their packets on the medium, multiple packets may be needed for a TResp. TReq packets are limited to a single packet. Multiple requests are sent as multiple TReqs. A TResp is allowed to be sequential packets. When the requester receives all the TResp packets in response to a TReq, the request is said to be complete and the message is delivered as one complete message to the client of ATP. ATP will assemble the message as one complete message and then deliver it to a client of ATP.

Every ATP packet has a sequence number in its header. This field is eight bits wide. This does not mean that it has 255 sequence numbers (counting in binary). Each bit is a sequence number. Therefore, if bit 0 is turned on, this is sequence number 0; if bit 1 is turned on, this is sequence number 1, and so forth.

In a TReq packet, it will indicate to the receiver the number of buffers available on the requester. When the receiver receives this packet, it will know how many response packets the requester is expecting.

When this field is set by a TResp packet, it is used as a sequence number (0 to 7). As stated before, each bit is not taken as a binary number; each bit represents an integer number (in the range of 0 to 7). This will indicate the number of the packet in the sequence of the response packets. Therefore, the sequencer on the requester can place the incoming packets in the appropriate buffer. As each good packet is received, the requester makes a log of this. For each packet not received in a sequence, the requester can make a request for a retransmit of this packet in the sequence. This is a selective reject type of retransmission (refer to Chap. 4 for more information on sequencing).

For example, if a TReq is transmitted and the sequence bitmap is set to (00111111), it is indicating to the destination that it has reserved six buffers for a response. In other words, the requester expects six packets of TResp in this response. As the TResp packets are sent back, the TResp packets will have a bit-map set to indicate the number of the packet that is being responded to. If the requester receives all packets except number 2 (00000010), it will send another TReq for the same information with its sequence bitmap set to (00000010). The destination should respond with that and only that TResp packet.

In the ATP data field (not the ATP header) would be an indication of what type of service is being requested, such as a read of a disk file on a file server. This type of information would be the session header and the application header, which will be discussed in a moment.

Printer Access Protocol

The Printer Access Protocol (PAP) uses connection-oriented transport (meaning a connection is established between the client and server before data is passed between the two). This is the protocol used to deliver data that a client workstation would send to a printing device on the network, enabling an application to write to the printer across the network.

PAP uses the services of NBP and ATP to find the appropriate addresses of the receiving end and to write data to the destination.

PAP provides five basic functions described in the following four steps:

1. Opening and closing a connection to the destination
2. Transferring data to the destination
3. Checking the status of a print job
4. Filtering duplicate packets

PAP calls NBP to get the address of the server's listening socket for printer services (the SLS). Data can be passed only after this socket is known. After the connection is established, data are exchanged. There is a 2-min timer to determine whether either end of the connection is closed. If the timer expires, the connection is closed. Either end of the connection may close the connection.

A workstation may determine the status of any print job. This may be executed with or without a connection being established. As for filtering duplicate packets, sequence numbers are assigned to each packet. A response from either end must include the original sequence number. Sequencing is fully explained in the section on ASP, which follows.

Figure 15-15 shows the block diagram of the PAP. Notice how the printer has the AppleTalk protocol stack in the printer software. This architecture is shown for a printer connected to the LocalTalk network. For EtherTalk or TokenTalk networks, the printer software is a spooler that resides in the Mac network server. The printer is directly attached to the Mac's printer port. The Mac uses AppleTalk to communicate between the Mac and the printer.

AppleTalk Session Protocol

The AppleTalk Session Protocol (ASP) does not care about the delivery of data. It does not care about the sequencing of data to ensure reliable delivery of the data. ASP's sole purpose is very straightforward. ASP, like

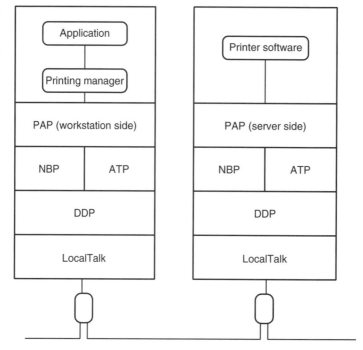

most session-layer protocols, opens, maintains, and closes sessions and sequences requests (not data) from upper-client software. It uses the lower-layer protocol ATP to sequence these requests.

ASP does provide a method for passing commands between a network station and its connected service. ASP provides delivery of these commands without duplication and makes sure the commands are received in the same order in which they were originally submitted. This orderly retrieval of commands is then responded to as the results become available. Like the other layers of the OSI model, the session-layer ASP communicates only with another ASP layer on another network station.

ASP is based on a client-server model. Upon initiation, a service (an NVE) on a server will ask ATP to assign it a socket number. ATP should respond to this request with a socket number received by DDP, and this number is the one used by the service on the server to let the network know about the service. The service will then notify NBP of the name and its socket number. This will be placed in the names table of the network station that offers this service, allowing any network station to find the service on the network. ASP will then provide a passive open connection on itself so that other stations on the Internet may connect to it. This is known as the *server listening socket* (SLS).

Before we continue, there are three types of named sockets that the text will be referring to:

1. Session Listening Socket (SLS): A service being offered listens on this socket.

2. Workstation Session Socket (WSS): This identifies the socket in the requesting workstation when a connection attempt to a service is being made.

3. Server Session Socket (SSS): Once a connection is made, this identifies to the workstation, the socket number to which to refer all future transactions.

Figure 15-16 depicts this interaction of sockets. The client side of ASP, on seeking a connection to a remote resource, must find the full internet address of the remote service (the NVE). ASP is said to place an active

Figure 15-16
Connection attempts and sessions using sockets.

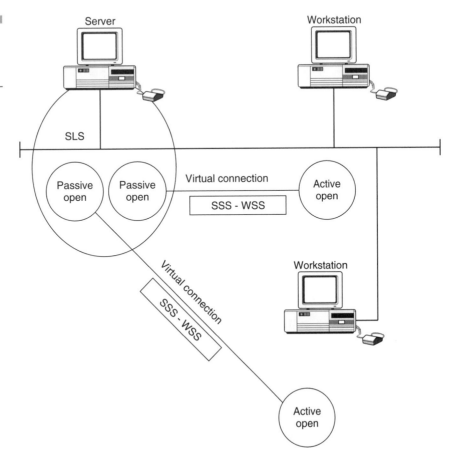

open connection request. In this, ASP will find the full internet address of the remote service and give itself a socket number, which will become the workstation session socket (WSS), and it will then request a connection from the remote resource. The server should have open sockets residing on DDP and will accept the connection request on one of these sockets. The virtual circuit (the connection or session) is then maintained between these two sockets (the WSS and the server socket, now known as the SSS). ASP builds upon ATP to provide a secondary level of transport service that is commonly required by client-workstation environments.

Since the most common application found on AppleTalk networks is the sharing of file and print services, the remainder of the text will deal with how a workstation and its file server communicate over the Internet.

Establishing a Session Two handshaking protocol steps must be performed before any session is established between a workstation and its file server.

1. The workstation and server must ask ASP to find out the maximum allowable command and reply sizes.

2. The workstation must find out the address of the server session listening socket (SLS) number by issuing a call to NBP.

The session listening socket, previously described, is an addressable unit to which the workstation will send all requests. This socket is opened by ASP, and the file server will make this socket well known on the network by broadcasting this socket to known entities through the use of NBP. This is the socket number for which the file server, for example, is listening for connection requests.

Refer to Fig. 15-17..

1. The workstation will ask ASP to open a session with a particular service on a file server.

2. ASP will transmit a session request packet to the file server. This packet will contain the workstation's socket number, to which the file server may respond. This WSS is an addressable unit so that the file server knows which process in the workstation to respond to in the workstation.

3. If the server accepts the session request, it will respond with the following:

Figure 15-17
Session requests.
(Courtesy Apple Computer)

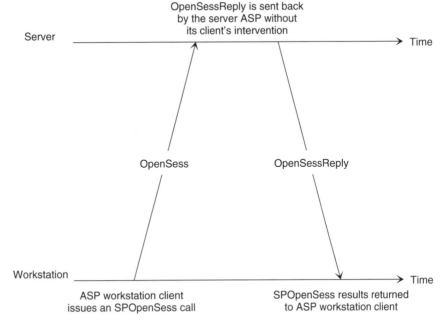

Figure 15-17
Session requests.
(Courtesy Apple Computer)

Acceptance indicator: An entity within the packet to indicate to the
   workstation that the request is accepted

Session identifier: An integer number that the server assigns to identify
   any particular session on the file server; Each unique workstation
   request is assigned an identifier

Address of the server session socket (SSS):   A new socket number that will
   uniquely identify that session on the server

The session is now considered established, and all communication will
continue between the two, using the WSS, the connection ID, and the
SSS.

Session Maintenance   Once the session is established, ASP must
maintain the session. There are maintenance commands specific to the
management of any AppleTalk session. Three are to be discussed here:

1. Tickling

2. Command sequencing

3. Discarding duplicate requests

Once a session is established, it will remain active until one end of the connection decides to quit the session. One other possible condition for session termination occurs when the path has become unreliable or one end has become unreachable. There will be times when the session will not be passing data or commands from one end of the connection to the other. For example, you may establish a connection, start up an application, and then not enter any data for a while. When this occurs, each end of the circuit does not know whether any circumstance may have occurred that would make the session inoperable. To find out whether the other side is operating normally, AppleTalk employs a protocol known as *tickling*. With this, each network station involved in a connection will periodically send a packet to the other to ensure the other end is functioning properly. (In other protocols, this is known as a keep-alive packet.) If either end does not receive this packet at certain time intervals, a timer will expire, and the session will be disconnected.

Command Sequencing If the same packet is received twice, the second (duplicate) request is discarded. This may happen when the originating station submits a packet and does not receive a response for it. The destination station of this packet may have been busy, or the packet may have been delayed while being forwarded through a busy router and the response packet was delayed. In any case, a timer will expire in the origination station and the packet will be retransmitted. Meanwhile, the server responds to the original packet and then receives the retransmission. The destination station must have a way of knowing that that particular packet was responded to. It knows this by sequence numbers. Each incoming packet is checked for the sequence number. If a packet received on the file server corresponds to a response sequence number, the duplicate packet will be discarded.

Data Transfer Once the connection is established and is operating properly, each end may send packets, which will read or write requests to the respective sockets. These reads and writes are actually data transfers between the two network stations.

ASP handles all requests from the workstation and the server. ASP is an initiator of requests and a responder to any requests on the network. Any of these commands will translate into ATP requests and responses. There are three formats for ASP:

1. Commands
2. Writes
3. Attention requests

As shown in Fig. 15-18*a*, ASP commands ask the file server to perform a function and to respond to these requests. Any command will translate into an ATP request of the session server socket (SSS). The reply will translate into an ATP response.

Figure 15-18
(*a*) Data transfer.
(*Courtesy Apple Computer.*) (*b*) ASP write.
(*Courtesy Apple Computer*)

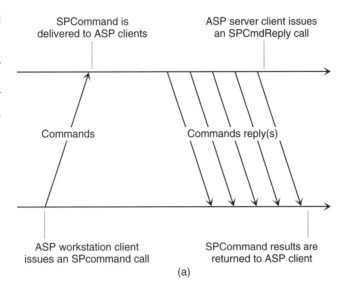

SPCommand is delivered to ASP clients

ASP server client issues an SPCmdReply call

Commands

Commands reply(s)

ASP workstation client issues an SPcommand call

SPCommand results are returned to ASP client

(a)

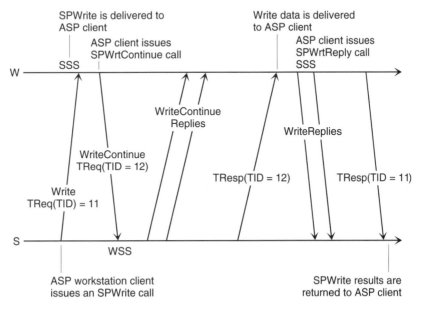

SPWrite is delivered to ASP client

ASP client issues SPWrtContinue call

Write data is delivered to ASP client

ASP client issues SPWrtReply call

W — SSS

SSS

WriteContinue Replies

WriteReplies

WriteContinue TReq(TID = 12)

Write TReq(TID) = 11

TResp(TID = 12)

TResp(TID = 11)

S —

WSS

ASP workstation client issues an SPWrite call

SPWrite results are returned to ASP client

W – Workstation
S – Server

(b)

A write is performed in Fig. 15-18 *b*. ASP on the server issues a single write command with its appropriate transaction ID. The workstation responds, stating that it is okay to continue writing on TID 12, and the server will do so. At a certain point in time, the workstation replies to the writes, ending with a TResp for TID 12. After all the write replies are sent to the server, the workstation acknowledges the end by a TResp on TID 11 for the server's TID.

If the workstation's command to its file server was to read part of a file, the response would be in multiple packets returned to the file server (if the file was larger than the largest packet size allowed on the network).

Attention requests are those requests that need immediate attention. No data will follow this packet.

Session End   Refer to Fig. 15-19*a* and *b*. The workstation or the server may close an established session. The workstation may close a session by sending a Close Session command to the same socket on the server (known as the server session socket, or SSS) in which the server was first established. The server may close a session by sending a command to the workstation socket (the workstation session socket or WSS) in which the session was first established. After a session is closed, the ASPs on both the server and the workstation must be notified, so that each may delete the proper information from its session table. All entries for those sessions in the tables are deleted.

The previous sections conclude the protocols needed for the AppleTalk network. These protocols accept data from the application layer and format them so that they may be transmitted or received on the network. A session may be started and maintained using the aforementioned protocols. Any application may be written to using those protocols. The largest application for AppleTalk is the AppleTalk Filing Protocol (AFP). This protocol allows network stations to share file and print services with each other.

AppleShare and the AppleTalk Filing Protocol

For a workstation to access files and other services of another workstation, the requesting station must invoke a high-level service known as Apple Filing Protocol. This is Apple's method of controlling server, vol-

Figure 15-19
(a) Workstation session close. (b) Server close.

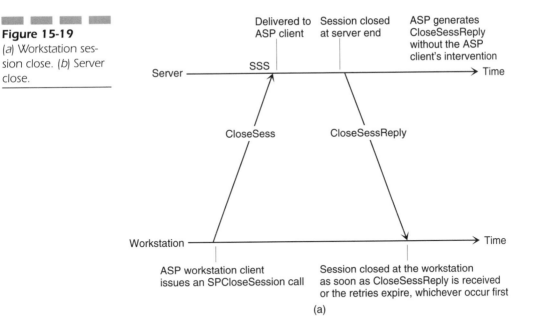

(a)

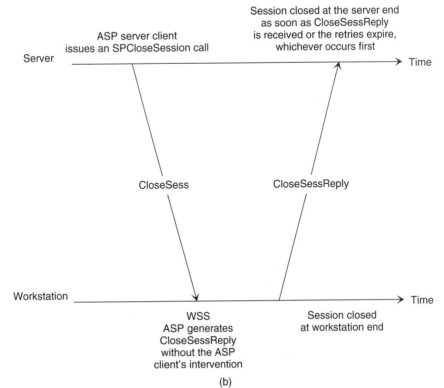

(b)

ume, directory, file, and desktop calls between a client and workstation on an AppleTalk network. For those familiar with other protocols, Microsoft's Server Message Block (SMB), Novell NetWare Control Protocol (NCP), and Sun's Network File System (NFS) are all similar in function to Apple's AppleShare and AFP. Apple decided not to follow these implementations, because they were in various stages of completion and SMB was written for DOS and NFS was written for UNIX. Neither could implement all of Apple's file structure easily. So Apple created a new remote file system using a protocol called AFP. The name given to use this is AppleShare. Simply stated, AFP is a service that uses the AppleTalk protocols to allow network stations to share data over the Internet.

Several entities within the AppleTalk network must be known before fully understanding the AppleTalk file/print network. These entities are:

File Servers. A file server is a device on the network that acts as a repository for files. These files can include applications as well as user data. File servers are usually the most powerful network station and contain a large-capacity hard disk. Multiple file servers are usually found on any network and can be segregated to file server, data server, and printer server.

Volumes. A volume is the top level of the Apple directory structure. Volumes may occupy an entire disk drive or may be many partitions on the same disk drive.

Directories and Files. These entities are stored in each volume. In any Apple machine, they are arranged in a branching tree structure. Directories are not accessed as files; they are addressable holding areas for files and may branch into other directories, because they carry a parent-child relationship.

File Forks. A file consists of two forks: a resource and a data fork. A resource fork contains system resources such as the icons and windows. A data fork contains the data in the file in unstructured sequence.

The protocol known as AppleShare allows an Apple computer to share its files with other users on the network. Every network station on the AppleTalk network may be set up to share files with any other station on the network. Stations may also be dedicated file servers on an internet, and all stations may have access to these servers if allowed by the network administrator. Each AppleShare folder has an owner and that owner will determine the access rights for that folder. The access rights—private, group, or all users—are self-explanatory. The ownership can be transferred to another owner.

Communication between a user's workstation and an AppleShare file server is accomplished using AFP. AFP runs as a client of ASP and is an upper-layer protocol to ASP. It will use ASP to enable a connection (known as a session) between a client and a file server. This session is a virtual connection that will allow protocol control information and user data to flow between the two network stations.

Prior to session establishment, a requesting station must know the address of the server it wishes to communicate with. Specifically, it must obtain the server's session listening socket (SLS). Servers will ask ATP for this, as was explained previously. After this information is returned, the server will use NBP to register the file server's name and type on the socket just received.

For a workstation to find a server, the workstation will place a lookup call to NBP, which should return the addresses of all the NVEs within the zone specified by the workstation. To find the names of all the file servers on a particular zone, the workstation would use: :AFPServer@*zone name*. NBP should respond to this lookup string with a listing containing all the active file servers specified in the zone name specified. Included in this listing will be the internet socket address (the SLS). The workstation will then choose a server. To get a particular server, the user could enter *server name*:AFPServer@*zone name*. This would return the SLS of the server, to which the user would then connect. These calls are all accomplished in the Chooser. It all sounds complicated, and it is—behind the user's screen, where the simplicity becomes involved. All the user has to do to connect to a file server is select Chooser under the Apple icon on the menu bar. Under Chooser, the user will then select the AppleShare icon. Chooser will then ask the user to select a zone. Once the user selects a zone, the Chooser will ask the user to choose a file server, all by using the mouse with windows and icons displayed on the user's screen.

The user would then log in to the server. Once the authentication is accomplished, the workstation would establish a session to the server, and communication would begin.

On the server side, once the server has opened the file server's socket and registered the file server's name, the server is ready to accept workstation requests, and workstations will call NBP to obtain the server's address and name. Once accomplished, the workstation will call ASP to open a session to that server.

All the resources are secured in three ways. Upon login, AFP provides user authentication (basically a password), volume password protection, and directory access controls based on the user authentication information.

Since its inception, AFP has continually undergone changes, mostly enhancements. Therefore, different versions of AFP will exist on an AppleTalk network. To determine which version the workstation will use with a server, the server (during the connection process) will tell the workstation which versions it supports. The workstation will then choose among the versions.

If the logon process is completed without error, the workstation will tell the server what it wants by issuing AFP calls to the server. Some of these calls are listed at the end of this chapter. When the workstation is done using the file services of a file server, the user should disconnect from that file server, allowing other connections to be established and freeing up memory and processing of other transactions that continue on that file server.

Once connected to a file server, the user can then select file volumes on the file server to attach to. These attached file volumes will show up as a network connection folder on the user's screen.

Figures 15-18 and 15-19 showed a workstation–file server connection. Once the workstation and file server have a session built, it is up to the workstation to determine which system calls are local and which are destined for a remote resource. The selected volumes will show up on the user's screen as a file folder with a LocalTalk connection to it. This is an icon that represents a volume, which is accessed just as the local hard disk is accessed. The user double-clicks on the networked file folder icon.

Apple's native file system commands are converted to AFP calls by being sent to the translator. These AFP calls are then sent to the file server. As shown in Fig. 15-20, a workstation program may send AFP calls directly, without having to use the translator. These are application programs that directly use a network, such as network mail programs.

AFP functions AFP operates on a series of calls, which are the commands that a workstation will submit to a file server. The AFP calls that are performed by AppleShare are listed in Table 15-7. These calls are completely transparent to the user.

AppleTalk

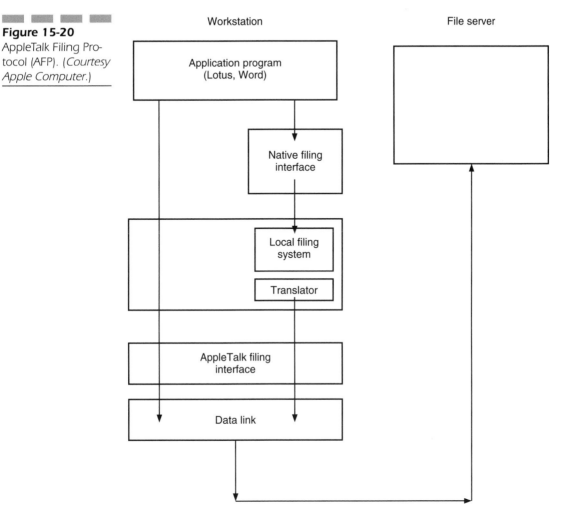

Figure 15-20
AppleTalk Filing Protocol (AFP). (*Courtesy Apple Computer.*)

TABLE 15-7 *AFP Calls*

Call	Function
Server calls	
FPGetSrvrInfo	Obtains descriptive information about the server
FPGetSrvrParms	Retrieves file server parameters
FPLogin	Establishes a session with a server, specifies the AFP version and user authentication to use
FPLoginCont	Continues the logon and user authentication process begun by FPLogin
FPLogOut	Terminates a session with a server
FPMapID	Maps a user ID or group ID to a user name or group name
FPChangePassword	Allows users to change their passwords
Volume calls	
FPOpenVol	Makes a volume available to a workstation
FPCloseVol	Informs a server that a workstation will no longer use a volume
FPGetVolParms	Retrieves parameters for a volume
FPSetVolParms	Sets the backup date for a volume
FPFlush	Writes to disk any modified data from a volume
Directory calls	
FPSetDirParms	Sets parameters for a directory
FPOpenDir	Opens a directory
FPCloseDir	Closes a directory
FPEnumerate	Lists the contents of a directory
FPCreateDir	Creates a new directory
File calls	
FPSetFileParms	Sets parameters for a file
FPCreateFile	Creates a new file
FPCopyFile	Copies a file from one location to another on the same file server
Combined directory-file calls	
FPGetFileDirParms	Retrieves parameters for a file or directory
FPSetFileDirParms	Sets parameters for a file or directory
FPRename	Renames a file or directory
FPDelete	Deletes a file or directory
FPMoveAndRename	Moves a file or directory to another location on the same volume and optionally renames it

TABLE 15-7 (Continued)

Call	Function
Fork calls	
FPGetForkParms	Retrieves parameters for a fork
FPSetForkParms	Sets parameters for a fork
FPOpenFork	Opens an existing file's data or resource fork
FPRead	Reads the contents of a fork
FPWrite	Writes to a fork
FPFlushFork	Writes to disk any of the fork's data that is in the server's buffer
FPByteRangeLock	Prevents other users from reading or writing data in part of a fork
FPCloseFork	Closes an open fork
Desktop database calls	
FPAddIcon	Adds an icon bitmap to the Desktop database
FPGetIcon	Retrieves the bitmap for a given icon
FPGetIconInfo	Retrieves an icon's description
FPAddAPPL	Adds mapping information for an application
FPRemoveAPPL	Removes mapping information for an application
FPGetAPPL	Returns the appropriate application to use for a particular document
FPAddComment	Stores a comment with a file or directory
FPRemoveComment	Stores a comment with a file or directory
FPGetComment	Retrieves a comment for a file or directory
FPOpenDT	Opens the Desktop database on a specific volume
FPCloseDT	Informs a server that a workstation no longer needs the volume's Desktop database

Digital Network
Architecture

Introduction

The original protocol of computer-to-computer communications, Digital Equipment Corporation's DECnet, was developed before any of today's LAN protocols. Currently, DECnet is about to enter its fifth phase, known as DECnet Phase V. This will be the version that incorporates Open Systems Interconnect (OSI) into the architecture.

History

DECnet Phase I was introduced in 1975 and ran on DEC's PDP-11 computers under the RSX operating system. It allowed for program-to-program communication, such as through file transfer, and remote file management between two or more computers. It supported asynchronous, synchronous, and parallel communication devices. It was simplex compared to today's topologies and protocols, but it did provide most of the same capabilities as today's internets. It was very powerful for its time.

In this and the next two releases, the DEC PDP-11 computer was the network. All computers communicated through the PDP-11. Ethernet was not available until the release of DECnet Phase IV. The data link for this network was a control message protocol, known as Digital Data Command Message Protocol, which ran on serial links between the PDP-11s.

DECnet Phase II was introduced in 1978 and provided several enhanced features over DECnet Phase I. Included in this release was support over more operating systems, TOPS-20 and DMS. Also enhanced were the remote file management and file transfer capabilities. DECnet Phase II could support a maximum of 32 nodes on a single network. Most important in this release was the modularity of the code. The code that provided DECnet was rewritten. With this and all subsequent DECnet releases, DEC kept the end-user interface the same. Only the underlying code changed.

DECnet Phase III was introduced in 1980. Included in this release was adaptive routing (the ability to find a link failure and route around it) and support for 256 nodes. The remote terminal, known as Remote Virtual Terminal capability, was introduced in this release. This allowed an end user to remotely log in to a remote processor as if the user were directly attached to that processor. The addition of downline loading operations, which allowed a program to be loaded and run in another (remote) computer, was included. Finally, this was the first phase that supported IBM's SNA architecture.

The release that is by far the most popular and most commonly found on Digital's products is DECnet Phase IV, introduced in 1982. With this release came support for Ethernet and further expansion of the number of nodes supported.

Since Digital was one of the originating companies to develop Ethernet into a standard (Xerox invented it and Digital, Intel, and Xerox developed it as a standard), they incorporated this into DECnet Phase IV. The second most noticeable improvement was the idea of area routing. This routing technique is similar to OSI's IS-IS routing. It remains the strongest link in the DECnet architecture.

The number of nodes (DEC terminology for network stations, workstations, routers, etc.) supported was increased to 64,449. An address is a 16-bit number with the first 6 bits assigned to the area and the remaining 10 bits allotted to the node number. Nodes are separated into distinct logical areas. The specification allows for 63 areas to be defined, with 1023 nodes in each of the 63 separate areas. This was shortsighted on Digital's part. This may seem like a lot of addresses, but it disallows the building of large Digital internets that interoperate. With DECnet, even personal computers need to be assigned a full node address.

Digital's routing is different from the routing of all other protocol types in that routers keep track of all the endnodes (user workstations) in an area. This greatly increases the size of a router's routing table. Separating nodes into areas not only reduces the traffic of updating routers, but it also decreases the size of routing tables per area. DECnet addressing and routing will be discussed later.

The network terminal concept introduced in Phase III was reinvented as the network virtual terminal in Phase IV. Also introduced with this release was X.25 packet-switched network support.

Excluded in later versions of this release was support for TCP/IP and Token Ring. Both of these network protocols became very popular throughout the 1980s, but Digital did not allow these protocols to run on DECnet Phase IV nodes. Around 1990, DEC finally delivered its version of TCP/IP for the VMS operating system after years of allowing a company known as Wollongong to provide this functionality.

Previous to the direct support of TCP/IP on VMS, DEC did provide TCP/IP functions for Ethernet through its Ultrix operating system. This was actually the Berkeley UNIX operating system adapted to run on VAX computers.

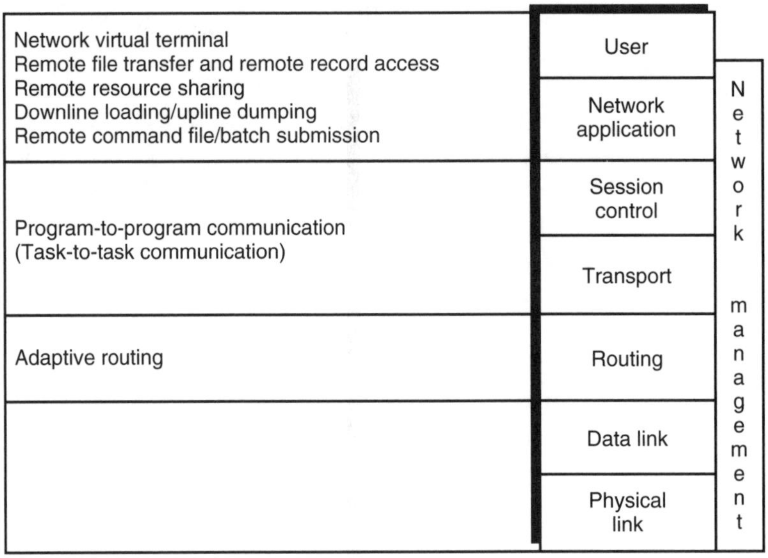

Figure 16-1
DECnet capabilities
and DNA layers.
(*Courtesy Uyless
Black.*)

Network virtual terminal Remote file transfer and remote record access	User	N e t w o r k m a n a g e m e n t
Remote resource sharing Downline loading/upline dumping Remote command file/batch submission	Network application	
Program-to-program communication (Task-to-task communication)	Session control	
	Transport	
Adaptive routing	Routing	
	Data link	
	Physical link	

As shown in Fig. 16-1, DECnet closely follows the seven-layer OSI model. The following text will discuss the protocol of Ethernet. DEC's support for Token Ring was released in 1993. The first two layers of Ethernet, physical and data link, were described in Chap. 2. Digital also includes support at the physical layer for

- X.21
- EIA-232-D
- CCITT V.24/V.28
- All wiring concepts for Ethernet

At the data-link layer, DNA supports not only Ethernet (IEEE 802.2/IEEE 802.3) but also a proprietary protocol known as Digital Data Communication Message Protocol (DDCMP). DDCMP is beyond the scope of this book. It has become a seldom-used protocol since the introduction of Phase IV support for Ethernet. Most companies have switched to Ethernet local and wide area networks for their Digital minicomputers. Figures 16-2 and 16-3 review the physical and data-link DNA layers. Figure 16-4 shows the network routing layer.

Digital's difference from other protocol implementations begin at the network layer. Here is where DECnet also becomes proprietary. This means that it is not an open protocol. Not just anyone can build this software protocol suite, because Digital holds the patents and copyrights to the architecture.

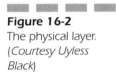

Figure 16-2
The physical layer.
(*Courtesy Uyless Black*)

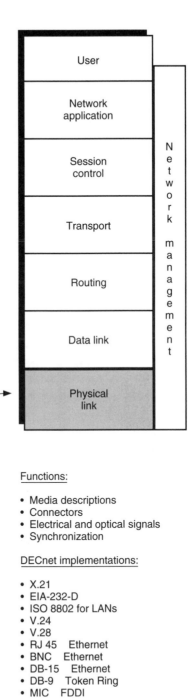

Functions:

- Media descriptions
- Connectors
- Electrical and optical signals
- Synchronization

DECnet implementations:

- X.21
- EIA-232-D
- ISO 8802 for LANs
- V.24
- V.28
- RJ 45   Ethernet
- BNC   Ethernet
- DB-15   Ethernet
- DB-9   Token Ring
- MIC   FDDI

Figure 16-3
The data-link layer.
(*Courtesy Uyless Black*)

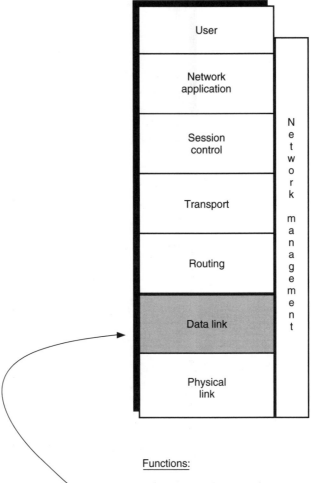

Functions:

- Dependable link operations
- Error checking
- Retransmissions
- Flow control
- Discerning data from "noise"

DECnet implementations:

- DDCMP
- HDLC
- ISO 8802 for LANs
- Ethernet
- LAPB for X.25 interfaces
- Token Ring
- FDDI

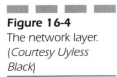

Figure 16-4
The network layer.
(*Courtesy Uyless
Black*)

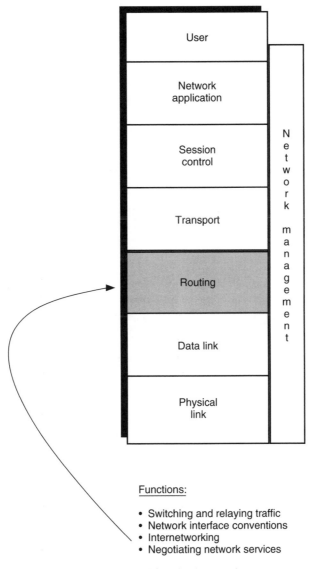

Figure 16-4
The network layer.
(*Courtesy Uyless
Black*)

Functions:

- Switching and relaying traffic
- Network interface conventions
- Internetworking
- Negotiating network services

DECnet implementations:

- X.25
- CLNS
- CLNP
- IP

The Routing Layer

Definitions

Before studying the DECnet Phase IV routing methods, we introduce specific terms will be used throughout the text, some of which have not been defined yet. The following list of terms is a reference for the text that follows.

Adjacency. According to the DECnet specification, an adjacency is a [circuit, nodeID] pair. For example, an Ethernet with n attached nodes (network stations) is considered as $n-1$ adjacencies (it does not include itself) for that router on that Ethernet. Basically, an adjacency is a node and its associated data-link connection. This could also be a synchronous connection to a router provided by a phone line.

Broadcast endnode adjacency (BEA). An endnode connected to the same Ethernet as a given node.

Broadcast router adjacency (BRA). A router connected to the same Ethernet as a given node.

Circuit. An Ethernet network (not an internet) or a point-to-point link.

Connectivity algorithm. The algorithm in the decision process whose function is to maintain path lengths (or hops); the routing function.

Cost. An integer number assigned to a router interface. This represents the cost of the port for routing. (See *Path cost.*)

Designated router. The router on the Ethernet chosen to perform duties beyond those of a normal router. This includes informing endnodes on the circuit of the existence and identity of the Ethernet routers. Also used for the intra-area routing for endnodes (discussed later).

Endnode. A node that cannot route.

Node. A device on the network. A router, bridge, minicomputer, and personal computer are examples.

Hop. The logical distance between two adjacent nodes. Nodes that are directly connected are said to be one hop away.

NBEA. Number of broadcasting endnode adjacencies.

NBRA. Number of broadcasting router adjacencies.

NN. The maximum node number in an area (less than 1024).

NA. A level 2 router parameter only, the maximum area number (less than 64).

Maxh. Maximum hops possible in a path to a reachable node in an area value (less than or equal to 30).

Maxc. The maximum path cost in an area (less than or equal to 1022).

Maxv. How high the Visits field may become before the packet is considered to be looping (less than or equal to 63).

AMaxh. Pertaining to level 2 routers only, (maximum path length to any area (less than or equal to 30).

AMaxc. Actual maximum path cost to any area (less than or equal to 1022).

Path. The route a packet takes from the source to the destination node. There are three types of paths:

- Endnode to endnode (direct routing)
- Endnode through level 1 router(s) in the same area
- Endnode through level 2 routers to an endnode destination in another area

Path cost. The sum of the circuit costs along a path between two nodes.

Path length. The number of hops between two nodes.

Point-to-point link. The link between two routers, usually a remote serial line connecting two routers.

Traffic assignment algorithm. The algorithm in the decision process that calculates the path costs in the routing database.

Functions of DECnet Routing

DECnet provides a first glimpse into hierarchical routing. Unlike other dynamic routing methods, routing tables are not distributed to all nodes on the entire internet. Routing tables are held to the local area routers, with endnodes (those that do not provide routing functions) reporting their status to their local routers.

The routing method of DECnet divides the Internet into areas. There are 63 areas allowed in a DECnet network. There may be up to 1023 addressable nodes in one area. Routers that provide routing functions between other routers in a local area are known as level 1 routers. Routers that provide routing between two areas are known as level 2 routers.

Upon startup, endnodes report their status to the router with the MAC destination multicast address "all routers." This is called the endnode hello packet. All level 1 routers will pick up these packets and build a level 1 database table of all known endnodes for the circuit (Ethernet, for example) they are attached to. These endnode multicasts will not be multicast across the router. This means that a router that receives an endnode hello will consume the packet. The router does not forward this packet.

Instead, level 1 routers will transmit their database tables (of all known endnodes) to other level 1 routers. In this way, all level 1 routers will be updated with all known endnodes in a particular area. Level 1 routers also maintain the address of the nearest level 2 router. To reduce the memory required to keep track of all end stations, DECnet routing is split into distinct areas. Level 1 routers track only the state of the endnodes in their respective areas.

Level 2 routers maintain a database of all known level 2 routers in the entire DECnet internet. Level 2 routers must perform level 1 functions as well as level 2 functions. Therefore, they adhere to all level 1 routing protocols, including receiving all local level 1 updates for their area. They provide the additional ability to route between areas.

Figure 16-5 shows the services and components used in DECnet routing. This figure is to be used as a reference for the text that follows.

Addressing

The addressing scheme used by DECnet Phase IV is based on area and node numbers. The total address is 16 bits long, with the area number being the first 6 bits and the node number the final 10 bits. To carry this addressing scheme from Phase III over to Phase IV, which supports Ethernet (Ethernet physical addresses are 48 bits long), an algorithm was derived. Whereas other protocols use a translator function to map a protocol address to a MAC address, DECnet literally maps its protocol address directly to the MAC address. A DECnet address is used in the DECnet packet header, but its translation to a MAC address overrides the burnt-in MAC address of the LAN controller. A table is not maintained to map a DECnet address to a MAC address.

Figure 16-5
Services and compo-
nents in DECnet rout-
ing. (*Courtesy Uyless
Black*)

DECnet Routing Layer Functions	
Service	Component
Packet paths	Determines path for packets if more than one path exists.
Topology changes	Alternate paths are used if a node or circuit fails; routing modules are changed to reflect changes.
Packet forwarding	Forwards packet to end communication layer at destination node or the next node if packet is not destined for the local node.
Node visits	Limits number of nodes that a packet can visit.
Buffer management	Manages buffers at nodes.
Packet return	Returns packets to end communication layer if packets are addressed to unreachable nodes (if requested by end communication layer).
Data-link monitoring	Monitors errors detected by the data-link layer.
Statistics	Gathers event data for network management layer.
Node verification	If requested by the network management layer, exchanges passwords with adjacent node.

Conversion Example   The DECnet address is 6.9 (area 6, node number
9). Put it into binary with 6 bits for the area number and 10 bits for the
node ID:

$$000110.0000001001$$

Split it into two 8-bit bytes:

$$00011000 \quad 00001001$$

Swap the two bytes:

00001001 00011000

Convert to hex:

0918

Adding the HIORD (Digital's constant high-order bytes, AA 00 04 00, explained next) gives AA 00 04 00 09 18 as the 48-bit physical address for a node residing in area 6 and a node ID of 9. Allowing 6 bits of area address and 10 bits for node ID gives a total of 65,535 possible addresses in a DECnet environment.

Figure 16-6 shows another example.

Address Constants The following addresses are reserved by Digital for the indicated usage:

HIORD (high-order bytes): The four bytes placed on the beginning of a 48-bit converted DECnet address. This is added after the address conversion is accomplished to complete the 48-bit DECnet address. These bytes are AA 00 04 00 xx xx.

All routers: The multicast address for packets that send out information pertaining to all the routers—AB-00-00-03-00-00.

All endnodes: A multicast address used by packets that contain information for all endnodes in an Ethernet segment (a circuit)—AB-00-00-04-00-00.

The protocol (Ethernet Type field) should be set to 60-03, which indicates a DECnet packet. DECnet uses the Ethernet frame format. DEC supports SNAP for Token Ring packets.

Areas

A DECnet network is split into definable areas (63 total). To accomplish area routing, there are two types of routers: level 1 and level 2. Level 1 routers route data within a single area and keep track of the state of all nodes in its area. Level 1 routers do not care about nodes that are outside their area. When communication takes place between two different areas, the level 1 routers send the data to a level 2 router.

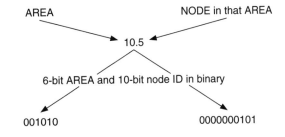

Figure 16-6
MAC address conversion for a DECnet address.

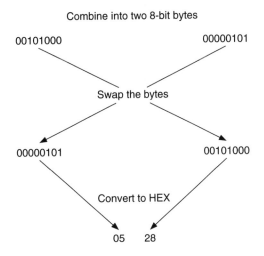

AREA   NODE in that AREA

10.5

6-bit AREA and 10-bit node ID in binary

001010   0000000101

Combine into two 8-bit bytes

00101000   00000101

Swap the bytes

00000101   00101000

Convert to HEX

05   28

Add the HIORD (4 Hi-order bytes assigned by DEC which are AA 00 04 00)

Gives the following 48-bit physical address
AA 00 04 00 05 28

Level 2 routers route traffic between two areas. Level 2 routers keep track of the least-cost path (not necessarily the fastest) to each area in the internetwork as well as the state of any of the nodes in the area.

Figure 16-7 shows three areas: 10, 8, and 4. Level 1 routers forward data within their own area. If a data packet is destined for another area, the level 1 router will forward the packet to its nearest level 2 router. A level 2 router automatically assumes the function of a level 1 router in that area. This router will ensure that the packet makes it to the final area. Then a level 1 router will forward the data to the final destination node.

Figure 16-7
DECnet areas. Level 2
routers can be
Ethernet-to-Ethernet
as well as Ethernet-
to-serial connections,
as shown here.

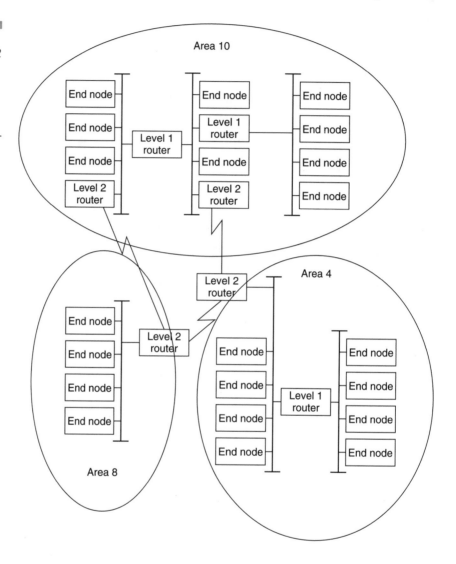

A level 2 router performs both level 1 and level 2 routing functions in the same router.

Figure 16-7 shows level 2 routers connected through serial lines. This is not always the case. Level 2 routers may connect two segments of Ethernet cable. Each Ethernet segment would be defined as a different area. Unless there are multiple routers defined with different area numbers connected to the same cable segment, there is only one area per cable segment. Unlike TCP/IP or AppleTalk, DECnet supports only one area per cable segment.

The Routing Database

There are two types of nodes in DECnet areas:

1. Endnodes are stations that do not have a routing capability. This means that they cannot receive and forward messages intended for other nodes. An endnode can send data (packets) to another, adjacent node (i.e., one on the same LAN, whether that is an endnode or another router). They cannot route a packet for another endnode or router. Examples of endnodes are personal computers and VAX minicomputers (with routing turned off).

2. A full-function node (a routing node) can send packets to any node on the network. This includes sending messages to local nodes as well as nodes that are in a different area (through a router). The computer equipment that Digital manufactures possesses the ability to act not only as a host, but also as a host that can provide network functionality and user application functionality. This text will discuss functionality of the network routing specification, not the dual functionality of a host providing application functionality plus the network routing functions. In other words, to explain the function of routing, the network is considered to have two entities: an endnode and a single device known as a router. A VAX mini and an external router can both provide full-function routing.

Declarations

Before we define the function of the DECnet routers, some declarations should be made.

- Router database tables are constructed from update routing messages received from adjacent nodes. These adjacent nodes could be other level 1 or 2 routers or endnodes in the local area (the same area number that is configured by the router). The information contained in these tables will be shown later. Since a DECnet internet is divided into areas, there are two types of tables: level 1 and level 2 tables.

- The function of level 1 routers is to store a distance based on cost and hop information. This distance is calculated from the router to any destination node within the local area. Level 1 routers receive database updates from adjacent level 1 routing nodes. This means that once a level 1 router builds a table containing a listing of endnodes that the router knows about, it sends this information to other level 1 routers (known as *adjacencies*) on its network. In this way the level 1 routing database table is updated periodically by the update routing messages received from adjacent level 1 routing nodes. Included in this message is the router's listing of known active endnodes in the area.

■ To update other level 1 routers, a routing node sends update routing messages to its adjacent routing nodes with cost and hop information. All level 1 routers will transmit this information to all other level 1 routers. This allows all level 1 table information to be eventually propagated to all routing nodes. Level 1 routers will not propagate their information to other areas.

■ Level 1 routers can function on Ethernet (broadcast circuit) or point-to-point networks (router to router through a serial line).

■ Level 2 routers perform level 1 routing functions as well as level 2 routing. They store cost and hop values from the local area to any other destination area. This is the main function of the level 2 routers—to route information to other areas on the DECnet internet. Level 2 routers receive database updates periodically from adjacent level 2 routing nodes.

■ Level 2 routers can function on Ethernet (broadcast circuit) and point-to-point networks (router to router across a serial line).

■ DECnet routers are adaptive to topology changes. If the path to a destination fails, the DECnet routing algorithm will dynamically choose an alternative path (if available). If, at any time, a physical line in the network goes down, all paths affected by that circuit will be recalculated by each routing node. Any time these algorithms reexecute, the contents of the databases are revealed to all adjacent nodes on each respective network.

■ DECnet routers are adaptive to different circuit types, including X.25, DDCMP, serial links, and Ethernet.

■ Event-driven updates mean that other routers are informed immediately of any changes in the network that would affect routing of a packet.

■ Periodic updates are timed intervals in which routing updates, test messages, and hello messages are sent. These timers are settable.

- An unknown address or network-unreachable packet will be returned (if requested).
- The number of nodes a packet has visited is tracked to keep the packet from endlessly looping in the network.
- Node verification—password protection is performed.
- Information is gathered for network management purposes.

Forwarding of Packets

DECnet uses not only adaptive routing, but also area routing. This means that DECnet routing is hierarchical. In other words, there is more than one layer in the routing (see Fig. 16-7). The routing techniques used by DECnet are not the same as those used in most networks today. When a source station tries to send data to a destination station, the path it takes to transmit the data is called the *path length* and is measured by the number of hops. A path between the source and destination station may never exceed the maximum number of hops for the network.

An example of hierarchical, or area, routing is the phone number system for calls within the United States, Mexico, or Canada. A 10-digit system is used to identify any phone. The first number is an area code, which identifies the area (according to the phone system) you live in. For example, 703 is the code for the northern part of Virginia. The second three digits are called the exchange number. This identifies which subpart of the area you live in. The last four digits indicate the phone itself. Therefore, it you wanted to call within the local area, you would not have to use all 10 digits. Using just the last seven digits, the phone company switches (analogous to data routers) will identify the exchange number and switch (route) the call. For a call to another area, the first three digits are used, and the phone company's switch will recognize this and switch the call to a switch in that area. The long-distance call would then be routed to the exchange number and, finally, to the individual phone.

Although DECnet uses hops, it is not based on a distance-vector algorithm. (For more information on distance-vector, see Chaps. 3 and 4.) DECnet routes are based on a different "cost" factor. The path with the lowest cost is chosen as the path a packet will take. This allows multiple paths to a destination to exist. The network manager may also assign a cost to each circuit (connection to the data link) on each station. The cost is an arbitrary number that is used to determine the best path for a data transfer. The total cost is the cumulative number between a source and destination station.

A router, not an endnode, will pick a path for data travel based on the least cost involved. A path may contain the highest number of hops but still have the least cost. Refer to Fig. 16-8. The path to node D can take many different routes. Before data are sent, node A must decide which path to node D has the lowest cost. As shown in the figure, A to B to C to D offers the lowest cost, so it is chosen as the path. Notice that this path has more hops than option 1. Hops are used in DECnet to determine the diameter of the network and to determine when a packet is looping, among other things. Assigning a cost is a much more efficient way of routing. If we based a path on the number of hops, option 1 would have been taken. If the path from B to D is a 9600-baud sync line and the paths from B to C and C to D are T1 serial lines, the hop algorithm will select the 9600-baud line, for it has the lowest hop count. Assigning costs allows the network administrator to determine the best path. One exception to cost-based routing is that when a node can be reached in two different ways and each path has the same cost, the router will arbitrarily select the route.

The most cost-effective route is the only one kept in the routing table. The one exception to this is when there are two paths to the same destination and the costs are equal.

Figure 16-8
DECnet least-cost routing. (*Courtesy Uyless Black*)

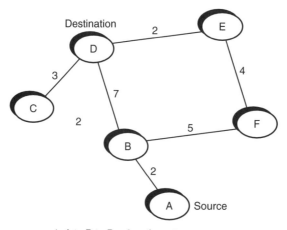

1. A to B to D = 9 path cost
2. A to B to C to D = 7 path cost
3. A to B to F to E to D = 13 path cost

Therefore, choose route 2

Cost numbers for circuits are arbitrarily assigned by a network administrator. There is not a strict standard used in choosing them. The DECnet architecture does have an algorithm for assigning them, though. The general principle behind it is, the slower the circuit (the link), the higher the cost. For example, a 9600-baud line will have a much higher cost than a T1 line. Table 16-1 shows the cost numbers most commonly assigned.

Depending on its level, a routing node in a DECnet network maintains at least one database. For a level 1 router, the database contains entries for path length and path cost to every node (workstation, host, etc.) in its area. It also maintains an entry for the whereabouts of a level 2 router in its area. Level 2 routers add a second database, known as an area routing table, which determines the least-cost path to other area routers in the network (not only the local area).

Node reachability means that a node can be accessed if the computed cost and the hops it takes to get there are do not exceed the number that is configured for that router.

A lot of parameters may be configured for a DECnet router to allow for efficient utilization of memory. If the DECnet network is small and has only a few nodes in a few areas, there is no need to configure the parameters with large entries. The memory will never be used. Most DECnet routers allocate memory space based on the configuration parameters. It is important to know the DECnet topology before config-

TABLE 16-1

Possible Cost
Parameters Based
on Line Speed

Speed	Cost	Speed	Cost
100 Mb/s	1	64 Kb/s	14
16 Mb/s	2	56 Kb/s	15
10 Mb/s	3	38.4 Kb/s	16
4 Mb/s	5	32 Mb/s	17
1.54 Mb/s	7	19.2 Kb/s	18
1.25 Mb/s	8	9.6 Kb/s	19
833 Kb/s	9	7.2 Kb/s	20
625 Kb/s	10	4.8 Kb/s	21
420 Kb/s	11	2.4 Kb/s	22
230.4 Kb/s	12	1.2 Kb/s	25
125 Kb/s	13		

uring a DECnet router. The memory space is not allocated dynamically based on what the router finds on the network.

Remember, in a DECnet environment, routers are not necessarily separate boxes that perform routing functions only. Routers may be contained in a Micro VAX, a VAX, or any other node on the network.

Routing the Data

There are two types of messages a router node may receive: data and control messages (network overhead, which has nothing to do with data).

Data messages contain user data that are transferred between two communicating stations, specifically at the end-to-end communication layer (OSI transport layer). The process of routing adds a route header to the packet that will enable the routers to forward the packet to its final destination. Control messages are exchanged by the routers that initialize, maintain, and monitor the status between the routers or endnodes.

Figure 16-9a shows the end station hellos from nodes 5.3, 5.2, and 5.1, respectively. These are multicast packets that are received by the router 5.4. This router is simply a level 1 router, for it has no other area to talk to. It will transmit level 1 routing updates from its Ethernet port.

Figure 16-9b shows two types of update packets that the routing function uses. One is for Ethernet and the other is for non-Ethernet (i.e., serial lines). The routers are level 2 routers. As shown in Fig. 16-9b the packet types used in broadcast networks (i.e., Ethernet) are

- Ethernet endnode hello message—picked up by level 1 and level 2 routers
- Ethernet router hello message—sent to all routers on the Ethernet by the router
- Level 1 routing message—sent by the level 1 router and containing the endnode database
- Level 2 routing message—sent by the level 2 router and containing level 2 routing table information

The packets transmitted between the routers on the serial lines (nonbroadcast networks) use the following routing messages only:

- Initialization message
- Verification message
- Hello and test messages

Figure 16-9 (a)
Level 1 routing
updates.

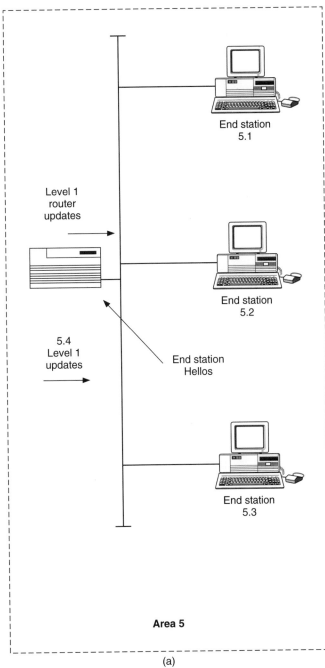

Figure 16-9 (a)
Level 1 routing
updates.

End station
5.1

Level 1
router
updates

End station
5.2

5.4
Level 1
updates

End station
Hellos

End station
5.3

Area 5

(a)

Figure 16-9 (b)
Routing updates

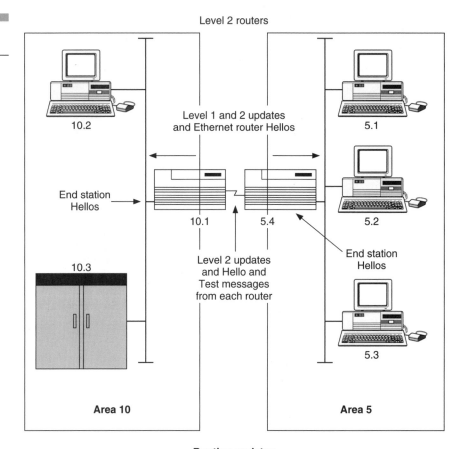

Level 2 routers

Level 1 and 2 updates
and Ethernet router Hellos

10.2

5.1

End station
Hellos

10.1   5.4

5.2

Level 2 updates
and Hello and
Test messages
from each router

10.3

End station
Hellos

5.3

Area 10

Area 5

Routing updates
Routers build databases based on
1) End station Hellos
2) Other level 1 routers
3) Other level 2 routers

(b)

■ Level 1 routing message
■ Level 2 routing message

The six types of routing control messages that DECnet Phase IV uses are defined as follows:

1. The routing message contains information that is used for updating the routing database of an adjacent node. Inside this message are the path cost and path length values for specific destinations.

2. The Ethernet router hello message is used to initialize and monitor routers on an Ethernet circuit. This is a multicast packet containing a list of all routers on the Ethernet circuit that the sending node has recently learned of via the Ethernet router hello messages. This packet is multicast periodically to all other stations (routers and endnodes alike) on the same Ethernet circuit so that all routers are updated with the most current status of other routers on the circuit. With transmission of this message, the designated router (DR) is also selected on that Ethernet circuit. Once the DR is selected, it will remain so until it is taken offline for whatever reason (other routers must hear from the designated router within a certain time period). One purpose of the DR is to assist end stations in discovering that they are on the same circuit. This is accomplished by setting a special intra-Ethernet bit in a packet that is forwarded from one endnode to another (more on this when endnodes are discussed). This enables the endnodes on the same Ethernet segment to communicate without using the router.

Specifically, the Ethernet router hello messages contain the sending router's ID, a timer called T3 (the timer for sender periodic hello messages), and the sending router's priority. This message also contains a listing of other routers that the transmitting router has heard from. The list of other routers will contain their node numbers and priorities. A new router will be added to a router's table provided that two variables—number of routers and number of broadcasting routing adjacencies—are not exceeded. A router will broadcast this message to all level 1 routers (destination address AB-00-00-03-00-00) and to all end stations (AB-00-00-04-00-00).

A router will not declare itself the designated router until a certain time has passed (according to the DECnet specification, this time defaults to 5 s).

An empty Ethernet router hello message indicates that the router is being brought down and other routers should modify their tables accordingly.

As stated before, every router is assigned a priority number. This number could allow it to become the designated router. One other use for the priority number of a router is that when routing tables become full (the maximum number of routers is reached), the router with the lowest priority is deleted from the table. If multiple routers have the same priority, the router with the lowest ID (area.node ID) will be deleted. An event will be transmitted indicating that the router is down.

3. The Ethernet endnode hello message is used to allow routers to initialize and monitor endnodes on an Ethernet circuit. This packet is transmitted as a multicast packet—all-routers multicast—by each active endnode on an Ethernet circuit to allow all routers to find out about all active endnodes on the circuit. The routers use this to monitor the status of the endnodes on their Ethernet. The routers make an entry in their tables for each endnode.

Once a router receives this packet, it should receive hellos from this adjacency at certain timer periods. If it does not, it will delete this entry from its table and generate an event that the adjacency is down. This period of waiting is set to three times the value of the router's T3 timer (the hello timer for the router).

4. The hello and test message tests the operational status of an adjacency. This type of packet is used on non-Ethernet (nonbroadcast) circuits when no messages are being transmitted across that line. For example, on a serial line interconnecting two DECnet routers, when there are no valid messages to transmit, this message is transmitted periodically so that the opposite end will not think the circuit is down. If this or any other message is not received within a certain time, the routing layer will consider the circuit to be down.

5. An initialization message appears when a router is initializing a non-Ethernet circuit. Contained in this message is information on the type of node, the required verification, maximum message size, and the routing version.

6. Finally, a verification message is sent when a node is initializing and a node verification must accompany the initialization.

Updating the Routers   Routing messages are propagated through the DECnet internet. Any node on the network can send a routing message to an adjacent routing node. When the adjacent routing node receives that message, it will compare the routing information in the message to its existing routing database table. If the information in the routing message contains new information, the table is updated. A new routing message containing the new information is then generated by that routing node and sent to all other adjacent routers on the Internet. Those routers will update and send new messages to their adjacent nodes, and so on. This is known as propagating the information. These messages are shown in Fig. 16-9 b.

If there are multiple areas on an internet (indicating the use of level 2 routers), routers must discard messages that do not pertain to them. For example, when a level 1 router receives an endnode or a router hello message, the first check accomplished is that the incoming message's ID (area.node ID) is the same as the router's ID (the areas should match). If they do not match, the packet is discarded. This means that a level 1 router will not keep a table of any adjacencies for other areas (for Ethernet, it will not keep track of endnodes in other areas). A level 1 router will keep endnode tables only for endnodes that have the same area ID as the router's.

Level 2 routers must keep a table of adjacencies to other level 2 routers, in addition to the adjacencies (level 1 routers and endnodes) in its own (single) area. It will discard any Ethernet endnode hello messages it receives from areas other than its own or any level 1 routing messages from other areas. When it receives a level 2 update message, it will include that router's ID in its update table. The router will also include that ID its next Ethernet router hello message. A level 2 update will contain an area number and an associated hop count and cost number to get to that area. These routing messages are broadcast using the all-routers broadcast (AA-00-00-03-00-00) in the destination address field of the MAC header.

When a router broadcasts its information to the network, and there is more information in the table than the Ethernet allows for a maximum-size packet, multiple packets are transmitted. This is similar to other routing protocols.

Router Operation   As shown in Fig. 16-10, the routing layer is split into two sublayers:

1. The routing initialization sublayer performs only the initialization procedures, which include initialization of the data-link layer. This involves setting up the drivers and controlling the Ethernet, X.25, and DDCMP. This is the layer that the routing layer of DECnet must talk to.

2. The routing control sublayer performs the actual routing, any congestion control, and packet lifetime control. This layer controls five different processes as well as the routing database.

The routing control sublayer controls the following five processes:

1. *Decision.*   Based on a connectivity algorithm that determines path lengths and another algorithm that determines path costs, this process will select a route for a received packet. Those algorithms are executed when the router receives a routing message (not a routable data message). The forwarding database tables are then updated based on the outcome of the invoked algorithms. The decision process will select the least-cost path to another node (level 1 routing) or the least-cost path to another area router (level 2 routing).

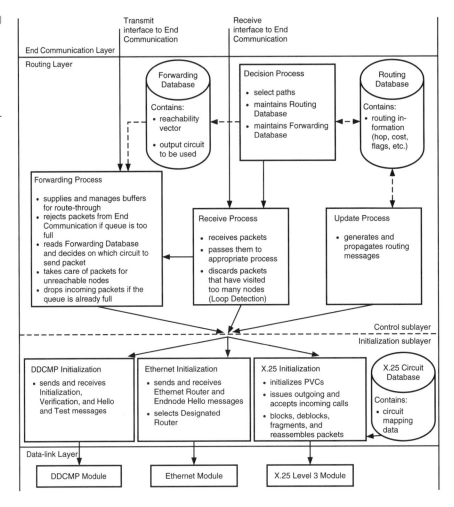

Figure 16-10
Routing layer components and their functions. *(Copyright © 1982 Digital Equipment Corporation)*

2. *Update.* This process is responsible for building and propagating the routing messages. These routing messages contain the path cost and path length for all destinations. Based on the decision process, these messages are transmitted when required. This process also monitors the circuits that it knows about by periodically sending routing messages to adjacent nodes. If the router is a level 1 router, it will transmit level 1 messages. If it is a level 2 router, it will transmit level 1 and level 2 routing messages.

- Level 1 routing packets are sent to adjacent routers within the router's area.
- Level 2 packets are sent to other level 2 routers.
- Level 1 routing packets contain information on all nodes in the router's home area.
- Level 2 routing packets contain information about all areas.
- Packets containing routing information are event-driven with periodic backup.

3. *Forwarding.* This process looks in a table to find a circuit (path) for forwarding a packet. If a path cannot be found, this process will return the packet to the sender or will discard the packet (depending on the option bits set in the packet route header, the RQR bit). This process also manages all the buffers required to support the tables.

4. *Receive.* This process inspects the packet's route header. Based on the packet type, this process gives the packet to another process for handling. Routing messages are given to the decision process, hello messages are given to the node listener process, and packets that are not destined for the router are given to the forwarding process.

5. *Congestion control.* This process uses a function known as transmit management. It handles the buffers, which are blocks of RAM memory set aside for storing (i.e., queuing) information until it can be sent. It limits the number of packets allowed in the queue. If this number is exceeded, the router is allowed to discard packets to prevent the buffer from overflowing.

Packet lifetime control prevents packets from endlessly looping in the network by discarding packets that have visited too many routers. This involves the loop detector, the node listener, and the node talker. The loop detector keeps track of how many times a packet has visited the node and removes the packet when it has visited the node a set amount of times. The node listener is the process that keeps track of adjacencies. It determines when a node has been heard from and if the identity of an adjacent node has changed. This is the process that determines whether an adjacency is declared down. In combination with this, the node talker is the process that allows hello packets to be transmitted. It places an artificial load on the adjacency so that failures can be detected.

Initialization and circuit monitoring are means by which the router obtains the identity of a neighboring node. For Ethernet circuits, the

Ethernet router hello and the Ethernet endnode hello messages perform these functions.

Forwarding of Data in a DECnet Environment

As stated before, data start at the application layer of network station and flow down through the OSI model towards the physical layer. As the data pass through each of the layers, additional information, known as header information, is added to the beginning to the packet. This information does not change the original information (the data from the application layer) in the packet; it is merely control information so that any node that receives this packet will know whether or not to accept it and how to process it. If the packet is not intended for that node, the header information will contain information on what to do with the packet.

A router must check its routing table to determine the path for a packet to reach its final destination. If the destination is in the local area, the router will send the packet directly to the destination node or to another router in that area to forward the packet to the destination node.

If the packet is for a node in a different area, the router will forward the packet to a level 2 router. That router will send it to another level 2 router in that area. The level 2 router in the other area will forward the packet to the destination node in the same manner as a level 1 router. Remember that level 2 routers act as both level 2 and level 1 routers.

Tables 16-2 and 16-3 show typical level 1 and 2 routing database tables of a level 2 router. This would be router A's (a level 2 router) forwarding routing database. Router A is shown in Fig. 16-11. From the table, you should notice that the number of reachable areas from this router is three, the number of reachable nodes is four for area 1 and two for area 11, and the number of adjacent routes is two. The table should look a little odd. In the level 1 database is information for two different areas. A level 1 router is not supposed to keep track of different area nodes— only the endnode IDs for its own area.

This is a multiport router that contains more than just two ports. Therefore, it can assign area and node IDs on a per-port basis. Therefore, this node is also a level 1 router for both area 1 and area 11. It must keep track of the nodes in both areas it connects to. In actuality, the router

TABLE 16-2

Level 1 Routing
Tables

Area 1							
Node	Port	Next Hop	Cost	Hops	BlkSize	Priority	Timer
1	1	1.1	3	1	1500	40	—
2	1	1.1	3	1	1500	40	40
15	1	1.1	3	1	1500	50	40
4	2	1.90	3	1	1500	40	30
6	2	1.90	3	1	600	40	40
9.0	2	1.90	3	1	1500	40	—
Area 11							
2	3	11.2	3	1	1500	40	40
30	3	11.2	3	1	1500	40	90
15	3	11.2	3	1	1500		
1	3	11.2	3	1	1500		

TABLE 16-3

Level 2 Routing
Table

Area	Port	Next Hop	Cost	Hops	Time
1	Local	—	—	—	—
11	3	11.2	3	1	60
50	3	11.2	6	2	40

would probably keep two tables—one for each area. The information is shown in the same table here for simplicity.

Definitions

LEVEL 1 TABLE DESCRIPTIONS

Forwarding destination of the packet (node and port). This contains the DECnet router address and the port number on the router (assuming the router has multiple ports) for possible packet destinations.

Next hop. The DECnet address (area and node address) that this router must send the packet to for it to be routed to its final destination.

Figure 16-11
Multiple DECnet
areas.

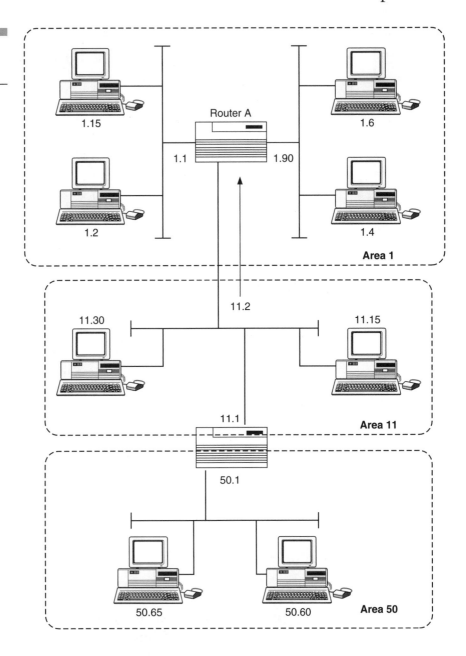

Cost.   A user-configurable number that the network administrator assigns to the circuit. Packets are routed using the path with the lowest cost.

Hops.   The number of hops (the number of routers) that a packet must traverse before reaching the destination.

BlkSize (block size).   The maximum size of a packet that can be sent to a destination node.

Priority.   The individual priority of a router. This is used to determine the designated router. The priority is also used in the following situation. If the number of the router's parameter is exceeded, then the routing database is updated as follows:

- The router with the lowest priority will be deleted from the database unless the new router has a lower priority than any other router in the database.

- If there are duplicate priority numbers, then the router with the lowest ID (48-bit Ethernet address) will be deleted from the database or the new router will not be added (if its Ethernet address is the lowest). Ethernet addresses are not duplicated. There will not be any decisions after this.

Time.   A parameter specifying how long this entry will remain in the table. This parameter is based on the configurable timer known as the hello time.

LEVEL 2 TABLE DESCRIPTIONS

Destination of the packet (area and port).   The DECnet area number (not node number) and the port number of the router to direct the packet to.

Next hop.   The DECnet address (area and node address) of the router to which a router will send the packet so that it may be routed to its final destination.

Cost.   Same as for level 1.

Hops.   Same as level 1 entry.

Time.   Same as level 1 entry.

It is important to note that a router has the possibility of transferring a packet through multiple routes. The decision on which route to take is based solely on the cost associated with a path to the final destination.

Endnode Packet Delivery

When an endnode wishes to communicate with another endnode on the same Ethernet, it may do so under the following conditions.

If the endnode has the destination station's address in its cache table, it may communicate directly with the endnode. If the destination station's address is not in the table, the endnode must send the directed packet to the designated router. Refer to Fig. 16-12. The designated router will then perform a lookup to see if the destination endnode is active. If the destination station's address is in the designated router's table, it will forward the packet to the endnode with the intra-area bit (in the route header) set to 1 (to indicate that this packet is destined for the local LAN). The destination station, upon receipt of that packet, will then send some type of response packet back to the originating station (adding that station's address to its local cache). The designated router will not be used for further communication.

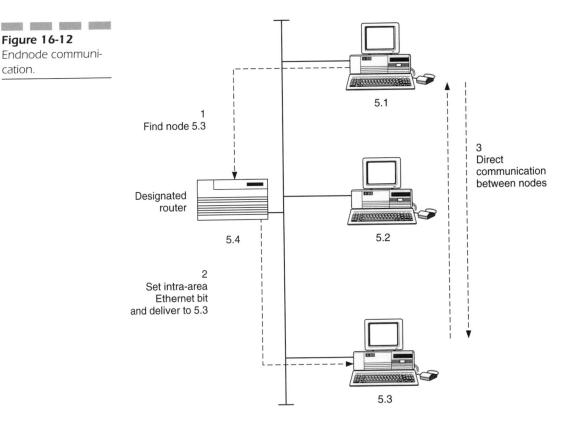

Figure 16-12
Endnode communication.

1
Find node 5.3

5.1

3
Direct
communication
between nodes

Designated
router

5.4

5.2

2
Set intra-area
Ethernet bit
and deliver to 5.3

5.3

The originating station, upon receipt of the destination station's response, will add the destination station's address to its cache and then communicate with it directly (no more help from the designated router). The endnode cache table may be aged out (the entry deleted when the station indicated in the cache table has not been heard from for a specified amount of time).

If the endnode does not have the address of the designated router (there is not an active router), it will try to send the packet to the destination station anyway.

A flowchart for DECnet routing is shown in Fig. 16-13.

A few final notes: As shown in Fig. 16-14a, there may be more than one area per Ethernet segment. The figure shows two areas on the same segment: area 5 and area 10. The only stipulation is that there must be two full-function nodes (or routers) on the same segment of Ethernet cable. Each of these routing nodes will have a different area. All of the rules stated in the previous paragraphs remain the same (Ethernet router hellos, level 1 updates, level 2 updates, etc.). Each router controls the nodes in its own area. A single router port may not support two areas.

Figure 16-14b shows a typical DECnet internet. It shows level 1 and level 2 routers, designated routers, end stations, and full-function nodes.

End Communication Layer: The DNA Transport Layer

The transport layer in the OSI model is used for reliable data transfer between two network stations, no matter where the two reside on the network. This is accomplished by setting up a session, establishing sequence numbers between the source and destination stations, and then passing data over the virtual circuit.

The DECnet transport-layer protocol that accomplishes this is called the network services protocol (NSP). It provides this as a service to the upper-layer protocols. Figure 16-15 summarizes the functions of the DNA transport layer.

The NSF provides the following functions:

■ Creates, maintains, and terminates logical links (virtual circuits)

■ Guarantees the delivery of data and control messages in sequence to a specified destination by means of an error control mechanism

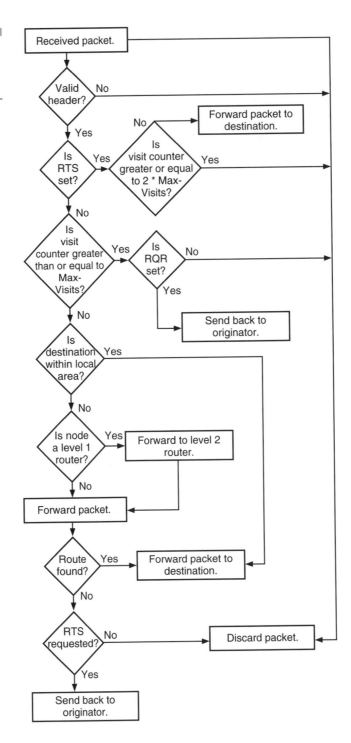

Figure 16-13
DECnet routing
(*Courtesy 3Com Corporation*)

Figure 16-14
(*a*) Multiple areas per
segment.

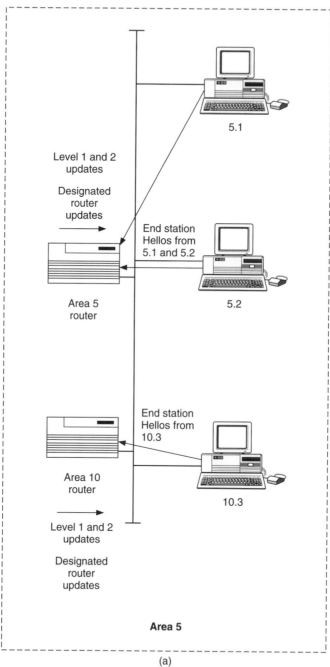

(a)

Figure 16-14
(b) Typical DECnet
internet.

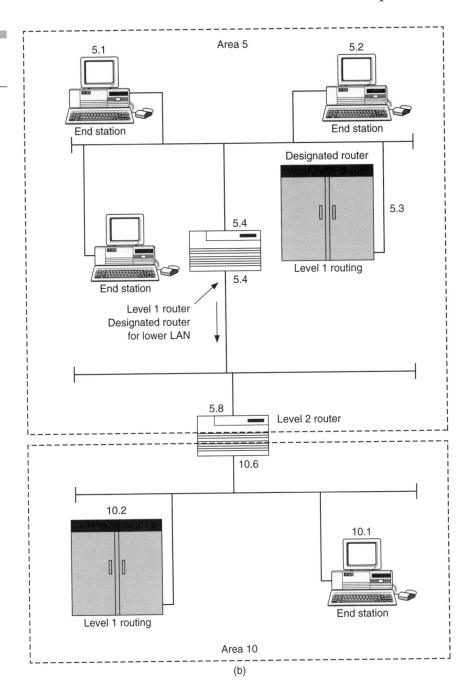

(b)

Figure 16-15
Functions of the
transport layer. (Courtesy Uyless Black)

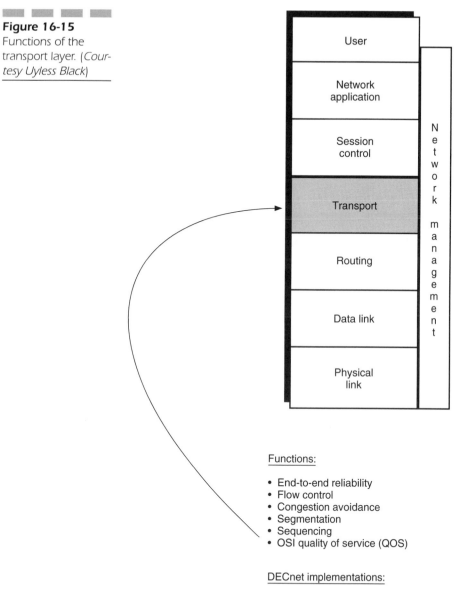

Functions:

- End-to-end reliability
- Flow control
- Congestion avoidance
- Segmentation
- Sequencing
- OSI quality of service (QOS)

DECnet implementations:

- Network Services Protocol (NSP)
- OSI transport protocol,
  classes 0, 2, and 4

- Transfers data into and out of buffers
- Fragments data into segments and automatically puts them back together at the destination
- Provides flow control
- Provides error control

To provide for logical-link service, flow control, error control, and other functions, there are three types of NSP messages:

1. Data

2. Acknowledgment

3. Control

Table 16-4 expands upon the functions of each type.

A logical link is a full-duplex logical channel that data may pass over. This is also known as a virtual circuit. It should be stated here that the session layer of NSP only requests these connections; it is the responsibility of the end communications layer to perform the work. These processes are completely transparent to the user. NSP sets up a connection, manages it (provides sequencing and error control), and then terminates the link when requested to do so.

There may be several logical links between two stations or many stations. Figure 16-16 shows the typical exchange between two end stations, A and B, having a logical link.

The exchange of information data between nodes A and B can be carried over two types of channels:

1. *Data channel:* Used to carry application data

2. *Other data subchannel:* Used to carry interrupt messages, data request messages, and interrupt request messages

The transport layer also fragments data into a size that the routing layer can carry. There is a maximum number of bytes the routing layer will handle, and it is up to NSP to fragment the data and give them to the routing layer. Each fragment contains a special number and other control information in the data segment message, which the receiving NSP layer interprets to understand how to put the data back together again in the order in which they were sent. Only data that are transmitted over the data channel are fragmented.

To provide error control, each end of the virtual link must acknowledge data that it received. Data received too far out of sequence will be

TABLE 16-4 NSP Messages

Type	Message	Description
Data	Data Segment	Carries a portion of a Session Control message. (This has been passed to Session Control from higher DNA layers and Session Control has added its own control information, if any.)
Data (also called Other Data)	Interrupt	Carries urgent data, originating from higher DNA layers. It also may contain an optional Data Segment acknowledgment.
	Data Request	Carries data flow control information and, optionally, a Data Segment acknowledgment (also called a Link Service message).
	Interrupt Request	Carries interrupt flow control information and optionally a Data Segment acknowledge (Link Service message.).
Acknowledgment	Data Acknowledgment	Acknowledges receipt of either a Connect Confirm message or one or more Data Segment messages and, optionally, an Other Data message.
	Other Data Acknowledgment	Acknowledges receipt of one or more Interrupt, Data Request, or Interrupt Request messages and, optionally, a Data Segment message.
	Connect	Acknowledges receipt of a Connect Initiate message.
Control	Connect Initiate and Retransmitted Connect Initiate	Carries a logical link Connect request from a Session Control module.
	Connect Confirm	Carries logical link Connect acceptance from a Session Control module.
	Disconnect Initiate	Carries logical link Connect rejection or Disconnect request from a Session Control module.
	No Resources	Sent when a Connect Initiate message is received and there are no resources to establish a new logical link (also called a Disconnect Confirm message).
	Disconnect Complete	Acknowledges the receipt of a Disconnect Initiate message (also called a Disconnect Confirm message).
	No Link	Sent when a message is received for a nonexistent logical link (also called Disconnect Confirm message).
	No Operation	Does nothing.

▬ ▬ ▬ ▬

Figure 16-16

Typical message
exchange between
two implementations
of NSP. (*Copyright ©
1982 Digital Equip-
ment Corporation*)

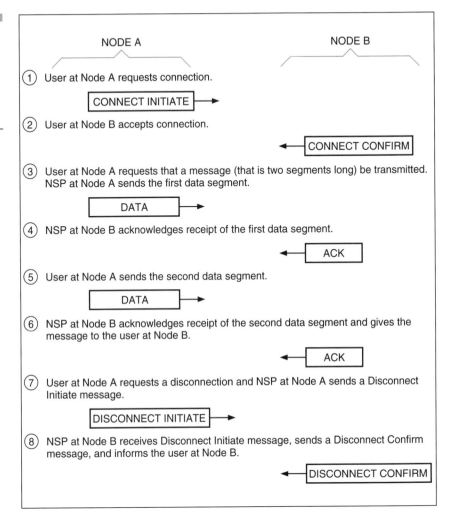

discarded, and a negative acknowledgment will be sent to the originator.
Any station sending data will not discard them (it will keep a copy of
the data in local RAM memory) until it has received a good acknowledgment from the recipient of the data.

Flow Control

NSP provides flow control to ensure that memory is not overrun (buffer overflow). Both types of data (normal and interrupt) can be flow-
controlled. When the logical link is first established, the two sides of the

link tell each other how the flow of data should be handled. There are two options:

1. No flow control.

2. The receiver will send the transmitter the number of data segments it can accept.

In addition to these rules, each receiving end of the link may at any time tell the opposite side (the transmitter) to stop sending data until further notice.

Figures 16-17 and 16-18 show the flow control and acknowledgment operations.

To provide for efficiency, any data type—whether a control data type or an application data segment—may contain the positive acknowledgment returned for previously received data. This means that there is not necessarily a separate ACK packet for data that are being acknowledged.

The Session Control Layer

Refer to Fig. 16-19. The session control layer protocol resides at the fifth layer of the OSI model and provides the following functions:

Mapping of node names to node addresses. This layer maintains a node name mapping table that provides a translation between a node name and a node address or its adjacency. This allows the session layer to select the destination node address or channel number for outgoing connect requests to the end communication layer. For any incoming connection requests, this allows the session control layer to properly identify the node that is making the connect request.

Identifying end users. This function determines if a process requested by an incoming connection request exists.

Activating or creating processes. The session control layer may start up a process or activate an existing process to handle an incoming connect request.

Validating incoming connect requests. A validation sequence is used to find out if the incoming connect request should be processed.

These functions are divided into five actions:

1. *Requests a logical link between itself and a remote node.* If the application desires a logical link between itself and a remote node, it will

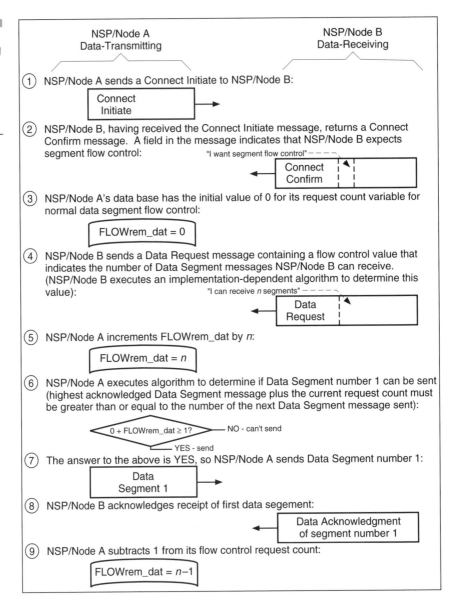

request the session layer for this. It will identify the destination
node address or a channel number for the end communication
layer by using the node name mapping table. It will format the
data for the end communication layer and issue a connect request
to this layer. It will then start an outgoing connection timer. If
this timer expires before the destination is heard from (accept or

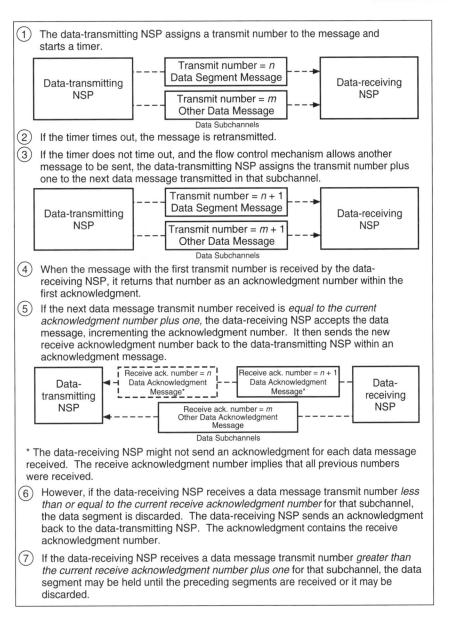

① The data-transmitting NSP assigns a transmit number to the message and starts a timer.

| Data-transmitting NSP | Transmit number = n Data Segment Message ----> | Data-receiving NSP |
| Transmit number = m Other Data Message ----> |

Data Subchannels

② If the timer times out, the message is retransmitted.

③ If the timer does not time out, and the flow control mechanism allows another message to be sent, the data-transmitting NSP assigns the transmit number plus one to the next data message transmitted in that subchannel.

| Data-transmitting NSP | Transmit number = $n + 1$ Data Segment Message ----> | Data-receiving NSP |
| Transmit number = $m + 1$ Other Data Message ----> |

Data Subchannels

④ When the message with the first transmit number is received by the data-receiving NSP, it returns that number as an acknowledgment number within the first acknowledgment.

⑤ If the next data message transmit number received is *equal to the current acknowledgment number plus one,* the data-receiving NSP accepts the data message, incrementing the acknowledgment number. It then sends the new receive acknowledgment number back to the data-transmitting NSP within an acknowledgment message.

| Data-transmitting NSP | <---- Receive ack. number = n Data Acknowledgment Message* ---- | Receive ack. number = $n + 1$ Data Acknowledgment Message* -- | Data-receiving NSP |
| <---- Receive ack. number = m Other Data Acknowledgment Message ------ |

Data Subchannels

* The data-receiving NSP might not send an acknowledgment for each data message received. The receive acknowledgment number implies that all previous numbers were received.

⑥ However, if the data-receiving NSP receives a data message transmit number *less than or equal to the current receive acknowledgment number* for that subchannel, the data segment is discarded. The data-receiving NSP sends an acknowledgment back to the data-transmitting NSP. The acknowledgment contains the receive acknowledgment number.

⑦ If the data-receiving NSP receives a data message transmit number *greater than the current receive acknowledgment number plus one* for that subchannel, the data segment may be held until the preceding segments are received or it may be discarded.

reject), it will cause a disconnect to be reported back to the requesting application layer.

2. *Accepts or rejects a connect from a remote node.* The end communication layer will tell the session control layer of a connection request from a remote (destination) node. The session control layer will check the incoming packet for source and destination end-

Figure 16-19
The session control
layer. (*Courtesy ©
Uyless Black.*)

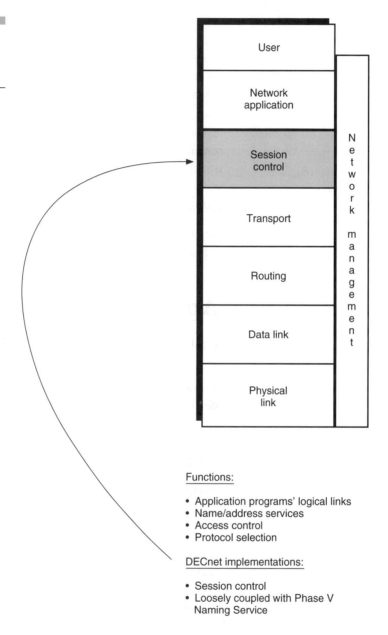

Figure 16-19
The session control
layer. (*Courtesy ©
Uyless Black.*)

Functions:

• Application programs' logical links
• Name/address services
• Access control
• Protocol selection

DECnet implementations:

• Session control
• Loosely coupled with Phase V
  Naming Service

user processes (socket or port numbers). With this it will identify,
create, or activate the destination end-user process. When the session control layer reads the incoming connection request packet,
it will enter a destination address with a destination name in its
table. After all this, it will deliver any end-user data to the destina-

tion application process. It will also validate any access control information.

3. *Sends and receives data across a valid logical link.*   Basically, the session control layer will pass any data between the application layer to the end communication layer to be delivered to the network.

4. *Disconnects or aborts an existing logical link.*   Upon notification from the application process, it will disconnect the session between two communicating nodes. Also, it will accept a disconnect request from the destination node and will deliver this to the application.

5. *Monitors the logical link.*

Figure 16-20 shows the relationship between the user processes, the session control layer, and the end communication layer.

Network Application Layer

DECnet defines the following modules at the network application layer (see Fig. 16-21):

■ The data access protocol (DAP) is the remote file access protocol. It enables file transfer between nodes on the DECnet network.

Figure 16-20
Session control model. (*Copyright © 1982 Digital Equipment Corporation.*)

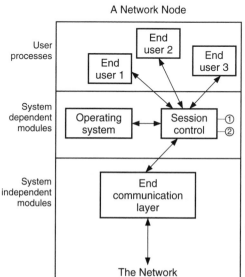

A Network Node

Comments

- End users are User, Network Application, and Network Management* modules.

User processes

End user 1, End user 2, End user 3

- Session Control is an interface to the End Communication layer for end user processes. It functions in conjunction with the operating system.
① and ② are data bases used by Session Control.

System dependent modules

Operating system — Session control —①—②

System independent modules

End communication layer

- The End Communication layer provides the logical link service to Session Control. Its functions are not dependent on individual operating systems.

* Network Management interfaces with Session Control in two ways: (1) to obtain access to the logical link service and (2) to monitor and control Session Control operations.

The Network

■■■ ■■■ ■■■ ■■■
Figure 16-21
The network applica-
tion layer. (*Courtesy
Uyless Black.*)

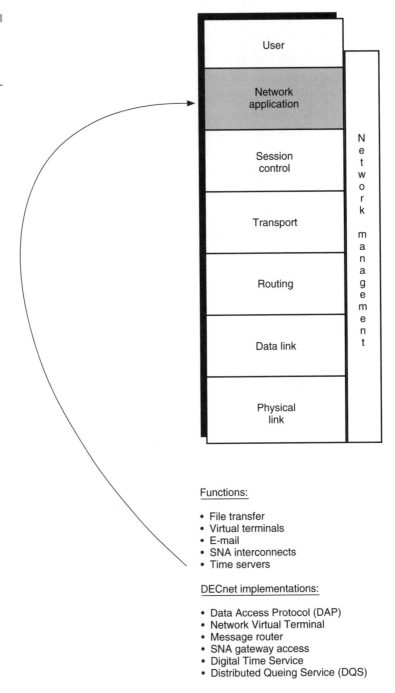

Functions:

- File transfer
- Virtual terminals
- E-mail
- SNA interconnects
- Time servers

DECnet implementations:

- Data Access Protocol (DAP)
- Network Virtual Terminal
- Message router
- SNA gateway access
- Digital Time Service
- Distributed Queing Service (DQS)

- The network virtual terminal (NVT) protocol allows remote terminals to act as if they were local to the processor—to "remote" to another host and act as if that terminal were locally attached.

- The X.25 gateway access protocol allows DECnet to interoperate over an X.25 link. This protocol is beyond the scope of this book.

- The SNA access protocol allows a DECnet network to interoperate with IBM's SNA. This protocol is beyond the scope of this book.

- The loopback mirror protocol tests logical links. This protocol is beyond the scope of this book.

Data Access Protocol

The data access protocol (DAP) is an application-level protocol. DAP permits remote file access within DNA environments. The following is a list of the functions of the DAP:

- Supports heterogeneous file systems

- Retrieves a file from an input device (a disk file, a card reader, a terminal, etc.)

- Stores a file on an output device (magnetic tape, a line printer, a terminal, etc.)

- Transfers files between nodes

- Supports deletion and renaming of remote files

- Lists directories of remote files

- Recovers from transient errors and reports fatal errors to the user

- Allows multiple data streams to be sent over a logical link

- Submits and executes remote command files

- Permits sequential, random, and indexed access of records

- Supports sequential, relative, and indexed file organization

- Supports wildcard file specification for sequential file retrieval, file deletion, file renaming, and command file execution

- Provides an optional file checksum to ensure file integrity

The DAP process, via the source and destination communicating stations, exchanges a series of messages. The initiation message contains information about operating a file system, buffer size, etc. These are negotiation parameters. Table 16-5 lists these messages.

TABLE 16-5 DAP Messages

Message	Function
Configuration	Exchanges system capability and configuration information between DAP-speaking processes. Sent immediately after a logical link is established, this message contains information about the operating system, the file system, protocol version, and buffering capability.
Attributes	Provides information on how data is structured in the file being accessed. The message contains information on file organization, data type, format, record attributes, record length, size, and device characteristics.
Access	Specifies the file name and type of access requested (read, write, etc.).
Control	Sends control information to a file system and establishes data streams.
Continue-transfer	Allows recovery from errors. Used for retry, skip, and abort after an error is reported.
Acknowledge	Acknowledges access commands and control messages used to establish data streams.
Access complete	Denotes termination of access.
Data	Transfers file data over the logical link.
Status	Returns status and information on error conditions.
Key definition attributes extension	Specifies key definitions for indexed files.
Allocation attributes extension	Specifies the character of the allocation when creating or explicitly extending a file.
Summary attributes extension	Returns summary information about a file.
Date and time attributes extension	Specifies time-related information about a file.
Protection attributes extension	Specifies file protection codes.
Name	Sends name information when renaming a file or obtaining file directory history.

Figure 16-22 shows a node-to-node file transfer using DAP, and Fig. 16-23 shows a file transfer over a DECnet network.

Network Virtual Terminal

There are times when a user connected to one host may need a connection to another host on the network. The Network Virtual Terminal

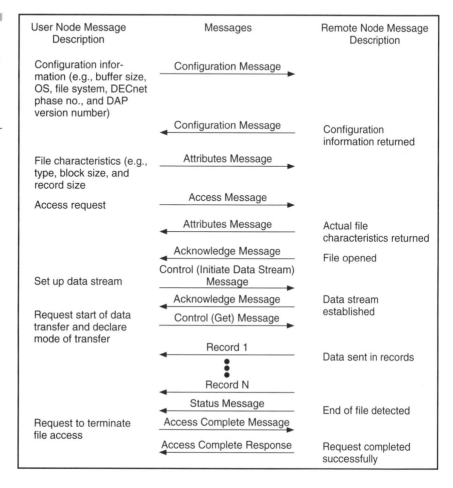

(NIT) protocol allows terminal communications to exist remotely as if the terminal were directly attached to the remote computer. Figure 16-24 shows the overall process for this. The following paragraphs explain the process.

The main protocol used for NVT is called the terminal communication protocol. It is the lower of the two sublayers (the other protocol is the command terminal protocol) of the NVTS, also known as the foundation layer. Its main goal is to establish and disconnect terminal sessions between applications and terminals. You may think of this as extending the session control layer to establishing and disconnecting sessions between endpoints (applications) that are specific to terminals. The endpoint in the host is called the *portal,* and the endpoint in the terminal end is called the *logical terminal.* A portal is a remote terminal identi-

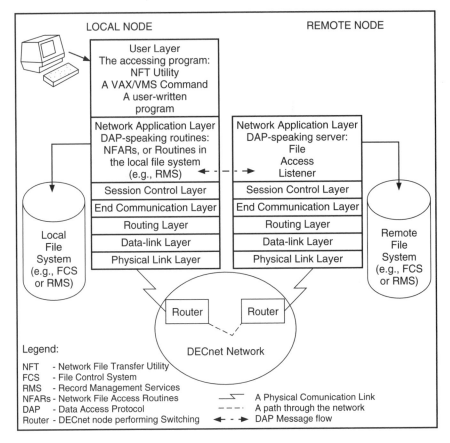

fier in the host, and a connection binds that identifier to an actual terminal. An NVT connection is called a *binding*.

Table 16-6 shows the messages that are exchanged between the terminal and its host using the terminal communication protocol.

The command terminal protocol is the second sublayer of NVT. It offers a set of functions mainly oriented toward command-line input. After a connection between a host and network terminal is made, this layer provides control over the network. The control messages that the terminal and host pass between them are shown in Table 16-7.

In a terminal session between a terminal and a host, the active terminal will request a binding to the destination host. This is shown in Fig. 16-24. The terminal management module in the server system requests a binding to the host system. To accomplish this, it invokes the services of the terminal communication service function (these actions are listed in

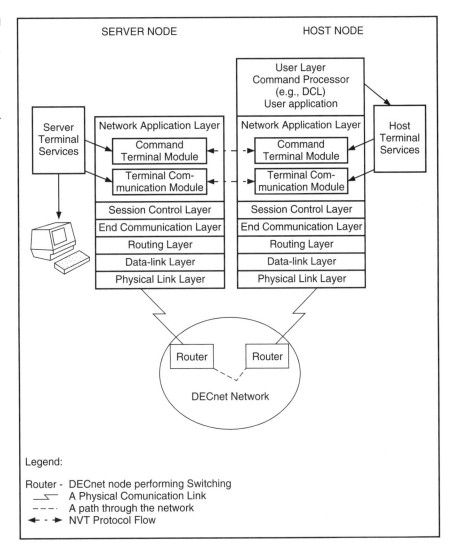

SERVER NODE   HOST NODE

Legend:

Router - DECnet node performing Switching

⎯⟋⎯ A Physical Comunication Link

– – – · A path through the network

◄ · ► NVT Protocol Flow

Table 16-6). Most DECnet users will know this as the SET HOST command entered at the terminal. SET HOST will allow a user at one host to connect to another host over the network. That is, a terminal connected to host A may now connect to a remote host B over the network (usually Ethernet).

The terminal communication service tries to connect to the remote system by requesting a logical link from the session control layer. At the destination host, the incoming logical link request is accepted and a portal is allocated. This is the beginning of the binding. The host module should accept the logical link. Once the logical link has been established,

TABLE 16-6

Terminal Communi-
cation Protocol
Messages

Message	Function
Bind request	Requests a binding; identifies version and type of sending system.
Rebind request	Requests a rebinding (reestablishes broken communications) for high-availability implementations.
Unbind	Requests that a binding be broken.
Bind accept	Accepts a bind request.
Enter mode	Requests entry of a new mode. (The only mode currently defined is command mode.) This selects the command terminal protocol as the higher-level protocol.
Exit mode	Request that the current mode be exited.
Confirm mode	Confirms the entry of a new mode.
No mode	Indicates that the requested mode is not available or confirms an exit mode request.
Data	Carries data (i.e., command terminal protocol information).

the Bind request is sent and, if accepted, a Bind Accept message is sent back to the requester. The binding has now been established.

Now that the binding has formed, the host will enter command mode (the command terminal protocol will now interact on the connection). This action takes place in the first terminal request from the login process. Both ends of the connection now enter into the command terminal protocol and remain with this protocol until the end of the connection. Requests and responses then take place as data are transferred across the link using the command terminal protocol.

The application program in the host system will send terminal service requests to the host operating system terminal services. The host terminal services issue corresponding requests to the host protocol module of a logout of the user from the host. The server protocol module reproduces those requests remotely and reissues them to the server terminal services.

Termination of this link usually comes from the application program running in the host system. This will happen by a logout by the host application With this, the host terminal communication services will send an Unbind request. The server will respond by releasing the link. Finally, the host disconnects the logical link.

Figure 16-25 shows NVT message exchanges during this connection.

TABLE 16-7

Command Terminal Protocol Messages

Message	Function
Initiate	Carries initialization information, as well as protocol and implementation version numbers.
Start read	Requests that a READ be issued to the terminal.
Read data	Carries input data from terminal on completion of a read request.
Out-of-band	Carries out-of-band input data.
Unread	Cancels a prior read request.
Clear input	Requests that the input and type-ahead buffers be cleared.
Write	Requests the output of data to the terminal.
Write complete	Carrier write completion status.
Discard state	Carries a change to the output discard state due to a terminal operator request (via an entered output discard character).
Read characteristics	Requests terminal characteristics.
Characteristics	Carries terminal characteristics.
Check input	Requests input count (number of characters in the type-ahead and input buffers combined).
Input count	Carries input count as requested with Check Input.
Input state	Indicates a change from zero to nonzero and vice versa in the number of characters in the input and type-ahead buffers combined.

Other Layers

Figure 16-26 shows the highest layer of the DNA model. This is the user layer. This layer consists of the users or user applications that were written or are being used in a networked environment. These could be specifically written network management, databases, or programs that are not part of the DNA model but use DNA for part of their operation.

Finally, Fig. 16-27 shows the network management layer used by DNA. It will not be explained here, but the figure shows the management functions of DNA.

Figure 16-25
Protocol message
exchange. (*Copyright
© 1982 Digital Equip-
ment Corporation*)

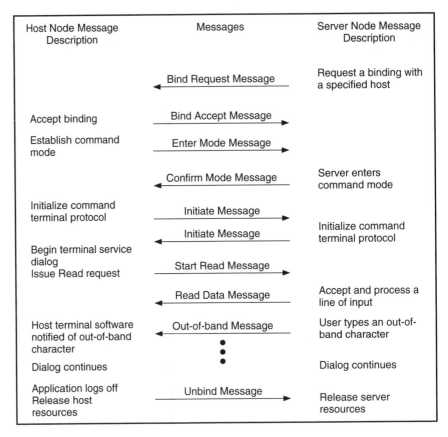

Host Node Message Description	Messages	Server Node Message Description
	← Bind Request Message	Request a binding with a specified host
Accept binding	Bind Accept Message →	
Establish command mode	Enter Mode Message →	
	← Confirm Mode Message	Server enters command mode
Initialize command terminal protocol	Initiate Message →	
	← Initiate Message	Initialize command terminal protocol
Begin terminal service dialog Issue Read request	Start Read Message →	
	← Read Data Message	Accept and process a line of input
Host terminal software notified of out-of-band character	← Out-of-band Message	User types an out-of-band character
Dialog continues	• • •	Dialog continues
Application logs off Release host resources	Unbind Message →	Release server resources

Digital Network Architecture

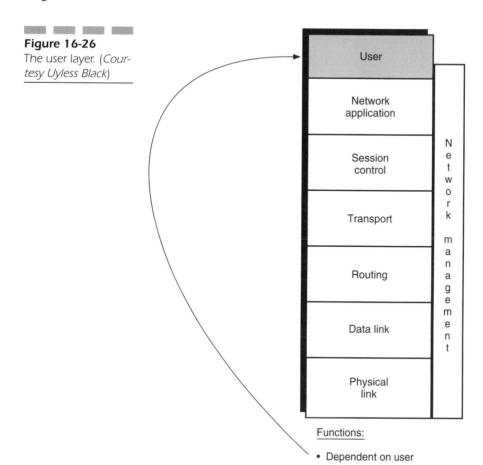

Figure 16-26
The user layer. (*Courtesy Uyless Black*)

Functions:

- Dependent on user

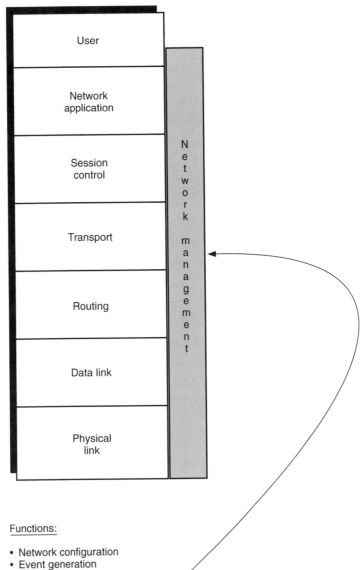

Figure 16-27
Network management. (*Courtesy Uyless Black*)

Functions:

- Network configuration
- Event generation
- Diagnostic operations
- Setting operational parameters
- Logging functions

DECnet implementations:

- DECmcc Director Product Family
- CMIP
- DECmcc TCP/IP SNMP Access
- DECmcc Extended LAN Manager
- TokenVIEW Plus

Local Area Transport

Introduction

The Local Area Transport protocol (LAT) is a Digital Equipment Corporation proprietary protocol whose primary function is to exchange data between a terminal server service (whether this service is a terminal emulation on a personal computer or a communications server) and its respective application host. These functions have been expanded to include direct host access and direct application access. It is primarily a session-layer protocol only and does not have a routing (network) layer involved. The layering of this protocol is shown in Figs. 17-1 and 17-2. LAT is sublayered as follows:

1. Service class sublayer
2. Slot sublayer
3. Virtual circuit sublayer

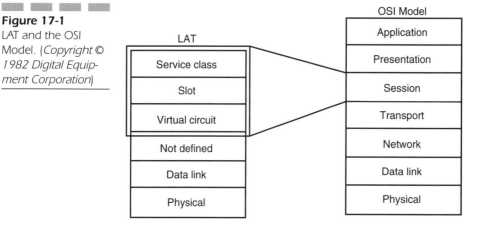

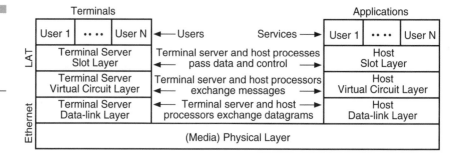

Figure 17-2 defines each of these sublayers. In these figures you should notice that LAT is defined at the OSI session layer. It does not provide network-layer services and therefore cannot be routed. It also is allowed to run on top of any data link that IEEE provides. Mainly, it is used on Ethernet/IEEE 802.3 LANs, but it is being migrated to IEEE 802.5 (Token Ring) LANs. The terms that are used in these figures will be discussed in detail after the following explanation on LAT node types. The LAT protocol may only be bridged to another network. It cannot be routed. LAT networks can be only extended by bridges.

Node Types on LAT LANs

There are two types of node on a LAT LAN:

1. *Terminal server:* An asynchronous terminal device that has multiple asynchronous connections on one end and a network connection on the other end. Users can access LAT services and initiate sessions on these devices.

2. *Load host:* A network station that can download the LAT software to a terminal server. This device is not always needed. Some terminal servers have a local boot device (a local floppy) that can load the LAT software into the terminal server.

The most common environment for the LAT protocol involves a device known as a terminal server. Another name for this device is a communication server. In this text, however, it will be called a *comm server.*

Comm servers allow any devices that primarily support the asynchronous communications protocol to attach to a network. Refer to Fig. 17-3, which shows the anatomy of a comm server. This device has multiple asynchronous connections on one end and one network connection. It represents one of the older data communications devices in the LAN arena. It allows asynchronous devices (such as modems, printers, terminals, and even some host ports) to connect to a LAN. It is an asynchronous multiplexor.

Comm servers today are still attaching to hosts, terminals, printers, and other serial devices, but with most terminals being replaced by personal computers, and mini- and mainframe computers having built-in network controllers, comm servers are providing communications connectivity in a different manner. They are being used in three basic areas:

Figure 17-3

Picture of a comm
server.

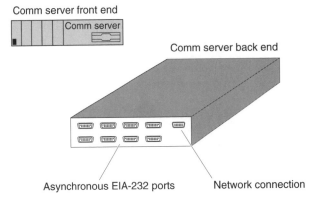

Comm server front end

Comm server

Comm server back end

Asynchronous EIA-232 ports   Network connection

predominantly in modem and printer pools, but also in the terminal
interface.

Figure 17-4 shows the aforementioned devices connected to a comm
server. At the bottom of the figure is a modem connection. Modem
pools are groups of modems clustered together for use by anyone who
has access to the comm server. By connecting modems directly to a
comm server, any user with access to the network can access one of the
modems (provided that no user has already established a connection to
that modem port on the comm server). Instead of a company buying
one modem for each user on the network, it can buy a few modems,
connect them to a comm server, and then have the users access the
modems through the network. This is extremely efficient, because few
modems are used 100 percent of the time. With connection to the
comm server, modems are now accessed by multiple users, thereby
increasing their use. Even if the modem port on the comm server is
busy, most networking companies have directed their comm server–to-
queue requests to a particular comm server port.

Comm servers can provide remote printing capabilities. For example,
one comm server may be connected to the host computer. Another
comm server may be placed somewhere on the network, with a printer
attached to it. A virtual circuit (a session) is established between the two
servers. Print jobs submitted on the host are printed to the comm server,
and that server sends the data over the network to the comm server with
the printer attached.

Devices are connected to comm servers with some type of cable, usually an EIA-232 cable. Comm servers are connected to the network and
provide a virtual connection (a software connection) over the LAN
between two devices. Some hosts with an internal network controller

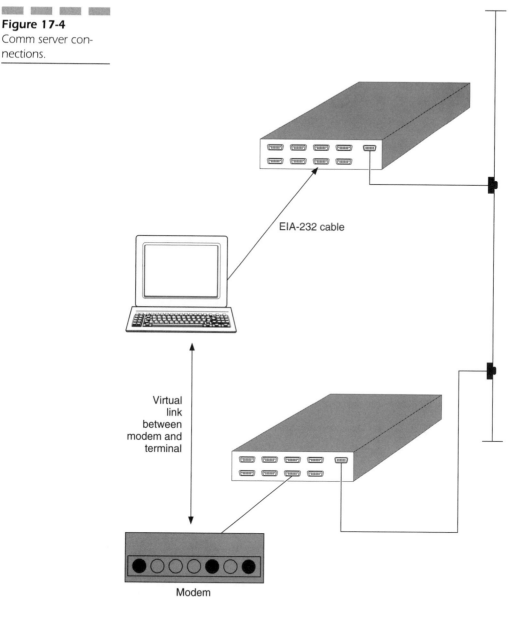

Figure 17-4
Comm server con-
nections.

EIA-232 cable

Virtual
link
between
modem and
terminal

Modem

(direct attachment to the LAN) are capable of sending data over a network without a comm server attached to it. A comm server can connect
to another comm server or to a directly attached network device.

There are two primary ways to create sessions using comm servers.
One is a user-initiated command, and the other is an application-
initiated command. With the user-initiated command (connected to a

comm sever), the user attaches a device to the comm server. Once the comm server is operational with its internal LAN operating software (in this case, LAT), the user enters a connect command to the comm server to indicate that a connection is wanted. This connect command is usually sent with a name in the form "connect VAX1," where VAX1 is the service name to which the user would like a connection (see Fig. 17-5). This connection is for a service between two comm servers. The virtual link is a software connection that has been established between the comm server and the host computer. The comm server can initiate multiple virtual connections with only one connection on the LAN segment. Once a connection is made by the two comm servers, users will never know that they are operating on a comm server. The connection emulates a direct connection between the terminal and the host.

For a host to connect to a service on a LAN, the setup process is similar, except the host makes the connection with user intervention.

Figure 17-5
Comm server links.

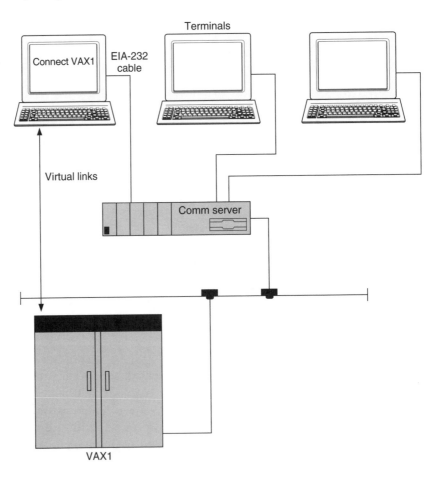

Comm servers can also connect directly to a service that is not another comm server. In this case, the remote connection has a direct connection to the LAN. The comm server can still access this remote service, just as if it were a comm server. The only requirement for all connections is that the LAN software that encapsulates the data to be transferred into a LAN packet must be the same on both sides of the connection.

The text of this chapter is based on the LAT protocol by Digital Equipment Corporation. LAT is an extremely fast and efficient protocol that allows data to be exchanged between two network stations. This protocol is a proprietary protocol written by DEC and was primarily intended to operate their comm servers and give them connections to services over the LAN. It is meant to operate on a highly reliable, high-bandwidth LAN. Because the protocol is proprietary, only its functions will be discussed in this text. Packet formats and protocol specifics will not be discussed.

LAT Topology

LAT network stations can communicate on a local LAN, or they may be separated by a bridge (Digital documentation refers to this as an *extended network*). Because LAT contains no network-layer protocols, it will not operate through a router unless that router also supports transparent bridging protocols. This type of router is known as a *multiprotocol router*. See Chap. 3 for more details on this type of router.

The LAT protocol assumes that once a connection is established between two network stations, one end or the other (not both) will maintain this connection. Furthermore, the LAT protocol assumes that the bandwidth of the medium (the LAN) is larger than the bandwidth needed for one LAT session. This is necessary because the LAT protocol can run over any access method, including serial lines.

This presents a serious problem of the LAT protocol. It is an extremely fast protocol, and a session may time out while waiting for a response from a remote session. This will not happen when the protocol is operating over a high-speed LAN such as Ethernet, but when the Ethernet LAN is extended through a device such as a bridge, and the remote bridge connection is a slow-speed serial line, the session may time out. Speeds needed to run LAT over a serial line are usually T1 and above. Speeds at 56 kbps and below will introduce delays long enough to cause time-outs to occur between two stations separated by a remote bridge. An easy algorithm to remember is to allocate 1 kB for each user across the WAN link. This means that a 56-kbps link can support 56 LAT users. Any more users across the link will cause delays.

LAT operates at the session layer of the OSI model. Three sublayers are interpreted at the session layer and are used for the LAT protocol. The other two OSI layers over which LAT can operate are the data-link layer and the physical layer. These last two layers transport data over the data link and were fully discussed in Chap. 2.

LAT uses the Ethernet frame format, and the Ethernet Type field for LAT is 6004. LAT was intended to operate over Ethernet and was built to enable terminals to be clustered on a comm server. The comm server has one connection to the Ethernet, and the other portion of the LAT protocol operates in the VAX hosts that are directly connected to the Ethernet LAN. The exception to this is that sessions can be established between two comm servers. For that matter, the LAT protocol can operate on any network station.

Because DEC holds the protocol proprietary, if any other company wants to copy it for their machines, they need the licensing from DEC. Because DEC supports the TCP/IP protocol (from Digital or from the company Wollongong), most companies are not going to do this. The TCP/IP protocol allows a more robust connection to devices on a LAN, and a lot of companies have reverted to it. This opens connections to VAX computers from many different types of computing devices. TCP/IP was designed with this in mind. Refer to Chaps. 7 through 12 for more information.

In order to operate LAT, there are entities that must exist on the LAN. These entities are

- LAT nodes, which are network stations containing the LAT software
- Network interfaces, which are the controller cards
- Token Ring (not yet standardized by DEC) or Ethernet LAN, which includes extending devices such as bridges or repeaters

The LAT protocol is defined at the session layer and contains the functions that follow.

Service Class Layer

The service class layer is the highest layer of the LAT architecture. It provides the offering and requesting of LAT network resources on the LAN, service announcements, directory functions, and groups (all will be explained in a moment). This layer uses a single module of protocol, known as Service Class 1, for interactive terminals. This type of service class is used by all LAT network stations for all service class functions. Service Class 1 functions include the following:

- Offers services
- Assigning service ratings
- Providing group codes to access services
- Sending service announcements and building service and service node directories using information in the announcements
- Maintaining a multicast time on service nodes that determines when service announcement messages are multicast
- Indicating flow control using the X-ON and X-OFF software flow control settings (for modem ports, the hardware signals of DTR/DSR and RTS/CTS can be used)

At the session layer, flow control is maintained by software X-ON and X-OFF parameters. As a buffer fills to near capacity, it tells the other side of the session to stop sending data by sending the X-OFF character. When it can begin to accept data again, it transmits an X-ON character. This can be initiated by the user typing CONTROL-S and CONTROL-Q on the keyboard, indicating to the service class layer that flow control should be enacted. This makes the slot layer control the flow of data by using slot flow control. Figure 17-6 depicts the flow control mechanism.

Figure 17-6

Flow control. (*Copyright © 1982 Digital Equipment Corporation*)

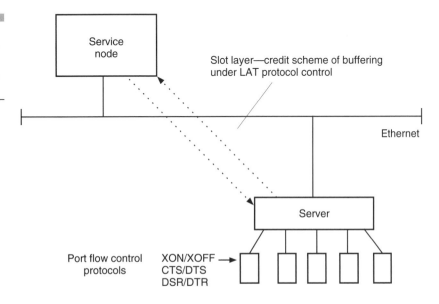

Basically, there are two types of flow control. One is user-initiated; the other, called slot flow control, is inherent to the LAT protocol. It is used more for congestion control and will be explained later.

Slot Layer

The slot layer establishes and maintains a session between two or more network stations. Its functions are as follows:

- Establishing and terminating sessions
- Providing for internal flow control for the session
- Providing for two full-duplex flow-controlled data channels and one non–flow-controlled data channel for each session
- Ending sessions when requested to do so by the service class layer

Following are definitions of the slots in the slot layer:

Start slot: Produced when a session is requested by a terminal server. The corresponding slot layer at the service node either accepts or rejects this start slot by responding with either a start slot (session accepted) or a reject slot (session rejected).

Reject slot: Transmitted by a service node to reject an incoming connection request.

Data slots: Used after a session has been accepted. While the session is active between the two network stations, data are transmitted using the data slot. There are three types of data slot:

The data-A slot—carries user data.

The data-B slot—carries port and session characteristics. It is also used to carry break, parity, and framing error indicators.

Attention slot—Indicates a special condition. It is not used in the normal data stream and is called the out-of-band slot. It can indicate the termination of output to a terminal.

Stop slot: Used to end a session. This slot is transmitted when a user disconnects from a service.

Flow control slot: Used to control the flow of data during the session. This ensures that data will not be lost during the session by making sure that each end of the session has space available to accept data. This will be covered in more detail later.

Virtual Circuit Layer

This layer transports data. It acts as either a master or a slave for any virtual circuit. A virtual circuit is a connection between two devices that are not cabled together. Network software enables these connections.

Following are the functions of the virtual circuit layer

- Initiating a virtual circuit between two network stations
- Maintaining a timer to determine when data is to be transferred between two network stations
- Transmitting the slots and credits for the flow control functions of the slot layer
- Controlling the multiplexing of data when a user has multiple sessions established
- Managing connections through the sequencing of data, determining when there is a transmission error, and controlling duplicated data messages
- Providing a timer for when data is not sent for a period of time (a keep-alive timer to ensure that the virtual circuit is still good even when there are no data to send)
- Terminating a virtual circuit when the session closes

This is the bottom of the LAT sublayers. The primary function of this layer is to establish and maintain a virtual circuit. It does not control a session—it controls a virtual circuit. Refer to Fig. 17-7. A session can be established between two stations. The LAT protocol allows not only one but multiple sessions to be established at the same time from or to the same network station. Up to 255 virtual circuits can be established between network stations.

These virtual circuits are different for a comm server than they are for a host. A session can be established between two comm servers or between a comm server and a host. (There is LAT software for PCs, but for the purposes of this chapter, we will use examples of comm servers and hosts.) For a session accepted by a host running the LAT software, up to 255 virtual circuits can be connected to it. For a comm server, one terminal port can have up to 255 virtual circuits.

The virtual circuit layer provides an interface for the slot layer for the transmission and reception of data or session control information. It is this layer that is responsible for transmitting messages between two net-

Figure 17-7
LAT service connec-
tions.

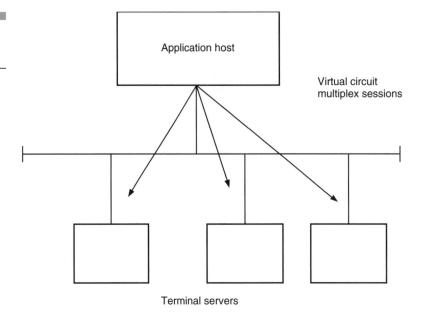

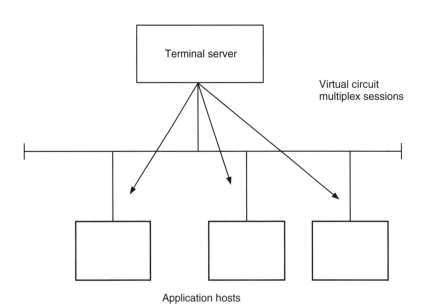

work stations and for providing the delivery of error-free data and the sequencing of these data.

Like all the other layers, the virtual circuit (VC) layer has flow control. With flow control, a network station cannot transmit data until the previous message sent has been acknowledged. If a message sent has not been acknowledged, the VC layer retransmits the message until it has been acknowledged or until the retransmit limit has been reached. Each of these protocols will be discussed in detail later in the chapter.

LAT Components

The components that make up the LAT network are *master nodes*, network stations that access services of a resource, and *Slave nodes*, network stations that provide services. A LAT network can consist of virtually any number of master and slave nodes, but there must be at least one of each. Furthermore, one network station or comm server can be both the master and slave.

LAT communication is the process of exchanging data between the master and the slave. This communication can be either data (user data) or commands to maintain the session. This process is truly peer-to-peer, which means that each network station that supports LAT can be either the master or the slave or both, and any network station can set up a session with another. There is a master and slave relationship, but any station can connect with any other. There is no specification that only certain network stations will be masters and others will be slaves.

Services

A service is a resource that has made itself available to a LAN (a host, a modem, a non-LAT host, etc.). Any network station that offers a LAT service is called a *service node*. These resources can include application programs, modems, printers, or any combination of hardware and software.

Each service resource is identified by a unique name. In order to establish data communication between two network stations using LAT, one end must connect to the service by using its name. A unique function of LAT is its capability of providing load balancing between service names. In other words, a service name can be used to identify multiple service nodes providing the same service. This feature includes an automatic failover mechanism in case one service node fails. In addition to this, one service node can provide multiple services.

Any network station that offers a LAT service broadcasts this information to the LAN by using a service-announcement message. Contained in these messages are the available services and identification information, such as the node name that has that service. All LAT stations on the LAN receive this information and, if their group numbers match, they build a table of available services.

Before transmitting a service-announcement message, a service node must calculate the service rating. The calculation of this rating includes the amount of activity on the node, the amount of available memory, and the processor type. This may be a static entry that is entered by the network administrator. For a terminal server, the rating also includes the number of available asynchronous ports and the amount of space available in the connection queue. A service rating is a number from 0 to 255 that describes the availability of the service. Each service in a service node has a rating. After calculating the service rating, the service node multicasts the service-announcement message.

The process ends with the terminal server building service and service node directories.

The Service-Announcement Process

With the information from the service-announcement message, the terminal server builds two directories in memory: a service directory and a service node directory.

The service directory contains the following information:

Service name

Service identification string (if available)

Node names of service nodes that offer each service

Current status of the service

Service rating of the service on each service node

The service node directory contains the following information:

Node name

Node identification

Node address

Node status

These directories are used in the process of connection attempts during the session establishment process. Service nodes do not maintain service directories, which is a reason that connection to LAT services is possible only for terminals connected to terminal servers.

In a LAT network, the object is for users to gain access to services that are available on the network. There are instances when security is needed, and not all users need access to all services on the network, so the group code was implemented to allow terminal servers to select which services on the LAT network they will make available to users.

Service announcements begin as soon as the LAT software is started. There is a timer that provides the interval at which these messages should be multicast. If a new service is created, this service is usually broadcast immediately; otherwise, this timer can be manually set.

Groups

When a LAT entity such as a terminal server receives a service-announcement message, it compares the service node group number with its known port group numbers. If there is a match, the terminal service processes the message. If there is not a match, the message is ignored.

Group numbers range from 0 to 255. They are assigned by the network administrator and are used for security purposes. All incoming messages to the LAT server are compared with a group number. They are assigned to a service, a terminal server, and even to an individual port on a terminal server. Each service node and the services it offers can have one or more group codes. These group codes are included in the service-announcement messages. During the process of reading these service-announcement messages, a terminal server decides whether it should process the data in the announcement or not.

When a terminal server reads the message, it first does a compare on the group code. If it finds a match, it puts the information from the message in the appropriate terminal server directories. If there are no matches with any of the group codes, the announcement message is discarded. The terminal server can conserve memory by discarding unneeded messages. In Fig. 17-8, terminal A can access the service of the modem and the service node (the host). It cannot access the service of the microVAX. Terminal B can access the service of the MicroVAX and the service node, but not the modem.

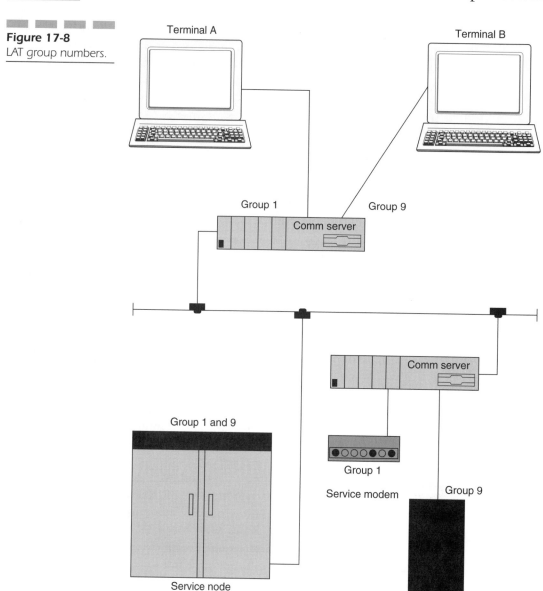

Figure 17-8
LAT group numbers.

Session Establishment

A session is an established communications path between two network stations. This path can be between any of the previously mentioned resources.

LAT offers two functions: the ability to establish a connection and the ability to offer a connection (or service). LAT services are usually held in host machines, but terminal servers can also offer services to the network, as shown in Fig. 17-9.

There are two types of connections: user-initiated and host-initiated. A user-initiated connection request is from a user trying to establish a connection with a service located somewhere on the LAN.

Host-initiated sessions are when application programs on the host initiate a connection request to a service on the LAN. This could be an application program establishing a LAT session with a terminal server on the LAN that has a printing service associated with it.

When a user at a terminal server issues a connect request, the terminal server must proceed through a series of steps to establish the connection. The steps are detailed subsequently.

If the connection request specifies only a service name, the server must select a service node before a session is attempted. This is a process of translating the service name into a node name. The server searches its service directory for the service name. If none is present, the connection attempt is aborted. If there is a service name present, the server selects a service node (more than one service node may provide the same service name) and checks to ensure that a group number is shared with it and with its address.

Once this procedure is accomplished, the terminal server verifies that a virtual circuit is available by sending a virtual circuit start message to the selected service node. Generally, this is accepted and the accepted node responds with a virtual circuit start message. A service node may reject the request if the service node has no available virtual circuits, if

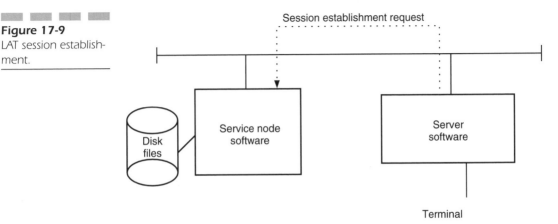

Figure 17-9
LAT session establishment.

there are insufficient resources to handle the request, or if an invalid message is received by the service node. Figure 17-10 shows the flowchart of this process.

After all this, a session is finally established. Once established, it must be maintained. The process of maintaining a session allows for the passing of data between the two stations, providing flow control, and disconnecting a session.

After a session is established, a virtual circuit enters into the run state, and data are finally transferred over the connection. Data transfer is very carefully controlled over the circuit.

Data Transfer

Data are transferred through a system of sublayer VC messages. Before data are transferred, a node must have a credit for every slot. A credit is a marker that allows the node that possesses it to transmit one slot of information to another node. A slot is a segment of data up to 255 bytes long. There is one credit per slot. The slot layer is the layer that establishes and maintains sessions for the service class layer. In turn, the slot layer uses the service layer to transmit and receive session information over virtual circuits. Information for a session is stored and transmitted using one or more slots. A slot is a message segment. This message segment contains information for a single session. The four slot types are start, data, reject, and stop.

The functions of the slot layer include establishing sessions between server ports and service nodes, providing two full-duplex flow-controlled data channels and a non–flow-controlled data channel for each and every session, and terminating sessions when requested by the service class layer.

Virtual circuit messages are of three types. They are named for the functions that they perform. Data transfer is shown in Fig. 17-11.

1. *Start messages:* These messages are exchanged between a server and a service node to start a virtual circuit. The service node responds to this request with a start message (sent from its slot layer), or it may send a stop message to reject the connection. Inside the start message is information to start the circuit. This includes the data-link frame size (which, in turn, will determine the size of the virtual circuit messages), the size of the slots, the size of the transmit and receive buffers, and the credits available for the session.

Figure 17-10
Service updates.
(Copyright © 1982
Digital Equipment
Corporation)

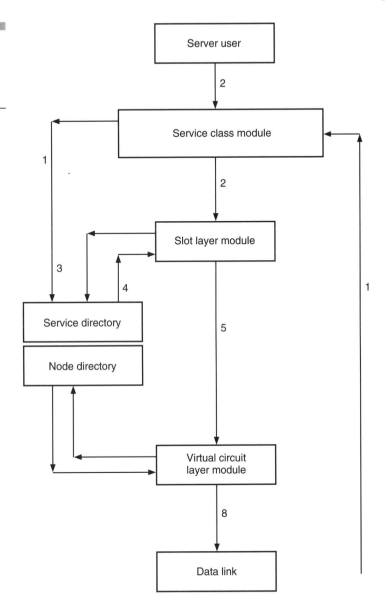

1. Service node multicast received—directory updated
2. User requests service connection
3. Slot layer looks up service name
4. Node offering is found
5. Slot layer requests virtual circuit layer to set up virtual circuit
6. Virtual circuit layer looks up node
7. Node address is found
8. Virtual circuit layer transmits the start message

Figure 17-11
LAT Communication
processes. (*Copyright
© 1982 Digital Equip-
ment Corporation*)

The slot layer establishes the first slot on a circuit.

The service node accepts the circuit request.

The circuit timer expires and the slot layer data (all that can
be contained in one RUN message) is transmitted.

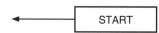

The service node acknowledges the receipt of the data and
optionally sends data to the terminal server.

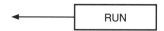

The keep-alive timer expires (due to lack of other data)
and the terminal server sends a keep-alive message.

The service node acknowledges receipt of the message.

The slot layer terminates the last slot on the virtual circuit.

2. *Run message:* A run message contains session information, which
   can include start session requests, credits for slot-layer flow control, user data, port status, or a stop session request. The virtual
   circuit layer at each network station exchanges run messages with
   each other until one end or the other sends a stop message. These
   messages are usually acknowledged by the other end. Slots are
   given to the virtual circuit layer by the slot layer to be transmitted in a virtual circuit run message. All sessions on the same virtual circuit share the same run message. The different types of
   message slots can be intermixed.

3. *Stop message.* This message is sent to indicate an end to a virtual
   circuit. Either end can initiate this message, but usually a server

sends it after the last session on a virtual circuit is disconnected. These are unacknowledged messages.

Upon reception of a frame, the VC layer hands the data to the slot sublayer. If the service node has anything to send, it builds and transmits the data.

Session Flow Control

Control of a circuit is simplex. A node may not transmit a run message until the previously transmitted run message has been acknowledged. The server continually transmits the message until it is acknowledged or until the retransmit timer expires.

Slot Flow Control

Unique to the LAT protocol is the idea of congestion control through the use of credits. Credits allow one end or the other to indicate how much buffer space is available on the other end of the connection. The slot layer stores information it receives in receive buffers. When a session is active, the receive buffers are filled by the user data and are routinely transmitted to the service node. These buffers may become full and not able to be emptied (for example, when the session is suspended by the user, i.e., the user switches sessions). In this case, the slot layer will not accept any more data until it can empty its buffers.

Flow of data is based on a credit exchange system between the two entities in a session. During session establishment, the two entities in the session negotiate the number of receive buffers on each end (usually, there are two buffers on each end). For each receive buffer, each end provides the other side of the session with a credit. A credit is simply a marker that allows the node that possesses it to transmit up to one slot (up to 255 bytes) of user data to the other end of the session. When the data are transmitted, the credit is transmitted with them. Therefore, the credit represents a free receive buffer on the other end of the session. Transmitting a credit ensures that there is room at the other end of the session for the data to be accepted.

Both ends of a session maintain a count of credits. When a partner's credit count is greater than 0, the partner can send data. When it runs out of credits, its credit will be 0, and it will not send any more data until the other end of the session sends more credits to it. When one end of the session empties its receive buffer, it returns a credit back to the other end of the session. This operation is shown in Fig. 17-12.

To make the protocol more efficient, data are not transferred on a keystroke-by-keystroke basis. They are transmitted on a timer basis. There is a circuit timer built into LAT that, when expired, transmits the data in its transmit buffers. There are usually two transmit buffers for

Figure 17-12
Data flow. (*Copyright © 1982 Digital Equipment Corporation*)

Terminal server   Service node

The terminal server user requests a service connection. This message includes initial flow control credit(s).

START

The service node accepts the connection request. The service node also includes initial flow control credit(s).

START

The terminal server user enters one or more characters which are stored in a data slot. If flow control credits are available, the slot is transmitted in a virtual circuit RUN message when the circuit timer expires. The terminal server decrements its flow control credit count for the session.

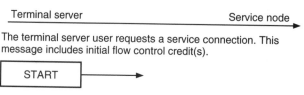
DATA

The service node processes the characters and possibly sends a response, usually with new flow control credit(s). The service node decrements its flow control credit count for the session. The terminal server increments its flow control count for the session if the service node sent additional credits.

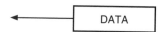
DATA

The service node sends some charactors. This step may not be possible if the service node's flow control credit count for the session is zero.

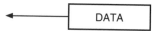
DATA

The terminal server processes the characters and optionally sends response data from the user as well as possible flow control credit(s). The server decrements its flow control credit count for the session. The service node increments its flow control if the terminal server sent additional credits.

DATA

The service node requests disconnection due to the user logging out of the service.

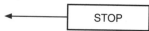
STOP

each established session. The terminal server sends data only when this timer expires.

If a terminal server has no data to send, it enters a balanced state. A circuit that has entered this state is known as *inactive*. It is the job of a terminal server to keep up with the state of the service node. With this state, a keep-alive timer is invoked. Each time the terminal server sends out a series of VC messages, it resets the keep-alive timer. There is a timer for each session.

If the timer expires, the terminal server transmits an empty message to the service node and expects a response. The service node responds with a data message or an empty message just to let the terminal server know the other side of the session is still active. If there is no response to the empty message, the terminal server retransmits it until it reaches its retransmit limit. If it reaches its retransmit limit, the terminal may respond with a circuit-down event process, in which the terminal server considers the service node to be unreachable and disconnects the session.

LAT can also perform a function known as *automatic failover*. This process is used with VAXclusters. VAXclusters are the physical grouping together of multiple VAX minicomputers on a LAN. By assigning a common service name to two or more cluster members, a user can reestablish communication sessions with the mini in the event that a previous session was aborted. With this, the user must log in again.

Session Termination

Finally, there are multiple ways to disconnect a session. First, the user may log out of the service node. The service node then exchanges messages with the terminal server to disconnect the session with the user's terminal server. Another way to disconnect a session is to type the disconnect command at the terminal server, which disconnects the user from the server.

Other LAT Services

Host-Initiated Requests

A host-initiated-request occurs when an application program on a general-purpose service node requests an application device. The service node transmits a multicast request for the address of the server offering the applications device and then attempts a connection to that server. Servers participate in this type of request by supplying their addresses and by processing host-initiated requests received from service nodes. This is really the reversal of the normal LAT communication, in which a server requests a service and attempts the connection to the service node.

There are times when it is necessary for a host to initiate a session. For example, when a host would like to send data to a terminal server that has a printer connected to it, the host would establish a session to the terminal server port with the printer attached, send the data, and then terminate the connection. This is all accomplished without any user intervention. This is shown in Fig. 17-13. This process involves three steps:

1. The host must translate the service name or the terminal server's name to its 48-bit physical address. A terminal server receives the message and, if the requested name matches, responds to the host. The terminal server's Ethernet address will be included in the response.

2. After the terminal server's Ethernet address is known, the service node (host) sends a message to that server. The message contains a request for a particular terminal server port, or, if the terminal server supports services, the host may request a service. During the process, group codes are checked. If there is not at least one match, the terminal server rejects the request.

3. The terminal server attempts to allow a session to be established. The terminal server either accepts the host request if a virtual circuit is available, queues the request if the port or the service is busy, or rejects the request.

If a terminal server must queue a request, it places it in a table that may have other connection attempts in it. Queues are maintained by the FIFO method. This means the first entry in is the first one back out (the first connection attempt that was queued will be the first in the queue to get a connection).

Session Management

Both servers and service nodes buffer user data, provide flow control between the server and its service node, and disconnect sessions. When multiple nodes offer a common service, servers can provide automatic failover for sessions to a service if a session is interrupted.

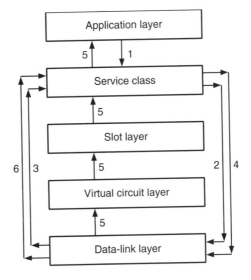

Figure 17-13
Host-initiated request.
(Copyright © 1982
Digital Equipment
Corporation)

1. An application process on a service node requests a connection to a port or service on a specific server.
2. The Service Class 1 module of the service node multicasts a solicit-information message to get the address of the server.
3. The response-information message received from the server goes directly from the service node's data-link layer to its Service Class 1 module.
4. The Service Class 1 module of the service node sends a command message directly through the data-link layer using the server's Ethernet address. This command message contains the host-initiated request.
5. If the requested port of service is available, the server responds with a virtual-circuit start or run message.
6. If the port of service is unavailable, the server responds with an error status message rejecting the request. An error status message goes directly from the data-link layer of the service node to its Service Class 1 module. If queuing were permitted by the service node, the server might queue a request rather than reject it.

Virtual Circuit Maintenance

Virtual circuit maintenance is performed through the use of keep-alive messages. A keep-alive message is transmitted over a session between a server and its service node when the circuit has been idle for a while (i.e., there is no user activity).

Connection Queue Maintenance

A server maintains its connection queue by transmitting status messages for each queued request directly to the node that requested the connection. Requests waiting in the queue for a connection can be removed by the manager of that server or by the individual that sent the request. Otherwise, when a service becomes available, the connection request is removed from the queue, and it begins the session establishment process, which is shown in Figs. 17-14 and 17-15.

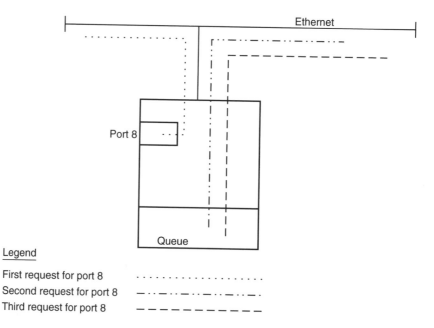

Figure 17-14

Queues. (*Copyright © 1982 Digital Equipment Corporation*)

Legend

First request for port 8	. . . .
Second request for port 8	— . . — . . — . . — . . — .
Third request for port 8	— — — — — — — — — —

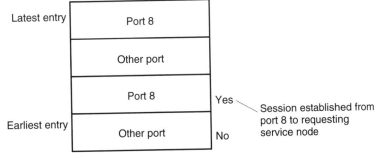

1. Queued requests when service becomes available

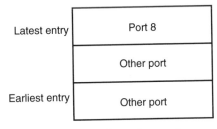

2. Queued requests after next connection made

Point-to-Point Protocol

Introduction

The Point-to-Point Protocol (PPP) provides a standard method of transporting multiprotocol datagrams across point-to-point links. In early implementations of data communication, communication between remote sites was accomplished using specially conditioned lines available from the phone company (AT&T, before the breakup). These lines were different from telephone lines in that they remained "up" full-time (because they had to carry digital data with a very low error rate). A lot of the lines that were used contained at least some segments of copper between the sites, and this allows noise (unwanted electrical signals) to permeate the cable and cause errors. We, as humans, can tolerate this, but computer communications cannot.

A leased line works similarly to having your phone off the hook full-time, with a continuous connection to the remote party. The lines were ordered from the phone company, which was informed that they were to be used for data. The telephone company patched a line together between the two sites that was "switched" permanently. They took special consideration to ensure that the lines were "clean" enough to be used with data.

Clean means that there is a very low bit error rate on the line, which is necessary if it is to be used for data rather than voice. Voice lines are not conditioned, and, in fact, the circuit is probably different every time you use the phone, even when dialing the same remote number.

Once a leased line was operational, this meant that basically the physical layer was operational. The data-link layer on the device attached to the phone line had to be worked out. It does not matter to a phone line what is transported over it—the main function of the device that attached to it was to make sure it could synchronize with the signals of the phone company. To the phone line, everything transmitted over it simply looks like a bunch of 0s and 1s, with special bits in front and back of this information called framing bits. These were used to interact with the phone company's equipment. The information that was transported over the phone lines was encapsulated with headers (just as with Ethernet or Token Ring). The information in the header was used by the remote end. In most cases the header information was very simple. It enabled the remote end to receive the data and give them to a special device handler for processing. However, each vendor added some tweaks to these headers, and the processing was just different enough between their products to cause interoperability problems, especially in the case of different network protocols (TCP, IPX, etc.).

In the early days of data communication this was not a problem, because both sides of the leased line used the same vendor's equipment. If one installed Cisco, then it was Cisco on both ends of the line; if it was 3Com, then it was 3Com on both ends. The reason for this is that with all the standards for local area networks, there was still no wide area network standard for communication between two nodes on a single link. As you know, there must be some data-link involvement in the transfer of data between two nodes. The data-link layer is responsible for addressing and framing (encapsulating data handed to it from the network layer) data so that they can be transferred as raw data by the physical layer. With no standard, no one vendor's equipment could talk to another's, and even if they could, each vendor had the right to change its product at any time, causing immediate pain to any other vendor trying to interoperate with them.

There is an architecture known as High-Level Data Link Control, but this is simply an architecture standard. Most vendors simply took that as the building block for their WAN point-to-point protocol encapsulation. There were many variations of this standard that included options for error handling, different protocol handling, frame sequencing, multiple line handling, etc. There are also multiple versions of HDLC that allowed for connection or connectionless delivery, and even one standard for connectionless delivery with acknowledgment. Each vendor implemented the HDLC architecture just a little differently for its own needs. (The most notable implementation of HDLC is the IBM SDLC protocol.) These different implementations resulted in incompatibility between vendors. Both ends of the link had to be equipped by the same vendor. The only other alternative was the X.25 standard (from the International Telecommunications Union [ITU], formerly known as the Consultative Committee for International Telephone and Telegraph [CCITT], which was the only public packet-switching standard at the time. Many customers opted for X.25 lines, and they worked just fine between most vendors. However, it did cause another party to get involved with the remote access of your equipment, and most X.25 lines were 19.2 kbps or slower.

The Point-to-Point Protocol (PPP) is based on HDLC as well, but it provides a standard method for transporting multiprotocol datagrams over point-to-point links. PPP comprises three main components:

1. A method for encapsulating multiprotocol datagrams

2. A link control protocol (LCP) for establishing, configuring, and testing the data-link connection

3. A family of network control protocols (NCPs) for establishing and configuring different network-layer protocols

Many vendors continued to support their own versions of HDLC until recently. Most did this for the increase in speed, but a few did it for the wrong reasons (single-vendor lock-in). PPP provides a standard method of transporting information over a serial line or over multiple serial lines (Multilink PPP, which we discuss near the end of the chapter).

There are many RFCs for PPP, which are listed throughout the chapter, but the two that I recommend reading are 1661 and 1990, which constitute the bulk of the protocol. There are other RFCs for the link control protocol and the network control protocol (one RFC for each protocol that will operate over the link, such as IPX, Bridge, XNS, AppleTalk, etc.). You should read these, as well, but do not try to memorize them like 1661 and 1990. The one other RFC that you should read is 1989, for link quality monitoring (LQM). These are all quick reading, and you really only need to read about the first 10 pages or so unless you are a software developer.

Details of PPP

The flowchart of the operation of PPP is shown in Fig. 18-1. It is a block diagram of how the protocol initializes and then maintains a link. As mentioned above, there are three functional components to PPP. The protocol originally starts in the DEAD state, from which the protocol uses LCP to initiate the link and enter into authentication (usually using two standard types of authentication, PAP and CHAP). After this is established the protocol enters into the NCP, which is used to control the protocol that will operate over PPP.

PPP Encapsulation

PPP encapsulation, shown in Fig. 18-2, is used to differentiate between the different protocols. Information handed to it is contained in the

Figure 18-1
PPP phases.

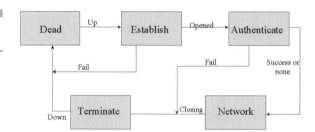

Figure 18-2
PPP encapsulation.

Protocol 8/16 bits	Information *	Padding *

Information field of the frame. The protocol field is either one or two bytes, and its purpose is to identify the type of data contained in the Information field. The Information field doesn't only contain data; it can contain control information, as well, that is used to establish and maintain a link.

The values used in the Protocol field are as follows:

0*xxx*–3*xxx*	Identify the network-layer protocol of specific packets
8*xxx*–b*xxx*	Identify packets belonging to the associated network control protocols (NCPs)
4*xxx*–7*xxx*	Used with protocols with low-volume traffic and have no associated NCP
c*xxx*–f*xxx*	Identify link-layer control protocols such as link control and authentication

This book does not list all the protocols—they are listed in the assigned numbers RFC (RFC 1700 as of this writing—check the ftp site listed in the RFC for the latest). The Protocol field simply identifies the data, which could be of control or data type.

PPP Operation

The operation of PPP between two points is very specific. A certain sequence of events must take place before the link is considered up and operational. It does not matter which vendor's equipment is being used—the events must take place exactly as stated, or the vendor is not in compliance.

The first phase of operation is the link dead phase. A link starts and ends in this phase. To "wake up" the link and enter the link establishment phase, an external event such as the detection of a carrier signal phone, the attached cable, or a network management event (such as dynamic configuration of the link) must take place. When this happens the link establishment phase is entered.

To start the link establishment phase, each end of a PPP link must send link control packets (LCPs) to each other. The purpose of this is to configure and test the link so that both sides agree on the configuration and protocols before bringing up the link. Both sides of the link exchange configuration packets.

Link Control Protocol

The link control protocol is indicated by a c021 in the Protocol field. This indicates that the packet contains link control protocol (LCP) information. The header that follows in the Information field is shown in Fig. 18-3. The Code field indicates one of the three types of LCP packet, which are described below. The Identifier field aids in matching requests with replies. The Length field indicates the length of the LCP packet, including the Code, Identifier, and Data fields.

There are three types of LCP packets: link configuration packets, link termination packets, and link maintenance packets. Link configuration packets are used to establish and configure a link using configure-request (code 1). A link that opens a connection transmits this type of LCP. Contained in the Information field are any options that the requestor would like to change from the set of defaults described in the RFC. This allows changes in the defaults, and the two sides must agree. The response to this packet can be one of the following:

- *Configure -ack (code 2):* Upon receipt of the configure-request, a response must be generated. This response is sent to the originator to indicate that the options that were sent are agreed to and can be supported by the remote node. Even if there were no options specified in the configure -ack, this packet can still be sent, indicating acceptance to the opening of a connection.

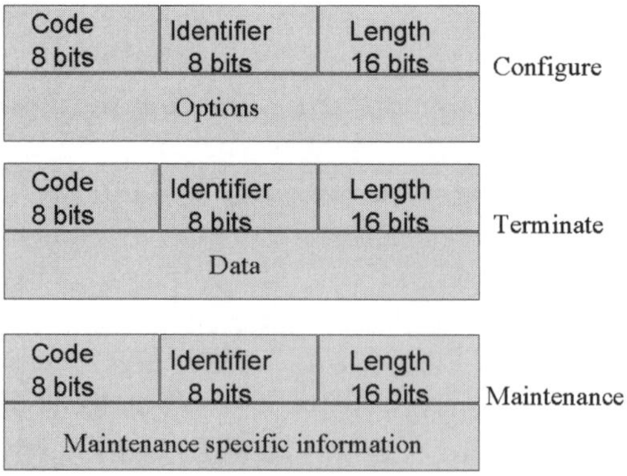

Figure 18-3
LCP packet description.

| Code 8 bits | Identifier 8 bits | Length 16 bits | Configure |
| Options | | | |

| Code 8 bits | Identifier 8 bits | Length 16 bits | Terminate |
| Data | | | |

| Code 8 bits | Identifier 8 bits | Length 16 bits | Maintenance |
| Maintenance specific information | | | |

- *Configure -nak (code 3):* This packet is sent in response to a configure-request to indicate that the options were recognizable, but they are not accepted. In the Options field of the packet are the options that are not accepted.

- *Configure-reject (code 4):* This packet is sent in response to those configure-request options that are not recognizable or are not acceptable for negotiation (which is like saying, "I don't want to talk about it."). This is different than the configure -nak in that the configure -nak is open for negotiation of an option, and this packet says it rejects it and it is not open for renegotiation.

The configuration options that are supported by RFC 1661 are as follows:

0   Reserved.

1   *Maximum-receive-unit:* Asks how large a frame will be supported. The default is 1500 bytes, but it can be configured by both sides if they agree on a larger or smaller size. It contains only the requested supporting length of the Information and Padding fields.

2   *Async-control-character map*

3   *Authentication protocol:* identifies the authentication protocol to be used Some implementations may request that a peer identify itself, and this field may ask the protocol that will be used for this security. It does not actually use the protocol here; it merely requests the type that will be used. Only one can be selected per link.

4   *Quality control:* Allows negotiation of the protocol used for link quality monitoring (LQM). By default, LQM is disabled. Sending this indicates that LQM is expected from the peer. Here, different protocols for LQM can be used in either direction.

5   *Magic number:* Allows detection of a loopback condition. This is a physical configuration in which the transmit leads are "looped back" to the receive leads so that a node can receive its own transmitted packets. Loopback conditions are used in modems for testing or to indicate that a line is down. In order to detect this, a node must select a number known as the magic number and place it in the Options field of the configure -request packet. The process for selecting these numbers is not specified, but common sources of numbers are the serial number of the node, the time of day, etc. You start with a "seed" number and "grow" the number with each packet sent. When a configure -request is received with this option, it compares it with the last configure -request sent with a magic number. If the two numbers are the same,

the node can assume that the link is looped back. It assumes, because it may not actually be looped back (if both sides picked the same magic number for instance).

6   Reserved

7   *Protocol field compression:*   Provides a method to negotiate the compression of the Protocol field. By default, the Protocol field is two bytes long, but it can be compressed into one byte. This can happen because a large quantity of the packets sent use Protocol field values of less than 256. This must be agreed upon by both sides—otherwise the fields will be misaligned for interpretation.

8   *Address and control field compression:*   Provides a method to negotiate the compression of the Address and Control fields. Because all packets between points on a point-to-point link have the same Address and Control fields, they are very easy to compress.

9   *FCS-alternatives*

10   *Self-describing-pad*

11   *Numbered-mode*

12   *Multilink-procedure*

13   *Callback*

14   *Connect-time*

15   *Compound-frames*

16   *Nominal-data-encapsulation*

17   *Multilink-MRRU*

18   *Multilink-short-sequence-number-header-format*

19   *Multilink-endpoint-discriminator*

20   *Proprietary*

21   *DCE-identifier*

Link termination packets terminate a PPP link using two codes:

- *Terminate-request (code 5):*   Used to signal to the other end that this node wishes to close the connection. This packet is continually sent until a terminate-ack is received or until it has been sent many times with no response.

- *Terminate-ack (code 6):*   Used to indicate to an originator of a terminate-request that the remote end agrees to the closing of the connection.

Link maintenance packets are used to manage and debug a link using the following codes:

- *Code-reject (code 7):* If an LCP packet is received that has a code that is not recognizable, it is sent to the other end of the link. It indicates a different version of LCP than the one that the receiving node is using.

- *Protocol-reject (code 8):* If an LCP packet is received, and it does not support the protocol indicated in the Protocol field, this packet is sent back. This packet type is usually sent when a new protocol is being attempted by the remote end to find out whether or not a particular protocol is supported.

- *Echo-request (code 9):* Sent to the remote end for many purposes such as debugging, link quality, etc.

- *Echo-reply (code 10):* Sent in response to an echo request packet.

- *Discard-request (code 11):* Used in maintenance of the link for debugging, performance testing, etc.

Once LCP has completed, the protocol considers itself to be in the open state, and the link quality monitor can be started at any time. In fact, this protocol can be run during authentication and network-layer protocol negotiation and during the passing of datagrams. The LQM measures the ability of a link to correctly transmit and receive datagrams and allow an implementation to switch to another line, if the protocol deems the link to be unsatisfactory.

The LQM measures data loss in units of packets, octets, and link-quality-reports. Each link quality monitoring implementation maintains counts of the number of packets and octets transmitted and successfully received and periodically transmits this information to its peer in a link-quality-report packet. These counters are always (at least they should be) ascending so that each time a report is received, the receiver can compare the number of packet, bytes, etc. sent with the number the remote end has received. This information is transmitted as a PPP packet of protocol type c025, which is reserved for link quality.

Link-quality-report packets provide a mechanism to determine the link quality, but it is up to each implementation to decide when the link is usable. The RFC does not state how to do this. It is assumed that it will be a *k* out of *n* decision, in that when *k* packets or bytes are considered bad or missing out of *n* packets transmitted, the line should be brought down or the session switched to another line. Using this approach allows the line to take "hits" but not be brought down every time a hit or a few hits occur.

Authentication

After LCP has brought a link up, through the link establishment phase, authentication is the next step. Authentication is, by default, turned off, but there are two standard methods of authentication used for PPP. One is the password authentication protocol, and the other is the challenge handshake authentication protocol (CHAP). For more information on the password authentication protocol (PAP), review RFC 1334. PAP is not as good as the CHAP authentication method, but it does provide simplex password checking. Passwords are sent in plain text and are therefore visible (no attempts are made to hide transmitted passwords), so there is no protection from playback or repeated trial-and-error attacks. The peer is in control of the frequency and timing of the attempts. This authentication method is most appropriately used where a plain text password must be available to simulate a login at a remote host. In such use, this method provides a similar level of security to the usual user login at the remote host.

When using a two-way handshake, PAP allows a nodes on a point-to-point connection to securely identify themselves. An ID/password pair is repeatedly sent by the peer to the authenticator until authentication is acknowledged or the connection is terminated.

The Code field is one octet and identifies the type of PAP packet. PAP Codes are assigned as follows:

1   *Authenticate-request:*   An ID/password pair is transmitted to the remote node.

2   *Authenticate-ack:*   Sent by the remote node to indicate that it recognizes and accepts the authenticate request.

3   *Authenticate-nak:*   Sent by the remote node to indicate that it does not accept the authenticate request.

For complete information on the challenge handshake authentication protocol (CHAP), review RFC 1994. CHAP is used to periodically verify the identity a peer by using a three-way handshake. This is done upon initial link establishment and may be repeated any time after the link has been established. This is accomplished as follows:

   1. After the link establishment phase is complete (during which there was a negotiation indicating CHAP would be used for authentication for at least one direction of the link), the authenticator sends a "challenge" message to the peer. CHAP can be used in neither, one, or both directions of the link. Each side of the link generates a challenge value, which can be anything previously agreed upon. The secret value is known to both sides but is not passed over the network.

2. Using the challenge value and the secret, a hash function is performed on them to create another value. The peer responds with this calculated value. Hashing yields a fixed-length value created mathematically to identify data uniquely. Refer to RFC 1321 for an example algorithm that is used with CHAP, called Message Digest Algorithm 5.

3. The authenticator checks the response against its own calculation of the expected hash value. If the values match, the authentication is acknowledged; otherwise the connection should be terminated. However, it may not be terminated—the RFC allows this to be implementer-dependent. One idea is to bring up only a subset of protocols when CHAP fails. Again, this is vendor-dependent.

4. At random intervals, the authenticator sends a new challenge to the peer and repeats steps 1 to 3.

What is unique about this authentication is that it is difficult to break the code. It is a better deterrent to hackers than the simple PAP protocol. CHAP provides protection against "playback attack" by a hacker through the use of an incrementally changing identifier and a variable challenge value. There is a secret that is shared between the two devices, but this secret is never transmitted over the link. The use of repeated challenges is intended to limit the time of exposure to any single attack. The authenticator is in control of the frequency and timing of the challenges. Each challenge is different because it depends on a variable. Therefore, if the challenge response were the same, even though it was based on a hash algorithm, the response would become predictable to an intruder—once one pass was done, the response would be the same for the next, and a hacker could simply capture the first response and then send it as its own.

After authentication is complete, the network layer protocol phase takes place. Here is where the two sides negotiate for the protocol types (IP, IPX, bridge, etc.) that each side can support. Each protocol is configured by using the appropriate control protocol for that network-layer protocol.

After the link has been established and optional facilities have been negotiated as needed by the LCP and authentication (optionally) has taken place, the next step in order for network packets to traverse a PPP link is for the link to negotiate the protocols it supports. Once each of the chosen network-layer protocols has been configured, datagrams from each network-layer protocol can be sent over the link.

There is one RFC per control protocol, and there is one control protocol per network protocol. For example, RFC 1332 describes the method for IP across a PPP link, which uses the same configuration format as LCP, but with a separate set of options. In the Option Type field are the options that are to be negotiated. Option values are assigned as follows:

1	*IP-addresses:*	Not used anymore—refer to RFC 1172.
2	*IP-compression-protocol:*	Compress the header using Van Jacobson compressions protocol.
3	*IP-address:*	Negotiate an IP address.

For the IPXCP (IPX Control Protocol) the following options are negotiated:

1	*IPX-network-number:*	Learn or determine a network number.
2	*IPX-node-number:*	Learn or inform your peer of your node number.
3	*IPX-compression-protocol:*	Negotiate a compression protocols such as Telebit Compressed IPX or Shiva Compressed NCP/IPX.
4	*IPX-routing-protocol:*	Negotiate the routing protocol of none RIP/SAP, or NLSP.
5	*IPX-router-name:*	Name your route.
6	*IPX-configuration-complete:*	Desired parameters are satisfied, end of negotiation.

There is not much for the control protocol to negotiate. It is a simplex safety mechanism to ensure that certain parameters are agreed upon by both sides before the link will pass the protocol's packets.

Multilink PPP

This protocol was originally set up to combine the B channels for ISDN (see Chap. 20), but it has progressed to support other communications lines and functions, as well, such as bandwidth on demand (in which, when a line is over capacity, a second line is brought up to the same destination) and load balancing (passing packets over both links) between the two links. Multilink PPP is a method for splitting, recombining, and sequencing datagrams across multiple logical data links. Basically, it combines multiple physical links into one logical link so that the total bandwidth of the multiple links is greater than that of any one of the links. The individual links do not have to be of the same type (frame relay, ISDN, leased, etc.), and they do not have to be of the same speed. However, they all must use the same path; for example, they cannot traverse different routers between the peers. They are still point-to-point. Refer to Fig. 18-4.

Using LCP (configuration option type 12), a link can indicate that it has more than one line to the peer and it would like to use both. When more than one line is used to the same peer, it is called a bundle. Packets are received over different physical link-layer entities, and each carries an identifier to indicate that it belongs to a certain bundle. These packets are then recombined and sequenced according to information in the MP header.

When a packet is received by the peer and by the protocol identifier, it indicates that it is MP. The packet is passed off to the MP functions in the peer for the packet to be properly placed into the stream. From here, the packet is handed off to its appropriate protocol for further processing.

The MP header format is shown in Fig. 18-5. This packet shows the long sequence number format. The other format for sequence numbers is 12 bits. The protocol is $0x00$-$0x3d$, which identifies an incoming packet as an MP packet. It will be handed off to the MP layer for processing. If the original packet has to be broken up into fragments to be sent across the link, the B and E bits are used. The B bit is the beginning fragment bit and indicates the beginning fragment, and the E bit indicates where

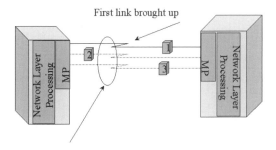

Figure 18-4
Multilink PPP topology.

Three physical links
one virtual link

Figure 18-5
Multilink PPP encap-
sulation.

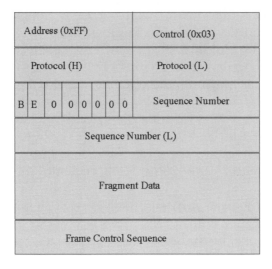

Address (0xFF)							Control (0x03)
Protocol (H)							Protocol (L)
B	E	0	0	0	0	0	Sequence Number
Sequence Number (L)							
Fragment Data							
Frame Control Sequence							

the fragment ends. This is different than the sequence number, which indicates where the packet lies in the stream of packets that were sent. In this way packets that were broken up can be put back together again to be handed off as a single packet to the appropriate protocol.

RFC 2125 (Bandwidth Allocation Protocol, or BAP) was written to manage the number of links in a multilink bundle. BAP defines datagrams to coordinate the addition and removal of individual links in a multilink bundle, as well as specifying which peer is responsible for which decisions regarding bandwidth management during a multilink connection.

Frame Relay

Introduction

This chapter is about the wide area network (WAN) protocol known as frame relay. It is not the intent of the following discussion to discuss frame relay from the provider's point of view. It will cover frame relay as it pertains to a customer's equipment. There are specialized engineers who maintain the provider's frame relay network, and that type of network is beyond the scope of this book. Frame relay on the provider's network can be of any technology, and if you ask about their internals you will get different answers from each provider. Different providers use ATM, FDDI, IP (with OSPF), or any combination of these technologies. This chapter will discuss frame relay as a WAN protocol to forward and receive network data between multiple CPE (customer premise equipment) sites.

A data-link layer protocol, frame relay is based on the ANSI standard known as T1.618. It is used to encapsulate and decapsulate data for transmission over a wide area network. This protocol can run on routers, bridges, remote terminal servers, and frame relay access devices (FRADs). It replaces the leased line connections that were once the norm for most wide area network services. The other alternative was X.25. The first public frame relay service was offered in 1991, and by 1997 there were over 100,000 ports (from the provider) installed. Frame relay is a streamlined WAN protocol that allows for speed and efficiency. It is assumed that the endpoints of the frame relay connection are intelligent and can correct for errors. Frame relay also assumes the use of very "clean" circuits. To this end, frame relay eliminated the requirement for stringent error and flow control that is found in X.25.

Before frame relay, the only standard packet-switched WAN protocol was X.25. However, X.25 was created when phone circuits were primarily constructed with copper and were noisy (contained unwanted electrical signals). While we as humans can tolerate noise on the lines, data cannot. Therefore, X.25 was very strict about error detection, and every X.25 hop performed error detection. Messages were also sent when the protocol discovered an error. There are three layers to the X.25 protocol (frame relay has two) providing for circuit setup, acknowledgments, and error detection. With all this, X.25 generally worked in the 9.6–19.2 kbps range. In those days, you had to. Frame relay does not have any of these features, and the protocol depends on cleaner telephone circuits and intelligent end devices for error control. As of this writing, frame relay provides speeds up to 45 Mbps. Table 19-1 provides a simple comparison

TABLE 19-1

Frame Relay and X.25

	X.25	Frame Relay
Common line speeds	9.6 kbps up to fractional T1	56 kbps up to T3
Overhead	High	Low
Line quality	Poor	Reliable
CRC checks	Each X.25 hop	End stations

between X.25 and frame relay. As you will see in the following text, frame relay is a WAN protocol, but it emulates a LAN environment.

Table 19-2 lists the standards by which frame relay is defined. There may be another set of standards that vendors claim they comply to, but they are basically based on the ones here. These standards are the product of a group known as the Frame Relay Forum. Their standards are preceded by the FRF insignia and can be found at *http://www.frforum.com*.

For the Internet, the standards are the RFCs. The beauty of RFCs is that they do not always pertain to the TCP/IP protocol. Frame relay is essentially a WAN protocol for TCP/IP, but there are a few RFCs for frame relay that include other protocols. The RFCs that concern frame relay are

- RFC 1490: Multiprotocol Interconnect over Frame Relay. This allows for multivendor interoperability in that it defines a standard encapsulation method for routed or bridged traffic. This is

TABLE 19-2

Frame Relay Standards

Standard	ANSI	ITU (CCITT)
Service description	T1.606	I.222/I233
Core aspects	T1.618	Q922
Access signaling	T1.617	Q933
PVC management	T1.617 Annex D	Q933 Annex A
Congestion management	T1.606 Addendum	I.370
Network-to-network interface (NNI)		I.372
Frame mode bearer service internetworking		I.555

also the RFC that is most associated with SNA and frame relay. However, it is not specifically written for the SNA protocol.

■ RFC 1293: Inverse ARP. Inverse ARP minimizes bandwidth utilization and simplifies or eliminates the manual configuration of data-link identifiers (DLCIs) and protocol (network) addresses. In a nutshell, it applies to frame relay stations that may have a DLCI, the frame relay equivalent of a hardware address, associated with an established permanent virtual circuit (PVC), but do not know the protocol address of the station on the other side of this connection. It also applies to other networks in a similar situation.

■ RFC 1315: Frame Relay MIB. This RFC defines the list of group and objects for the SNMP protocol.

Definitions

The following terms will be used throughout the text:

Committed information rate (CIR). The bandwidth defined for a virtual circuit.

Permanent virtual circuit (PVC). A predefined virtual circuit.

Virtual circuit (VC). The connection between two frame relay ports.

User-to-network interface (UNI). Specifies signaling and management functions between a frame relay network device and the end user's device.

Network-to-network interface (NNI). Specifies signaling and management functions between two frame relay networks (not discussed in this book).

Data Link Connection Identifier (DLCI). The "MAC address" for a frame relay circuit. It identifies the VC between the CPE (customer premise equipment) and the frame relay switch.

Mesh (full or partial). Refers to the frame relay topology and explains the connectivity between customer sites. For frame relay to work effortlessly, all sites should be connected to all other sites (full mesh). However, this is not efficient, so a partial mesh topology is most often used. A partial mesh allows all sites to connect to at least one or two common sites (hub and spoke).

There are two sides to frame relay. Figure 19-1a shows the overall topology view of a frame relay network. First there is the frame relay

Figure 19-1
(a) Overview of
frame relay topology.
(b) Frame relay
topology with
components.

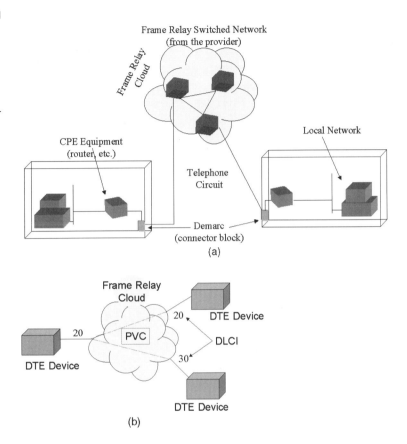

provider. This is not the router company. This is the telephone company
or an independent company that has built a frame relay service through
the telephone company. The frame relay provider enables the forwarding
of data from and to each of your sites that has frame relay connectivity.
They are the cloud, if you will. They provide the service to your door
and stop at a point known as the demarcation point, or demarc (pronounced "dee-mark") for short. This is where the other side of frame
relay comes into play: the frame relay vendor or CPE vendor.

To provide information to the frame relay service and be able to
receive frame relay information, a piece of equipment at the customer
site is needed. This is known as the customer premise equipment (CPE).
You need a device such as a router, bridge, or a frame relay access device
(FRAD) to "gateway" your information from the LAN to the WAN. The
frame relay service provider does not provide this. The provider takes
the information from the vendor device and transports it over the
frame relay cloud to the remote site. The vendor takes information from

the LAN, formats it for proper transmission, and gives it to the service provider to be forwarded across the WAN.

Another way to view frame relay is shown in Fig. 19-1 *b*. Here the frame relay topology is shown with all of its components. A physical connection is made between the CPE device and the first frame relay switch. The switch can accommodate many physical connections from different CPEs. Passed across each physical connection are the virtual circuits (VCs). A permanent VC (PVC) is the most common type used with frame relay. (The alternative is switched VC, which is a connection that is set up dynamically, or on demand. It is not discussed in this book.) Each VC is identified by a number, called the DLCI. All of this is fully explained in the following text.

Before frame relay, what did most customers do for high-speed links (higher than X.25)? This is shown in Fig. 19-2. A customer network was linked over the wide area via a leased line. This is a special line available from the telephone company that is "conditioned" for high-speed data use. This line is set up between two specified points and used only by the customer for its data. It is not shared, at least as far as the customer knows. The CPE uses a single hardware port for each circuit, and there is also one CSU/DSU (customer service unit/digital service unit, a device to condition the data for use on the circuit) for each leased line. This means that there is one serial port dedicated to each remote site. The more remote sites, the more CPE serial ports and the more CSU/DSUs. Not only is this expensive, but it also requires lots of space to house all

Figure 19-2
Leased line WAN.

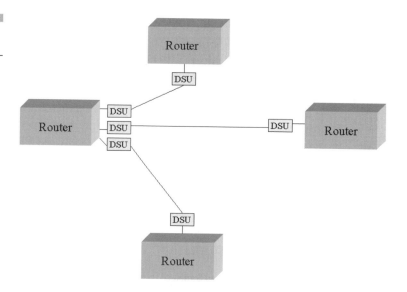

the equipment needed. Frame relay replaced 99 percent of the hardware with software functions.

The frame relay function encapsulates data for transmission. This encapsulation uses headers and trailers around the data for "routing" through the frame relay network. The frame relay frame is shown in Fig. 19-3. It contains the following fields.

- *Data link connection identifier (DLCI).*   This identifies the PVC that connects two points over the frame relay WAN. It has local significance only. This means that the number assigned as the DLCI has meaning only between the CPE and the first frame relay switch. Interestingly enough, the DLCI can be the same on both sides of the cloud. The frame relay cloud uses this connection number to look up the path to the other end of the connection and forward the data.

- *Command/response (C/R).*   This field is not used.

- *Extended addressing.*   Frame relay uses the ITU (International Telecommunications Union, formerly the CCITT) Q922 protocol, which specifies a three- or four-byte frame relay header that contains an extra address bit. However, this is not used in most frame relay implementations.

- *Explicit congestion notification fields (Forward and Reverse) (FECN, BECN).*   These fields allow for flow control.

- *Discard eligibility (DE).*   Set by the CPE device, this bit indicates that the this frame is eligible for discarding should the frame relay network deem it necessary. Frame relay allows for burst transmission rates above the committed rate. If the switch is capable of forwarding the frames, it will. If the switch cannot forward frames, it will look for those frames that have the DE bit set and discard those frames first.

Figure 19-3
Q.922 frame relay frame.

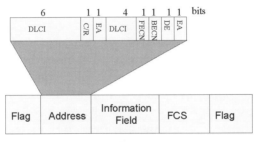

Frame Relay Structure

DLCI	Comment
0–15	Reserved
16–1007	Assignable (by provider)
1008–1023	Reserved

Using the range of DLCIs shown in Table 19-3, there are 991 PVCs that can be assigned, which is more than enough for any frame relay CPE device. Remember, DLCIs have only local significance, and two routers that are physically side by side can have duplicate PVCs as long as they connect to different frame relay ports on the switch side of the circuit.

The DLCI is the "MAC address" for frame relay. In the router, for instance, an ARP cache is kept that contains DLCI–to–IP address mappings for remote routers via PVCs. The first switch receives the data from the router, takes a look in its map table, and then forwards the data over its network to the final destination. Frame relay tries to emulate a LAN. The DLCI allows an address (hardware emulation) to be assigned to a PVC. The DLCI has only local significance in that it is used between the DTE and DCE interfaces. In other words, it will be assigned to a PVC between the router on the customer's site and the first switch. The first switch is "provisioned" (has a mapping of DLCIs) that indicate where the other end of the connection is (for a PVC). In this way, any data received on this interface will be forwarded through the provider's network and received properly at the other end of the PVC.

Refer to Fig. 19-4, which shows two PVCs with the same DLCI number on different parts of the map. Two identical DLCIs could not be assigned to the same router, however. The first switch has a mapping (a database table) that contains information on where to forward the traffic based on the received PVC number. The switch does not know about the protocol address (IP address or IPX address) of the routers at the customer site. Remember, frame relay is a protocol that operates at the data-link layer. Internally, the frame relay provider may run IP, but this would not interface with the IP of the routers within the customer site. The mapping simply contains a DLCI number and the next hop that the switch should forward the traffic to. If the provider is running IP as its forwarding method for its switched network, the whole packet would be encapsulated in another IP header from the provider's equipment and forwarded to the final destination as indicated by the map table.

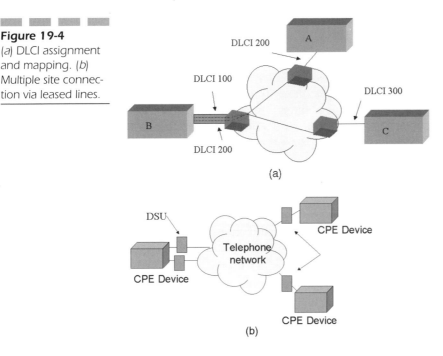

Figure 19-4
(a) DLCI assignment
and mapping. (b)
Multiple site connec-
tion via leased lines.

Many frame relay switch providers do not use IP internally and have constructed their own proprietary method for forwarding.

Address Resolution

Basically, there are two standardized methods used for address resolution:

- ARP
- Inverse ARP

ARP was explained in the TCP/IP chapter of this book, and its use is exactly the same with frame relay. Briefly, IP understands 32-bit addressing, and some of the underlying link layers have different addressing mechanisms. Therefore, a mapping of IP to link-layer addressing must be kept. This is what ARP does. Given the IP address, ARP requests the hardware address. The remote node should respond with the hardware address—in the case of frame relay, the DLCI. This is not the DLCI of the remote router, but the DLCI of the local router pertaining to that PVC. The encapsulation that is used is RFC 1490, "Multiprotocol over Frame Relay." It uses the SNAP header, and the remote router, if it's running RFC 1490 as well, should respond with the ARP reply. This allows the ARP cache to be built using the DLCI and the IP address, just like the MAC layer address–to–IP address mapping. Other protocols such as NetWare (IPX) do not have this problem, for the host address is the DLCI (or the MAC address if on a LAN, for example).

Inverse ARP is an extension to the ARP protocol that is protocol independent. A version of ARP runs with AppleTalk to determine addresses as well. This is called AARP, or AppleTalk ARP.

Inverse ARP can be used with any network layer protocol and enables the router to translate a given DLCI to a specific protocol address. Within the frame relay environment, new PVCs may be announced through the exchange of signaling messages between the frame relay switch and the router. These signaling exchanges provide and indication of the DLCI assigned to the PVC, but do not provide information regarding protocol addressing. Inverse ARP enables the router to discover the protocol address of the remote node (the other end of the individual PVC) associated with the newly announced DLCI. Previously, the only method for accomplishing this was to manually configure the router to statically indicate the protocol address of the remote node of the PVC. Again, this would not scale as the network grew. Inverse ARP is also extensible to other protocols, including protocols of LAN and WAN media, not just network protocols of NetWare (IPX).

Upon response from the remote end of the PVC, an ARP cache entry is built mapping the DLCI to the protocol address, just as in the ARP process for broadcast media (Ethernet, Token Ring, etc.). Again, the purpose of inverse ARP is to find the remote end of a PVC protocol address when a new DLCI is added.

A new DLCI is added by the provider. In configuring the switches to provide you with a PVC between two points, the job of the provider is simply that: to provide a packet-switched circuit between those two points. The LMI (local management interface, discussed later) will update the router to indicate what PVCs have been added, but this does not enable the IP addresses of the remote ends to be identified.

Frame relay reduces the amount of hardware that you need from both the router vendor and the telephone vendor. The number of hardware ports on the router and the number of DSUs required to connect to the phone company are both reduced. The number of circuits required from the phone company is less too. Frame relay provides many, many benefits for WAN connectivity. Even better, it provides for multivendor connectivity through the use of the protocol as a standard.

Circuits

Three circuits are currently defined for frame relay:

- Permanent: a prebuilt "always on" vitual circuit.

- Multicast: The ability to send information over a reserved DLCI to the frame relay switch, which in turn multicasts the information to specific endpoints (preconfigured) on the network. The advantage here is that the sending device sends the information only once and the frame relay network replicates and forwards it.

- Switched: This is the ability to dynamically build a VC when required. This is much like dialing a number from your phone. A VC can be set up at any time to a remote device and then torn down after use.

A PVC is not what you are probably thinking: a leased line. However, a PVC emulates a leased line through a software function. The circuit is permanently built between two points, but packet switching is used to transport packets. A PVC means that the circuit is built between two sites, and neither side can call any other site without a reprovision of the frame relay switch. It is called a virtual circuit because the two points are connected to each other, but only by means of the packet-switched network. Other packets will be the exact same ports and switches to be forwarded to their destinations. The packet is switched to its final destination, but there is no dedicated hardware pipe built between the two sites. Once the PVC is set up between two sites there is no need for setup each time data are transmitted. The circuit is continually there. The circuit is prearranged in the frame relay cloud. Remember, frame relay is defined only between the point on the frame relay cloud and the customer site. The methods and software/hardware in the cloud are not defined by frame relay. In fact, any combination of routing techniques can be used in the cloud, including ATM switching and IP routing.

The number of PVCs that can be supported (by Q.933 Annex A, explained in the frame relay management section) is limited by the maximum frame size that can be supported by the user device and the network on the bearer channel (e.g., when the maximum frame relay information field size is 1600 octets, then a maximum of 317 PVC status elements may be encoded in the status message).

An interesting aspect of frame relay is that the hardware port can support multiple PVCs. See Fig. 19-4, which uses a router to demonstrate this purpose. With a leased line (which is simply what frame relay emulates via software), every circuit is assigned a physical port. A DSU is connected to that port, and a line is hardware-constructed through the phone company to the destination. The inefficiency arises when the single router needs to connect to multiple sites. Multiple hardware ports and DSUs are needed to connect to each site. In fact, each site that needs connections to more than one remote site must repeat this.

Committed Information Rate

The ability to transmit data from the customer site to the switch is available using three methods:

- Committed data, which the frame relay network is obligated to transmit

- Excess data, which the frame relay network will attempt to transmit if the resources are available

- Any data beyond this, which the frame relay network will reject

The committed information rate (CIR) is the minimum transmission rate at which data are guaranteed to be transmitted and received on the PVC. The CIR is analogous to the speed of a leased line. However, the line speed of frame relay (the speed between the CPE and the frame relay switch) is usually much faster than the CIR. You could have a CIR of 64 kbps and a line speed (sometimes called the access or port speed) of 1.544 Mbps. If there are multiple PVCs on the same line, each has the capability of transmitting at the CIR if the other PVCs are idle. This means that the CIR is shared between the number of PVCs configured on the circuit. If no other PVCs are transmitting, a single PVC may use the entire CIR for transmission.

Frame relay provides an additional capability known as burst rate. There are three parameters here:

- *Committed burst rate (B_c):* The maximum amount of data allowed to be transmitted over a specified time interval.

- *Excess burst rate (B_e):* The maximum amount of data exceeding B_c that can be sent over a specified time interval.

- *Committed rate measurement interval (T_c):* The time interval during which the user can send only B_c data and B_e amounts of data.

Data can only be transmitted at an instantaneous rate based on the line speed, which is normally higher than the CIR. It is hardly reasonable to reject the second bit of a transmission because it arrived too fast, and so the CIR is determined by averaging over a period, typically one second (this is known as T_c). In general, the duration of T_c is proportional to the "burstiness" of the traffic. T_c is computed as $T_c = B_c/\text{CIR}$. T_c is not a periodic time interval. It is used only to measure incoming data, during which it acts like a sliding window. Incoming data triggers the T_c interval. T_c is configured on the frame relay switch by the provider when the circuit is provisioned (set up).

The committed burst size (B_c) is defined by the standards to reflect this and is the amount of data that the network is committed to accept in any (1 s) time period, regardless of the instantaneous rate. For example, if the line speed is 64 kbps and the CIR 32 kbps, the network will accept data at 64 Kbps for half a second, but it need not accept any more during that second.

The excess burst size (B_e) defines the rate up to which the network will attempt to transfer data received in excess of B_c.

While these rules seem relatively simple, their implementations by the equipment designers and network providers can be quite different. For example, in some networks the CIR really is committed and is underwritten with underlying bandwidth. In others it is provided on a statistically averaged basis, in a similar way to airline seat bookings, and thus may not always be available under high traffic loads.

The way excess data are handled varies even more widely, with the data transfer rates ranging from very little to almost all of the data offered to the network above the CIR. The way this is handled can be at least as important as the CIR value in determining PVC throughput, yet it is given little prominence and many customers do not even consider it when specifying a frame relay service.

These parameters are set up during the provisioning of the PVC on the provider network.

Congestion Control

Congestion can be caused by many different things on the network. Frame relay is a packet-switched environment in which there are multiple (meaning hundreds of thousands) of companies sharing the same frame relay (the cloud) equipment. It is not like a leased line, where the subscriber (the customer) owns the entire pipe and, based on the clock, the sender can send as much information down that pipe as desired. The pipe is completely open for two endnodes (routers, etc.) to communicate over, and no other packets are on that pipe except for the ones that they transmit.

Frame relay is different. This is a gross exaggeration, but basically the pipe is shared by all who have subscribed to that provider. This is true at the core (the cloud). The same equipment is used to forward data from multiple companies. You should be able to see that congestion will occur, possibly simply by one customer oversubscribing its link and constantly bursting its data.

Frame relay provides two very simple mechanisms to indicate to the edge routers that congestion is occurring on the switches:

- Discard eligibility (DE) bit
- FECN and BECN bits

The DE bit is set by the router and indicates to the receiving switch (any switch in the cloud) that this packet is allowed to be discarded if the switch is congested and needs to reduce the number of packets for forwarding. Packets with the DE bit set are the first packets to get deleted when the switch becomes congested. This bit can be set by the edge router based on various conditions; for example, the packet was sent while the router was bursting, the packet was sent from the low-priority queue in the router, or the CIR is set to zero, which means that all packets will have the DE bit set.

No need to worry about a packet being dropped by the switch. Frame relay assumes that the higher layer of the protocol (such as TCP) will time out and retransmit the packet. The frame switch does not provide indications that it dropped any particular packet. However, it can tell the edge routers that it is getting congested, which is an indication that frames are being discarded or dropped.

Two bits in the frame header allow for this notification: FECN and BECN (pronounced "fecken and becken"). These stand for forward explicit congestion notification and backward explicit congestion notification, respectively. They are one bit in length. Refer to Fig. 19-5. A BECN is sent upstream and a FECN is sent downstream. A BECN is sent to notify the sending edge devices (the routers) that congestion is occurring on the frame relay network. It does not indicate what or where, but simply that congestion is occurring. (Remember, that frame relay is fast and simple, with most error mechanisms stripped out to allow for speed and efficiency.)

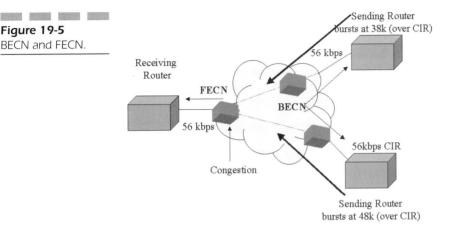

Figure 19-5
BECN and FECN.

What the routers do after receiving the BECN is up to the router vendor implementation. Usually, it is a *k* out of *n* procedure, which means that once *k* BECNs are received out of *n* frames, then some type of congestion control is invoked. Some implementations simply convert strictly back to the CIR. Others revert to 75 percent of the CIR (or some other percentage of the CIR until the BECN is no longer received) and slowly go back to the CIR and then start to burst again.

If a router continues to receive BECNs, it may be time to increase the remote router's access speed. Figure 19-5 indicates that the receiving router is at the same speed (CIR) as the two sending routers. The sending routers are set to 56k CIR, as is the one receiving router. However, after the two sending routers burst, congestion occurs on the one switch, causing BECNs and FECNs to be transmitted. Instead of having the sending routers slow down and the upper layer protocols having to retransmit, the CIR on the remote router should be increased. What does this mean?

Proper frame relay implementation indicates that a receiving router should have twice the port access speed of the sum of all the remotes. Refer to Fig. 19-6. Although this figure exaggerates the point, the receiving router must be at least the same speed as the sum of the remote routers, but two times this speed is better. This is best shown in a hub-and-spoke configuration.

Again, frame relay is a packet-switched WAN. Congestion can occur because too much data are coming into the network (from many sources), and the network is simply overloaded. It might not be only your data that have overloaded the network.

Figure 19-6
Port speeds.

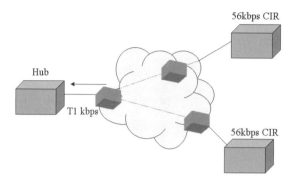

Multicast Data

Multicast is becoming a very popular method of transmitting real-time data (voice and video) or non–real-time data that must be received by multiple receivers.

Again, in a leased line environment, the forwarding of multicast data is no problem. When the router assigns one IP address to every PVC, multicast works fine. However, in the default configuration of frame relay, there are multiple PVCs that are assigned one IP address. These are commonly called group-mode PVCs. All PVCs are grouped together and have the same WAN IP address. Multicast does not work well in this configuration. In this case, there are two options:

- Multicast-enabled frame relay
- Router replication

Some of the DLCIs are reserved by switch providers for multicast replication. Any data received on these DLCIs are known to be multicast frames. The switch is preconfigured with a list of corresponding DLCIs that the frame should be forwarded to (see Fig. 19-7). Having the frame relay switch provide for the replication of multicast traffic is the best method for handling multicast. However, most providers do not support multicast.

In the absence of a multicast-enabled frame relay service, multicast packets can still be sent, although this method is not as efficient as having the switch perform the replication. As shown in Fig. 19-8, the router is capable of doing the replication and sending it out over each PVC. What is the problem with this? There are a few. First, all PVCs will

Frame Relay

Figure 19-7
Multicast-enabled switch.

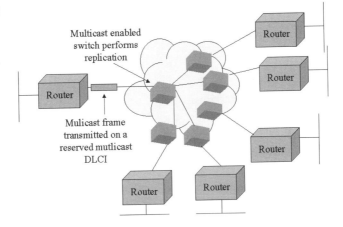

Multicast enabled switch performs replication

Mulicast frame transmitted on a reserved mutlicast DLCI

receive this information, even the ones that do not need it. There are specific algorithms to ensure that only those paths indicating that they need the multicast packet will receive it. With router replication, all PVCs will get the replicated packet and the remote router will discard it.

Second, having the frame relay provider perform the replication saves bandwidth on your trunk line. Otherwise, instead of one packet being sent, *n* packets are sent, where *n* is the number of PVCs. Also, the router is forced to replicate the packet *n* times, and this consumes buffer space and CPU cycles. The frame relay service providers are indicating that they will provide this service in the first half of 1998, but it remains to be seen whether this will happen.

Figure 19-8
Router replication of a multicast frame.

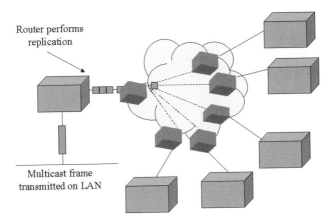

Router performs replication

Multicast frame transmitted on LAN

Frame Relay Management

Frame relay management encompasses two functions:

- *Consolidated link layer management (CLLM).* In-band congestion control is accomplished though the BECN and FECN bits in every frame. These bits are DLCI dependent. Therefore, they indicate congestion on a PVC-by-PVC basis.

  CLLM is an out-of-band congestion indicator. Instead of one PVC, it reports on a group of PVCs that are likely to have congestion. The CPE device, upon receiving a CLLM message, will take appropriate action on the group of PVCs (usually by slowing the transmission rate). The DLCI is 1007 for a two-byte-address PVC.

- *Local management interface (LMI).* A management entity in frame relay provides a method to insure that the link between the DTE and the frame relay network is fully operational. This interface allows for dynamic updates of changes in the status of a PVC (add, change, or remove). This management interface provides for simple XON/XOFF flow control. It also operates on a reserved DLCI, either 0 (ANSI/ITU-T standard) or 1023 (Frame Relay Forum [Group of Four]).

LMI Protocols

LMI is not exchanged end to end between routers; it is exchanged between the router and the local frame relay provider. As a result, the LMI type at one location does *not* have to match the LMI type at other locations. If you have a frame relay router in Sarasota, Florida, the LMI may be of one type and another router in Spokane, Washington, may use another type of LMI. LMI provides status and outage notification for frame relay PVCs.

Generally, LMI includes a few standards:

Generic—"Group of Four" enhancements to the Annex D specification

ANSI—ANSI Annex D T1.617-1991 specification

Q933a—CCITT Q933 Annex A

LMI specifications defined by the Group of Four (now known as the Frame Relay Forum) within the standard "Frame Relay Specifications with Extensions" are incompatible with both ITU-T and ANSI standards. This is the basis for the LMI standard commonly known as REV.1.

The channel used for LMI is 1023 under the Group of Four specification and 0 under ITU-T/ANSI. This prevents LMI from functioning with a Group of Four implementation on one side and an ITU-T or ANSI implementation at the other end of a frame relay link. In addition, the Group of Four has defined the unidirectional LMI procedures only, referred to as the heartbeat process. Running LMI on DLCI 1023 at one end of a circuit and DLCI 0 at the other end will not stop the frame relay circuit from operating, but it will prevent the LMI from functioning.

There are three protocols used for this management:

Status inquiry: A "keep alive" message is usually sent every 10 s and ensures a good connection between the CPE device and the frame relay switch. Most interfaces use a k out of n method to determine a nonusable interface. This means that if the CPE device misses k out of n messages, it will bring the frame relay line down.

Status: A complete update of the network is initiated by the periodic status message, sent every 30 s.

Status inquiry: A status inquiry message is sent by the switch any time there is a change in the network, such as an added, changed, or deleted PVC.

These messages are used to perform three procedures:

1. Link integrity verification (LIV)
2. Notification of PVC status
3. Notification of added or deleted PVCs

Periodically (based on a timer) the CPE device transmits a status inquiry message to the frame relay network. The frame relay network side responds with a status message containing the requested LIV information element (IE). Both sides maintain a send sequence number and a receive sequence number. The link integrity is checked by a sequence number generation and checking process.

The user side sends a status message containing a link integrity verification IE. The report type may either be full status or link integrity only. The link integrity IE contains the send sequence number of the sender (user side), which is incremented each time a status inquiry message is sent, and a receive sequence number, which is equal to the send sequence number contained within the LIV element of the last status message received from the network side.

As shown in Fig. 19-9, a LMI frame produces information about each PVC through information elements. These IEs are produced one per PVC, and their details are shown in Fig. 19-10. There are three types:

- *Report type.* This indicates whether the message is a full status request, LIV only (test the link), or a request for a single PVC. This is indicated by the record type.

- *Verification type.* This is for synchronization of sequence numbers.

- *PVC status.* This indicates the type of PVC. The N bit indicates that a new PVC has been added, the A bit indicates that the PVC is active and alive, and the S bit indicates that it is a spare PVC. A very misunderstood bit here is the A bit. At one time, many frame relay providers did not support the A bit. This bit allows a router to determine that a PVC is not active. It is still configured, but not active. This can cause havoc in non–full mesh topologies, for a link may appear to be up and operational when it really is not. The A bit indicates that a PVC is dead for now, so no traffic should be routed over this PVC.

The network side must respond by sending a status message containing a LIV with the send sequence number of the sender (network side), which is incremented each time a status message is sent, and a receive sequence number equal to the send sequence number contained within

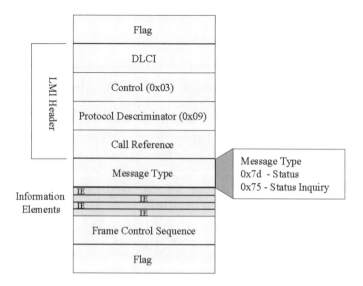

Figure 19-9
LMI frame layout.

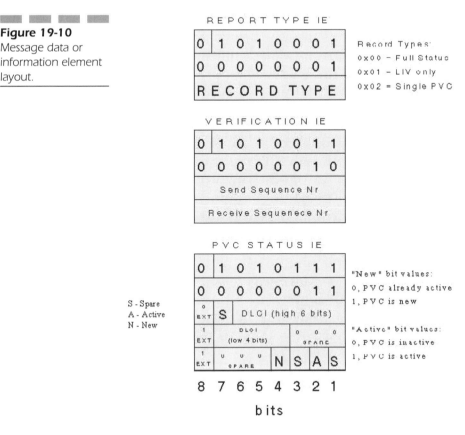

Figure 19-10
Message data or information element layout.

the LIV IE of the last status inquiry message received from the user side. The network side maintains a timer that is restarted each time a status message is sent.

For link integrity, there are no methods in the specification that indicate how to correct the error. Therefore, error correction is implementation dependent.

The PVC status message tells the CPE site whether existing PVCs are active or inactive or whether they have been added or deleted. PVCs are added or deleted by the provider, and the CPE follows along. A report type of full status is sent in the status inquiry message by the CPE device. The frame relay switch should respond with a status message with a report type of full status along with PVC status IEs for each existing PVC.

A PVC status of active indicates to the CPE device that the PVC is available for use. As indicated earlier, if a provider does not support this bit, it can create havoc on your network. Your router, for instance, will think that the PVC is active and will continue to try to use that PVC, but the switch knows it is down and will not forward data across it. With this message, the router will know that the PVC is not active and can converge to find a new route.

For a new PVC, the frame relay switch will set the N bit on a PVC status IE in addition to the active/inactive status indicator if the PVC has been added since the last status message containing full PVC status. A PVC is indicated as deleted simply by the omission of a PVC status IE in the status message. If the CPE device receive a status message with the N bit set, and the device believes that PVC already exists, then the CPE device will understand that the PVC was deleted and then added back. Another condition that could take place is the CPE device receiving a status message about a PVC it did not know about (it did not have the N bit set). The CPE device should log this as an error.

Using LMI, a router and switch can communicate information. This information is very simple and generic and basically covers the integrity of the line and the PVCs. No real statistical data are contained in LMI. Again, it is assumed that the end stations or CPE devices are very intelligent, and the frame relay circuit is simply needed to carry data over a WAN.

Frame Relay Topologies

There are multiple ways to configure a frame relay interface. The names for these configurations are not standard, so the most useful names are given here to describe the functions. Be aware that these configurations are given different names on the vendors' devices.

Group PVC (Fig. 19-11): All PVCs are grouped together as a single interface, and the single interface is assigned one IP (or network protocol) address.

Direct assignment PVC (Fig. 19-12): Each PVC is assigned one IP (or network protocol) address even though there is still only one physical address. For both group and direct PVC there is one network number for all connections in the frame relay cloud. Each port will have a different host number but the same network number.

Hybrid PVC: A combination of the two modes above. Some of the PVCs are grouped together and configured as group PVCs, and other PVCs are individually configured.

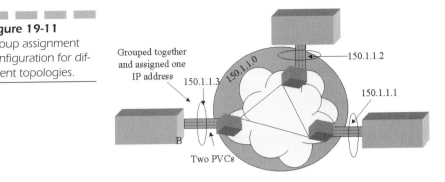

Figure 19-11
Group assignment
configuration for dif-
ferent topologies.

Figure 19-11
Group assignment
configuration for dif-
ferent topologies.

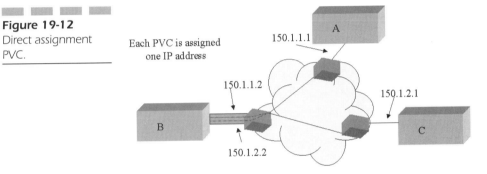

Figure 19-12
Direct assignment
PVC.

Group PVCs are grouped on a single hardware port of the edge device
(i.e., a router). The individual PVCs are each identified by the DLCI. The
DLCI indicates to the first switch where the packet should be forwarded
by means of a table that is configured on the switch by the frame relay
provider. In this way, the data will be forwarded to the correct destination and other destinations will not see the data. However, a routable
protocol such as IP still sees that interface as a single interface and not as
multiple interfaces. Therefore, on the edge device the port is assigned
one protocol (i.e., IP) address.

In the direct assignment PVC configuration, each PVC is assigned one
protocol address, enabling each PVC to appear as a separate circuit.
Although this allows each PVC to act as its own circuit, it requires more
resources on the router to maintain each PVC. This can become cumbersome when scaling (growing the network). However, it is the best configuration for some topologies that include the routing protocol of OSPF
and provide for dial backup.

In the hybrid PVC configuration, some of the PVCs are configured
into a group and assigned one group address, but for bridge protocols

they are treated as direct assignment PVCs. This is very useful when you have a combination of routable and nonroutable protocols such as IP and bridge.

These configurations are valid only for configuration on the router. The switch does not have any idea that these configuration are in place. Broadcast-oriented protocols such as IPX for NetWare and IP for TCP/IP are not well suited for the typical frame relay environment. Therefore, the router vendors devised these methods in order to run the existing suite of protocols across a frame relay environment.

It is important to know the topology of a frame relay. Common frame relay topologies are shown in Fig. 19-13 and include the full mesh, partial mesh, and star (i.e., hub and spoke). Another consideration in frame relay is bridged networks.

In the full-mesh topology, the frame relay provider set up PVCs between all frame relay CPE devices. As shown in the figure, routers A, B, and C all have PVCs to each other. This is the topology that most emulates a LAN topology for WAN. This is the most desirable topology for frame relay, with one caveat. It requires a lot of PVCs, and as your network expands, the number of PVCs expands at the rate of $n(n-1)/2$, where n is the number of nodes to be connected. As your number of routers connected to the frame relay increases, the network becomes very inefficient and very expensive. However, this is the best topology for routing protocols such as RIP to work with. All routers and networks can see each other in a full-mesh environment.

Remember that frame relay emulates leased lines through software (PVCs). In a leased line environment, every connection basically has a

Figure 19-13
Frame relay topologies.

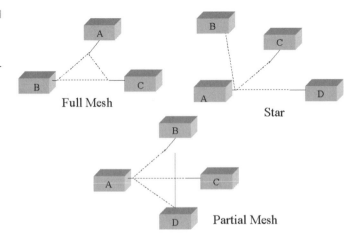

Full Mesh

Star

Partial Mesh

connection to every other connection, no matter what the topology. This is because multiple hardware ports are used, one for each connection, and therefore, every port has its own IP address assignment. Routing update protocols are able to be forwarded to all other routers, and, no matter what the topology, all networks are seen by all routers.

However, this is not true for the other topologies. Since frame relay emulates the hardware circuits via software, a network protocol still sees the port as one physical port. And by definition of protocols such as split horizon, a routing update may be forwarded only to ports that the router did not receive the update from. This presents a problem for frame relay topologies other than full mesh. Take IP/RIP, for example. In the star topology, if router C updates the hub router A with its network number for an Ethernet port on the router, router A is not allowed to rebroadcast this information back out the frame relay port. The IP service on router A still sees the port as one network number, and not as multiple PVCs each with its own network number. This is shown in Fig. 19-14.

Another problem is bridged networks. Yes, they still exist out there, and in some configurations they work very well for the customer. However, bridging over frame relay presents its own set of problems. Remember that bridges will forward broadcast and multicast packets to all active ports except for the one that they received the packet on. Again, since frame relay is emulating hardware ports via software, a broadcast packet received on a frame relay interface cannot be forwarded back out that interface. This keeps other frame relay routers from receiving the much-needed broadcast or multicast traffic. The forwarding of broadcast and multicast packets in a bridged network is a necessary event.

Figure 19-14
Non-full-mesh connectivity scenario.

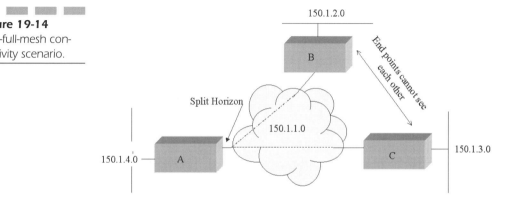

Frame Relay and IP

A frame relay network must be carefully designed. The following are some helpful thoughts.

- *Static routes.* For a star or partial-mesh configuration, static routes offer a good alternative for frame relay. The endpoint routers need a default route back to the hub. The hub router may or may not need static routes; this depends on the manufacturer of the router at the hub site.

- *Network summaries.* Try to design your network with an IP addressing scheme that is contiguous, so that you can summarize networks in the update packets over the router frame relay. This leads to less traffic across the frame relay.

- *OSPF.* Most router vendors now support an attribute known as "point-to-multipoint," which allows OSPF to understand the topology as a star. Also, in this environment the frame relay routers should have their priority set to 0 to ensure that they do not become the designated or backup designated routers on the frame relay.

- Make sure that you are using inverse ARP; this eliminates the problem of static entries for mappings.

- Make sure your CPE device is RFC 1490 compliant. This makes for easier work, such as ARP, and facilitates interoperation with multiple CPE vendors (e.g., Cisco, Bay, and 3Com).

With the star topology, the endpoints cannot see each other without some help. In Fig. 19-13 router B cannot see router C or D. If a node off of router B wanted to communicate with a node off of router C or D, a special route would have to be placed into router B, known as a default route, pointing to router A. Router B would forward any packets for unknown networks to router A, which would forward them to one of the other routers.

While this may not seem like a big deal, it does not scale. If there are 40 routers behind router B, all 40 would have to have know of the default route, and if there are loops behind router B, all would have to be carefully managed in the event of failure of a router in the looped network.

There is a solution for this that router vendors are supporting. Using the same hub-and-spoke or star topology, router A can configure each

PVC to have its own IP address, thereby allowing the IP interface to think that there are multiple hardware ports on the interface, each with its own IP address. Using this method, router A is allowed to update router C with the network number of the Ethernet on router B, for the IP interface thinks there are multiple IP interfaces on the single hardware port. Also, router B can be updated with the network numbers from router C, and the same for router D.

One other solution is to turn off split horizon at the hub router. This enables the router to re-advertise a route that it learned back through the port that it learned it from. Yes, this can be dangerous in that it can allow loops to form, and most router vendors allow only split horizon to be disabled for IP. But without assigning each PVC a separate address, it is the simplest method. In simple configurations, such as true hub and spoke, it works very well.

Frame Relay and IPX

IPX is known as a "chatty" protocol. This simply means that it talks a lot (lots of overhead packets), and some of the things it has to say do not need to be rebroadcast everywhere. A great example of this is the SAP protocol. This protocol advertises the services available on a server. However, it does not just advertise for one protocol; every server advertises its services. The minimum number that a server will advertise is two; one for file service and one for print service. The problems associated with the RIP protocol are also found with the SAP protocol. The update function is the same as for RIP. Therefore, it will have the same problems on a frame relay network as IP does.

One suggestion is simply to turn off some of the advertising. This is primarily accomplished by setting filters in the routers. For example, PowerChute™ UPS and HP JetDirect™ are two services that probably do not need to be advertised throughout a network, especially over the WAN.

Another solution is to increase the periodic advertisement timer for SAP. Most services are available full time, do not go away, and therefore do not require updates every 60 s (the default periodic timer). Therefore you may want to reset this timer to every 10 minutes or even higher to reduce the amount of bandwidth consumed for IPX SAP traffic.

A final solution is to turn off SAP broadcasts altogether and configure static SAPs in the routers.

Other Tuning Suggestions for Frame Relay Installations

Frame relay is a great protocol and is the standard for WAN connectivity. Its speed ranges from 9.6 (sometimes lower) to over 45 Mbps per port rate. Most configurations are software tunable and can be reconfigured in a matter of seconds (however, getting the order through the provider is a different matter). Reconfiguring a frame relay circuit is really a matter of reconfiguring the frame relay switch or the router. There generally is no need to replace hardware or software on either end, although there are instances where this may be necessary, such as configuring a device from T1 to 45 Mbps. But for most purposes, changes are accomplished through software.

RFC 1490: Multiprotocol Interconnect over Frame Relay

RFC 1490 describes the a standard method used for carrying network data over a frame relay backbone. This is also the RFC that enables frame relay and SNA to interoperate in that IBM supports frame relay directly out of the front-end processor. This is discussed in detail later.

Two types of data will travel across a frame relay backbone: routed and bridged frames. Therefore, the RFC covers both routable and non-routable (bridging) protocols. All protocols must encapsulate their packets within a Q922 Annex A frame. Refer to Fig. 19-15.

Q922 addressing is usually two bytes in length. Q922 allows for three- and four-byte addressing, but this is discouraged and is not usually implemented by frame relay providers. The control field is always set to 0x03, but it can be negotiated to another value (this is beyond the scope of this book). The PAD field is used to allow for alignment to a two-byte boundary.

The NLPID (network level protocol ID) is administered by ISO and CCITT. It contains identifiers to allow frame relay to recognize IP, CLNP (ISO OSI), and IEEE SNAP encapsulated packets. In other words, this field tells the receiver what protocol is encapsulated.

The size of the encapsulated frames varies. The minimum size is five bytes between the flags (frame relay header and trailer). This is expanded to six bytes for three-byte DLCI and to seven bytes for four-byte

Figure 19-15
RFC 1490 encapsulation.

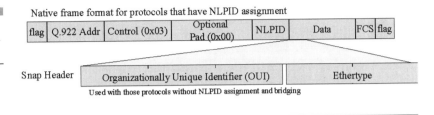

DLCI (extended addressing is discouraged, though). Some frame relay implementations support only a maximum of 262-byte frames. Others support 1600 bytes or larger. This is really up to the frame relay provider. Therefore, frame sizes are configurable.

For those protocols that do not have an assigned NLPID, the SNAP header is used. However, the NLPID is used with the SNAP header. What this means is that if the NLPID is 0x80 (0x means hex), then a SNAP header follows. RFC 1490 simply identifies the encapsulation when frame relay is used as a backbone. However, IP is assigned a NLPID of 0xCC. This is a direct assignment of an NLPID, and therefore SNAP is not used.

One problem in encapsulating bridge packets is the MAC header of the encapsulated packet. For example, you could be using IEEE 802.3 (Ethernet), IEEE 802.4 (Token Bus), IEEE 802.5 (Token Ring), or FDDI. Each has a different MAC header. For example, Ethernet does not have the Frame Status field that the IEEE 802.5 frame has. When we encapsulate a bridged frame, we literally take the whole frame and encapsulate it; we do not strip off the headers and trailers first. Therefore, there are different encapsulations for each MAC type. Table 19-4 indicates the PID types for each, and Fig. 19-16 shows some of the encapsulations. To encapsulate bridge frames over frame relay, we use the IEEE 802.1 reserved OUI (organizationally unique identifier) of 0x00-80-C2.

Something else should have clicked here. With a standard for encapsulation of network-layer protocol data across frame relay, you should be able to have two different vendors communicate across a frame relay cloud. This is correct. Once RFC 1490 was established as a standard, dif-

TABLE 19-4

PID Values for OUI 0x00-80-C2 (Bridge Encapsulation)

With Preserved FCS	Without Preserved FCS	Medium
0x00-01	0x00-07	IEEE 802.3/Ethernet
0x00-02	0x00-08	IEEE 802.4
0x00-03	0x00-09	IEEE 802.5
0x00-04	0x00-0A	FDDI
	0x00-0B	IEEE 802.6
	0x00-0E	Spanning tree BPDU

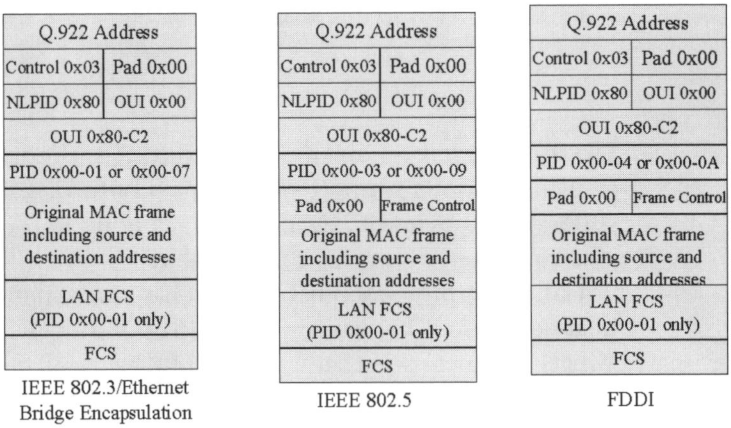

Figure 19-16

RFC 1490 bridge encapsulation.

Q.922 Address	
Control 0x03	Pad 0x00
NLPID 0x80	OUI 0x00
OUI 0x80-C2	
PID 0x00-01 or 0x00-07	
Original MAC frame including source and destination addresses	
LAN FCS (PID 0x00-01 only)	
FCS	

IEEE 802.3/Ethernet
Bridge Encapsulation

Q.922 Address	
Control 0x03	Pad 0x00
NLPID 0x80	OUI 0x00
OUI 0x80-C2	
PID 0x00-03 or 0x00-09	
Pad 0x00	Frame Control
Original MAC frame including source and destination addresses	
LAN FCS (PID 0x00-01 only)	
FCS	

IEEE 802.5

Q.922 Address	
Control 0x03	Pad 0x00
NLPID 0x80	OUI 0x00
OUI 0x80-C2	
PID 0x00-04 or 0x00-0A	
Pad 0x00	Frame Control
Original MAC frame including source and destination addresses	
LAN FCS (PID 0x00-01 only)	
FCS	

FDDI

ferent vendors could communicate over frame relay, for each side would know exactly how to decapsulate and therefore process the packet. This means that a Cisco router and a Bay Networks router can communicate across a frame relay cloud.

Configuration Challenges

As discussed previously, are some challenges in using frame relay. First, it emulates a LAN environment and makes assumptions based on this. It expects to have full-mesh connectivity. Full mesh means that each router has a PVC to every other router in the WAN. This can get messy and expensive. Depending on how you configure your router, this can consume resources in the form of memory and CPU. It can get expensive in that you have to pay for all those PVCs and get them all configured correctly. A better topology is the star or the partial mesh, which offers advantages in both resources and cost. Accomplishing this does not require changing anything about frame relay. The CPE vendors (i.e., those that make routers, bridges, FRADs, etc.) modified their configurations and software to accommodate frame relay. Use the routers to their fullest extent.

Many routers today have useful options to enable any network to become very efficient using frame relay. The ability to place data into different queues is a very powerful technique to control the bandwidth. Priority Queues as well as protocol prioritization allow you to set up multiple queues (simply areas of memory) that will hold packets until an algorithm chooses to forward the packets.

Currently most routers operate under a simple queue function known as FIFO, or first in first out. Cisco, however, uses its own proprietary mechanism, known as *weighted fair queuing*. It is a very fair method for forwarding non–time-sensitive data, and because the Internet or intranet forwards mostly text data, this system works very well. Text (or any data-like information, including graphics) can tolerate delays and dropped packets are retransmitted. However, data transfer of voice and video cannot tolerate delays or dropped packets. This is known as real-time forwarding, and different methods are being devised to facilitate this new type of forwarding.

A queue is nothing more than a holding area. As packets arrive, the front-end filter reads certain user-configurable contents of the packet. Based on a match, the filter will then place the traffic into a queue. As a simple example, the queues can be named high, normal, and low. The normal queue is one that was used in the FIFO calculation. This has priority over the low queue but not the high queue. This is where all traffic comes if there is no filter match. It may also be the queue used for traffic that is very bursty. For example, suppose received IPX SAP information is to be forwarded out a WAN port, but the hardware cannot forward the traffic fast enough, and the packets need a place to stay until the hardware can get to it. The traffic that is delay sensitive (cannot tolerate delays) is placed in the high queue. This packets in this queue are transmitted more often than those in another queue. The traffic that is not delay sensitive and can be sent to the final destination at any time is placed in the low queue. This is the least-looked-at queue for forwarding. Traffic such as email and FTP could be placed in this queue. So manage your network by managing the routers. Queues and filters help out tremendously in a frame relay environment.

Another suggestion is try not to bridge. I know that some protocols that have to be bridged (LAT for DEC terminal and printer traffic and NetBIOS). However, most routers allow NetBIOS to be encapsulated in IP for transmission over a WAN. This may work at your site.

The last thing to monitor is the BECN and FECN bits. A substantially large number of these bits indicate trouble. Use the CIR to control the discard option.

Alternative Frame Relay Implementations

Frame relay has come a long way since its first implementation in 1991. Back then, it simply carried multiprotocol data over PVCs. Since then it has been enhanced and modified to support SNA, voice, and switched virtual circuits (SVCs). Figure 19-17 shows a generic SNA topology and frame relay SNA topology.

Frame relay access devices (FRADs, also known as frame relay assemblers/disassemblers) allow those connections to attach to a frame relay network without the need to have frame relay directly on the equip-

Figure 19-17
Traditional SNA and frame relay SNA.

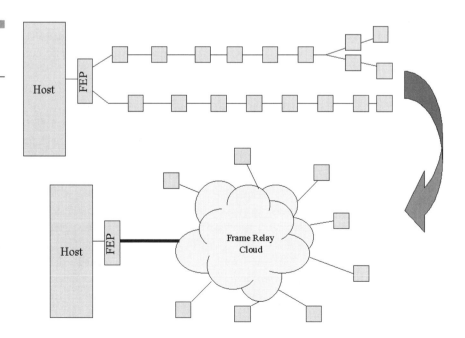

ment itself. For SNA, one end of the connection can be a FRAD and the other a direct connection into the front-end processor (FEP) itself.

Interestingly enough, a FRAD could be a router, based on the definition above. A FRAD does not have to be associated with SNA, although this is usually the case. The key to all SNA or non-SNA frame relay is RFC 1490 encapsulation. This is what makes it possible for SNA, non-SNA, or a mixed environment using FRADs to operate. Another interesting point is that there are no polls involved here. This is not SDLC conversion. This is known as switched-service SNA in that the FEP does not poll and the subsystem does not respond. The subsystem sends data when it has data to send, and the FEP understands this using the switched-service functions.

There are three methods for connection of SNA devices over a frame relay network. Refer to Fig. 19-18. There are many vendors making FRADs, which can lead to interoperability problems. The interoperability standard is that the device must be RFC 1490 compliant. This is

Figure 19-18

SNA over frame relay technologies.

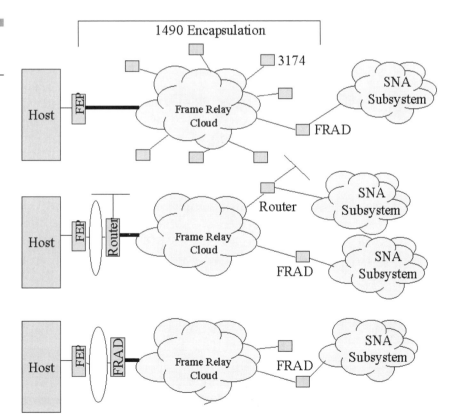

key. Without this, interoperability is impossible to ensure. SNA extensions must be added to the device as well. Remember that SNA is not a FRAD's only function. A FRAD enables devices that do not have frame relay built in with the ability to access frame relay networks. Also, that the 3174 subsystem controller has the capability to connect directly to frame relay as well. This is shown at the top of Fig. 19-18.

The lower part of Fig. 19-18 shows routers connecting SNA devices over frame relay. The routers are running RFC 1490 and Data Link Switching (DLSw). A FRAD could communicate over frame relay to the router using RFC 1490 encapsulation. Also, DLSw can communicate over the frame relay in that DLSw encapsulates data in an IP datagram, which is then sent out of the frame relay circuit, decapsulated by the router, and then forwarded to the FEP.

This is how one could be led to believe that routers can act as FRADs. But if you think about it, any device that implements DLSw, which allows SNA data to be forwarded through the use of IP, in turn supports IP over frame relay and therefore performs FRAD functions for LAN SNA devices. This has many advantages in that LLC2 devices are locally polled (instead of sending the polls over the frame relay network), eliminating possible time-outs, and FRAD-to-FRAD messages are guaranteed. This, along with running RIP or OSPF as the routing protocol, allows rerouting to take place if necessary. An added advantage of this is the interoperability of FRADs, routers, and SNA devices over frame relay. Again, this is due to RFC 1490 specifying a standard encapsulation for both bridging and routing, so that any 1490 packet should be able to be interpreted.

The bottom illustration shows end-to-end FRAD devices transporting SNA data from equipment that does not have the capability of connecting to the frame relay network directly.

Frame relay continues to displace the leased line environment, but as of this writing leased lines still account for over 60 percent of the WAN market. But frame relay is making a steady and progressive forward path.

One environment that may deter the advance of frame relay is not a new WAN protocol, but rather the Internet service provider (ISP). Yes, ISPs give you access to the Internet, but they are starting to provide much more than that. They are enabling their networks to carry all of a corporation's data. This means that the ISP can provide the WAN service for a corporation.

How this is accomplished is very similar to the frame relay topology. Frame relay providers give WAN access to many customers, but they all basically use the same equipment (the very definition of packet switching). ISPs are providing connections to customer sites through their own POPs (points of presence, spaces where customers connect to the provider and the provider connects to the phone company for transmission to other locations).

Does this mean that ISPs are becoming small phone companies but mainly for data? Yes, in a sense you can think of it in this way in terms of topology. However, the ISPs are not phone companies, and in fact, they need the phone company lines to communicate over their networks. This can get very complicated and it is very involved, especially since telecommunications deregulation took place.

The best way to think about it is that ISPs are providing WAN backbones to corporations so that corporations do not have to provide this internally. This means that the ISPs will have the frame relay connections, and the growth will be (for frame relay) within the ISP environment. A corporation could have a leased or dial (ISDN) line to the first POP of the ISP. The ISP will then route that data over its backbone to the intended destination POP. This could be over frame relay or any other WAN technology or any combination of WAN technologies. This is accomplished using the technology of virtual private networks (VPNs), a function not covered in this book.

But ISPs are young and they are not proven. They do not come close to the coverage of a phone company such as MCI, BellSouth, etc. They provide ubiquitous service not only locally, but internationally as well.

CHAPTER 20

ISDN

Adapted from Gary C. Kessler and Peter Southwick, *ISDN: Concepts, Facilities, and Services,* signature edition, New York: McGraw-Hill, 1998

I. Telecommunications Background

Before discussing any details of ISDN or B-ISDN technology, standards, protocols, or implementations, it is necessary for the reader to have some baseline understanding of certain aspects of telecommunications. Here we review some of the data and telecommunications background topics relevant to this understanding and provide a broad overview rather than in-depth analyses or motivation.

- *Communications basics.* Introduces terms and concepts such as analog and digital signaling, amplifiers and repeaters, and passband and bandwidth, as well as the structure of the U.S. telephone network, the telephone local loop, and multiplexing.
- *Digital telephony.* Discusses why the telephone network has migrated from an analog network to a digital one and describes how human voice is digitized, how digital signals are carried on the local loop, and how full-duplex communication is accomplished on a digital local loop. The digital carrier hierarchies are also introduced.
- *Types of switched networks.* Defines, compares, and differentiates among circuit switching, packet switching, and fast packet switching technologies.
- *Open Systems Interconnection (OSI) Reference Model.* The OSI model describes a framework for network communications protocols. The OSI model will be introduced and defined, as will be the protocol architecture for packet switching and X.25.

Communications Basics

Analog and Digital Signals

One of the most important concepts for our discussion of telecommunications is that of a *signal.* Signals are the representation of information. In today's communications systems, signals are usually an electrical current or voltage, where the current or voltage level is used to represent data. Communications systems can employ either analog or digital signals (Fig. 20-1).

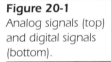

Figure 20-1
Analog signals (top)
and digital signals
(bottom).

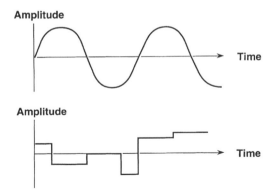

An *analog signal* is one that can take on a continuous set of values within some given range, to directly represent information. Examples of analog signals are human voice, video, and music. Analog signals are also sometimes referred to as modulated signals.

A *digital signal* is one that may take on only a discrete set of values within a given range, such as a battery that can supply either 13 or 23 volts (V). Binary signals, in particular, are digital signals that may take on only two values, 0 or 1. Digital signals are sometimes referred to as unmodulated signals.

The distinction between analog and digital signals is important. Sound is produced when air is set in motion by a vibrating source, such as the vocal cords or the strings of a guitar. When the vibrating source pushes out, the air particles around it are compressed; when the source pulls back, the air particles are pulled apart. This compression and expansion of air particles causes a chain reaction through the air away from the original source, generating a sound wave. Sound waves are analog signals, where each in-and-out vibration is called a *cycle*. The frequency of the signal is measured in cycles per second, or hertz (Hz). A sound wave with a frequency of 261.63 Hz, for example, is the musical note middle C.

An analog signal can be transported over an electrical circuit if the sound waves can be made to alter the characteristics (voltage, frequency, or amperage) of the electrical circuit so as to represent the analog signal. This function is performed by a microphone, similar to the one found in your telephone.

Human voice comprises a particular type of analog signal, namely, a mixture of sinusoidal (sine) waves. The telephone network has been designed specifically to handle analog, human voice signals. An ISDN will only carry digital signals, although we will still want to send human voice through the network. Later discussion in this section describes how human voice is carried in a digital network.

Amplifiers and Repeaters

The analog telephone network contains amplifiers to boost the signals so that they can be carried over long distances. Unfortunately, all copper media (e.g., twisted pair and coaxial cable) act as an antenna to pick up electrical signals from the surrounding environment. This noise can come from many sources, such as fluorescent lights, electric motors, and power lines. An amplifier, by the nature of its function, boosts the signal (e.g., the human voice) and background noise equally. The effects of noise, then, are additive in an analog network; the noise boosted by one amplifier becomes input to the next amplifier.

Amplifiers in the analog network are poorly suited for digital signals. Digital signals are represented as square waves, although they leave the transmitter in a rounded-off form, looking much like an analog signal. An amplifier, then, would accept a poorly formed digital signal plus any noise on the line and output a louder, degraded signal.

Digital networks use signal regenerators (or repeaters) instead of amplifiers. A signal regenerator accepts the incoming digital signal and then creates a new outgoing signal. Repeaters are typically placed every 6000 feet [ft, or 2 kilometers (km)] or so in the long distance digital network. Since the signal is regenerated, the effects of noise are not additive from repeater to repeater.

Structure of the Telephone Network

The telephone network structure in the United States has an interesting history that is beyond the scope of this book. Nevertheless, the evolution of the network over the last 15 years is worth examining because of its impact on user services during the 1980s and beyond.

The Predivestiture Network   Figure 20-2a shows the major components of the public switched telephone network (PSTN) in North America prior to 1984. The implementation of this hierarchical network was started in the 1930s and essentially completed in the 1950s; it has undergone continual modifications as technology and population demographics have changed and the effects of divestiture have been felt. The 10-digit telephone numbers commonly used in North America expedite call routing within this switching hierarchy; in the simplest of implementations the first three digits identify the area code, the next three digits identify the end office, and the last four digits identify the end user.

ISDN

Figure 20-2a
Telephone network switching hierarchy in the United States (predivestiture).

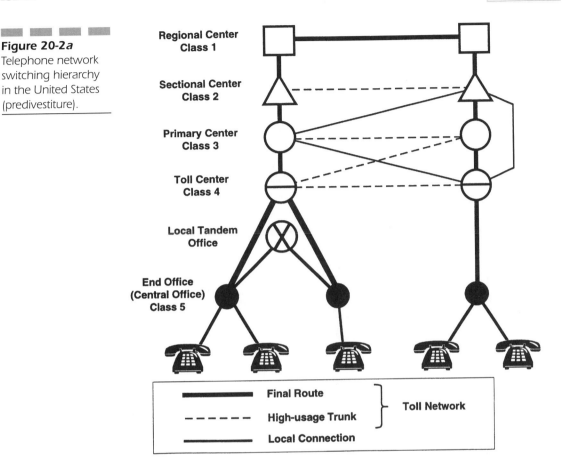

Regional Center
Class 1

Sectional Center
Class 2

Primary Center
Class 3

Toll Center
Class 4

Local Tandem
Office

End Office
(Central Office)
Class 5

———— **Final Route**	⎫
– – – – – **High-usage Trunk**	⎬ **Toll Network**
———— **Local Connection**	⎭

An end user, such as a residential or business customer, is directly connected to a telephone network end office, also called a class 5 or C.O. Users are typically connected to the C.O. over a single twisted wire pair called the *local loop*. In today's telephone networks, the number of customers requesting service has often outnumbered the capacity of the end offices. Rather than invest in larger and larger switching systems, network providers often install multiple switches in a single C.O. or wire center. This has led to some confusing terminology. For the purpose of this book, the building which houses the switching equipment is called the C.O. and the equipment providing the switching service is called the *local exchange* (LE). For large networks, a single C.O. can contain multiple LEs.

Class 4 offices served two functions. As a toll switch, a class 4 office was part of the long distance (toll) network. Alternatively, a class 4 office could act as a tandem switch, to interconnect those class 5 offices not having sufficient interoffice traffic to justify direct trunks. Local tandem office switches also handled overflow traffic on direct trunks between end offices. The distinction between toll and tandem became particularly important in the United States after the 1984 divestiture of AT&T resulted in the separation of communications resources within the network.

Routing between C.Os uses the fewest number of intermediate switching offices to minimize the cost of carrying the traffic. The actual route selected depends on such factors as the distance between the two C.Os, the current network traffic level, and time of day.

A connection between two users who are physically connected to the same C.O. requires only the involvement of that single switching office. Where two subscribers were connected to different C.Os and the two class 5 offices were attached to the same class 4 office, that toll center would make the connection. When the C.Os were further apart, other switching offices were used, although it was not necessary that class 5, 4, or 3 offices always connected through the next higher level of switch. A higher-class switch could perform lower-class switching functions; a class 5 office, for example, could be served by a class 4, 3, 2, or 1 office.

The final (primary) route structure shown in Fig. 20-2a was supplemented by an alternate routing structure. To minimize heavy traffic loads at the higher levels and signal degradation when the route involved many trunks and switching offices, high-usage trunks were used between any switching offices where economically justified.

By 1980, there were over 19,000 class 5 offices and 200 million local loops in the Bell System in the United States. There were also about 900 class 4 offices, 204 class 3 offices, 63 class 2 offices, and 10 class 1 offices.

The Modification of Final Judgment (MFJ), signed by Judge Harold H. Greene in August 1982, represented the settlement of the U.S. Government's 1974 antitrust suit against AT&T. The MFJ was the basis of the breakup of AT&T and the Bell operating telephone companies.

AT&T's Plan of Reorganization, filed in December 1982, provided the new structure and organization of the U.S. telephone industry after January 1, 1984. According to the plan, AT&T would retain long distance communication services and communications equipment manufacturing businesses, as well as other assets such as Bell Laboratories. Local telephone service, provided by 22 Bell operating companies (BOCs), were organized into seven regional Bell holding companies (RBHCs), or regional BOCs (RBOCs). All Bell System assets were assigned to either AT&T or an RBOC, since the MFJ disallowed joint ownership of switching equipment. As a result, AT&T kept the toll network switches (class 1 to 4 offices and some

Figure 20-2b
Telephone network
switching hierarchy
(postdivestiture).

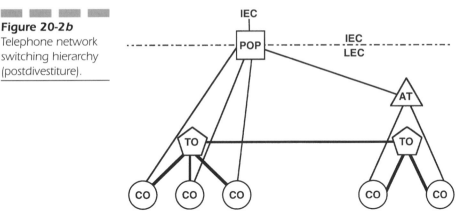

Figure 20-2b
Telephone network
switching hierarchy
(postdivestiture).

tandem switches) and the RBOCs kept local switching equipment (class 5
and some tandem offices).

The Postdivestiture Network   Figure 20-2b shows the telephone switching hierarchy that resulted from divestiture. Local telephone service areas
were redrawn by divestiture into approximately 250 local access and transport areas (LATAs). The RBOCs, of course, are not the only local telephone
companies. There are several hundred independent telephone companies
(ITCs) in the United States, which are non-Bell-system providers such as
Cincinnati Bell, Frontier Telephone,[1] General Telephone (GTE), Southern
New England Telephone (SNET),[2] and others. The LATAs were defined
ostensibly to ensure a fair and equal market to the RBOCs and ITCs. Since
the BOCs and ITCs are limited to transporting traffic only within a LATA
(intraLATA), they are called local exchange carriers (LECs). The LECs maintain C.O.s and tandem offices for high-volume inter-C.O. traffic.

Transport of traffic between LATAs (interLATA) is carried by the
interexchange carriers (IECs), such as AT&T, MCI, Sprint, and WorldCom.[3]
There are a few exceptions to this rule; several high-traffic corridors have
been defined, such as in the New York City area, where the LEC provides
some interLATA service. The interLATA toll network is owned by the
IECs. The IECs maintain point-of-presence (POP) switches to carry inter-

[1]Formed after the purchase of Rochester Telephone by Long Distance North.
[2]Now affiliated with SBC.
[3]Formed by the merger of LDDS and WilTel.

LATA traffic. C.Os may connect directly to the POP or via an access tandem (AT) switch.

IntraLATA traffic has typically been carried by the LEC. This market has been opened for competition, so now a customer will be served by the LEC and can choose an IEC for interLATA traffic and another carrier for intraLATA toll calls. This environment has created a very large number of service providers and complicates the orderly network structure of the old Bell system.

The Telecommunications Act of 1996 is again changing the structure described above. Where LECs were once limited to providing local service only (and limited to a given geographic region) and IECs to long distance service, we will soon find LECs and IECs competing equally for both local and long-haul service. In addition, we will see an increasing number of competitors in the local access market (once a monopolistic haven for the LEC). These competitive local exchange carriers (CLECs) will offer local dial tone and other voice and data services.

Passband and Bandwidth

Before analyzing the requirements for transmitting human voice in a digital form, we must first define the bandwidth associated with voice and the telephone local loop.

Recall that the frequency of an analog signal is the number of complete sine waves (or vibrations) sent every second and is measured in cycles per second, or hertz. The *passband* of a channel is the range of frequencies that can be carried by that channel; the *bandwidth* is the width of the passband. For example, while one television channel may use the 470.5- to 476.5-megahertz (millions of hertz, MHz) passband and another channel uses the 800- to 806-MHz passband, both channels have a bandwidth of 6 MHz.

Human voice can produce sounds in the approximate frequency range of 30 to 10,000 Hz (10 kilohertz, or kHz), for a bandwidth of 9.97 kHz. The ear can hear sounds in the 20- to 20,000-Hz frequency range (19.98-kHz bandwidth).

The Telephone Local Loop

The telephone local loop has a 4-kHz-band channel with the frequency range from 0 to 4000 Hz. This channel actually carries human voice in the frequency range of roughly 300 to 3400 Hz. This may be surprising con-

Figure 20-3
Average speech sig-
nal energy.

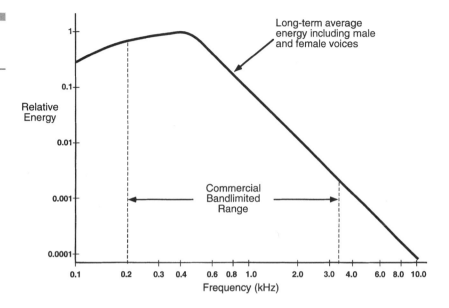

Figure 20-3
Average speech signal energy.

sidering that the human voice produces sounds between 30 and 10,000 Hz. How can a channel with a bandwidth of 3.1 kHz carry the information content of a source with a 9.97-kHz bandwidth?

In fact, the local loop is not meant to carry just any analog signal but is optimized for human voice. Figure 20-3 shows that the major portion of the relative energy of the human voice signal is in the passband from about 200 to 3500 Hz. This is the frequency range where the bulk of the power, intelligibility, and recognizability of the voice signal occurs. Thus, the 300- to 3400-Hz passband is adequate for acceptable-quality human voice transmissions.

Note that a channel with a bandwidth of 3.1 kHz cannot carry all of the information in the voice's frequency range. Voice can be limited to a 3.1-kHz band because the ear can obtain most of the necessary information for conversation (namely, intelligibility and recognition) from that narrow band.

Consider the case of music, however, which is intended to be pleasing over a larger frequency spectrum than is required in normal voice conversation. A transmission facility carrying music must use a larger bandwidth than voice. Think about what happens when someone plays a musical instrument over the telephone; it sounds flat and tinny because it is missing all frequency components below 300 Hz and above 3400 Hz.

The primary reason that the telephone network uses the narrow 3.1-kHz band rather than the entire 10 kHz of voice is that the narrow band allows more telephone conversations to be multiplexed over a single physical facility. This is particularly important for the facilities connecting telephone switching offices.

Multiplexing

Multiplexing in a network allows a single resource to be shared by many users. In particular, multiplexers in the telephone network allow many voice conversations to be carried over a single physical communications line.

Analog communications facilities typically use frequency division multiplexing (FDM) to carry multiple conversations. FDM divides the available frequency among all users and each user has an assigned channel for as much time as necessary (Fig. 20-4). In the case of voice, each conversation is shifted to a different passband with a bandwidth of approximately 4 kHz (3.1 kHz for the voice signal and 900 Hz for guard bands to prohibit interchannel interference). Since the bandwidth is held constant, the integrity of the user's information is maintained even though the passband has been altered.

FDM is a scheme with which we are all familiar. Television stations, for example, each require a 6-MHz passband and all TV channels simultaneously share the available bandwidth of the air. The TV set in our house, then, acts as a demultiplexer to tune in only the passband (i.e., the channel) that we want to watch. This is also the same principle used for cable TV and radio channels.

Digital signals are typically multiplexed on a communications facility using time division multiplexing (TDM). Whereas FDM provides each user with part of the frequency spectrum for all of the time that the user requires, a TDM scheme provides each user with the entire frequency spectrum for a small burst of time (Fig. 20-5). In the figure, time slots are granted on a round-robin basis to the five users who share the channel.

Figure 20-4
Frequency division multiplexing.

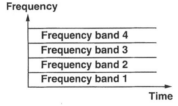

Figure 20-5
Time division multi-
plexing.

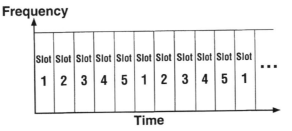

Digital Telephony

The Move to a Digital Telephone Network

Prior to the early 1960s, the United States telephone network was an integrated analog network, meaning that all devices, facilities, and services were analog in nature. In fact, plain old telephone service (POTS) was all that was offered; the telephone network merely carried analog utterances from one point to another.

While the telephone local loop today is still predominately analog, the rest of the network has been migrating toward using digital facilities for over 30 years. In the United States, AT&T started investigating the use of digital transmission facilities and switches in the 1950s. The first digital lines used in North America were T1 carriers, introduced into the network in 1962 for traffic between telephone switching offices. T1 carriers use two wire pairs to separate the transmit and receive function, as opposed to the local loop's single pair.

The first digital switch, AT&T's Number 4*ESS* (Electronic Switching System), was introduced into the toll network in 1976 and Nortel[4] introduced their digital toll switch, the DMS-200, in early 1980. Throughout the 1980s, digital switches were introduced into C.O.s as well, notably Nortel's DMS-100 and AT&T's No. 5*ESS* switch, introduced in 1981 and 1982, respectively. The majority of C.O.s in the United States now use digital switches.

The introduction of digital switches and digital carrier facilities allowed portions of the telephone network to operate more efficiently and meant that new types of communications services could be offered to some customers. Even if all switching offices and interoffice trunks in

---

[4]Formerly Northern Telecom International (NTI).

the network were digital today, the network still contains analog local loops. The "analog" local loop, then, is the weakest link in the "end-to-end digital" chain, and POTS continues to be the primary service available to most end users.

The ultimate goal is to build an IDN, which is a network where all switches, interoffice trunks, local loops, and telephones are digital.

There are a number of reasons for converting network facilities from analog to digital, but the overriding one is economy. Digital facilities and digital devices are less expensive to design, build, and maintain than comparable analog devices. Indeed, the microprocessor revolution of the last 25 years has caused digital devices to propagate to all facets of life and to plummet in price, while the cost of analog devices has remained relatively stable. Another reason for this conversion is that the digital equipment results in less noise being transmitted along with the information signals. This means that a digital facility will provide "cleaner" communications paths. Finally digital devices are less prone to mechanical failure.

Digital C.Os are easier to design because the computer *is* the switch. Since most digital C.Os use some form of TDM, the speed of a *digital switch* can be increased by defining more time slots. The limiting factor is the speed at which signals can be turned on and off within the digital devices. As computer processors become faster and relatively less expensive, it becomes easier to build faster and larger digital switches.

Once the network has been completely converted to digital, it is a natural conceptual step to observe that many types of different services may be carried by this network if all these services are delivered in a digital form.

Digitizing Voice and Pulse Code Modulation

To carry human voice in a digital form, the voice signal is sampled 8000 times per second. This sampling rate is based upon Harry Nyquist's Sampling Theorem, which shows that to be able to accurately reproduce an analog signal from a series of samples, sampling must occur at twice the highest frequency of the signal. The local loop passband is taken to be between 0 and 4000 Hz (the total bandwidth of an FDM voice channel including both the voice signal and guard bands). The maximum frequency, 4 kHz, requires a sample rate of 8000 times per second, corresponding to a sample interval of 125 microseconds (μs).

Each sample of the voice signal is converted to a digital bit stream. The process of converting the analog sample to a bit stream is pulse code mod-

Figure 20-6
Pulse code modulation.

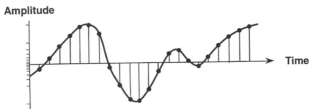

ulation (PCM) and is performed by a device called a CODEC (COder-DECoder). The CODEC may be located in a digital switch, in which case the local loop between the telephone and switch carries analog signals. Alternatively, the CODEC may be placed in the telephone set, in which case the local loop carries digital signals.

Figure 20-6 shows the voice digitization scheme. The voice signal is sampled once every 125 μs, or once every 1/8000 s. This sampling, called pulse amplitude modulation (PAM), results in an analog level corresponding to the signal at that moment. The amplitude of the PAM sample is mapped to a discrete value on the amplitude axis; this digital encoding is the PCM step.

Note the nonlinearity of the PCM amplitude scale. The amplitude levels are defined to be closer together at lower volumes and farther apart at higher volumes; this is called *companding* (compression expanding). There are two main companding algorithms used in digital telephony; the μ-law is used primarily in the United States, Canada, and Japan, while the A-law is used throughout most of the rest of the world.

The PCM companding rules define 255 (μ-law) or 256 (A-law) amplitude levels; therefore, each voice sample is coded as an 8-bit word. Since 8000 samples are taken each second, the bit rate of a single voice channel is 64,000 bps. This is sometimes referred to as digital signaling level 0, or DS-0.

It should be noted that the analog voice signal cannot be mapped exactly onto the digital amplitude scale. Thus, the digitized signal is not an exact replica of the original signal; the difference is called the *quantization error*. The quantization error has an additive effect, so the error becomes greater each time the signal undergoes an analog-to-digital (A/D) or digital-to-analog (D/A) conversion (which was the case when switches were analog and trunks were digital). Due to the large number of amplitude levels and the use of companding, the quantization error is minimized, and what little remains is easily compensated for by the listener.

Companding reduces the effect of the quantization error by utilizing a nonlinear scale on the amplitude axis. By concentrating on the lower-amplitude (volume) signals as shown in Fig. 20-6, PCM is able to achieve 9-bit (512 levels) accuracy while only actually employing an 8-bit code. For the comfort of most humans, PCM is designed to catch the subtleties and nuances when people are talking softly rather than when they are yelling. PCM with companding (A-law or μ-law) meets the quality standards of the analog telephone network and is considered "toll quality."

The Digital TDM Hierarchy

T1 carriers were the first digital carriers employed in the United States. A T1 carrier multiplexes 24 voice channels over a single transmission line using TDM. The basic unit of transmission is a *frame*, which contains one PCM sample from each of the 24 channels. Since a sample is represented by 8 bits, a single frame contains 192 bits of user data. Each frame is preceded by a single framing bit; thus, a single T1 frame contains 193 bits (Fig. 20-7). Since each frame contains one sample from each voice channel, there must be 8000 frames per second on the T1 channel. This yields a bit rate of 1.544 Mbps, which is also known as digital signaling level 1 (DS-1). Since 8000 bps are for framing, the actual user data rate is 1.536 Mbps.

The T1 carrier multiplexes 24 voice channels using TDM. Several T1 carriers, in turn, can be multiplexed using TDM to form even higher-speed carrier channels. The number of channels multiplexed together is defined by the digital TDM hierarchy. The T-carrier system in the United States, with its associated digital signal (DS) level designator, follows the AT&T digital TDM hierarchy (Table 20-1). A portion of this hierarchy is also used in Canada, Japan, Taiwan, and South Korea.

The other widely used TDM hierarchy is based upon the Conference of European Postal and Telecommunications (CEPT) administrations standard (Table 20-1). The first level of the CEPT digital hierarchy multiplexes 32 time slots (each with 8 bits, yielding 64-kbps channels), yielding a frame with 256 bits (Fig. 20-8) and a bit rate of 2.048 Mbps. One of the 32 time slots is used for signaling, one is used for frame alignment, and the remaining 30 are used for actual user data, resulting in a user data rate of 1.920 Mbps. This is referred to as the CEPT level 1, or E1, frame format.

"T" and "E" carriers predate digital switching. Therefore, in the early days of digital carriers, the switches were still analog. That meant that a CODEC had to be placed on both ends of every digital carrier between

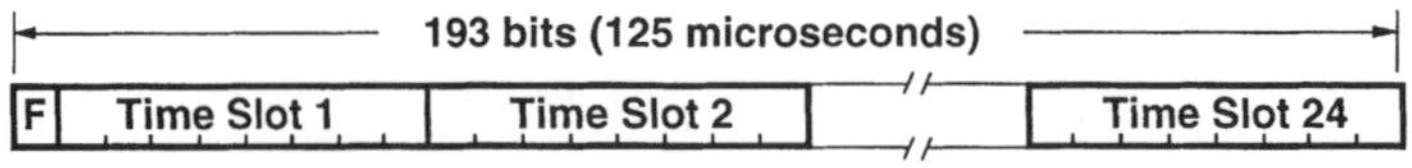

Figure 20-7
T1 frame format.

193 bits (125 microseconds)

| F | Time Slot 1 | Time Slot 2 | // | Time Slot 24 |

TABLE 20-1

TDM Hierarchy
Used in North
America, Europe,
and Japan

Digital Multiplexing Level	Number of Equivalent Voice Channels	Bit Rate (Mbps)		
		N. America	Europe	Japan
DS-0/E0/J0	1	0.064	0.064	0.064
DS-1/J1	24	1.544		1.544
E1	30		2.048	
DS-1C/J1C	48*	3.152		3.152
DS-2/J2	96	6.312		6.312
E2	120		8.448	
E3/J3	480		34.368	32.064
DS-3	672	44.736		
DS-3C	1344*	91.053		
J3C	1440*			97.728
E4	1920		139.264	
DS-4	4032	274.176		
J4	5760			397.200
E5	7680		565.148	

*Intermediate multiplexing rates.

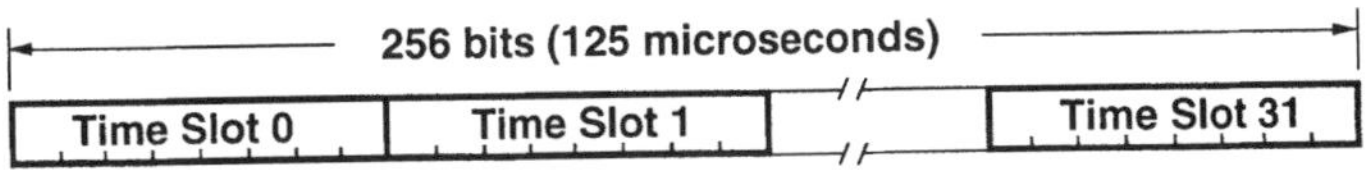

Figure 20-8
E1 frame format.

every pair of offices. Thus, one telephone connection might be routed through several switches, requiring that the coding/decoding process be performed several times. In a fully digital network, there will be a single coding and single decoding step, since the CODEC will be part of the end-user equipment. Even if the local loop is analog, the CODEC will be placed in the C.O., which would still only require a single coding and decoding operation.

The telecommunications industry realized many years ago that higher speeds could be economically achieved by using an optical fiber medium.

To ensure international compatibility at higher speeds, work began in the mid-1980s to define a single digital hierarchy based on fiber and able to incorporate the "low-speed" copper-based digital hierarchies. The digital hierarchy for optical fiber is known in North America as the Synchronous Optical Network (SONET).

The SONET optical hierarchy is based upon building blocks in increments of 51.84 Mbps, roughly corresponding to the DS-3 line rate. The 51.84-Mbps rate is called the Synchronous Transport Signal level 1 (STS-1) when referring to an electrical signal or Optical Carrier level 1 (OC-1) when referring to an optical signal. Standards already define the format for rates from 51.84 Mbps (OC-1) to 9953.28 Mbps (OC-192), as shown in Table 20-2.

SONET not only defines high-speed communications over fiber but also a consistent multiplexing scheme. With SONET, an OC-n line rate is exactly n times 51.84 Mbps, so an OC-n transmission is formed by byte-multiplexing n OC-1 frames. This results in the very straightforward design of SONET multiplexers. It is also very different from the T1 and CEPT hierarchies where different levels use different multiplexing and framing schemes.

The SONET standard is specified for North America; its international counterpart is known as the Synchronous Digital Hierarchy (SDH). The main format difference between the two is that the basic SDH rate is 155.52 Mbps, designated Synchronous Transport Module level 1 (STM-1), which is equivalent to SONET's OC-3/STS-3. SDH rates are also shown in Table 20-2.

TABLE 20-2

SONET Optical Carrier (OC) and SDH Synchronous Transport Module (STM) Levels

Line Rate (Mbps)	SONET Level	SDH Level
51.840	OC-1	
155.520	OC-3	STM-1
466.560	OC-9	
622.080	OC-12	STM-4
933.120	OC-18	
1244.160	OC-24	STM-8
1866.240	OC-36	STM-12
2488.320	OC-48	STM-16
4976.640	OC-96	STM-32
9953.280	OC-192	STM-64

Digital Signals on the Local Loop

A major stumbling block to sending digital signals between the C.O. and customer site is today's local loop. The local loop comprises a twisted pair of 22-26 gauge copper wire. The twisting of the wire pair reduces the crosstalk and interference from multiple pairs within the same bundle of wires within a cable. The average length of a local loop in the United States is about 18,000 ft (18 kft, or 5.5 km).

Load coils, which are induction loops, counteract the buildup of capacitance created by long runs of twisted pair. Load coils are placed on the local loop to reduce the distortion of voice frequencies in the wire pair. While the load coils ensure that the voice signal is recognizable after traveling the distance between the customer site and the C.O., they effectively limit the voiceband to frequencies below 4000 Hz. *Bridged taps* are also present on the local loop; they reduce installation time for new customer connections but also negatively affect digital transmission on the loop by attenuating the transmitted signals. In the United States, loops longer than 18,000 ft (5.4 km) have load coils on the loop every 6000 ft (1.8 km), starting 3000 ft (900 m) from the central office.

In the United States today, the composition of local loops is

- Nonloaded loops no more than 18 kft in length (70 percent).
- Loaded loops greater than 18 kft in length (15 percent).
- So-called *derived loops,* comprising nonloaded loops up to 12,000 ft in length connected to remote wiring distribution equipment connected to the central office via a fiber or digital carrier (15 percent).

The problem with the analog local loop is that while 3.1 kHz is sufficient for carrying human voice signals, it is not sufficient for carrying the frequencies required to represent high-speed digital data. Square waves are composed by combining sine waves of different frequencies. Stable, recognizable square waves require frequency components much higher than 4000 Hz, making the loaded local loop inadequate for digital communication.[5]

The 4000-Hz bandwidth limitation of the local loop is imposed by the network architecture and not by the physical medium itself. In fact,

---

[5]Fourier's theorem states that any repeating, periodic waveform can be approximated by the sum of a (possibly infinite) set of sine waves. A 1000-Hz square wave, for example, would require sine wave components with frequencies 500, 1500, 2500, 3500, ... Hz. Even this relatively low-speed square wave would be severely degraded in the local loop's 300- to 3400-Hz passband.

twisted pair may be used in analog telephony applications with a bandwidth up to 250 kHz, which requires amplifiers every 16.5 to 19.8 kft (5 to 6 km), and in digital applications with bit rates over 6 Mbps, which requires repeaters approximately every 9 kft (3 km). Therefore, digital communication over twisted pairs is possible once load coils are removed and bridged taps are accounted for on the line. This unloaded local loop, if used for digital applications, is called a digital subscriber line (DSL). Chapter 23 of *ISDN* covers the family of DSL standards and protocols.

It should be noted that some current LAN products and standards utilize unshielded twisted pair media at data rates above 100 Mbps; SONET/ SDH speeds (155 Mbps) can also be achieved. The length of the wire, however, is limited to a hundred or so meters, well short of the local loop requirements of several miles. This distance limitation is due to the attenuation of frequencies of the signal. As data transfer rates increase, the usable distances decrease, if all other factors remain the same. As an example, a 20-kHz signal will travel, and be recognized, twice as far as a 40-kHz signal.

Full-Duplex Communication on the Local Loop

A nonloaded local loop can carry digital transmissions. The next step is to accomplish simultaneous, two-way (full-duplex) communication over a digital loop.

Today's analog local loops carry sounds in both directions at the same time. The voice signals from both parties are on the local loop simultaneously; in addition, bridged taps, wire gauge changes, and splices can cause echo of the signal back to the transmitter. Full-duplex communication over the local loop is not a problem in analog applications. When people talk over the telephone, the brain filters out their words when they are echoed back. For data applications, modems typically split the bandwidth of the local loop in half to achieve full-duplex communication; the originating modem will usually transmit in the lower half of the passband, and the answering modem will usually transmit in the upper half.

Splitting the bandwidth in half is not possible in a digital environment, since digital signals cannot be confined to a given passband. Alternatively, the T1 approach could be adopted; two pairs of wire could be used, one for transmit and one for receive. This solution, however, is not a viable one for the local loop since hundreds of millions of miles of new cable would have to be installed in the United States alone. Instead, two other approaches are used to achieve full-duplex digital communication over a two-wire DSL.

The first method is called time-compression multiplexing (TCM). TCM works as follows: If we wish a facility to operate at x-bps full-duplex, we can simulate that by operating the facility at $2x$-bps half-duplex, where each data stream travels in opposite directions over the shared facility at different times. TCM requires facilities at both ends of the communications channel to constantly and quickly turn the line around, an operation called *ping-ponging*.

The half-duplex facility, in fact, really has to operate at a rate somewhat above $2x$ bps to accommodate facility turnaround time and propagation delay. In those systems employing TCM, most 56- and 64-kbps full-duplex signals are carried on a 144-kbps half-duplex channel, a ratio of 2.57:1 and 2.25:1, respectively. TCM was developed in the early 1980s for AT&T's Circuit Switched Digital Capability service and is used today over two-wire facilities in AT&T's Switched 56 and Nortel's Datapath services. It was proposed for ISDN local loops but is not used due to the restrictions in cable length (9000 ft).

The second approach to achieving full-duplex communication over the DSL is to use a device called a *hybrid with echo canceler*. The hybrid circuit mixes and separates the transmit and receive signals onto a single twisted pair. An echo canceler does exactly what its name implies; the transmitter remembers what it is sending and "subtracts" the appropriate signal from the incoming transmission, thus eliminating the returning echo. This requires complex algorithms but, in fact, is the method of choice for use in high-speed modems and on ISDN local loops. Echo cancelers are discussed in more depth in Chap. 5 of *ISDN*.

Types of Switched Networks

To fully understand and appreciate ISDN services, it is necessary to understand circuit, frame, and packet switching (*circuit-, frame-,* and *packet mode,* respectively, in the ISDN vernacular). These switching techniques are in common use today and are supported by an ISDN. Before discussing these types of switching, it is useful to examine the characteristics of voice and data calls.

Voice calls typically have the following characteristics:

■ *Delay sensitive.*   Silence in human conversation conveys information, so the voice network cannot add (or remove) periods of silence.

■ *Long hold time.*   Telephone calls usually last for a relatively long time compared to the time necessary to set up the call; while it may take 3 to 11 s to set up a telephone call, the average call lasts for 5 to 7 minutes (min).

■ *Narrow passband requirement.*   As we have already seen, a 3.1-kHz passband is sufficient for human voice. Furthermore, increasing the bandwidth available for the voice call does not affect the duration of the call.

Data calls have different characteristics, including:

■ *Delay insensitive.*   Most user data does not alter in meaning because of being delayed in the network for a few seconds; a packet containing temperature information from the bottom of Lake Champlain, for example, will not change in meaning due to a short delay in the network.

■ *Short hold time.*   Most data traffic is bursty (i.e., the bulk of the data is transmitted in a short period of time, such as in interactive applications). A 90/10 rule is often cited to demonstrate this: 90 percent of the data is transmitted in 10 percent of the time. Since data transmission will tend to be very fast, long call setup times yield inefficient networks.[6]

■ *Wide passband utilization.*   Data can use all of a channel's available bandwidth; if additional bandwidth is made available for a data call, the duration of the call can decrease.

Figure 20-9 shows the general structure of a switched network. Hosts (end users) are connected to a network to gain communications pathways to each other. Nodes are switches within the network. In a switched network, the path between a pair of hosts is usually not fixed. Therefore, host 1 might connect to host 3 via nodes A, C, E, D or via nodes A, B, D.

In the telephone network, telephones could be considered hosts while LEs are nodes. The network is switched because two end users do not have a permanent, dedicated path between them. Instead, a path is established on request and released at the end of the call. One primary difference between types of switching networks is whether the communications channel allocated between nodes for a given call is shared with other calls or not.

Circuit Switching

Circuit switching is the most familiar type of switching to most people; the telephone network provides an excellent example of this type of network. In

---

[6]For the purpose of this discussion, a data call is defined as a single interactive session between hosts. This is not to be confused with the amount of time spent on line by a person (i.e., surfing the net).

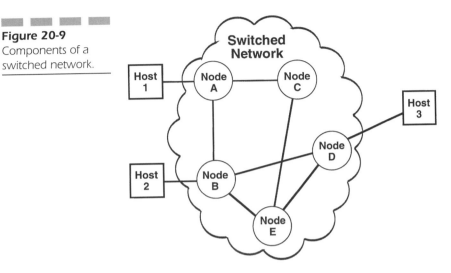

Figure 20-9
Components of a
switched network.

a circuit switched network, the communications pathway between two users is fixed for the duration of the call and is not shared by other users. During the call, the circuit is equivalent to a physical pair of wires connecting the two users. Although several users may share one physical line by use of FDM or TDM, only one user is assigned to a single channel at any given time.

In circuit switching, a connection is obtained between two users by establishing a fixed pathway through the network. The route is established after the calling party initiates the call setup procedure by telling the network the address of the called party (i.e., after the user dials the telephone number).

Circuit-switched connections are well suited for voice traffic. The dedicated pathway is required due to the delay-sensitive nature of voice calls. Also, the long call setup time is compensated for by the relatively long call hold time.

For similar reasons, circuit-mode connections are not as well suited for data calls. The bursty nature of data means that a long call setup procedure wastes time for a short transmission of data. Since the circuit switched network is optimized for human voice, all channels have a narrowband passband (4 kHz or 64 kbps); again, this means that data calls will have a longer duration. If the connection is maintained between sessions, dedicating the channel means that the channel is idle much of the time. While data may be (and is) carried over circuit switched facilities, it is an inefficient use of those facilities and not optimal for data transmission.

It should be noted that a new set of digital circuit-mode services have entered the market. These services offer wideband circuit-switched channels at rates to 45 Mbps. These dial-up wideband services are typically employed as backup facilities for disaster recovery operations where the value of the connection more than compensates for the data's inefficiencies.

Packet Switching

Packet switching was first described for data communications in the early 1960s. With packet switching, there is no dedicated end-to-end physical connection between two users during information exchange; instead, users submit their messages to the network for delivery to another user. The end-to-end connection between users, then, is logical rather than physical. Since internode channels are not dedicated to a specific end-to-end connection, they may be shared by many end-to-end logical connections. In this way, packet switching optimizes use of network resources by allowing several applications to share transmission facilities so that physical channels are never idle except in the total absence of traffic. Packet switching is suitable only for delay-insensitive traffic because variations in traffic loads can cause queuing delays.

In packet switched networks (PSNs), user messages are subdivided into units for transmission called *packets*. While packets may vary in size, they have a fixed maximum, such as 128 or 4096 octets. The receiver has the responsibility to reassemble the original message from the incoming packets.

A packet switched connection defines a logical pathway between two hosts through the packet network but does not dedicate any physical facilities to that connection. In this way, several packet switched connections can share the physical resources, optimizing use of the network resources. When packets are received by a node, they are placed in buffers and sent on to the next node in the path at the next available opportunity. Having multiple users share a physical resource on an as-needed basis is a type of *statistical TDM*.

A potential problem with statistical TDM is that some transmissions will be delayed. For example, if two packets are ready for transmission on the same physical line at the same time, the node will send one of them and buffer (store) the other one. The short delay is not a problem for many data applications, however, since the most common data applications are delay-insensitive. Delay-sensitive applications,[7] however, cannot use this scheme.

---

[7]Some data applications, such as stock market information and real-time monitoring systems, are, in fact, delay-sensitive and cannot tolerate excessive network delays.

paschoolchoice.org

DNS record

```
paschoolchoice.org.      IN   NS    ns1.internet-dns.net.
paschoolchoice.org.      IN   NS    ns2.internet-dns.net.


paschoolchoice.org.      IN   A     xxx.xxx.xxx.xxx.
mail.paschoolchoice.org. IN   A     xxx.xxx.xxx.xxx
mail2.paschoolchoice.org. IN  A     xxx.xxx.xxx.xxx


www.paschoolchoice.org.    IN CNAME paschoolchoice.org.


paschoolchoice.org.     IN  MX  10  mail.paschoolchoice.org.
paschoolchoice.org.     IN  MX  50  mail2.paschoolchoice.org.
```

Packets are sent to a network node by the user (host) and are forwarded through the network from node to node until delivered to the destination host. As we observed above, the transmitting node must store the packet until it can forward it to the next node. For this reason, packet switching is called a *store-and-forward* strategy.

Packet networks can operate in one of two connection modes, namely *connection-oriented* or *connectionless.* In a connection-oriented, or *virtual circuit (VC)* network, an end-to-end logical connection must be established between two hosts before data exchange can take place. Connection-oriented networks are somewhat analogous to the telephone network: before people can talk to each other, a call must be established. The route of the connection is set up exactly once; and the network can handle error and flow control since the route is usually fixed for the duration of a connection. With connectionless, or *datagram* network, no logical connection is needed prior to the exchange of data. These networks are analogous to the postal system: a datagram is transmitted whenever there is data to send; each datagram is individually routed (and, therefore, may arrive at the destination out of sequence); and neither error nor flow control can efficiently be provided by the network.

Two other terms are used to describe the services of a packet network. *Reliable service* means that the network guarantees sequential delivery of packets and can notify senders when packets have been lost. An *unreliable service* refers to one where the network does not guarantee packet delivery and does not notify the sender when packets are lost or discarded.

Traditional packet switched public data networks (PSPDNs), offer a reliable, VC service. When two hosts wish to communicate, they establish a VC between them defining the logical host-to-host connection. A reliable VC connection means that while packets are guaranteed to be delivered to the destination and to be delivered in sequence, no physical lines are dedicated to the connection between the two hosts. Even though all packets associated with a VC probably follow the same route through the network, no user "owns" a physical line. For example, in Fig. 20-9, a VC between host 1 and host 3 and a VC between host 2 and host 3 will certainly share the physical path between nodes B and D and the physical path between node D and host 3.

The Internet, as a particular example of a packet network, offers an unreliable datagram service for its users. Datagram packets are sent into the network, each having its own address and each finding the "best" route to the destination. The network does not guarantee delivery of packets, much less sequential delivery. The Internet depends on protocols at the end-communicating hosts for a "reliable" end-to-end connection. This networking strategy might seem to be the weaker of the packet networking alternatives, but datagram systems are very popular because of the simplicity of the underlying network and the relative reliability of the facilities. Datagram networks also display a better recovery in the face of mode outages because each packet is routed independently.

Fast Packet Technologies

The concepts behind packet switching have yielded new high-performance packet-mode services called *fast packet switching*. Fast packet services and technologies are characterized by an assumption that the network infrastructure is a low-error-rate, high-speed digital network and depends on end-user systems for error correction (and some error detection). Today's fast packet services are, in fact, unreliable; data units with errors are discarded by the network and end users are not notified of such data loss. Fast packet services have two forms, *frame relay* and *cell relay*.

Frame relay is conceptually similar to VC packet switching. Frames can be of varying size, much like packets in a PSN. Hosts on a frame relay network establish a VC prior to exchanging frames, and the network discards frames with errors. The difference is that the hosts are responsible for end-to-end reliable communication. Frame relay is an additional packet-mode service for ISDN and is described in more detail in Chaps. 15 through 17 of *ISDN*.

Cell relay, unlike frame relay and packet switching, uses a fixed-size transmission entity called a *cell*. Utilization of a fixed-size cell allows many optimizations to be made in network switches and has better statistical multiplexing capabilities, allowing the concurrent transport of many traffic types, including voice, video, graphics, and data. Cell relay is currently being offered in both connection-based (VC) and connectionless (datagram) forms. A connection-oriented cell relay technology is the basis of asynchronous transfer mode (ATM). Connectionless cell relay technology is the basis of SMDS. (See Chaps. 18—21 of *ISDN* for further discussion of these technologies.)

Open Systems Interconnection Reference Model

During the 1960s and 1970s, companies such as Burroughs, Digital Equipment Corporation (DEC), Honeywell, and IBM defined network communications protocols for their computer products. Because of the proprietary nature of the protocols, however, the interconnection of computers from different manufacturers, or even between different product lines from the same manufacturer, was very difficult.

In the late 1970s, the International Organization for Standardization (ISO) developed the Reference Model for Open Systems Interconnection. The OSI model comprises a seven-layer architecture which is the basis for open network systems, allowing computers from any vendor to communicate with each other.

The goals of the OSI model are to expedite communication between equipment built by different manufacturers. The layering of the OSI model provides transparency; that is, the operation of a single layer of the model is independent of the other layers.

The OSI model is described here because it provides an excellent reference with which to compare and contrast different protocols and functionality. Implementations of OSI are few and far between, however, and it could be argued that the Transmission Control Protocol/Internet Protocol (TCP/IP) is the best implementation so far of an open systems protocol suite.

OSI Layers

The OSI model specifies seven functional protocol layers (Fig. 20-10). Peer layers across the network communicate according to *protocols*; adjacent layers in the same system communicate across an *interface*. Network architectures (such as ISDN and B-ISDN) specify the function of the layers, the protocol procedures for peer-to-peer communication, and the communication across the interface between adjacent protocol layers. Actual implementations and algorithms are not typically specified.

Figure 20-10
Reference model for
OSI.

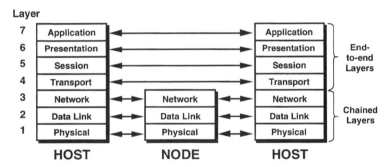

The lower three layers of the OSI model are:

- *Physical Layer* (*layer 1*).   Specifies the electrical and mechanical characteristics of the protocol used to transfer bits between two adjacent devices in the network; common examples include EIA-232-E (formerly RS-232-C), EIA-530, High-Speed Serial Interface (HSSI), V.24, V.28, and V.35.

- *Data Link Layer* (*layer 2*).   Specifies the protocol for error-free communication between adjacent devices across the physical link; common examples include IBM's Binary Synchronous Communications (BISYNC) and Synchronous Data Link Control (SDLC) protocols, DEC's Digital Data Communication Message Protocol (DDCMP), ISO's High-level Data Link Control (HDLC), and ITU-T's Link Access Procedures Balanced (LAPB), Link Access Procedures on the D-channel (LAPD), and Link Access Procedures to Frame Mode Bearer Services (LAPF).

- *Network Layer* (*layer 3*).   Specifies protocols for such functions as routing, congestion control, accounting, call setup and termination, and user-network communications. Examples include IP, ISO's Connectionless Network Protocol (CLNP), and ISDN's call control procedures (Q931 and Q2931).

These three layers are called the *chained layers* and comprise procedures for host-to-node and node-to-node communication. End users (hosts), as well as all switching devices (nodes) along the route between the hosts, must implement these protocol layers.[8]

The upper four layers of the OSI model are:

- *Transport Layer* (*layer 4*).   Specifies the functions and classes of service for error-free communication between hosts across the subnetwork. Examples include TCP and ISO's Transport Protocol (TP).

- *Session Layer* (*layer 5*).   Specifies process-to-process communication, error recovery, and session synchronization.

- *Presentation Layer* (*layer 6*).   A set of general, non-application-specific user services, such as encryption, authentication, and text compression.

- *Application Layer* (*layer 7*).   Specifies the user interface to the network and a set of specific user applications. Sample applications and protocols include TCP/IP's Simple Mail Transfer Protocol (SMTP) and ITU-T X.400 for e-mail, X.500 for directory services, TCP/IP's Telnet and

---

[8]Exceptions are the switches in fast packet networks. These devices, while using three layers for VC establishment (frame relay and ATM), use only a two layer stack for normal data transfers.

ISO's VT protocol for remote login and virtual terminals, TCP/IP's File Transfer Protocol (FTP) and ISO's File Transfer Access Method (FTAM) for file transfers, TCP/IP's Simple Network Management Protocol (SNMP) and ISO's Common Management Information Protocol (CMIP) for network management, and the Hypertext Transfer Protocol (HTTP) for the World Wide Web.

These four layers are called the *end-to-end layers* since they are implemented only in hosts. End-to-end information is transparent to the chained layers. The network nodes that deal with the chained layers generate higher-layer protocol traffic specific to their applications; a switching node, for example, could generate network management traffic using the SNMP protocol, but this would not affect the operation of the chained layers or the network nodes.

The ITU-T standard ISDN protocols define a user-network interface that comprises only the chained layers. While an ISDN itself can provide many types of services using many types of protocols, the ISDN user-network interface is designed to be a common set of protocols for user access to the network, regardless of the required service.

Packet Switching and X.25

ITU-T Recommendation X.25 defines the interface between a user and a PSPDN. User hosts are called *data terminal equipment* (DTE) and the network nodes are called *data circuit-terminating equipment* (DCE). X.25 is very important to ISDNs; packet switching will be supported by ISDNs, and X.25 is the most widely used packet switching protocol today. Recommendation X.25 defines three protocol layers corresponding to a user-network interface:

- *Layer 1.* Exchanges bits between the DTE and DCE; the Physical Layer specified by X.25 is based on Recommendations X.21 and X.21 *bis.*[9]

- *Layer 2.* Ensures error-free communication between the DTE and DCE; the X.25 Data Link Layer protocol is LAPB.

- *Layer 3.* Provides rules for the establishment of virtual calls and the ability to have several simultaneous virtual calls on a single physical

---

[9]X.21 is not implemented in public X.25 networks in North America; the common customer premises interface is EIA-232, EIA-530, or V.35, which are options under X.21 *bis.*

channel between the DTE and DCE; this protocol is called the Packet Layer Protocol (PLP).

ITU-T Recommendation X.75 is similar to X.25. Originally written for internetworking between PSPDNs, it has taken on a more general role for internetworking many types of packet networks, including PSPDNs and ISDNs.

Protocol Architectures

ISDN does not specifically discuss OSI, packet switching, or details of Recommendation X.25. A basic understanding of the OSI model, however, will enhance understanding of the discussion of ISDN and B-ISDN protocol suites. In addition, an understanding of the X.25 protocol architecture and packet switching will enhance understanding of ISDN packet-mode operations and fast packet services.

II. Terms, Definitions, Standards

An ISDN is a digital network that can provide many types of services to a user. The real thrust of the ISDN standards is not how the network operates but how the user communicates with the network and accesses network services. ISDN standards, then, define the interface between the user and the network. This interface is in the form of a set of protocols, including a message set used to request services.

Many terms associated with ISDN, such as D-channel, B-channel, 2B+D, 23B+D, basic rate, primary rate, NT1, TA, bearer services, ITU-T, ANSI are familiar . Section II introduces and defines many of the terms used in the standards, literature, and vendor's ISDN product and service descriptions.

The use of these terms herein is not intended to confuse or intimidate the reader. On the contrary, the terms have rather precise meanings and facilitate discussion about the network, its components, and its services. The concepts of ISDN are actually quite straightforward except for the new language that has been introduced to discuss them.

ISDN cannot succeed as a global telecommunications strategy without international standards. Section II concludes with an introduction to the organizations responsible for creating the ISDN standards. It is impossible to understand any of the ISDN compatibility issues without some knowledge of the players in the ISDN standards game.

The international standards provide a common framework for all national service providers. In the United States, we do not have a single service or equipment provider. Hence, an additional set of standards is necessary for consistent implementation of an ISDN. These definitions are described in the National ISDN (NI) standards, a set of implementation documents that define the services and signaling for the United States.

ISDN Channels

In data communications, a *channel* is a unidirectional conduit through which information flows. A channel can carry digital or analog signals comprising user data or network signaling information. In ISDN and other digital TDM environments, a channel generally refers to a time slot on a transmission facility and is full-duplex (bidirectional).

In today's telephone network, the local loop connection between the user and C.O. provides a single analog channel, for different types of information. First, the loop is used to carry signals between the user's equipment and the network. The telephone, for example, places a short circuit on the line to indicate that the handset has been taken off-hook. A dial tone from the network signals the user to enter the telephone number. Pulses or tones representing the dialed digits, busy signals, and ringing signals also appear over the local loop. Second, after the call is established, the loop carries user information, which may be voice, audio, video, or data, depending upon the application. These two types of usage represent two logical channels, one for signaling and one for user services.

In an ISDN, the local loop carries only digital signals and comprises several channels used for signaling and user data. The different channels coexist on the local loop using TDM. There are three basic types of channels defined for user communications in an ISDN, differentiated by their function and bit rate (Table 20-3):

- *D-channel.* Carries signaling information between the user and the network; may also carry user data packets
- *B-channel.* Carries information for user services, including voice, audio, video, and digital data; operates at the DS-0 rate (64 kbps)
- *H-channel.* Same function as B-channels but operates at bit rates above DS-0

These channels are described in more detail below.

Channel	Function	Bit Rate
B	Bearer services	64 kbps
D	Signaling and packet-mode data	16 kbps (BRI)
		64 kbps (PRI)
H_0	Wideband bearer service	384 kbps
H_1	Wideband bearer services	
H_{10} (23B)*		1.472 Mbps
H_{11} (24B)		1.536 Mbps
H_{12} (30B)		1.920 Mbps
N × 64	Variable bandwidth bearer services	64 kbps to 1.536 Mbps in 64-kbps increments

*An H_{10}-channel is defined by ANSI, but not by the ITU-T.

The D-Channel

All ISDN devices attach to the network using a standard physical connector and exchange a standard set of messages with the network to request service. The contents of the service-request messages will vary with the different services requested; an ISDN telephone, for example, will request different services from the network than will an ISDN television. All ISDN equipment, however, will use the same protocol and same set of messages. The network and user equipment exchange all service requests and other signaling messages over the ISDN D-channel. Typically a single D-channel will provide the signaling services for a single ISDN interface (access point). It is possible for a single ISDN device (e.g., a PBX) to be connected to the network with more than one ISDN interface. In this scenario, it is possible for the D-channel to provide signaling information for many ISDN interfaces. This capability saves channel and equipment resources by consolidating all signaling information on one channel; it is only available on the T-carrier ISDN interface, as discussed below.

Although the D-channel's primary function is for user-network signaling, the exchange of these signaling messages is unlikely to use all of the available bandwidth. Excess time on the D-channel is available for user's packet data and, indeed, the transport of packet-mode data is the secondary function of the D-channel. The excess time is deemed to be great enough to allow service providers to offer user data services at rates up to 9.6 kbps on the D-channel. This is a bargain for users because the full 16 kbps of the D-channel is typically available. User-network signaling messages always have priority over data packets.

The D-channel operates at either 16 or 64 kbps, depending upon the user's access interface.

The B-Channel

Signals exchanged on the D-channel describe the characteristics of the service that the user is requesting. For example, an ISDN telephone may request a circuit-mode connection operating at 64 kbps for the support of a speech application. This profile of characteristics describes what is called a *bearer service*. Bearer services (described in more detail in Chap. 3 of *ISDN*) are granted by the network, allocating a circuit-mode channel between the requesting device and the destination. At the local loop, the B-channels are designated to provide this type of service.

The primary purpose of the B-channel, then, is to carry the user's voice, audio, image, data, and video signals. No service requests from the user are sent on the B-channel. B-channels always operate at 64 kbps, the bit rate required for digital voice applications.

The B-channel can be used for both circuit-mode and packet-mode applications. A circuit-mode connection provides a transparent user-to-user connection, allowing the connection to be specifically suited to one type of service (e.g., television or music). In the circuit mode, no protocols above the physical layer (64 kbps) are defined by the ITU-T for the B-channels; each user of a B-channel is responsible for defining the protocols to be used over the connection. It is also the responsibility of the users to assure compatibility between devices connected by B-channels. Packet-mode connections support packet switching equipment using protocols such as X.25 or frame relay. The ISDN can provide either an internal packet-mode service or provide access to an existing PSPDN for packet service. In either case, the protocols and procedures of the PSPDN must be adhered to when requesting packet-mode service.

B-channels can be used on an on-demand basis, as described above, or on a permanent basis. If a B-channel is provisioned for permanent service, no D-channel signaling is required for the operation of the B-channel. A sample application might be the provisioning of a permanent B-channel for high-speed (64-kbps) PSPDN access or frame relay access.

The most important point to remember is the relationship between B- and D-channels. The D-channel is used to exchange the signaling messages necessary to request services on the B-channel.

H-Channels

A user application requiring a bit rate higher than 64 kbps may be obtained by using wideband channels, or H-channels, which provide the bandwidth equivalent of a group of B-channels. Applications requiring bit rates above 64 kbps include LAN interconnection, high-speed data, high-quality audio, teleconferencing, and video services.

The first designated wideband channel is an H_0-channel, which has a data rate of 384 kbps. This is equivalent to logically grouping six B-channels together.

An H_1-channel comprises all available time slots at a single user interface employing a T1 or E1 carrier. An H_{11}-channel operates at 1.536 Mbps and is equivalent to 24 time slots (24 B-channels) for compatibility with the T1 carrier. An H_{12}-channel operates at 1.920 Mbps and is equivalent to 30 time slots (30 B-channels) for compatibility with the E1 carrier.

ANSI has designated an H_{10}-channel, operating at 1.472 Mbps and equivalent to 23 time slots on a T1 interface. This channel was defined by ANSI to support a single wideband channel and a D-channel on the same T1 access facility; with an H_{11}-channel, a D-channel and wideband channel cannot coexist on the same T1 interface.

Another set of ISDN channels has been defined for variable bit rate applications, called an $N \times 64$ channel. This channel is similar in structure to the H-channels except it offers a range of bandwidth options from 64 kbps to 1.536 Mbps in increments of 64 kbps. When a user requests an $N \times 64$ channel for a given call, the service request contains the type of channel ($N \times 64$) and the value of N (1 to 24). A benefit to users of an $N \times 64$ channel is that they do not require inverse multiplexing equipment on the premises since the network maintains time slot sequence integrity between the N 64-kbps time slots. Another advantage of the $N \times 64$ channel is the ability to customize the bandwidth requirements to the application.

Access Interfaces

An *access interface* is the physical connection between the user and the ISDN that allows the user to request and obtain services. The concept of an access interface is a familiar one to users of today's networks. Most residences, for example, have a single-line telephone and, accordingly, a single connection to the local C.O. This single local loop can be said to comprise two logical channels, as described earlier, one for user-network signals (on- and off-hook) and one for user data (voice and tones).

As the number of simultaneous users increases at a customer location, so does the requirement for the number of physical resources to handle those users. A second local loop, for example, can provide a second telephone line, while multiple trunk circuits can provide multiple lines between a customer's PBX and the C.O. Access to other networks and/or network services (e.g., a packet network or the Internet) can be provided by bringing additional lines to the customer's premises. It is not uncommon for a business location to have many individual lines connecting it to the C.O. for such services as telephony, fax, a point-of-sale terminal, and remote security.

ISDN access interfaces differ somewhat from today's telephone network access interfaces. First, one goal of ISDN is to provide all services over a single network access connection (physical resource), independent of the equipment or service type. Second, ISDN access interfaces comprise a D-channel for signaling multiplexed with some number of B-channels for user data. This design allows multiple information flows simultaneously on a single physical interface.

ISDN recommendations from the ITU-T currently define two different access interfaces, called the *basic rate interface* (BRI) and *primary rate interface* (PRI). These access interfaces specify the rate at which the physical medium will operate and the number of available B-, D-, and H-channels (Table 20-4).

Bellcore documents use a slightly different set of terms, namely, *basic rate access* (BRA) and *primary rate access* (PRA). The use of this terminology stems from the separation of the service access from the physical interface; a BRA, for example, could be physically delivered to a location in a form other than a single two-wire interface.

TABLE 20-4	Interface	Structure*	Total Bit Rate	User Data Rate
ISDN Access Interface Structures	Basic rate interface (BRI)	$2B+D_{16}$	192 kbps	144 kbps
	Primary rate T1 interface (PRI) E1	$23B+D_{64}$† $30B+D_{64}$	1.544 Mbps 2.048 Mbps	1.536 Mbps 1.984 Mbps

*The D-channel operates at 16 kbps in the BRI and at 64 kbps in the PRI.
†This is one possible PRI configuration, and the most common today. Other configurations are also possible, such as 24B.

Basic Rate Interface

The BRI comprises two B-channels and one D-channel, and is designated 2B+D. The BRI D-channel always operates at 16 kbps.

The BRI will typically be used in one of two ways. First, it can provide ISDN access between a residential or business customer and the ISDN LE. Alternatively, it can provide ISDN access between user equipment and an ISDN-compatible PBX in a business environment. As a tariffed offering, the BRI can be ordered in configurations other than 2B+D, and other nomenclature may be encountered. If the BRI is to be used only for telephony and no data will be sent on the D-channel, the configuration is sometimes called *2B+S* (the D-channel is for signaling only). If only a single B-channel is required, a *1B+D* or *1B+S* arrangement may be ordered; packet data is allowed on the D-channel in the former and not in the latter.

Finally, if only low-speed (9.6-kbps) packet data is required, a *0B+D* configuration can be ordered. These configurations allow ISDN to be customized for customer applications and are priced differently based on the number of active channels. It should be noted that in all of these configurations, the interface's physical characteristics are the same; the only difference is in which channels have been activated by the LE and what type of traffic is allowed on the D-channel.

The user data rate on the BRI is 144 kbps (2×64 kbps + 16 kbps), although additional signaling for the physical connection requires that the BRI operate at a higher bit rate. The specific rates of the interface and the overhead associated with those rates are discussed in Chap. 5 of *ISDN*.

Primary Rate Interface

The PRI also has a number of possible configurations. The most common configuration in North America and Japan is designated 23B+D, meaning that the interface comprises 23 B-channels plus a single D-channel operating at 64 kbps. Optionally, the D-channel on a given PRI may not be activated, allowing that time slot to be used as another B-channel; this configuration is designated 24B.[6] This PRI description is based on the T1 dig-

---

[6]The presence of a D-channel at the user-network interface is essential in ISDN for the exchange of signaling information to control services. Therefore, at least one PRI at a customer interface must be configured as 23B+D. Since the D-channel on one PRI may control other physical interfaces, subsequent PRIs may be configured as 24B. In these cases, a second PRI is often configured with a backup D-channel.

ital carrier. It operates at a bit rate of 1.544 Mbps, of which 1.536 Mbps are user data.

A 30B+D PRI is also defined that comprises 30 B-channels and 1 D-channel. Based on the E1 digital carrier, it operates at 2.048 Mbps, of which 1.984 Mbps are user data.

The PRI contains more channels than a typical end-user device will use. The PRI is, in fact, primarily intended to provide access to the network by some sort of customer premises switching equipment, such as a remote access server PBX, multiplexer, or host computer.

When a wideband application requires more throughput than that provided by a B-channel, the PRI can be configured to provide H-channel access. When this configuration is used, the number of available B-channels will decrease by the number of time slots used by the H-channel(s). An example would be a videoconferencing system needing 384 kbps (an H_0-channel) for a call. The supporting PRI would have extra bandwidth available for a D-channel and 17 B-channels. If the video system needed an H_{11}-channel, no B- or D-channel time slots would be available. This flexibility allows the PRI to act as a wideband access system and a narrowband access system, depending on the application active at any time. The same bandwidth (time slots) can be configured for different types of channels on demand.

Functional Devices and Reference Points

Several different devices may be present in the connection between CPE and the network to which the CPE is attached. Consider the relatively simple example of a customer's connection to the telephone network. All of the subscriber's telephones are connected with inside wiring to a junction box in the customer's building; the local loop provides the physical connection between the junction box and the LE. As far as the customer is concerned, the CPE is communicating directly with the exchange; the junction box is transparent.

Other equipment may also be present. If a PC is attached to the telephone network, for example, a modem will replace the telephone. In a PBX environment, the telephones and modems are attached to the PBX, which will provide on-site switching; the PBX is, in turn, connected to the LE.

Protocols describe the rules governing the communication between devices in a network. With all of the devices mentioned here, questions might arise as to which protocols are to be used where and who is responsible for defining the protocols. The telephone, for example, uses a familiar protocol that is specified by the network; certain current represents the off-hook signal, special pulses or tones represent the dialed digits, etc.

A modem follows the same protocol as a telephone on the side that connects to the telephone network. It uses a different protocol, however, on the side that connects to a PC; EIA-232-E and the Hayes AT-command set, for example, are commonly used between a PC and external modem. The modem acts as a signal converter so that digital signals output from the PC will be suitable for the analog telephone network.

The presence of a PBX adds another layer of complexity. A telephone connected to a PBX follows protocols specified by the PBX manufacturer, which is why many PBX-specific telephones are not usable on the public telephone network. The PBX, in turn, must use network-specified protocols for the PBX-to-network communication.

In today's communications environment it is often difficult to separate the devices from the functions they perform. The case of the PC communicating over the telephone network is an example. Above we described three devices, a PC, a modem, and a network; each has a specific function and is governed by a set of protocols. What would happen if the PC had an internal modem? The number of functional devices would remain the same, but the number of physical devices would be reduced to two, the PC and the network. In this example the number of functional devices and the number of actual physical devices differ due to the packaging of the devices.

These same ideas are extended to ISDN. The ISDN standards define several different types of devices. Each device type has certain functions and responsibilities but may not represent an actual physical piece of equipment. For that reason, the standards call them *functional devices.*

Since the ISDN recommendations describe several functional device types, there are several device-to-device interfaces, each requiring a communications protocol. Each of these functional device interfaces is called a *reference point.*

The paragraphs below describe the different functional devices and reference points, which are shown in Fig. 20-11.

ISDN Functional Devices

The network device that provides ISDN services is the *LE.* ISDN protocols are implemented in the LE, which is also the network side of the ISDN

Figure 20-11
ISDN functional
devices and reference
points.

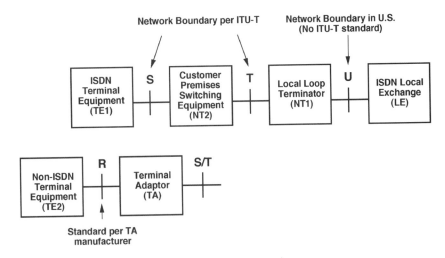

local loop. Other LE responsibilities include maintenance, physical interface operation, timing, and providing requested user services.

Some ISDN exchange manufacturers further break down the functions of the LE into two subgroups called *local termination* (LT) and *exchange termination* (ET). The LT handles those functions associated with the termination of the local loop, while the ET handles switching functions. For simplicity and generality, this book will usually refer only to the LE and avoid specific references to LT or ET except where necessary. Also included in the LE is equipment specialized to support the ISDN services. These have to do with the signaling used in ISDN and the incorporation of packet-mode or frame-mode data in the ISDN list of services. The first is a packet handler (PH). This device is responsible for the decoding of all ISDN signaling packets passed between the LE and the ISDN subscriber. It is also used to distinguish user X.25 data on the D-channel from signaling data and routes the user data toward its destination on the associated PSPDN. The second device (or devices) is the network signaling system employed for the ISDN. In today's environment this signaling system is Signaling System No. 7 (SS7); it is described in *ISDN*. The SS7 device is responsible for the creation and interpretation of the signaling messages used between LEs in the ISDN. The final device is a frame handler (FH), which has a function similar to the PH but supports frame relay user traffic rather than ISDN signaling and X.25 traffic. The FH, like the PH, can be an integral part of the LE or an adjunct processor attached to the LE.

Network termination type 1 (NT1), or local loop terminator equipment, represents the termination of the physical connection between the customer site and the LE. The NT1's responsibilities include line performance monitoring, timing, physical signaling protocol conversion, electrical conversion, and power transfer.

Network termination type 2 (NT2) equipment are those devices providing customer site switching, multiplexing, and/or concentration. This includes PBXs, multiplexers, routers, host computers, terminal controllers, and other CPE for voice and data switching. An NT2 will be absent in some ISDN environments, such as residential or Centrex ISDN service. NT2s distribute ISDN services to other devices that are attached to it. In this role, the NT2 might perform some protocol conversion functions as well as distribution functions. One of the primary distribution functions is the network signaling on behalf of the attached terminals. The NT2 is responsible for all signaling to the network. As an example, a PBX might terminate an analog telephone and allow access to an ISDN PRI for a connection to other subscribers. In this case the PBX is providing protocol conversion from the analog voice to the ISDN digital voice and is collecting the dialed digits from the telephone and creating a signaling message for the LE.

Terminal equipment (TE) refers to end-user devices, such as an analog or digital telephone, X.25 data terminal equipment, ISDN workstation, or integrated voice/data terminal (IVDT). *Terminal equipment type 1* (TE1) are those devices that utilize the ISDN protocols and support ISDN services, such as an ISDN telephone or workstation. *Terminal equipment type 2* (TE2) are non-ISDN-compatible devices, such as the analog telephones in use on today's telephone network.

A *terminal adapter* (TA) allows a non-ISDN device (TE2) to communicate with the network. TAs have particular importance in today's ISDN marketplace; nearly every device in use in today's data and telecommunications environment is TE2. TAs allow analog telephones, X.25 DTEs, PCs, and other non-ISDN devices to use the network by providing any necessary protocol conversion.

The reader should note that a single physical piece of equipment can take on the responsibilities of two or more of the functional devices defined here. For example, a PBX might actually perform NT1 (local loop termination) and NT2 (customer site switching) functions; this combination is sometimes referred to as NT12. In the same theme, an ISDN telephone can be purchased that has a TA and an NT1 built-in; this combination is referred to as a bargain.

ISDN Reference Points

The ISDN reference points define the communication protocols between the different ISDN functional devices. The importance of the different reference points is that different protocols may be used at each reference point. Four protocol reference points are commonly defined for ISDN, and are called R, S, T, and U.

The R reference point is between non-ISDN terminal equipment (TE2) and a TA. The TA will allow the TE2 to appear to the network as an ISDN device, just like a modem allows a terminal or PC to communicate over today's telephone network. There are no specific ISDN standards for the R reference point; the TA manufacturer will determine and specify how the TE2 and TA communicate with each other. Examples of R reference point specifications include EIA-232-E, V.35, and the Industry Standard Architecture (ISA) bus.

The S reference point is between ISDN user equipment (i.e., TE1 or TA) and network termination equipment (NT2 or NT1). The T reference point is between customer site switching equipment (NT2) and the local loop termination (NT1). ISDN recommendations from the ITU-T, the primary international standards body for ISDN (described further on), specifically address protocols for the S and T reference points. In the absence of the NT2, the user-network interface is usually called the S/T reference point.[7]

One of the more controversial and pivotal aspects of ISDN, at least in the United States, was the transmission standard across the local loop between the NT1 and the LE, called the U reference point. The ITU-T considers the physical NT1 device to be owned by the network administration; that makes the local loop part of the network. Therefore, the ITU-T views the S or T reference points as the user-network boundary. ITU-T recommendations do not address internal network operations, so they have no standard for transmission across the local loop (U reference point).[8]

The U.S. Federal Communications Commission (FCC), however, does not adopt this same view. Since the NT1 is on the customer's site, the FCC considers it to be customer owned. Since network equipment (the LE) is on one side of the U reference point and user equipment (the NT1) is on the other, it is clearly the local loop that represents the user-network boundary according to the FCC. Furthermore, operation across the user-network boundary in the United States must be described by a public standard

---

[7]The ambivalence of what to call this bus results in different providers referring to the user-network interface as the S, T, or S/T interface in the absence of the NT2. We will use the term S/T because it is somewhat more "complete." It is an S interface from the viewpoint of the TE and a T interface from the viewpoint of the NT1.

[8]In fact, the ITU-T standards do not actually refer to the U reference point; they mention the R, S, and T reference points and the *transmission line.*

and, in fact, is the subject of a U.S. national standard from the American National Standards Institute (ANSI, described below).

Although not shown in Fig. 20-11 or described further in this chapter, some manufacturers define a V reference point between LEs in an ISDN. This reference point identifies the network node interface and is transparent to the user.

B-ISDN Channels, Functional Devices, and Reference Points

For reference, it should be noted that ITU-T B-ISDN standards use the same functional device and reference point designations as defined above. The notation for devices and protocols with broadband capability, however, includes the letter *B*. A B-ISDN terminal, for example, will be designated B-TE1, and the B-ISDN T reference point will be designated T_B.

Standards Organizations

The ITU-T

The organization primarily responsible for producing international ISDN standards is the International Telecommunication Union Telecommunication Standardization Sector (ITU-T), formerly known as the International Telegraph and Telephone Consultative Committee (CCITT, or Comité Consultatif International Télégraphique et Téléphonique). Although the ITU has been an agency of the United Nations (U.N.) since 1948, the ITU's formal beginning dates back to 1865, and it is the world's oldest intergovernmental agency.

The ITU-T produces standards describing access to public telecommunications networks and the services to be offered by those networks but does not specify the internal operation of the networks. It is for this reason that the ITU-T does not define the U reference point; the ITU-T views the local loop as part of the network and, therefore, an internal matter.

Since the 1960s, CCITT standards, called *recommendations*, were formally adopted at plenary sessions held every 4 years. The recommendations were published in a set of books referred to by the color of their cover; 1988 recommendations are contained in the Blue Book, 1984 in the Red Book, 1980 in the Yellow Book, and 1976 in the Orange Book, for example.

This schedule and publishing scheme fell out of favor during the 1980s due to the time lost waiting for the next plenary and the tremendous amount of paper that was required to publish recommendations, including republishing those recommendations that had not changed; the Red Book, for example, was 9900 pages in length and the Blue Book, just 4 years later, was about 20,000 pages. At the 1988 Plenary Session, the CCITT passed a resolution that each new or revised recommendation would be published as soon as it was finalized. This not only responded to environmental concerns but also expedited availability of completed works.

The CCITT's sister standards organization is the International Radio Consultative Committee (CCIR). The CCIR concentrates on specifications for radio communications, including radio- and satellite-based ISDN. In March 1993, the ITU underwent a significant reorganization and reassignment of responsibilities. One result was that both the CCITT and CCIR were renamed to the ITU-T and ITU Radiocommunication Standardization Sector (ITU-R), respectively. The quadrennial CCITT Plenary Assembly is now known as the World Telecommunication Standardization Conference.

There are five classes of membership within the ITU-T:

A. *Administration members* represent a country's telecommunications administration and act as the official voting representative. The Postal, Telephone, and Telegraph (PTT) administration is typically a country's Class A member; since the United States does not have a PTT, the State Department is the U.S. Class A member.

B. *Recognized Private Operating Agencies* (RPOA) are private or government organizations that provide a public telecommunications service, such as Lucent, MCI, Sprint, and BT Tymnet.

C. *Scientific and industrial organization members* are any other commercial organization with an interest in the ITU-T's work, such as Alcatel, Lucent, DEC, IBM, Nortel, and Siemens.

D. *International organization members* include other international organizations with an interest in the ITU-T's work, such as ISO.

E. *Specialized treaty agencies* are agencies organized by treaty whose work is related to the ITU-T's, such as the World Health Organization and World Meteorological Organization.

Although only Class A members can officially vote at the sessions, all members can participate at the study group (SG) and working group level. As the official U.S. representative, the State Department coordinates all U.S. participation with the ITU-T.

The work of the ITU-T is performed by 15 SGs and other committees (Table 20-5). ITU-T recommendations are identified by a letter followed by a number, where the letter indicates the general topic of the recommendation series. Notable topics include:

- *E-series.* Telephone network and ISDN
- *G-series.* International telephone connections and circuits
- *I-series.* ISDN
- *Q-series.* Telephone switching and signaling networks
- *V-series.* Digital communication over the telephone network
- *X-series.* Public data communication networks

ITU involvement with ISDN dates back over 20 years. In 1968, a CCITT SG meeting convened to discuss the integration of switching and transmission. The meeting resulted in the formation of a special SG devoted to

TABLE 20-5

Study and Advisory Groups of the ITU Telecommunication Standardization Sector (1993–1996)

SG 1—Service Definition

SG 2—Network Operation

SG 3—Tariff and Accounting Principles

SG 4—Network Maintenance

SG 5—Protection Against Electromagnetic Environment Effects

SG 6—Outside Plant

SG 7—Data Networks and Open System Communications

SG 8—Terminals for Telematic Services

SG 9—Television and Sound Transmission

SG 10—Languages for Telecommunication Applications

SG 11—Switching and Signaling

SG 12—End-to-End Transmission Performance of Networks and Terminals

SG 13—General Network Aspects

SG 14—Modems and Transmission Techniques for Data, Telegraph, and Telematic Services

SG 15—Transmission Systems and Equipment

Telecommunication Standardization Advisory Group (TSAG)

this topic (Study Group Special D), which later became SG XVIII. ITU-T SG 11 (formerly CCITT SG XVIII) has responsibility for digital networks including ISDN. Among other things, they are responsible for writing the I-series recommendations defining ISDN and specifying appropriate services and protocols. Figure 20-12 shows the general organization of the I-series recommendations. ITU-T I-series recommendations are listed in App. B of *ISDN.*

Study Group D dealt with IDNs and integrated services networks (ISNs); the former is a network that contains digital switches and transmission facilities from end to end, and the latter is a network that offers many types of services from a single network. In 1971, another CCITT meeting was convened to discuss these terms. Several countries wanted the word *digital* added to the ISN term since integrated services can, conceivably, be offered by an analog network, which was not the subject of SG XVIII's work. Other countries felt that integration of services was only feasible in a digital environment, thus adding the word was redundant. The compromise was to coin the term *integrated services (digital) network.* Even-

Figure 20-12

Organization of ITU-T I-series (ISDN) recommendations. (*From ITU-T Recommendation I.110*)

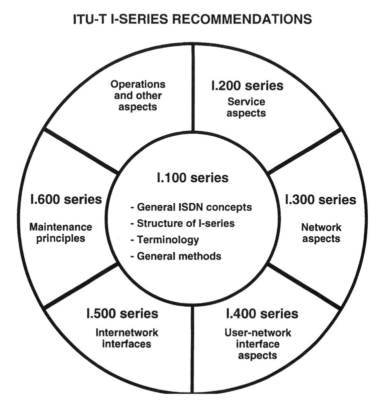

ITU-T I-SERIES RECOMMENDATIONS

Operations and other aspects

I.200 series
Service aspects

I.600 series
Maintenance principles

I.100 series
- General ISDN concepts
- Structure of I-series
- Terminology
- General methods

I.300 series
Network aspects

I.500 series
Internetwork interfaces

I.400 series
User-network interface aspects

tually the parentheses were dropped, yielding the term that we have today. Note that this activity was occurring well before digital switches were widely introduced into the telecommunications network.

Under the direction of SG XVIII, CCITT leadership in organizing and developing ISDN standards grew. The first set of ISDN recommendations were formally published in 1984 (Red Book), followed by updated specifications in 1988 (Blue Book) and almost continuously since then.

Other SGs participate in the ISDN standards process by virtue of the recommendations that they prepare, which often overlap. Recommendations for public data networks (X-series), ISDN and telephone network switching and signaling (Q-series), and addressing and numbering plans (E-series), for example, can all pertain to ISDN. For this reason, several I-series recommendations are also assigned Q-, X-, or other series' recommendation numbers. Support of X.25 terminals on an ISDN, for example, is described in Recommendations I.462 and X.31; these two standards are identical. Similarly, ISDN signaling procedures on the D-channel are listed as both Recommendation I.451 and Q931. In the Red Book, dual-listed recommendations were often published in both locations; this changed in the Blue Book so that recommendations were published in only one location and referenced in the others. In this book, ITU-T recommendations will be referred to by the series and number where they are actually published. An I-series or other alternate designation, if any, will be given in parentheses.

The American National Standards Institute

ANSI is the primary standards-setting body in the United States. Formed in 1918, ANSI is a nonprofit, nongovernmental organization supported today by more than a thousand trade organizations, professional societies, and corporations. ANSI itself does not create standards per se, but rather coordinates and sanctions the activities of appropriate organizations that do write standards.

ANSI Standards Committee T1 is responsible for producing U.S. national telecommunications standards. T1-series standards include T1 and other digital carrier specifications,[9] ISDN, the U.S. version of SS7, SONET, and frame relay.

---

[9]The designation of the T1 carrier and the T1 committee is coincidental. The *T-carrier* designation for digital carriers was made by AT&T in the 1960s, and the *T1* assignment for telecommunications was made by ANSI in the 1980s.

The secretariat for the T1 Committee is the Alliance for Telecommunications Industry Solutions (ATIS). ATIS was originally incorporated as a not-for-profit association in 1983 and called the Exchange Carriers' Standards Association (ECSA). Renamed in 1993, ATIS comprises members of the telecommunications industry to address exchange access, interconnection, and other technical issues that have resulted from divestiture. ATIS supports a number of industry forums on topics such as ordering and billing, network operations, bar code specifications, electronic data interchange, open network architecture, network reliability, and electrical protection.[10]

The work of the T1 committee, formed in 1984, is handled by six subcommittees, each dealing with different aspects of telecommunications. The various T1 subcommittees and working groups are listed in Table 20-6, and T1 standards related to ISDN and B-ISDN are listed in App. B of *ISDN*.

Recognizing the hierarchical organization of ANSI will help in understanding and appreciating the standards-making process. Each committee and working group meets approximately three to six times each year, and sometimes more. Each group comprises individuals representing companies that have an interest in the progression of the standard, including vendors, manufacturers, users, and service providers.

Suppose, for example, that the T1S1.1 task group produces a new ISDN specification for adoption as a standard. All members of the task group must reach a consensus before the specification advances to the draft standard phase. When the group is ready to forward the standard, it goes to the T1S1 subcommittee. That level, too, must approve the document before it is forwarded; any negative comments are referred to the task group. After the subcommittee approves the document, it is forwarded to the T1 committee, where the document enters a public comment period during which time *anyone* can make a comment on the draft standard. Again, negative comments must be resolved by the task group. Subsequent drafts again go out for public comment. Only after all of these stages are passed can a standard be formally adopted. Due to the schedule of meetings and the required time period for comments, even a relatively noncontroversial draft standard can take a year or more for formal adoption.

ANSI is playing a significant role in the development of ISDN standards. In particular, the local loop operation (U reference point) for the BRI is standardized in the United States only by ANSI, since it is not the topic of any ITU-T recommendation. ANSI is also actively creating other ISDN, B-ISDN, and related standards for the United States.

---

[5]ATIS also sponsors the ANSI O5 Committee, developing standards for wood poles and wood products for utility structures.

TABLE 20-6

Subcommittees and Working Groups of the ANSI T1 Committee (July 1995)

T1A1—Performance and Signal Processing

    T1A1.2—Network Survivability Performance

    T1A1.3—Performance of Digital Networks and Services

    T1A1.5—Multimedia Communications Coding and Performance

    T1A1.7—Signal Processing and Network Performance for Voiceband Services

T1E1—Interfaces, Power and Protection of Networks

    T1E1.1—Analog Access

    T1E1.2—Wideband Access

    T1E1.4—Digital Subscriber Loop (DSL) Access

    T1E1.5—Power Systems—Power Interfaces

    T1E1.6—Power Systems—Human and Machine Interfaces

    T1E1.7—Electrical Protection

    T1E1.8—Physical Protection

T1M1—Internetwork Operations, Administration, Maintenance and Provisioning

    T1M1.1—Internetwork Planning and Engineering

    T1M1.3—Testing and Operations Support Systems and Equipment

    T1M1.4—Administrative Systems (inactive)

    T1M1.5—OAM&P Architectures, Interfaces and Protocols

T1P1—Systems Engineering, Standards Planning and Program Management

    T1P1.1—Program Management and Standards Planning

    T1P1.2—Systems Engineering for Personal Communications Networks and Service Aspects

    T1P1.3—Systems Engineering for Wireless Access and Terminal Mobility

    T1P1.4—Wireless Interfaces

T1S1—Services, Architectures and Signaling

    T1S1.1—ISDN Architecture and Services

    T1S1.3—Common Channel Signaling

    T1S1.5—Broadband ISDN

T1X1—Digital Hierarchy and Synchronization

    T1X1.3—Synchronization and Tributary Analysis Interfaces

    T1X1.4—Metallic Hierarchical Interfaces

    T1X1.5—Optical Hierarchical Interfaces

ANSI maintains a liaison to the ITU-T and ISO. ANSI standards are often forwarded to the international standards community for adoption worldwide or to form the basis of an international standard.

Bellcore

After the breakup of AT&T, Bellcore was formed as the research and development arm of the seven RBOCs in the United States. Bellcore is an active participant in the national and international standards process.

Bellcore is responsible for defining implementation standards and service requirements for the RBOCs. Historically, Bellcore produced a series of documents. Technical Advisories (TAs) were draft specifications which, after industry review, become Technical References (TRs) for implementation; Framework Advisories (FAs) provided general technology guidance and overview; and Special Reports (SRs) provided general information that was outside the scope of implementation or technology specifications. In 1994, Bellcore adopted a new scheme whereby most technical specifications are called Generic Requirements (GRs).

In 1996, Bellcore was sold to Science Applications International Corporation (SAIC). Prior to that time, Bellcore was funded by the RBOCs. Bellcore has taken a lead role within the United States for defining many aspects of services to be offered by local telephone companies, including ISDN, B-ISDN, frame relay, SONET, metropolitan area networks (MANs), switching technology, operations technology, network management, and billing systems. In addition to producing a large number of ISDN-related reports and specifications, they have developed National ISDN, a family of implementation specifications providing guidance to the industry to ensure compatibility between ISDN switches, services, and CPE from multiple vendors. National ISDN is discussed in Chap. 9 of *ISDN* and Bellcore's ISDN-related specifications are listed in App. B of *ISDN*.

The European Telecommunications Standards Institute

Anticipating the approach of a single-market economy in the late 1980s, the European Commission published a report that argued that the harmonization of Europe required a pan-European telecommunications network infrastructure based upon standard equipment and services. From this, the European Telecommunications Standards Institute (ETSI) was founded in 1988 to accelerate the development of such standards for the European Union (then called the European Community). ETSI is an independent, self-funded organization, headquartered in France.

ETSI's membership composes over 400 equipment manufacturers, public network service providers, users, and research organizations from over two dozen European countries. About a dozen non-European countries, such as Australia, Canada, Israel, Japan, and the United States, have observer status in ETSI, while the Commission of the European Communities and the European Free Trade Association Secretariat each has a special Counselor status.

ETSI's technical work is performed under the auspices of a Technical Assembly comprised of 11 technical committees (Table 20-7) and over 60 subtechnical committees, project teams, and working groups. Like most standards organizations, ETSI works on a consensus basis. Subcommittee's draft documents require technical committee approval before going out for public comment. Technical committees review the document after the public inquiry phase, followed by voting by the national standards organizations of ETSI member countries. Once forwarded by a subtechnical committee, a standard may take nearly a year before it is finally adopted, unless accelerated procedures are adopted. In addition, absolute consensus

TABLE 20-7

ETSI Technical Committees (1995)

Business TeleCommunications (BTC)*
Equipment Engineering (EE)
Human Factors (HF)
Methods for Testing and Specification (MTS)†
Network Aspects (NA)
Radio Equipment and Systems (RES)‡
Satellite Earth Stations and Systems (SES)
Signaling Protocols and Switching (SPS)
Special Mobile Group (SMG)
Terminal Equipment (TE)
Transmission and Multiplexing (TM)

*Formerly BT.
†Formerly called Advanced Testing Methods (ATM).
‡Includes work of the former Paging Systems (PS) committee.

is not required for adoption of a standard; a document can be approved by a 71 percent weighted vote of the members. Formally approved ETSI specifications are called a European Telecommunications Standard (ETS), while an Interim ETS (I-ETS) is an approved specification that must be discarded or converted to an ETS within 2 to 5 years. Compliance with ETSI standards is voluntary. ETSI's ISDN- and B-ISDN-related standards are listed in App. B of *ISDN*.

Other Standards Organizations

There are many other standards organizations that affect computer networking and telecommunications, representing government agencies, professional organizations, and industry. Among these other organizations creating ISDN-related standards are:

- *ISO*.[11]   Formed in 1947, ISO is a nongovernmental organization with a charter to promote the development of worldwide standards to facilitate international cooperation, communication, and commerce. ISO's networking standards include the OSI Reference Model, the HDLC bit-oriented protocol, international LAN standards, and OSI protocols. ISO has over 100 members, each representing a national standards organization; ANSI is the U.S. representative to ISO.

- *International Electrotechnical Commission (IEC)*.   Created in 1906, the IEC is one of the world's oldest international standards organizations. Focusing on the areas of electronics and electricity, the IEC and ISO closely coordinate their activities and many of their standards are jointly adopted.

- *National Institute for Standards and Technology (NIST)*.   Formerly known as the National Bureau of Standards (NBS), NIST has taken a lead role in defining ISDN applications. In particular, NIST has formed the North American ISDN Users' Forum (NIUF) to identify ISDN applications and to guide and encourage manufacturers in developing those applications that are of most interest to users. The NIUF is discussed in Chap. 11 of *ISDN*.

- *United States Telephone Association (USTA)*.   Formerly the U.S. Independent Telephone Association (USITA), the USTA is a national trade association representing local exchange carriers in the United States.

---

[11]The term *ISO* is not an acronym, but comes from the Greek term *isos* meaning *equal*.

Originally founded in 1897, at the time when Alexander Graham Bell's original patents were expiring, the USITA represented the non-Bell, or ITCs. After the breakup of AT&T in 1984, their charter changed to represent all LECs, so the word *independent* was dropped from the name. The USTA provides a broad range of services, including representing the industry before governmental and legislative bodies, hosting an annual trade show, providing training, and producing specifications to enhance network and vendor interoperability.

Other standards bodies focusing on other types of networks affect ISDN as well. The Institute of Electrical and Electronics Engineers (IEEE), an international professional society headquartered in the United States, is accredited by ANSI to produce standards in several areas; in particular, the IEEE 802 Project creates LAN and MAN standards. The Internet Engineering Task Force (IETF) is responsible for producing standards related to the Internet, the international network interconnecting thousands of subnetworks around the world using the TCP/IP protocol suite. Both the IEEE and IETF have generated ISDN-related internetworking specifications.

In the United States, AT&T has long published standards for manufacturers wishing to attach their equipment to the public telephone network. Although local telephone service is not currently offered by AT&T, they manufactured the majority of C.O. switches used by the local telephone companies in the United States today. AT&T technical bulletins and other publications remain industry standards for both local telephone service from the C.O. and long distance services via AT&T's toll network. AT&T continues to play an important role in ANSI and ITU-T standards development. In 1995, the manufacturing portion of AT&T was divested and was incorporated under the name Lucent Technologies. Lucent carries on AT&T's role in standards and equipment advancements.

Industry Consortia

The major standards organizations mentioned above—including ITU, ANSI, and ISO—create *de jure* (by law) standards, meaning that their specifications are mandated by some form of legislation or treaty and provide certain guarantees and/or protections to those who adhere to the standards. Unfortunately, the de jure standards process, as described above, can be very slow. This sometimes leads to standards being adopted that are incomplete or out of date upon publication, leading to a confused marketplace rather than a cohesive one.

In the last several years, a number of important *de facto* (by fact) "standards" organizations have evolved, particularly in the areas of fast packet switching technologies and B-ISDN. These groups were formed as industry consortia to promote and accelerate development of a single technology. All of the consortia involve service providers, equipment vendors, and users and are international in their scope. All include committees working on technical aspects of the technology, as well as public education and awareness. These groups form *implementation agreements* rather than standards; these agreements specify the way in which member companies have agreed to implement these technologies and services but do not provide de jure standards, although the implementation agreements are invariably based on existing de jure standards or are forwarded to appropriate standards organizations.

The most important of the industry consortia for our study of ISDN and related technologies are:

- The ADSL Forum, formed in 1994 to help telephone service providers and equipment vendors realize ADSL's market potential. The ADSL Forum focuses on network, protocol, architectural, and marketing issues related to ADSL, VDSL, and other xDSL technologies.

- The ATM Forum (ATMF), formed in 1991 to address specific implementation issues for ATM services that were not covered in ITU recommendations.

- The Frame Relay Forum (FRF), formed in 1991 to address technical frame relay issues that were beyond the scope of the existing ANSI and ITU frame relay standards.

- The SMDS Interest Group (SIG), formed in 1991 to promote Bellcore's SMDS specifications. It was started specifically to address technical issues that were beyond the scope of the SMDS user-network interface and to promote awareness of the service. The SIG was disbanded in 1997, ostensibly because it had fulfilled its charter and the service was stable from a technical perspective.

These groups are discussed in detail elsewhere in *ISDN*. Appendix B of *ISDN* lists relevant implementation agreements.

Summary

ISDN terminology is different from today's telecommunications terminology and is, therefore, sometimes cumbersome. Nevertheless, the terms have precise meaning and provide a common language platform with which to discuss ISDN issues.

Similarly, there appear to be an overabundance of standards and organizations that produce them. Each, however, has its own charter and responsibilities. Increasingly, when charters overlap, standards groups cooperate to eliminate redundant standards. However, national and international standards will continue to exist; the international standards bodies will focus on basic operations, procedures, and internetworking while the national standards bodies will focus on issues specific to their country's networks (and which, in some cases, would never be addressed or resolved by an international body).

REFERENCES

ANSI Std.802.2, 1985. *ISO/DIS 8802/2: Local Area Networks Logical Link Control*, 1984, 1996.—ISBN 0-471-82748-7.

Black, Uyless. *Computer Networks Protocols: Standards and Interfaces*, Prentice-Hall, 1987.—ISBN 0-13-165754-2 025.

Black, Uyless. *OSI—A Model for Computer Communications Standards*, Prentice-Hall, 1991.—ISBN 0-13-637133-7.

Black, Uyless. *TCP/IP and Related Protocols*, McGraw-Hill, 1992.—ISBN 0-07-005553-X.

CCITT X.21. *Specifications for Accessing Circuit-Switched Networks*.

CCITT X.25. *Specifications for Connecting Data Terminal Equipment to Packet-Switched Networks*.

Chappell, Laura. *NetWare LAN Analysis*, Sybex and Novell Press, 1993.—ISBN 0-78-211143-2.

Comer, Douglas. *Internetworking with TCP/IP*, vol. 1, 2d ed., Prentice-Hall, 1991.—ISBN 0-13-468505-9.

Dalal, Yogen K, and Prentice, Robert S. *Forty-Eight-Bit Absolute Internet and Ethernet Host Numbers*, Xerox Corporation Office Systems Division, 1981.

Digital Equipment Corporation. *Data Access Protocol Functional Specification (DAP), Version 5.6.0.* October 1980. Order number-AA-K177A-TK.

Digital Equipment Corporation. *DECnet Digital Network Architecture, Phase IV: General Description*, May, 1982. Order number AA-N149A-TC.

Digital Equipment Corporation. *DECnet Digital Network Architecture, Phase IV: NSP Function Specification*, December 1983. Order number AA-X439A-TK.

Digital Equipment Corporation. *DECnet Digital Network Architecture, Phase IV: Routing Layer Function Specification, Version 2.0.0*, December 1980. Order number AA-X435A-TK.

Digital Equipment Corporation. *DECnet Digital Network Architecture, Phase IV: Network Management Function Specification*, December 1983. Order number AA-X437A-TK.

Digital Equipment Corporation. *DECent Digital Network Architecture, Phase IV: Session Control Function Specification, Version 1.0.0,* November 1980. Order number AA-K182A-TK.

Digital Equipment Corporation. *Networks—Communications Local Area Transport (LAT): Network Concepts,* June 1987. Order number AA-HY66A-TK.

Digital Equipment Corporation. *LAT Network Concepts,* October 1990. Order number AA-LD848-TK.

Digital Intel and Xerox Corporations. *The Ethernet—A Local Area Network: Data-Link Layer and Physical Layer Specifications, Version 2.0* (also known as the *Blue Book* when ordering from Xerox Corporation), November 1982.

IBM Token Ring Network Architecture Reference, 3d ed., September 1989. Order number SC30-3374-02.

IEEE Document Standard 802.1d. May 2, October 1991.

IEEE Document Standard 802.1H. *MAC Layer Bridging of Ethernet in IEEE 802 LANs.*

IEEE Document Standard P802.5M-D6. *Unproved Draft of Media Access Control (MAC) Bridges,* P802.1D, 1991.

IEEE Std. 802.5-1989. *Local Area Networks 802.5 Token Ring Access Method,* Institute of Electrical and Electronic Engineers, 1989. ISBN 1-55937-012-2.

ISO 8072.OSI Transport Layer service definitions.

ISO 8073.OSI Transport Layer protocol specifications.

ISO 8208.Network Layer protocol specification for connection-oriented service based on CCITT X.25 specifications.

ISO 8326.OSI Session Layer service definitions including transport classes 0, 1, 2, 3, and 4.

ISO 8473.Network Layer protocol and addressing specification for connectionless network service.

ISO 8649/8650.Common application and service elements (CASE) specifications and protocols.

ISO 8822/23/24.Presentation Layer specifications.

Martin, James, and Leben, Joe. *DECnet Phase V: An OSI Implementation,* Prentice Hall, 1992,—ISBN 1-S5568-076-9.

Miller, Mark A. *LAN Protocol Handbook,* M&TBooks, 1990.—ISBN 1-55851-099-0.

Miller, Mark A., *Troubleshooting Internetworks*, M&TBooks, 1991.—ISBN 1-55851-236-5.

Naugle, Matthew G., *Local Area Networking*, McGraw-Hill, 1991.—ISBN 0-07-046455-3.

Novell NetWare Multiprotocol Router-Basic, Version 1.0: User Guide, March 1992. 123-000194-001.

Novell NetWare System Interface Technical Overview, Addison Wesley, 1990.—ISBN 0-201-57027-0.

Oppen, Derek C., and Dalal, Yogen K., *The Clearinghouse: A Decentralized Agent for Locating Named Objects in a Distributed Environment*, October 1991. Order number OPD-T8103.

Request for Comment 768 User Datagram Protocol (UDP).

Request for Comment 791 Internet Protocol.

Request for Comment 792 Internet Control Message Protocol ICMP.

Request for Comment 793 Transmission Control Protocol (TCP).

Request for Comment 1058 Routing Information Protocol (RIP).

Rose, Marshall T. *The Simple Book: An Introduction to Management of TCP/IP-Based Internets*, Prentice-Hall, 1991.—ISBN 0-13-812611-9.

Sidhu, Gursharan S., Andrews, Richard F., and Oppenheimer, Alan B. *Inside Apple Talk*, 2d ed., Addison Wesley, 1993.—ISBN 0-201-55021-0.

Stallings, William. *Handbook of Computer Communications Standards*, vol.1: *The Open Systems Interconnection (OSI) Model and OSI-Related Standards, 1st ed., Prentice-Hall, 1988.—ISBN 0-672-22664-2.*

Stevens, Richard W. UNIX Network Programming, Prentice-Hall, 1990.—ISBN 0-13-949876-1.

Tanenbaum, Andrew S. *Computer Networks*, 2d ed., Prentice-Hall, 1989. ISBN 0-13-162959-X.

3Com Corporation. *Education Services Internet Packet Exchange (IPX) Protocol Student Guide*, June 1991.

3Com Corporation. *NetBuilder Bridge/Router Operation Guide*, manual number 09-0251-000, April 1991.

Turner, Paul, *NetWare Communications Process* (application notes), Novell, September 1990.

Xerox Corporation. *Clearinghouse Protocol*, April 1984. Order number XNSS 078404.

Xerox Corporation, *Courier: The Remote Procedure Call Protocol*, December 1981. Publication number XNSS 038112.

Xerox Corporation *Internet Transport Protocols,* January 1991. XNSS 029101. (This book covers the network and transport layers for the XNS protocol.)

Xerox Corporation. *Sequenced Packet Protocol Connection Parameter Negotiation, November 1990. Publication number XNSS 339011.*

Xerox Corporation, Xerox Network Systems Architecture: General Information Manual, April 1985. XNSG 068504.

INDEX

Index

Index

ABOUT THE AUTHOR

Matthew Naugle is Director of Engineering at Zipcom, Raleigh, NC. He has also held senior positions at 3Com Corporation, Proteon, and Bay Networks. Naugle is the author of McGraw-Hill's *Local Area Networking, 2/e.*